Engineering Cyber-Physical Systems and Critical Infrastructures

7

Series Editor

Fatos Xhafa, *Departament de Ciències de la Computació, Technical University of Catalonia, Barcelona, Spain*

The aim of this book series is to present state of the art studies, research and best engineering practices, real-world applications and real-world case studies for the risks, security, and reliability of critical infrastructure systems and Cyber-Physical Systems. Volumes of this book series will cover modelling, analysis, frameworks, digital twin simulations of risks, failures and vulnerabilities of cyber critical infrastructures as well as will provide ICT approaches to ensure protection and avoid disruption of vital fields such as economy, utility supplies networks, telecommunications, transports, etc. in the everyday life of citizens. The intertwine of cyber and real nature of critical infrastructures will be analyzed and challenges of risks, security, and reliability of critical infrastructure systems will be revealed. Computational intelligence provided by sensing and processing through the whole spectrum of Cloud-to-thing continuum technologies will be the basis for real-time detection of risks, threats, anomalies, etc. in cyber critical infrastructures and will prompt for human and automated protection actions. Finally, studies and recommendations to policy makers, managers, local and governmental administrations and global international organizations will be sought.

D. Jude Hemanth · Tuncay Yigit · Utku Kose · Ugur Guvenc
Editors

4th International Conference on Artificial Intelligence and Applied Mathematics in Engineering

ICAIAME 2022

Editors
D. Jude Hemanth
Department of ECE
Karunya University
Karunya Nagar, Tamil Nadu, India

Utku Kose
Department of Computer Engineering
Suleyman Demirel University
Isparta, Türkiye

Tuncay Yigit
Department of Computer Engineering,
Faculty of Engineering
Süleyman Demirel University
Isparta, Türkiye

Ugur Guvenc
Electric-Electronic Department
Duzce University
Düzce, Türkiye

ISSN 2731-5002 ISSN 2731-5010 (electronic)
Engineering Cyber-Physical Systems and Critical Infrastructures
ISBN 978-3-031-31958-7 ISBN 978-3-031-31956-3 (eBook)
https://doi.org/10.1007/978-3-031-31956-3

This Springer imprint is published by the registered company Springer Nature Switzerland AG
The registered company address is: Gewerbestrasse 11, 6330 Cham, Switzerland

Foreword

The recent state of technological improvements shows that intense use of Artificial Intelligence is an essential factor to ensure better data processing in problem solutions. Furthermore, effective solutions result to better user experience, which has been a vital factor in terms of personalized technology use trend of the new century. Such personalized user experience is also a sign for the improved capabilities of Artificial Intelligence. Improved capabilities in terms of Artificial Intelligence have been provided best via Deep Learning and hybrid intelligent systems of multiple Machine Learning techniques. At the final, the end user, who is consumer and responsible for the developed intelligent technologies, is the critical actor in terms of the momentum seen in the scientific literature.

Because the scientific literature is very active, it is an urgent need to have idea about how Deep Learning as well as hybrid intelligent systems are applied in different fields. We know that the problems of biomedical have a long-time strong relation with the literature of Artificial Intelligence. But it appeared already that the flexible, data-oriented Artificial Intelligence techniques are too active in all other fields such as electronics, mechatronics, geology, geophysics, chemistry, communication, education and even fine arts. Especially creative outcomes by Deep Learning models in fine arts and robotic automation replacing human jobs take many of recent interests but almost all kinds of technical applications have been using Artificial Intelligence methodologies for a long time period. So, it is a need to keep informed about the latest advancements in order to have idea about how the literature of Artificial Intelligence is advancing and building the future world. That may be done effectively by examining the recent collection of research works appeared at scientific events.

This book including the research works presented at the 4th International Conference on Artificial Intelligence and Applied Mathematics in Engineering (ICAIAME 2022) is one of the latest reference works, which enable readers to track the latest literature in terms of Artificial Intelligence. As the coverage of the event is with not only engineering fields but also Applied Mathematics, the collection provided in the book reports about how different engineering fields and mathematical methodologies have been interacting with the revolutionary techniques of Artificial Intelligence. The provided content is strong in terms of having a wide coverage of Artificial Intelligence applications and reporting the latest comparative results in order to see a competitive state of different techniques targeting similar problem topics. Outcomes provided inside the book are useful to have more idea about how the human–machine interaction is advancing and how the humankind may take critical steps to have the best from Artificial Intelligence-based solutions in different areas. I think researchers as well as degree students will effectively use this reference work in their research projects. The book will be useful in even courses given for Bachelor, Master and Doctorate degrees. The whole content of the book shows that there were efforts to choose the best quality research to be presented at ICAIAME 2022 and included in the book after careful reviews.

It was a pleasure for me to write a foreword for such a remarkable book work for the ICAIAME event series. For their valuable efforts and kind invitation, I would like to thank dear editors Dr. Hemanth, Dr. Yigit, Dr. Kose and Dr. Guvenc. It was also a pleasure for me to read about the latest research outcomes provided by the chapter authors of this book. I believe that all readers will enjoy the content and benefit greatly from this timely book to organize their further research plans.

Deepak Gupta

Preface

Artificial Intelligence has been effectively changing the way of engineering and triggering critical advancements in the context of industrial fields. With the wind of the industry 4.0, the future is already being structured over advanced smart tools, which are already common objects of daily life. It is important that the required computational background of such smart tools is built with the advanced technologies, which are result of the scientific contributions by Artificial Intelligence. As this paradoxical run is actually a recursive mechanism within the history of technology, the synergy between hardware and software seems too strong in 2020s so far. So, industrial outcomes of engineering solutions are in the front of everybody. This is an indicator that there is a great research flow in different engineering fields.

Gathering the latest, cutting-edge research by 2022, this book is the new volume of the International Conference on Artificial Intelligence and Applied Mathematics in Engineering (ICAIAME). As held in Baku, Azerbaijan, the 4th ICAIAME has gained again a remarkable interest from international researchers, who are working with Artificial Intelligence to improve quality of outcomes of different engineering fields and Applied Mathematics. It is critical that the current trend of Artificial Intelligence-based research is still with revolutionary contributions by Deep Learning, alternative applications of Data Science methods and even intelligent optimization solutions for the modeled real-world problems. Of course, the role of solutions by Applied Mathematics has a major role in shaping the mathematical outcomes regarding technical aspects of engineering problems. All these facts are reflected in this volume of ICAIAME 2022 in the context of wide diversity of application areas resulting findings, which are effective to understand current and future potentials in technological rise. It is also critical that the outcomes of this book will give idea to learn more about current collaborative relation between human and the machine.

As it was done in the previous volumes of the ICAIAME, both independent reviewers and the scientific committee of the new ICAIAME 2022 had valuable efforts to review the content of submitted-presented research works and gave their views to improve their quality in terms of technical quality, organization and readability. The organizing committee of the ICAIAME thinks that the findings of the research presented in this event are among remarkable results shaping the future of collaboration between the Artificial Intelligence and technical fields. It is also thought that the influence of societal aspects of Artificial Intelligence and advanced technology will be discussed more year by year.

The editors of this book: 4th International Conference on Artificial Intelligence and Applied Mathematics in Engineering—ICAIAME 2022 would like to send thanks to all participants, who took part in the event. As it has been always done, the special thanks are for all keynote speakers as well as the session chairs, who shared their critical scientific and academic views for latest research outcomes. The ICAIAME 2022 would not be successful without intense efforts by the event staff. So, all the best compliments are for the ICAIAME 2022 staff. While moving towards the new year 2023, all ICAIAME committees and the staff wish all the best for the next upcoming events.

D. Jude Hemanth
Tuncay Yigit
Utku Kose
Ugur Guvenc

International Conference on Artificial Intelligence and Applied Mathematics in Engineering 2022

Web: http://www.icaiame.com
E-Mail: info@icaiame.com

Briefly About

ICAIAME 2022 is meeting with the latest requirements of the Academic Promoting Programme applied in Turkey.

4th International Conference on Artificial Intelligence and Applied Mathematics in Engineering (ICAIAME 2022) will be held within 20–21–22 May 2022, in Baku, the capital city of the brother country Azerbaijan. The main theme of the conference, which will be held at Azerbaijan Technical University with international participations along a three-day period, is solutions of Artificial Intelligence and Applied Mathematics in engineering applications. Acceptance of abstracts for ICAIAME 2022 papers is in Turkish, English and Azerbaijani. Full text acceptance will be in Turkish and English. Abstracts submitted in Azerbaijani language must be translated into Turkish or English

The main theme of ICAIAME 2022 is "Artificial Intelligence and Mathematics Based Cyber Security". However, applicants are encouraged to submit original research work in Engineering Sciences and Applied Mathematics. High-quality, accepted and presented English papers will be published in the Springer Series: Engineering Cyber-Physical Systems and Critical Infrastructures.

Conference Scope/Topics (as not limited to): In Engineering Problems:

- Machine Learning Applications
- Deep Learning Applications
- Intelligent Optimization Solutions
- Robotics/Softrobotics and Control Applications
- Hybrid System-Based Solutions
- Algorithm Design for Intelligent Solutions
- Image/Signal Processing-Supported Intelligent Solutions
- Data Processing-Oriented Intelligent Solutions
- Cyber Security Intelligent Solutions
- Real-Time Applications in Cyber-Critical Infrastructures

- Security Protocols based on Intelligent Systems
- Intelligent Solutions in Intrusion Detection/Prevention Systems
- Prediction and Diagnosis Applications
- Linear Algebra and Applications
- Numerical Analysis
- Differential Equations and Applications
- Probability and Statistics
- Cryptography
- Operations Research and Optimization
- Discrete Mathematics and Control
- Nonlinear Dynamical Systems and Chaos
- General Engineering Applications
- General Topology
- Number Theory
- Algebra Analysis
- Applied Mathematics and Approximation Theory
- Mathematical Modelling and Optimization
- Intelligent Solutions in Civil Engineering
- Graph Theory
- Kinematics
- Cryptography

Conference Posters

Springer
ICAIAME
2022
International Conference
on Artificial Intelligence and
Applied Mathematics
in Engineering
20-21-22 May 2022
Baku, Azerbaijan
(Azerbaijan Technical University)
Şuşa
2022
more information
www.icaiame.com
info@icaiame.com

General Committees

Honorary Chairs

İlker Hüseyin Çarıkçı	Rector of Süleyman Demirel University, Turkey
Vilayet Veliyev	Rector of Azerbaijan Technical University, Azerbaijan
Şahin Bayramov	Rector of Mingachevir State University, Azerbaijan
İbrahim Diler	Rector of Applied Sciences University of Isparta, Turkey
Musa Yıldız	Rector of Gazi University, Turkey

General Chair

Tuncay Yiğit	Süleyman Demirel University, Turkey

Conference Chairs

Cemal Yılmaz	Mingachevir State University, Azerbaijan
Yusuf Sönmez	Azerbaycan Teknik Üniversitesi, Azerbaycan
Hasan Hüseyin Sayan	Gazi University, Turkey
İsmail Serkan Üncü	Isparta Applied Sciences University, Turkey
Utku Köse	Süleyman Demirel University, Turkey
Mevlüt Ersoy	Süleyman Demirel University, Turkey

Organizing Committee

Mehmet Gürdal	Süleyman Demirel University, Turkey
Anar Adiloğlu	Süleyman Demirel University, Turkey
Şemsettin Kılınçarslan	Süleyman Demirel University, Turkey
Kemal Polat	Bolu Abant İzzet Baysal University, Turkey
Okan Bingöl	Applied Sciences University of Isparta, Turkey
Cemal Yılmaz	Gazi University, Turkey
Ercan Nurcan Yılmaz	Mingachevir State University, Azerbaijan
Hamdi Tolga Kahraman	Karadeniz Technical University, Turkey
M. Ali Akcayol	Gazi University, Turkey

Jude Hemanth	Karunya University, India
Uğur Güvenç	Düzce University, Turkey
Akram M. Zeki	International Islamic University Malaysia, Malaysia
Asım Sinan Yüksel	Süleyman Demirel University, Turkey
Bogdan Patrut	Alexandru Ioan Cuza University of Iasi, Romania
Halil İbrahim Koruca	Süleyman Demirel University, Turkey
Erdal Aydemir	Süleyman Demirel University, Turkey
Ali Hakan Işık	Mehmet Akif Ersoy University, Turkey
Muhammed Maruf Öztürk	Süleyman Demirel University, Turkey
Osman Özkaraca	Muğla Sıtkı Koçman University, Turkey
Bekir Aksoy	Isparta Applied Sciences University, Turkey
Mehmet Kayakuş	Akdeniz University, Turkey
Gürcan Çetin	Muğla Sıtkı Koçman University, Turkey
Murat İnce	Süleyman Demirel University, Turkey
Gül Fatma Türker	Süleyman Demirel University, Turkey
Ferdi Saraç	Süleyman Demirel University, Turkey
Cevriye Altıntaş	Isparta Applied Sciences University, Turkey
Hamit Armağan	Süleyman Demirel University, Turkey
Recep Çolak	Isparta Applied Sciences University, Turkey

Secretary and Social Media

Çilem Koçak	Isparta Applied Sciences University, Turkey

Accommodation and Registration/Venue Desk

Recep Çolak	Isparta Applied Sciences University, Turkey
Cem Deniz Kumral	Isparta Applied Sciences University, Turkey

Travel/Transportation

Hamit Armağan	Süleyman Demirel University, Turkey

Web/Design/Conference Session

Ali Topal	Isparta Applied Sciences University, Turkey

Scientific Committee

Tuncay Yiğit (General Chair)	Süleyman Demirel University, Turkey
Utku Köse (Committee Chair)	Süleyman Demirel University, Turkey
Ahmet Bedri Özer	Fırat University, Turkey
Akram M. Zeki	Malaysia International Islamic University, Malaysia
Ali Keçebaş	Muğla Sıtkı Koçman University, Turkey
Ali Öztürk	Düzce University, Turkey
Anar Adiloğlu	Süleyman Demirel University, Turkey
Aslanbek Nazıev	Ryazan State University, Russia
Arif Özkan	Kocaeli University, Turkey
Aydın Çetin	Gazi University, Turkey
Ayhan Erdem	Gazi University, Turkey
Cemal Yılmaz	Mingachevir State University, Azerbaijan
Çetin Elmas	Azerbaijan Technical University—Vice Rector, Azerbaijan
Daniela Elena Popescu	University of Oradea, Romania
Eduardo Vasconcelos	Goias State University, Brazil
Ekrem Savaş	Rector of Uşak University, Turkey
Ender Ozcan	Nottingham University, England
Ercan Nurcan Yılmaz	Mingachevir State University, Azerbaijan
Erdal Kılıç	Samsun 19 May University, Turkey
Eşref Adalı	İstanbul Technical University, Turkey
Gültekin Özdemir	Süleyman Demirel University, Turkey
Hamdi Tolga Kahraman	Karadeniz Technical University, Turkey
Hüseyin Demir	Samsun 19 May University, Turkey
Hüseyin Merdan	TOBB Economy and Technology University, Turkey
Hüseyin Şeker	Birmingham City University, England
Igbal Babayev	Azerbaijan Technical University, Azerbaijan
Igor Lıtvınchev	Nuevo Leon State University, Mexico
İbrahim Üçgül	Süleyman Demirel University, Turkey
İbrahim Yücedağ	Düzce University, Turkey
İlhan Koşalay	Ankara University, Turkey
Jose Antonio Marmolejo	Panamerican University, Mexico
Jude Hemanth	Karunya University, India
Junzo Watada	Universiti Teknologi PETRONAS, Malaysia
Kemal Polat	Bolu Abant İzzet Baysal University, Turkey
Marat Akhmet	Middle East Technical University, Turkey

Barış Akgün	Koç University, Turkey
Bekir Aksoy	Isparta Applied Sciences University, Turkey
Cevriye Altıntaş	Isparta Applied Sciences University, Turkey
Deepak Gupta	Maharaja Agrasen Institute of Technology, India
Dmytro Zubov	The University of Information Science and Technology St. Paul the Apostle, Macedonia
Enis Karaarslan	Muğla Sıtkı Koçman University, Turkey
Esin Yavuz	Süleyman Demirel University, Turkey
Fatih Gökçe	Süleyman Demirel University, Turkey
Ferdi Saraç	Süleyman Demirel University, Turkey
Gül Fatma Türker	Süleyman Demirel University, Turkey
Gür Emre Güraksın	Afyon Kocatepe University, Turkey
Gürcan Çetin	Muğla Sıtkı Koçman University, Turkey
Iulian Furdu	Vasile Alecsandri University of Bacau, Romania
Mehmet Kayakuş	Akdeniz University, Turkey
Mehmet Onur Olgun	Süleyman Demirel University, Turkey
Murat İnce	Süleyman Demirel University, Turkey
Mustafa Nuri Ural	Gümüşhane University, Turkey
Okan Oral	Akdeniz University, Turkey
Osman Palancı	Süleyman Demirel University, Turkey
Paniel Reyes Cardenas	Popular Autonomous University of the State of Puebla, Mexico
Remzi Inan	Applied Sciences University of Isparta, Turkey
S. T. Veena	Kamaraj Engineering and Technology University, India
Serdar Biroğul	Düzce University, Turkey
Serdar Çiftçi	Harran University, Turkey
Ufuk Özkaya	Süleyman Demirel University, Turkey
Veli Çapalı	Süleyman Demirel University, Turkey
Vishal Kumar	Bipin Tripathi Kumaon Institute of Technology, India
Anand Nayyar	Duy Tan University, Vietnam
Hamit Armağan	Süleyman Demirel University, Turkey
Recep Çolak	Isparta Applied Sciences University, Turkey
Simona Elena Varlan	Vasile Alecsandri University of Bacau, Romania
Ashok Prajapatı	FANUC America Corp., USA
Katarzyna Rutczyńska-Wdowiak	Kielce University of Technology, Poland
Mustafa Küçükali	Information and Communication Technologies Authority
Özkan Ünsal	Süleyman Demirel University, Turkey
Tim Jaes	International Project Management Association, USA

Keynote Speaks

1. Prof. Dr. Çetin Elmas (Gazi University, Turkey) **"Democracy, Artificial Intelligence and Threats"**
2. Prof. Dr. Ender Özcan (University of Nottingham, England) **"An Extended Classification Of Selection Hyper-Heuristics"**
3. Prof. Dr. Hüseyin Seker (University of Birmingham, England) **"The Power of Data and The Things It Empowers: What have we learned in the age of Covid-19"**
4. Prof. Dr. Jude Hemanth (Karunya Institute of Technology and Sciences, India) **"Why Deep Learning Models Over Traditional Machine Learning Models?"**
5. Prof. Dr. M. Ali Akcayol (Gazi University, Turkey) **"Deep Network Architecture for Image Colorization"**
6. Prof. Dr. Sergey D.Bushuyev (Kyiv National University, Ukraine) **"Development Of Intelligence Society On The Basis Of Global Trends"**
7. Dr. Mladen Vukomanovi (Vice President, International Project Management Agency – IPMA, Croatia) **"Application Of Machine Learning To Construction Projects: A Case Of Bexel"**

Acknowledgement

As the editors, we would like to thank Lect. Cem Deniz Kumral (Isparta University of Applied Sciences, Turkey) for her valuable efforts on pre-organization of the book content, and the Springer team for their great support to publish the book.

Contents

Sentiment Analysis in Turkish Using Transformer-Based Deep Learning Models 1
Oktay Ozturk and Alper Ozcan

Mathematical Modeling of an Antenna Device Based on a T-Shaped Waveguide of the Microwave Range 16
Elmar Z. Hunbataliyev

Higher-Order and Stable Numerical Scheme for Nonlinear Diffusion System via Compact Finite Difference and Adaptive Step-Size Runge-Kutta Methods 30
Shodijon Ismoilov, Gurhan Gurarslan, and Gamze Tanoğlu

Explainable Artificial Intelligence (XAI) for Deep Learning Based Intrusion Detection Systems 39
Mehmet Sevri and Hacer Karacan

A Color Channel Based Analysis on Image Tessellation 56
Turan Kibar and Burkay Genç

Modeling of Shear Strength of Basalt Fiber Reinforced Clay (BFRC) Soil Using Artificial Neural Network (ANN) 73
Mehmet Fatih Yazıcı, Ahmetcan Sungur, and Sıddıka Nilay Keskin

Change of the Internet with Blockchain and Metaverse Technologies 82
Ismet Can Sahin and Can Eyupoglu

A Language-Free Hate Speech Identification on Code-mixed Conversational Tweets 102
Pelin Canbay and Necva Bölücü

Decision Trees in Causal Inference 109
Hulya Kocyigit

The Resilience of Unmanned Aerial Vehicles to Cyberattacks and Assessment of Potential Threats 122
Ahmet Ali Süzen

TensorFlow Based Feature Extraction Using the Local Directional Patterns 130
Hamidullah Nazari and Devrim Akgun

Ensuring the Invariance of Object Images to Linear Movements for Their Recognition 140
Rahim Mammadov, Elena Rahimova, and Gurban Mammadov

AI-Based Network Security Anomaly Prediction and Detection in Future Network 149
Gunay Abdiyeva-Aliyeva and Mehran Hematyar

Mathematical Modeling of the Antenna Devices of the Microwave Range 160
Islam J. Islamov, Mehman H. Hasanov, and Elmar Z. Hunbataliyev

Design and Simulation of the Miniaturized Dual-Band Monopole Antenna for RFID Applications 176
Kayhan Çelik

Exploring the Driven Service Quality Dimensions for Higher Education Based on MCDM Analysis 186
Aleyna Sahin, Mirac Murat, Gul Imamoglu, Kadir Buyukozkan, and Ertugrul Ayyildiz

Deep Learning-Based Traffic Light Classification with Model Parameter Selection 197
Gülcan Yıldız, Bekir Dizdaroğlu, and Doğan Yıldız

Investigation of Biomedical Named Entity Recognition Methods 218
Azer Çelikten, Aytuğ Onan, and Hasan Bulut

Numerical Solutions of Hantavirus Infection Model by Means of the Bernstein Polynomials 230
Şuayip Yüzbaşı and Gamze Yıldırım

Instance Segmentation of Handwritten Text on Historical Document Images Using Deep Learning Approaches 244
Umid Suleymanov, Vildan Huseynov, Ilaha Manafova, Asgar Mammadli, and Toghrul Jafarov

Synthetic Signal Generation Using Time Series Clustering and Conditional Generative Adversarial Network 254
Nurullah Ozturk and Melih Günay

Transfer Learning Based Flat Tire Detection by Using RGB Images 264
Oktay Ozturk and Batuhan Hangun

Cyber Threats and Critical Infrastructures in the Era of Cyber Terrorism 274
Zeynep Gürkaş-Aydin and Uğur Gürtürk

Secure Data Dissemination in Ad-Hoc Networks by Means of Blockchain 288
Cansin Turguner, Engin Seven, and Muhammed Ali Aydin

Battery Charge and Health Evaluation for Defective UPS Batteries via Machine Learning Methods 298
Mehmetcan Çelik, İbrahim Tanağardıgil, Mehmet Uğur Soydemir, and Savaş Şahin

Covid-19: Automatic Detection from X-Ray Images Using Attention Mechanisms 309
Cemil Zalluhoğlu and Cemre Şenokur

Lexicon Construction for Fake News Detection 320
Uğur Mertoğlu and Burkay Genç

A Transfer Learning Approach for Skin Cancer Subtype Detection 337
Burak Kolukısa, Yasin Görmez, and Zafer Aydın

Competencies Intelligence Model for Managing Breakthrough Projects 348
Sergey Bushuyev, Igbal Babayev, Natalia Bushuyeva, Victoria Bushuieva, Denis Bushuiev, and Jahid Babayev

Secure Mutual Authentication Scheme for IoT Smart Home Environment Using Biometric and Group Signature Verification Methods 360
Hisham Raad Jafer Merzeh, Mustafa Kara, Muhammed Ali Aydın, and Hasan Hüseyin Balık

Fuzzy Method of Creating of Thematic Catalogs of Information Resources of the Internet for Search Systems 374
Vagif Gasimov

Scattered Destruction of a Cylindrical Isotropic Thick Pipe in an Aggressive Medium with a Complex Stress State 383
Sahib Piriev

Estimating the Resonance Frequency of Square Ring Frequency Selective Surfaces by Using ANN 396
Mehmet Yerlikaya and Hüseyin Duysak

Application of Kashuri Fundo Transform to Population Growth and Mixing Problem 407
Haldun Alpaslan Peker and Fatma Aybike Çuha

Attack Detection on Testbed for Scada Security 415
Esra Söğüt and O. Ayhan Erdem

Conveyor Belt Speed Control with PID Controller Using Two PLCs and LabVIEW Based on OPC and MODBUS TCP/IP Protocol 423
Arslan Tirsi, Mehmet Uğur Soydemir, and Savaş Şahin

AI-Powered Cyber Attacks Threats and Measures 434
Remzi Gürfidan, Mevlüt Ersoy, and Oğuzhan Kilim

A QR Code-Based Robust Color Image Watermarking Technique 445
Gökhan Azizoğlu and Ahmet Nusret Toprak

Repairing of Wall Cracks in Historical Bridges After Severe Earthquakes Using Image Processing Technics and Additive Manufacturing Methods 458
Pinar Usta, Merdan Özkahraman, Muzaffer Eylence, Bekir Aksoy, and Koray Özsoy

Analysis Efficiency Characteristics Multiservice Telecommunication Networks Taking into the Account of Properties Self-similar Traffic 465
Bayram G. Ibrahimov, Mehman H. Hasanov, and Ali D. Tagiyev

Research and Analysis Methods for Prediction of Service Traffic Signaling Systems Using Neural Network Technologies 473
Bayram G. Ibrahimov, Cemal H. Yilmaz, Almaz A. Aliyeva, and Yusif A. Sonmez

An Image Completion Method Using Generative Adversarial Networks 483
Eyyüp Yıldız, Selçuk Sevgen, and M. Erkan Yüksel

Construction 4.0 - New Possibilities, Intelligent Applications, Research Possibilities 490
Krzysztof Kaczorek, Nabi Ibadov, and Jerzy Rosłon

Tone Density Based Sentiment Lexicon for Turkish 500
Muazzez Şule Karaşlar, Fatih Sağlam, and Burkay Genç

User Oriented Visualization of Very Large Spatial Data with Adaptive Voronoi Mapping (AVM) 515
Muhammed Tekin Ertekin and Burkay Genç

Comparative Analysis of Machine Learning Algorithms for Crop Mapping Based on Azersky Satellite Images 537
Sona Guliyeva, Elman Alaskarov, Ismat Bakhishov, and Saleh Nabiyev

Defect Detection on Steel Surface with Deep Active Learning Methods on Fewer Data 549
Bahadır Gölcük and Sevinç İlhan Omurca

Determining Air Pollution Level with Machine Learning Algorithms: The Case of India 560
Furkan Abdurrahman Sari, Muhammed Ali Haşıloğlu, Muhammed Kürşad Uçar, and Hakan Güler

A New Epidemic Model with Direct and Indirect Transmission with Delay and Diffusion 582
Fatiha Najm, Radouane Yafia, Ahmed Aghriche, and M. A. Aziz Alaoui

Machine Learning-Based Biometric Authentication with Photoplethysmography Signal 595
Bahadır Çokçetn, Derya Kandaz, and Muhammed Kürşad Uçar

DDOS Intrusion Detection with Machine Learning Models: N-BaIoT Data Set 607
Celil Okur, Abdullah Orman, and Murat Dener

Research and Analysis of the Efficiency Processing Systems Information Streams of Telematic Services 620
Bayram G. Ibrahimov, Gulnar G. Gurbanova, Zafar A. Ismayilov, Manafedin B. Namazov, and Asif A. Ganbayev

An Artificial Intelligence-Based Air Quality Health Index Determination: A Case Study in Sakarya 630
Salman Ahmed Nur, Refik Alemdar, Ufuk Süğürtin, Adem Taşın, and Muhammed Kürşad Uçar

Anomaly Detection in Sliding Windows Using Dissimilarity Metrics in Time Series Data 640
Ekin Can Erkuş and Vilda Purutçuoğlu

Operating a Mobile Robot as a Blockchain-Powered ROS Peer: TurtleBot Application 652
Mehmed Oğuz Şen, Fatih Okumuş, and Adnan Fatih Kocamaz

Opportunities and Prospects for the Application of Intelligent Robotic Devices in the Agricultural Sector 662
A. Mustafayeva, E. Israfilova, E. Aliyev, E. KHalilov, and G. Bakhsiyeva

Detection of Diabetic Macular Edema Disease with Segmentation of OCT Images 671
Saliha Yeşilyurt, Altan Göktaş, Alper Baştürk, Bahriye Akay, Derviş Karaboğa, and Özkan Ufuk Nalbantoglu

Innovative Photodetector for LIDAR Systems 680
K. Huseynzada, A. Sadigov, and J. Naghiyev

FastText Word Embedding Model in Aspect-Level Sentiment Analysis of Airline Customer Reviews for Agglutinative Languages: A Case Study for Turkish 691
Akın Özçift

Prediction of Electric Energy in Hydroelectric Plants by Machine Learning Methods: The Example of Mingachevir Dam 703
Almaz Aliyeva, Mevlüt Ersoy, and M. Erol Keskin

Technology and Software for Traffic Flow Management 713
Kahramanov Sebuhi Abdul

Comparision of Deep Learning Methods for Detecting COVID-19 in X-Ray Images 723
Hakan Yüksel

Rule-Based Cardiovascular Disease Diagnosis 740
Ayşe Ünlü, Derya Kandaz, Gültekin Çağil, and Muhammed Kürşad Uçar

Author Index 751

Sentiment Analysis in Turkish Using Transformer-Based Deep Learning Models

Oktay Ozturk[1(✉)] and Alper Ozcan[2]

[1] School of Computing, Wichita State University, Wichita, KS 67260, USA
oxozturk1@shockers.wichita.edu

[2] Department of Computer Engineering, Akdeniz University, Antalya, Turkey
alperozcan@akdeniz.edu.tr

Abstract. Social media, e-commerce, review, and blogging websites have become important sources of knowledge as information and communication technology has advanced. Individuals can share their thoughts, complaints, feelings, and views on a wide range of topics. Because it seeks to identify the orientation of the sentiment present in source materials, sentiment analysis is a key field of research in natural language processing. Sentiment analysis is a natural language processing (NLP) task that received the attention of many researchers and practitioners. The majority of earlier studies in sentiment analysis mainly focused on traditional machine learning (i.e., shallow learning) and, to some extent, deep learning algorithms. Recently, transformer-based models have been developed and applied in different application domains. These models have been shown to have a huge potential to advance text classification and, particularly, sentiment analysis research fields. In this paper, we investigate the performance of transformer-based sentiment analysis models. The case study has been performed on four datasets that are in Turkish. First, preprocessing methods were used to remove links, numerals, unmeaningful, and punctuation characters from the data. Unsuitable data was eliminated after the preprocessing phase. Second, each data set splitted into two parts; 80% for training, 20% for testing. Finally, transformer-based BERT, ConvBERT, ELECTRA, traditional deep learning, and machine learning algorithms have been applied to classify sentences into two or three classes, which are either positive, neutral, or negative. Experimental results demonstrated that transformer-based models could provide superior performance in terms of F-score compared to the traditional machine learning-based and deep learning models.

Keywords: BERT · ELECTRA · Deep learning · Transformer-based models · Natural Language Processing · Sentiment analysis · Turkish

1 Introduction

There is a tremendous amount of text on the internet, and more is on the way. Every second, more than 9K tweets and 2.9M e-mails are sent [1]. In 2020, 283K

D. J. Hemanth et al. (Eds.): ICAIAME 2022, ECPSCI 7, pp. 1–15, 2023.
https://doi.org/10.1007/978-3-031-31956-3_1

new pages were created in Turkish Wikipedia [2]. Making sense of this pile of text data has some practical implications for both researchers and companies by obtaining meaningful data. For example, valuable information can be extracted from job advertisements for analyzing current required competencies [34], from medical texts for detecting the symptoms [21], from tweets for brand management [26], Moreover, from online reviews for managing customer satisfaction [39]. One of the methods that have been used in this inference process with the widest application area is sentiment analysis.

Texts can be classified as positive, negative, or neutral via sentiment analysis. The application area of sentiment analysis is quite wide. It can benefit in fields such as politics, [3], health [5], finance [51], and psychology [31]. Hence, sentiment analysis becomes one of the popular research fields for natural language processing. Although deep learning [4] and machine learning-based [53] methods have been getting most of the attention recently. Various hybrid [7] and lexicon-based [44] approaches are also available in the literature. Besides classifying sentences as a whole, studies are available to detect the term that the sentiment is about. It helps to clarify sentences that consist of sentiments for two or more aspects at the same time. For instance, in "Service was good but desserts were expensive," we can obtain that opinion on "service" is positive while opinion on "desserts" is negative. Several SemEval tasks [35,45] include aspect-based sentiment analysis.

The morphologically rich structure of Turkish such as being an agglutinative language that makes Turkish words to be formed with the attachment of morphemes to roots raises some challenges. For instance, an English phrase consisting of several words (such as "when you came") can be represented in Turkish with only one word (as "geldiğinde"). Hence, it causes some difficulties in Turkish sentiment analysis. Sentiment analysis studies in Turkish mostly focus on tweet classification [18,41], movie review analysis [15,49], and customer review analysis [15,38]. Various methods have been applied in these studies such as lexicon-based [49], machine learning-based [9,28] and deep learning-based [15,18], including hybrid ones [22,41]. Also, there are some aspect-based sentiment analysis studies [23,24] which aim to identify different sentiments in a text together with the target terms they relate to.

In this study, a transformer based comprehensive sentimental analysis are presented for Turkish language. To the best of our knowledge, it is the first study that applies the ELECTRA model in Turkish sentiment analysis research. For the proper evaluation and bench marking different transformer based, LSTM based, BERT based, Naïve Bayes based model results are compared. For this purpose, four different data sets have been used in training and testing phases, including movie reviews, 3 or 2 class hotel reviews, and product comments.

The rest of the paper is organized as follows. After discussing related sentiment analysis studies2, we explained data sets that are used in 3.1. 3, provides a brief knowledge about different sentiment analysis methods used in this research. I 4, describe our experimental setup, and we shared the results of the experiments in 5. Finally, we conclude our study in 6.

2 Related Works

Sentiment analysis has been applied in various languages and fields, especially in English, using many methods. In studies conducted for Turkish, different approaches have been used accordingly, and the success of these methods in different domains has been measured. However, it can be said that some cutting-edge methods, especially transformer-based ones, have not been studied sufficiently yet. Here, we investigate related studies with different approaches.

Early sentiment analysis researches in Turkish focuses on lexicon-based methods [49] propose a framework by translating an existing lexicon to Turkish. They evaluate the framework with a corpus of movie reviews. Additionally, [17] introduce a comprehensive polarity lexicon called SentiTurkNet for Turkish sentiment analysis. They suggest that SentiTurkNet outperforms the direct translation method [27] analyzes Turkish tweets with the help of lexicon-based approach to classify them as positive, negative, or neutral, reaching a success rate of 80%. Also, a polarity calculation approach is proposed [56], considers linguistic features of Turkish in sentiment analysis. Sentiment analysis also got its share from the spread of machine learning to various fields. [9] Use word embeddings with various machine learning algorithms and test them with domain-specific tweets. In addition, [33] investigates effects of the 36-word embedding based representation consisting of different word embedding methods, and they suggest that in Turkish sentiment analysis, the representation that includes word2vec method outperforms other representations. [29,36] implement machine learning algorithms such as Naïve Bayes, random forest, support vector machine and k-nearest neighbors. In addition to them, [46] suggest that logistic regression algorithm outperforms them with accuracy of 77.35%. Also, [13] investigate the success of active learning in sentiment analysis. [40] compares various artificial intelligence methods with a lexicon-based approach and reports that random forest method performs best in Turkish Twitter data compared to support vector machines, maximum entropy, and decision tree models with the success rate of 88.5%.

Later, as machine learning algorithms could not consider features such as word order in a sentence, various deep learning models have been utilized in sentiment analysis. [15] implement recurrent neural networks with Long Short-Term Memory units and then test them with customer and movie reviews. [18] apply a deep learning technique for sentiment analysis utilizing a neural network model with six dense units and three dropout layers to show that sentiment analysis can be used for disaster management. [54] compares traditional methods with deep neural networks with data sets in five different topics in Turkish. Additionally, while observing the contribution of different training data sizes to the success of a system, [42] apply BERT model for Turkish sentiment analysis.

Apart from that, various hybrid methods have been proposed. [22] combines lexicon-based approach with machine learning-based techniques such as Naïve Bayes, support vector machines, and J48. [41] hierarchically combines random forest and support vector machine methods for Turkish sentiment analysis. Also,

[32] emphasize positive effects of the preprocessing in Turkish sentiment analysis researches.

3 Methods

In the past, different sentimental analysis techniques have been used for Turkish language. The performance and architecture of Transformers-based BERT and ELECTRA models are compared with different machine learning and deep learning models in the below part of this section. Moreover, datasets used for evaluation are also described below.

3.1 Data Sets

Turkish sentiment analysis studies are mostly based on tweets or reviews of movies, hotels, and products. Accordingly, five present sentiment analysis data sets, including tweets, movie critiques, and product comments, have been used for this research. These data sets will be examined in the following sections.

Turkish Movie Critique Dataset. Beyazperde, similar to IMDb is an online platforms that contains user's reviews for movies, TV shows, biographies, and ratings. The studied dataset which is released by [19], Contains 10662 Turkish movie critiques with their respective class scale from 1 to 5. The ratings of critiques are divided into 2 classes: 1 and 2-star reviews are treated as negative, while 4 and 5-star reviews are treated as positive. 3-star reviews are have been removed from the dataset since they are vague.

Turkish Product Comments Dataset. Another data set used compose user comments toward various phones and the sentiments of those comments released in Kaggle by Bahar Yılmaz [55]. While the data set in question contains 937 unique product comments, 739 are positive, and the remaining 237 are negative.

Turkish Customer Feedback Dataset. Turkish customer feedback dataset consists of 8485 feedbacks made for different online stores and sentiments of these reviews. Positive and negative samples have been evenly distributed in this dataset, and it is released by [10]. Besides, 2938 neutral comments were added to this dataset, and 3-class and 2-class versions have been used.

3.2 Machine Learning Models

Machine learning uses statistical and mathematical methods to make inferences from previously or currently obtained data and make predictions about unknown data due to inferences. Machine learning is divided into three categories: supervised learning, unsupervised learning, and reinforcement learning. Supervised machine learning is divided into two as classification and regression. The classification method is given a result between classes taught to the machine (black or white, right or wrong, right or left, etc.). Another hand, unsupervised machine learning uses the clustering method to make predictions on a data set whose structure is unknown. The clustering method is defined as the grouping of similar data according to defined criteria. In unsupervised machine learning, there is no human operator or class label. This method is generally used in cases where there are extensive data sets and has the ability to make decisions about these data by the clustering method. On the other hand, supervised machine learning is taught with specified data sets and then tries to make predictions about new incoming data. Unlike the supervised and unsupervised machine learning methods, reinforcement learning search for such intelligent actions that an agent could take according to the environment to maximize the performance.

There are many supervised and unsupervised machine learning algorithms for text classification and sentiment analysis tasks. This paper uses Support Vector Machines (SVM), Random Forest, and Naïve Bayes supervised algorithms to classify sentences as negative, positive, or neutral.

Term Frequency-Inverse Document Frequency (TFIDF). TFIDF, short for term frequency-inverse document frequency, is a numerical metric in information retrieval that is meant to represent how relevant a word is to a document in a list or dataset. It is generally used as a weighting factor in text mining, text classification, and sentiment analysis. The tf-idf value increases respectfully to the number of times a word occurs in the document and is balanced by the count of documents in the corpus that contain the word, which helps to adapt to the fact that certain words usually appear more frequently. In order to classify sentences as positive, negative, or neutral, we use TFIDF as features. Term frequency in order the term t is defined as in Eq. 1.

$$\mathrm{tf}(t,d) = \frac{f_{t,d}}{\sum_{t' \in d} f_{t',d}} \tag{1}$$

where ft, d is the raw count of a term in a document, i.e., the number of times that term t occurs in document d. The second step for calculating TFIDF is to calculate inverse document frequency (IDF). IDF is defined as Eq. 2.

$$idf(t,D) = \log \frac{N}{|\{d \in D : t \in d\}|} \tag{2}$$

where N: represents total number of documents in the corpus, $N = |D|$ and $|\{d \in D : t \in d\}|$: denotes the total appearance of the term t in the document. Finally, TFIDF defined as Eq. 3.

$$tfidf(t,d,D) = tf(t,d) \cdot idf(t,D) \tag{3}$$

Word Embedding Methods. Machine learning-based algorithms have been used in sentiment analysis researches frequently. It uses statistical and mathematical methods to make inferences from previously or currently obtained data automatically via learning it from existing training data or grouping obtained data according to defined criteria. Occasionally, for sentiment analysis, word vectors are used with labeled data to train models via these algorithms. In this paper, we use support vector machines, random forest, and Naïve Bayes algorithms to create models trained with mentioned data sets and classify sentences as positive or negative.

3.3 Deep Learning Models

Artificial neural networks (ANNs) are an information processing technology that has the capacity to learn, memorize, and reveal the relationship between data. ANNs consist of many processing units linked together in a weighted manner, making them a prominent solution for some natural language processing problems such as sentiment analysis. It can provide additional aspects for data to investigate, such as the order of the words in a sentence, which machine learning methods lack. To give additional insight into Turkish sentiment analysis studies, ANN models such as LSTM, BiLSTM, and CNN have been utilized to classify sentences into positive, negative, or neutral.

Recurrent Neural Networks. It is a modified version of feed-forward network used for sequential data. It takes an input and change the hidden state and tries to predict the timestamp of the data. The hidden states consist of multiple layers combined with nonlinear functions to better integrate the information in different timestamps allowing it to make better predictions. Even if each unit's non linearity is very easy, iterating it over time results in quite rich dynamics [43]. The RNN's recurrent structure can be seen in 1.

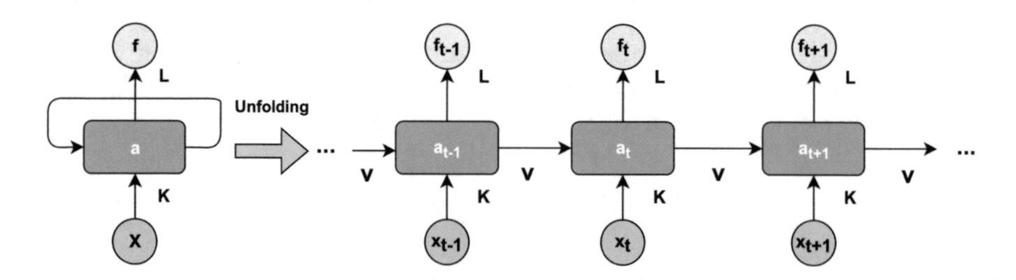

Fig. 1. RNNs recurrent structure. Bottom represents input state, middle represents hidden state, and top represents output state. The network's weights are K, L, and V. [8]

LSTM and BiLSTM Neural Networks. Some problems in the training of basic RNNs have led to the emergence of other deep learning structures. For example, the vanishing gradient problem occurs when we try to distribute a large input space into 0–1 intervals. In vanishing gradient, small changes are observed in the output when huge changes are applied in input due to distribution of input space, which results in small derivative and gradients of the loss function approaches to zero. Hence, it makes learning very difficult. To overcome RNNs' these problems, Long Short-Term Memory networks are proposed. With its additional gates (2), LSTMs carry out important information throughout the sequence as remembering them much longer than RNNs and make predictions accordingly. Also, Bidirectional Long Short-Term Memory networks consist of two LSTMs and working forward and backward, providing additional information to the network.

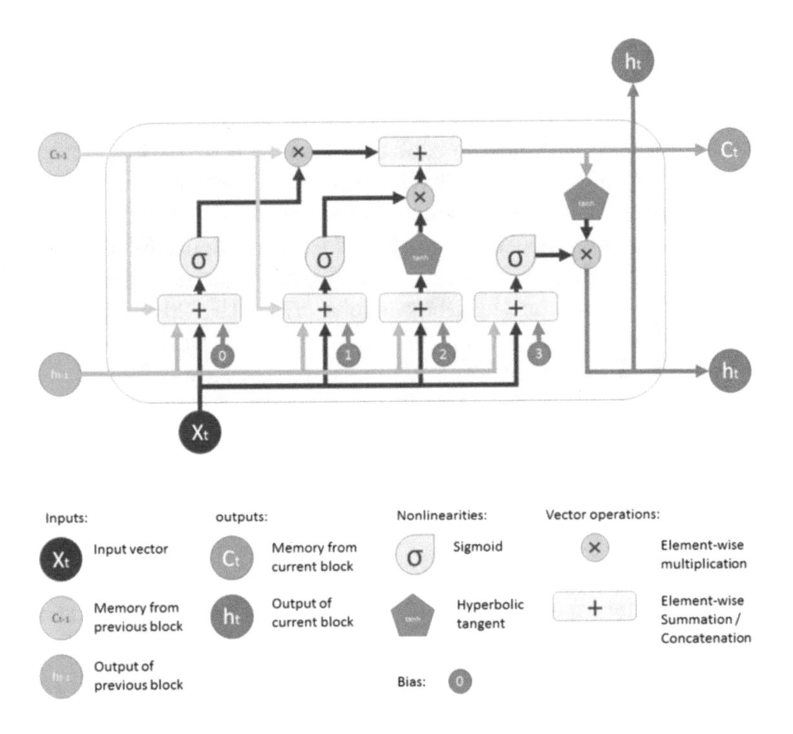

Fig. 2. The representation of a LSTM block [52].

BERT. Transformer is a deep learning algorithm developed in 2017, mostly used in the natural language processing field [47]. Like recurrent neural networks (RNNs), transformers are designed to process natural language in tasks such as translating and text classification. Unlike RNNs, it is not necessary to use transformers with an ordered data sequence. For instance, if the given sentence is a natural language term, we do not need to address its origin before terminating the transformer. Because of this feature, transformer allows more comparisons than RNNs, which reduces training time [47].

Transformer, replacing former recurrent neural network models like long short-term memory, has easily become the standard of choice for various NLP tasks [50], such as text classification and document generation. Since transformer boost parallelization amid preparing, it has empowered preparing on bigger data sets than it was presented. This has the drive to the improvement of pre-trained models such as BERT [20] and GPT-3 [12] which have been trained with large data sets and can be fine-tuned to various natural language problems.

BERT. Bidirectional Encoder Representations from Transformers (BERT) is a transformer-based language model developed [20] for downstream natural language processing tasks. What distinguishes BERT from its predecessors is its bidirectional structure, which can be fine-tuned easily. There are too many parameters in the BERT; the structure requires high computational hardware and a large amount of training data for the training process. For these reasons, the model is trained to learn language structure with large data sets. In specific tasks, fine-tuning with less data for different tasks can be performed.

The main English version of BERT comes with two pre-trained models of the following types: BERT_{BASE} model, which contains 12 heads, 12 layers, 768 hidden, 110M parameter neural network architecture, and BERT_{LARGE} model with 24 layers, 1024 hidden, 16 heads, 340M parameter neural network architecture; both are trained on the BookCorpus [57] with 800 million words and English Wikipedia corpus with 2.5 million words. The workflow of the BERT can be seen in Fig. 3. Since it is easy to fine-tune, there are other adaptations of BERT for a language other than English, such as Arabic, that [6], Persian [25], Dutch [48], and Turkish. BERTurk [37] have been used in this article for sentiment analysis in Turkish. It is a fine-tuned BERT model trained with various Turkish corpora.

ELECTRA. While BERT is more effective than conventional language models with its bidirectional structure, it still has a considerable cost of training. To eliminate this, a method called "Efficiently Learning an Encoder that Classifies Token Replacements Accurately" (ELECTRA) has been proposed [16]. Like BERT, ELECTRA can be fine-tuned for downstream NLP tasks. According to [16], ELECTRA has less training time and outperforms BERT on downstream NLP tasks.

ELECTRA model operates on masking and tokenization. Initially, a random sequence of tokens are generated and replace with [MASK] tokens. Afterward, the generator predicts the original form of these [MASK] tokens and output is discriminated in ELECTRA based on the value either it is original or replaced by the generator. Figure 4 illustrates the general workflow of ELECTRA model. The hyper parameters of basic ELECTRA and BERT_{BASE} models are same while the difference is generated in designing the large models. BERT being a popular model contains various adaptations. In the similar way, ELECTRA

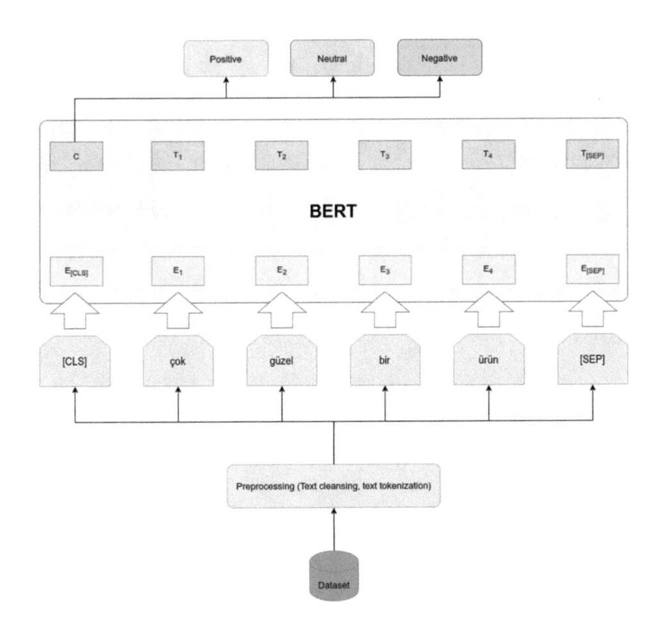

Fig. 3. BERT model for Turkish sentiment analysis task.

contains different adaptations for language like German [14] and Turkish. For sentimental analysis using ELECTRA the BERTTurk corpora is used for training the model.

4 Evaluation

For training and testing of the machine learning and deep learning models Google Colab [11] environment is used in Ubuntu 18.04 with GPU. Machine learning and deep learning libraries such as Tensorflow, scikit-learn, and python is used for the development of the model. To evaluate the ELECTRA and BERT model for sentimental analysis real movie reviews and hotel reviews dataset is used. Both of the proposed model performance is compared to the traditional sentimental classifier methods like SVM, Decision Trees, and Random Forest. Moreover, to train the transformer based models, fine tuning of pre-trained BERTurk and Turkish ELECTRA model detail is described in Sect. 3.1. The testing of these models is performed on the unseen dataset and performance metrics such as accuracy, precision, recall, and F1 scores are calculated.

5 Results and Discussion

The obtained results of proposed and base models are shown in Table 1 for 3 class customer feedback datasets in Table 2 for 2 class customer feedback datasets in Table 3 for the product comments dataset, and in Table 4 for the movie critique

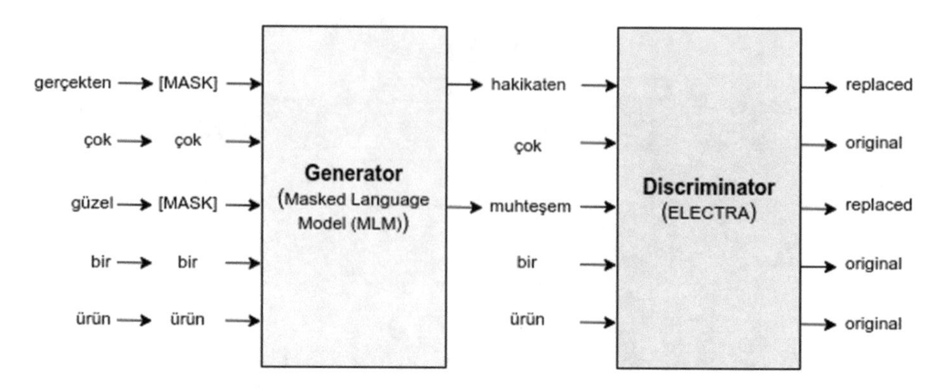

Fig. 4. ELECTRA model for Turkish sentiment analysis task.

dataset. Accuracy, precision, recall, and F1 measure scores are have been provided for each method. In general, transformer-based sentiment analysis methods like BERT and ELECTRA performed better than traditional machine and deep learning models. For the movie review dataset, ELECTRA performs best. However, other methods except decision trees do not perform poorly either. For the customer feedback dataset, when the number of words per sentence increases according to the movie review dataset, there is an increase in the performance of all methods. However, this time, BERT outperforms ELECTRA and other methods. Another remarkable issue at this point is that the performance of SVM gets closer to BERT.

In conclusion, when various sentiment analysis methods are tested with data sets that have different characteristics, such as the number of words per sentence, we observe that transformed-based methods have notable success for Turkish.

Table 1. Sentiment Analysis Results on 3 Class Customer Feedback Dataset

Algorithm	Accuracy	Precision	Recall	F1 Score
Decision Trees	54.395%	54.371%	54.588%	54.307%
Random Forest	65.343%	65.320%	64.850%	64.717%
SVM	69.030%	69.020%	69.266%	69.126%
Naïve Bayes	67.668%	67.754%	67.753%	67.461%
LSTM	61.315%	61.194%	61.465%	61.304%
BiLSTM	60.805%	60.736%	60.944%	60.829%
BERT	**74.084%**	**74.117%**	**74.113%**	**74.061%**
ConvBERT	76.344%	73.642%	73.571%	73.588%
ELECTRA	73.107%	73.307%	73.162%	73.224%

Table 2. Sentiment Analysis Results on 2 Class Customer Feedback Dataset

Algorithm	Accuracy	Precision	Recall	F1 Score
Decision Trees	79.292%	79.203%	79.692%	79.184%
Random Forest	86.312%	86.292%	86.337%	86.303%
SVM	89.734%	89.775%	89.845%	89.732%
Naïve Bayes	90.560%	90.551%	90.566%	90.556%
LSTM	86.497%	86.518%	86.520%	86.497%
BiLSTM	88.738%	88.770%	88.714%	88.728%
BERT	93.550%	93.555%	93.564%	93.550%
ConvBERT	**93.580%**	**93.581%**	**93.580%**	**93.580%**
ELECTRA	92.497%	92.503%	92.501%	92.497%

Table 3. Sentiment Analysis Results on Product Comments Dataset

Algorithm	Accuracy	Precision	Recall	F1 Score
Decision Trees	74.736%	74.711%	74.777%	74.711%
Random Forest	88.421%	88.430%	88.430%	88.421%
SVM	91.578%	91.578%	91.578%	91.578%
Naïve Bayes	92.631%	92.553%	93.636%	92.579%
LSTM	84.210%	84.358%	84.175%	84.182%
BiLSTM	90.526%	90.666%	90.558%	90.522%
BERT	95.180%	95.468%	94.769%	95.059%
ConvBERT	**96.938%**	**95.238%**	**96.530%**	**95.861%**
ELECTRA	94.936%	94.444%	95.744%	94.836%

Table 4. Sentiment Analysis Results on Movie Critique Dataset

Algorithm	Accuracy	Precision	Recall	F1 Score
Decision Trees	78.949%	78.928%	78.964%	78.935%
Random Forest	86.169%	86.155%	86.181%	86.162%
SVM	89.123%	89.136%	89.129%	89.123%
Naïve Bayes	89.123%	89.133%	89.125%	89.122%
LSTM	84%	84.031%	84.027%	83.999%
BiLSTM	82.875%	82.870%	82.878%	82.872%
BERT	92.604%	92.607%	92.605%	92.604%
ConvBERT	92.638%	92.686%	92.640%	92.636%
ELECTRA	**92.651%**	**92.665%**	**92.650%**	**92.651%**

6 Conclusion

This paper proposed a comprehensive benchmark of Transformer-based BERT and ELECTRA models in Turkish with various machine learning and deep learning approaches, using four different data sets, including movie reviews, customer reviews, and tweets. Up to our knowledge, this is the first study that uses ELECTRA and ConvBERT for sentiment analysis in Turkish.

Additionally, we point out the untapped potential of transformer based models in the Turkish sentiment analysis. For future work, other transformer-based models and different variations should be considered, such as RoBERTa [30].

Acknowledgements. We thank Stefan Schweter for providing fine-tuned Turkish BERT model for the community.

References

1. Internet live stats - internet usage; social media statistics. https://www.internetlivestats.com/. Accessed 6 Oct 2022
2. Wikistats - statistics for wikimedia projects. https://stats.wikimedia.org/. Accessed 6 Oct 2022
3. Abercrombie, G., Batista-Navarro, R.: Parlvote: a corpus for sentiment analysis of political debates. In: LREC (2020)
4. Al-Smadi, M., Talafha, B., Al-Ayyoub, M., Jararweh, Y.: Using long short-term memory deep neural networks for aspect-based sentiment analysis of Arabic reviews. Int. J. Mach. Learn. Cybern. **10**(8), 2163–2175 (2018). https://doi.org/10.1007/s13042-018-0799-4
5. Alamoodi, A., et al.: Sentiment analysis and its applications in fighting covid-19 and infectious diseases: a systematic review. Expert Syst. Appl. 114155 (2020). https://doi.org/10.1016/j.eswa.2020.114155, http://www.sciencedirect.com/science/article/pii/S0957417420308988
6. Antoun, W., Baly, F., Hajj, H.: Arabert: transformer-based model for Arabic language understanding. ArXiv abs/2003.00104 (2020)
7. Appel, O., Chiclana, F., Carter, J., Fujita, H.: A hybrid approach to the sentiment analysis problem at the sentence level. Knowl.-Based Syst. **108**, 110–124 (2016). https://doi.org/10.1016/j.knosys.2016.05.040, http://www.sciencedirect.com/science/article/pii/S095070511630137X. New Avenues in Knowledge Bases for Natural Language Processing
8. Arisoy, E., Sethy, A., Ramabhadran, B., Chen, S.: Bidirectional recurrent neural network language models for automatic speech recognition. In: 2015 IEEE International Conference on Acoustics, Speech and Signal Processing (ICASSP), pp. 5421–5425 (2015). https://doi.org/10.1109/ICASSP.2015.7179007
9. Ayata, D., Saraçlar, M., Özgür, A.: Turkish tweet sentiment analysis with word embedding and machine learning. In: 2017 25th Signal Processing and Communications Applications Conference (SIU), pp. 1–4 (2017)
10. Bilen, B., Horasan, F.: Lstm network based sentiment analysis for customer reviews. Politeknik Dergisi, 1–1 (2021). https://doi.org/10.2339/politeknik.844019
11. Bisong, E.: Google Colaboratory, pp. 59–64. Apress, Berkeley (2019). https://doi.org/10.1007/978-1-4842-4470-8_7

12. Brown, T., et al.: Language models are few-shot learners. ArXiv abs/2005.14165 (2020)
13. Cetin, M., Amasyali, M.F.: Active learning for Turkish sentiment analysis. In: 2013 IEEE INISTA, pp. 1–4 (2013)
14. Chan, B., Schweter, S., Möller, T.: German's next language model. ArXiv abs/2010.10906 (2020)
15. Ciftci, B., Apaydin, M.: A deep learning approach to sentiment analysis in Turkish. In: 2018 International Conference on Artificial Intelligence and Data Processing (IDAP), pp. 1–5 (2018)
16. Clark, K., Luong, M.T., Le, Q.V., Manning, C.D.: Electra: pre-training text encoders as discriminators rather than generators. ArXiv abs/2003.10555 (2020)
17. Dehkharghani, R., Saygin, Y., Yanikoglu, B.A., Oflazer, K.: Sentiturknet: a Turkish polarity lexicon for sentiment analysis. Lang. Resour. Eval. **50**, 667–685 (2016)
18. Demirci, G.M., Keskin, S., Doğan, G.: Sentiment analysis in Turkish with deep learning. In: 2019 IEEE International Conference on Big Data (Big Data), pp. 2215–2221 (2019)
19. Demirtas, E., Pechenizkiy, M.: Cross-lingual polarity detection with machine translation. In: Proceedings of the 2nd International Workshop on Issues of Sentiment Discovery and Opinion Mining, WISDOM'13. Association for Computing Machinery, New York, NY, USA (2013). https://doi.org/10.1145/2502069.2502078
20. Devlin, J., Chang, M.W., Lee, K., Toutanova, K.: Bert: pre-training of deep bidirectional transformers for language understanding (2019)
21. Dreisbach, C., Koleck, T.A., Bourne, P.E., Bakken, S.: A systematic review of natural language processing and text mining of symptoms from electronic patient-authored text data. Int. J. Med. Inform. **125**, 37–46 (2019). https://doi.org/10.1016/j.ijmedinf.2019.02.008, http://www.sciencedirect.com/science/article/pii/S1386505618313789
22. Erşahin, B., Aktas, Ö., Kilinç, D., Ersahin, M.: A hybrid sentiment analysis method for Turkish. Turk. J. Electr. Eng. Comput. Sci. **27**, 1780–1793 (2019)
23. Çetin, F.S., Eryigit, G.: Türkçe hedef tabanlı duygu analizi İçin alt görevlerin İncelenmesi - hedef terim, hedef kategori ve duygu sınıfı belirleme (2018)
24. Çetin, F.S., Yildirim, E., Özbey, C., Eryigit, G.: Tgb at semeval-2016 task 5: Multi-lingual constraint system for aspect based sentiment analysis. In: SemEval@NAACL-HLT (2016)
25. Farahani, M., Gharachorloo, M., Farahani, M., Manthouri, M.: Parsbert: transformer-based model for Persian language understanding. ArXiv abs/2005.12515 (2020)
26. Greco, F., Polli, A.: Emotional text mining: customer profiling in brand management. Int. J. Inform. Manag. **51**, 101934 (2020). https://doi.org/10.1016/j.ijinfomgt.2019.04.007, http://www.sciencedirect.com/science/article/pii/S0268401218313598
27. Karamollaoğlu, H., Dogru, I., Dörterler, M., Utku, A., Yildiz, O.: Sentiment analysis on Turkish social media shares through lexicon based approach. In: 2018 3rd International Conference on Computer Science and Engineering (UBMK), pp. 45–49 (2018)
28. Kaya, M., Fidan, G., Toroslu, I.H.: Sentiment analysis of Turkish political news. In: 2012 IEEE/WIC/ACM International Conferences on Web Intelligence and Intelligent Agent Technology, vol. 1, pp. 174–180 (2012)
29. Kırelli, Y., Arslankaya, S.: Sentiment analysis of shared tweets on global warming on twitter with data mining methods: a case study on Turkish language. Comput. Intell. Neurosci. **2020** (2020)

30. Liu, Y., et al.: Roberta: a robustly optimized Bert pretraining approach. ArXiv abs/1907.11692 (2019)
31. Maryame, N., Najima, D., Hasnae, R., Rachida, A.: State of the art of deep learning applications in sentiment analysis: psychological behavior prediction. In: Bhateja, V., Satapathy, S.C., Satori, H. (eds.) Embedded Systems and Artificial Intelligence, pp. 441–451. Springer Singapore, Singapore (2020)
32. Mulki, H., Haddad, H., Ali, C.B., Babaoglu, I.: Preprocessing impact on turkish sentiment analysis. In: 2018 26th Signal Processing and Communications Applications Conference (SIU), pp. 1–4 (2018)
33. Onan, A.: Sentiment analysis in Turkish based on weighted word embeddings. In: 2020 28th Signal Processing and Communications Applications Conference (SIU), pp. 1–4 (2020)
34. Pejic-Bach, M., Bertoncel, T., Meško, M., Živko Krstić: Text mining of industry 4.0 job advertisements. Int. J. Inform. Manag. **50**, 416–431 (2020). https://doi.org/10.1016/j.ijinfomgt.2019.07.014, http://www.sciencedirect.com/science/article/pii/S0268401218313677
35. Pontiki, M., Galanis, D., Pavlopoulos, J., Papageorgiou, H., Androutsopoulos, I., Manandhar, S.: Semeval-2014 task 4: aspect based sentiment analysis. In: COLING 2014 (2014)
36. Rumelli, M., Akkuş, D., Kart, Ö., Isik, Z.: Sentiment analysis in turkish text with machine learning algorithms. In: 2019 Innovations in Intelligent Systems and Applications Conference (ASYU), pp. 1–5 (2019)
37. Schweter, S.: Berturk - bert models for turkish (2020). https://doi.org/10.5281/zenodo.3770924
38. Seyfioglu, M.S., Demirezen, M.: A hierarchical approach for sentiment analysis and categorization of turkish written customer relationship management data. In: 2017 Federated Conference on Computer Science and Information Systems (FedCSIS), pp. 361–365 (2017)
39. Sezgen, E., Mason, K.J., Mayer, R.: Voice of airline passenger: a text mining approach to understand customer satisfaction. J. Air Transp. Manag. **77**, 65–74 (2019). https://doi.org/10.1016/j.jairtraman.2019.04.001, http://www.sciencedirect.com/science/article/pii/S0969699718304873
40. Shehu, H., Sharif, M.H., Uyaver, S., Tokat, S., Ramadan, R.: Sentiment analysis of Turkish twitter data using polarity lexicon and artificial intelligence (2020)
41. Shehu, H.A., Tokat, S.: A hybrid approach for the sentiment analysis of Turkish twitter data. In: Hemanth, D.J., Kose, U. (eds.) Artificial Intelligence and Applied Mathematics in Engineering Problems, pp. 182–190. Springer International Publishing, Cham (2020)
42. Sigirci, I.O., et al.: Sentiment analysis of Turkish reviews on google play store. In: 2020 5th International Conference on Computer Science and Engineering (UBMK), pp. 314–315 (2020)
43. Sutskever, I., Martens, J., Hinton, G.E.: Generating text with recurrent neural networks. In: L. Getoor, T. Scheffer (eds.) Proceedings of the 28th International Conference on Machine Learning, ICML 2011, Bellevue, Washington, USA, 28 June - 2 July (2011), pp. 1017–1024 (2011). https://icml.cc/2011/papers/524_icmlpaper.pdf. Accessed 6 Oct 2022
44. Taboada, M., Brooke, J., Tofiloski, M., Voll, K.D., Stede, M.: Lexicon-based methods for sentiment analysis. Comput. Linguist. **37**, 267–307 (2011)
45. Toh, Z., Su, J.: Nlangp at semeval-2016 task 5: Improving aspect based sentiment analysis using neural network features. In: SemEval@NAACL-HLT (2016)

46. Uslu, A., Tekin, S., Aytekin, T.: Sentiment analysis in Turkish film comments. In: 2019 27th Signal Processing and Communications Applications Conference (SIU), pp. 1–4 (2019)
47. Vaswani, A., et al.: Attention is all you need (2017)
48. de Vries, W., van Cranenburgh, A., Bisazza, A., Caselli, T., van Noord, G., Nissim, M.: Bertje: a dutch bert model. ArXiv abs/1912.09582 (2019)
49. Vural, A., Cambazoglu, B.B., Senkul, P., Tokgoz, Z.O.: A framework for sentiment analysis in Turkish: application to polarity detection of movie reviews in Turkish. In: ISCIS (2012)
50. Wolf, T., et al.: Transformers: State-of-the-art natural language processing. In: Proceedings of the 2020 Conference on Empirical Methods in Natural Language Processing: System Demonstrations, pp. 38–45. Association for Computational Linguistics (2020). https://doi.org/10.18653/v1/2020.emnlp-demos.6
51. Xing, F.Z., Malandri, L., Zhang, Y., Cambria, E.: Financial sentiment analysis: an investigation into common mistakes and silver bullets. In: COLING (2020)
52. Yan, S.: Understanding LSTM and its diagrams (2017). https://blog.mlreview.com/understanding-lstm-and-its-diagrams-37e2f46f1714. Accessed 6 Oct 2022
53. Yang, P., Chen, Y.: A survey on sentiment analysis by using machine learning methods. In: 2017 IEEE 2nd Information Technology, Networking, Electronic and Automation Control Conference (ITNEC), pp. 117–121 (2017)
54. Yildirim, S.: Comparing deep neural networks to traditional models for sentiment analysis in Turkish language (2020)
55. Yılmaz, B.: Product comments dataset (2020). https://www.kaggle.com/baharyilmaz/product-comments-dataset Accessed 6 Oct 2022
56. Yurtalan, G., Koyuncu, M.: Çigdem Turhan: a polarity calculation approach for lexicon-based Turkish sentiment analysis. Turk. J. Electr. Eng. Comput. Sci. **27**, 1325–1339 (2019)
57. Zhu, Y., et al.: Aligning books and movies: towards story-like visual explanations by watching movies and reading books (2015)

Mathematical Modeling of an Antenna Device Based on a T-Shaped Waveguide of the Microwave Range

Elmar Z. Hunbataliyev(✉)

Department of Radioengineering and Telecommunication, Azerbaijan Technical University, H. Javid ave. 25, AZ 1073 Baku, Azerbaijan
elmarzulfugar59@gmail.com

Abstract. In this work, mathematical modeling of an antenna device based on a T-shaped waveguide of the microwave range is carried out. The dependence of the coupling coefficient and the level of the branched power (points) on the height and length of the inhomogeneity, as well as the distribution of the heights of the inhomogeneities, the amplitude distribution of the antenna device with direct and reverse excitation, the calculated radiation patterns of the antenna device with direct and reverse excitation, the radiation patterns of the antenna model with equidistant the location of inhomogeneities, the amplitude-phase distribution of the electromagnetic field with an equidistant location of inhomogeneities. The structure of the distribution of inhomogeneities in the waveguide obtained as a result of the work carried out makes it possible to implement antennas with two radiation patterns, which are formed depending on the direction of excitation. The properties of the obtained radiation patterns allow us to consider this antenna as one of the main elements in the development of a two-channel phased antenna array of radar stations with an increased rate of coverage. A technique for mathematical modeling of an antenna device based on a T-shaped waveguide in the microwave range has been developed. On this basis, the possibilities of optimizing the characteristics of a two-channel phased antenna array have been studied. Problems of parametric synthesis are solved and technical solutions are found with optimal electrodynamic and design-technological characteristics for a two-channel phased antenna array for radar stations.

Keywords: Antenna · T-shaped · Waveguide · Microwave Devices · Radiation Pattern

1 Introduction

Antenna arrays based on a T-waveguide can be used in radio communications and radar, both as an independent antenna and as an element of a flat or conformal antenna array.

From the prior art linear antenna arrays [1], manufactured by the method of printing technology, providing high repeatability of electrical characteristics at a relatively low manufacturing cost. Due to design features, as a rule, vibrator-type radiators are used as radiating elements in such antennas.

D. J. Hemanth et al. (Eds.): ICAIAME 2022, ECPSCI 7, pp. 16–29, 2023.
https://doi.org/10.1007/978-3-031-31956-3_2

Also known are linear arrays of slot emitters, which are a set of slots cut in the screen, excited by a symmetrical or asymmetrical strip line by crossing the slot with a strip line [2, 3]. Also known are linear antenna arrays of collinear slot radiators [4, 5], cut in the wall of the waveguide, which are powered by a waveguide wave excited in the waveguide using a system of pins, closed [4] or not closed [5] to one of the walls of the waveguide, connected to the output lines of a strip or microstrip power divider. The closest claimed technical essence, i.e. the prototype is a linear antenna array [4]. Similar to the proposed technical solution, this antenna, adopted as a prototype, contains a power distributor on a symmetrical strip line, collinear slot radiators and conductive side walls, closing the screens of a symmetrical strip line between themselves to localize waveguide waves that excite the slots. However, the prototype antenna contains paired radiating slots located on opposite sides of the side surfaces of the waveguide, designed to form isotropic radiation patterns (RP) in the cross-sectional plane of the waveguide and cannot serve as an element of a flat or conformal array.

The technical problem lies in the possibility of creating a linear antenna array, which is an integrated structure consisting of a power division system based on a strip line and a system of emitters, while the polarization plane of such an antenna is perpendicular to the RP plane. The technical result consists in solving this technical problem. The specified technical result is provided in a strip linear antenna array containing collinear slot radiators and a power divider on a symmetrical strip line, the output strip conductors of which are closed by conductive jumpers to one of the screens of the strip line, and two conductive walls closing the screens of the strip line between themselves, forming narrow walls of a rectangular waveguide, the wide walls of which are formed by strip line screens, characterized in that the slot radiators are cut in the first conductive wall located near the rectilinear edge of the strip line screens, parallel to the conductive walls and the axial line of the slot radiators, and the output strip conductors pass through discontinuities in the second conductive wall, contain T-shaped strip branching in each radiator and close to one of the screens of the strip line inside a rectangular waveguide near the edges of the slot radiator. An additional feature is that the conductive wall containing the slot emitters is covered with a dielectric on the outer side of the waveguide. An additional feature is that the grating can be made using the technology of multilayer printed circuit boards in the form of metallized layers on the surface of two dielectric sheets stacked together, and the conductive walls forming the narrow walls of a rectangular waveguide and the shorting of the strip conductors to the screen are made in the form of metallized holes in said dielectric sheets, wherein the slot emitters are formed by gaps between said metallized holes. An additional feature is that the distance between the narrow walls of the rectangular waveguide is less than half the wavelength, taking into account the dielectric constant of the medium between the screens.

On Fig. 1 shows a linear antenna array based on a T-shaped waveguide. The proposed antenna contains T-shaped strip elements with shorts, where one T-shaped element is used to excite two adjacent radiators. Such a constructive solution in the antenna leads to in-phase and equal-amplitude excitation of a pair of adjacent radiators, which, when forming radiation patterns of a special shape, requiring the use of an unequal-amplitude and out-of-phase distribution of the exciting field along the antenna opening, leads to the appearance of spurious switching lobes in the RP. The proposed antenna is free from this disadvantage.

The main requirements that will be imposed on such a device include the maximum possible separation angle of the maxima of the RP of the main and auxiliary "mirror" beam from the normal (at least $20° \pm 1°$) while maintaining a low level of side lobes below minus 30 dB (by power).

According to [6–10], for the deviation of the RP maximum by the angle θ, the distance between the inhomogeneities must satisfy the condition

$$\sin\theta = \frac{\lambda}{\lambda_g} - \frac{\lambda}{2d}, \tag{1}$$

where λ is the wavelength, λ_g is the wavelength in the waveguide, and d is the spacing of inhomogeneities (Fig. 1). However, as the deviation angle θ of the pattern increases, the value of d increases until the linear lattice no longer satisfies requirement $\frac{d}{\lambda} \leqslant \frac{1}{1+\sin\theta}$ [11], which leads to the appearance of side maxima of the pattern.

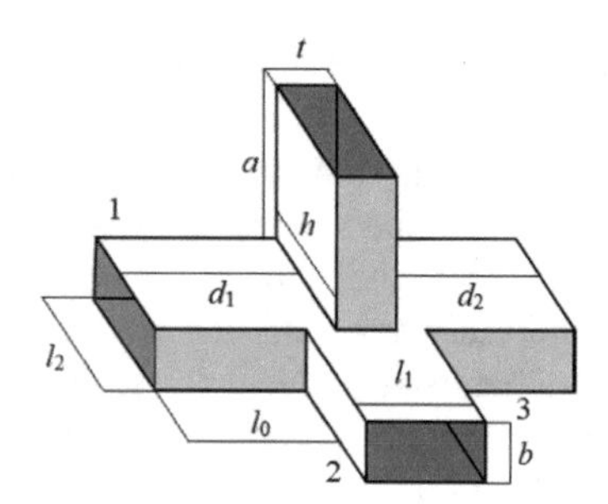

Fig. 1. Location of inhomogeneities at the bottom of the antenna: 1 – profile of a two-channel *T*-waveguide, 2 – first exciter, 3 – second exciter.

At the same time, for negative values of θ, the value of d decreases, and at the length of inhomogeneities $l = \lambda_g/2$ [3, 4], their mutual overlap occurs, violating the asymmetry condition, which reduces the level of radiated power. The length of the radiating element in this case is selected from the consideration of matching its wave impedance with the main line [12–14].

2 Obtaining RP with the Required Parameters

Thus, to solve the problem, namely, to obtain a RP with the required parameters, it is necessary to consider the case when the length of the inhomogeneities is defined as $l = \lambda_g/p$, provided that the shortening factor $p \geq 2$.

It is known that the amplitude of the radiated wave of a single inhomogeneity placed in the profile of the *T*-waveguide (Fig. 1) is determined by the function of the coupling coefficient [15]

$$\alpha(h, l) = 1 - \exp\left(-2k(h)\frac{l}{\lambda}\right), \tag{2}$$

where $k(h)$ is the per unit attenuation coefficient of a wave passing over an inhomogeneity of height h and length l:

$$k(h) = \pi^2 \frac{q^2}{\sqrt{1-q}} \frac{1 - \delta_t \frac{q}{\sqrt{1-q^2}}}{(1+\delta_t)^4} \frac{\left(\frac{2hq}{\lambda}\right)^2}{\left(1 - \frac{2hq}{\lambda}\right)\left(1 - 2\frac{2hq}{\lambda}\right)}, \tag{3}$$

where $q = \frac{\lambda}{\lambda_c}$, $\lambda_c = 4a + 8\frac{b}{\pi}\ln 2$,

$$\delta_t = \frac{4q}{\pi}\frac{b}{\lambda}\left[2\ln\left(\frac{2 - \frac{t}{2b}}{2\left(1 - \frac{t}{2b}\right)}\right) - \frac{t}{2b}\ln\left(\frac{\frac{t}{2b}\left(2 - \frac{t}{2b}\right)}{\left(1 - \frac{t}{2b}\right)^2}\right)\right]. \tag{4}$$

The wavelength in the waveguide $\lambda_g(h) = \lambda/\gamma(h)$ is determined through the wave deceleration coefficient

$$\gamma(h) = \sqrt{1-q^2}\left(1 + \frac{\delta_t q^2}{1-q^2}\right) - \frac{q^2}{\sqrt{1-q^2}} \frac{\frac{2hq}{\lambda}}{1 - \frac{2hq}{\lambda}} \frac{1 - \frac{\delta_t q^2}{1-q^2}}{(1+\delta_t)^2}. \tag{5}$$

Since the application of this formula is confirmed only in the case of inhomogeneity length $l = \lambda_g/p$, where $p = 2$, verification of its applicability area for cases $p > 2$ is required.

To do this, using the HFSS software package, we built a model (Fig. 2) of an T-waveguide according to the design described in [16–18], operating at a frequency $f_0 =$ 5 GHz with dimensions $a = 40$ mm, $b = 20$ mm, $t = 2$ mm, and the analysis of the radiating properties of the inhomogeneity located in it was carried out.

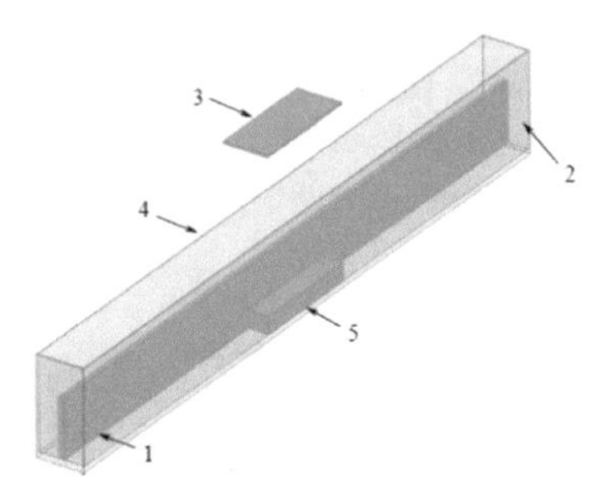

Fig. 2. Model of a T-waveguide with a single inhomogeneity.

The model includes an excitation port - 1, a load port - 2, a receiving port - 3 with a length of $\lambda/2$, located at a distance λ from the radiating edge of the T-waveguide profile - 4 with an inhomogeneity - 5. All elements of the radiating T-waveguide are made of aluminum, the rest of the space is filled with air. During the simulation, the height h and the length of the inhomogeneity $l(h) = \lambda_g(h)/p$ vary.

Figure 3 shows the obtained dependence of the branched power level and the values of the coupling coefficients calculated by formula (2) on the height of the radiator inhomogeneity.

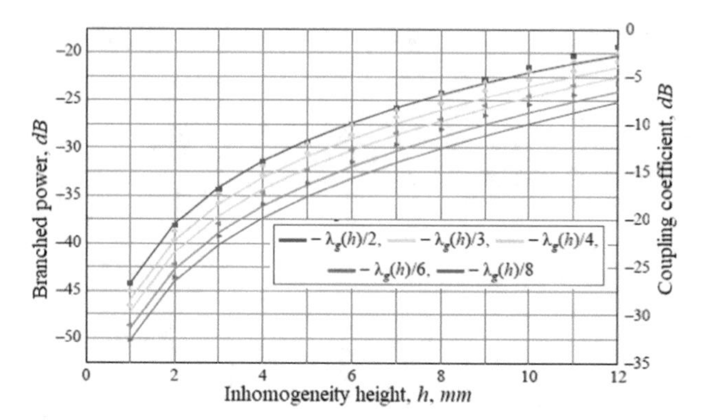

Fig. 3. Dependence of the coupling coefficient (solid line) α and the branched power level (points) on the height and length of the inhomogeneity.

Since the quantities under consideration are directly dependent on each other, we will take as the coefficient of their proportionality the value of the branched power of a known inhomogeneity of length λ_g (1 mm)/2 and compare them on one graph.

According to the obtained results, the application of formula (2) is permissible only in the case of $p = 2$. To calculate the coupling coefficients for $p \geq 2$, we introduce empirically selected correction factors into (2):

$$\alpha'(h, l) = \left(1 - \frac{p-2}{10}\right)\alpha\left(h\left(1 + \frac{p-2}{10}\right), l\right), \tag{6}$$

after which the dependence takes the form as in Fig. 4.

With this in mind, we will calculate and simulate a two-channel linear antenna with an amplitude distribution of $I_n = 0,08 + 0,02 \sin \frac{\pi n}{N}$ for the level of side lobes of minus 40 dB and the deviation angle of the main lobe equal to minus 20°.

According to (1), in order to deviate the RP by an angle $\theta = -20°$, the distance between the elements should be

$$d = \frac{\lambda}{2} \frac{1}{\frac{\lambda}{\lambda_g(0) - \sin\theta}} = 33,7\,\text{mm}.$$

With this value of d, 75 inhomogeneities can be placed on a profile with a length of 34 λ. In this case, in order to fulfill the condition of the absence of mutual overlap, their length should be $l = \lambda_g(0)/4 = 24{,}3$. At this stage, the calculation takes the wavelength in an empty waveguide, because h_n values are not yet known.

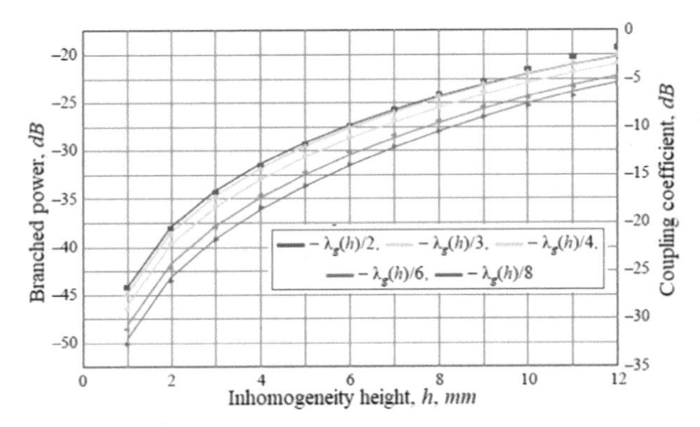

Fig. 4. Dependence of the coupling coefficient (solid line) α' and the branched power level (points) on the height and length of the inhomogeneity.

To calculate the heights of inhomogeneities we use the formulas from [3]:

$$h_n = \frac{\lambda_c}{4} \frac{\sqrt{k_n^2 + 4\pi^2 \frac{q^2}{\sqrt{1-q^2}} \frac{1-\delta_t \frac{q^2}{\sqrt{1-q^2}}}{1+\delta_t^4} k_n - 3k_n}}{\pi^2 \frac{q^2}{\sqrt{1-q^2}} \frac{1-\delta_t \frac{q^2}{\sqrt{1-q^2}}}{1+\delta_t^4} - 2k_n}, \tag{7}$$

where

$k_n = -\frac{\ln(1-\alpha_n')}{2\frac{l}{\lambda}}$, $\alpha_n' = \frac{I_n^2}{P_{in}-P_n}$, $P_{in} = \frac{1}{\eta}\sum_{i=0}^{N} I_i^2$, $P_n = \sum_{i=0}^{n-1} I_i^2$, $\eta = 95\%$.

In the obtained discontinuity height distribution $h_n \leq 8$ mm (Fig. 5), while the maximum error α' does not exceed 0,2 dB (Fig. 4).

3 Determination of the Amplitude Distribution of the Antenna

Knowing the height distribution, it is possible to restore the amplitude distribution, including when excitation of the emitter from the opposite side (Fig. 6):

$$I_{P12n} = \sqrt{\alpha'^2(h_n, l_n)\prod_{i=0}^{n-1}\left(1-\alpha'^2(h_i, l_i)\right)}, \tag{8}$$

$$I_{P12n} = \sqrt{\alpha'^2(h_{N-n}, l_{N-n})\prod_{i=0}^{n-1}\left(1-\alpha'^2(h_{iN-}, l_{N-i})\right)}. \tag{9}$$

From Fig. 6 shows that excitation from the opposite side distorts the amplitude distribution, reducing the contribution of the latter emitters. At the same time, the radiation pattern predicted in this case (Fig. 7) is characterized by an acceptable side-lobe level and width, which can be used in radar stations.

Using the obtained values of h_n, d, l, a model of a two-channel linear radiator was constructed (Fig. 7), the excitation of which is carried out through the end by port P1, the wall of the opposite end acts as a load specified by port P2. To obtain a second radiation pattern, it is necessary to change the direction of radiation of the ports. To determine the parameters of the amplitude and phase of radiation in the near zone, the radiation boundary is set in the form of a line placed above the radiating part of the T-waveguide at a height λ.

On Fig. 8 shows the pattern of the model in comparison with the calculated data. As can be seen, the radiation pattern is characterized by a high level of side lobes, and the position of the main forward and reverse lobes is shifted from the expected by 5°, which is caused by phase distortion (Fig. 9), which increases as the wave propagates through the inhomogeneities. In this case, the obtained amplitude distribution agrees quite accurately with the calculated data, despite the accepted simplification in calculating l.

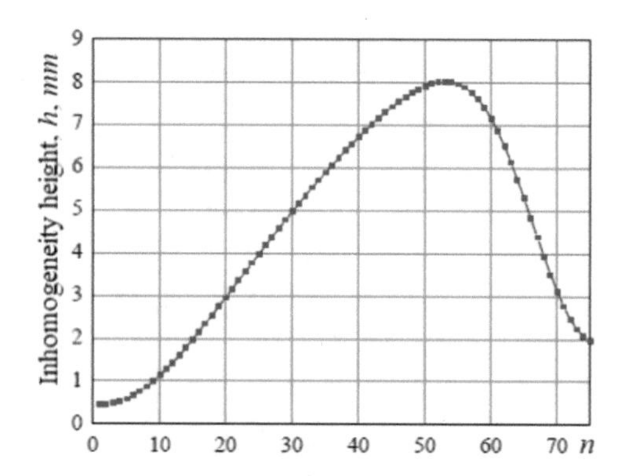

Fig. 5. Distribution of inhomogeneity heights h_n.

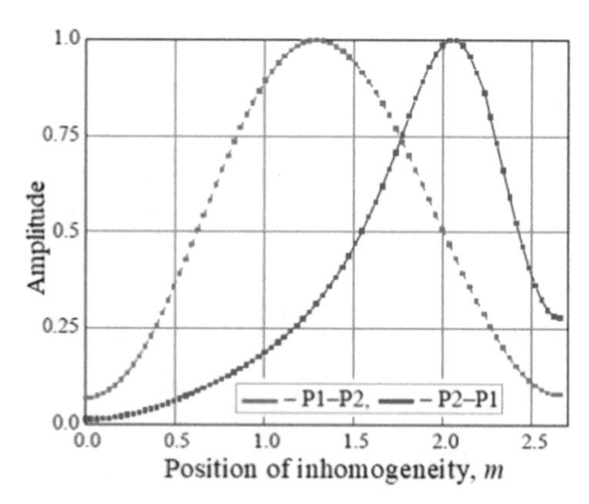

Fig. 6. Amplitude distribution of the linear emitter with forward (P1–P2) and reverse (P2–P1) excited.

The issue of compensating for phase distortions in the aperture of such a linear radiator was raised in [3]. Based on the consideration that at each point where a traveling

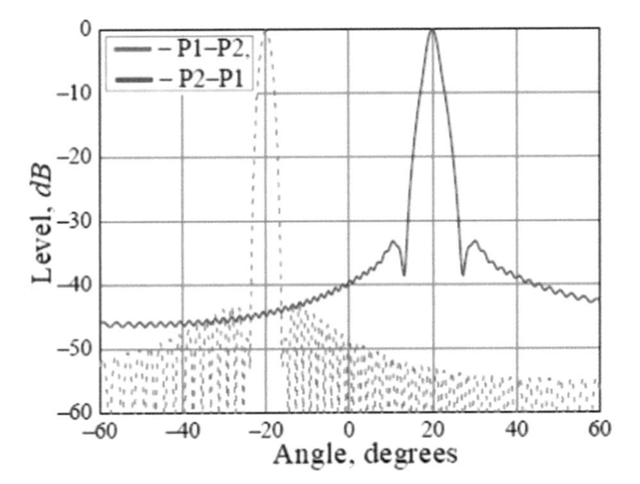

Fig. 7. Calculated RP of a linear radiator with a straight torque (P1–P2) and reverse (P2–P1) excitation.

waves branches off along the waveguide, there are incident, transmitted, and branched waves that satisfy the condition for complex amplitudes $E = E_1 + E_2$, the phase distortion introduced by each inhomogeneity is represented as a combination of two types of distortions $\delta F_n = \delta F_n^{rad} + \sum_{i=0}^{n-1} \delta F_i^{tr}$, the terms of which can be represent by linear functions of the height of the inhomogeneity h:

$$\delta F^{rad}(h) = vh; \delta F^{tr}(h) = uh. \tag{10}$$

Thus, at a fixed length of inhomogeneities $l = \lambda_g\ (h)/2$, phase distortions can be compensated for by their non-equidistant placement, which is determined by ratio:

$$d_n = \frac{\lambda}{2} \frac{\frac{\lambda_g(h_{n-1}) + \lambda_g(h_n)}{2\lambda_g(0)} - \frac{u\sum_{i=0}^{n-1} h_i + v(h_n - h_{n-1})}{\pi}}{\gamma(0) - \sin\theta_0}, \tag{11}$$

where $\gamma(0)$ is the deceleration coefficient of an empty T-waveguide, u, v are the coefficients calculated on the basis of the experimental phase distribution [3].

Let us determine the value dn for the case $l = \lambda_g(h)/p$ for $p \geq 2$. To do this, we represent the phase of the radiated wave of the zero and first inhomogeneities by the following expressions:

$$F_0^{rad} = \frac{\pi}{p} + \delta F_1^{rad}, \tag{12}$$

$$F_1^{rad} = \frac{2\pi}{p} + \delta F_0^{tr} + \left(\frac{\pi}{p} + \delta F_1^{rad} - \pi\right) + \frac{2\pi}{\lambda_g(0)}\left(d_1 - \frac{l_1 + l_0}{2}\right), \tag{13}$$

where π/p and $2\pi/p$ are the phase incursion to the center and over the entire length of the zero inhomogeneity; F_0^{tr} – distortion of the phase of the wave passing through

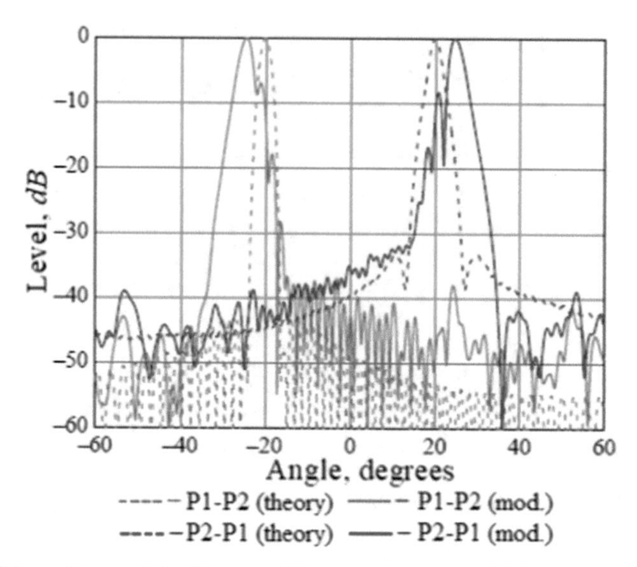

Fig. 8. RP of the emitter model with an equidistant arrangement of inhomogeneities: solid line - the result of modeling, dotted line – calculated.

the zero inhomogeneity; F_0^{rad}, F_1^{rad} – distortion of the phase of the wave emitted by the corresponding inhomogeneity; π – taking into account the phase reversal on the opposite side of the T-waveguide crest [4]; $\frac{2\pi}{\lambda_g(0)}\left(d_1 - \frac{l_1+l_0}{2}\right)$ – phase incursion in the empty section of the waveguide.

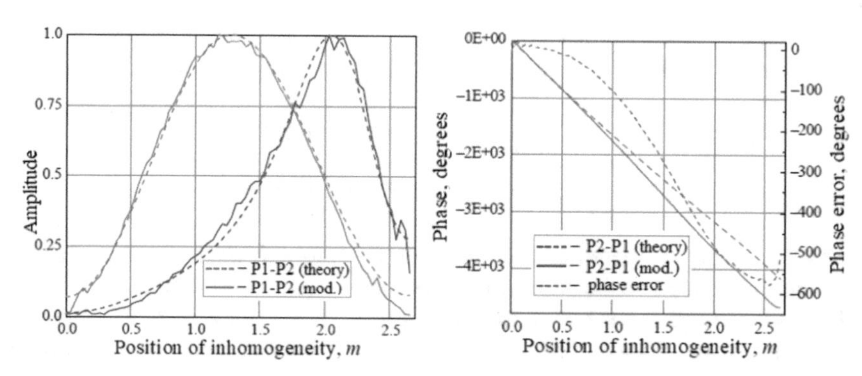

Fig. 9. Amplitude-phase distribution of the field with an equidistant arrangement of inhomogeneities.

To deflect the beam through the angle θ, condition $F_1^{rad} - \delta F_0^{rad} = d_1 \sin\theta$, must be satisfied, from which we obtain:

$$d_1 = \frac{\lambda}{2} \frac{\frac{\lambda_g(h_1)+\lambda_g(h_0)}{p\lambda_g(0)} - \frac{\delta F_1^{rad} - \delta F_0^{rad} + \delta F_0^{tr}}{\pi} - \left(\frac{2}{p} - 1\right)}{\gamma(0) - \sin\theta_0}, \quad (14)$$

taking into account (6), the interelement distance between inhomogeneities can be written as:

$$d_n = \frac{\lambda}{2} \frac{\frac{\lambda_g(h_{n-1})+\lambda_g(h_n)}{p\lambda_g(0)} - \frac{u\sum_{i=0}^{n-1} h_i + v(h_n - h_{n-1})}{\pi} - \left(\frac{2}{p} - 1\right)}{\gamma(0) - \sin\theta}. \quad (15)$$

Thus, the resulting expression allows us to calculate the distribution d_n to compensate for the phase distortion introduced by inhomogeneities of any length, defined as $l_n = \lambda_g(h)/p$. In particular, for $p = 2$ we obtain expression (11).

To compensate for phase distortions in functions (10), it is necessary to determine the coefficients u and v, for which you can use the expressions [3]:

$$u = -\frac{\sum_{n=1}^{N} [(c_2(h_n - h_0) - \delta F_n)(H_n + c_1(h_n - h_0))]}{\sum_{n=1}^{N} [H_n + c_1(h_n - h_0)]^2}, \quad (16)$$

$$v = -\frac{N(u\overline{H} - \overline{F})}{H_N - Nh_0}, \quad (17)$$

where $c_1 = -\frac{-N\overline{H}}{H_N - Nh_0}$, $c_2 = -\frac{-N\overline{F}}{H_N - Nh_0}$, $H_n = \sum_{i=0}^{n-1} h_i$, $\overline{H} = \frac{1}{N}\sum_{n=1}^{N} H_n$, $\overline{F} = \frac{1}{N}\sum_{n=1}^{N} \delta F_n$.

In this case, they are equal to $u = -35{,}181\ \text{m}^{-1}$, $v = 56{,}488\ \text{m}^{-1}$. The resulting distribution d_n is shown in Fig. 10.

Since the distribution of the positions of the inhomogeneities dn has changed, the values of hn must also change. In turn, the length of the inhomogeneities must be changed in accordance with the relation $l_n = \lambda_g(h_n)/p$.

As a result of the correction, a decrease in the magnitude of the phase error is observed, which improves the parameters of the pattern (Figs. 11 and 12), however, the correspondence to the fully set target is still not achieved.

When repeating the phase correction operation according to the described method, a further decrease in the influence of phase distortion was achieved (Figs. 13 and 14), while the obtained RP are characterized by the correct deviation angle and a low sidelobe level not exceeding minus 30 dB.

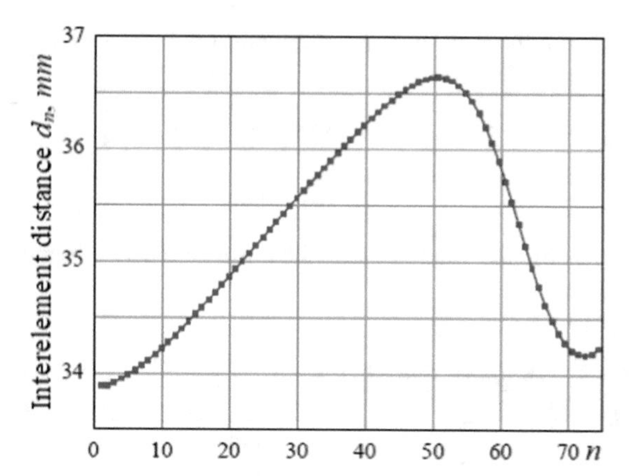

Fig. 10. Distribution of distances between inhomogeneities after compensation of phase distortions.

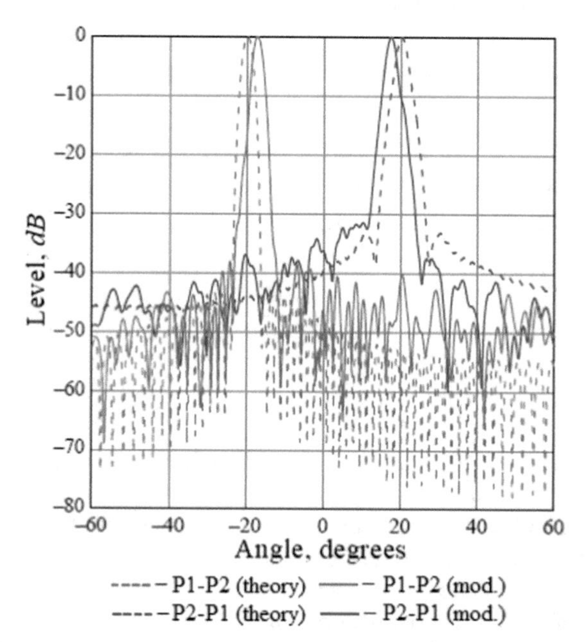

Fig. 11. RP of the emitter model after the first correction tions: solid line – simulation result, dotted line – calculated.

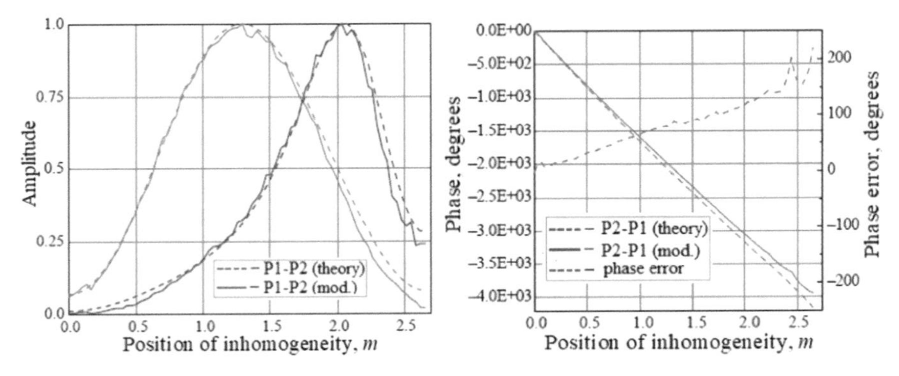

Fig. 12. Amplitude-phase distribution of the field after the first correction.

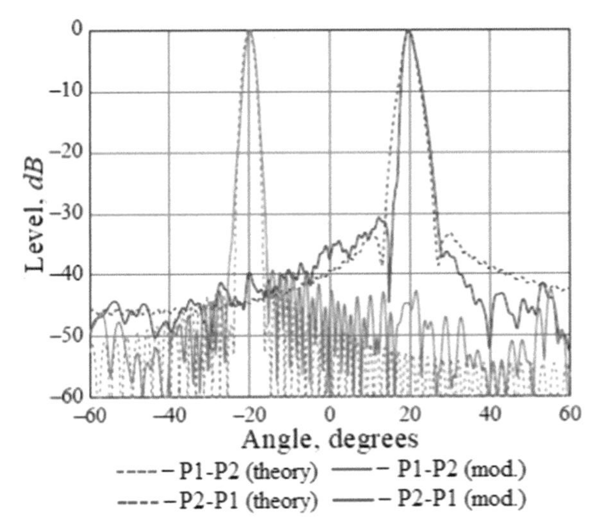

Fig. 13. Radiator model RP after the second correction: solid - modeling result, dotted line – calculated.

The existing trend towards a decrease in phase distortion persists with further iterations to a lesser extent and eventually converges to the same values that differ little from those obtained, which is explained by the presence of a nonlinear component of the distortion described in [3]. However, a strict account of these corrections using the technique given in [3] is impossible due to a decrease in the length of inhomogeneities, which will be the subject of further research. Nevertheless, in practice, already at the second iteration, the result fully satisfies the criteria of the task.

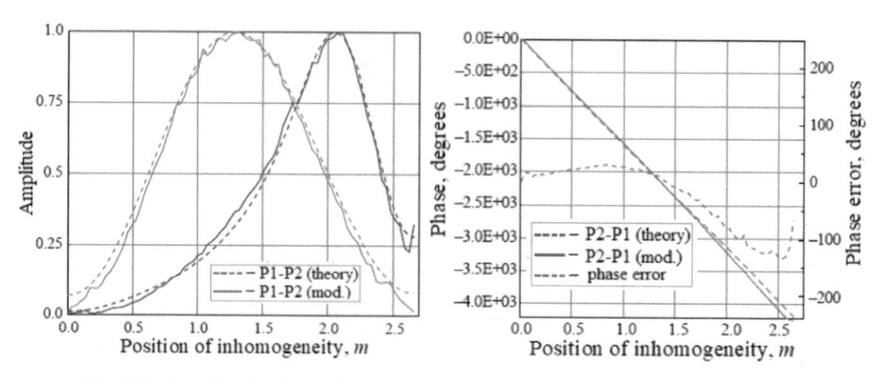

Fig. 14. Amplitude-phase distribution of the field after the second correction.

4 Conclusion and Recommendation

The structure of the distribution of inhomogeneities at the bottom of a semi-open grooved waveguide obtained as a result of the work carried out makes it possible to implement antennas with two radiation patterns, which are formed depending on the direction of excitation. The properties of the obtained RP make it possible to consider this emitter as one of the main elements in the development of two-channel phased antenna arrays of radar stations with an increased scan rate.

References

1. Rahimzadeh, R.M., Lamecki, A., Sypek, P., Mrozowski, M.: Residue-pole methods for variability analysis of S-parameters of microwave devices with 3D FEM and mesh deformation. J. Radio. Eng. **29**(1), 10–20 (2020)
2. Sun, D.Q., Xu, J.P.: Real time rotatable waveguide twist using contactless stacked air-gapped waveguides. IEEE Microwave Wirel. Compon. Lett. **27**(3), 215–217 (2017)
3. Li, F., et al.: Design and microwave measurement of a Ka-band HE11 mode corrugated horn for the Faraday rotator. IET Microw. Antennas Propag. **11**(1), 75–80 (2017)
4. Liu, Z.Q., Sun, D.Q.: Transition from rectangular waveguide to empty substrate integrated gap waveguide. Int. J. Electron. Lett. **55**(11), 654–655 (2019)
5. Yousefian, M., Hosseini, S.J., Dahmardeh, M.: Compact broadband coaxial to rectangular waveguide transition. J. Electromagn. Waves Appl. **33**(9), 1239–1247 (2019)
6. Menachem, Z.: A new technique for the analysis of the physical discontinuity in a hollow rectangular waveguide with dielectric inserts of varying profiles. J. Electromagn. Waves Appl. **33**(9), 1145–1162 (2019)
7. Singh, R.R., Priye, V.: Numerical analysis of film-loaded silicon nanowire optical rectangular waveguide: an effective optical sensing. Micro Nano Lett. **13**(9), 1291–1295 (2018)
8. Taghizadeh, H., Ghobadi, C., Azarm, B., Majidzadeh, M.: Grounded coplanar waveguide-fed compact MIMO antenna for wireless portable applications. Radio. Eng. **28**(3), 528–534 (2019)
9. Islamov, I.J., Ismibayli, E.G., Gaziyev, Y.G., Ahmadova, S.R., Abdullayev, R.Sh.: Modeling of the electromagnetic feld of a rectangular waveguide with side holes. Prog. Electromagn. Res. **81**, 127–132 (2019)

10. Islamov, I.J., Shukurov, N.M., Abdullayev, R.Sh., Hashimov, Kh.Kh., Khalilov, A.I.: Diffraction of electromagnetic waves of rectangular waveguides with a longitudinal. In: IEEE Conference 2020 Wave Electronics and its Application in Information and Telecommunication Systems (WECONF). INSPEC Accession Number: 19806145 (2020)
11. Khalilov, A.I., Islamov, I.J., Hunbataliyev, E.Z., Shukurov, N.M., Abdullayev, R.Sh.: Modeling microwave signals transmitted through a rectangular waveguide. In: IEEE Conference 2020 Wave Electronics and its Application in Information and Telecommunication Systems (WECONF). INSPEC Accession Number: 19806152 (2020)
12. Islamov, I.J., Ismibayli, E.G.: Experimental study of characteristics of microwave devices transition from rectangular waveguide to the megaphone. IFAC-PapersOnLine **51**(30), 477–479 (2018)
13. Ismibayli, E.G., Islamov, I.J.: New approach to definition of potential of the electric field created by set distribution in space of electric charges. IFAC-PapersOnLine **51**(30), 410–414 (2018)
14. Islamov, I.J., Ismibayli, E.G., Hasanov, M.H., Gaziyev, Y.G., Abdullayev, R.Sh.: Electrodynamics characteristics of the no resonant system of transverse slits located in the wide wall of a rectangular waveguide. Prog. Electromagn. Res. Lett. **80**, 23–29 (2018)
15. Islamov, I.J., Ismibayli, E.G., Hasanov, M.H., Gaziyev, Y.G., Ahmadova, S.R., Abdullayev, R.Sh.: Calculation of the electromagnetic field of a rectangular waveguide with chiral medium. Prog. Electromagn. Res. **84**, 97–114 (2019)
16. Islamov, I.J., Hasanov, M.H., Abbasov, M.H.: Simulation of electrodynamic processes in the cylindrical-rectangular microwave waveguide systems transmitting information. In: Aliev, R.A., Kacprzyk, J., Pedrycz, W., Jamshidi, M., Babanli, M., Sadikoglu, F.M. (eds.) ICSCCW 2021. LNNS, vol. 362, pp. 246–253. Springer, Cham (2022). https://doi.org/10.1007/978-3-030-92127-9_35
17. Islamov, I.J., Hunbataliyev, E.Z., Zulfugarli, A.E.: Numerical simulation of characteristics of propagation of symmetric waves in microwave circular shielded waveguide with a radially inhomogeneous dielectric filling. Cambridge Univ. Press: Int. J. Microwave Wirel. Technol. **9**, 1–7 (2021)
18. Islamov, I.J., Hunbataliyev, E.Z., Abdullayev, R.Sh., Shukurov, N.M., Hashimov, Kh.Kh.: Modelling of a microwave rectangular waveguide with a dielectric layer and longitudinal slots. In: Durakbasa, N.M., Gençyılmaz, M.G. (eds.) Digitizing Production Systems. LNME, pp. 550–558. Springer, Cham (2021). https://doi.org/10.1007/978-3-030-90421-0_47

Higher-Order and Stable Numerical Scheme for Nonlinear Diffusion System via Compact Finite Difference and Adaptive Step-Size Runge-Kutta Methods

Shodijon Ismoilov[1(✉)], Gurhan Gurarslan[2], and Gamze Tanoğlu[1]

[1] Izmir Institute of Technology, Urla, Izmir, Turkey
ismoilovshodi@gmail.com, gamzetanoglu@iyte.edu.tr
[2] Pamukkale University, Denizli, Turkey
gurarslan@pau.edu.tr

Abstract. In this study, an efficient numerical method is proposed for the numerical solution of a class of two-dimensional initial-boundary value problems governed by a non-linear system of partial differential equations known as a Brusselator system. The method proposed is based on a combination of higher-order Compact Finite Difference (CFD) scheme and stable time integration scheme which is known as adaptive step-size Runge-Kutta method. The performance of adaptive step-size Runge-Kutta (RK) formula of third-order accurate in time and Compact Finite Difference scheme of sixth-order in space are investigated. The proposed method has been compared with the studies in the literature. Several test problems are considered to check the accuracy and efficiency of the method and reveal that the method is an efficient and reliable alternative to approximate the Brusselator system.

1 Introduction

In characterization of biological and chemical reactions a process which plays a significant role is diffusion. Mathematical modelings involving this process are of high importance to study a wide range of patterns of chemical species [1]. One of the reaction-diffusion systems describing such patterns is so-called Brusselator system. Initially, the Brusselator model was proposed by Prigogine and R. Lefever in 1968 [2–4]. The system consists of two variables interrelated with reactant and product chemicals whose concentrations are controlled [5]. Certain processes that these equations model can be seen in plasma and laser physics in multiple coupling between modes, in enzymatic reactions, in formation of turing pattern on animal skin and in formation of ozone by atomic oxygen through a triple collision [6, 12]. Since, the analytical solutions of these equations are not found yet, they are of interest from the numerical point of view.

D. J. Hemanth et al. (Eds.): ICAIAME 2022, ECPSCI 7, pp. 30–38, 2023.
https://doi.org/10.1007/978-3-031-31956-3_3

The general reaction-diffusion Brusselator system is the non-linear system of partial differential equations

$$\begin{aligned} \frac{\partial u}{\partial t} &= \alpha - (\beta + 1)u + u^2 v + \gamma \nabla^2 u(x,t) \\ \frac{\partial v}{\partial t} &= \beta u - u^2 v + \gamma \nabla^2 v(x,t) \end{aligned} \tag{1}$$

with initial conditions and Neumann boundary conditions

$$u(x,0) = f(x), \qquad v(x,0) = g(x), \qquad x \in \hat{\mathcal{D}} \tag{2}$$

$$\frac{\partial u(x,t)}{\partial n} = \frac{\partial v(x,t)}{\partial n} = 0, \qquad (x,t) \in \partial\hat{\mathcal{D}} \times [0,T]. \tag{3}$$

The model has been considered by a lot of researchers throughout the years. They have proposed different approaches for one-dimensional and two-dimensional form of the equation to get numerical simulations of the model. Different types of meshfree algorithms have been developed, the approach that Al-Islam and his group [5] have taken is by combining radial basis multiquadric functions and first-order finite difference method while the meshfree algorithm developed by Kumar and his coworkers [6] is based on radial basis multiquadric functions combined with differential quadrature technique to get numerical solution of this model. Several methods based on B-spline functions are studied throughout the years. A modified trigonometric cubic B-spline functions coupled with differential quadrature method has been applied to this model by Alqahtani in [7]. Two different modified cubic B-spline based on differential quadrature algorithm have been studied in [8,9]. Onarcan and his team [10] have also developed a cubic trigonometric B-spline interrelated with Crank-Nicholson method in time. A numerical technique based on Lucas and Fibonacci polynomials coupled with finite difference method and polynomial based differential quadrature method are studied to approach the solution of the model in [11,12], respectively. Some of the popular approaches are based on fractal-fractional differential operators. In [13] the classical differential operators has been replaced by fractal-fractional differential operators due to the power law, exponential decay, and the generalized Mittag-Leffler kernels and the results are fairly accurate. Jena and his colleagues [14] have developed a semi-analytical technique called fractional reduced differential transform method characterized by the time-fractional derivative to approximate the solution of Brusselator model.

In this work, we have presented a different approach where we applied a higher-order compact finite difference schemes to estimate the solution of two-dimensional Brusselator system. The method of lines is used to reduce the system of non-linear partial differential equation into a system of non-linear ordinary differential equation. The obtained system is then solved by an adaptive step-size Runge-Kutta method. A famous test problem is chosen from the literature to demonstrate the efficiency and accuracy of the method.

2 Numerical Methods

In this section, a brief introduction to Compact Finite Difference and adaptive step-size Runge-Kutta schemes are given. To increase the accuracy in the solution, CFD formulae include the derivatives of neighbouring nodes in the calculation. Since, neighbouring derivatives are also unknowns, we get more than one unknown per equation. However, when the scheme is applied to all the nodes sufficient equations are produced to find all the unknowns [15]. As for time integration, adaptive step-size Runge-Kutta method is used, that is an ODE integrator having some adaptive control in each step-size along its progress. The usual purpose of such integrators is to achieve a desired accuracy in the solution with minimum computational cost [16]. That is obtained by passing quickly through the interval where the solution changes slowly and passing carefully through the interval where the solution changes rapidly [17].

2.1 Spatial Discretization

In the present work, the implicit compact schemes are used to approximate the spatial derivatives where we compute all the estimated derivatives along a grid line and solve a linear system of equations. We introduce a family of fourth order schemes for interior nodes. For the nodes near the boundary suitable lower order schemes are introduced.

2.1.1 CFD Schemes for the Second Derivative

The family of schemes to be introduced has three stencils at the left-hand side, i.e., it produces three unknowns per equation. For simplicity we consider a uniformly spaced mesh, i.e., $x_1, x_2, \ldots, x_{i-1}, x_i, x_{i+1}, \ldots, x_N$ and given the values of a function at the nodes $u_i = u(x_i)$. The mesh size is denoted by $\Delta x = x_{i+1} - x_i$. The second derivatives can be given at interior nodes as follows [18]

$$\hat{\alpha} u''_{i-1} + u''_i + \hat{\alpha} u''_{i+1} = b\frac{u_{i+2} - 2u_i + u_{i-2}}{4(\Delta x)^2} + a\frac{u_{i+1} - 2u_i + u_{i-1}}{2(\Delta x)^2} \tag{4}$$

which gives a $\hat{\alpha}-$family of fourth-order schemes where

$$a = \frac{4}{3}(1 - \hat{\alpha}), \qquad b = \frac{1}{3}(-1 + 10\hat{\alpha}) \tag{5}$$

by choosing proper values explicit or implicit compact schemes can be created.

Surprisingly, setting $\hat{\alpha} = \frac{2}{11}$ we get a scheme of sixth-order for the second derivative as follows

$$\frac{2}{11} u''_{i-1} + u''_i + \frac{2}{11} u''_{i+1} = \frac{12}{11}\frac{u_{i+2} - 2u_i + u_{i-2}}{4(\Delta x)^2} + \frac{3}{11}\frac{u_{i+1} - 2u_i + u_{i-1}}{2(\Delta x)^2} \tag{6}$$

this particular choice of $\hat{\alpha}$ is deduced from the first term of truncation error, for more details see [15,18]. We use this scheme for internal nodes, i.e., $i = 3, 4, \ldots, N-2$.

2.1.2 Formulae Near the Boundary for the Second Derivative

Problems with non-periodic boundary conditions require special treatment so does our case. Thus, we introduce new formulae for the second derivatives for the near boundary nodes [15,18]. For the nodes $i = 1$ and $i = N$ the approximations are called non-central or one-sided and are given respectively:

$$u_1'' + 11u_2'' = \frac{1}{(\Delta x)^2}(13u_1 - 27u_2 + 15u_3 - u_4), \tag{7}$$

$$11u_{N-1}'' + u_N'' = \frac{1}{(\Delta x)^2}(13u_N - 27u_{N-1} + 15u_{N-2} - u_{N-3}). \tag{8}$$

For the nodes $i = 2$ and $i = N-1$ we used Eq. (4) again with $\hat{\alpha} = \frac{1}{10}$ and $a = \frac{12}{5}$, $b = 0$:

$$\frac{1}{10}u_{i-1}'' + u_i'' + \frac{1}{10}u_{i+1}'' = \frac{12}{5}\frac{u_{i+1} - 2u_i + u_{i-1}}{2(\Delta x)^2}. \tag{9}$$

After combining all the given schemes, i.e., (6)–(9), it forms a tridiagonal matrix as follows

$$\begin{bmatrix} 1 & 11 & & & & & \\ \frac{1}{10} & 1 & \frac{1}{10} & & & & \\ & \frac{2}{11} & 1 & \frac{2}{11} & & & \\ & & & \ddots & & & \\ & & & \frac{2}{11} & 1 & \frac{2}{11} & \\ & & & & \frac{1}{10} & 1 & \frac{1}{10} \\ & & & & & 11 & 1 \end{bmatrix} \begin{bmatrix} u_1'' \\ u_2'' \\ u_3'' \\ \vdots \\ u_{N-2}'' \\ u_{N-1}'' \\ u_N'' \end{bmatrix} = \begin{bmatrix} \hat{\beta}_1 \\ \hat{\beta}_2 \\ \hat{\beta}_3 \\ \vdots \\ \hat{\beta}_{N-2} \\ \hat{\beta}_{N-1} \\ \hat{\beta}_N \end{bmatrix} \tag{10}$$

where $(u_1'', \ldots, u_N'')^T$ are unknowns and the right-hand side $\hat{\beta}_i$ is the linear combination of known values of function u_i at some nodes as shown in the schemes above. The method of choice to solve this matrix is Thomas algorithm. We denote this whole package, i.e., Eq. (10), with "CFD6" (compact finite difference scheme of sixth order) for the rest of the work.

2.2 Temporal Discretization

The adaptive step-size Runge-Kutta method that we use in this work deserves a brief explanation. This method uses two ordinary Runge-Kutta formulas from the family of embedded Runge-Kutta formulas to compare the results [19,20]. The advantage of these types of formulas is that they share the same evaluation points as we shall see. Therefore, the amount of computational cost is minimized [21].

Assuming the governing equation is

$$\frac{\partial u}{\partial t} = \mathcal{L}u \tag{11}$$

where in our case $\mathcal{L}$ is a non-linear operator independent of time. We start the integration from t_0 to $t_0 + \Delta t$ (step $k+1$) using 3rd order Runge-Kutta formula

$$\begin{aligned} f_1 &= \mathcal{L}u_k \\ f_2 &= \mathcal{L}(u_k + \Delta t f_1) \\ f_3 &= \mathcal{L}(u_k + \frac{\Delta t}{4}(f_1 + f_2)) \\ u_{k+1} &= u_k + \frac{\Delta t}{6}(f_1 + 4f_3 + f_2). \end{aligned} \tag{12}$$

With the above f_1 and f_2, we can use the 2nd order Runge-Kutta method to get a less accurate solution at t_{k+1}:

$$u^*_{k+1} = u_k + \frac{\Delta t}{2}(f_1 + f_2) \tag{13}$$

taking the difference between u_{k+1} and u^*_{k+1} we get

$$e = \|u_{k+1} - u^*_{k+1}\| = \frac{\Delta t}{3}\|f_1 - 2f_3 + f_2\| \tag{14}$$

according to this difference e scales as $(\Delta t)^3$, we need to bound e within the desired accuracy ϵ, as the computations in [17] shows the new step-size should satisfy

$$\Delta t_{new} < \Delta t(\frac{\epsilon}{e})^{1/3} \tag{15}$$

to satisfy the above inequality, we use

$$\Delta t := 0.9\Delta t(\frac{\epsilon}{e})^{1/3} \tag{16}$$

to update our step-size. Now, if e is larger than ϵ, the equation determines how much to decrease the step-size when this failed step is retried, if e is smaller than ϵ, then the step is successful, furthermore, the equation determines how much increase in the step-size can be made safely for the next step. More information about these types of methods can be found in [16,17,22,23].

2.3 Applications to Brusselator Model

In this section, the implementation of the given schemes to the two-dimensional Brusselator model is shown. The system with its initial and boundary conditions can be deduced from Eq. (1).

Let us define the uniform mesh as $\{(x_i, y_j) : i = 1, 2, \dots, N,\ j = 1, 2, \dots, M\}$ with $\Delta x = x_{i+1} - x_i$ and $\Delta y = y_{j+1} - y_j$. Also, let us denote the semi-discrete form of u and v, respectively, by

$$\begin{aligned} U(t) &= \{u(x_i, y_j, t) : i = 1, 2, \dots, N,\ j = 1, 2, \dots, M\}, \\ V(t) &= \{v(x_i, y_j, t) : i = 1, 2, \dots, N,\ j = 1, 2, \dots, M\}, \end{aligned} \tag{17}$$

i.e., U and V are both matrices of size $M \times N$. Now, let U_{xx} denote the approximation of the second derivative of u in the direction of x where U_{xx} is a matrix of columns

$$\{(u''_{1,j}, u''_{2,j}, \dots, u''_{N,j})^T : j = 1, 2, \dots, M\}, \tag{18}$$

i.e., M number of Eq. (10)'s are produced to solve. And let U_{yy} denote the approximation of the second derivative of u in the direction of y where U_{yy} is a matrix of rows

$$\{(u''_{i,1}, u''_{i,2}, \dots, u''_{i,M}) : i = 1, 2, \dots, N\}, \tag{19}$$

i.e., N number of Eq. (10)'s are produced to solve. Similarly, the approximations of derivatives of v are denoted by V_{xx} and V_{yy}. Using these matrices a semi-discrete form of Eq. (1) for two-dimensional version of the model can be written as

$$\begin{aligned} \frac{\partial U}{\partial t} &= \alpha - (\beta + 1)U + U^2V + k(U_{xx} + U_{yy}) \\ \frac{\partial V}{\partial t} &= \beta U - U^2V + k(V_{xx} + V_{yy}). \end{aligned} \tag{20}$$

From here on, the adaptive step-size Runge-Kutta method, in Sect. 2.2, is applied for time integration. Neumann boundary conditions are approximated by classical finite difference methods.

3 Numerical Simulations

In this section, a numerical solution of the Brusellator's equation is estimated for the given problem to test the current numerical schemes. Fortunately, we have the exact solution for this problem to demonstrate the effectiveness and accuracy of the method. The results are compared to some results found in the literature. The numerical computation is performed using uniform grids. All computations were done by codes produced in MatLab.

Test Problem. Consider the two dimensional Brusselator equation over the domain $\mathcal{D} = \{(x, y) \mid 0 \leq x \leq 1,\ 0 \leq y \leq 1\}$ with the exact solution given as follows [6]

$$\begin{aligned} u(x, y, t) &= \exp(-x - y - 0.5t) \\ v(x, y, t) &= \exp(x + y + 0.5t) \end{aligned}$$

where the initial and boundary data is extracted from this solution. The results are approximated with the proposed methods using parameters $\alpha = 0$ $\beta = 1$, $k = 0.25$ at $t = 2$ with desired accuracy $\epsilon = 10^{-4}$ in time integration. In Table 1 the estimated results of ours and [5] at various times are listed and compared with the exact solution. As for the accuracy of the presented method L_∞ norms of the test problem are presented in Table 2. In addition, two different L_∞ norms obtained by [6,11] are brought for comparison purpose. The illustrations shown in Figs. 1 and 2 demonstrates that u and v have opposite behaviours, u dissipates while v increases as times goes.

Table 1. Comparison of numerical results produced by CFD6 with the results of [5] at the point (0.40, 0.60) at different times.

t	u			v		
	CFD6	[5]	Exact	CFD6	[5]	Exact
0.30	0.3166	0.3174	0.3166	3.1582	3.158	3.1582
0.60	0.2725	0.2732	0.2725	3.6693	3.668	3.6693
0.90	0.2346	0.2351	0.2346	4.2631	4.262	4.2631
1.20	0.2019	0.2024	0.2019	4.9530	4.952	4.9530
1.50	0.1738	0.1742	0.1738	5.7546	5.754	5.7546
1.80	0.1496	0.1499	0.1496	6.6859	6.685	6.6859

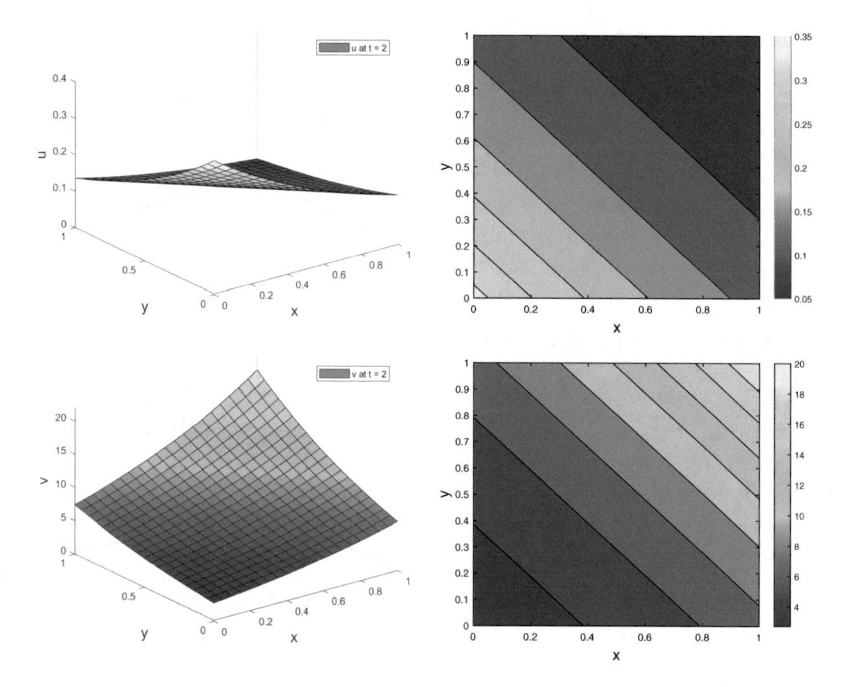

Fig. 1. The physical behaviour of concentration of u and v in 3D and contour of the given test problem at times $t = 2$.

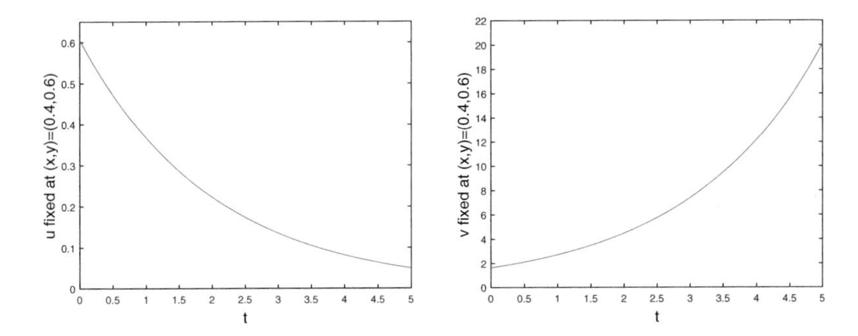

Fig. 2. The solution profile of u and v of the given test problem fixed at $(x, y) = (0.4, 0.6)$ over $0 < t \leq 5$.

Table 2. L_∞ norms of the problem at $t = 2$ with the accuracy tolerance $\epsilon = 10^{-4}$.

N	[6]		[11]		CFD6	
	u	v	u	v	u	v
10×10	1.6947×10^{-06}	8.5877×10^{-05}	2.7094×10^{-05}	1.7571×10^{-03}	6.4521×10^{-07}	1.1565×10^{-05}
15×15	1.5364×10^{-06}	7.9857×10^{-05}	2.3714×10^{-05}	2.3714×10^{-04}	9.0706×10^{-07}	1.2555×10^{-05}
21×21	1.3452×10^{-06}	1.017×10^{-06}	1.3115×10^{-05}	1.4635×10^{-03}	1.7645×10^{-06}	2.0922×10^{-05}

4 Conclusion

In this work, a numerical approach for two-dimensional Brusselator system was taken under consideration. We investigated the performance of a sixth-order compact finite difference scheme for space combined with a third order time integrator called adaptive step-size Runge-Kutta method. The proposed method was capable of producing highly accurate results with minimal computational cost, not to mention its easy implementation. The results obtained were compared with the best results obtained in [6]. The comparison tables has shown that the present results are highly accurate. In addition, several test problems especially in [1,6,12] were also considered and compared to check the accuracy and efficiency of our method and reveal that the method is very efficient and reliable. Therefore, it is safe to conclude that the presented method is a very efficient and reliable alternative for solving Brusselator system.

References

1. Mittal, R.C., Kumar, S., Jiwari, R.: A cubic B-spline quasi-interpolation algorithm to capture the pattern formation of coupled reaction-diffusion models. Eng. Comput. **38**, 1375–1391 (2021). https://doi.org/10.1007/s00366-020-01278-3
2. Prigogine, I., Lefever, R.S.: Symmetry breaking instabilities in dissipative systems. II. J. Chem. Phys. **48**(4), 1695–1700 (1968)
3. Prigogine, I., Nicolis, G.: Self-organisation in nonequilibrium systems. In: Dissipative Structures to Order through Fluctuations, pp. 339–426 (1977)

4. Lefever, R.S., Nicolis, G.: Chemical instabilities and sustained oscillations. J. Theor. Biol. **30**(4), 267–284 (1971)
5. Al-Islam, S., Ali, A., Al-Haq, S.: A computational modeling of the behavior of the two-dimensional reaction-diffusion Brusselator system. Appl. Math. Model. **34**(12), 3896–3909 (2010)
6. Kumar, S., Jiwari, R., Mittal, R.C.: Numerical simulation for computational modelling of reaction-diffusion Brusselator model arising in chemical processes. J. Math. Chem. **57**, 149–179 (2019)
7. Alqahtani, A.M.: Numerical simulation to study the pattern formation of reaction-diffusion Brusselator model arising in triple collision and enzymatic. J. Math. Chem. **56**(6), 1543–1566 (2011)
8. Jiwari, R., Yuan, J.: A computational modeling of two dimensional reaction-diffusion Brusselator system arising in chemical processes. J. Math. Chem. **52**(6), 1535–1551 (2011)
9. Mittal, R.C., Rohila, R.: Numerical simulation of reaction-diffusion systems by modified cubic B-spline differential quadrature method. Chaos, Solitons Fractals **92**, 9–19 (2016)
10. Onarcan, A.T., Adar, N., Dag, I.: Trigonometric cubic B-spline collocation algorithm for numerical solutions of reaction-diffusion equation systems. J. Math. Chem. **37**(5), 6848–6869 (2018)
11. Haq, S., Ali, I., Nisar, K.S.: A computational study of two-dimensional reaction-diffusion Brusselator system with applications in chemical processes. Alex. Eng. J. **60**(5), 4381–4392 (2021)
12. Mittal, R.C., Jiwari, R.: Numerical solution of two-dimensional reaction-diffusion Brusselator system. Appl. Math. Comput. **217**, 5404–5415 (2011)
13. Saad, K.M.: Fractal-fractional Brusselator chemical reaction. Chaos, Solitons Fractals **150**(12), 111087 (2011)
14. Jena, R.M., Chakraverty, S., Rezazadeh, H., Domiri, G.D.: On the solution of time-fractional dynamical model of Brusselator reaction-diffusion system arising in chemical reactions. Math. Methods Appl. Sci. **43**(7), 3903–3913 (2010)
15. Hoffmann, K.A., Chiang, S.T.: Computational Fluid Dynamics, vol. 3, 4th edn, pp. 117–137. A Publication of Engineering System^{TM} (2000)
16. Press, W.H., Teukolsky, S.A.: Adaptive Stepsize Runge-Kutta integration. Comput. Phys. **6**, 188 (1992). https://doi.org/10.1063/1.4823060
17. Lu, Y.Y.: Numerical Methods for Differential Equations, Department of Mathematics City University of Hong Kong Kowloon, Hong Kong, pp. 13–16
18. Lele, S.K.: Compact finite difference schemes with spectral-like solution. J. Comput. Phys. **103**, 16–42 (1992)
19. Cicek, Y., Gucuyenen, K.N., Bahar, E., Gurarslan, G., Tanoğlu, G.: A new numerical algorithm based on Quintic B-Spline and adaptive time integrator for coupled Burger's equation. Comput. Methods Differ. Equ. **11**, 130–142 (2022)
20. İmamoğlu, K.N., Korkut, S.Ö., Gurarslan, G., Tanoğlu, G.: A reliable and fast mesh-free solver for the telegraph equation. Comput. Appl. Math. **41**(5), 1–24 (2022)
21. Bahar, E., Gurarslan, G.: B-spline method of lines for simulation of contaminant transport in groundwater. Water **12**(6), 1607 (2020)
22. Dormand, J.R., Prince, P.J.: A family of embedded Runge-Kutta formulae. J. Comput. Appl. Math. **6**(1) (1980)
23. Dormand, J.R., Prince, P.J.: High order embedded Runge-Kutta formulae, Department of Mathematics and Statistics, Teesside Polytechnic, Middlesbrough, Cleveland, TS1 3BA, U.K

Explainable Artificial Intelligence (XAI) for Deep Learning Based Intrusion Detection Systems

Mehmet Sevri[1(✉)] and Hacer Karacan[2]

[1] Computer Engineering Department, Faculty of Engineering and Architecture, Recep Tayyip Erdogan University, Rize, Turkey
mehmet.sevri@erdogan.edu.tr
[2] Computer Engineering Department, Faculty of Engineering, Gazi University, Ankara, Turkey
hkaracan@gazi.edu.tr

Abstract. In recent years, attacks on computer networks and web applications have become increased and sophisticated, and detecting these attacks has become a major challenge. Using deep learning techniques has remarkably increased the performance of intrusion detection systems due to their high accuracy levels and effectiveness. However, with their complex architectures, deep learning models are black-boxes since their working mechanism for making predictions cannot be explained transparently. In critical systems such as intrusion detection systems, only having high accuracy performance is not enough for the machine learning models to be used in real-time. Additionally, it is important that the model is transparent, and predictions of the model can be interpreted. In this study, explainable artificial intelligence (XAI) techniques that can be used to explain deep learning-based intrusion detection systems and interpret model predictions are examined in detail. The study proposes a new framework for interpreting deep learning models trained on various benchmark anomaly datasets using LIME (Local Interpretable Model-Agnostic) and SHAP (Shapley Additive Explanations) explainers. The proposed framework provides the local interpretations for individual predictions of deep learning-based intrusion detection models with visuals. In addition, by determining the most effective feature vectors for all inputs, it is ensured that the operations of the black-box models can be explained globally. With this study, novel and successful deep learning-based intrusion detection models that can detect anomalies in both computer networks and web applications are developed, and the explanations of the models and the interpretations of individual predictions are provided to ensure real-time usage.

Keywords: Intrusion Detection System (IDS) · Deep Learning · Explainable Artificial Intelligence (XAI) · Web Security · Network Security

1 Introduction

Cyber security has become a critical and essential concept for individuals, companies, institutions, and states with the widespread use of internet technology. Thanks to the new

D. J. Hemanth et al. (Eds.): ICAIAME 2022, ECPSCI 7, pp. 39–55, 2023.
https://doi.org/10.1007/978-3-031-31956-3_4

attack tools and broad attack surfaces, hackers can carry out very sophisticated attacks in a short time and significant economic gains can be obtained. The economic gains that can be achieved with cyber-attacks increase the attackers' desire and cause significant increases in the number of attacks and victims' economic losses. The concept of cyber security consists of different branches such as network security, application security, critical infrastructure security, Internet of Things (IoT) security, and cloud security. These branches are also divided into sub-branches within themselves and constitute the whole of the cyber security concept. The scope of this study consists of network security and application security areas. Network security includes the processes of detecting and preventing threats, attacks, and damage to computer networks from outside or inside. Network security is generally aimed at analyzing network packets at OSI layer 4 and detecting the attack patterns contained. On the other hand, application security concerns the analysis of application-level traffic in OSI layer 7 and the detection of attacks to the application level. Application security covers different areas such as ransomware, malware, virus, web application security, and mobile application security. In this study, the field of web application security was handled. Brief information about the areas of network security and web application security discussed in the study was presented in below.

Network-based intrusion detection systems (NIDS) are used to detect and prevent network-based attacks by analysing in the packet level at the network layer. Similarly, web application firewalls (WAF) are used for intrusion detection and prevention for web applications. WAF can analyze web traffic at the application layer. Both NIDS and WAF development use similar methodologies. The development of intrusion detection systems uses two fundamental strategies: signature-based and anomaly-based (IDS) [1]. Signature-based intrusion detection systems are based on a database that contains known attack signatures and vulnerability patterns. A footprint is dedicated to each attack pattern that can be hosted in network packets or web traffics. Signature-based IDS can detect and prevent attacks by matching the footprints of attacks with the records in the database. One important disadvantage of this approach is that it can only prevent known attack footprints. Maintaining the database and updating anomalous signatures is a major challenge. Anomaly-based intrusion detection systems detect anomalous activities by learning the normal behaviours of the network or applications. The development of anomaly-based IDS is usually carried out by training from pre-recorded normal network packets or application requests. A trained anomaly-based IDS can classify a request as an anomaly if it is outside of known instances. The most significant advantage of anomaly-based systems is their ability to detect new vulnerabilities derived from known attacks. On the other hand, these systems has a major disadvantage of generating alarms with a high false-positive (FP) rate. Different machine learning-based models can be used in the development of anomaly-based systems. Deep learning (DL)-based algorithms have recently become popular due to their outstanding performance. However, lots of machine learning methods such as deep learning are not transparent since the model's behaviours and predictions are not fully explainable. Therefore, the models created by these methods are called black-box models. In general, the transparency and explainability of AI systems are inversely proportionate to their intrusion detection performance. While the detection performance of anomaly-based IDS models with a sophisticated structure such as deep learning increases, their explainability decreases. The model's high performance alone does not provide the essential assurance to deploy it in vital systems such as health, finance, human resources, and intrusion detection.

When prior intrusion detection research is examined, it is seen that the most successful anomaly-based intrusion detection systems are generally black-box models. In addition, in related studies, it is seen that model explainability and interpretation of predictions are not addressed. In general, the explainability part of artificial intelligence-based intrusion detection systems is insufficient for trust in real-time usage. In order to enable real-time use of anomaly-based models, it is an important requirement in terms of trust and transparency in the model that knowing significant inputs and parameters in the attack type predictions. Furthermore, current academic research reveals that anomaly-based models can be misled by simple manipulations. To determine whether the developed black-box IDS is affected by such manipulations, it is necessary to interpret the system's predictions. In the scope of this study, it is aimed to provide the explainability and interpretability of NIDS and WAF systems based on different deep learning algorithms. In this study, an XAI framework based on two different explanation approaches has been developed to interpret the model predictions of the IDSs based on the black-box deep learning models. While the predictions of individual inputs are interpreted in the first approach (local explainability), global explanations are generated in the second method, allowing the determination of the most important features in the model classification. Local Interpretable Model-Agnostic Explanations (LIME) and SHapley Additive exPlanations (SHAP) approaches were used to provide explanations of black-box models.

The main contributions of this study are presented below;

- Deep learning-based IDS models have been developed that can detect attacks at both the network layer and the application layer with a high detection rate. In this regard, NIDS and WAF models based on successful CNN and DNN algorithms have been developed.
- An XAI framework has been developed to explain and interpret the predictions of black-box deep learning-based IDS models. Thus, while it may be explained which input values were used to generate the model predictions, it can also be interpreted as how the model performs in general on the whole entire dataset.
- The effectiveness of the proposed XAI framework was demonstrated on different anomaly datasets (network and web application). In this way, the robustness of the XAI framework has been demonstrated.
- To contribute to similar cyber security studies in the future and reproducibility, the source codes and datasets used in the study will be shared publicly.

The study consists of four chapters, including the introduction, and brief information about the organization of the study was presented below. Preliminary information about explainable approaches and deep learning models is presented in the second section of the study. In the third part, the performance of the deep learning-based IDS systems and the explanations of anomaly models based on the proposed XAI framework are presented. Conclusions and future studies are presented in the last section of the study.

2 Related Works

Explainable artificial intelligence is one of the current trending fields of study. There are few studies that include XAI in anomaly-based intrusion detection studies. Wang et al. [2], carried out a study about the explanation of intrusion detection models based on general machine learning algorithms with SHAP. The authors developed artificial neural network-based multi-classifier and One-Vs-All classifier models to classify attack types. To the best of the authors knowledge, their work is the first explainable machine learning study in the field of cybersecurity.

Sarhan et al. [3] proposed a structure that can explain black-box models based on Random Forest (RF) and Deep Feed Forward (DFF) trained on different IoT network anomaly datasets. The authors used the SHAP approach as the explanation method. Similarly, in the study [4], anomaly models were developed using decision tree (DT) and RF algorithms that the authors trained on three different IoT network anomaly datasets. The authors proposed an XAI approach with the heatmaps created based on the SHAP explanation approach.

In the study [5], the authors proposed a decision tree-based IDS model and explanation structure to increase confidence in anomaly-based IDS systems. The authors aimed to explain the model with rule inferences based on the nodes and leaves of the simple decision tree trained on the KDD dataset.

In the study [6], the authors proposed an explanation approach based on features generalizer in order to explain the network anomaly detection models. They developed IDS models based on traditional machine learning algorithms such as support vector machine (SVM), RF, Extra Tree and Naive Bayes (NB). The feature selection and their significant analysis provide information about the dataset. However, there is no direct connection between feature selection and the black-box model explanation process since the model predictions were not used in the explanation process. Although the correlation between features and class labels is important, feature analysis independent of model predictions is insufficient in terms of the model trust.

In the study [7], the authors proposed an explanation structure based on genetic algorithms to explain their deep learning-based network traffics classification models. The authors used the Resnet model as the classifier model. The authors developed an explainability method based on the dominant feature selection of the genetic algorithm by grafting the outputs of the Resnet model to the chromosomes in the genetic algorithm structure.

Mariano et al. [8] proposed an XAI method based on using the adversarial approach to explain the misclassified predictions of DNN based network anomaly model. The authors used the NSL-KDD network anomaly dataset. The proposed XAI approach is based on the adversarial method that attempts to make true classification by changing the minimum numbers of feature values in the inputs that the DNN model misclassifies.

When the above studies are examined, it is seen that cyber security studies involving XAI are generally network-based ones. The number of studies dealing with the explanation of web anomaly is limited. Our study differs from other studies in terms of providing explainability at both the network layer and application layer levels. The XAI framework proposed in this study is important in terms of providing models based on different deep learning algorithms and explaining traffics in different OSI layers. In addition, the XAI framework proposed within the scope of the study is designed to support IDS models developed with all other deep learning models, as well as DNN and CNN used in the study.

3 Preliminaries About the DL Algorithm and Explainable AI (XAI) Approaches in the Study

It has recently been widely used in research on the development of intrusion detection systems since deep learning shows high performance in many different problems. Depending on the problem, different deep learning approaches could be preferred. When the studies in the cyber security field are examined, it is seen that DNN and CNN-based deep learning approaches are frequently preferred. Since the main purpose of this study is to develop an XAI framework that provides the explanation of deep learning-based IDS models, it has mainly focused on the explainable AI in the study. In addition, the basic structures of DNN and CNN architectures are briefly summarized.

3.1 Deep Learning Architectures in the Development of IDS Models

DL is a sub-field of artificial intelligence and has algorithms and methods that are widely used today. DL is one of the trending topics in computer science, which is frequently used in many machine learning areas such as image processing, sound processing, natural language processing. Especially with the increase in computing power with GPUs, important classification performances have begun to be achieved by creating multi-layered (deep) and multi-neural networks. DL enables the transformation of high-dimensional data in terms of features and volume into lower-dimensional and meaningful outputs. It does this with DL algorithms that contain many parameters (the source of the depth concept). The computational operations are performed in the single-layer structure of traditional machine learning algorithms. These operations are performed gradually with different abstraction levels and layers with many parameters in DL architectures. The outputs calculated in each layer are transferred to the next layers. The model outputs are then obtained, with nonlinear functions representing the most important features of the data. At each transition between layers, the attributes that best represent the data are transferred and an architecture is provided in which the targeted output can be represented with the best features [9]. Deep learning techniques are not new methods, but with the development of CPU and GPU systems and the increase in computing power, more successful models have emerged on large amounts of data than traditional machine learning algorithms.

3.1.1 Deep Neural Network (DNN)

The most fundamental deep learning architecture is DNN. DNN consists of many layers, each with many neurons [10]. Neuron connections provide transitions between layers, and as the number of deep and neuron connections increases, so does the number of parameters that need to be optimized [11]. Because of their simple structure, DNN has the advantage of being easier to optimize and working faster than other DL architectures [10]. DNN architectures work successfully in the development of intrusion detection systems as well as in the application of a wide range of problems [12].

3.1.2 Convolutional Neural Network (CNN)

The CNN method is one of the most frequently used and most striking deep learning architectures in the literature. It has achieved significant success in image processing with CNN-based models. The CNN architecture is typically composed of a series of convolution layers, an optional pooling layer, a fully connected layer, and a classification layer. CNN architecture can create auto-filters from the given images, similar to the filters used in image processing before, by using convolution and pooling layers and determining the most significant features. CNN architecture tries to converge to the minimum error value by continuously updating the parameters with network feedbacks [13].

3.2 The Explainer AI (XAI) Approaches Used in the Study

In order to use machine learning-based models in critical systems, the model should be transparent and its predictions should be interpretable by system administrators. The interpretation of machine learning models can be performed in three phases based on the time the model was created. The interpretation can be performed before the model training (pre-modelling explainability), while the model is running (explainable modelling) or after the model is created (post-modelling explainability). Pre-modelling explainability can be performed with exploratory data analysis (PCA, t-SNE), summarizing the data set, and interpretable feature extraction techniques. In addition, feature selection can be performed with feature importance models such as Generalized Linear Models (GLM) or extra trees classifier, Random Forest (RF). Secondly, due to their transparent structures, some AI models such as decision trees, linear regression, and k-nn can be interpreted while using models that provide interpretability by themselves. AI models that do not have self-explanatory structures are called black-box models as explained above. The interpretation of deep learning models becomes impossible due to the large number of layers and the number of parameters that can reach millions. In order to interpret and explain black-box models, post-hoc models and post-modelling explainability techniques should be used after model training. There are two main methods that are frequently used in ensuring the explanations of deep learning models: post-hoc local explanation and feature relevance. In addition, model simplification, text explanation, model visualization, and feature relations estimation methods can be used to support the explanation of black-box models [14].

There are two post-hoc approaches for interpreting AI models: global explanations and local explanations. Global explanations aim to show the general behaviour of the model over the entire dataset. In this context, the aim is to measure the contribution of model behaviour subcomponents as well as the contribution of input features to model predictions in general. In global explainability, it provides a general interpretation of the model by considering all inputs in the dataset or inputs of the same class together. On the other hand, local explanations aim to interpret the predictions of individual instances and to reveal the contribution of the parameters in the input to the model prediction. As a result of the local explanations, it is possible to interpret how the model generates the prediction for each individual instance.

In the related literature, it is seen that post-hoc explanation methods are generally used in the interpretation of black-box models. The most commonly used explanation methods are LIME (Local Interpretable Model-agnostic Explanations), SHAP (SHApley Additive exPlanations), and decision tree rules-based methods. In addition, the Layer-wise Relevance Propagation (LRP) method can be used for artificial neural network models. The sections that follow explain the general structures and processes of post-hoc explainability methods used in the study for the interpretation of anomaly-based intrusion detection systems (IDS).

3.2.1 Local Interpretable Model-Agnostic Explanations (LIME)

The LIME approach is a model agnostic explainer AI approach that provides the explanation of discrete individual predictions of black-box models through surrogate models. LIME provides local explanations of individual predictions of a pre-trained black-box model. LIME generates subset permutations based on the absence of features in an individual input, creates predictions using an explainable model, and works by determining the most significant feature values. The LIME method generates a new sub-dataset by permuting based on the absence of features in the desired input. Predictions of the original black-box model of the instances in the sub-dataset are added as labels to the sub-dataset. A self-explanatory model (linear regression, Lasso, decision tree) is trained using the sub-dataset and the labels. Discrete linear explanations are generated based on the sub-dataset predictions that are closest to the original model's targeted input prediction. LIME aims to use the least number of features in the targeted input to explain prediction of model. By using these discrete linear explanations, Lasso error minimization based on the desired K features is performed and the weight values of the features are calculated (see Eq. 1) [15]. LIME can be applied to tabular, textual and image datasets. The calculations of the explainability values based on the input features of the LIME approach were shown in Eq. 1 and Eq. 2.

$$\xi(x) = argmin_{g \in G} L(f, g, \pi x) + \Omega(g) \tag{1}$$

$$L(f, g, \pi x) = \sum_{z, z' \in Z} \pi x(z) \left(f(z) - g(z')\right)^2 \tag{2}$$

The working flow of how LIME performs the interpretation of prediction of individual inputs is listed below [16].

- Firstly, an instance wanted to make explanation is handled and the black-box model prediction of this input is generated.
- A sub-dataset is created with permutations based on the absence of features in the input. Black-box model predictions are generated for inputs in the synthetically generated sub-dataset.
- The newly created instance in the sub-dataset are weighted by measuring the distance from the original input. This process is carried out in LIME by simply taking the weight value as the ratio of the number of features in the new input to the number of features in the original input.
- A self-interpretable model is trained with this sub-dataset based on the permutations of features in the targeted input. Self-interpretable model is linear regression by default in the LIME, but it can be used instric interpretable models such as decision tree or Lasso.

- The black box model prediction for targeted input can be explained using the explanations provided by the self-interpretable model.

LIME's main advantages are that it works with various data types, including textual, visual, and tabular data, and that it has a clear and simple explanation structure. Furthermore, because of its model-agnostic structure, it can explain various AI models and continue to explain even if the black-box model changes. However, the need for a field expert to perform the LIME kernel settings and application is a significant challenge, and explanations will be insufficient if the kernel settings are not done correctly.

3.2.2 SHapley Additive exPlanations (SHAP)

SHAP [17] provides global explainability for the black-box model as well as providing local explainability for individual inputs. The SHAP approach performs global explainability and can identify the features that contribute the most to a black-box model due to its game theory-based computation method. The SHAP explanation approach, like LIME, is model agnostic and post-hoc. It uses a method very similar to LIME for local explanations. SHAP uses self-interpretable surrogate models (linear regression, decision tree) to explain the black-box models. The calculation of the contribution of the features to the prediction is carried out by calculating the Shapley values based on game theory. SHAP uses coalitional game theory when calculating the contribution of features to prediction. As a result, it is using a method for calculating each feature's individual contribution to a multi-player game score, either individually or as a group. While SHAP calculates the effects on prediction by changing or removing the values of features in input, it uses game theory to evaluate the impact of multiple features on model prediction at the same time. While SHAP generates coalition vectors based on the presence of feature values, it assigns different feature values at random based on the marginal distribution instead of the original feature value in the input. KernelSHAP, the basic SHAP structure, is used for local explainability. One of the most significant disadvantages of KernelSHAP is its slowness. It requires high computational power to calculate shap values. The mathematical equations used in calculating the explainability values based on the input features in the KernelSHAP approach are seen in Eq. 3 and Eq. 4.

$$g(z') = \phi_0 + \sum_{j=1}^{M} \phi_j z'_j \tag{3}$$

$$L(f, g, \pi_x) = \left[f\left(h(z')\right) - g(z')\right]^2 \pi_x(z') \tag{4}$$

4 DL Based IDS Models and Dataset

In this section, the structures of the developed NIDS and WAF models and the network and web anomaly datasets used in training are explained. Since the main motivation of the study was to ensure the explainability of deep learning-based IDS models, performance optimization and high performance of deep learning models were not directly targeted. Deep learning models used in intrusion detection systems previously developed by us were used as NIDS and WAF models in this study [1, 18, 19]. NIDS and WAF models were created using DNN and CNN-based models. Within the scope of the study, binary-class anomaly detection models that can classify "normal" and "abnormal" traffic were developed. Multi-class models in which attack types are detected were not included in this study. However, all the applied methods and techniques in the study are also applicable in multi-class classification models. In the study, a total of four deep learning-based IDS systems, two of which are NIDS and two are WAF, were developed. NSL-KDD dataset was used in the training of NIDS models. GAZI-HTTP web anomaly dataset, which was created by us, was used in the development of WAF models.

Table 1. The architecture of the DNN classifier model

Layer (Type)	Input Shape	Output Shape	Activation Function
Embedding	(m, n)	$(m, n, 8)$	–
Flatten	(m, n, 8)	(m, n * 8)	–
Dense-1	$(m, n * 8)$	$(m, 64)$	tanh
Dropout-1	$(m, 64)$	$(m, 64)$	–
Dense-2	$(m, 64)$	$(m, 64)$	tanh
Dropout-2	$(m, 64)$	$(m, 64)$	–
Dense	$(m, 64)$	$(m, 2)$	Softmax

Table 2. The architecture of the CNN classifier model

Layer (Type)	Input Shape	Output Shape	Activation Function
Embedding	(m, n)	$(m, n, 16)$	–
Reshape	(m, n, 16)	(m, n, 16, 1)	–
Conv2D – 1	$(m, n, 16, 1)$	$(m, n, 1, 512)$	ReLU
Conv2D – 1	$(m, n, 1, 512)$	$(m, n - 1, 1, 512)$	ReLU
MaxPooling2D - 1	$(m, n, 1, 512)$	$(m, 1, 1, 512)$	–
MaxPooling2D - 2	$(m, n - 1, 512)$	$(m, 1, 1, 512)$	–
Concatenate	$(m, 1, 1, 512)$	$(m, 1, 2, 512)$	–
Flatten	$(m, 2, 1, 512)$	$(m, 1024)$	–
Dense	$(m, 1024)$	$(m, 2)$	Softmax

DNN and CNN-based models were trained on each dataset. The structures of the developed models and datasets are described below. The architectural structure of the DNN and CNN models for binary-class attack detection is shown in Tables 1 and 2, respectively. NSL-KDD dataset [20] was used in the training of NIDS models. NSL-KDD dataset is a well-known benchmark dataset that has been used frequently in network anomaly studies for many years. The NSL-KDD dataset contains network-based tabular data belonging to different attack types. The NSL-KDD dataset is suitable for developing both binary-class and multi-class models.

As mentioned above, GAZI-HTTP web anomaly dataset was used in the development of WAF models in the study. This dataset contains eight different types of web attacks and one "normal" label for a total of nine distinct classes. The dataset consists of web requests collected from various web applications and it is suitable for developing both binary-class and multi-class models. In this study, attack types were taken under a single "abnormal" label and used in binary-class model training. The distribution of the GAZI-HTTP dataset was shown in Table 3. As can be seen, it has an imbalanced distribution of cases.

Table 3. Distributions of GAZI-HTTP dataset

Type	Total Size	Distribution
Valid	73529	67.74%
SQLi	10803	9.95%
XSS	9139	8.42%
Command Injection	3235	2.98%
LFI	2885	2.66%
Open Redirect	2725	2.51%
CRLF	2610	2.40%
SSI	1963	1.81%
XXE	1656	1.53%
Total	108545	100%

5 Explanations of IDS and WAF Models

In this section, local and global explanations of the developed NIDS and WAF models are presented. While the NSL-KDD dataset that is used to train NIDS models contains tabular data, the GAZI-HTTP dataset, which is used to train WAF models, contains textual web requests. SHAP was used in realizing the global explanations of the DL models. SHAP can identify the significance features for the black-box model by measuring the contribution of the features in a dataset to the model. Shapley values calculated based on coalition game theory were used in realizing global explanations. The contribution of a feature to the model is directly proportional to the Shapley value. In this context, the feature with the highest Shapley value is the most significant feature for the model. The mean Shapley value of a feature is obtained by calculating the average of the Shapley values calculated for each instance in the dataset or subset. The SHAP charts given below contain the average Shapley value calculated over the whole dataset unless otherwise stated.

The 20 most significant features that make the highest contribution to the DNN and CNN based NIDS models trained with the NSL-KDD dataset, according to the average Shapley values, are shown in Figs. 1 and 2, respectively. As can be seen from these figures, the most significant feature for NIDS models was specified as the "dst_bytes" (number of data bytes from destination to source). The second most important feature is determined as "src_bytes" (number of data bytes from source to destination). In this case, we can say DL-based NIDS models developed on NSL-KDD were mostly affected by the amount of data.

The effects of the 20 features based on the Shapley values to class-based in classifying of "normal" and "abnormal" for DNN based NIDS models trained on NSL-KDD can be seen in Figs. 3 and 4, respectively. Each point in these figures represents a single SHAP value for a prediction and feature. The red color indicates a higher value for a feature, whereas the blue color indicates a lower value for a feature.

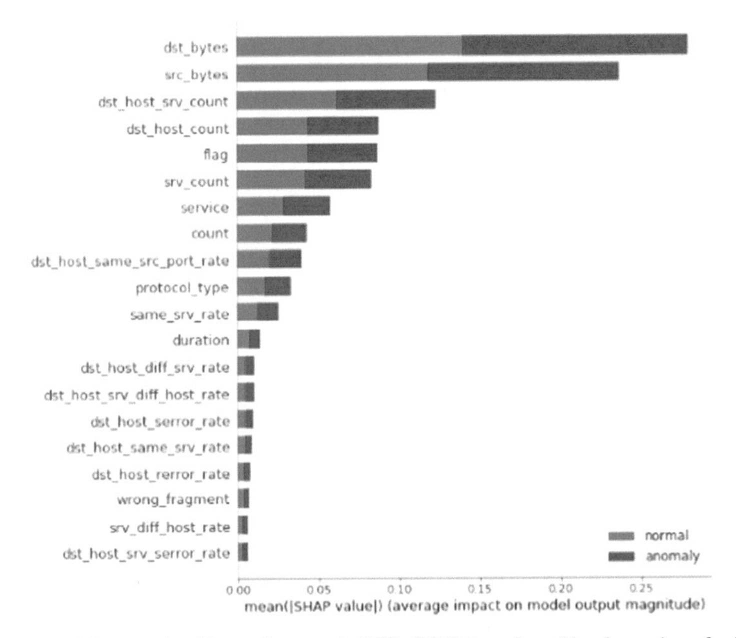

Fig. 1. The 20 most significant features in NSL-KDD based on Shapley values for DNN

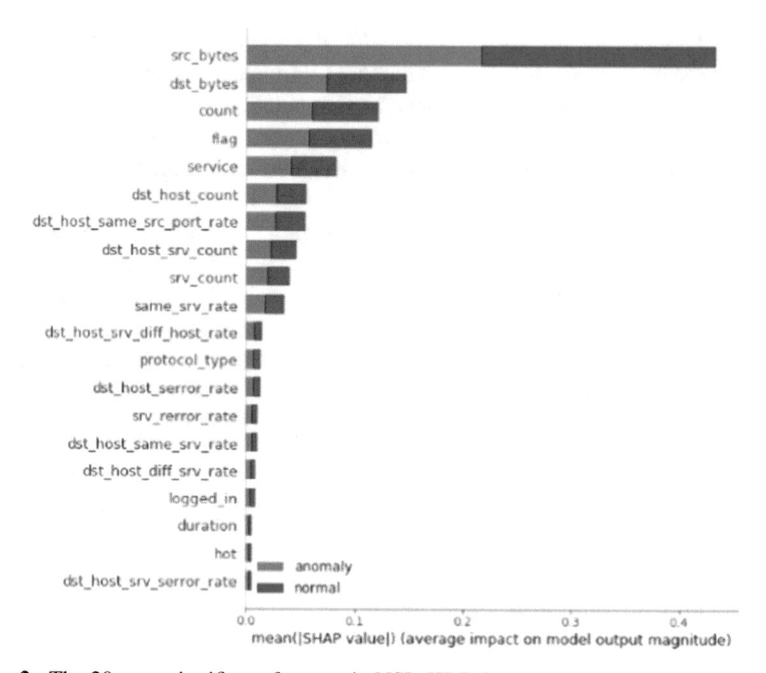

Fig. 2. The 20 most significant features in NSL-KDD based on Shapley values for CNN

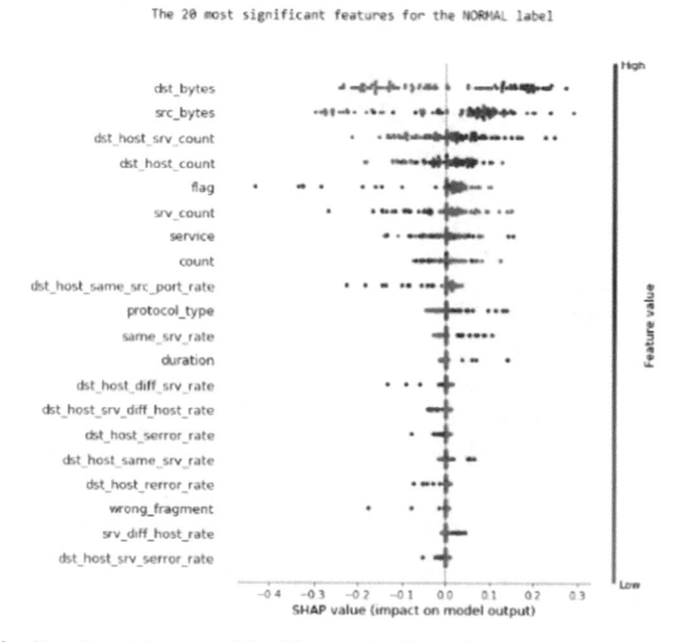

Fig. 3. Class-based impacts of the 20 most significant features of a normal instance

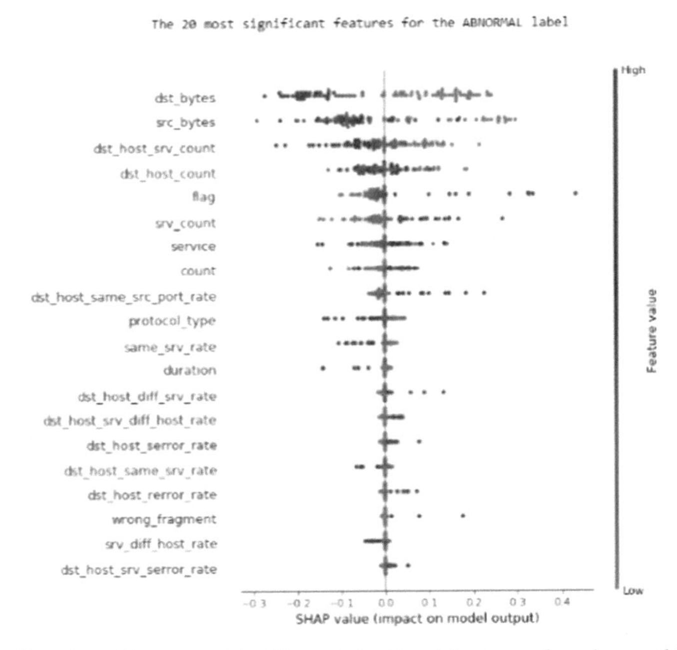

Fig. 4. Class-based impacts of the 20 most significant features of an abnormal instance

Local explanations of individual inputs can also be performed with the SHAP. The local explanations based on SHAP of normal and abnormal labelled instances on the NSL-KDD dataset were shown in Fig. 5, respectively.

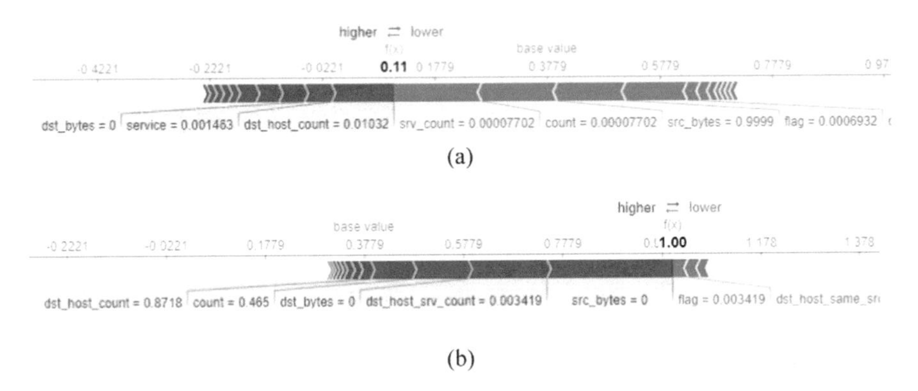

Fig. 5. Local explanations based on SHAP of normal (a) and abnormal (b) instances

LIME is used to create local explanations of individual inputs. The explanations of two sample instances classified as normal and abnormal by the DNN model trained on the NSL-KDD dataset were shown in Figs. 6 and 7, respectively. LIME performs explanations with a linear model based on the absence of features in individual inputs. Although LIME has a simple structure, it has high explanatory power. The weight values of features that show impacts of the features on the classification process for an attack instance are shown in Fig. 7. For these instances, the effect values were calculated over the 15 most significant features, and the totals constitute the prediction probabilities of the class labels.

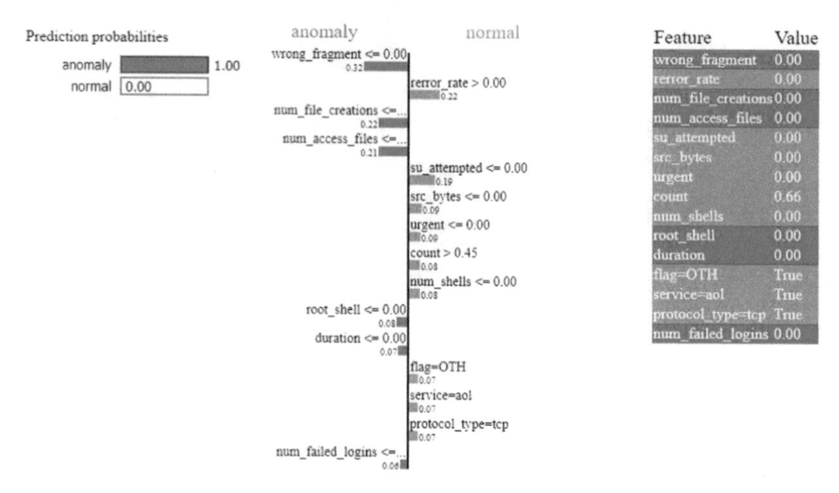

Fig. 6. Explanation of the DNN model prediction for a normal instance with LIME

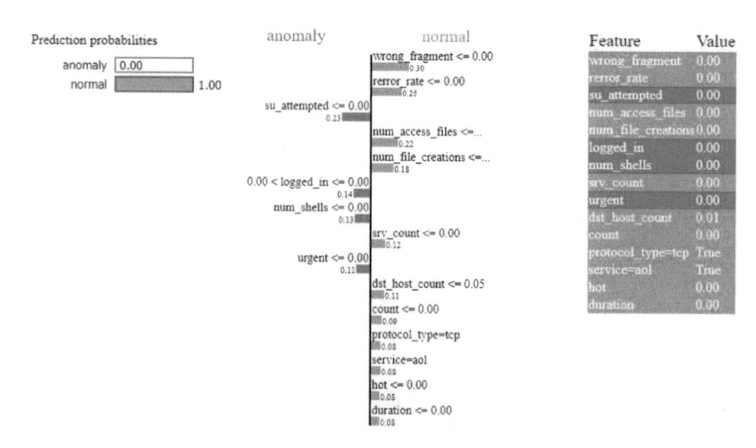

Fig. 7. Explanation of the DNN model prediction for an abnormal instance with LIME

The explanation of the DNN model prediction of a web request that contains SQLi attack footprints in the GAZI-HTTP dataset by LIME is shown in Fig. 8. It can be seen from Fig. 8 that LIME pointed out the correct words for SQLi attack. Thanks to the structure created with LIME in the study, the power of textual graphics was used in the explanation of web requests. Similarly, the explanation of an Open Redirect (ORED) attack with LIME is shown in Fig. 9.

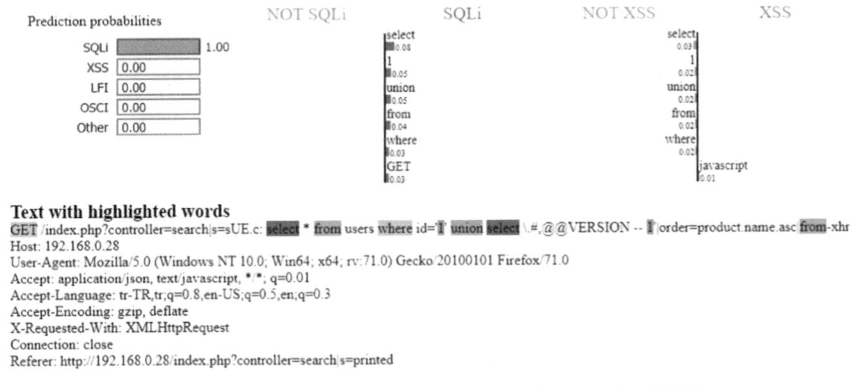

Fig. 8. Explanation of a SQLi anomaly instance by LIME

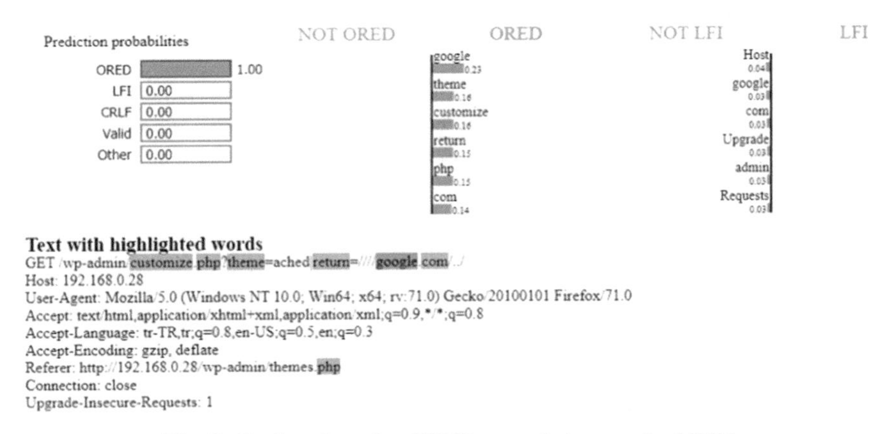

Fig. 9. Explanation of an ORED anomaly instance by LIME

6 Conclusions and Future Work

For confidence in IDS, anomaly-based intrusion detection systems should be transparent and have interpretable decisions. In this study, the explanations of deep learning-based NIDS and WAF models were aimed and an XAI framework was developed. Intrusion detection models were developed at both the network layer and application layer levels in the study. SHAP and LIME-based structures were created to provide global and local explanations of the IDS models developed based on DNN and CNN. The mean Shapley values were used to explain the most significant features affecting the models. Feature weight values generated by LIME were used in addition to the Shapley value in the local explanations. The study is important in terms of considering both web application security and network security together. In addition, the proposed structure has been created to support all deep learning and even state-of-the-art machine AI algorithms, as well as the DNN and CNN models presented in the study. The study demonstrated the most effective features of various attack types, as well as a visual explanation of the model predictions. Local explainability was achieved by two different methods with

LIME and Shap. In web requests, property values are words in the request; in network packets, explanations are provided with properties containing continuous or categorical values. Global explainability was achieved with Shap values based on coalition game theory. As a result, the most significant features and relationships in the datasets that affect black box model predictions were highlighted. In future studies, it is planned to develop a new domain-specific explanation approach that can be used to explain deep learning-based intrusion detection models instead of existing approaches. Thus, it will be possible to make more specific and stronger explanations according to the anomaly types.

References

1. Karacan, H., Sevri, M.: A novel data augmentation technique and deep learning model for web application security. IEEE Access **9**, 150781–150797 (2021)
2. Wang, M., Zheng, K., Yang, Y., Wang, X.: An explainable machine learning framework for intrusion detection systems. IEEE Access **8**, 73127–73141 (2020)
3. Sarhan, M., Layeghy, S., Portmann, M.: An explainable machine learning-based network intrusion detection system for enabling generalisability in securing IoT networks. arXiv e-prints, arXiv preprint arXiv:2104.07183 (2021)
4. Le, T.T.H., Kim, H., Kang, H., Kim, H.: Classification and explanation for intrusion detection system based on ensemble trees and SHAP method. Sensors **22**(3), 1–28 (2022)
5. Mahbooba, B., Timilsina, M., Sahal, R., Serrano, M.: Explainable artificial intelligence (XAI) to enhance trust management in intrusion detection systems using decision tree model. Complexity **2021**, 1–14 (2021)
6. Islam, S.R., Eberle, W., Ghafoor, S.K., et al.: Domain knowledge aided explainable artificial intelligence for intrusion detection and response. arXiv preprint arXiv:1911.09853 (2019)
7. Ahn, S., Kim, J., Park, S.Y., Cho, S.: Explaining deep learning-based traffic classification using a genetic algorithm. IEEE Access **9**, 4738–4751 (2021)
8. Marino, D.L., Wickramasinghe, C.S., Manic, M.: An adversarial approach for explainable AI in intrusion detection systems. In: 44th Annual Conference of the IEEE Industrial Electronics Society, pp. 3237–3243 (2018)
9. Goodfellow, I., Bengio, Y., Courville, A.: Deep Learning. MIT Press, Cambridge (2016)
10. Schmidhuber, J.: Deep learning in neural networks: an overview. Neural Netw. **2015**(61), 85–117 (2015)
11. Dellaferrera, G., Woźniak, S., Indiveri, G., Pantazi, A., Eleftheriou, E.: Introducing principles of synaptic integration in the optimization of deep neural networks. Nat. Commun. **13**(1), 1–14 (2022)
12. Bengio, Y.: Learning Deep Architectures for AI. Now Publishers Inc., Delft (2009)
13. Guo, Y.M., Liu, Y., Oerlemans, A., et al.: Deep learning for visual understanding: a review. Neurocomputing **187**, 27–48 (2016)
14. Arrieta, A.B., Díaz-Rodríguez, N., Del Ser, J., et al.: Explainable artificial intelligence (XAI): concepts, taxonomies, opportunities and challenges toward responsible AI. Inf. Fusion **58**, 82–115 (2020)
15. Ribeiro, M.T., Singh, S., Guestrin, C.: Why should I trust you? Explaining the predictions of any classifier. In: Proceedings of the 22nd ACM SIGKDD International Conference on Knowledge Discovery and Data Mining, pp. 1135–1144 (2016)
16. Molnar, C.: Interpretable machine learning (2020). https://lulu.com
17. Lundberg, S., Lee, S.I.: A unified approach to interpreting model predictions. arXiv preprint arXiv:1705.07874 (2017)

18. Sevri, M., Karacan, H.: Two stage deep learning based stacked ensemble model for web application security. KSII Trans. Internet Inf. Syst. (TIIS) **16**(2), 632–657 (2022)
19. Sevri, M., Karacan, H.: Deep learning based web application security. In: 2nd International Conference on Advanced Technologies, Computer Engineering and Science (ICATCES), pp. 349–354 (2019)
20. Tavallaee, M., Bagheri, E., Lu, W., Ghorbani, A.: A detailed analysis of the KDD CUP 99 Data Set. In: IEEE Symposium on Computational Intelligence for Security and Defense Applications (CISDA), pp. 1–6 (2009)

A Color Channel Based Analysis on Image Tessellation

Turan Kibar[1(✉)] and Burkay Genç[2]

[1] Institute of Informatics, Hacettepe University, Ankara, Turkey
kibarturan@gmail.com
[2] Computer Engineering Department, Hacettepe University, Ankara, Turkey
burkay.genc@hacettepe.edu.tr

Abstract. In this study, it is investigated whether separate triangulation of the RGB components that make up the image would be more efficient in terms of size and quality than direct triangulation of the main image. Different tessellation, point selection and coloring techniques were used for the research, and which technique was better at which point and the advantages it provided were investigated. Image channels are generally used in areas such as improving underwater photographs, identifying disease in computer aided diagnosis, and cryptography, but the advantages of transmitting and storing image data have not been adequately investigated. In our research, it has been shown experimentally that instead of keeping the vertex coordinates and color of all triangles forming the triangulation, it is sufficient to keep one-third of it, and it is more advantageous in terms of sizing to keep the color of a certain number of clusters instead of keeping the colors of all triangles.

Keywords: image triangulation · image tessellation · RGB components

1 Introduction

The importance of image storage is increasing day by day. Semiconductor part producers store photos of wafers and different components for production functions, in conjunction to meet compliance and liability necessities. Insurance companies store vehicle and structural images to confirm damage claims. Businesses and law industries store images of contracts and several other types of documents. Online shopping services require to store massive amounts of product photos.

The focus of this paper is on whether the separate triangulation of the RGB components that make up the image will be more efficient in size and quality than the triangulation of the main image directly. It will also be studied whether the use of different layering algorithms, different point selection algorithms and different color selection techniques would make a difference between the output and the original image. Difference will be compared quantitatively by calculating the Structural Similarity Index (SSIM) for each case. In order to do this, one of the tessellation techniques, Delaunay Triangulation, will be used, and unlike previous studies, RGB components of the image will be triangulated

D. J. Hemanth et al. (Eds.): ICAIAME 2022, ECPSCI 7, pp. 56–72, 2023.
https://doi.org/10.1007/978-3-031-31956-3_5

separately instead of the main image, and while the final image is being created, these created triangulations will be combined and the image will be reconstructed. How different the result from the original image will be found by changing the parameters used while creating the triangulation. Since it will be possible to determine the depth of the triangulation to be made by the user, it will provide the user with a choice in cases where image quality is important, or speed is a factor.

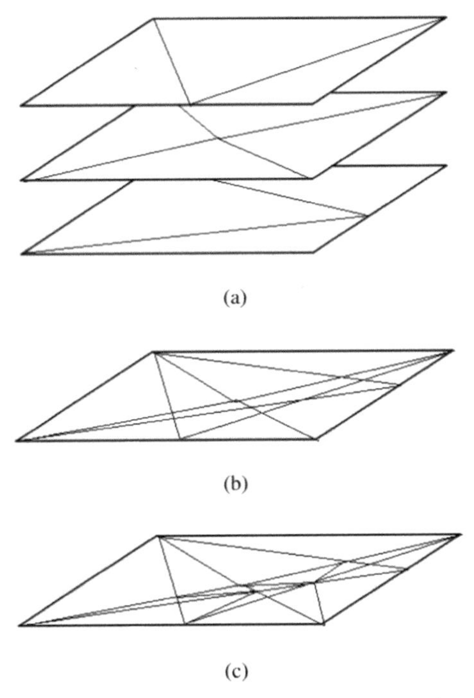

Fig. 1. RGB Channels and superposed result (a) RGB Channels to be tessellated (b) Channels superimposed (c) Resultant triangles

As shown in Fig. 1a, when the RGB layers are triangulated separately, a total of 10 ($3 + 4 + 3$) regions are created. Then these layers are superimposed as in Fig. 1b, 15 regions are formed (1 pentagon, 3 quadrilaterals, 11 triangles). When the polygons in these regions are converted into triangles, 20 triangles are formed as shown in Fig. 1c.

In this study, 3 images were reconstructed in 18 different ways, using 3 different tessellation techniques mentioned above, 3 different point selection techniques and 2 different coloring techniques. Total of 54 experiments are run. These techniques are summarized below:

Tessellation Methods:

- Direct Image Tessellation
- Tessellation of Superposed Channels
- Superposition of Tessellated Channels

Point Selection Methods:

- Randomization
- Local Variance
- Entropy

Coloring Methods:

- Full Coloring
- Cluster Coloring

Combinations of these methods were run and the result images were written to disk and all results were poured into the spreadsheet.

Tessellation is done by Delaunay Triangulation. Delaunay Triangulation is used because output triangles are the most uniform triangles out of other triangulation methods and triangles are close to regular simplices. Similarity results are measured by SSIM (Structural Similarity Index Measure). K-means clustering is the technique used for determining clusters of colors. For the simulations we used the Python programming language (Python Software Foundation, https://www.python.org/).

Method devised in this paper provides the opportunity to transmit and store images in less size without sacrificing much on quality. In our method, we achieve this by employing tessellation of RGB channels instead of the tessellating the main image. By employing channel triangulation, 3.46 times less storage is achieved with a marginal quality tradeoff and it is shown that Tessellation of Superposed Channels method can achieve similarity ratios close to 1 with ease and cluster coloring may be preferred over full coloring when storage, encoding and transmission of the image over Internet are the main concerns.

2 Related Works

Manish et al. (2014), proposed a method for an RGB image encryption and decryption using discrete wavelet transform and using RGB channels in the encode and decode phases [1].

Selimovic et al. (2021), proposed a new method of steganography based on a combination of Catalan objects and Voronoi–Delaunay triangulation. Steganography is the practice of hiding a secret message inside of something that is not secret. The main goal of his paper is to transmit a message via the Internet (or some other medium) using an image so that the image remains uninterrupted. Some classified message hidden in the picture sent by using encrypted Delaunay Triangulation hiding the message in the vertices of the Delaunay triangulation [2].

Image channels are frequently used in computer aided disease and cancer analysis. Peng et al. (2012) (as cited in Xingyu (2017)), studied on finding out whether the patients have prostate cancer or not. He divides images into channels, takes the red channel from the RGB color space, the yellow channel from the CMYK color space and the Hue channel from the HSV color space and examines the color content and other features of these channels. He diagnoses the diseased prostate by comparing a normal prostate image with the images obtained from different channels according to the mean and standard deviations in terms of color content. Xingyu also made an extensive table of which color spaces and channels are used to diagnose which disease in her study [3, 4].

Carrasco et al. (2020), used the Red Channel of RGB color palette and Delaunay triangulation in his study to determine the stomatal cell spatial distribution on the leaves of the plants [5].

Lou Beaulieu et al. (2020), used subtraction of green channel from red channel of Chinese hamster ovary cell microscopic image to retrieve the signal to be worked on in order to use it for the study of a cellular phenomenon called mechano-transduction involved in sensing mechanical pain in biological organisms [6].

Ghate et al. (2021) (as cited in Shaik (2018)), since the green channel is preserved in underwater photography and the red channel with longer wavelength is more attenuated, the blurry image is enhanced by transferring some of the content in the green channel to the red channel to strengthen the red channel signal [7, 8].

Song et al. (2019), Delaunay triangulation and RGB channels are used for constructing 3D underwater mapping by iteratively applying underwater optical model to each depth pixel of each channel for correcting refraction effects. Delaunay triangulation is used to determine which triangle ray hits first to choose the intersection point with the minimum depth [9].

Wang et al. (2021), RGB Color channels of images are used which are taken from Unmanned Air Vehicles in order to determine the rice cluster distribution uniformity of mechanized paddy field transplanting in south China. He used binarization of the image with respect to the intensity values of pixels in Red and Green channels. Image is binarized if red or green channel value of an individual pixel is greater than a certain threshold value in order to identify seedlings correctly and to separate it from the soil. Later, he used Delaunay triangulation to obtain the polygonal area that includes the rice clusters which will be measured for uniformity [10].

Toivanen et al. (1999), developed a method using Delaunay triangulation to compress the image, using the "Distance Transform on Curved Space (DTOCS)" technique [11].

Morell et al. (2014), developed a method that creates planes from a 3-dimensional point cloud, compresses the points on these planes with Delaunay triangulation, and allows the colors of these points to be transmitted and reconstructed by preserving them with the color partitioning phase [12].

Prasad et al. (2016), perform uniform color triangulation after detecting edges in the image and divide the image into homogeneous regions in terms of color. Side by side and similarly colored triangles are then converted to polygons [13].

Pal et al. (2018), stored password in the image format by segregating the text data into the 3-channelled RGB image to overcome the dependency of key used to encrypt the information which is exposed to either both or one of the parties which creates a vulnerability [14].

3 Method

3.1 Direct Image Tessellation

In this technique, 3 different point selection methods are used to obtain vertices and then Delaunay triangulation is applied on these vertices to obtain simplices. After that colors of simplices are determined and image is recreated. Example outputs are shown in Fig. 2.

Fig. 2. Lena image and its various triangulations. (a) Input image (b) 100 triangles (c) 1000 triangles (d) 2000 triangles

3.2 Superposition of RGB Channels After Triangulation and Coloring

In this technique steps are as follows as seen as Fig. 3:

- Separation of the input image into red, green and blue channels
- Tessellation of each channel by Delaunay Triangulation
- Coloring each tessellation
- Superposition of channels.

When Superposition of Tessellated Channels algorithm is compared to Direct Image Tessellation algorithm, Direct Image Tessellation algorithm gives better similarity results. In case of quality of the tessellations between these two algorithms, on average 5% of decrease is measured. Decrease stems very likely from the observation that if triangles on all layers do not coincide they create a low-pass filter effect and reduce each other's brightness because the triangles in individual channels are bigger than triangles formed in Direct Image Tessellation algorithm in size. Since it represents a larger area, a more general color is assigned as a representative color because of its large size.

When output image sizes are compared as per Table 1, on average 3.46 times smaller storage is achieved by Superposition of Tessellated Channels algorithm with respect to Direct Image Tessellation. Instead of storing 2000 triangles, it is sufficient to keep 300 triangulations (100/100/100) total. (When 3 triangulations are superposed, approximately 21.62 times the individual layer polygons occur on the final image as per calculations.) A great advantage of this technique over other image storage techniques is that it is independent of the size of the input image. The size of the image does not matter, as we only store the coordinates, faces and colors of the simplices.

3.3 Coloring After Superposition of Triangulated RGB Channels

In this technique, instead of tessellating and coloring channels and recombining them afterwards, channels are triangulated and superposed first and resultant tessellation is colored. Instead of adding RGB colors of each layer on top of each other, this technique superpositions triangulations and forms polygons first, then determines the contours of these polygons and infer color of that polygon by retrieving all pixel colors that form the polygon and finding median color of it from the input image and recreating it.

Example outputs for the figures can be seen on Fig. 4. In this technique steps are as follows as seen on Fig. 5.

- Separate the input image into red, green and blue channels
- Triangulate each channel by Delaunay Triangulation
- Get point indices and coordinates for the triangles formed
- Fill each triangle with different colors channel by channel
- Add channels (Fig. 6)
- Find the unique colors in the result, each will correspond to a polygon and its value will tell which two initial triangles intersect there.
- For every unique polygon find contour (Fig. 7)
- Get the average color of the pixels composing the contours from input image (Fig. 8)
- Recreate the final image using these colors

Although the polygons are a few pixels wide, SSIM that are very close to 1.0 can be reached with this method easily (Tessellation of Superposed Channels), but the other methods saturate after a point. For example, some artifacts started to appear in Lena image after 6000 triangles (3000 points) in Direct Image Tessellation method, but there was no such problem in this technique.

Very high number of polygons on the output can be achieved such as 97870, 361718 polygons. Also, high SSIM (similarity) results can be achieved with this technique.

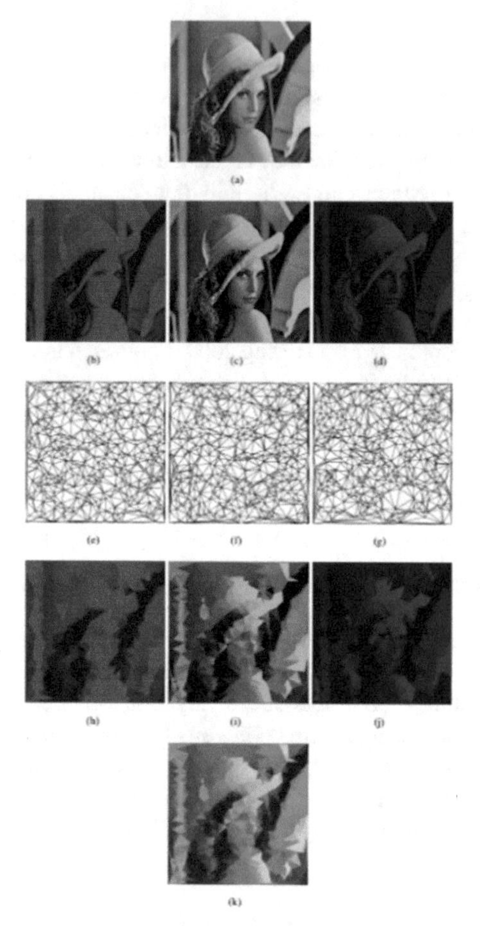

Fig. 3. Steps of Superposition of Tessellated Channels Algorithm (a) Lena image (b) Red channel (c) Green channel (d) Blue channel (e) Red channel triangulation (1000 triangles) (f) Green channel triangulation (1000 triangles) (g) Blue channel triangulation (1000 triangles) (h) Red channel output (i) Green channel output (j) Blue channel output (k) Output image is a superposition of 3(h), 3(i) and 3(j)

Highest SSIM result achieved by Tessellation of Superposed Channels method on shapes image using Entropy point selection. This resulted in 0.985 SSIM on 62848 polygons.

In Tessellation of Superposed Channels method, not only triangles, but also polygons such as quadrilaterals and pentagons and other various polygons can be formed. Also, when polygon count is small, sharp edges can form much more frequently in this technique than other two techniques.

3.4 Point Selection and Coloring Techniques Used

Randomization, Local Variance and Entropy is used as point selection techniques. Full Coloring and Cluster Coloring are used as coloring techniques.

Table 1. Comparison of output file sizes between Direct Image Tessellation Method and Superposition of Tessellated Channels Method (Gain)

Point Selection Method	Average Final Image Polygon Count	Size (kB) in Technique 1 (Direct Image Tesselation)	Size (kB) in Technique 2 (Superposition of Tesselated Channels)	Gain (File Size for Technique 2 / File Size for Technique 1)
Randomization	200	6,67	2,18	3,06
Randomization	2000	70,00	17,73	3,95
Randomization	6000	222,30	55,52	4,00
Randomization	200	6,53	2,16	3,02
Randomization	2000	68,49	17,58	3,90
Randomization	6000	217,70	55,04	3,95
Randomization	200	7,33	4,54	1,62
Randomization	2000	68,38	20,14	3,39
Randomization	6000	215,27	57,33	3,75
Local Variance	200	5,53	1,95	2,83
Local Variance	2000	56,73	14,59	3,89
Local Variance	6000	171,24	46,35	3,69
Local Variance	200	5,69	1,94	2,94
Local Variance	2000	54,35	15,15	3,59
Local Variance	6000	145,19	46,64	3,11
Local Variance	200	6,80	4,36	1,56
Local Variance	2000	62,39	18,34	3,40
Local Variance	6000	192,35	51,90	3,71
Entropy	200	5,72	1,94	2,95
Entropy	2000	60,74	14,45	4,20
Entropy	6000	196,04	46,86	4,18
Entropy	200	5,69	1,92	2,96
Entropy	2000	60,36	15,11	3,99
Entropy	6000	193,30	47,73	4,05
Entropy	200	6,69	4,36	1,54
Entropy	2000	62,08	18,34	3,39
Entropy	6000	196,85	51,90	3,79

Point selection per local variance and entropy steps are as follows:

- find the local variance/entropy of the input image as per standard deviation formula (mean of the square of image minus the square of the mean of image)
- choose a point is where the variance/entropy is highest
- apply a gaussian mask to reduce the values around the previously selected point to prevent the next point to be selected from being too close to the previously selected point.
- choose another point where the variance/entropy is highest

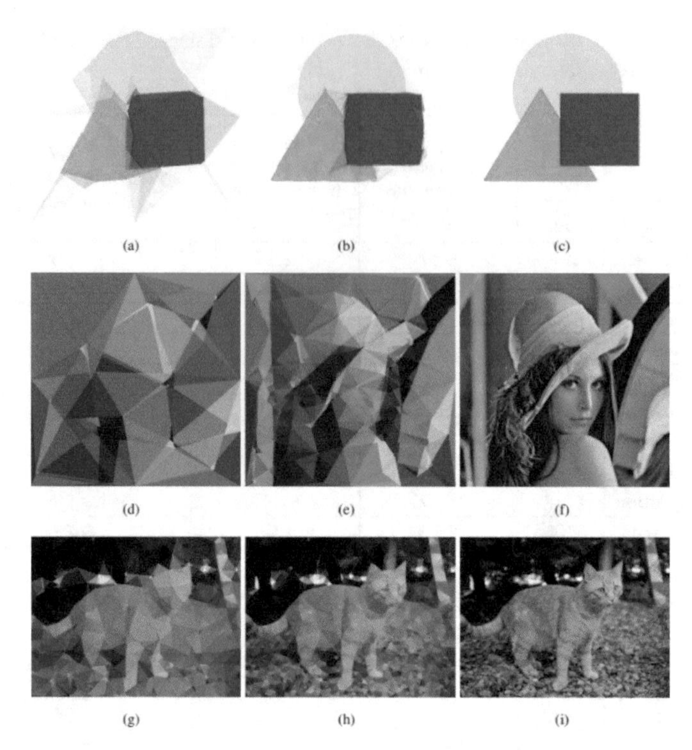

Fig. 4. Examples of Tessellation of Superposed Channels Algorithm. (a) 36 triangles, 227 polygons (b) 116 triangles, 991 polygons (c) 6016 triangles, 50680 polygons (d) 36 triangles, 234 polygons (e) 116 triangles, 1090 polygons (f) 8016 triangles, 64812 polygons (g) 216 triangles, 2465 polygons (h) 1016 triangles, 11120 polygons (i) 40016 triangles, 361718 polygons

Examples of point selection per local variance and entropy are given in Fig. 9 and Fig. 10.

For Full Coloring, steps are:

- generate pixel coordinate list with respect to the shape of the image
- find which triangle which pixel belongs to
- create a table of all triangles, pixels and pixel colors
- group pixel colors by triangles and then find median of the colors which represents the triangle color

For Cluster Coloring, steps are:

- desired number of clusters is given to the method and the most dominant colors in the image is found
- Delaunay triangulation is applied on image
- median color of each triangle found
- full colors are compared with the cluster colors and the triangle colors are replaced with these cluster colors

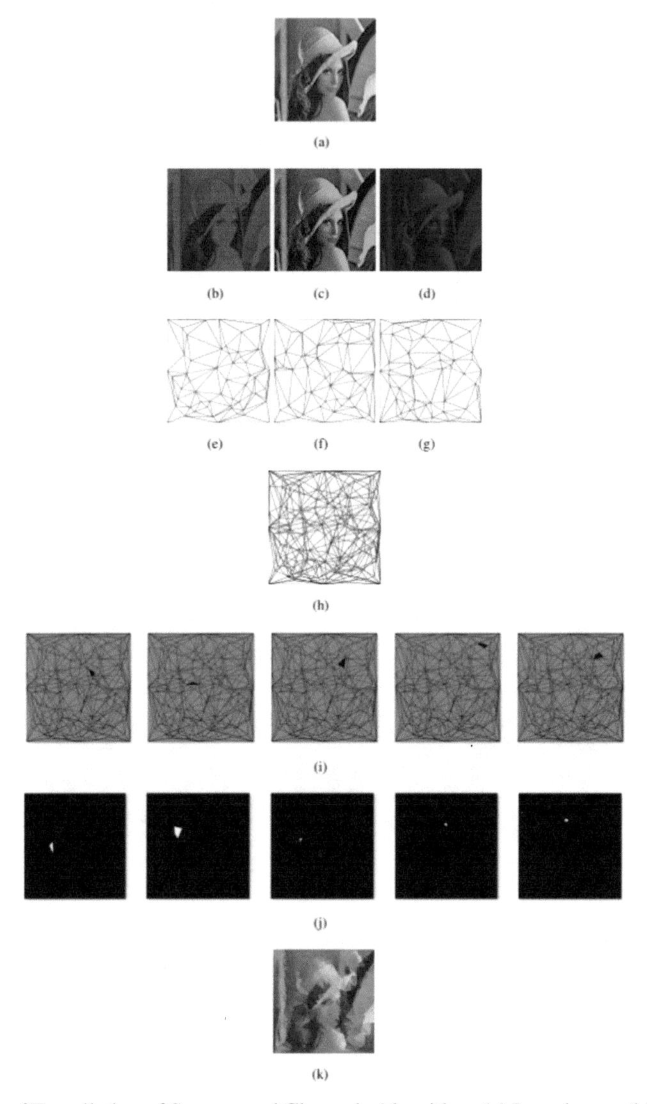

Fig. 5. Steps of Tessellation of Superposed Channels Algorithm. (a) Lena image (b) Red channel (c) Green channel (d) Blue channel (e) Red channel triangulation (100 triangles) (f) Green channel triangulation (100 triangles) (g) Blue channel triangulation (100 triangles) (h) Superposed triangulations (1221 polygons) (i) Resultant polygons (5 arbitrary samples taken) (j) Polygons prepared for contour retrieval (5 arbitrary samples taken) (k) Output

Comparison is done with respect to the Euclidean distance between the cluster colors and the triangle color. The cluster color with the lowest Euclidean distance is assigned instead of the triangle color. This is done for all triangles.

Examples of cluster coloring are given in Fig. 11 and Fig. 12.

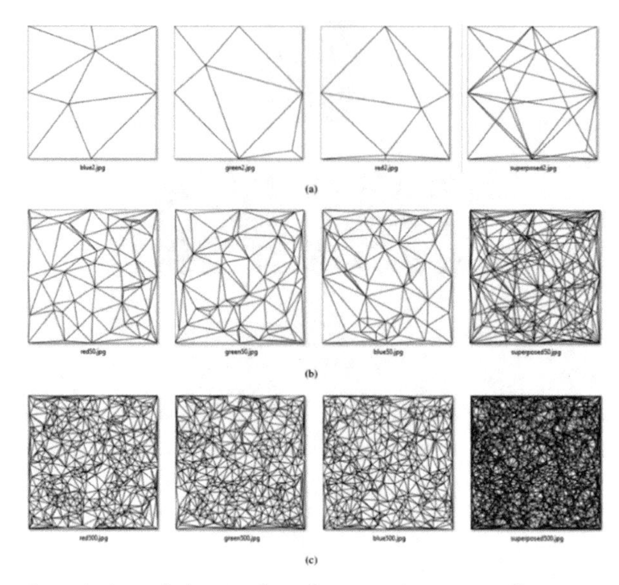

Fig. 6. Various channel triangulations on Lena image and corresponding superposed outputs (a) 2 points on each channel (Final wireframe: 57 Polygons) (b) 50 points on each channel (Final wireframe: 1177 Polygons) (c) 500 points on each channel (Final wireframe: 11608 Polygons)

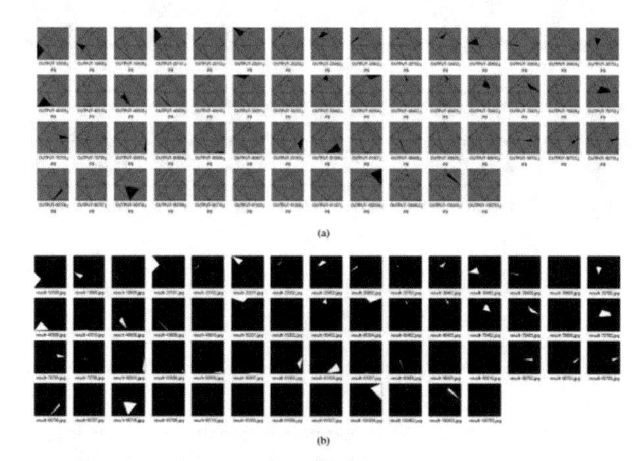

Fig. 7. Retrieving Contours from Superposed Triangulations (2 points triangulated on each channel, total of 57 Polygons) (a) Polygons of superposed wireframe (b) Isolated Polygons

4 Results

The biggest improvement by employing channel triangulation is that we gain 3.46 times less storage for %5 quality tradeoff (Superposition of Tessellated Channels method). Rather than storing 2000 triangle data, it's sufficient to store 300 triangulation data totals. (100 on each channel). Or instead of storing 6000 triangles, it will be enough to store 900 triangle's data (300 on each channel). This is a big advantage for minimizing

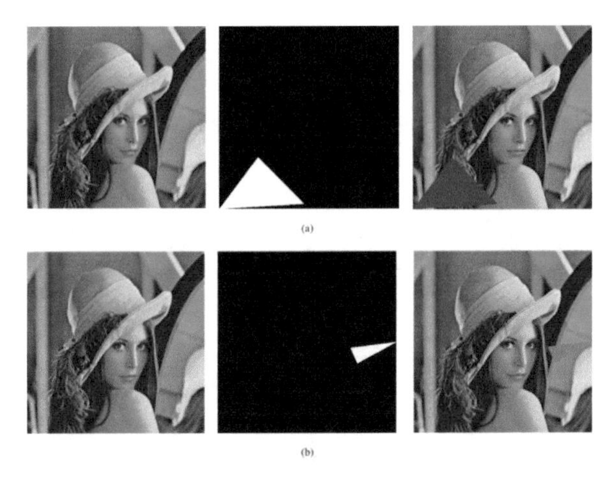

Fig. 8. Retrieving median Color of a Triangle from Main Image (a) An arbitrary contour (b) Another arbitrary contour

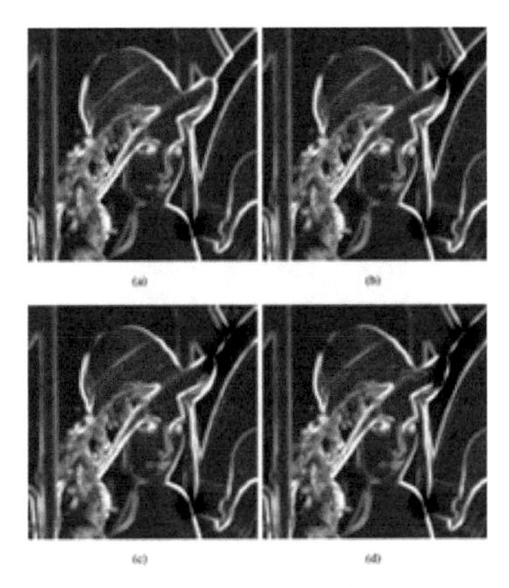

Fig. 9. Outputs of Local Variance Algorithm and Gaussian Mask for Lena image (a) 1st point chosen (b) 1st point masked (top right) (c) 2nd point masked (d) 3rd point masked

storage. Also, size of the image does not matter, as we only store the coordinates, faces and colors of the triangles.

As per Tessellation of Superposed Channels method, when 3 triangulations are superposed, approximately 21.62 times the individual layer polygons occur on the final image. The biggest advantage of this method is that the similarity ratio can approach to 1 with very large polygon numbers with ease. In other methods, increasing the number of polygons at a certain point contributes very little to the similarity (reaches saturation) and some artifacts start to appear, but in this method, the similarity can be very high since the polygons can be even a few pixels in size. As per the disadvantages, triangles or quadrilaterals or other polygons with sharp corners may form in triangulations when

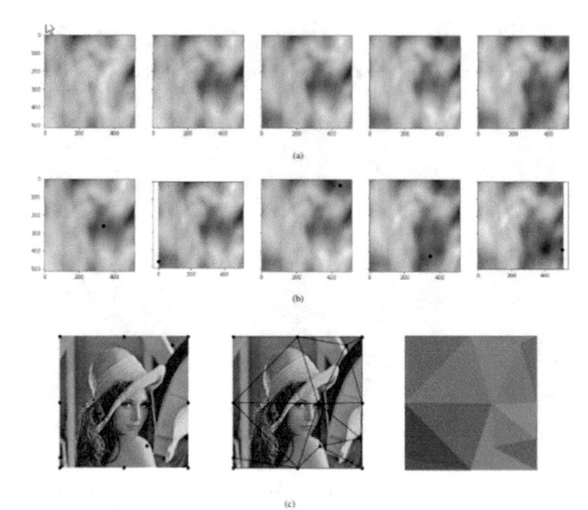

Fig. 10. Five-point (without edge points) Entropy Algorithm for Lena image (a) Entropy map before Gaussian Masking (b) Entropy map after Gaussian Masking around selected point (c) Points, Triangulation, and Tessellated Lena images

Fig. 11. Various outputs of clustering algorithm, Lena image, Superposition of tessellated channels method (a) Red channel (400 Triangles, 2 Clusters), Green Channel (400 Triangles, 2 Clusters), Blue Channel (20 Triangles, 50 Clusters) superposed (b) Red channel (10K Triangles, 10 Clusters), Green Channel (2K Triangles, 2 Clusters), Blue Channel (400 Triangles, 2 Clusters) superposed (c) Red channel (10K Triangles, 10 Clusters), Green Channel (10K Triangles, 50 Clusters), Blue Channel (10K Triangles, 50 Clusters) superposed

triangulated in a small number of points. On the other hand, when the number of points is increased, this disadvantage disappears.

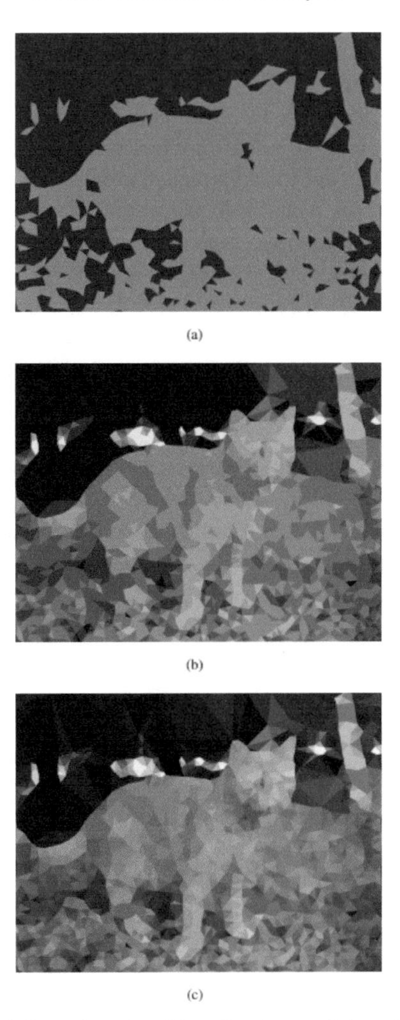

Fig. 12. Various outputs of clustering algorithm, Cat image, Direct Image Tessellation method (a) 2K triangles, 2 Clusters (b) 2K triangles, 10 Clusters (c) 2K triangles, 50 Clusters

In Tessellation of Superposed Channels method, it was necessary to find the polygons of image after superimposition process. There are some suggested polygon clipping algorithms for this. Like the Sutherland-Hodgeman algorithm or the Weiler-Atherton algorithm. We have considered using these algorithms, but these algorithms generally serve to determine which polygons are formed by intersecting specific two polygons so not practical for our case. In our case, we had to produce our own algorithm because there were 3 channels. The polygon determination process we use, first assigns an arbitrary color to the triangulations in each channel, and then the colors are summed. Then, pixels with the same value are determined and it is assumed that pixels with the same value form a polygon. The increment value of automatic color assignment should be arranged in order to prevent duplicates. This algorithm is not limited by the number of channels.

The number of channels can be increased. This algorithm is used for determination of the polygons in final superimposed image and it is very accurate.

When Point Selection methods are examined, it is seen that the best algorithm is the entropy method. On average, the entropy method is % 6,6 better than the random method and % 4,4 better than the local variance method.

When coloring algorithms examined, as the number of clusters increases, higher similarity rates are achieved, but when full coloring is done instead of cluster coloring, an expected increase is not observed. For example, when full coloring is used instead of 35 cluster colors, there is an average increase in SSIM of only % 1,7. (Refer to Fig. 13). For 10K triangles, we should store 10K particular colors however with a % 1.7 tradeoff of the image quality, we only need to store 35 values of RGB colors. This shows us that it may not be worth the time to compute and store the colors of all the triangles when doing triangulation. Cluster coloring may be preferred over full coloring. It is thought that this will provide an advantage over full coloring in the storage, encoding and transmission of the image over Internet. In addition, it is found experimentally that median color of the pixels is giving better results than finding mean color of the pixels to determine the color of the triangle. (Refer to Fig. 14).

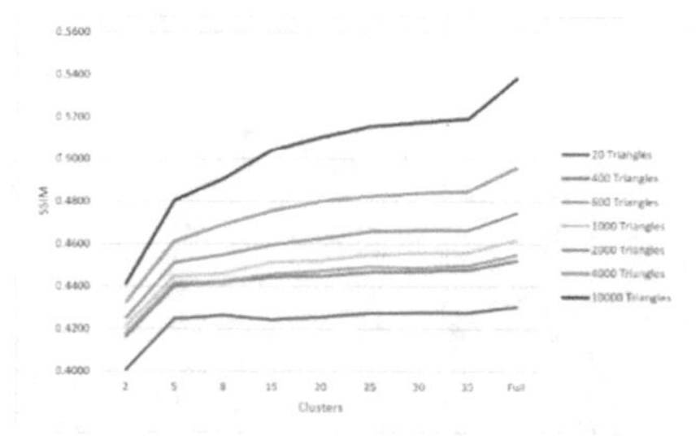

Fig. 13. K-Means Cluster Coloring image for various Clusters and Similarity Measure (Direct Image Tessellation Method, Entropy Point Selection, Cat Image)

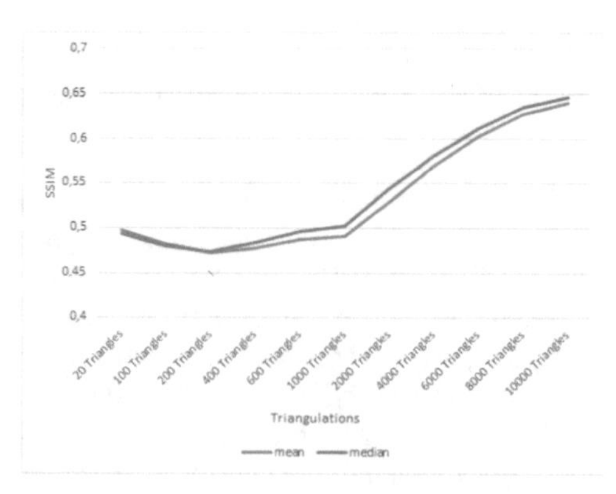

Fig. 14. SSIM results of triangulations for mean and median determination of average pixel color of triangle.

5 Future Work

As triangulation technique, Delaunay triangulation can be refined to triangulate recursively on regions where non-homogeneity occurs more. First image will be divided into equal windows. In each window, according to the similarity criterion desired in the original image, recursive triangulation can be made until similarity criterion is reached. Initially, points are selected in that window, then points are triangulated, the colors of the triangles are determined and the similarity ratio is calculated. If the desired similarity criterion is reached, the triangulation process stops. If the similarity criterion is not reached, this process is repeated recursively. The quality of the picture will change according to the size of the window to be selected here and the desired similarity value.

If we look at the point selection methods, point selection is indeed a problem of locating areas where non-homogeneities occur more e.g. as color changes, transitions and variations. Three point selection techniques are used in this paper but there are other techniques such as Super Pixel algorithm, using the weak texture from noise level estimation from a single image [15], 2D wavelet transform and methods which look for the features for homogeneity and use them to find homogenous regions and then select the inverse. These techniques can be used for future studies on point selection part on this topic.

References

1. Kumar, M., Mishra, D.C., Sharma, R.K.: A first approach on an RGB image encryption. Opt. Lasers Eng. **52**, 27–34 (2014)
2. Selimović, F., Stanimirović, P., Saračević, M., Krtolica, P.: Application of Delaunay triangulation and Catalan objects in steganography. Mathematics **9**(11), 1172 (2021)
3. Li, X.: Circular probabilistic based color processing: applications in digital pathology image analysis. Doctoral dissertation. University of Toronto, Canada (2017)
4. Peng, Y., Jiang, Y., Yang, X.J.: Computer-aided image analysis and detection of prostate cancer: using immunostaining for Alpha-Methylacyl-CoA Racemase, p63, and high-molecular-weight cytokeratin. In: Machine Learning in Computer-Aided Diagnosis: Medical Imaging Intelligence and Analysis, pp. 238–256. IGI Global (2012)
5. Carrasco, M., Toledo, P.A., Velázquez, R., Bruno, O.M.: Automatic stomatal segmentation based on Delaunay-Rayleigh frequency distance. Plants **9**(11), 1613 (2020)
6. Beaulieu-Laroche, L., et al.: TACAN is an ion channel involved in sensing mechanical pain. Cell **180**(5), 956–967 (2020)
7. Ghate, S.N., Nikose, M.D.: Recent trends and challenges in Image Enhancement Techniques for Underwater Photography. NVEO-NATURAL VOLATILES ESSENTIAL OILS J.—NVEO **8**, 12272–12286 (2021)
8. Shaik, M., Meena, P., Basha, S., Lavanya, N.: Color Balance for underwater image enhancement. Int. J. Res. Appl. Sci. Eng. Technol. **6**, 571–581 (2018)
9. Song, Y., Köser, K., Kwasnitschka, T., Koch, R.: Iterative refinement for underwater 3D reconstruction: application to disposed underwater munitions in the Baltic sea. ISPRS-Int. Arch. Photogram. Remote Sens. Spat. Inf. Sci. **42**, 181–187 (2019)
10. Wang, X., Tang, Q., Chen, Z., Luo, Y., Fu, H., Li, X.: Estimating and evaluating the rice cluster distribution uniformity with UAV-based images. Sci. Rep. **11**(1), 1–11 (2021)

11. Toivanen, P.J., Vepsäläinen, A.M., Parkkinen, J.P.: Image compression using the distance transform on curved space (DTOCS) and Delaunay triangulation. Pattern Recogn. Lett. **20**(10), 1015–1026 (1999)
12. Morell, V., Orts, S., Cazorla, M., Garcia-Rodriguez, J.: Geometric 3D point cloud compression. Pattern Recogn. Lett. **50**, 55–62 (2014)
13. Prasad, L., Skourikhine, A.N.: Vectorized image segmentation via trixel agglomeration. Pattern Recogn. **39**(4), 501–514 (2006)
14. Pal, S.K., Anand, S.: Cryptography based on RGB color channels using ANNs. Int. J. Comput. Netw. Inf. Secur. **10**(5), 60–69 (2018)
15. Liu, X., Tanaka, M., Okutomi, M.: Noise level estimation using weak textured patches of a single noisy image. In: 2012 19th IEEE International Conference on Image Processing, pp. 665–668. IEEE (2012)

Modeling of Shear Strength of Basalt Fiber Reinforced Clay (BFRC) Soil Using Artificial Neural Network (ANN)

Mehmet Fatih Yazıcı[1](✉), Ahmetcan Sungur[2], and Sıddıka Nilay Keskin[1]

[1] Civil Engineering Department, Suleyman Demirel University, Isparta, Turkey
{mehmetyazici,nilaykeskin}@sdu.edu.tr

[2] Department of Civil Engineering, Graduate School of Natural and Applied Sciences, Suleyman Demirel University, Isparta, Turkey

Abstract. Multilayered ANN is widely used as a modeling tool in geotechnical engineering because it successfully predicts the nonlinear relationship between different parameters and because the network can be trained using only input-output data pairs without simplifying the problem or making assumptions. This study aims to develop an ANN model using the results of 39 direct shear tests conducted on clay soil reinforced with basalt fiber. For this purpose, an ANN-based predictive model, in which the normal stress (σ) applied during the direct shear test, fiber length (L), and fiber content (F_c) are used as input parameters, was created and the shear strength (τ) of the reinforced soil was predicted. A feed-forward network structure consisting of one hidden layer was preferred and a backpropagation (BP) algorithm was used in the training phase. A trial and error procedure was used to determine the number of neurons in the hidden layer. 20% of the data was used for testing and 80% for training. The mean squared error (MSE) and coefficient of determination (R^2) indicators were chosen to measure the success of the network in predicting the actual output values. When using a linear transfer function (LTF) in the output layer and the tangent sigmoid transfer function (TSTF) and two neurons in the hidden layer, the predictions that provide the best agreement with the measured values were obtained.

Keywords: artificial neural network · back-propagation algorithm · soil reinforcement · basalt fiber

1 Introduction

Soil reinforcement techniques are widely used to improve the shear strength properties of low bearing capacity and/or high compressibility soils. Recently, there has been an increased interest in stabilizing soils with fiber materials as an alternative to traditional soil stabilization techniques (cement, lime, geosynthetic material, etc.), as they are sustainable and environmentally friendly materials with low cost and high tensile strength [1]. To determine the shear strength of fiber-reinforced soils in the laboratory, triaxial or unconfined compression and direct shear tests are generally conducted. However,

D. J. Hemanth et al. (Eds.): ICAIAME 2022, ECPSCI 7, pp. 73–81, 2023.
https://doi.org/10.1007/978-3-031-31956-3_6

laboratory tests used to determine shear strength can consume a lot of time depending on the type and number of tests, and some types of tests can be quite costly. For this reason, soft computing tools have been developed that give better results than traditional methods to use time efficiently [2]. ANN, which is one of the soft computing methods, is frequently used in geotechnical engineering as in many other disciplines [3]. Some of the studies carried out with ANN in geotechnical engineering are; the prediction of the axial capacity of piles [4, 5], the prediction of unconfined compressive strength of soil reinforced with waste material [3], the prediction of differential and maximum settlement of a piled raft foundation and moment on the raft foundation [6], the prediction of compaction properties, permeability and shear strength of the soil [7], the prediction of undrained shear strength of a cohesive soil [2], the prediction of cohesion and internal friction angle values of soils [8–10], the prediction of liquefaction potential of a gravelly soil [11], the prediction of displacement of cantilever wall [12]. However, there is no study on the prediction of the shear strength of basalt fiber reinforced soils by ANN. In this study, an ANN-based predictive model was developed using the study's experimental data conducted by Sungur et al. [13] on the shear strength of BFRC soil. For this purpose, the shear strength of fiber-reinforced clay soil was predicted by creating an ANN-based predictive model using the fiber length, fiber content, and normal stress values applied during the direct shear test as input parameters. To determine the model that gives the best agreement between the measured and the prediction values, the structure of the network was changed, and the best ANN-based predictive model was tried to determine.

2 Material and Method

2.1 Dataset

In this study, using the data from the experimental study by Sungur et al. [13], a model ANN was developed to investigate the effects of adding basalt fibers on the shear strength of clay soil with low plasticity. In the study conducted by Sungur et al. [13], basalt fibers with 9 mm, 15 mm, and 22 mm lengths and 9–23 μm diameter were added to the soil in the amounts of %0.5–1.0, %0.5–2.0, and %0.5–1.5 by dry weight of the soil respectively. Samples with dimensions of 60 mm × 60 mm × 20 mm were prepared from fiber-soil mixtures and were sheared at a displacement rate of 1 mm/min under 60, 120, and 240 kPa normal stress in the direct shear test. As a result, shear strength values of 39 different samples were obtained (Fig. 1). Table 1 shows the descriptive statistics parameters of the study data. The stress values given in this table are calculated based on the corrected area.

2.2 ANN

Multilayered ANN is widely used as a modeling tool due to its success in predicting the nonlinear relationship between different parameters and because the network can be trained with input-output data pairs only, without making assumptions or simplifying the problem [14]. ANNs are soft computing tools inspired by the information processing

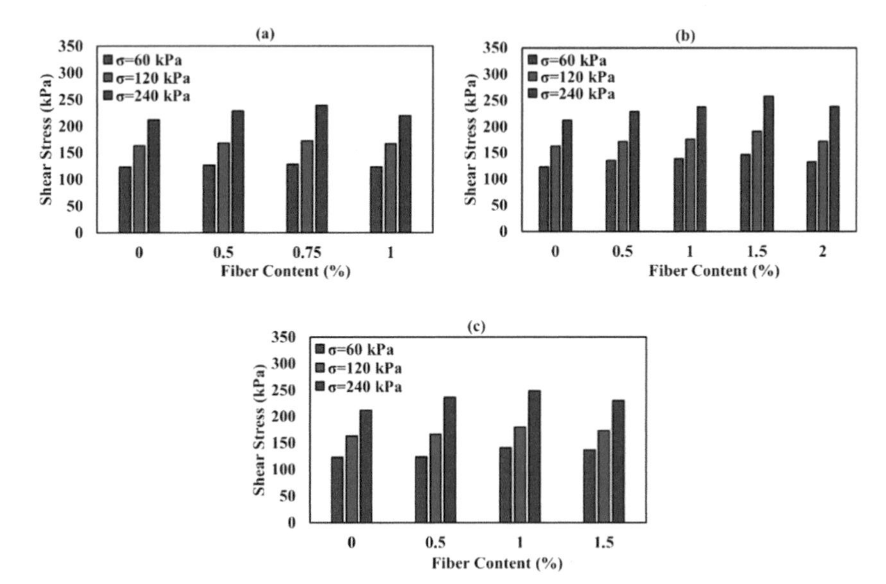

Fig. 1. Shear strength values of clay soil reinforced with basalt fiber in the length of (a) 9 mm, (b) 15 mm and (c) 22 mm [13]

technique of the human brain. The ANN modeling philosophy is like that employed in the development of traditional statistical models. In both situations, the purpose of the model is to capture the relationship between inputs and corresponding outputs [5]. The most popular form of ANNs that use a feedforward network structure to account for the nonlinear relationship between input and output neurons is the multilayer perceptron (MLP). In practice, the BP algorithm, which is one of the supervised learning methods, is generally used to train the MLP network [5, 6, 10, 15, 16]. A multilayered ANN is composed of an input layer and an output layer with several processing units (also called neurons or nodes) interconnected by weighted connections, and one or more hidden layers between these layers. The hidden layer(s) generate(s) an output that propagates forward until it arrives at the output layer, where the prediction value of the network is achieved. In each neuron, the input values (X_i) from the neurons of the previous layer are summed by multiplying the connection weights (w_i), and a bias value (b) is added to the obtained value. Then, this calculated value is passed through a transfer function (f) and the output value (O_n) of the relevant neuron is produced (Fig. 2). Finally, the predicted value of the network is compared with the actual measured output value, and the difference between these values is considered an error. The errors are propagated backward using the gradient descent rule, and weights and biases in the network are updated for all units [7, 17]. The output value (O_n) of a neuron whose mathematical model is presented in Fig. 2 is calculated with the following equation.

$$O_n = f\left(\sum_{i=1}^{n} (w_i X_i) + b\right) \tag{1}$$

Table 1. Descriptive statistical values of the parameters

Symbol	Unit	Min.	Max.	Mean	SD*	SC*	KC*
F_c	–	0	2	0.788	0.603	0.345	−0.655
L	cm	0.9	2.2	1.531	0.510	0.113	−1.344
σ	kPa	57.402	243.398	138.334	74.863	0.390	−1.450
T	kPa	122.625	258.525	177.632	42.475	0.305	−1.260

**The SD, SC, and KC values in Table 1 represent the standard deviation, skewness, and kurtosis coefficients, respectively.*

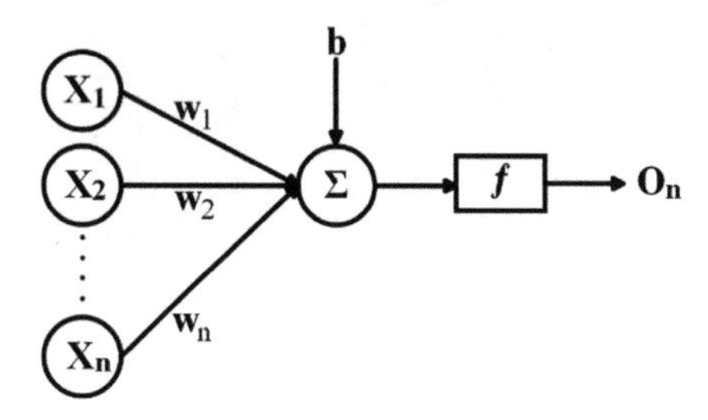

Fig. 2. Mathematical model of a neuron

In a feedforward ANN model, the number of neurons in the input and output layers is determined according to the structure of the problem, while the structure of the hidden layer is created by a trial-and-error procedure. As the number of hidden layers increases, the computation time increases and the overfitting problem occurs, which causes the network to be weak in predicting the data that was not included in the training. The size of the training set and the number of weights (number of observations), which are directly related to the number of hidden layers and neurons, determine the probability of overfitting [18, 19]. The greater the number of weights concerning the size of the training set, the larger the ability of the network to memorize features of the training data. Due to overfitting, generalization for the test and validation dataset is disappeared and the network gives predictions that deviate from the true values [20]. In the literature, it is seen that the best ANN-based predictive model is generally obtained when one hidden layer is used [3, 4, 11, 12, 21, 22]. For this reason, one hidden layer is preferred for the MLP network in this study. Apart from the heuristic methods and suggestions are given in the literature for determining the optimal number of neurons, there is no clear statement. That's why the optimum number of neurons is generally determined experimentally. The heuristic formulas proposed in the literature to find the optimal number of neurons in the hidden layer are given in Table 2. N_i and N_0 represent the number of nodes in the input and output layers, respectively (Table 2). In studies conducted within the scope of civil engineering, 2 to 10 neurons are usually used in one hidden node, and using more than 10 neurons in the hidden node has no remarkable effect on optimizing the performance of the network [23]. Taking into account the above-mentioned literature recommendations for determining the number of neurons, the MLP network structure which consists of one hidden layer with 1–10 neurons was used in this study. The input layer consists of 3 neurons (L, F_c, and σ), and the output layer consists of a single neuron (τ). The analyzes were conducted in two stages, and the TSTF [21] was used in the entire network in the first stage. In the second stage, the TSTF in the hidden layer and the LTF in the output layer is preferred [3]. The Levenberg-Marquardt algorithm [4, 15, 24, 25] was utilized to train the network. 80% of the data was utilized for training and the rest for testing. MSE and R^2 indicators, which are widely preferred to interpret the performance of the network, were preferred.

Table 2. Proposed formulas for selecting the number of neurons in the hidden layer [4]

Formula	Reference
$\sqrt{N_i N_o}$	[19]
$\frac{(N_i+N_O)}{2}$	[27]
$\frac{2+N_O N_i+0,5N_O(N_O^2+N_i)-3}{N_i+N_O}$	[28]
$2N_i$	[29]

3 Results and Discussion

The R^2 and MSE values showing the success of each network with different structures in predicting training and testing data are shown in Table 3. When Table 3 is examined, it is seen that the R^2 values are very close to 1. This means that the success of all ANN models in this study is acceptable. However, it has been observed that the best performance is obtained when 2 neurons and TSTF are used in the hidden layer and the LTF is used in the output layer. The R^2 and MSE values obtained from this model are 0.963, 68.670 (training phase) and 0.978, 41.455 (testing phase), respectively. Although the MSE values for the training phase are lower for some of the networks listed in Table 3, it can be seen that these models provide much higher MSE values when predicting test data. This situation is thought to be caused by overfitting. The structure of the network selected as the best predictive model is given in Fig. 3.

The regression plots showing the relationship between the output values of the best predictive model and the actual values are shown in Fig. 4. As shown in Fig. 4, the R^2 is equal to 0.963, 0.978, and 0.966, and MSE values equal to 41.455, 68.670, and 63.088 for the training, testing, and all data, respectively, show the applicability of the ANN-based predictive model. Mean absolute error (MAE) and root mean square error (RMSE) values in addition to MSE, are also given in Fig. 4.

In Fig. 3, $\varphi_1(x)$ and $\varphi_2(x)$ represent the TSTF and LTF, respectively. Also, b_1 and b_2 represent the bias values.

To better understand the performance of the ANN, the predictions of the network and the corresponding measured values were compared in both the training and testing phases (Fig. 5). Figure 5 shows that the network's outputs are generally very close to the

Table 3. Performance of ANN models

Model No.	N_H	ANN Results							
		S_1*				S_2**			
		Training		Testing		Training		Testing	
		R^2	MSE	R^2	MSE	R^2	MSE	R^2	MSE
1	1	0.939	144.502	0.979	78.387	0.962	68.109	0.976	45.549
2	2	0.970	55.875	0.965	62.845	0.963	68.670	0.978	41.455
3	3	0.980	36.618	0.953	88.670	0.964	89.443	0.939	117.320
4	4	0.993	15.085	0.976	135.338	0.876	262.891	0.848	281.883
5	5	0.914	176.088	0.971	74.016	0.970	55.011	0.937	135.938
6	6	0.976	43.131	0.963	108.102	0.964	65.626	0.944	119.407
7	7	0.935	118.371	0.857	274.466	0.870	209.017	0.958	121.411
8	8	0.960	93.247	0.849	296.388	0.989	22.074	0.877	258.992
9	9	0.918	170.350	0.950	120.184	0.988	23.105	0.959	112.733
10	10	0.954	95.974	0.945	131.947	0.948	95.586	0.831	364.195

N_H=Number of nodes in the hidden layer

** The TSTF is used throughout the network (S_1)*

*** The TSTF and LTF were used in the hidden and output layers, respectively (S_2).*

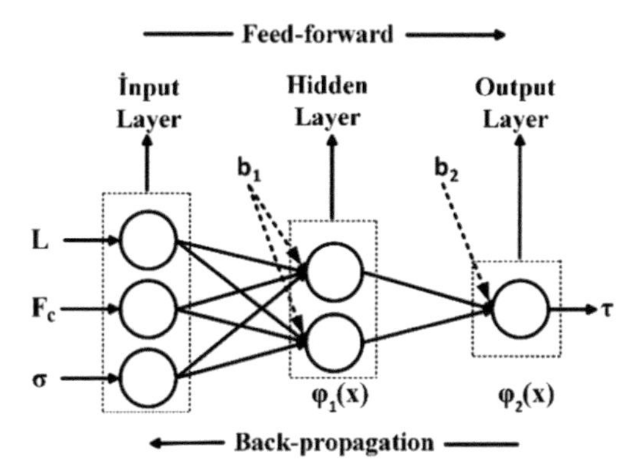

Fig. 3. Structure of the ANN model that gives the best predictions

measured values in both the training and testing phases. Especially in the testing phase, the excellent agreement between the predicted values and the measured values confirms the generalization ability of the proposed ANN model. As a result, it can be said that the ANN-based predictive model, which uses fiber length, fiber content, and normal stress values applied in the direct shear test as input parameters, can be successfully applied to predict the shear strength of BFRC.

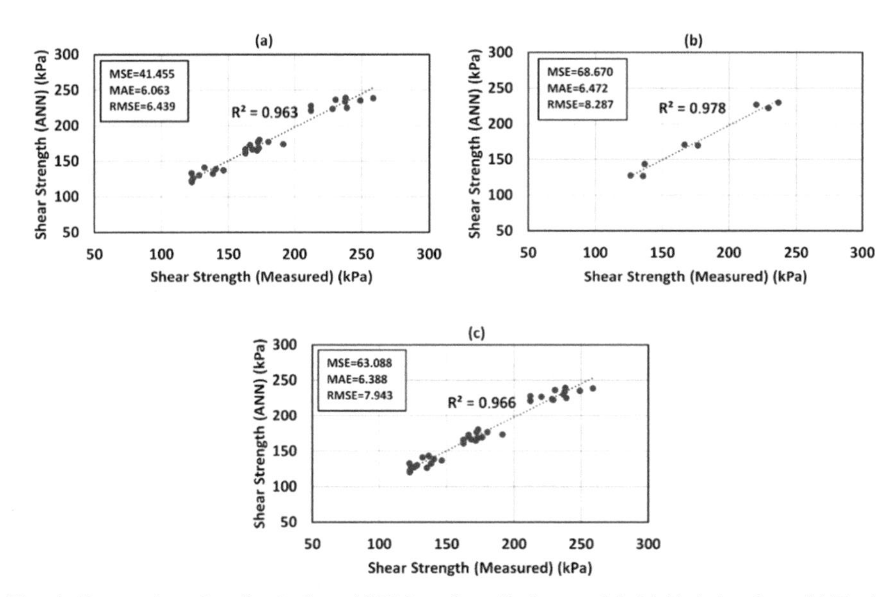

Fig. 4. Regression plots for the best ANN-based prediction model, (a) Training data, (b) Testing data, and (c) All data

Overall, as seen in Figs. 4 and 5, the success of the network in producing results close to the measured values indicates that ANN is a method that requires no prior assumptions and can provide a relatively reliable solution for evaluating the shear strength of BFRC soils. Determining the shear strength of fiber-reinforced soils by laboratory tests may require a lot of time depending on the test type, test variables, and therefore the number of tests, and some test types can be quite costly. By using the ANN model structure proposed in this study, the number of experiments required for the prediction of the shear strength of the BFRC soil can be reduced, time can be used more efficiently and research can be completed at a lower cost.

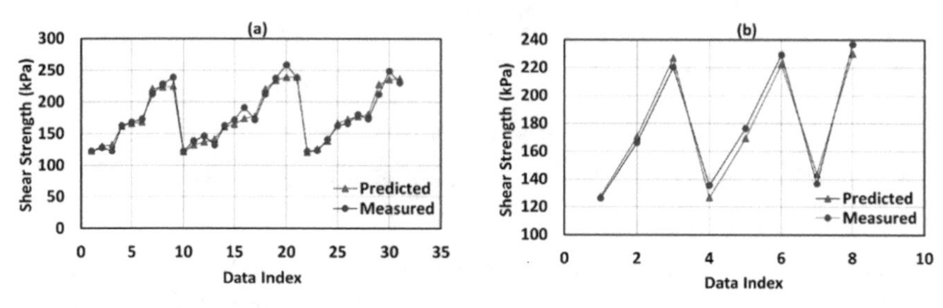

Fig. 5. Comparison of predicted and measured shear strength values; (a) Training data, (b) Testing data

4 Conclusion

In this study, an ANN model was developed using fiber length, fiber content, and normal stress applied in the direct shear test as input parameters to predict the shear strength of BFRC soil. Until a network structure that produced outputs close to the actual data was obtained, the number of neurons in the hidden layer was changed from 1 to 10. R^2 and MSE indicators were used to evaluate the performance of the network. When two neurons and the TSTF are used in the hidden layer, and the LTF is used in the output layer, the predictions that perfectly agree with the measured values are obtained. As a result, the ANN method can be used as an excellent numerical soft computing tool in predicting the shear strength of BFRC soil. Thus, the shear strength of BFRC soils can be calculated with a smaller number of experiments and therefore cost.

References

1. Yazıcı, M.F., Keskin, S.N.: Review on soil reinforcement technology by using natural and synthetic fibers. Erzincan Univ. J. Sci. Technol. **14**(2), 631–663 (2021)
2. Richard, J.A., Sa'don, N.M., Karim, A.R.A.: Artificial neural network (ANN) model for shear strength of soil prediction. Defect Diffusion Forum **411**, 157–168 (2021)
3. Park, H.I., Kim, Y.T.: Prediction of strength of reinforced lightweight soil using an artificial neural network. Eng. Comput.: Int. J. Comput.-Aided Eng. Softw. **28**(5), 600–615 (2011)
4. Momeni, E., Nazir, R., Armaghani, D.J., Maizir, H.: Application of artificial neural network for predicting shaft and tip resistances of concrete piles. Earth Sci. Res. J. **19**(1), 85–93 (2015)
5. Shahin, M.A.: Frontiers of structural and civil engineering. Can. Geotech. J. **47**, 230–243 (2010)

6. Rabiei, M., Choobbasti, A.J.: Innovative piled raft foundations design using artificial neural network. Front. Struct. Civ. Eng. **14**(1), 138–146 (2020)
7. Tizpa, P., Chenari, R.J.: ANN prediction of some geotechnical properties of soil from their index parameters. Arab. J. Geosci. **8**, 2911–2920 (2015)
8. Ghoreishi, B., et al.: Assessment of geotechnical properties and determination of shear strength parameters. Geotech. Geol. Eng. **39**, 461–478 (2021)
9. Venkatesh, K., Bind, Y.K.: ANN and neuro-fuzzy modeling for shear strength characterization of soils. In: Proceedings of the National Academy of Sciences, India Section A: Physical Sciences, pp. 1–7 (2020)
10. Kiran, S., Lal, B.: Modelling of soil shear strength using neural network approach. Electron. J. Geotech. Eng. **21**(10), 3751–3771 (2016)
11. Xu, Q., Kang, F., Li, J.: A neural network model for evaluating gravel liquefaction using dynamic penetration test. Appl. Mech. Mater. **275**, 2620–2623 (2013)
12. Huang, Z., Zhang, D., Zhang, D.: Application of ANN in predicting the cantilever wall deflection in undrained clay. Appl. Sci. **11**(20), 9760 (2021)
13. Sungur, A., Yazıcı, M.F., Keskin, S.N.: Experimental research on the engineering properties of basalt fiber reinforced clayey soil. Eur. J. Sci. Technol. **28**, 895–899 (2021)
14. Shahin, M.A., Jaska, M.B., Maier, H.R.: Artificial neural network applications in geotechnical engineering. Aust. Geomech. **36**(1), 49–62 (2001)
15. Ataseven, B.: Forecasting by using artificial neural networks. Oneri J. **10**(39), 101–115 (2013)
16. Dao, D.B., et al.: A spatially explicit deep learning neural network model for the prediction of landslide susceptibility. CATENA **188**, 104451 (2020)
17. Alzo'Ubi, A.K., Ibrahim, F.: Predicting the pile static load test using backpropagation neural network and generalized regression neural network-a comparative study. Int. J. Geotech. Eng. **15**(2), 1–13 (2018)
18. Baum, E.B., Haussler, D.: What size net gives valid generalization. Neural Comput. **6**, 151–160 (1989)
19. Masters, T.: Practical Neural Network Recipes in C++. Academic Press, New York (1993)
20. Kaastra, I., Boyd, M.: Designing a neural network for forecasting financial and economic time series. Neurocomputing **10**, 215–236 (1996)
21. Moayedi, H., Rezaei, A.: An artificial neural network approach for under-reamed piles subjected to uplift forces in dry sand. Neural Comput. Appl. **31**, 327–336 (2019)
22. Wang, Y., Cong, L.: Effects of water content and shearing rate on residual shear stress. Arab. J. Sci. Eng. **44**, 8915–8929 (2019)
23. Lai, J., Qiu, J., Feng, Z., Chen, J., Fan, H.: Prediction of soil deformation in tunnelling using artificial neural networks. Comput. Intell. Neurosci. **2016**, 16p (2016)
24. Ramasamy, M., Hannan, M.A., Ahmed, Y.A., Dev, A.K.: ANN-based decision making in station keeping for geotechnical drilling vessel. J. Mar. Sci. Eng. **9**(6), 596 (2021)
25. He, S., Li, J.: Modeling nonlinear elastic behavior of reinforced soil using artificial neural networks. Appl. Soft Comput. **9**, 954–961 (2009)
26. Acar, R., Saplıoğlu, K.: Detection of sediment transport in streams by using artificial neaural networks and ANFIS methods. Nigde Omer Halisdemir Univ. J. Eng. Sci. **9**(1), 437–450 (2019)
27. Ripley, B.D.: Statistical Aspects of Neural Networks, Networks and Chaos-Statistical and Probabilistic Aspects. Chapman & Hall, London (1993)
28. Paola, J.D.: Neural network classification of multispectral imagery. Master thesis. The University of Arizona, USA (1994)
29. Kanellopoulas, I., Wilkinson, G.G.: Strategies and best practice for neural network image classification. Int. J. Remote Sens. **18**, 711–725 (1997)

Change of the Internet with Blockchain and Metaverse Technologies

Ismet Can Sahin[1](✉) and Can Eyupoglu[2]

[1] Department of Computer Engineering, Hezârfen Aeronautics and Space Technologies Institute, National Defence University, 34149 Istanbul, Turkey
ismetcnsahin@gmail.com

[2] Department of Computer Engineering, Turkish Air Force Academy, National Defence University, 34149 Istanbul, Turkey

Abstract. A new trend that has started to be talked about again in the technology world, especially in recent years, is Web 3.0. As it is said, the concept of Web 3.0 was also being talked about in 2004. In those years, we were feeling the transition to Web 2.0 and were waiting for the transition to Web 3.0 on the one hand. Somehow that drastic change didn't happen. Bitcoin has created a change in the financial world, and Etherium has brought it back to the agenda. Can the blockchain technology on which the coins are based also completely change the internet? Can Web 3.0 do this by passing the Web 2.0 class and protecting our data privacy? These questions are not ridiculous, if the change takes place, it will not be limited only to the Internet, and it will also intersect with the Metaverse. This study reveals the impact of the blockchain on the transition from Web 2.0 to Web 3.0 that you need a decentralized structure to protect data privacy. Besides, this study provides an insight into the contributions of this change that we will also see in the Metaverse.

Keywords: Big Data Privacy · Web 3.0 · Blockchain · Metaverse

1 Introduction

New technology is rapidly reshaping the way we live our lives. As a result, the trackable digital breadcrumbs we leave behind can paint not just an accurate picture of what we do and where we go but also our identity and even uncover more personal information such as medical conditions and social security numbers. These tracking methods may also lead to people coming to know highly sensitive details like our bank account numbers, credit card info, and even places you frequent regularly putting your safety at risk. This highlights the importance of using anonymous location services to ensure your data is concealed and out of reach of all prying eyes. A growing and newly relevant topic in computer science is social engineering. It refers to an act of manipulating people into performing actions or divulging confidential information and has now become easier due to these smart devices. In addition, some people are worried about the privacy of others when it comes to information being provided on the Internet regarding government agencies recording the people who travel in and out of a state or country using

D. J. Hemanth et al. (Eds.): ICAIAME 2022, ECPSCI 7, pp. 82–101, 2023.
https://doi.org/10.1007/978-3-031-31956-3_7

license plate scanners that protect citizens from terrorism. This type of data is gathered via surveillance cameras for future reference so that law enforcement can look into a certain situation but more than often such images may be stored on servers somewhere else as this company's privacy policy states, therefore, putting both its consumers and producers at risk [1].

Big data refers to a meaningful and trackable representation of information obtained from a variety of sources including social media posts, search engine requests, photos, videos, blog posts, and log files. Currently, these types of data are stored in structural databases and many companies are required to process terabytes of data per day. Traditional infrastructure is inadequate for processing all this information with the help of analytics and turning them into usable analytical data. When it comes to making decisions, companies who rely on relational databases use their knowledge of the data they obtained by either employing database scalability solutions to analyze big-data big data storing it all. This is a standing solution but not a viable one however as these types of databases are struggling to store increasing amounts of big data. Furthermore, when it comes to discovering and preserving personal privacy, scaling up a high volume of unstructured data introduces new obstacles for traditional databases. Many database anonymization solutions will help make sure that the stakes won't get too steep when dealing with personal data using traditional database solutions. While large-scale data management solutions have only just begun to grow exponentially, the number of innovative applications that manage more than single terabytes is quickly growing. Most organizations rely on various kinds of data anonymization (de-identification) techniques to ensure security and privacy. Verifiable verbal or written guarantees are the most widespread solutions for ensuring confidentiality and integrity. However, it has been found that these solutions have not been very fruitful. Passwords, with their high level of security and multi-factor authentication; data dynamic and distributed data systems where information is shared by consents and combined in one place for safekeeping and security; and practices that are regularly applied to secure data integrity and confidentiality; technical solutions like encryption that are also used at low levels are essential yet flawed. However, more advanced solutions including cryptography are developing rapidly [2].

Nowadays, we can all agree that the web has been transformed into something much bigger and better than we could have ever imagined. Providing us with dozens of amazing features, it has cultivated a new, faster way to communicate whether it be through social media or the availability of global sharing and easy access to the most up-to-date information and news. While it can enhance our knowledge in many aspects it also runs the risk of initiating new problems such as hindering democracy if too much power is given to one centralized company. Another problem would be Internet censorship or security issues since valuable information is collected by accessing servers which leads us to growing security threats as well due to personal data being stored. Before blockchain technology, data was centralized, and no protocol would support the decentralization of data. As a result, this led to all of our data being collected by mega-platforms and owned by them. Through the Decentralized Web, this issue will be solved. The token model will make sure people invest in protocols not only in the applications on top but also in the development infrastructure as well [3].

Blockchain is the primary technology behind Bitcoin. It uses a decentralized system that stores information in different blocks which can then be updated by other users on the distributed ledger using anonymous public-private key encryption. This means that all transactions are logged, and any changes made to the underlying code can be identified and traced back, making blockchain an effective method of issuing smart contracts without a need for trust [4].

The Metaverse is a revolutionary system designed to help you improve your management and organizational skills through various features that are both easy to use and easy to understand. In it, you will be able to interact with 3D models, practical applications, virtual items, and the Metaverse team itself - all via our proprietary software developed by experienced professionals in that very field. Metaverses, now a household word, make their way into the private lives of millions every day. This wasn't always the case. Neal Stephenson (a brilliant science fiction writer) wrote about the modern metaverse in his 1992 book Snow Crash. This concept has been around for a long time, though; The Cave Allegory by Plato explains how we viewed the notion of loneliness before it became synonymous with the "modern-convenience" of online socializing. Metaverses have recently started to populate our homes, starting as little games designed to keep kids quiet, but resulting in full-length graphic novels that sometimes reach blockbuster status. Recently, they have made news in the technology world with some big players announcing major investments, and ambitious plans for the development of entirely new and futuristic metaverses. Among them are Microsoft2, Epic Games, 3, and Meta4, the tech holding in which Facebook was (uncoincidentally) rebranded. Mark Zuckerberg (Meta and Facebook CEO) envisioned the metaverse as the next evolution of the Internet. Mark Zuckerberg, CEO of Facebook and Meta said earlier this year: "I think, we're building this so that someday you can use it. I mean eventually, you'd want to be able to put on these glasses or contact lenses and just see a bunch of stuff happening around you." What is metaverse? It's simply "Internet 3.0." Akin to the mobile Internet, the metaverse views both our physical and digital networking as evolution, so rather than venturing off into some "galaxy far far away" it fully intends on setting foot in reality. The fascinating aspect of this comes from being able to experience your digital self-alongside everyone else and be who you want to be. For that to even happen, let alone take off, there will have to be technological advancements that allow for the building of all the instances. And that is where: Biggest up-and-coming technologies: Virtual Reality (VR), Augmented Reality (AR), and the Digital Twin concept. VR will help to create 3D immersive spaces AR will integrate the virtual and physical worlds, allowing for real-world data to be included in virtual spaces. Wearable sensors on avatars in visualized digital environments will mimic the movements of people within the real world and feed additional real-world data into these environments. The metaverse will also feature a ready market where both physical and virtual items can be purchased via non-fungible tokens. These purchasable tokens result from the next generation of networking technologies that allow for increased connectivity through augmented reality. While incorporating the latest technologies, social media, big data, and advanced algorithms into our metaverse program and as we are monitoring these developments vigilantly for any potential problems, you know that certain things could go wrong. So we do everything we can to ensure that these challenges don't become too difficult for us to face. In this context, it seems evident that security might be a bigger challenge than ever before, plus it'll require some adjusting [5].

In this study, we provide various contributions. We first review the concepts of big data privacy, web 3.0, blockchain, and then the formation of the latest Metaverse concept. Next, we describe the current background of web 2.0. In this new context, we highlight the current security and privacy issues that have been raised like never before. Therefore, it is necessary to switch from web 2.0 to web 3.0. As we move from Web 2.0 to Web 3.0, we are reviewing our new virtual worlds and existing Metaverses in the last section.

2 Big Data Privacy

Our digital universe is growing faster than ever. According to the latest data (Fig. 1), by the year 2020 our accumulated data will reach approximately 40,000 exabyte. The term big data is used to describe large sets of information. According to Accenture research, in the next four years, it's predicted that 6.59 million terabytes of data will come from sensors, mobile devices, and machines, which is twice the amount of information Facebook collects every day. The collection of all this data is affecting businesses around the world in ways they never imagined possible. For example, big data has allowed Apple to predict how many units they need to make for each holiday season offering as well as local grocery stores to predict what days throughout the week produce sells best. Although this may sound like a nightmare for some small businesses because increased storage costs can pile high, imagine what opportunities arise when you're able to make more informed decisions about your future [3].

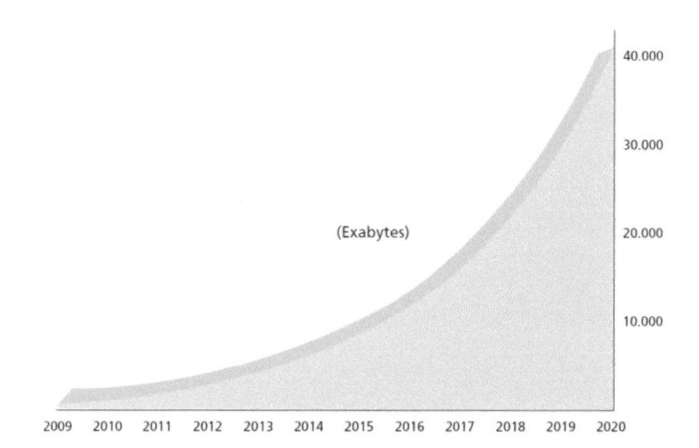

Fig. 1. Expected Data Growth, 2010–2020 [3].

Big data has opened the doors to a better future by allowing artificial intelligence to improve itself. This innovation helps solve problems in useful ways, which is good for everybody except those who might be fearful about the end of mankind. Big data is also sold to marketing companies that use the data to keep viewers informed about different products, parts, and other things that may be relevant to them and thus help motivate users to buy something they haven't bought before. However, since big data can be used for malicious intentions, such as invading their privacy or making false promises about things that are not true, it is important to be very careful when entering your information online, and this is the occurrence of a loss of democracy. Lanier notes that presently, "Big data and artificial intelligence systems are primarily economic and political entities that disenfranchise most people." What might seem like a good bargain can often become an illusion that costs so much more than one could have imagined. Take for example the trade-off between privacy and profit. By swapping the value of our data, users in fact lordship over it and most importantly profit from what they generate with their own time. Democracy suffers as well since without getting paid by one's fork, a majority of users deprive themselves of the income and financial freedom that goes along with earning profits. And it doesn't stop there either: when we make our data available to social media platforms, which are private businesses against which we're voluntarily entering into a contract, we give up the right to any related profits when these platform owners sell us out to third parties (which happens all the time). Not only do these tech behemoths harm the businesses they compete with, but they also prevent new companies from accessing the same data, and thus a true playing field of competition cannot exist. Rarely are two mega-platforms alike and thus only one of them can own that piece of data leading to further centralization. This policy has the effect of creating more monopolies rather than allowing for multiple choices for each willing consumer [3].

Big data may contain sensitive personal information that is vulnerable to being accessed by the wrong person. Privacy is highly important when dealing with big data. The biggest challenge in handling big data is preserving an individual's privacy because it might contain either personally identifiable information or other types of personally identifiable information like GPS coordinates and social security numbers, etc. It should be taken into consideration how to share sensitive information on an unsafe network such as the Internet of Things (IoT). One way of solving that problem is to ensure privacy-preserving data mining and publishing to keep the identities of individuals, who have been involved with certain sensitive operations, private all along [6].

As technology advances, customer demands evolve as well. One recent development in technology is the unmanned aerial vehicle (UAV), which is a plane that doesn't need to be piloted by a human. The popularity of UAVs will continue to rise and so will the number of applications for them and that means there's a lot more data being created than ever before. That's why it's become increasingly important to make sure all of your company's collected data remains protected at all times [7].

In the healthcare industry, data is being generated continuously. We see this in our everyday lives; for example, when we visit our primary physician or visit a specialist. Other times this occurs when we decide to go to the hospital, or if we need a surgery that requires an operation. In addition to all of this, there's also the matter of wearables and other smart devices that we use daily. Information from these devices is being sent over to our primary care providers and even insurance companies, who then use it to calculate how much they're going to pay for certain procedures based on what's been incurred so far using this information. A major concern across the health industry is patient privacy and security. Every health provider has files on patients, but how much do they know about us? Some of us might need to provide basic information, like if we have a chronic illness. Other types of data, though like our genetic information are often kept entirely separate from health records. Many patients are concerned about how easy it would be to steal their data. If hackers manage to steal or lose private patient information, that could spell bad news for everyone involved. From both a legal and financial standpoint (a breach can cost billions) and from a practical one (a breach can lead to worsened care quality), not only hospitals but also other facilities must take such threats very seriously [8].

3 Web 1.0, Web 2.0, Web 3.0

Sir Tim Berners-Lee created the World Wide Web in 1989. He was working at CERN in Geneva, Switzerland when he came up with the idea for a system that allowed researchers to stay in touch via bulletin board systems. These digital noticeboards allowed users to share information quickly, but only left messages made from text files whereas Berners-Lee wanted to allow links between documents on the internet. This led him to develop HTML (Hypertext Markup Language), which became the standard for creating web pages. In recent years, he has advocated the vision of a semantic web [9].

There are three qualities of the Web, which we are going to talk about today. These three characteristics are Web 1.0 as a web of cognition, Web 2.0 as a web of human communication, and lastly Web 3.0 as a web of cooperation. We aren't talking in technical terms here so don't get confused. The three characteristics we mentioned before are just indications or tips that you can use to describe and realize privacy on the internet based on our observations and experiences [10].

3.1 Web 1.0

What we call the web is the world wide web, the network that surrounds the world. This is where this definition of this concept was first made. This screen on Tim Berners-Lee's computer at the Cern laboratory in Switzerland is the world's first web page. Two important points on this page attract our attention. The first is to see the words world wide web written somewhere for the first time, which shows the vision. And the second one says a word called hypermedia. It refers to the method of the topic, that is, it will consist of information connected by Internet links. This connection concept is the core of Web 1.0 (Fig. 2). Until then, the information that is singular in computers is being linked together. Web 1.0 is very important for us, but it was insufficient. There weren't many content producers back then. There were static pages created by a small number of people. People who connected to the Internet were mostly reading and consuming. 10 years have passed since 1999, Web 2.0 has started to be dec with hoarse voices. The form of the connection is visible in Web 1.0.

One of the first incarnations of the web we know today, as envisioned by Tim Berners-Lee, was called 1.0; this is exactly what most people wanted for their websites during that time. The goal for having a site at that time was simply to establish an online presence; people wanted their sites to be available all over the entire world so anyone could check them out whenever they choose to do so [12].

Fig. 2. WWW or Web 1.0 [11].

3.2 Web 2.0

The evolution of the Internet is now making waves, and everyone from IT professionals to businesses to Web users is taking note. The second phase in the Internet's evolution has been dubbed Web 2.0 or, more poetically, the "wisdom Web", "people-centric Web", "participative Web", and "read/write Web" [13].

Now we can say that the period when the actions of those who enter the Internet to produce writing on the one hand, instead of passive actions of consuming reading, began (Fig. 3). Static websites have been replaced by dynamic ones. The search mechanisms that are similar to dec indexes we use to access information have disappeared, and search engines consisting of only one search box have appeared. With the decimation of the search engine in 2004, this change accelerated, and the Internet was no longer connected to computers, to mobile devices. It has become easier to write and produce content. Websites began to be fed with content created by users.

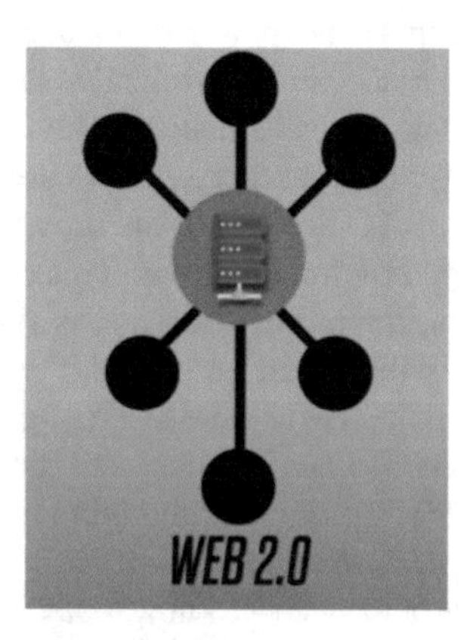

Fig. 3. Web 2.0 [11].

Most of the internet brands that we have heard about as world giants were born during this period during the transition from Web 1.0 to Web 2.0 and have developed and grown at a great pace. Most of those who could not make the transition here has disappeared.

Web 2.0 platforms such as YouTube, Twitter, and Facebook have gained much popularity recently because they allow users to exchange huge amounts of data easily. With the help of these sites, people can share their photos, videos, and ideas with other internet users around the world. To achieve this, the sites will collect data on various areas including user movements, history of resources they have visited, and interactions occurring between different users on a particular network. The discussion on privacy in web 2.0 is therefore significant because it brings up an important issue that we should be aware of the fact that personal data of users is collected without our consent for a variety of reasons including the promotion of third-party products and services, personalization of advertisements and also analysis for security purposes among others [14].

With this exciting wind of change, we have started to observe the ways of web 3.0 on the horizon. We were ready to surf in the waves that he would create. Interestingly, the inventor of the web was also waiting for this web 3.0 and called it the semantic web. But this transition did not happen for many reasons.

The Web 2.0 phenomenon is generating a lot of interest in the business and marketing communities, but it is also attracting a lot of criticism. Experts in the field are concerned about privacy. They believe that there is too much focus on collecting and disseminating personal information [15].

The average college student on Facebook uses the site for 1 h and a half each day, with updated status messages every six minutes. They share more personal information every single time they log in than any other type of user and their privacy settings are often forgotten or neglected [16].

Also, we propose that the challenge that exists lies in allowing users to easily express usage policies for their data. Since privacy is not well understood by Web 2.0 users, as a result of unintended consequences many teenagers accept that their posted data may unintentionally identify them. However, when compared with teen bloggers from 2004, teens today do understand the nuances of oversharing and self-expression on social media. Both privacy and security demand explicit study and regulation through both federal and private bodies [17].

3.3 Web 3.0

To answer the question "What is Web 3.0?", different Internet experts have approached web 3.0 from different angles and with their ideas of what it entails. Some consider it to mean a semantic web and some see it as being about personalization. The technical writer Conrad Wolfram for example sees Web 3.0 as a computer world where information-generating computers will assist human thinking instead of humans doing all the work themselves [18].

We would like you to pay attention to one detail, as you can see in the photos above, both in Web 1.0 and in Web 2.0. There are common characteristics. The middle point. That's where all the connections go. A centralized structure is needed for the Internet. Let's talk in web 1.0 language you have made a website and you are hosting it on the server. If we talk about the Web 2.0 language, you can open an account on Twitter or Facebook and communicate with the whole world. This time the server was replaced by platforms. We need centralized structures and intermediaries in both structures. Isn't the Internet what we call connecting two computers? Why can't I connect directly to my friend's computer? This is one of the things that Web 3.0 wants to accomplish. The tool allows you to make connections without the need for platforms. It is Web 3.0 to be able to set up a decentralized internet that communicates from peer to peer (Fig. 4).

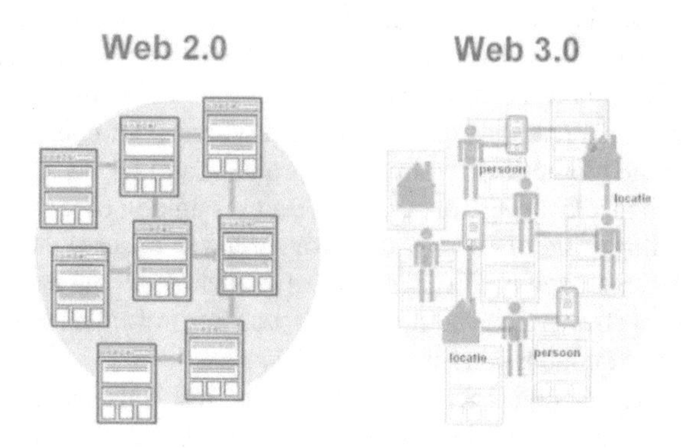

Fig. 4. Comparison of Web 2.0 and Web 3.0 [19].

When we hear the word decentralized, crypto coins come directly to our minds. Because they are digital currencies that can be transferred directly from user to user. If it is possible to transfer material value to each other in this way, why not transfer information?

In Web 2.0, it's as if we keep our identity in a centralized database. So we rely on them to secure our information [20]. However, this is expected to change with the Blockchain infrastructure in Web 3.0.

4 Blockchain

Many cryptographic technologies have been introduced over recent years with some proving groundbreaking and others proving flawed. The main promising cryptographic technology behind Bitcoin has been the Blockchain. The Blockchain is an open immutable ledger that records in a chronological chain of blocks all transactions involving cryptocurrencies including Bitcoin. This allows for a transparent, decentralized system to be created along with a high level of security thanks to cryptography which leads to new possibilities presented by these features as seen within various applications developed thanks to the Blockchain such as voting systems, IoT applications, supply chain management, banking and even healthcare [20].

A blockchain is a peer-to-peer network that consists of nodes. Each node has a copy of the ledger, and all the copies are synchronized and updated to ensure a single version of the truth. While all nodes have access to the same information, each node operates independently of one another without giving up any data until it's requested by someone who can validate who has the right to use it.

Once validated, the transaction is signed off on by both parties, who then record it in a block where it becomes permanently embedded into the chain where everyone in that cryptocurrency network can see it. In blockchain transactions, there is no centralized authority. Each node on the network needs to verify that this request is from an authentic source. This makes digital signatures a necessity for sending messages through the system to check their authenticity. Also, since the message is used to create a signature this means that the integrity of the message is ensured since no one can alter the message without invalidating the signature. In addition to assuring that the message cannot be modified, using a message for a signature also helps users avoid disclosing their private keys for their wallets (the key which is required to unlock your bitcoins) because anyone who has it would be able to access all of your funds. Elliptic Curve Digital Signature Algorithm ECDSA is what Bitcoin uses to verify the authenticity as well as provide security in transactions [3].

Blockchain, the technology that began with bitcoin in 2008 is a way of facilitating transactions, moving funds, or anything of value. Whereas typically one would have to put his/her trust in a third party to oversee the process and facilitate it, cryptocurrencies allow people to manage their own money and keep control over their finances. The first blockchain was built around "proof-of-work", wherein anyone could add blocks but only one block could be "chosen". Over the years, however, newer technologies have been developed which prevent centralization by allowing groups to decide on what chains are included in the main chain and which ones stay separate. One of the most common methods for reaching consensus on a blockchain is through Proof of Work (PoW), also known as "mining". This is because it's the fairest way to ensure that all nodes play by the same rules. The trouble, however, is that there will always be dishonest miners who try to use loopholes in the system to validate their blocks to re Bitcoin or another cryptocurrency for themselves. What this means, though, is that PoW can potentially cause various forks and splits due to some nodes recognizing 1 block while other nodes recognize another. If 2 entirely separate blocks are mined on top of each other at nearly the same time, then a blockchain will fork into 2 distinct chains. We've seen this happen with Ethereum Classic and Ethereum. Blockchain is developing and evolving. As per the data mentioned in Table 1, Blockchain 1.0 was a period of innovation in the financial industry with the advent of Bitcoin, which was innovative at its time because it attempted to create a single financial system online that could be used on a global level. In this environment, Bitcoin strived to become decentralized because blockchain core values were focused on decentralization and decentralization. Blockchain 2.0 is an era where smart contracts are being adopted as shown by Ethereum smart contracts, which made it possible to execute contracts with their legal effects [21].

Table 1. Blockchain Paradigm Evolution Direction [21].

Blockhain 1.0	Blockhain 2.0	Blockhain 3.0
Currency transfer, Digital Payment System, Remittance, Cryptocurrency	Decentralized autonomous organization, Bonds, Mortgages, Smart property, A smart contract, Stock, Loans	AI, Government, Science, IoT, Culture, Art, Big Data, Health, Public

One of the things that makes blockchain so secure is that it takes away the need to trust individual network nodes. There's no central authority that you need to trust, because security is provided by cryptography and distributed consensus [22].

There are several important properties of blockchain that make it an ideal tool for security and trust in complex and unknown operation environments [23]:

- The distributed ledger, often found in blockchains, is a continuously growing list of records. This provides a history of all interactions and transactions that were ever committed. Because the ledger is maintained by a large number of nodes, data loss protection is provided [23].
- A blockchain is a structure of data blocks that are chained together using cryptographic hashes. As shown in Fig. 5, each block contains a block identifier, timestamp, cryptographic hash value, and the cryptographic hash value of the previous block. This chaining of transaction blocks makes it very difficult to change anything in the history of the blockchain [23].
- The security of the network is not just about trusting the individual nodes. Transactions are verified by multiple nodes in the network through a consensus algorithm. This allows the network to keep functioning even if some of the nodes fail or act maliciously. Consequently, this provides a high degree of assurance for all data recorded in the ledger [23].

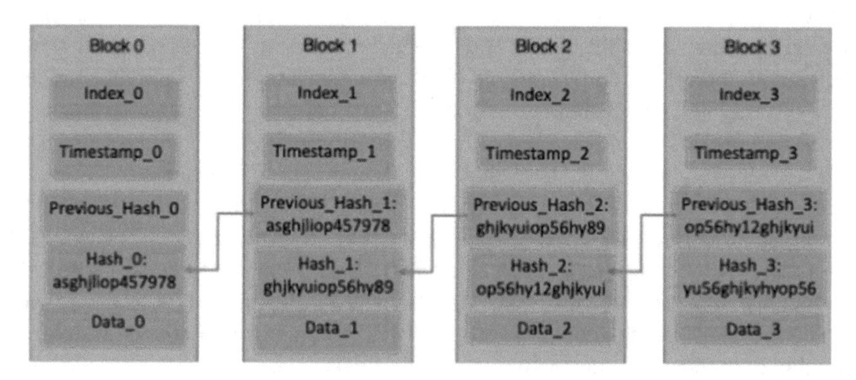

Fig. 5. Blockchain Structure [23].

4.1 Consensus in Distributed Systems and Blockchain

Consensus is a vital part of distributed systems, not just blockchain. It's needed anytime multiple processes or nodes need to share a common data item. There are two main types of blockchain: permissionless and permission. In a permissionless blockchain, nodes are anonymous. This means that a new, tampered transaction block can be added, resulting in a fork. A fork occurs when a valid transaction doesn't match an invalid one. The goal of a consensus algorithm is to get all of the nodes to agree on a single value. This is important in a permissioned blockchain because the nodes are known entities and not anonymous [24].

When it comes to blockchain technology, it's important to achieve consensus among the nodes of the network, even though they might not be entirely trustworthy. This is depicted in Fig. 6. In general, consensus is something to consider in distributed systems where the nodes of a network are faulty or the communication is unreliable. Consensus can be achieved in distributed systems in spite of some failures. Additionally, when defining a consensus system, the communication model (synchronous or asynchronous) is also considered relevant [24].

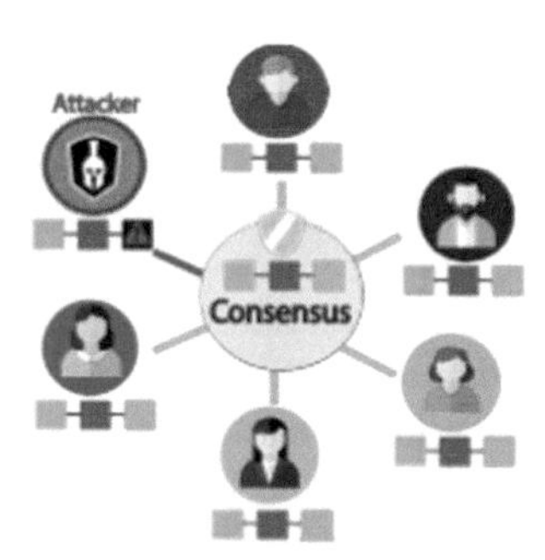

Fig. 6. Consensus in Distributed Systems [24].

Despite concerns over the safety of the digital sphere and the personal data management available on the Web 3.0 blockchain network, cryptography is making it possible to encrypt information, giving full control back to the individual. It's increasingly important for us all to have access to our data and therefore we should use a distributed ledger or decentralized database that features private transactions. Consumers can also benefit by being in control of where their data is stored and how it's used - beyond what they were able to do before. This enables us to approve or reject third-party requests for access and any needed changes related to the processing of that data. No matter if it is permission, notice, or other requests about data, the new smart contract solution will handle each request individually in a way that supports every stakeholder [25].

Thanks to blockchain, a new era of networked and decentralized systems can be built in which information is kept more secure, economic activity is made easier, personal data is managed with higher privacy guarantees and personal identity itself becomes interoperable.

4.2 Decentralized Protocols

The Hypertext Transfer Protocol (HTTP) which was originally developed by an individual at CERN named Tim Berners-Lee allows for decentralized publishing. However, the Internet we have today is centralized even though HTTP was once created for decentralization. This means that HTTP is a stateless protocol, which meant that the connection between browser and server would be dropped once a webpage had been loaded into the browser's cache. This issue lead to the need for a data layer that would retain the state of connections, and this data layer was provided by companies like Google and Facebook. The Internet is centralized because of its inability to keep track of stored data and how they propagate. Internet applications were designed so that they could function individually, with little concern with how the data was passed back and forth from one to the other. Using blockchains as the network is a way for decentralized applications to share everything openly. Additionally, blockchains provide a shared transmission layer for their users that eliminates the need for centralized data centers. One of the many benefits of blockchain protocols is that it takes out the middleman when transactions between users are carried out. For example, you might use an application like Uber which coordinates with your location and an available driver nearby to pick you up and take you to your destination. But what if this data could be shared between multiple applications without a centralized server? One of the layers on top of blockchain protocols like Bitcoin, Neo, or Ethereum can serve as the shared data layer that different dApps can rely on outside of the platform they were built with. This way there aren't any differences in how different applications are related to each other [3].

There is a monetary equivalent to your knowledge, even your identity. How do you think companies that turn into giants with Web 2.0 get them to use it for free? Receiving your data, which has a monetary value, consists of mining data on storage and selling it. There is a cash equivalent for the hours you spend on these platforms. Even if you use it for free, you are investing your time, not your cash. You are investing and it should be worth it. It provided us with access to information resources, these platforms introduced us to each other and fused us. He also realized that he can create algorithms that will allow these people to come to me more and spend more time on my platform. Facebook innocently added a like button. Some were surprised at what to do to click on the like button more. They started coming. They were jealous of those who were watching them because they were tempted. Companies that noticed these behavioral changes began to highlight more of the content they said. Moreover, they kept the algorithms they used to do this from people. Web 2.0 has taken us to the past instead of moving humanity to the future with very big promises.

5 Data Monarchy (Server), Data Democracy (P2P Networks)

Data monarchies began to be used when the Internet was going to equalize conditions and democratize information (Fig. 7).

Suppose you have made an application and you want to share it, you are expected to apply to the relevant monarchy and pay at least 30% tax after publication if the application is considered suitable for consideration (Fig. 8).

Of course, a price should be paid for a service, but in the case where we are coming from, there are several points of control over this. The decision-makers at that point were given too much authority and power, and we gave them these powers often without realizing it. If the Web 3.0 current becomes stronger, then something may change our contribution to this study. In this way, the ownership of the information will have changed hands. His control is returning to us, to individuals (Fig. 9).

We are not just talking about them theoretically, but in practical applications, the Apps in Web 2.0 are called dApps in Web 3.0. It is called a Decentralized App (Fig. 10).

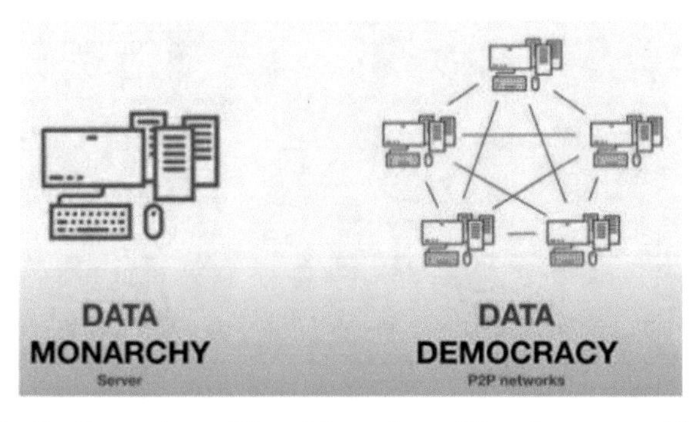

Fig. 7. Comparison of Data Monarchy and Data Democracy [26].

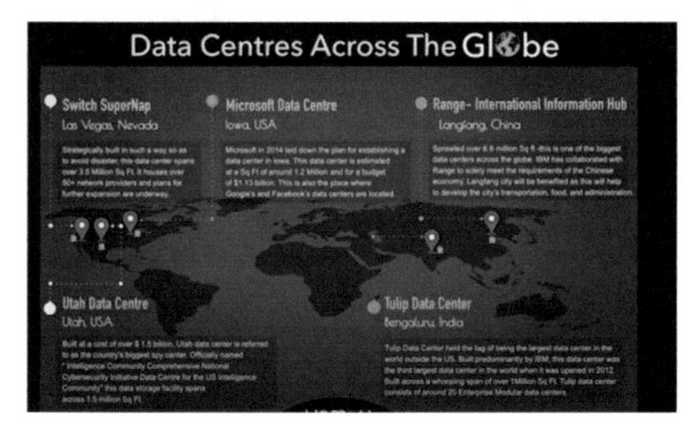

Fig. 8. Data Centers [27].

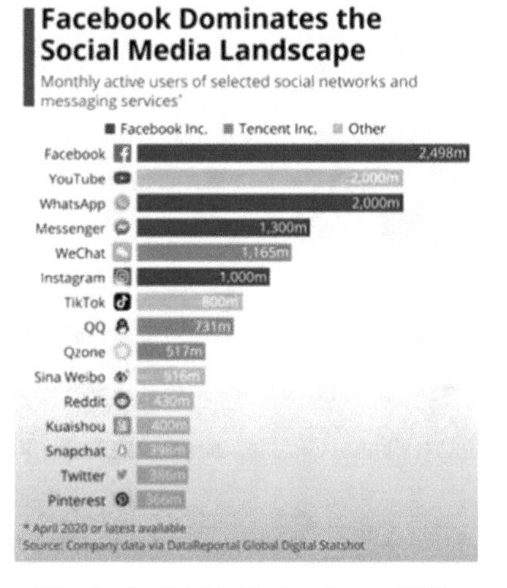

Fig. 9. Social Media Landscape [28].

Fig. 10. App > DApp [29].

For example, instead of the browsers that we use to use on the Internet, it is recommended to use dApps that work with wallet logic (Fig. 11). Thus, every behavior, every information, and even your interest on the Internet is transformed into a value. I can share these activities with those who deserve them while accumulating these values. If I use content, I pay directly to the manufacturer.

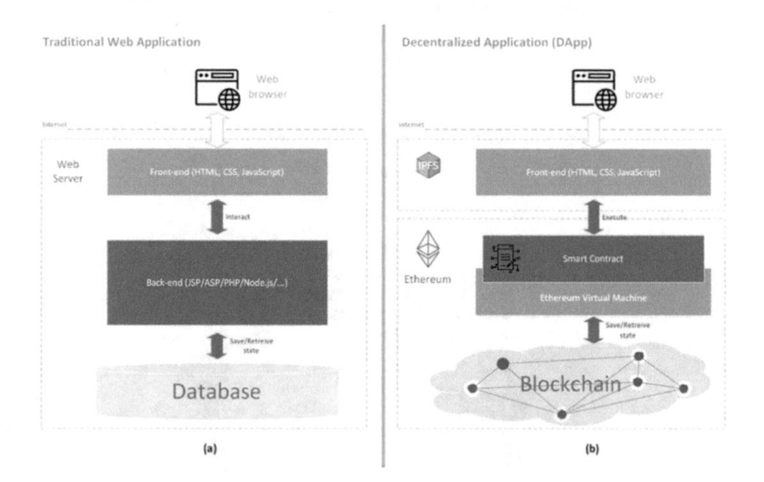

Fig. 11. The architecture of Blockchain [30].

The artist prepares his work, singer's music, and author's writing and delivers it directly to his target audience. The middleman disappears, and we don't give our data to large companies. In them, we don't develop algorithms using our data and we don't play with our emotions.

At this point, a concept called NFT has already emerged to regulate such copyright-related activities. Thousands of cryptocurrencies are also available for material purchases. In other words, blockchain-based technologies can inflate the sails of Web 3.0.

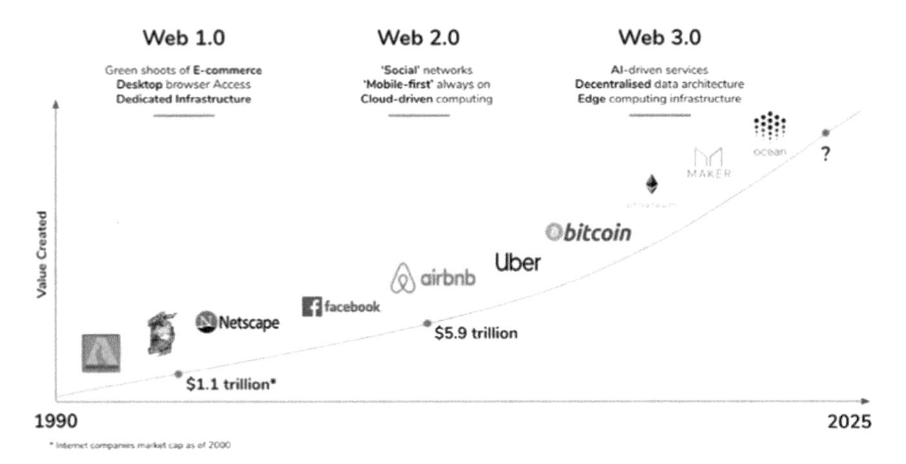

Fig. 12. The Evolution of the Web [31].

The topic has a relationship with the metaverse, we'll talk about it in the next section. In addition, you will be able to continue communicating with your family and friends after you die with data that will be generated on the Metaverse platform while you are alive. In short, it will be shared that our data should be in a safe area to continue to be used from the platform, while we live and after we die.

6 Metaverse

The virtual world (and even IRL) is growing more rapidly than ever since the arrival of the 4th industrial revolution. The real world has been converted from physical space into data, and 3D modeling programs have led what was once physical into the digital marketplace. Here, we ask if data in the VR space is being tracked accurately enough by using our trust technology to ensure that all insights are transparent moving forward because in virtual reality trust technology like blockchain for example can be considered one of the most promising options to come out. With blockchain technology, what will be its applications that have the potential to create a more immersive and enjoyable experience for users in the Metaverse environment? [21].

Meta comes from the Greek language, and it's a prefix that means more all-encompassing or transcending. Verse is an abbreviated form of universe, and it signifies a space/time container. So, when you put these two words together, you get metaverse: a brand-new word that represents a digital living space where traditional social systems are novel and changed. To build a metaverse, state-of-the-art technologies like virtual reality (VR), digital twin, and blockchain are used to map everything in our real world to a parallel universe. For example, users can play, work, and live with friends from any place in the metaverse. In 1992, Neal Stephenson proposed the initial conception of the metaverse in his famous science fiction novel Snow Crash. In the metaverse, people use digital avatars to control and compete with each other to upgrade their status. However, the metaverse is still in its conceptional stage; there are common standards and very few real implementations are available [32].

The metaverse is a combination of various cutting-edge technologies such as 6G, artificial intelligence (AI), VR and digital twins. The basic technologies necessary for the metaverse are [32]:

- Extended reality technology, including augmented reality (AR) and virtual reality (VR), is critical for the metaverse. AR can superimpose digital information on the physical world, while VR makes users feel like they're experiencing the digital world vividly. These techniques are key to developing the metaverse, which creates digital space where users can interact just as they would in the real world [32].
- Digital twin technology is used to create a virtual copy of a real-world object. This is done by using real-world data to predict the expected behavior of the object. In the metaverse, digital twins can mirror the real world into the virtual world. This means that the metaverse can be used to find trial solutions to unsolved issues in the real world [32].
- Blockchain technology is essential to the metaverse for two primary reasons. First, blockchain can be used as a repository to store data anywhere within the metaverse. Second, blockchain provides a complete economic system that can link the virtual world of the metaverse with the real world. This is done primarily through NFTs, which allow virtual goods to become physical objects. In other words, blockchain allows users to trade virtual items in the same way as they would in the real world. Consequently, blockchain becomes the bridge between the real world and the metaverse [32].

There are two important trends about the future of the Internet, Metaverse and Web 3.0, the Web 3.0 infrastructure represents the Metaverse superstructure. The universes surrounding it express the user experience. Just as mobile devices have accelerated our transition to Web 2.0, Metaverse tools and VR helmets, smart glasses that are expected to become widespread soon, can accelerate our transition to Web 3.0 on other wearable devices. Many new companies have already been created for such an Internet-based blockchain, which continues to grow every day (Fig. 12). Alternative solutions are being developed to replace the old companies that have taken a cornerstone in every field.

6.1 As a Metaverse, Live Forever

The Czech-based company announced that it has created a virtual universe that can allow people to continue to live and talk to their deceased relatives. The company claims they have successfully created a virtual space where users will be able to exist forever through avatars, which will carry the data of their personality and conversations with friends or family members even long after they pass away (Fig. 13).

It is well known that the world is increasingly incorporating technological advances. With the concept of the metaverse, parents acknowledge a future where people spend their time interacting in virtual worlds with help from wearable technologies as opposed to interacting in real-life environments. There's a new metaverse that is currently under development by a VR game company called Somnium Space. The VR platform is similar to Facebook but only differs in one aspect which is that this VR allows you to communicate with your deceased relatives and relatives.

The company provides artificial intelligence to create automated avatars of real people based on their recorded movements, body language, and conversations within the application - and ultimately restore them as hyperrealistic characters in an "alternate reality". With augmented reality technology, Somnium aims to back up its servers via user-generated data. Especially when someone passes away without saving your data backup, their relatives will always be able to attempt to communicate with them again via this "other world".

Fig. 13. Avatars are created with data obtained from their images, voice, and movements [33].

The security and privacy of this received big data, the fact that it will be stored on Somniun's servers reveals that Blockchain is needed for living people and deceased people, and the transition to Web 3.0 is required.

7 Conclusion

Will the improvements made to replace the old companies in every field to be able to own our data to protect data privacy continue? Or is it possible that it will be possible shortly, we still don't know yet if we will be talking about Web 3.0 after 10 years? We think that the rapid acclimatization of people to digitalization with the pandemic will perhaps increase the changes and use in this area. We are even excited about the possibility of the adaptability of several principles introduced by Blockchain to the Internet as well. If these principles are supported by smart contracts and certain systems, we can turn them into a democracy instead of that data monarchy on the internet. Then the transition to a decentralized autonomous organization can be made in the real world, after democracy, algocracy can begin to emerge. There are a few steps that have been taken in this regard, some designs related to the cities that are planned to be established in the real world have appeared, and we will be examining this in the next study. We mentioned hypermedia in Web 1.0. It changed the way we read the information first, and the way we produce content, and it changed the world. If we want to switch to Web 3.0, we must add a new one to this special word. Now the connection alone is not enough, we should also add the word trust. In the future, the Internet should not only connect information and people but also be able to create mutual trust between us. We have put forward this study to create trust in the real and virtual universes where we will live. We hope that the word trust will also be added to the development of the Internet and the formation of virtual universes with this work.

References

1. Falchuk, B., Loeb, S., Neff, R.: The social metaverse: battle for privacy. IEEE Technol. Soc. Mag. **37**, 52–61 (2018). https://doi.org/10.1109/MTS.2018.2826060
2. Eyupoglu, C., Aydin, M.A., Sertbas, A., Zaim, A.H., Ones, O.: Büyük Veride Kişi Mahremiyetinin Korunması, pp. 177–184 (2017). https://doi.org/10.17671/gazibtd.309301
3. Alabdulwahhab, F.A.: Web 3.0: the decentralized web blockchain networks and protocol innovation, pp. 1–3 (2018). https://doi.org/10.1109/CAIS.2018.8441990
4. Guo, L., Xie, H., Li, Y.: Data encryption based blockchain and privacy preserving mechanisms towards big data, pp. 2–6 (2020). https://doi.org/10.1016/j.jvcir.2019.102741
5. Pietro, R.D., Cresci, S.: Metaverse: security and privacy issues, pp. 2–7 (2021). https://doi.org/10.1109/TPSISA52974.2021.00032
6. Eyupoglu, C., Aydın, M.A., Zaim, A.H., Sertbas, A.: An efficient big data anonymization algorithm based on chaos and perturbation techniques, pp. 1–3 (2018). https://doi.org/10.3390/e20050373
7. Lv, Z., Qiao, L., Hossain, M.S., Choi, B.J.: Analysis of using blockchain to protect the privacy of drone big data, pp. 45–46 (2021). https://doi.org/10.1109/MNET.011.2000154
8. Bhuiyan, M.Z.A., Zaman, A., Wang, T., Wang, G., Tao, H., Hassan, M.M.: Blockchain and big data to transform the healthcare, pp. 62–68 (2018). https://doi.org/10.1145/3224207.3224220
9. Naik, U., Shivalingaiah, D.: Comparative study of Web 1.0, Web 2.0 and Web 3.0, pp. 2–3 (2009). https://doi.org/10.13140/2.1.2287.2961
10. Fuchs, C., Hofkirchner, W., Schafranek, M., Raffl, C., Sandoval, M., Bichler, R.: Theoretical foundations of the web: cognition, communication, and co-operation. Towards an understanding of Web 1.0, 2.0, 3.0, pp. 1–3 (2010). https://doi.org/10.3390/fi2010041
11. Web 3.0 Nedir. https://www.isdoyazilim.com/nedir/web-3.0-n
12. Getting, B.: Basic Definitions: Web 1.0, Web. 2.0, Web 3.0 (2017). http://www.practicalecommerce.com/articles/464/Basic-Definitions-Web-10-Web-20-Web-30/
13. Murugesan, S.: Understanding Web 2.0, pp. 34–41 (2007). https://doi.org/10.1109/MITP.2007.78
14. Fuchs, C.: Web 2.0, presumption, and surveillance, pp. 289–305 (2010). https://doi.org/10.24908/ss.v8i3.4165
15. Smith, H.J., Dinev, T., Xu, H.: Information privacy values, beliefs and attitudes: an empirical analysis of Web 2.0 privacy, pp. 989–991 (2011). https://doi.org/10.2307/41409970
16. Gross, R., Acquist, A.: Information revelation and privacy in online social networks (the Facebook case), pp. 72–74 (2005). https://doi.org/10.1145/1102199.1102214
17. Cormod, G., Krishnamurthy, B.: Key differences between Web 1.0 and Web2.0, pp. 1–4 (2008). https://doi.org/10.5210/fm.v13i6.2125

18. Nath, K., Iswary, R.: What comes are Web 3.0? Web 4.0 and the future, pp. 1–2 (2015). https://www.researchgate.net/publication/281455061_What_Comes_after_Web_30_Web_40_and_the_Future
19. Web 3.0. https://empowermenttech12.wordpress.com/2017/06/24/the-current-state-of-ict-technologies/
20. Saraf, C., Sabadra, S.: Blockchain platforms: a compendium, pp. 4–5 (2018). https://doi.org/10.1109/ICIRD.2018.8376323
21. Jeon, H., Youn, H., Ko, S., Kim, T.: Blockchain and AI meet in the metaverse. In: Advances in the Convergence of Blockchain and Artificial Intelligence, pp. 2–3 (2021). https://doi.org/10.5772/intechopen.99114
22. Angin, P., Mert, M.B., Mete, O., Ramazanli, A., Sarica, K., Gungoren, B.: A blockchain-based decentralized security architecture for IoT. In: Georgakopoulos, D., Zhang, L.-J. (eds.) ICIOT 2018. LNCS, vol. 10972, pp. 3–18. Springer, Cham (2018). https://doi.org/10.1007/978-3-319-94370-1_1
23. Angin, P.: Blockchain-based data security in military autonomous systems, pp. 363–364 (2020). https://doi.org/10.31590/ejosat.824196
24. Chaudhry, N., Yousaf, M.M.: Consensus algorithms in blockchain: comparative analysis, challenges and opportunities, pp. 58–59 (2019). https://doi.org/10.1109/ICOSST.2018.8632190
25. Kolain, M., Wirth, C.: Privacy by blockchain design: a blockchain-enabled GDPR-compliant approach for handling personal data, pp. 1–2 (2018). https://doi.org/10.18420/blockchain2018_03
26. The Future of Web 3 is Here. https://njkhanh.com/the-future-of-web-3-is-here-p5f3131393237
27. T Data Centres across the Globe – Infographics. https://www.vembu.com/blog/data-centres-across-the-globe-infographics/
28. Facebook Inc. Dominates the Social Media Landscape. https://www.statista.com/chart/5194/active-users-of-social-networks-and-messaging-services/
29. T How web 3.0 blockchain Would Impact Businesses. https://appinventiv.com/blog/web-3-0-blockchain-impact-on-businesses/
30. Farnaghi, M., Mansourian, A.: Blockchain, an enabling technology for transparent and accountable decentralized public participatory GIS, pp. 1–2 (2020). https://doi.org/10.1016/j.cities.2020.102850
31. Evolution of Web. https://dev.to/pragativerma18/evolution-of-web-42eh
32. Gadekallu, T.R., et al.: Blockchain for the metaverse: a review, pp. 4–5 (2020). https://doi.org/10.48550/arXiv.2203.09738
33. Live Forever' mode seeks to erase death using the metaverse. https://www.digitaltrends.com/computing/vr-company-somnium-space-live-forever-in-metaverse/

A Language-Free Hate Speech Identification on Code-mixed Conversational Tweets

Pelin Canbay[1](✉) and Necva Bölücü[2]

[1] Department of Computer Engineering, Sutcu Imam University, Kahramanmaras, Turkey
pelincanbay@ksu.edu.tr

[2] Department of Computer Engineering, Hacettepe University, Ankara, Turkey
necva@cs.hacettepe.edu.tr

Abstract. Hate speech is rapidly increasing on social media, which has a negative impact on society. Timely detection of hate speech in these media is crucial. Currently, researchers are working intensively in this field, especially in English. Since social media is a platform with no language restriction and anyone can share content in a different language, there is a need for language-free solutions in hate speech identification. In this paper, we present a solution submitted by our team PC1 for Subtask 2 – Identification of Conversational Hate Speech in Code Mixed Languages (ICHCL) of the Hate Speech and Offensive Content Identification (HASOC-2021) shared task. The dataset provided by HASOC for Subtask-2 includes the Twitter comments with the content of the parent comments, which are labeled as HOF (Hate and Offensive) or NOT. The tweets are from different languages and different scripts. Here, we proposed a language-free solution to identify hate speech from code-mixed and script-switched texts. Our solution uses ASCII based script normalization on the dataset to translate all language-specific or emotional characters into proper ASCII format. Character n-gram features of the formatted texts are used with their tf-idf values in a neural network based classification model in our solution. Our work's test score in macro precision is 0.6672, and macro F1 score is 0.6537 (ranked 10th).

Keywords: ASCII transformation · char-grams · hate speech · script normalization

1 Introduction

Social media provides convenient communication services to users regardless of social and demographic boundaries. People can easily make their voices heard by a large audience on these media. Because of this convenience, unfortunately, these voices are also easy to spread when they sound hateful. Hateful and offensive speech can be highly damaging to societies [1, 2]. If these speeches are identified in time, some measures can be taken in advance, and societal health can be saved from harm. Considering this situation, there is an increasing need for studies on the detection of hate speech in order to take the necessary measures.

D. J. Hemanth et al. (Eds.): ICAIAME 2022, ECPSCI 7, pp. 102–108, 2023.
https://doi.org/10.1007/978-3-031-31956-3_8

Timely identification of hate speech on social media is a necessity today, but this process is fraught with many difficulties. With Subtask-2, HASOC tackled two significant challenges together in 2021; identification of hate speech in code-mixed languages and identification of hate speech within a conversation. Social media is an interactive platform, so texts in these media should not be considered standalone [3]. Since texts in social media are contextual, they should be treated together with their parents' comments. In multilingual societies, social media streams may contain texts with mixed codes and even mixed scripts. This diversity makes identifying hate speech more challenging than assessing standalone and monolingual texts. To identify hate speech in the face of these challenges, we proposed a language-free solution that includes script normalization to transform mixed-script texts into a standard representation. Moreover, n-gram features are extracted to learn the specific characteristics of tweets based on hate speech identification. For identification, we used tf-idf values of n-grams in a neural network based classification model.

2 Dataset Description

HASOC is an organization that has held hate speech identification contests in 2019 [4], 2020 [5], and 2021 [6]. In 2021, the organization announced two subtasks and released datasets for these subtasks. Subtask-2 is a brand new problem offered that year. Subtask-2 aims to classify the tweets with their parent comments into Hate and Offensive (HOF) or Not categories. The major challenge of this subtask is that the provided dataset contains English, Hindi, and code-mixed Hindi tweets, and also some tweets are script-mixed. The dataset includes 5740 individual tweets with their parents and is provided to the participants to create a classifier that accurately identifies hate speeches with their parent posts. The details of the dataset and the overview of the HASOC Subtask-2 can be found in [7].

3 Proposed System Description

The variety of texts used in social media reduces the success of identifying whether they are hate speech or not. Since the provided dataset is code-mixed, we developed a language-free classifier that can accept texts from any language and translate them into a standard format. For this purpose, we perform data cleaning on the texts and then decode them into the standard ASCII format. With the formatted texts, we extracted character n-gram features of all texts. Finally, we used the tf-idf values of the n-gram features for neural network based classification to obtain the identification accuracy. The overview of the proposed system is shown in Fig. 1. The details of each methodological module are explained in the subsections.

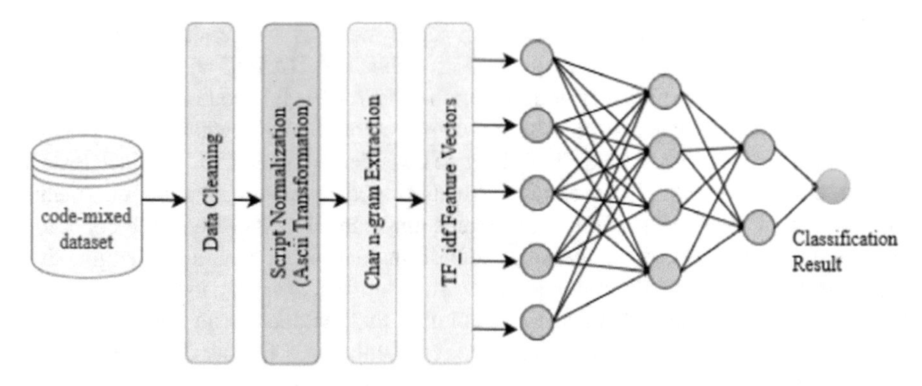

Fig. 1. The Proposed System Overview

3.1 Data Cleaning

In line with our goal of creating an utterly language-free solution for hate speech identification, we did not use language-dependent data cleaning processes such as stop word extraction or stemming. However, texts on social media contain many noises, like usernames or URLs. In this phase, in addition to converting lowercase letters, we extracted some noisy or context irrelevant parts from the texts before the script normalization. These parts are URLs (http://….), special tags (@USER, @RT,….), and extra spaces.

3.2 Script Normalization

Script normalization converts linguistic units from any language or special characters, such as emojis, into a smaller standard character set. In multilingual societies, people use different languages together in speaking and writing. The transition from one language to another in an utterance is called code-mixing or code-switching [8]. People may prefer to use the specific script and character set of the code in question in code-mixing, which is called script-switching. In this case, processing the code-mixed text requires language-dependent preprocessing for each language. On the other hand, script-level normalization not only reduces the processing cost but also makes the texts more applicable for foreign researchers.

For script-level normalization, we used the ASCII transformation. ASCII is a subset of Unicode, the universal character set containing more than 140.000 characters. ASCII is the most compatible character set, consisting of 128 characters composed of English letters, digits, and punctuation. To convert a wide range of Unicode characters into a more straightforward ASCII representation, we used the anyascii[1] library written in Python. The library converts characters that are not ASCII to an appropriate ASCII based on their meaning or appearance. The changes of the texts in our solution on an example are given respectively in Table 1.

[1] https://pypi.org/project/anyascii/ Last visited: 23-09-2021.

Table 1. A samples of text transformation.

Method	Output
Original Tweet	Just realised that the #TwitterBan news has made many #tukdetukde gang become pro Modi. Suddenly they are worried abt his wellbeing and how concerned are they and also not wanting him to suffer repercussions.... Such love for Modi was unseen before. **अभी आप की दुकान बंद!** https://t.co/EbU4yfn0W6 @ashokepandit Picking one quote out of many Not cool @brother
Cleaned Tweet	Just realised that the #TwitterBan news has made many #tukdetukde gang become pro Modi. Suddenly they are worried abt his wellbeing and how concerned are they and also not wanting him to suffer repercussions.... Such love for Modi was unseen before. **अभी आप की दुकान बंद!** Picking one quote out of many Not cool
Normalized Tweet	Just realised that the #TwitterBan news has made many #tukdetukde gang become pro Modi. Suddenly they are worried abt his wellbeing and how concerned are they and also not wanting him to suffer repercussions.... Such love for Modi was unseen before. abhi apki dukane bnd ! Picking one quote out of many :yawning_face: Not cool
Char 4-grams	just ust st r t re rea real eali alis lise ised sed ed t d th tha that hat at t t th the the he # e #t #tw #twi twit witt itte tter terb erba rban ban an n n ne new news ...

3.3 Classification Model

The feature vector used to feed the neural model in our solution contains character 4-g with tf-idf weights. Our classification model is based on neural network architecture. The architecture of our model and its properties are shown in Table 2.

We split the training dataset provided by HASOC for Subtask-2 into 80% train dataset and 20% validation dataset to find the optimal hyperparameters. A grid search algorithm was used to find the optimal hyperparameters and the layer properties of the model. Depending on the problem and the dataset, the most accurate results are obtained by the neural network properties given in Table 2. The best parameters for training the model are; Adam optimization [9] (learning rate = 0.001), epochs = 50 and batch size = 16. Using these parameters, obtained accuracy from the train and validation sets shown in Fig. 2.

Table 2. Classification model and its properties

Layer	Unit	Activation Function
Dense	256	Rectified Linear Unit
Dropout	0.1	
Dense	64	Rectified Linear Unit
Dense	16	Hyperbolic Tangent
Dense	1	Sigmoid

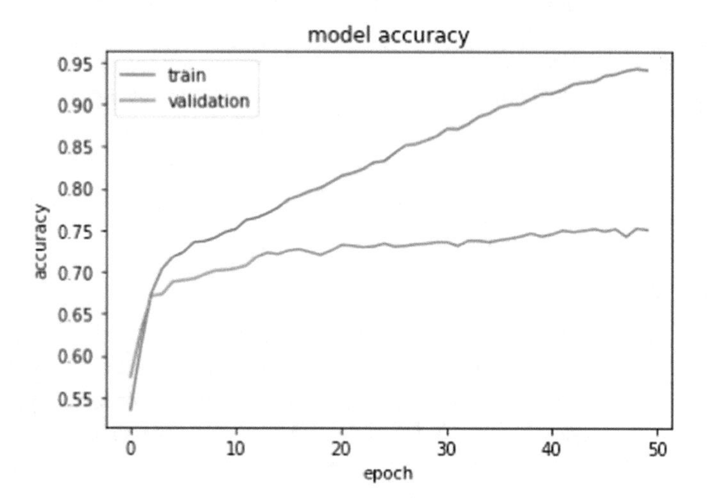

Fig. 2. Train and Validation Set Accuracy

Train and validation accuracy increase together until 50 epochs. After 50 epochs, the accuracy of training continues to increase while the accuracy of validation decreases. This shows that after 50 epochs, the system starts to overfit the train set. For this reason, the model training was stopped after 50 epochs.

4 Results and Discussion

As shown in Fig. 1, our solution consists of methodological steps. We used the Logistic Regression Classifier to observe the differences in classification success within each methodological step. Logistic regression is one of the most commonly used conventional classifiers used in various text classification studies [10–12]. Since our study involves a phase-wise solution, the classification reports of each phase are presented separately in Table 3.

The results shown in Table 3 are the classification reports obtained from the validation set. The first three results (Original Tweets, Cleaned Tweets, and Cleaned ASCII Tweets) are taken from the Logistic Regression classifier on tf-idf of word unigrams. The result of

Cleaned ASCII N-gram Tweets is also obtained from the Logistic Regression classifier on tf-idf of character 4-g features. Among the other n values (2, 3, 4, 5, 6), 4-g gave a better result. We obtained the highest validation accuracy from our neural network based classification model, explained in the Classification Model section. Our model is more accurate than logistic regression because many possibilities were tried to select the most appropriate neural network parameters and properties for the problem. A great success was achieved by optimizing the artificial neural networks' parameters and adaptable properties. The results obtained by our team on the test dataset of Subtask-2 are 0.6672 for macro precision and 0.6537 for macro F1 score (ranked $10th$).

Table 3. Classification report of each phases

Text	Accuracy	Class	Precision	Recall	F1_score
Original Tweets	0.72	HOF NOT	0.73 0.71	0.70 0.74	0.72 0.73
Cleaned Tweets	0.72	HOF NOT	0.73 0.71	0.70 0.75	0.72 0.73
Cleaned ASCII Tweets	0.72	HOF NOT	0.73 0.71	0.70 0.74	0.72 0.72
Cleaned ASCII N-gram Tweets	0.73	HOF NOT	0.74 0.71	0.70 0.75	0.72 0.73
Cleaned ASCII N-gram Tweets (Our Model)	0.75	HOF NOT	0.77 0.74	0.72 0.78	0.74 0.76

5 Conclusion and Future Work

Identifying hate and offensive speech in a conversation plays a vital role in social activities. When the conversation is in code-mixed language, identification becomes more difficult. This study presents a language-free solution based on script normalization and n-gram feature extraction. Script normalization is performed using the ASCII transformation and can be performed in all different script texts to achieve a standard representation of all languages. Script normalization normalizes mixed script texts and provides a default representation of the ASCII character set for other special characters such as emojis. Although using a standard code set might not make a more considerable difference in identifying hate speech, it reduces processing complexity and also makes code-mixed texts more understandable and applicable to foreign researchers.

In the future, we plan to apply our model to different code-mix datasets and use other classification models to improve the accuracy of hate speech identification.

References

1. Fortuna, P., Nunes, S.: A survey on automatic detection of hate speech in text. ACM Comput. Surv. **51**, 1–30 (2018)
2. Kovács, G., Alonso, P., Saini, R.: Challenges of hate speech detection in social media. SN Comput. Sci. **2**, 1–15 (2021)
3. Modha, S., Majumder, P., Mandl, T., Mandalia, C.: Detecting and visualizing hate speech in social media: a cyber watchdog for surveillance. Expert Syst. Appl. 161 (2020)
4. Mandl, T., et al.: Overview of the hasoc track at fire 2019: Hate speech and offensive content identification in Indo-European languages. In: Proceedings of the 11th Forum for Information Retrieval Evaluation, FIRE'19, Association for Computing Machinery, New York, NY, USA, pp. 14–17 (2019)
5. Mandl, T., Modha, S., Kumar, A., Chakravarthi, M.B.R.: Overview of the hasoc track at fire 2020: Hate speech and offensive language identification in Tamil, Malayalam, Hindi, English and German. Forum for Information Retrieval Evaluation, FIRE 2020, Association for Computing Machinery, New York, NY, USA, pp. 29–32 (2020)
6. Modha, S., et al.: Overview of the HASOC Subtrack at FIRE 2021: Hate Speech and Offensive Content Identification in English and Indo-Aryan Languages and Conversational Hate Speech. FIRE 2021: Forum for Information Retrieval Evaluation, Virtual Event, ACM (2021)
7. Satapara, S., Modha, S., Mandl, T., Madhu, H., Majumder, P.: Overview of the HASOC Subtrack at FIRE 2021: Conversational Hate Speech Detection in Code-mixed language. Working Notes of FIRE 2021 – Forum for Information Retrieval Evaluation, CEUR (2021)
8. Rudra, K., Sharma, A., Bali, K., Choudhury, M., Ganguly, N.: Identifying and analyzing different aspects of English-Hindi code-switching in twitter. ACM Trans. Asian Low-Resour. Lang. Inf. Process. **18**, 1–28 (2019)
9. Kingma, D.P., Ba, J.: Adam: a method for stochastic optimization. arXiv preprint arXiv:1412.6980(2014)
10. Malmasi, S., Zampieri, M.: Challenges in discriminating profanity from hate speech. J. Exp. Theor. Artif. Intell. **30**, 187–202 (2018)
11. Canbay, P., Sezer, E.A., Sever, H.L.: Binary background model with geometric mean for author-independent authorship verification. J. Inf. Sci. (2021)
12. Modha, S., Majumder, P., Mandl, T.: An empirical evaluation of text representation schemes to filter the social media stream. J. Exp. Theor. Artif. Intell. 1–27 (2021)

Decision Trees in Causal Inference

Hulya Kocyigit(✉)

Karamanoglu Mehmetbey University, Karaman, Turkey
hk20902@gmail.com

Abstract. Nonrandomized studies are commonly employed in medical science to quantify the effects of treatments on outcomes. There may be differences between the treated and untreated groups due to the dearth of random treatment assignments of participants in observational research. Thus, predictions of the treatment effect may be distorted as a result of these discrepancies. Statistical procedures such as propensity score are required in this case to eliminate or limit the effects of confounding variables. Even though a main factor logistic regression is widely used to estimate the propensity score, using machine learning algorithms in observational studies have considerably progressed in recent decades. This paper examines two fundamental problems: estimating propensity score and determining covariate balance among the treatment groups. So, this study evaluates the effectiveness of propensity score matching rely on the random forest and decision trees methods option (i.e., recursive partitioning CART, bagged CART, boosted CART, pruned CART, conditional inference trees, and C5.0 model) to logistic regression that could achieve the comparable results with fewer assumptions and higher consistency. In order to compare various decision trees methods in PS matching for estimating average treatment effects (ATEs) in terms of bias and mean squared error (MSE), we conduct a Monte Carlo simulation under the three different scenarios. In the framework of PS approaches, it is also looked into the necessity of tweaking machine learning settings. According to the results, the application of these adjusted decision trees methods to estimate the PS may better the robustness of PS matching estimators. Thus, causal inference has been a favorable method that is frequently used in many disciplines such as medicine, epidemiology, economics, public health to know causality between treatment and control group.

Keywords: Nonrandomized study · propensity score · decision trees · matching · overlap

1 Introduction

In principle, statistical inference is concerned with measuring aspects of a population's distribution by relying on a sample. Scientists are frequently interested in learning more than just the associations within a population's joint likelihood distribution. Researchers might be curious about how the distribution changes when it is operated upon: does A give rise to B?. More than merely looking at a probability distribution is required to understand causal links. The approach of causal inference emphasizes how to describe and respond to observed associations. Although causal inference theory provides several techniques

D. J. Hemanth et al. (Eds.): ICAIAME 2022, ECPSCI 7, pp. 109–121, 2023.
https://doi.org/10.1007/978-3-031-31956-3_9

for avoiding logical errors in arguments, others say that it exaggerates mathematics and deduction at the expense of research description and contextual synthesis, both of which are necessary for reliable real-world inference. Pearl [1] states that any relationship that can be expressed in terms of a joint distribution of observable factors is an associational notion, while any relationship that cannot be described from the distribution simply is a causal concept. Thus, causal inference has been a favorable method that is frequently used in many disciplines such as medicine, epidemiology, economics, and public health to know the causality between treatment and control groups. In literature, different types of data such as randomized control trials, and observational or experimental design data can be preferred to express cause-effect relationships. In literature, different types of data such as randomized control trials, and observational or experimental data can be preferred to express cause-effect relationships. Random control trials are comfortable tools for scientists to solve predictive problems regarding causation between treated and untreated groups in the future. Randomized experiments perform unbiased treatment effect estimators and make it easier to apply simple and classic statistical procedures. The probable outcomes paradigm, which goes beyond established approaches, provides intuitive and straightforward claims regarding causation in a random experiment [2–4].

Nevertheless, such tests might be either prohibitively priced, inappropriate, or unattainable. As a result, extracting causal effects from pseudo-observational data generated by analyzing a system without subjecting it to interferences is beneficial [3, 5]. Austin and Stuart [6] proposed that treatment selection bias, in which treated participants differ markedly from untreated participants, can occur in observational studies investigating the effect of treatment on outcomes. Consequently, treatment effects cannot be evaluated simply by assessing outcomes between treatment arms. This problem can be addressed in part by incorporating knowledge on measurable confounders into the research design or treatment effect estimation. Classical techniques of adjustment are frequently constrained by the fact that they can only adapt to a small number of factors. This restriction does not apply to propensity scores, which offer a scalar summary of the covariate information [3]. Rosenbaum and Rubin [7] developed the propensity score in the 1980s which can be described as the probability of treatment provided by observed given variables. Logistic regression is usually utilized to estimate propensity scores for a binary treatment. Parametric models, on the other hand, necessitate assumptions about variable selection, covariates' distribution, and defining interaction among covariates. If parametric assumptions are met, covariate balance between treated and untreated groups may not be attained, resulting in a skewed impact estimate.

This paper seeks at using some decision tree approaches as an option for logistic regression. Unlike statistical techniques for modeling, which presuppose a data model with variables determined from the data, machine learning employs a learning technique to extract the link between an outcome and a treatment without using an a priori data model. This study is organized as follows: Sect. 2 presents a framework of causal inference, looks over the estimation of propensity score based on types of decision tree models, and describes matching method depending on the propensity score. Section 3 proceeds over a variety of Monte Carlo simulations to see how well these approaches for predicting treatment effects perform. In Sect. 4, I offer a simulation study in which I assess the effectiveness of several matching methods in evaluating the influence of different treatment scenarios. So, a summary of the findings is presented and put in the perspective of previous research. Finally, I wrap things up with a summary and some reflections in Sect. 5.

2 Methods

2.1 Framework and Overview of Causal Effect

For this work, I utilize the appropriate notation. Let T represent a binary variable that represents the treatment level (i.e., $T_i = 1$ $T_i = 1$ for treated group vs $T_i = 0$ $T_i = 0$ for the untreated group). Also, X constitutes n-dimensional an array of observed baseline variables. Let Y refer to a continuous outcome variable. So, the potential two possible outcomes, which denote (0) and (1)Y_i $Y_i Y_i$ Y_i outcomes for untreated and treated groups, are considered when given under identical conditions. So, one potential outcome is written as $Y_i = T_i Y_i(0) + (1 - T_i) Y_i(1)$ $Y_i = T_i Y_i(0) + (1 - T_i) Y_i(1)$. Shortly, Y_i Y_i is expressed to be equivalent as Y_i Y_i (0) when $T_i = 0$and $T_i = 0$and beside, Y_i Y_i is occurred to be equal as Y_i Y_i (0) when $T_i = 1$. $T_i = 1$. Average causal treatment effect (ATE) is expressed by $E[Y_i(1)] - -E[Y_i(0)]$ $E[Y_i(1)] - -E[Y_i(0)]$. . Average causal treatment effect in treated (ATT) is represented difference between the outcome groups: $E[Y_i(1)] - E[Y_i(0)|T = 1]$ $E[Y_i(1)] - E[Y_i(0)|T = 1]$.

2.2 Estimation of Propensity Score

The propensity score is expressed as a probability of obtaining treatment depending on a set of variables, any model associating a set of variables and any group of confounder variables might be utilized to estimate propensity score [8]. It means that the two treatment arms are practically similar for a specified propensity score. Evaluating the two treatments for each propensity score and then aggregating the treatment effect throughout all propensity scores would be a simple implementation [9]. A logistic regression model is used to estimate the propensity score, in which the binary variable representing treatment status is regressed on measured baseline variables. Logistics regression implies that variables are additive, sometimes high-order interaction on the log-odds scale. LR as a parametric model with choose large interaction terms and polynomial terms illustrate high performance. Nevertheless, choosing appropriate high-order and interaction terms between variables has been highly crucial.

Machine learning techniques have been offered as an option to parametric models (i.e., logistics regression, covariate balance propensity score model, etc.) because of being easy construction of models. Even though various ML methods are discussed about estimating treatment effect and reduce bias between treatment groups, there is limited studies under umbrella of decision tree models. So, I focus on five decision tree-based techniques to be employed. A propensity score is described as Pr (T=t, X), where T stands for receiving treatment and X indicates a set of observable variables. The following approaches are applied to estimate propensity scores.

Recursive Partitioning: When applying recursive partitioning to estimate propensity scores, the data is divided into groups with similar treatment levels depending on the levels of covariates or a threshold employed for continuous covariates. Interactions and nonlinear relationship terms are not necessarily applied in fitting the recursive partitioning method whereas logistic regression, which is a parametric approach, demands that interactions and nonlinear relationships between baseline covariates be provided before constructing the model. Recursive partitioning algorithms have the added feature of automatically managing missing data in variables by evaluating all possible circumstances for each split [10].

***C5.0 Algorithms*:** One of the new family members of decision trees based on Machine Learning Algorithms is the C5.0 algorithm [11] which was created as an improved form of the well-known and frequently employed C4.5 classifier.C5.0 approach produces several decision trees and then merged to advance the predictions. There are some advantages to using this method such as reducing expense for covariate misclassification, allowing signs for missing or non-appropriate cases, provide sampling and cross-validation [12, 13].

The Conditional Inference Tree (CIT) is the machine learning technique with the fewest requirements among the many data mining approach. Propensity scores might be estimated using CIT by computing the percentage of trees that categorized each observation as a treatment [14, 15].

***CART Prune*:** CART is a binary recursive partitioning algorithm that can treat both continuous and nominal properties as objectives and predictors. No binning is necessary or advised because the data is managed in its pure form. The data is separated into two subgroups starting at the root node, and each of the subgroups is divided into more small groups. Without the use of a halting rule, trees are expanded to their maximum size; in essence, the tree-growing process comes to a halt when no more splits are feasible due to a shortage of data. The tree is then trimmed back to the root using the innovative cost-complexity pruning strategy. The next split to be pruned is the one that has the least impact on the tree's overall performance on training data. The CART mechanism is designed to generate a series of nested pruned trees, each of which is a contender for being the best tree [16, 17].

***Boosted Model*:** Friedman (2001) [18] proposed gradient boosted regression trees that are a kind of machine learning technique relying on tree averaging. Boosted model, despite training a large number of entire high variance tress that are then averaged to minimize overfitting, incorporates little trees one by one, each with a high bias. So, the new tree that is built in each iteration specifically on the documents that are accountable for the present remaining regression error [19]. When dealing with data sets with binary data or categorical data with fewer than tens of categories, the boosted trees model performs. Boosted tree models, unlike linear models, may represent non-linear interactions between characteristics and the targets. One thing to keep in mind is that tree-based models aren't meant to function with very few features.

2.3 Propensity Score Matching

Propensity score matching is the way of forming matched sets of treated and untreated individuals with similar propensity score values [20]. There are numerous matching techniques, as is widely known such as matching on the propensity score, kernel matching, nearest-neighbor matching, or caliper matching. According to [21], one-to-one or pair matching is the most typical application of propensity score matching, in which pairings of treatment and control individuals are generated so that matched subjects have identical propensity score values. Only changes in a nonlinear function of the variables, namely the predicted propensity score, are considered in this study. There are two reasons behind this decision. First of all, balancing for variations in propensity scores between treatment and control groups reduces all major biases related to observed variable differences. Secondly, finding similar matches on the propensity score is easier than finding close matches on all factors together. Let e(x) represents the propensity score, and $l(x) = \ln e(x)$ $l(x) = \ln e(x)$ denotes the linearized propensity score (lps), or call as odds ratio logarithm. To be more explicit, we'll employ the squared difference in lps as the metric [22]:

$$\mathrm{d_l}\left(\mathrm{x}, x'\right) = (\mathrm{l(x)} - \mathrm{l}(x'))^2 = \left(\ln\left(\frac{\mathrm{e(x)}}{1 - \mathrm{e(x)}}\right) - \ln\left(\frac{\mathrm{e}(x')}{1 - \mathrm{e}(x')}\right)\right)^2 \tag{1}$$

3 Simulation Study

This paper describes the framework of the Monte Carlo simulations used to look into the potential bias of different propensity score methodologies for computing conditional treatment effects. Monte Carlo simulations are based on results from a variation of [23] 's simulation design that compared the ability of several propensity score models to balance observed variables across treatment and control subjects. This study created 15 variables ($\mathrm{X_1} - \mathrm{X_{12}}$) as a variety of continuous and binary covariates for each simulation. The binary covariates were standard normal random variables that had been dichotomized (at cutoff value 0), whereas the continuous variables were generated based on the standard normal variables. The first four covariates (i.e., $\mathrm{X_1}-\mathrm{X_4}$) are related with both of treatment (T) and outcome (Y), three covariates (i.e., $\mathrm{X_5}-\mathrm{X_7}$) are associated with only treatment, three covariates (i.e., $X_8 - X_{10}$ $X_8 - X_{10}$) are connected with only outcome and finally, five covariates (i.e., $\mathrm{X_{11}}, \mathrm{X_{12}}, \mathrm{X_{13}}, \mathrm{X_{14}}, \mathrm{X_{15}}$) are not associated with neither treatment nor outcome. Also, correlation between some of covariates ranging from 0.2 to 0.9 were used such as corr ($\mathrm{X_1} - \mathrm{X_5}) = 0.2$, $\mathrm{corr}(\mathrm{X_2} - \mathrm{X_6}) = 0.9$, $\mathrm{corr}(\mathrm{X_3} - \mathrm{X_8}) = 0.2$, and $\mathrm{corr}(\mathrm{X_4} - \mathrm{X_9}) = 0.9$. So, $\mathrm{X_1}, \mathrm{X_3}, \mathrm{X_5}, \mathrm{X_6}, \mathrm{X_8}, \mathrm{X_9}, \mathrm{X_{11}}, \mathrm{X_{14}}$ covariates are generated as dichotomous variables, while $\mathrm{X_2}, \mathrm{X_4}, \mathrm{X_7}, \mathrm{X_{10}}, \mathrm{X_{12}}, \mathrm{X_{13}}, \mathrm{X_{15}}$ covariates are based on the continuous normal random variables. This study created 500 datasets, each with 500 individuals, for each of the simulation scenarios. After generating covariates, four different treatment scenarios are created as:

Treatment A: $Pr[T = 1|X_i] = (1 + \exp\{-(\theta_0 + \theta_1 * X_1 + \theta_2 * X_2 + \theta_3 * X_3 + \theta_4 * X_4 + \theta_5 * X_5 + \theta_6 * X_6 + \theta_7 * X_7)\})^{-1}$
Treatment B: $Pr[T = 1|X_i] = (1 + \exp\{-(\theta_0 + \theta_1 * X_1 + \theta_2 * X_2 + \theta_3 * X_3 + \theta_4 * X_4 + \theta_5 * X_5 + \theta_6 * X_6 + \theta_7 * X_7 + \theta_2 * X_2 * X_2 + \theta_4 * X_4 * X_4 + \theta_7 * X_7 * X_7)\})^{-1}$
Treatment C: $Pr[T = 1|X_i] = (1 + \exp\{-(\theta_0 + \theta_1 * X_1 + \theta_2 * X_2 + \theta_3 * X_3 + \theta_4 * X_4 + \theta_5 * X_5 + \theta_6 * X_6 + \theta_7 * X_7 + \theta_1 * 0.5 * X_1 * X_3 + \theta_2 * 0.7 * X_2 * X_4 + \theta_3 * 0.5 * X_3 * X_5 + +\theta_4 * 0.7 * X_4 * X_6 + \theta_5 * 0.5 * X_5 * X_7 + \theta_1 * 0.5 * X_1 * X_6 + \theta_2 * 0.7 * X_2 * X_3 + \theta_3 * 0.5 * X_3 * X_4 \theta_4 * 0.5 * X_4 * X_5 + \theta_5 * 0.5 * X_5 * X_6)\})^{-1}$
Treatment D: $Pr[T = 1|X_i] = (1 + \exp\{-(\theta_0 + \theta_1 * X_1 + \theta_2 * X_2 + \theta_3 * X_3 + \theta_4 * X_4 + \theta_5 * X_5 + \theta_6 * X_6 + \theta_7 * X_7 + \theta_1 * 0.5 * X_1 * X_3 + \theta_2 * 0.7 * X_2 * X_4 + \theta_3 * 0.5 * X_3 * X_5 + +\theta_4 * 0.7 * X_4 * X_6 + \theta_5 * 0.5 * X_5 * X_7 + \theta_1 * 0.5 * X_1 * X_6 + \theta_2 * 0.7 * X_2 * X_3 + \theta_3 * 0.5 * X_3 * X_4 \theta_4 * 0.5 * X_4 * X_5 + \theta_5 * 0.5 * X_5 * X_6 + \theta_2 * X_2 * X_2 + \theta_4 * X_4 * X_4 + \theta_7 * X_7 * X_7)\})^{-1}$

where $\theta_i = (0, 0.8, -0.25, 0.6, -0.4, -0.8, -0.5, 0.7)$. All treatment scenarios as in above, which determined the degree of complexity in the relationship between covariates and treatment assignment, were used in $\text{Prob}(T = 1) = \frac{1}{\{1+\exp(-\text{treatment scenario}-\xi*\tau\}}$ to be generated the final form of treatment.

An outcome version is considered as:

$$Pr[Y = 1|X_i, T_i] = \left(1 + \exp\left\{-\begin{pmatrix}\alpha_0 + \alpha_1 * X_1 + \alpha_2 * X_2 + \alpha_3 * X_3 + \alpha_4 * X_4 \\ +\alpha_5 * X_8 + \alpha_6 * X_9 + \alpha_7 * X_{10}\end{pmatrix}\right\}\right)^{-1}$$

where $\alpha_i = (-3.85, 0.3, -0.36, -0.73, -0.2, 0.71, -0.19, 0.26, -0.4)$. The outcome scenario is used into $\text{Prob}(Y = 1) = \frac{1}{\{1+\exp(-\text{outcome scenario}-\vartheta*\zeta)\}}$ that generates the final version of outcome. In here, true treatment effect defines as –0.4. After complete generating all 12 covariates, 4 different treatment scenarios and an outcome in this simulation, I construct different propensity estimation strategies.

In real-world circumstances, researchers choose which variables to include in estimating techniques, and these approaches are modeled after some of those choices. The variables that were taken into account as:

Model 1: only variables $(X_1 - X_4$ $(X_1 - X_4)$ are associated with either treatment and outcome.
Model 2: variables $(X_1 - X_7$ $(X_1 - X_7)$ are related to treatment individuals.
Model 3: variables $(X_1 - X_4$ and $X_8 - X_{10})$ $(X_1 - X_4$ and $X_8 - X_{10})$ are associated with the outcome individual.
Model 4: All covariates $(X_1 - X_{10}$ $(X_1 - X_{10})$ have a direct or indirect relationship with both the outcome and the treatment.
Model 5: All possible covariates $(X_1 - X_{15}$ $(X_1 - X_{15})$ are included in the model.

The average of estimated SE, RMSE, absolute bias and relative bias are presented to evaluate this simulation's performance. Absolute bias was computed by $\left|\widehat{\beta_1} - \beta_1\right|$, where β_1 is true treatment effect (i.e. defined by –0.4). Relative bias is formulated as $100*\frac{\text{Absolute Bias}}{\beta_1}$ $100*\frac{\text{Absolute Bias}}{\beta_1}$. This paper computed the average of SE based on taking average of the SE of treatment effects estimates for each simulated data. Lastly,

RMSE is defined as represented taking square root of means square error for each estimator. It's formulated by $\sqrt{\frac{1}{n}\sum_{i=1}^{n}(\beta_i - \beta)^2}$. I generated 500 datasets, each involving of sample size $n = 500$ individuals. All results were carried out on Mac OS X platform with R 3.3.0 statistical packages.

4 Simulation Results

Table 1–Table 4 summarizes the results of simulation studies that are conducted. All tables below display the absolute bias, relative bias, average estimated SE, and RMSE of estimated treatment effect for scenarios A-D in assessment utilizing propensity score-matched data. In other words, all tables illustrate the properties of result metrics after matching on propensity scores. The absolute bias of boosted models tends to be higher than the rest of the four decision trees techniques meanwhile C5.0 model seems to be yielded less absolute bias and relative bias across all model scenarios. When model scenarios (i.e., model 1, model 2, model 3, model 4, and model 5) are assessed in the same model estimators in Table 1, Model 2 using C5.0 estimators produced less absolute bias and relative bias. So, it means that the true propensity score model (i.e., model 2 was included by covariates) produced less bias and balanced between treatment and outcome groups when considering the C5.0 model to estimate propensity score values. In other words, adding confounders variables in propensity score models intended to be created remarkable bias and RMSE values. Besides, absolute relative bias over boosted estimator model were 0.219 for Model 1,0.245 for Model 2,0.226 for Model 3, 0.255 for Model 4 and 0.282 for Model 5 (Table 1). Thus, these results indicated us that boosted model might not be a good option to estimate propensity scores and set up a good balance between treatment arms under no matter considering which variables are involved in model scenarios. The averages of estimated SEs were smallest in the C5.0 method and then in CIT for the model 2 scenario, reflecting different model scenarios of covariates after matching (Table 2). For boosted models' method, the absolute bias ranged from 0.219 to 0.282 while for CIT it ranged from 0.186 to 0.196. Adding instrument variables in the propensity score estimate (options Model 2, 4 and 5) resulted in greater relative bias for all five estimators throughout the board.

The bias in the recursive partitioning approach is 0.201, 0.216, 0.204, 0.218, and 0.223, in which they seem to be close to each other, from model 1 to model 5. However, the values at models reflect the opposite with boosted method producing a remarkable larger bias and relative bias. For evaluations where the difference between estimator approaches' range of differences is lower, the range was limited to values that highlight the major variations; for RMSE, I presented 0.250 to 0.463 across all models for the boosted model (Table 3). As expected, the C5.0 method was associated with the lowest positivity bias and absolute relative bias across Model 1, Model 2 and Model 4 compared to other methods in Table 3. Those results indicated that if I added instrument variables in models or removed some variables, which are related to treatment, C5.0 methods performed increased bias and largest RMSE. Even the CIT method throughout all models has been following close results in terms of metrics with the C5.0 method. In other words,

Table 1. Absolute bias, relative bias, average estimated SE and RMSE of treatment A on outcome Y based on the Recursive partitioning, Prune CART, Boosted CART, Conditional tree and C5.0 models for five different model scenarios

Estimators	Model Scenarios	Absolute Bias	Relative Bias(%)	Ave. Estimated SE	RMSE
Recursive Partitioning	*Model 1*	0.205	51.28	0.138	0.256
	Model 2	0.204	51.07	0.139	0.256
	Model 3	0.196	49.10	0.138	0.252
	Model 4	0.208	52.07	0.139	0.258
	Model 5	0.206	51.68	0.139	0.263
CART Prune	*Model 1*	0.188	47.23	0.138	0.239
	Model 2	0.199	49.93	0.140	0.246
	Model 3	0.191	47.90	0.138	0.236
	Model 4	0.194	48.55	0.140	0.240
	Model 5	0.191	47.89	0.141	0.237
Boosted	*Model 1*	0.219	54.81	0.138	0.275
	Model 2	0.245	61.44	0.139	0.306
	Model 3	0.226	56.53	0.138	0.285
	Model 4	0.255	63.96	0.139	0.324
	Model 5	0.282	70.69	0.140	0.354
Conditional Tree	*Model 1*	0.186	46.60	0.138	0.231
	Model 2	0.196	49.11	0.140	0.249
	Model 3	0.187	46.75	0.139	0.229
	Model 4	0.197	49.43	0.140	0.249
	Model 5	0.195	48.98	0.140	0.247
C5.0	*Model 1*	0.185	46.26	0.141	0.232
	Model 2	0.177	44.32	0.139	0.223
	Model 3	0.230	57.67	0.142	0.250
	Model 4	0.174	43.62	0.139	0.213
	Model 5	0.179	44.88	0.140	0.215

Remark: Model 1: $X_1 - X_4\, X_1 - X_4$, Model 2: $X_1 - X_7\, X_1 - X_7$, Model 3: $X_1 - X_4 and X_8 - X_{10}$ $X_1 - X_4 and X_8 - X_{10}$, Model 4: $X_1 - X_{10}\, X_1 - X_{10}$, Model 5: $X_1 - X_{15}\, X_1 - X_{15}$.

I observed that the CIT and C5.0 methods are preferable than the rest of the methods. Overall, in Table 3, the C5.0 method has illustrated better performance generally.

Also, in Table 4, while the CIT method with all types of models obtained slightly better bias, boosted method produced the worst bias, relative bias, and the largest RMSE especially in more complex treatment scenarios among of five methods. The CIT method

Table 2. Absolute bias, relative bias, average estimated SE and RMSE of treatment B on outcome Y based on the Recursive partitioning, Prune CART, Boosted CART, Conditional tree and C5.0 models for five different model scenarios

Estimators	Model Scenarios	Absolute Bias	Relative Bias(%)	Ave. Estimated SE	RMSE
Recursive Partitioning	*Model 1*	0.190	47.52	0.138	0.247
	Model 2	0.214	53.52	0.139	0.270
	Model 3	0.193	48.48	0.138	0.245
	Model 4	0.218	54.55	0.139	0.273
	Model 5	0.225	56.33	0.140	0.284
CART Prune	*Model 1*	0.189	47.38	0.138	0.233
	Model 2	0.189	47.45	0.139	0.242
	Model 3	0.183	45.98	0.138	0.231
	Model 4	0.191	47.75	0.140	0.246
	Model 5	0.198	49.55	0.140	0.251
Boosted	*Model 1*	0.216	54.20	0.138	0.271
	Model 2	0.259	64.92	0.141	0.333
	Model 3	0.228	57.17	0.138	0.294
	Model 4	0.284	71.08	0.141	0.360
	Model 5	0.300	75.16	0.142	0.376
Conditional Tree	*Model 1*	0.175	43.79	0.140	0.222
	Model 2	0.166	41.50	0.138	0.209
	Model 3	0.245	61.45	0.140	0.262
	Model 4	0.175	43.95	0.139	0.214
	Model 5	0.181	45.41	0.139	0.216
C5.0	*Model 1*	0.174	43.53	0.140	0.218
	Model 2	0.155	38.95	0.138	0.196
	Model 3	0.241	60.32	0.141	0.259
	Model 4	0.174	43.53	0.139	0.211
	Model 5	0.184	46.14	0.139	0.219

Remark: Model 1: $X_1 - X_4$ $X_1 - X_4$, Model 2: $X_1 - X_7$ $X_1 - X_7$, Model 3: $X_1 - X_4 and X_8 - X_{10}$ $X_1 - X_4 and X_8 - X_{10}$, Model 4: $X_1 - X_{10}$ $X_1 - X_{10}$, Model 5: $X_1 - X_{15}$ $X_1 - X_{15}$.

appears to have increased its performance in Table 4 according to its yield in resulting of Table 3 in terms of measurements. In other saying, the difference in their performance between CIT and C5.0 have remarkably grown when all treatment is generated from linear treatment to complex treatment scenario. It is clear that when all tables are evaluated from treatment A- treatment D scenarios, boosted method tends to blow up the bias, relative bias, and values of RMSE for all models, respectively. That indicates that when complex treatment scenarios lead to produce more bias for boosted method. Another remarkable point is that the CART prune method consistently resulted in less bias when the treatment variable is considered less complex or linear (i.e., treatment A and B) but the trend conversed for more complex treatment scenarios (i.e., treatment C and D). In addition, Recursive partitioning and CART prune methods have produced parallel and close measurement values when no matter which treatment scenarios are considered according to Table 3 and Table 4. To put it plainly, changing your treatment scenario does not make sense for both recursive partitioning and CART prune methods under different circumstances models.

Table 3: Absolute bias, relative bias, average estimated SE and RMSE of treatment C on outcome Y based on the Recursive partitioning, Prune CART, Boosted CART, Conditional tree and C5.0 models for five different model scenarios

Estimators	Model Scenarios	Absolute Bias	Relative Bias(%)	Ave. Estimated SE	RMSE
Recursive Partitioning	*Model 1*	0.201	50.32	0.138	0.253
	Model 2	0.216	54.41	0.139	0.276
	Model 3	0.204	51.14	0.138	0.258
	Model 4	0.218	54.60	0.139	0.280
	Model 5	0.223	55.98	0.140	0.286
CART Prune	*Model 1*	0.201	50.35	0.139	0.252
	Model 2	0.206	51.53	0.140	0.262
	Model 3	0.197	49.39	0.139	0.252
	Model 4	0.204	51.19	0.140	0.256
	Model 5	0.202	50.63	0.141	0.257
Boosted	*Model 1*	0.307	76.95	0.202	0.382
	Model 2	0.328	82.24	0.204	0.415
	Model 3	0.313	78.46	0.203	0.391
	Model 4	0.334	83.69	0.205	0.424
	Model 5	0.364	91.23	0.208	0.463
Conditional Tree	*Model 1*	0.198	46.60	0.138	0.250
	Model 2	0.208	49.11	0.140	0.262
	Model 3	0.202	46.75	0.139	0.253
	Model 4	0.215	49.43	0.140	0.271

(continued)

Table 3: (*continued*)

Estimators	Model Scenarios	Absolute Bias	Relative Bias(%)	Ave. Estimated SE	RMSE
	Model 5	0.212	48.98	0.140	0.266
C5.0	*Model 1*	0.193	48.44	0.143	0.243
	Model 2	0.195	48.87	0.140	0.244
	Model 3	0.214	53.57	0.145	0.238
	Model 4	0.161	40.25	0.141	0.198
	Model 5	0.169	42.35	0.142	0.209

Remark: Model 1: $X_1 - X_4$ $X_1 - X_4$, Model 2: $X_1 - X_7$ $X_1 - X_7$, Model 3: $X_1 - X_4 and X_8 - X_{10}$ $X_1 - X_4 and X_8 - X_{10}$, Model 4: $X_1 - X_{10}$ $X_1 - X_{10}$, Model 5: $X_1 - X_{15}$ $X_1 - X_{15}$.

Table 4. Absolute bias, relative bias, average estimated SE and RMSE of treatment D on outcome Y based on the Recursive partitioning, Prune CART, Boosted CART, Conditional tree and C5.0 models for five different model scenarios

Estimators	Model Scenarios	Absolute Bias	Relative Bias(%)	Ave. Estimated SE	RMSE
Recursive Partitioning	*Model 1*	0.196	49.19	0.138	0.249
	Model 2	0.216	54.06	0.140	0.276
	Model 3	0.198	49.56	0.138	0.255
	Model 4	0.221	55.32	0.140	0.281
	Model 5	0.227	56.88	0.140	0.291
CART Prune	*Model 1*	0.197	49.37	0.139	0.245
	Model 2	0.207	51.88	0.140	0.261
	Model 3	0.191	47.84	0.139	0.242
	Model 4	0.217	54.41	0.140	0.274
	Model 5	0.208	52.10	0.140	0.266
Boosted	*Model 1*	0.305	76.37	0.203	0.384
	Model 2	0.351	87.96	0.207	0.437
	Model 3	0.314	78.56	0.204	0.395
	Model 4	0.374	93.72	0.208	0.474
	Model 5	0.406	101.71	0.211	0.502
Conditional Tree	*Model 1*	0.185	46.60	0.138	0.237
	Model 2	0.199	49.11	0.139	0.249

(*continued*)

Table 4. (*continued*)

Estimators	Model Scenarios	Absolute Bias	Relative Bias(%)	Ave. Estimated SE	RMSE
	Model 3	0.180	46.75	0.138	0.231
	Model 4	0.192	49.43	0.139	0.241
	Model 5	0.190	48.98	0.139	0.242
C5.0	*Model 1*	0.182	45.41	0.142	0.230
	Model 2	0.173	43.31	0.139	0.225
	Model 3	0.231	57.95	0.143	0.252
	Model 4	0.176	44.00	0.140	0.215
	Model 5	0.177	44.42	0.141	0.215

Remark: Model 1: $X_1 - X_4$ $X_1 - X_4$, Model 2: $X_1 - X_7$ $X_1 - X_7$, Model 3: $X_1 - X_4 and X_8 - X_{10}$ $X_1 - X_4 and X_8 - X_{10}$, Model 4: $X_1 - X_{10}$ $X_1 - X_{10}$, Model 5: $X_1 - X_{15}$ $X_1 - X_{15}$.

5 Conclusion

Propensity score approaches have proven a significant instrument in many fields where the investigation of observation research data is widespread. It has been noted that propensity scores are computed utilizing logistics regression or sometimes have limitations using parametric methods [23–25]. For all methods across all different propensity score models considered from very simple to extremely complex treatment scenarios (i.e., from Treatment-A scenario in Table 1 to Treatment-B scenario in Table 2), all methods were assessed in terms of bias, relative bias, an average of SE, and RMSE metrics. This paper examines tens of simulated datasets to show when recursive partitioning, CART prune, boosted, conditional tree, and C5.0 methods techniques outperform or close the results to each other under any circumstances. Even though CIT and C5.0 methods tend to produce close results for each treatment scenario, C5.0 appears to work exceptionally well under all model scenarios across all treatment scenarios. The flexibility of boosted model for all models, on the contrary hand, is likely to offer the worst results when the full propensity score model scenario (i.e., model 5) has unimportant variables or more complex treatment. So, it means that the type of treatment scenario that matters in propensity score models affect the comparative performance of the boosted techniques. I hope that all recommendations throughout this paper will assist a researcher in selecting the optimal instrument and empathize with which decision tree methods are chosen for the job.

References

1. Pearl, J.: The foundations of causal inference. Sociol. Methodol. **40**(1), 75–149 (2010)
2. Rubin, D.B.: Estimating causal effects of treatments in randomized and nonrandomized studies. J. Educ. Psychol. **66**(5), 688 (1974)

3. D'Agostino, J.R., Ralph, B.: Propensity score methods for bias reduction in the comparison of a treatment to a non-randomized control group. Stat. Med. **17**(19), 2265–2281 (1998)
4. Trojano, M., et al.: Observational studies: propensity score analysis of non-randomized data. Int. MS J. **16**(3), 90–97 (2009)
5. Austin, P.C.: The use of propensity score methods with survival or time-to-event outcomes: reporting measures of effect similar to those used in randomized experiments. Stat. Med. **33**(7), 1242–1258 (2014)
6. Austin, P.C., Stuart, E.A.: Moving towards best practice when using inverse probability of treatment weighting (IPTW) using the propensity score to estimate causal treatment effects in observational studies. Stat. Med. **34**(28), 3661–3679 (2015)
7. Rosenbaum, P.R., Rubin, D.B.: The central role of the propensity score in observational studies for causal effects. Biometrika **70**(1), 41–55 (1983)
8. McCaffrey, D.F., Ridgeway, G., Morral, A.R.: Propensity score estimation with boosted regression for evaluating causal effects in observational studies. Psychol. Meth. **9**(4), 403 (2004)
9. He, H., et al. (ed.): Statistical Causal Inferences and their Applications in Public Health Research. Springer (2016)
10. Porro, G., Iacus, S.M.: Random recursive partitioning: a matching method for the estimation of the average treatment effect. J. Appl. Economet. **24**(1), 163–185 (2009)
11. Pang, S., Gong, J.: C5. 0 classification algorithm and application on individual credit evaluation of banks. Syst. Eng. Theory Pract. **29**(12), 94–104 (2009)
12. Bujlow, T., Tahir, R., Pedersen, J.M.: A method for classification of network traffic based on C5. 0 Machine Learning Algorithm. In: 2012 International Conference on Computing, Networking and Communications (ICNC). IEEE, pp. 237–241 (2012)
13. Ciampi, A., et al.: Recursive partition: a versatile method for exploratory-data analysis in biostatistics. In: Biostatistics, pp. 23–50. Springer, Dordrecht (1987)
14. Han, H., Kwak, M.: An alternative method in estimating propensity scores with conditional inference tree in multilevel data: a case study. J. Korean Data **30**(4), 951–996 (2019)
15. Pandya, R., Pandya, J.: C5. 0 algorithm to improved decision tree with feature selection and reduced error pruning. Int. J. Comput. Appl. **117**(16), 18–21 (2015)
16. Therneau, Terry M., et al. An introduction to recursive partitioning using the RPART routines. Mayo Foundation: Technical report, 1997
17. Steinberg, D.: CART: classification and regression trees. In: The Top Ten Algorithms in Data Mining, pp. 193–216. Chapman and Hall/CRC (2009)
18. Friedman, J.H.: Greedy function approximation: a gradient boosting machine. Ann. Stat. 1189–1232 (2001)
19. Mohan, A., Chen, Z., Weinberger, K.: Web-search ranking with initialized gradient boosted regression trees. In: Proceedings of the Learning to Rank Challenge, pp. 77–89. PMLR (2011)
20. Rosenbaum, P.R., Rubin, D.B.: Assessing sensitivity to an unobserved binary covariate in an observational study with binary outcome. J. Roy. Stat. Soc. Ser. B (Methodol.) **45**(2), 212–218 (1983)
21. Austin, P.C.: An introduction to propensity score methods for reducing the effects of confounding in observational studies. Multivar. Behav. Res. **46**(3), 399–424 (2011)
22. Imbens, G.W., Rubin, D.B.: Causal inference in statistics, social, and biomedical sciences. Cambridge University Press (2015)
23. Setoguchi, S., et al.: Evaluating uses of data mining techniques in propensity score estimation: a simulation study. Pharmacoepidemiol. Drug Saf. **17**(6), 546–555 (2008)
24. Cannas, M., Arpino, B.: A comparison of machine learning algorithms and covariate balance measures for propensity score matching and weighting. Biom. J. **61**(4), 1049–1072 (2019)
25. Lee, B.K., Lessler, J., Stuart, E.A.: Improving propensity score weighting using machine learning. Stat. Med. **29**(3), 337–346 (2010)

The Resilience of Unmanned Aerial Vehicles to Cyberattacks and Assessment of Potential Threats

Ahmet Ali Süzen(✉)

Department of Computer Engineering, Isparta University of Applied Sciences, Isparta, Turkey
ahmetsuzen@isparta.edu.tr

Abstract. An Unmanned Aerial Vehicle (UAV) is an aircraft that does not have a pilot and passenger and is used in military and civilian activities. UAVs can be managed with autonomous or remote-control systems. It is used for observation and tactical planning in missions such as image processing, scanning, and destruction, especially in military areas. Recently, they have been widely used in military operations within the framework of national security. The fact that UAVs are architecturally composed of intelligent electronic systems and artificial intelligence-based software poses cyber threats. In this study, the resilience of UAVs against cyber-attacks and their potential threats in the airspace were assessed. To analyze potential risks and the current situation, Distributed Denial of Service (DDOS), De-authentication, Jamming, Sniffing, Coding, and Man-in-the-middle (MITM) attack surfaces were examined in UAV network architectures. As a result of the tests performed on the discovered attack matrices, assessments were made to increase the level of potential resilience and prevent potential threats.

Keywords: Unmanned Aerial Vehicle · Cyber Security · Threats and Attacks

1 Introduction

Reconnaissance, surveillance, and real-time image transmission are critical, especially in the military. In these sensitive processes, Unmanned Aerial Vehicle's (UAV) has been developed due to the need for both minimizing human errors and faster results [1]. UAVs are basically non-human vehicles that can be remotely controlled, autonomously self-directed, or both [2, 3]. Today, UAV systems are used in many tasks such as intelligence gathering, reconnaissance, long-term surveillance, target detection and tracking, laser marking, and monitoring of border activities such as terrorism and smuggling [4, 5]. In addition, the gradual development of technology reduces the cost of accessing this technology. As a result, UAV applications have become widespread in many civilian areas such as the service sector, agriculture, and photography, apart from the military field.

UAV is classified as non-operator, acting remotely or autonomously, and being able to carry deadly or non-lethal payloads according to its usage activities [5]. UAV weights can be evaluated with many techniques such as flight altitudes and times, propulsion

D. J. Hemanth et al. (Eds.): ICAIAME 2022, ECPSCI 7, pp. 122–129, 2023.
https://doi.org/10.1007/978-3-031-31956-3_10

systems, payloads they carry, and areas of use [6]. While designing UAV systems, they must meet specific criteria according to their usage area and the purpose of the activity. These criteria are stated as follows [4].

- The suitability of the platform design for the task,
- The precision and reliability of the guidance systems,
- Reliability of automatic control/guidance software,
- Performance of mission payloads,
- Reliability and safety of the communication system,
- Mission planning, implementation, monitoring, evaluation capabilities and ease of use/training of ground control systems,
- Connection features to C4I (Command, Control, Communications, Computers and Intelligence) systems,
- Compliance with atmospheric and electromagnetic environmental conditions,
- Reliability, backup, readiness features,
- Having open system architecture and common system solutions.

The possibility that controls and data transmission in UAVs may be open to external interference as a result of the use of the electromagnetic spectrum is one of the critical weaknesses of the system. In retrospect; It is seen that there are events such as the capture of real-time images transmitted from the UAV by the attacker and the infection of the computer with malicious software in the mission command centers. For this reason, in our study, the potential risks of UAVs were examined to evaluate their resilience against cyber-attacks. The attack vectors in the UAV network model were used in the evaluation process. Despite the risks that may arise as a result of this, evaluations were made to prevent potential threats.

2 Examining UAV Architecture

UAVs can vary in system complexity depending on the type of mission. The scope of the study is based on the components of a basic UAV for the assessment of potential risks. As seen in Fig. 1, the UAV is grouped into three components: the UAV segment, the communication segment, and the ground segment [7]. Risk-free UAV architectures are not considered, as this study focuses on assessing the potential risks of UAVs from the framework of attack vectors.

2.1 UAV Segment

The unmanned aerial vehicle model includes Airframe, Actuators, and Power systems. The airframe represents the mechanical structures of UAVs. Actuators are responsible for the transmission and movement of control within the system to mechanical parts. The Power System is the necessary component to store and distribute the energy necessary for the operation of the system. Aircraft division mainly consists of mechanical and transmission-based components [8]. Therefore, the potential risk is not foreseen and cannot be evaluated.

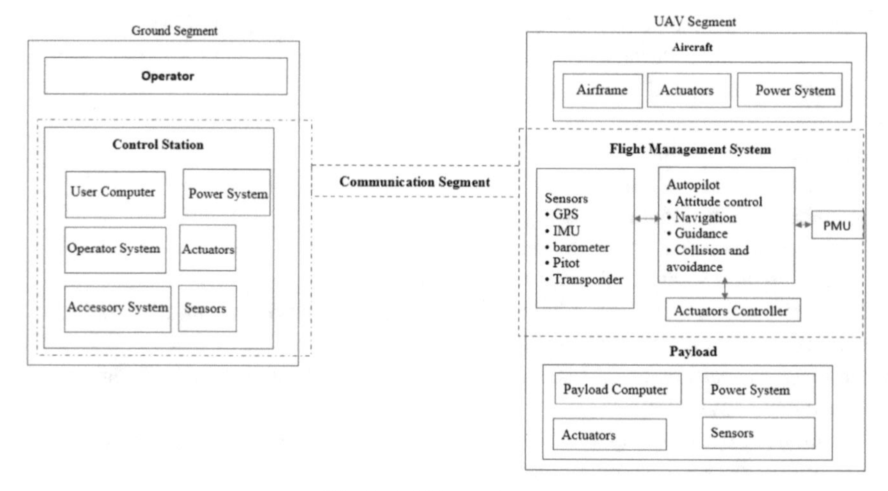

Fig. 1. UAV Architecture

2.1.1 Flight Management System

The Flight Management System (FMS) is the hardware board that monitors the status of the vehicle and controls actuators/motors to perform a safe and automatic flight. The basic elements of this module are; autopilot, basic sensors, power management unit, actuator controller, and embedded communication modules [9]. FMSs are one of the key points that can be manipulated the most in terms of cyber-attack vectors. In particular, the fact that autopilot is both software and hardware-based and is in communication with all components of the FSM increases the risk. As an autopilot capability, it performs the flight status or programmed flight plan of the UAV. The autopilot processes the data coming from the sensors, especially for a smooth and targeted flight [10]. At the beginning of these sensors are the Global Position System (GPS) and pressure. Problems and threats experienced in the simultaneous transmission of such critical data lead UAVs to critical errors.

2.2 Communication Segment

This layer includes communication systems that provide remote control or data collection during operation of UAVs both autonomously and with a command center. Transmission of telemetry, flight control and configuration data are critical, especially in situations such as performing a safe flight and monitoring the flight [11]. The communication segment can vary greatly in complexity of communication systems, depending on the connectivity needs of the intended operation. For simultaneous operation, the communication system can provide WiFi, RF, 4G/LTE, satellite links between the UAV and the Ground Control Station (GCS) [12]. This also creates communication-based attack risks.

3 Attack Vectors and Potential Risk Assessments

It is seen that data transmission between UAVs or from UAV to GCS is bidirectional when traditional UAV architecture and network structures are examined. Data transmission of UAVs works wirelessly in open environment in both categories. The most important point in the realization of wireless data transmission is the establishment of a secure connection. The next part of the study includes the evaluation of the potential risks of the UAV components and the presentation of possible solutions.

3.1 Flying Ad-Hoc Network (FANET)

FANET is one of the most important communication designs in multi-UAV systems as shown in Fig. 2. FANET is a model based on communication between UAVs without the need for a communication point [13]. In FANET network structures, at least one UAV must be connected to a ground or satellite-based hub. FANET systems are used in multiple UAV architectures such as search and rescue, border surveillance, and simultaneous data acquisition [14]. Communication between UAVs takes place within the field of view and is vulnerable to eavesdropping. In case of possible eavesdropping, the simultaneous data in the network can be eavesdropped on by the attackers. The use of bidirectional antennae in FANET networks provides data transmission to the desired UAV and reduces the risk of eavesdropping. FANET is insufficient for the internal communication of UAVs and communication with the command center due to the lack of central control and fixed topology. New generation UAVs should prefer Software Defined Networking (SDN) to provide simultaneous communication and central control. Software-based networks called SDN are preferred for a secure network topology [15]. SDN basically has application, control, and infrastructure layers. Thanks to these layers, setting up and configuring the network is more flexible and simpler. SDN-based UAV communication is anticipated to be a solution that reduces potential risks, especially in multi-tasking UAV systems.

3.2 Global Position System (GPS)

GPS is an essential component of UAVs, especially in automatic mapping, and flight and location reporting. Instant information about the UAV location is processed via the satellite with GPS [16]. Military (P-Code) signals are encrypted, but not civilian signal data, because satellites are used for different purposes. The signals from the satellite can be extremely weak in some cases. GPS receivers are vulnerable to attacks due to both the lack of encryption of data and signal-level weaknesses. In such a scenario, it can perform a jamming attack (GPS jamming), in which attackers can interface the original GPS signals with higher strength signals. Likewise, the GPS receiver may be subject to GPS spoofing attacks involving fake locations. As a result of this attack, it allows the UAV to become completely dysfunctional or to detect its location.

Especially considering that it undertakes critical tasks in UAV systems, the GPS component must be resistant to cyber-attacks [17]. Signal processing, Spatial processing, and Data processing methods are recommended against the potential risks of GPS jamming and GPS spoofing. Spatial and signal processing measures require the development of

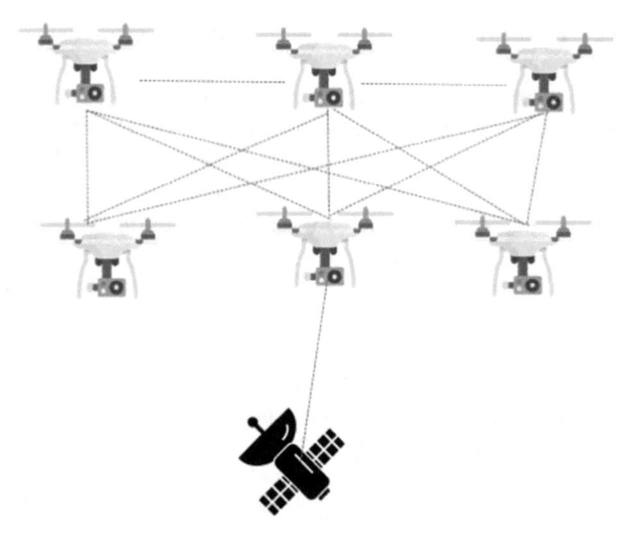

Fig. 2. FANET Communication Architecture

specific UAV-specific GPS receivers (hardware and software) [15]. Here, cryptographic machine learning methods come to the fore despite GPS spoofing attacks. In Data Processing, it can also be applied to existing GPS receivers in the market. The important thing in this method is to make the receiver data more complex with methods such as Receiver Autonomous Integrity Monitoring (RAIM) algorithms.

3.3 Communication

Technologies such as WIFI, RF, 4G/LTE, and satellite are preferred for communication between UAVs and with the center. Although a wired connection is used for the communication of UAV components and sensors, the main communication takes place via wireless connections. Bluetooth, Zigbee, and WIFI are widely used for short-distance wireless communication of UAVs [18]. WiMAX and Cellular technologies are preferred for long-distance communication. Satellite communication is generally recommended when networks such as WiMAX and Cellular are not available [19].

The UAV uses sensors (camera, temperature, GPS, etc.) that provide many data streams according to its task. Where data is of critical importance, it is monitored simultaneously from central systems. Attacks to this wireless data stream cause serious data losses. In addition, data manipulation during data transfer creates risky scenarios. The most common potential risk in UAV wireless communications is jamming attacks [20]. Jamming attacks aim to disrupt the communication of the data flow with interference or conflict methods [21]. Distributed Denial-of-service (DDoS) attacks are the most preferred method in jamming attacks. DDOS attacks send bogus requests to the UAV wireless data communication, resulting in the communication being interrupted or corrupted. In addition, a De-authentication attack can be made to a UAV within the communication network. De-authentication attacks are when a UAV within the network sends a false request to leave the network communication. With a de-authentication attack, a UAV is dropped from the network, posing a critical risk for multiple UAV missions. Resistance to jamming attacks can be increased by cutting too many requests coming from the same address, or by setting a window to control collision rates. Firewalls or authentication systems that control the UAV network architecture according to certain rules are preferred to prevent such attacks.

3.4 Ground Control Station (GCS) and Autopilot

The Ground Control Station (GCS) is a land-based or sea-based control center for the control and simultaneous data transmission of UAVs [23]. GCSs are software-based and wirelessly connected to UAVs. When UAVs are deployed autonomously, they achieve the target with autopilot management [22]. An autopilot is a structure that provides a command to the engine by predicting the flight situation. In some cases, Autopilot can be connected to central systems for purposes such as downloading flight data, obtaining an update package, or reconfiguring it. GCS software is installed on computers, smartphones, or tablet hardware. For this reason, threats in the operating system affect the GCS software [24]. Man in the middle (MITM) attacks are seen on computers belonging to GCS software. MITM is a technique where an attacker monitors the data between the UAV and the GCS, and in the attack, there is access to sensitive data without the consent of the users. Many MITM attacks such as aggressive eavesdropping and URL manipulation can occur with GCS software uses being vulnerable to other computer activities.

It makes it vulnerable to malware infection or unauthorized access as a result of potential risks in UAV communication with GCS software. The activities of malicious software can transfer or change critical data such as flight parameters, flight plan, sensor data to malicious attackers. The following precautions can be implemented to protect Autopilot and GCS systems from potential risks.

- In data transmission between UAV and GSC, bidirectional control should be provided with an authentication mechanism.
- In data transmission between the UAV and the GSC, the integrity of the data must be ensured by hash checking and the identity must be authenticated.
- In data transmission between UAV and GSC, the data must be sent encrypted to the destination.
- In data transmission between UAV and GSC, log records should be kept for examining potential risks.
- Intrusion detection techniques need to be applied to software systems at GSC.

4 UAV Attack Assessment

It is seen that data transmission points pose potential risks when looking at UAV network architectures and components. The UAV performs its missions in the airspace with wireless communication. We can talk about cyber-attacks where there is wireless communication. UAV attack vectors are addressed to sensors, network architectures, UAV communications, and GCS software that generate data. Attacks on UAVs and their target sources are given in Table 1.

Table 1. Attacks on UAVs and their target sources

Type of Cyber-Attack	UAV Source
Spoofing	Sensors
Jamming	Sensors
De-authentication attack	Communication
Zero Day Vulnerabilities	GCS, Autopilot
Malware	GCS
DDOS	Autopilot, Communication
Code Injection, Code Modification	GCS
MITM	GCS
Packet Sniffing	Sensors, Communication

5 Conclusions and Future Work

Recently, the use of UAVs has been increasing in military and civilian areas where human activities are limited. UAVs are used as single and multiple to perform specific tasks. Some are autonomous and some are managed by GSC. The realization of all these systematic managements with wireless and smart sensors also brings potential cyber-attacks. In this study, the potential risks of UAV network architectures and components are examined. The resilience of UAV systems against cyber-attacks and solutions against possible risks were made. In addition, in the next phase of the study, it is foreseen to develop an SDN-based solution for establishing a secure UAV network architecture. Thus, it is aimed to provide a secure structure for multiple UAV communication and simultaneous data transmission.

References

1. Gupta, L., Jain, R., Vaszkun, G.: Survey of important issues in UAV communication networks. IEEE Commun. Surv. Tutorials **18**(2), 1123–1152 (2015)
2. Yao, H., Qin, R., Chen, X.: Unmanned aerial vehicle for remote sensing applications—A review. Remote Sens. **11**(12), 1443 (2019)
3. Konar, M., Kekeç, E.T.: İnsansız Hava Araçlarının Uçuş Süresinin Termal Hava Akımları Kullanılarak Arttırımı. Avrupa Bilim ve Teknoloji Dergisi **23**, 394–400 (2021). https://doi.org/10.31590/ejosat.874809
4. Caner, E.: İnsansız Hava Araçları. Millî Güvenlik ve Askerî Bilimler Akademik Dergisi **1**(1), 213–240 (2014)
5. Kwon, C., Liu, W., Hwang, I.: Analysis and design of stealthy cyber attacks on unmanned aerial systems. J. Aerosp. Inform. Syst. **11**(8), 525–539 (2014)
6. Arjomandi, M., Agostino, S., Mammone, M., Nelson, M., Zhou, T.: Classification of unmanned aerial vehicles. Report for Mechanical Engineering class, pp. 1–48. University of Adelaide, Adelaide, Australia (2006)

7. Cummings, M.L., Bruni, S., Mercier, S., Mitchell, P.J.: Automation architecture for single operator, multiple UAV command and control. The Int. Command Control J. **1**, 1–24 (2007)
8. Liu, X.-F., Guan, Z.-W., Song, Y.-Q., Chen, D.-S.: An optimization model of UAV route planning for road segment surveillance. J. Cent. South Univ. **21**(6), 2501–2510 (2014). https://doi.org/10.1007/s11771-014-2205-z
9. Ghazi, G., Botez, R.M.: Identification and validation of an engine performance database model for the flight management system. J. Aerosp. Inform. Syst. **16**(8), 307–326 (2019)
10. Neretin, E.S., Budkov, A.S., Ivanov, A.S., Ponomarev, K.A.: Research on modernization directions of the human-machine interface of flight management system for future civil aircrafts. J. Phys.: Conf. Ser. **1353**(1), 012007 (2019)
11. Bithas, P.S., Michailidis, E.T., Nomikos, N., Vouyioukas, D., Kanatas, A.G.: A survey on machine-learning techniques for UAV-based communications. Sensors **19**(23), 5170 (2019)
12. Luo, C., Miao, W., Ullah, H., McClean, S., Parr, G., Min, G.: Unmanned aerial vehicles for disaster management. In: Durrani, T.S., Wang, W., Forbes, S.M. (eds.) Geological Disaster Monitoring Based on Sensor Networks. SNH, pp. 83–107. Springer, Singapore (2019). https://doi.org/10.1007/978-981-13-0992-2_7
13. Singh, K., Verma, A.K.: Flying adhoc networks concept and challenges. In: Mehdi Khosrow-Pour, D.B.A. (ed.) Advanced Methodologies and Technologies in Network Architecture, Mobile Computing, and Data Analytics, pp. 903–911. IGI Global (2019). https://doi.org/10.4018/978-1-5225-7598-6.ch065
14. Vijitha Ananthi, J., Subha Hency Jose, P.: A review on various routing protocol designing features for flying ad hoc networks. In: Shakya, S., Bestak, R., Palanisamy, R., Kamel, K.A. (eds.) Mobile Computing and Sustainable Informatics. LNDECT, vol. 68, pp. 315–325. Springer, Singapore (2022). https://doi.org/10.1007/978-981-16-1866-6_23
15. Zhao, L., Yang, K., Tan, Z., Li, X., Sharma, S., Liu, Z.: A novel cost optimization strategy for SDN-enabled UAV-assisted vehicular computation offloading. IEEE Trans. Intell. Transp. Syst. **22**(6), 3664–3674 (2020)
16. Kwak, J., Sung, Y.: Autonomous UAV flight control for GPS-based navigation. IEEE Access **6**, 37947–37955 (2018)
17. Khan, S.Z., Mohsin, M., Iqbal, W.: On GPS spoofing of aerial platforms: a review of threats, challenges, methodologies, and future research directions. PeerJ Comput. Sci. **7**, e507 (2021)
18. Abdalla, A.S., Marojevic, V.: Communications standards for unmanned aircraft systems: the 3GPP perspective and research drivers. IEEE Commun. Stand. Mag. **5**(1), 70–77 (2021)
19. Zolanvari, M., Jain, R., Salman, T.: Potential data link candidates for civilian unmanned aircraft systems: a survey. IEEE Commun. Surv. Tutorials **22**(1), 292–319 (2020)
20. Fouda, R.M.: Security vulnerabilities of cyberphysical unmanned aircraft systems. IEEE Aerosp. Electron. Syst. Mag. **33**(9), 4–17 (2018)
21. Arthur, M.P.: Detecting signal spoofing and jamming attacks in UAV networks using a lightweight IDS. In: 2019 international conference on computer, information and telecommunication systems (CITS), pp. 1–5. IEEE (2019)
22. Tran, T.D., Thiriet, J.M., Marchand, N., El Mrabti, A., Luculli, G.: Methodology for risk management related to cyber-security of Unmanned Aircraft Systems. In: 2019 24th IEEE International Conference on Emerging Technologies and Factory Automation (ETFA), pp. 695–702. IEEE (2019)
23. Jung, J., Nag, S.: Automated management of small unmanned aircraft system communications and navigation contingency. In: AIAA Scitech 2020 Forum, p. 2195 (2020)
24. Muniraj, D., Farhood, M.: Detection and mitigation of actuator attacks on small unmanned aircraft systems. Control. Eng. Pract. **83**, 188–202 (2019)

TensorFlow Based Feature Extraction Using the Local Directional Patterns

Hamidullah Nazari[1(✉)] and Devrim Akgun[2]

[1] Computer and Informatics Engineering Department, Institute of Natural Science and Technology, Sakarya University, Esentepe Campus, 54050 Serdivan, Sakarya, Turkey
hamidullahnazary22@gmail.com

[2] Software Engineering Department, Faculty of Computer and Information Sciences, Sakarya University, Esentepe Campus, 54050 Serdivan, Sakarya, Turkey
dakgun@sakarya.edu.tr

Abstract. Python is an interpreted high-level programming language and is especially popular in realizing machine learning and data science applications since it provides various conveniences and rich libraries for developers. TensorFlow is a well-known open-source machine learning library with many utilities to develop machine learning applications. The TensorFlow framework accelerates computation processes due to multi-core computing units such as CPUs and GPUs and various compiled algorithms implementing machine learning and deep learning applications. TensorFlow operators defined for tensors usually work in the GPU device and thus perform the computations faster than the sequential implementations. In this study, the Local Directional Pattern (LDirP) algorithm, a feature extraction method in computer vision, was developed with the TensorFlow library. Therefore, the new LDirP algorithm written using the TensorFlow operators can benefit from the GPU hardware acceleration without low-level parallel programming. The new algorithm written using the TensorFlow operators can benefit from the GPU hardware acceleration without low-level parallel programming. The proposed implementation was evaluated using various sizes of images, and the speed-up performance of the TensorFlow-based algorithm was presented using comparative evaluations. The results show that implementing the LDirP method using TensorFlow provides significant acceleration ratios over the naïve Python equivalent of the algorithm.

Keywords: Local Directional Patterns · TensorFlow · Feature Extraction · Machine Learning · GPU Acceleration

1 Introduction

Feature extraction methods are commonly utilized in deep learning and machine learning applications. This type of feature extraction can be used as an interlayer

D. Akgun—These authors contributed equally to this work.

D. J. Hemanth et al. (Eds.): ICAIAME 2022, ECPSCI 7, pp. 130–139, 2023.
https://doi.org/10.1007/978-3-031-31956-3_11

in machine learning and deep learning network models. The LDirP algorithm is one of the crucial computer vision algorithms used to extract image features. Its purpose is to help improve the learning process and reduce the size of the dataset. In 2010, T. Javid et al. [1] proposed the LDirP algorithm as a more stable method, less impacted by image noise. This method provides considerable refinements in texture analysis. It is widely utilized in a variety of image applications and research. For example, significant successes have been in facial recognition [2], facial expression [3,4], and other computer vision problems. For cloud video data object recognition, the LDirP approach is employed [5], facial emotion recognition [6], iris recognition [7,8], fingerprint recognition [9], gait gesture recognition [10], gender classification detection with face images [11], illumination robust optical flow [12], approximate age of individuals detection and classification [13], hypertrophic cardiomyopathy diagnosis index [14], detection of malignant masses using digital mammography [15], plant species recognition [16], texture based informal settlement classification [17], tissue characterization in X-Ray and ultrasound images [18], kinship verification [19] and others, it gives a positive result. TensorFlow is a high performance numerical computing open source software framework. Because of its modular architecture can run on a wide range of platforms (CPUs, GPUs, TPUs) and devices, including PCs, server clusters, mobile, and edge devices. Users can define complex computational operations with TensorFlow and create custom layers in deep learning networks [20]. Previously, Local Binary Pattern (LBP) and Local Derivative Pattern (LDP) algorithms which are fundamental feature extraction algorithms have implemented using TensorFlow operators [21,22]. According to evaluations on various sizes of images and batch-size it has been observed that TensorFlow provides significant acceleration ratios. In this study, the LDirP algorithm is developed using TensorFlow operators. The proposed algorithm can be used as image preprocessing and an intermediate layer in deep learning. Various image and random number dimensions were tested on 28×28, 56×56, 112×112, 224×224, and 448×448 to measure algorithm performance. Several batch sizes were also tested for each image and the number of dimensions in the experimental evaluations. Organization of the paper is as follows; in the following section details of the LDirP algorithm have been presented. In the third section the details of the TensorFlow Based algorithm have been explained. Performance evaluations of the method have been presented using comparative evaluations in the fourth section. Finally, brief conclusions about the overall paper have been given.

2 Background

The LDirP algorithm is one of the crucial feature extraction methods, and it's called novel feature extraction. It is a method that requires intensive operation because it operates on each input image pixel. The LDirP algorithm generates an eight-bit binary code assigned to the pixel of each image to be applied. This approach applies the Kirsch mask filters [23] shown in equation 1 to the image to be used, with a frame size of 3×3.

$$\begin{bmatrix} -3 & -3 & 5 \\ -3 & 0 & 5 \\ -3 & -3 & 5 \end{bmatrix} \begin{bmatrix} -3 & 5 & 5 \\ -3 & 0 & 5 \\ -3 & -3 & 5 \end{bmatrix} \begin{bmatrix} 5 & 5 & 5 \\ -3 & 0 & -3 \\ -3 & -3 & -3 \end{bmatrix}$$

$$\begin{bmatrix} 5 & -3 & -3 \\ 5 & 0 & -3 \\ 5 & -3 & -3 \end{bmatrix} \begin{bmatrix} -3 & -3 & -3 \\ 5 & 0 & -3 \\ 5 & 5 & -5 \end{bmatrix} \begin{bmatrix} -3 & -3 & -3 \\ -3 & 0 & -3 \\ 5 & 5 & 5 \end{bmatrix}$$

Kirsch Mask Filter

A Kirsch mask is a nonlinear filter operation used to find edge images. It is like a compass and consists of eight directions. By picking a 3×3 frame with the Kirsch mask filter, the LDirP algorithm encodes each pixel of the image to be applied. Each frame size consists of 8 bits. Based on the middle pixel of the 3×3 frame size selected in the x and y directions of the image, the corresponding response value is calculated by applying Kirsch mask values in the (m0, ..., m7) eight directions. Given the pixels of an image in the (x, y) directions, I and the corresponding mi mask are used for each I direction. The response value of the i-th direction is calculated as follows.

$$M_i = \sum_{k=-1}^{1} \sum_{l=-1}^{1} M_i(k+1, l+1) * I(x+k, y+l) \tag{1}$$

In equations two and three, the LDirP code is generated by determining the most significant bit of the corresponding response values.

$$LDirP_{i,y}\,(m_0, ..., m_7) = \sum_{i=1}^{7} s(m_i, m_k) 2^i \tag{2}$$

$$s(x) = \begin{pmatrix} 1, \; x => 0 \\ 0, \;\; x < 0 \end{pmatrix} \tag{3}$$

In the 2nd and 3rd equations, the most significant bit is found and subtracted from each (m0, ..., m7), then one is taken if the extracted value is greater than 0 and zero if it is less (Figs. 1).

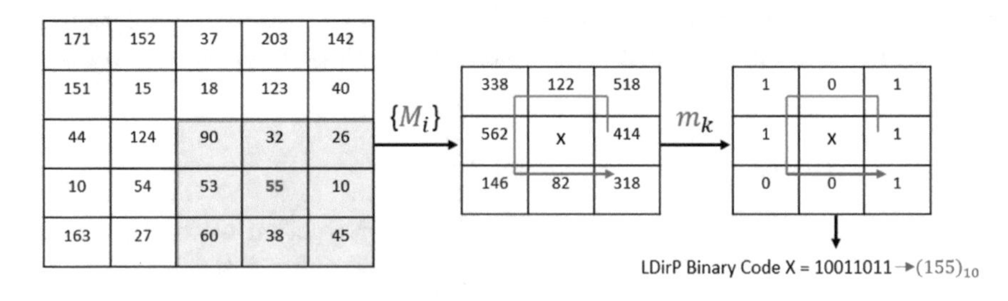

Fig. 1. LDirP Code for a Central Pixel 55

Fig. 2. (a) and (c) are Sample Images from ImageNet [1], (b) and (d) are Their LDirP Transform Results

3 Implementation of the LDirP Algorithm with TensorFlow

TensorFlow is an open-source, end-to-end modern framework for implementing machine learning and deep learning applications. It was originally developed for large numerical computations. The purpose of its fast running is that it works at both the CPU and GPU levels. TensorFlow can execute on many platforms, such as local machines, cloud clusters, 64-bit Linux, macOS, Windows, Android, and iOS. It has APIs for various high level programming languages, such as Python, C++, Java, JavaScript, and Go.

TensorFlow uses tensors to perform mathematical operations. A tensor consists of an n-dimensional array. TensorFlow can run computation operations on the GPU unit without requiring the user to write parallel code at the GPU level. Most of the essential functions, such as *add, multiple, matmul,* etc., are realized with the support of the GPU.

According to the LDirP method, all pixels in the input image are calculated independently. The implementation of the LDirP algorithm with a tensor is performed as shown in Fig. 2. Using a 3×3 matrix, all image pixels to be applied are multiplied by Kirsch mask filter directions from m0 to m7. For example, all aspects of the selected matrix for m0 are taken on a row basis, multiplied by the Kirsch mask values, then added, and continue until m7. Then, k = 3 (MSB - Most Significant values) are taken, and the calculation process occurs (Fig. 3).

```
1   def tf_ldirp(x):
2           NW_tf = tf.constant([[5, 5, -3],
3                                [5, 0, -3],
4                                [-3, -3, -3]],dtype=tf.int32) # m0
5
6           N_tf = tf.constant([[5, 5, 5],
7                               [-3, 0, -3],
8                               [-3, -3, -3]],dtype=tf.int32) # m1
9
10          NE_tf = tf.constant([[-3, 5, 5],
11                               [-3, 0, 5],
12                               [-3, -3, -3]],dtype=tf.int32) # m2
13
14                              # m3,m4, m5,m6
15
16          W_tf = tf.constant([[5, -3, -3],
17                              [5, 0, -3],
18                              [5, -3, -3]],dtype=tf.int32) # m7
19          #Select Elements within the Mask
20          y00=y[:,0:M-2, 0:N-2]; y01=y[:,0:M-2, 1:N-1]; y02=y[:,0:M-2, 2:N  ]
21
22          y10=y[:,1:M-1, 0:N-2]; y11=y[:,1:M-1, 1:N-1]; y12=y[:,1:M-1, 2:N  ]
23
24          y20=y[:,2:M, 0:N-2];   y21=y[:,2:M, 1:N-1];    y22=y[:,2:M, 2:N ]
25
26          # Row 0
27          tf_m0 = tf.multiply(tf.cast(y00,dtype='int32'), NW_tf[0,0] )
28          #--------------------------------------------------
29          tmp=tf.multiply(tf.cast(y01,dtype='int32'), NW_tf[0,1] )
30          tf_m0 =tf.add(tf_m0,tmp)
31          #--------------------------------------------------
32          tmp=tf.multiply(tf.cast(y02,dtype='int32'),NW_tf[0,2] )
33          tf_m0 =tf.add(tf_m0,tmp)
34          # Row 1 ---------------------------------------------
35          tmp=tf.multiply(tf.cast(y10,dtype='int32'), NW_tf[1,0] )
36          tf_m0 =tf.add(tf_m0,tmp)
37          #------------------------------
38          tmp=tf.multiply(tf.cast(y11,dtype='int32') , NW_tf[1,1] )
39          tf_m0 =tf.add(tf_m0,tmp)
40          #------------------------------
41          tmp=tf.multiply(tf.cast(y12,dtype='int32') , NW_tf[1,2] )
42          tf_m0 =tf.add(tf_m0,tmp)
```

Fig. 3. TensorFlow Based Calculation LDirP Algorithm

4 Experimental Results

The execution times of the sample states of LDirP were measured using the time command from the Python time library, as shown in snippet 1. All measurements were repeated 30 times and used to calculate the mean of the experimental results. The algorithm has been tested using several images defined in batch size and several image dimensions (Fig. 4).

```
Random Test Numbers :
[[ 95   4 143 241  92  11 165]
 [238 159 133 246 250 119 140]
 [  5  31 109 188 218 226  31]
 [106  98 131 227 233 110  82]
 [148  15 127 190 125 144  71]
 [ 88  50 245  68 121  93 165]
 [207 198  79  79 109  75 101]]
```

(a)

```
Tensorflow Computation Result:
[[ 13  70 200  76  98 196  22]
 [ 35  21  49  11  97  67  26]
 [ 97 194  19  49 145  25 145]
 [137 208  25  49 200 145  49]
 [ 97  26 161 104 196 161  50]
 [ 13 200 168 140  70 164 145]
 [ 35  76 140 196 100 100 196]]
```

(b)

```
Python Computation Result :
[[ 13  70 200  76  98 196  22]
 [ 35  21  49  11  97  67  26]
 [ 97 194  19  49 145  25 145]
 [137 208  25  49 200 145  49]
 [ 97  26 161 104 196 161  50]
 [ 13 200 168 140  70 164 145]
 [ 35  76 140 196 100 100 196]]
```

(c)

Fig. 4. (a) For a Given Random Matrix, (b) and (c) Examples Verification Results Using TensorFlow and Python

```
1 # Initial running time
2 start_time  =  time.time()
3
4 #Realization of LDirP with TensorFlow for given batch of images.
5 tf_result  =  tf_LDirP(batch_of_images).numpy()
6
7 #Given terminate time and find out elapsed time
8 elapsed_time = time.time() - start_time
```

Code snippet 1. TensorFlow Implementation of LDirP Transform

Experimental results were performed on Windows 10 using Python 3.9.0 and TensorFlow 2.8.0. Computer hardware tested using Intel (R) Core (TM) i5-

9300H CPU@2.40 GHz, 8 GB of memory, NVIDIA GeForce GTX 1650, 896 CUDA cores, and 56 texture units.

The test images were resized to various dimensions such as 28 × 28, 56 × 56, 112 × 112, 224 × 224, 448 × 448. Image content does not affect computation times. Table 1 shows the Python implementation results in seconds compared to TensorFlow. According to the results, CPU running time varies with different image sizes.

Table 1. Running Times for the LDirP Algorithm in Python (seconds)

Image Size					
Batch-size	28 × 28	56 × 56	112 × 112	224 × 224	448 × 448
1	0.1419	0.5610	2.2750	9.1460	37.7049
2	0.3280	1.1249	4.6010	18.5197	77.2780
4	0.5610	2.3319	9.6339	37.5741	150.2489
8	1.1279	4.5120	18.1270	73.7650	299.3680
16	2.4780	10.0979	37.1329	147.0970	594.3374
32	4.5960	18.3220	73.8619	299.5495	1189.7717
64	9.7819	33.1159	155.6256	631.6109	2547.9083

When the image size increases, the run times also increase. Because LDirP performs pixel based processing, as the image size increases, the number of pixels also increases, so the running times increase. For example, the processing time

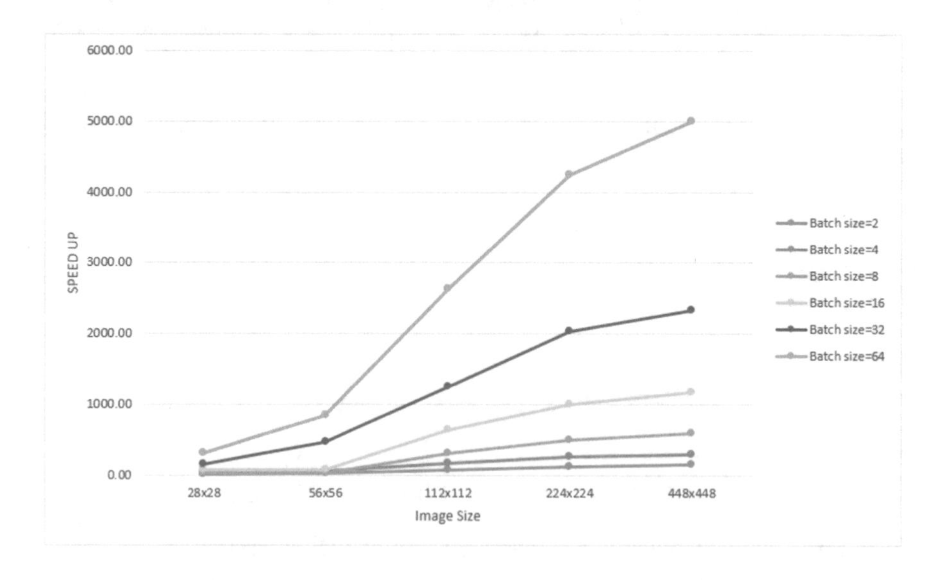

Fig. 5. TensorFlow speed-up on Python calculations.

for 112×112 image size is 2.2750 s, and the processing time for 448×448 images is 37.7049 s. As the image size and batch size increase, the CPU running times also increase, as shown in Fig. 5.

Table 2. Running Times for the LDirP Algorithm in TensorFlow (seconds)

Image Size					
Batch-size	28 × 28	56 × 56	112 × 112	224 × 224	448 × 448
1	0.0301	0.0370	0.0574	0.1448	0.5075
2	0.0302	0.0371	0.0575	0.1449	0.5077
4	0.0308	0.0373	0.0576	0.1467	0.5079
8	0.0310	0.0376	0.0578	0.1471	0.5080
16	0.0312	0.0390	0.0582	0.1474	0.5082
32	0.0314	0.0392	0.0588	0.1477	0.5084
64	0.0316	0.0394	0.0591	0.1489	0.5095
128	0.0319	0.0396	0.0597	0.1496	0.5098
256	0.0321	0.0398	0.0602	0.1501	0.5108

Table 2 shows the results of the processing times for the various image sizes on the GPU in TensorFlow. As shown in Table 2, the speed-up of executing LDirP with TensorFlow are due to the employment of GPU and multicore CPU.

Figure 5. Graph representation of TensorFlow's application of the LDirP algorithm to images while utilizing the CPU.

5 Conclusion

This study presents a method using TensorFlow operators to speed up LDirP transformation calculations. The LDirP algorithm, which extracts features by applying the Kirsch mask filter to the images in the eight directions depending on pixel based operations, is handy for machine learning and computer vision applications. As the experimental result shows, these algorithms are frequently expensive to build, such as Python scripts. The LDirP method has been specified in terms of matrix operations, so TensorFlow operators are able to be utilized efficiently. TensorFlow uses tensors that consist of n-dimensional arrays; Using this kind of tensor entitles the creation of custom functions. These custom functions may be used as image preprocessing or as custom layers of deep learning. The LDirP method performs independent pixel-based calculations on images; when running time is compared, it has been observed that TensorFlow is significantly faster according to the Python script implementation. The reason is that TensorFlow functions are compiled with optimization and support multicore CPUs and GPU devices. According to the results, it has been observed that the acceleration with TensorFlow increased when the image size and batch size increased. Also, the acceleration results may differ depending on computer hardware and software versions. Implementing LDirP in TensorFlow and comparing various images and batches gives a good idea of TensorFlow acceleration.

References

1. Jabid, T., Kabir, M.H., Chae, O.: Local directional pattern (LDP)-a robust image descriptor for object recognition. In: 2010 7th IEEE International Conference on Advanced Video and Signal Based Surveillance, pp. 482–487. IEEE (2010)
2. Zhong, F., Zhang, J.: Face recognition with enhanced local directional patterns. Neurocomputing **119**, 375–384 (2013)
3. Jabid, T., Kabir, M.H., Chae, O.: Local directional pattern (LDP) for face recognition. In: 2010 Digest of Technical Papers International Conference on Consumer Electronics (ICCE), pp. 329–330. IEEE (2010)
4. Zhou, J., Xu, T., Gan, J.: Feature extraction based on local directional pattern with SVM decision-level fusion for facial expression recognition. Int. J. Bio-science Bio-technology **5**(2), 101–110 (2013)
5. Nayagam, M.G., Ramar, K.: Reliable object recognition system for cloud video data based on LDP features. Comput. Commun. **149**, 343–349 (2020)
6. Basu, A., Dsouza, G., Regi, R., Saldanha, A., Guide, R.C.: Facial emotion recognition using LDP with SVM. St. Francis Institute of Technology, Technical report (2016)
7. Hezil, N., Hezil, H., Boukrouche, A.: Robust texture analysis approche for no-ideal iris recognition
8. Madhuvarshini, N.: Iris recognition using modified local line directional pattern
9. Kumar, R., Chandra, P., Hanmandlu, M.: Local directional pattern (LDP) based fingerprint matching using SLFNN. In: 2013 IEEE Second International Conference on Image Information Processing (ICIIP-2013), pp. 493–498. IEEE (2013)
10. Uddin, M.Z., Khaksar, W., Torresen, J.: A robust gait recognition system using spatiotemporal features and deep learning. In: 2017 IEEE International Conference on Multisensor Fusion and Integration for Intelligent Systems (MFI), pp. 156–161. IEEE (2017)
11. Jabid, T., Kabir, M.H., Chae, O.: Gender classification using local directional pattern (LDP). In: 2010 20th International Conference on Pattern Recognition, pp. 2162–2165. IEEE (2010)
12. Mohamed, M.A., Rashwan, H.A., Mertsching, B., García, M.A., Puig, D.: Illumination-robust optical flow using a local directional pattern. IEEE Trans. Circuits Syst. Video Technol. **24**(9), 1499–1508 (2014)
13. Hu, M., Zheng, Y., Ren, F., Jiang, H.: Age estimation and gender classification of facial images based on local directional pattern. In: 2014 IEEE 3rd International Conference on Cloud Computing and Intelligence Systems, pp. 103–107. IEEE (2014)
14. Gudigar, A., et al.: Novel hypertrophic cardiomyopathy diagnosis index using deep features and local directional pattern techniques. J. Imaging **8**(4), 102 (2022)
15. Abdel-Nasser, M., Rashwan, H.A., Puig, D., Moreno, A.: Analysis of tissue abnormality and breast density in mammographic images using a uniform local directional pattern. Expert Syst. Appl. **42**(24), 9499–9511 (2015)
16. Hirasen, D., Viriri, S.: Plant species recognition using local binary and local directional patterns. In: 2020 2nd International Multidisciplinary Information Technology and Engineering Conference (IMITEC), pp. 1–9. IEEE (2020)

17. Shabat, A.M., Tapamo, J.-R.: A comparative study of the use of local directional pattern for texture-based informal settlement classification. J. Appl. Res. Technol. **15**(3), 250–258 (2017)
18. Abdel-Nasser, M., et al.: Breast tissue characterization in X-ray and ultrasound images using fuzzy local directional patterns and support vector machines. In: VISAPP (1), pp. 387–394 (2015)
19. Chergui, A., Ouchtati, S., Telli, H., Bougourzi, F., Bekhouche, S.E.: LPQ and LDP descriptors with ml representation for kinship verification. In: The Second Edition of the International Workshop on Signal Processing Applied to Rotating Machinery Diagnostics (SIGPROMD 2018), pp. 1–10 (2018)
20. Andrade-Loarca, H., Kutyniok, G.: tfShearlab: the TensorFlow digital shearlet transform for deep learning. arXiv preprint arXiv:2006.04591 (2020)
21. Akgün, D.: A TensorFlow implementation of local binary patterns transform. MANAS J. Eng. **9**(1), 15–21 (2021)
22. Akgün, D.: A TensorFlow based method for local derivative pattern. Mugla J. Sci. Technol. **7**(1), 59–64 (2021)
23. Venmathi, A., Ganesh, E., Kumaratharan, N.: Kirsch compass kernel edge detection algorithm for micro calcification clusters in mammograms. Middle-East J. Sci. Res. **24**(4), 1530–1535 (2016)

Ensuring the Invariance of Object Images to Linear Movements for Their Recognition

Rahim Mammadov[1], Elena Rahimova[1], and Gurban Mammadov[2(✉)]

[1] Azerbaijan State Oil and Industry University, Azadliq av. 16/21, Baku AZ1010, Azerbaijan
rahim1951@mail.ru

[2] Azerbaijan State Scientific Research Institute for Labor Protection and Occupational Safety, Tabriz st.108, Baku AZ1008, Azerbaijan
qurban_9492@mail.ru

Abstract. In automated production, automated systems for recognizing two-dimensional images of these products are often used to control and diagnose the quality of manufactured products. However, in this case, difficulties arise with the reliability of recognition of these images due to the orthogonal displacement of object images, rotation of the image around the center of gravity, and rescaling of the image of the object. These destabilizing factors lead to significant methodological errors in estimating the proximity measure between the checked and standard objects. Therefore, the process of object image recognition must be invariant to changes in the position of the image. In the existing methods of pattern recognition, either invariant topological and geometric features are chosen for these changes (these features are less reliable), or non-invariant geometric features are selected with the subsequent provision of invariance to linear changes in images (they become more informative). Therefore, the problem is to ensure the invariance of the geometric features of images to linear changes. Existing methods cannot fully solve this problem. Therefore, an algorithm is proposed to solve this issue. The issue is solved by the fact that when an image with a random position appears, the angle of rotation of the image around the axis of inertia of the image is found. Then, the image is virtually returned to its original standard position. Thus, the problem of invariant recognition of two-dimensional images is solved. In the proposed algorithm, the contour points of two-dimensional images of objects are given by coordinates in the Cartesian coordinate plane. Therefore, for the correct recognition of such images, it is necessary to ensure that the image points are invariant to displacement and rotation. For the image to be invariant to orthogonal displacement, the coordinate system must be moved to the center of gravity of the image being defined. Then, by the moments of inertia about the coordinate axes of the image, the angle of its rotation relative to the initial position of the reference object is determined. After evaluating the angle of rotation of the image, the coordinates of the contour points are found by rotating the reference image in the computer memory by this angle. Then the coordinates of the contour points of the current image are compared with the coordinates of the contour points of the rotated reference image. Thus, the proposed algorithm makes it possible to invariantly recognize two-dimensional images of objects. The proposed algorithm was simulated on a computer and positive results were obtained.

D. J. Hemanth et al. (Eds.): ICAIAME 2022, ECPSCI 7, pp. 140–148, 2023.
https://doi.org/10.1007/978-3-031-31956-3_12

Keywords: robotic complexes · vision systems · invariance · linear displacement · image rotation

1 Introduction

In adaptive robots and flexible industrial systems, there are problems of image recognition of objects subjected to control and diagnostics in technological lines in order to check the quality of manufactured products. The efficiency of recognition of such images of objects largely depends on the comparison of the parameters of the images of the controlled and standard objects and the adoption of the necessary decision on image recognition based on the results obtained. These processes occur in the presence of destabilizing factors, due to which there are errors in the measurement of image parameters, which creates certain difficulties in obtaining correct solutions. The accuracy of decision-making in pattern recognition is in many respects close between recognizable and reference images, according to which a decision is made about the belonging of incoming information to one or another information set [1, 4]. When recognizing images of objects, methodological errors arise associated with their linear changes (changing the scale of images, rotating the image around the center of gravity and changing the orthogonal position in the coordinate plane). This is due to the fact that the products after manufacturing lie on the conveyor line randomly, since their strict orientation in space is a complex and costly task. This leads to the loss of information about the number, place and absolute value of features, random errors in measuring the values of features of objects, etc. These and other destabilizing factors reduce the reliability of pattern recognition, to eliminate which recognition must be invariant to the above linear changes in the image [2, 3]. To ensure invariance to image rotation around the center of gravity and to changing the position of the image, various methods and means have been proposed [14, 16, 18]. However, these methods either cannot provide complete invariance of pattern recognition or cannot provide high reliability of pattern recognition. For example, topological and some geometric features are invariant to linear changes in images, but due to their high integration, they do not provide sufficient recognition reliability [2, 6, 9, 11]. The analysis of these works showed that these methods are not perfect and cannot be used for high reliability and invariant image recognition. Therefore, research aimed at finding methods and means to achieve the invariance of pattern recognition with high reliability remains relevant. In this paper, we propose to use the solution of this issue using statistical moments as reference points of images and, based on this principle, develop an image recognition technique that is invariant to their linear transformations.

2 Problem Statement

When recognizing images of objects, there are certain difficulties associated with rotating its image around its axis of gravity and changing its scale, which leads to the loss of information about the number, place and absolute value of features, random errors in measuring the values of features of objects, etc. These and other destabilizing factors reduce the reliability of image recognition, to eliminate which recognition must be invariant to linear image changes [5, 7, 8, 13, 20].

To ensure invariance to image rotation around its center of gravity and large-scale changes in the image, various methods and means were proposed [12, 17]. However, these methods cannot provide the greatest invariance of pattern recognition. Therefore, research aimed at finding the best methods and means to achieve the invariance of image recognition remains relevant.

3 Comparative Analysis

Let's take a brief look at the recognition methods that allow you to use a system to combine the solution of the problem of image detection and measurement of its parameters. Of all the features in the drawings, we will consider only the geometric features. There are several ways to achieve sensitivity to transformation of the recognition system, in particular, two groups of approaches can be distinguished among the most studied transformation transitions. The methods of the first group include spatially insensitive properties (for example, the Moments method, the Fourier method of images).

Proponents of the alternative approach work with object models and try to combine objects observed and used in training by selecting parameters.

The method based on the analysis of the amplitudes of the individual harmonics of the Fourier spectrum of images has a number of advantages, such as a small number of important features for recognition, an unambiguous relationship between the rotation of the image or the corresponding rotation and the scale of the spectrum. The harmonic displacement of the spectrum can be used to measure the corresponding displacement in the image [10, 15].

Mellin and Fourier-Mell transformations also reduce the number of features such as Fourier transforms, which simplifies the recognition scheme. The double-scale invariance of the latest method allows to stabilize the verticality of the statistical properties of the measurement descriptions based on them, thereby increasing the accuracy of the measurement. These methods are performed only in coherent optical systems, and require a sufficiently complex image analyzers to achieve a certain degree of invariance.

The secant method is used to recognize images that are large enough to be contoured using segments or straight sequences. For this method it is not enough to draw the contours of the images, but also to divide the angular area of the image into segments, which can contain several or more objects. The most important condition when using this method is the stability of the visible shape of the object [19] (Table 1).

After comparing different algorithms and schemes for solving the problems of recognition and identification of known objects, it can be concluded that among the most promising schemes are the characteristics of the object determined (controlled) by synchronous detection of the center of the image and the geometric moments of its image. Its Fourier transform is used, or one of the methods of determining the position of the main maximum of the correlation function of an object description, which correlates schemes using a priori synthesized discriminant functions. However, when there are sufficiently arbitrary and a priori unknown changes in geometric parameters (properties), for example, the scale and shape of its description, for example, the use of the recognized methods of consideration is not effective enough.

Table 1. Mutual analysis of the methods used

The method of obtaining the main features of the image		Fourier transform (analysis of harmonic Fourier spectrum)	Transformation of Mellina	Fourier-Mellina transformation	Secant method	Geometric moments method	Method of geometric moments of space-frequency spectra of the image	Optical correlation method
Early signs		Harmonic amplitude	Harmonic amplitude	Harmonic amplitude	Distribution of secant lengths and angles between them	Geometric moments of the image	Geometric moments of the individual harmonic Fourier transform of the image	The basic maximum of the correlation function and its position
Acquired invariance	displacement	+ (+/-)	-	+ (+/-)	+	+ (+/-)	+ (+/-)	+ (+/-)
	Rotation	-	-	-	+(length) - (corners)	+ (+/-)	+ (+/-)	+ (+/-)
	To change the scale	-	+	+	+ corners) - length)	+ (+/-)	+ (+/-)	+ (+/-)

4 Problem Solving

Let's assume that information about the object is stored in memory and they can be displayed in the XOY coordinate system on the monitor screen as an image – 1 (I1). If the object is participating in the movement (parallel movement by Δx, Δy in the direction of the coordinate axes x, y, rotation around its center of gravity by an angle φ), then its new image (I2) on the monitor screen is accompanied by a change in linear dimensions (Fig. 1). Therefore, in order to recognize an object from images, it is necessary to solve the problem of achieving the invariance of image recognition to linear image transformations.

With the help of the AutoCAD system, the image of the object after the movement (taking into account the scale number k) will be represented as I2.

Determine the coordinates of the center of gravity (point C1) of the image-1 stored in the memory of the known coordinates A_i ($X_{I;}$ Y_i) of the object

$$X_{ci} = \frac{\sum_{i=1}^{n} X_i}{n}; \; Y_{ci} = \frac{\sum_{i=1}^{n} Y_i}{n} \tag{1}$$

We draw a rectangular coordinate system $X'C1\,Y'$ and calculate the new coordinates of the image points -1 in this coordinate system:

$$\begin{aligned} X_i' &= X_i - X_{C1} \\ Y_i' &= Y_i - Y_{C1} \end{aligned} \tag{2}$$

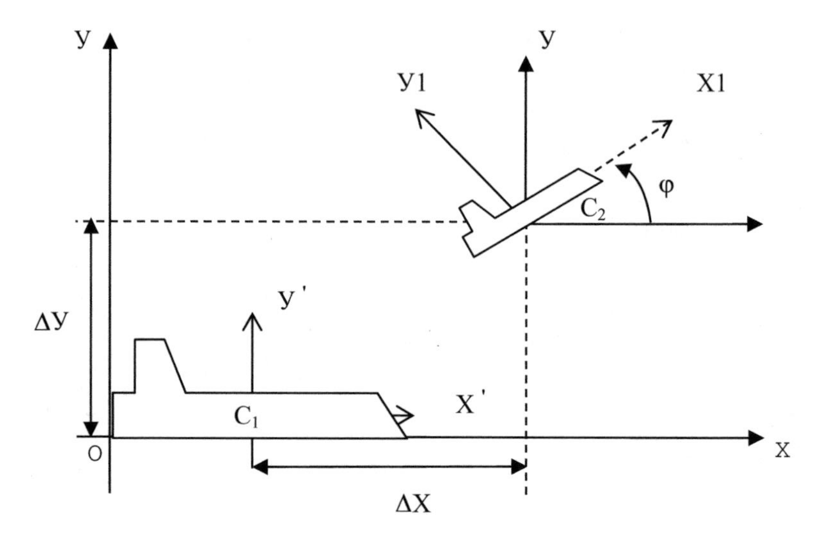

Fig. 1. Calculation of moments during the linear displacement of the object

We calculate the moments of inertia of the image – 1 relative to the X′C1 $\mathcal{Y}'$ system

$$
\begin{aligned}
J'_x &= \sum_{i-1}^{n} (y'_i)^2 \Delta s \\
J'_y &= \sum_{i=1}^{n} (x'_i)^2 \Delta s \\
J'_{x'y'} &= \sum_{i=1}^{n} x'_i y'_i \Delta s
\end{aligned}
\tag{3}
$$

When image-2 is rotated around the center of gravity C2 at an angle φ, the coordinate system X′C1,$\mathcal{Y}'$ goes into the coordinate system X ~ C2,$\mathcal{Y}$ ~. Taking into account the scale number k, the new coordinates Ai (Xi ~;$\mathcal{Y}$i ~), elementary area Δs ~, moments of inertia Jx ~, Jy ~, Jx ~ y ~ take the form:

$$
\begin{aligned}
&\mathrm{x}^{\sim} = \mathrm{kx}'; \mathrm{y}^{\sim} = \mathrm{ky}'; \Delta s^{\sim} = \mathrm{k}^2\Delta \mathrm{s}' \\
&\mathrm{J}_{\mathrm{x}}^{\sim} = \sum\nolimits_{i=1}^{n} (x_i^{\sim})^2 \Delta s^{\sim} = k^4 J'_x \\
&\mathrm{J}_{\mathrm{y}}^{\sim} = \sum\nolimits_{i=1}^{n} (y_i^{\sim})^2 \Delta s^{\sim} = k^4 J'_y \\
&\mathrm{J}_{\mathrm{x}}^{\sim}\mathrm{J}_{\mathrm{y}}^{\sim} = \sum\nolimits_{i=1}^{n} x_i^{\sim} y_i^{\sim} \Delta s = k^4 J_{x'y'}
\end{aligned}
\tag{4}
$$

Through the point C2 we draw the coordinate axes C2X″, C2Y″, parallel, respectively, to the axes OX, OY. We choose the length of the coordinate grid step equal to the length of the grid step of the system X~C2,$\mathcal{Y}$~, Δs″ = Δs~ = k2Δs′.

We determine the coordinates of the points Ai (X''_i; y″) in the X″C₂,Y″ system and calculate the moment of inertia $J_{x''y''}$.

$$J_{x''y''} = \sum x''_i y''_i \Delta s \tag{5}$$

It is known that when the coordinate axes are rotated by φ, the moments of inertia $J_{x''y''}$, $J_x^{\sim}$, $J_y^{\sim}$ and $J_x^{\sim}{}_y$ are related by the dependence

$$J_{x''y''} = \frac{J_y^{\sim} - J_x^{\sim}}{2} sin2\phi + J_{xy}^{\sim} cos2\phi \tag{6}$$

Hence, taking into account (4), we have:

$$2J_{x''y''} = k^4\left(J'_y - J'_x\right)\sin2\varphi + 2k^4 J_{x'y'}\cos2\varphi \tag{7}$$

We introduce the following notation:

$$\begin{aligned} a &= J_{y''} - J_{x'} \\ b &= 2J_{x'y'}; \\ c &= \frac{2J_{x''y''}}{k^4} \end{aligned} \tag{8}$$

Then (6) takes the form:

$$c = a \cdot \sin2\varphi + b \cdot \cos2\varphi \tag{9}$$

Easily solved Eq. (8) is an equation for determining the angle of rotation of image-2 around its center of gravity C_2.

Table 2. In object recognition Coordinates and recognition percentage after rotation of the object

Object coordinate data in screen coordinate system		Object image coordinates in screen coordinate system (after movement)		Recognizable calculated object coordinates		Relative errors of object recognition (%)	
X	Y	X*	Y*	XT	YT	Px%	Py%
2,5	2,5	92,5	93,6	2,5	2,5	0,04	0,02094
7,5	2,5	94,7	94,8	7,5	2,5	0,01	0,02094
12,5	2,5	96,9	96,1	12,5	2,5	0,0047	0,02094
17,5	2,5	99,0	97,3	17,5	2,5	0,00212	0,02094
22,5	2,5	101,2	98,6	22,5	2,5	0,00067	0,02094

(continued)

Table 2. (*continued*)

Object coordinate data in screen coordinate system		Object image coordinates in screen coordinate system (after movement)		Recognizable calculated object coordinates		Relative errors of object recognition (%)	
X	Y	X*	Y*	XT	YT	Px%	Py%
27,5	2,5	103,4	99,8	27,5	2,5	−0,0002	0,02094
32,5	2,5	105,5	101,1	32,5	2,5	−0,0009	0,02094
37,5	2,5	107,7	102,3	37,5	2,5	−0,0014	0,02094
42,5	2,5	109,9	103,6	42,5	2,5	−0,0017	0,02094
47,5	2,5	112,0	104,8	47,5	2,5	−0,002	0,02094
2,5	7,5	91,3	95,8	2,5	7,5	0,04127	0,00405
7,5	7,5	93,5	97,0	7,5	7,5	0,01082	0,00405
12,5	7,5	95,6	98,3	12,5	7,5	0,00473	0,00405
17,5	7,5	97,8	99,5	17,5	7,5	0,00212	0,00405
22,5	7,5	100,0	100,8	22,5	7,5	0,00067	0,00405
27,5	7,5	102,1	102,0	27,5	7,5	−0,0002	0,00405
32,5	7,5	104,3	103,3	32,5	7,5	−0,0009	0,00405
37,5	7,5	106,4	104,5	37,5	7,5	−0,0014	0,00405
42,5	7,5	108,6	105,8	42,5	7,5	−0,0017	0,00405
7,5	12,5	92,2	99,2	7,5	12,5	0,01082	0,00067
12,5	12,5	94,4	100,4	12,5	12,5	0,00473	0,00067
7,5	17,5	91,0	101,3	7,5	17,5	0,01082	−0,0008

X and Y – Cartesian coordinates of the initial image in the table;
X* and Y* – Cartesian coordinates of the image after movement in the table;
XT and YT Recognizable calculated object coordinates;
Px% and Py % Relative errors of object recognition (%).

The initial coordinates of the object (X, Y) using the X*, Y* image coordinates, rotation angle, scale factor, parallel translation parameters and coordinates of the object's center of gravity are determined by the formulas

$$
\begin{aligned}
X &= [k * X_{c1} + (x - \Delta x) * \text{Cos}\varphi + (y - \Delta y) * \text{Sin}\varphi]/k \\
Y &= [k * Y_{c1} - (x - \Delta x) * \text{Sin}\varphi + (y - \Delta y) * \text{Cos}\varphi]/k
\end{aligned} \tag{10}
$$

5 Conclusion

Thus, the proposed algorithm allows invariant recognition of two-dimensional images. The proposed algorithm was simulated on a computer and positive results were obtained. To test the proposed methodology, as an example, the data indicated in Table 2 was used. According to the coordinates in the first two columns of the table, the area (So = 550), the length of the contour (Lo = 140) and the coordinates of the center of gravity (X_{c1} = 21,818; $У_{c1}$ = 6,136) of the object were determined. Using the AutoCAD system, the images of the object were moved parallel to the coordinate axes $\Delta x = 100$, $\Delta y = 100$ rotated around their center of gravity by $\varphi = 300$, and the image dimensions were reduced 2 times. We measured the image coordinates relative to the screen coordinate system, the area (Si = 137.5) and the length of the contour (Pi = 70) of the image.

With the help of the proposed technique, after the corresponding calculations, the following results were obtained: $\varphi = 30°$, k = 0.5. The table shows that the maximum relative error of object recognition in this case did not exceed 0.04%.

References

1. David Forsyth, A.: 'Computer Vision' First Indian Edition. Pearson Education (2003)
2. Liu, J.: 2D shape matching by contour flexibility. IEEE Trans. Pattern Anal. Mach. Intell. **19**, 1–7 (2008)
3. Mammadov, R.G.: Correction of estimation errors of the measure of affinity between objects at recognition patterns for intellectual systems. In: Proceedings of the International Scientific Conference «Проблемы кибернетики и информатики», – Baku: – October 23–25, pp. 21–24 (2006)
4. Шапиро, Л. Компьютерное зрение [Электронный ресурс] / Л. Шапиро, Дж. Стокман. – Москва: БИНОМ. Лаборатория знаний.- 763с (2015)
5. Senthil Kumar, K.: Object recognition using shape context with canberra distance. Asian J. Appl. Sci. Technol. (AJAST) **1**(2), 268–273 (2017)
6. Волосатова, Т.М., Козов, А.В.: Особенности методов распознавания образов в автоматической системе управления поворотом мобильного робота // Мехатроника, автоматизация, управление. 2018. Т. 19. N 2. С. 104–110
7. Lowe, D.G.: Distinctive image features from scale-invariant keypoints. Int. J. Comp. Vision **60**(2), 91–110 (2004)
8. Nantogma, S., Xu, Y., Ran, W.: A Coordinated air defense learning system based on immunized classifier systems. Symmetry **13**, 271 (2021)
9. Duda, R.O., Hart, P.E., Stork, D.G.: Pattern Classification (2nd edn.), 738 p. Wiley-Interscience (2001)
10. Mammadov, R., Rahimova, E., Mammadov, G.: Increasing the reliability of pattern recognition by analyzing the distribution of errors in estimating the measure of proximity between objects. pattern recognition and information processing (PRIP'2021). In: Proceedings of the 15th International Conference, 21–24 Sept. 2021, Minsk, Belarus. – Minsk : UIIP NASB, pp. 111–114 (2021)
11. Yrehab, F., Rabab, M., Fayez, W., Emad, E.-S.: Rotation-invariant neural pattern recognition system using extracted descriptive symmetrical patterns. Int. J. Adv. Comput. Sci. Appl. **3**(5), 151–158 (2012)
12. Jiang, H., Yu, S.X.: Linear solution to scale and rotation invariant object matching. In: IEEE Conference Computer Vision and Pattern Recognition, pp. 2474–2481 (2009)

13. Liu, B., Hang, W., Weihua, S., Zhang, W., Sun, J.: Rotation-invariant object detection using sector-ring hog and boosted random ferns. Vis. Comput. **34**(5), 707–719 (2018)
14. Mammadov, R.G., Aliyev, T.Ch., Mammadov, G.M.: 3D object recognition by unmanned aircraft to ensure the safety of transport corridors. In: İnternational conference on problems of logistics, management and operation in the East-West transport corridor (PLMO), pp. 209–216. Baku, Azerbaijan 27–29 Oct 2021
15. Mammadov, R.G., Rahimova, E.G., Mammadov, G.M.: Reducing the estimation error of the measure of proximity between objects in pattern recognition. In: The International Conference on Automatics and Informatics (İCAİ'2021), pp. 76–81. IEEE, Varna, Bulgaria 30 Sept.–2 Oct. 2021
16. Mammadov, R.G., Aliev, T.Ch.: Difinition of orientation of objects by the system of technical vision. In: The Third International Conference "Problems of Cybernetics and Informatics PCI '2010", vol. 1, pp. 259–262. Baku (2010)
17. Хачумов, М.В. Инвариантные моменты и метрики в задачах распознавания графических образов // – Россия: Современные наукоемкие технологии, – 2020. №4, Ч.1. – с.69 – 77
18. Ejima, T., Enokida, S., Kouno, T.: 3D object recognition based on the reference point ensemble. In: International Conference on Computer Vision Theory and Applications, pp. 261–269. Portugal (2014)
19. Ha, V.H.S., Moura, J.M.F.: Afine-permutation invariance of 2-D shape. IEEE Trans. Image Process. **14**(11), 1687–1700 (2005)
20. Cortadellas, J., Amat, J., de la Torre, F.: Robust normalization of silhouettes for recognition application. Pattern Recognition Lett. **25**, 591–601 (2004)

AI-Based Network Security Anomaly Prediction and Detection in Future Network

Gunay Abdiyeva-Aliyeva[1(✉)] and Mehran Hematyar[2]

[1] Laboratory of Control in Emergency Situations (Doctorate of Science/Postdoc), Institute of Control Systems of the Ministry of Science and Education of the Republic of Azerbaijan, Baku, Azerbaijan
gunay.abdiyeva@isi.az

[2] Azerbaijan Technical University, Huseyn Javid Street, 25, Baku, Azerbaijan
mehran@aztu.edu.az

Abstract. Anomalies are observed in non-compliance with specific rule sets that manifest as certain arrangements that deviate from a commonly placed concept of custom type. It should be said that every system has a propensity to show anomalous traits, further highlighting the significance of the concept of Anomaly Detection. A crucial point of data analysis for identification of anomalous data contained in a specified data set is seen in Anomaly Detection. Intrusion detection mechanisms, an important technology in network security, systematically track suspicious activities or patterns in a system and indicate if any events are vulnerable to an attack. Intrusion detection methods are divided into two categories: misuse detection and anomaly detection. With little or no correspondence to classification problems, measuring the efficiency of anomaly detection algorithms is more difficult. On the one hand, since true anomalies are uncommon in nature, the ground reality of anomalies remains unknown. Anomaly detection algorithms, on the other hand, often generate an anomalous score for each item. If the anomalous scores of objects are greater than a certain threshold, they are called anomalies.

Furthermore, anomaly detection is synonymous with the prediction of anomalies, changes observed in functionality, and tracking unprecedented patterns in the distribution of data in networks. Anomaly detection could be a potential, crucial tool in various sectors, ranging from financial security, medicine, natural sciences, manufacturing firms to electricity firms; the array is infinite. Incorporating, or embedding the concept of Artificial Intelligence into the detection of anomalies is justified because of the fact that it possesses special skill sets in the mold of accuracy/precision, automation, performing evaluations in real time, and self-learning (with streams of training data, of course) to bolster the efficacy with which anomaly detection is implemented.

Keywords: IDS · KNN · Rock curve · Anomaly detection · Naïve Bayes · NIDS

1 Introduction

As a consequence of the rapid spread of information technology across different fields in daily activities, it is now imperative that network security be improved on, and strengthened. However, to begin with, anomalies are observed in noncompliance with specific

D. J. Hemanth et al. (Eds.): ICAIAME 2022, ECPSCI 7, pp. 149–159, 2023.
https://doi.org/10.1007/978-3-031-31956-3_13

rulesets, that manifest as certain arrangements that deviate from a commonly placed concept, of custom type. It should be said that every system has a propensity to show anomalous traits, further highlighting the significance of the concept of Anomaly Detection. A crucial point of data analysis for identification of anomalous data contained in a specified data set is seen in Anomaly Detection. In a similar manner, it is researched to a broad extent in machine learning, and statistics as well; the concept is otherwise, in some cases, referred to as deviation detection, exception mining, novelty detection and outlier detection, as stated by Gunay, Mehran, and Sefa [1]. According to Gunay, Mehran, and Sefa [1], Network Anomaly Detection (abbreviated as NAD) plays an instrumental role in the field of Network Intrusion Detection Systems (NIDSs), and has been of great significance in detection and thwarting of novel attacks in the past three decades; in addition, it is domain specific. To offer any form of validation to the myriad of studies which have been carried out within the domain, it is sternly proposed that sophisticated, all-inclusive models be divided into vised, in order to deal with data/records increasing at an exponential rate with respect to types and nature of attacks (existing/zero-day-attack), frequency of attacks, and suchlike.

One of the major cruxes in Intrusion Detection Systems (IDSs) is the detection of anomalies with great precision, together with reduced costs of computation from the enormous volume of data to be processed. In conjunction with prerequisite assistance from data mining methodologies and algorithms of machine learning, picking out powerful attributes from data is bound to lessen the high computational costs, and in the end, see to it that anomaly detection is conducted with greater precision.

2 Intrusion Detection

Intrusion detection mechanisms, an important technology in network security, systematically track suspicious activities or patterns in a system and indicate if any events are vulnerable to an attack, according to Hu et al. [2]. Intrusion detection methods are into two categories: signature-based/misuse detection and anomaly detection. Signature-based misuse detection can only identify proven attacks by comparing the actions of incoming intrusions to historical information and predefined laws. It can simply detect known attacks that follow the particular patterns that have been integrated into the system, but falls short when it comes to identifying unfamiliar/new undefined attacks [3]. However, by explicitly computing anomalies, anomaly detection establishes a natural behavior for the structures and subsequently detects incoming intrusions. Anomaly intrusion detection system functions by identifying exceptional patterns that are not in conformity with normal network traffic [4]. It is capable of detecting new threats, but it is also prone to raising false alarms [5].

3 Algorithms and Methodologies Used in AI-Based Anomaly Detection

3.1 AI Based Network Traffic Flows Classification

Several common techniques, such as the probability-based naive Bayes classifier, hyperplane-based support vector machines, instance-learning-based k nearest neighbors, the sigmoid function-based logistic regression method, and rule-based classification like decision trees, have been used to construct a data driven predictive model [6, 7]. A large number of researchers have used the machine-learning classification techniques described above in the area of cybersecurity, specifically for detecting intrusions or cyber-attacks. Li et al. [3], for example, used the hyperplane-based support vector machine classifier with an RBF kernel to classify predefined attack categories such as DoS, Probe or Scan, U2R, R2L, and normal traffic using the most common KDD'99 cup dataset used the J48 and Naive Bayes algorithms to classify the UNSW-NB15 dataset, and discovered that the J48 algorithm, with 18 features, gave an accuracy rate of 18% based on gain ratio (GR) evaluation technique. Patel [8] designed a model that entailed a feature selection algorithm developed on attribute values, central points and association rule mining. The UNSW-NB15 and NSL-KDD datasets were then tested on the model, with Naive Bayes technique and Expectation-maximization clustering used in detection anomalies, and the model was observed to return a significant level of accuracy within a relatively short processing time. In an attempt to come up with a more efficient system, Zheng et al. [9] used a least-squares support vector machine classifier to train the model using large data sets. Hu et al. [2] used a version of the support vector machine classifier to identify the anomalies in their study. Dsir C et al. [10] adopted a broader approach involving the RepTree algorithm and protocols in a two-phase approach – the model was evaluated using the NSL-KDD and UNSW-NB15 datasets. In the first phase, network traffic flows were classified into 'UDP', 'TCP' and 'OTHER', after which they were distinguished into anomaly and normal. The second phase involved the use of a multi class algorithm to classify the anomalies that were detected. The analysis showed that the model had an accuracy of 89.85% and 88.95% for NSL-KDD and UNSW-NB15 respectively. In research carried out by Wagner et al. [11] used a one-class support vector machine classifier to detect anomalies and various forms of attacks, including NetBIOS scans, DoS attacks, POP spams, and Secure Shell (SSH) scans. Clustering is a machine learning methodology that divides power consumption data into different units, commonly referred to as clusters, and thus aids in the classification of unlabelled datasets as regular or abnormal (even with many dimensions). Because of its simplicity, this anomaly detection technique has sparked interest in a variety of research topics, including network intrusion detection, Internet of things (IoT), sensor networks, suspicious activity detection in video surveillance, anomalous transaction detection in banking systems, and suspicious account detection on online social networks. Ghori et al. [12] proposed a back propagation technique, wherein training was done in tandem with generated input and equivalent targets that were eventually implemented into the network, to detect anomalies revolving around DoS, U2R, R2L and Probe attacks. Their model did not, however, show a very promising result as the detection rate was below 80% for each of the intrusion types. Liu et al. [13] employed the feed forward and back propagation learning algorithms to detect victim end DOS anomalies, with an unsupervised correlation technique being used in actualizing feature selection. Upon testing this model on both UNSW-NB15 and NSL-KDD datasets, the authors discovered that it returned highly satisfactory results in terms of detection time and training time. Clustering techniques have been found to possess a prospective capability to learn and identify anomalies from consumption time-series without the need for explicit explanations, as stated by Pastore et al. [4]. One-class classification, otherwise referred to as one-class learning (OCL), divides initial power consumption patterns into two groups: positive (normal) and negative (abnormal), and then classifies them accordingly. As provided in a 2016 study by Shi et al. [5], while the negative party attempts to design classification algorithms can be missing, incorrectly sampled, or ambiguous. As a result, OCL is a difficult classification problem that is more difficult to solve than OCL. Problems of traditional classification, which attempt to distinguish between data from two or more groups based on data on training consumption apply to all of the classes. Various approaches and methodologies for detecting anomalous consumption footprints based on OCL have been proposed in the literature under-listed below. The one-class support vector machine (OCSVM) was used in a 2011 study conducted by Hock [14] to find the smallest hypersphere that encompasses all of the power observations. To identify power consumption occurring at abnormal rates, a kernel-based one-class neural network (OCNN) is proposed in the research of Chalapathy, Menon & Chawla [7]. It combines deep neural networks' (DNN) ability to derive progressive rich representations of power signals with OCL, forming a tight envelope around typical power consumption patterns. Two approaches to one-class convolutional neural networks (OCNN) were proposed in separately conducted but connected research by Oza & Patel [8], and Zheng et al. [9]. They both employ the use of a zero-centred Gaussian noise in the latent space as the pseudo-negative class and train the model using the cross-entropy loss to learn an accurate representation of the considered class as well as the decision boundary. In addition to the foregoing, when labelled data are missing, one-class random forest (OCRF) is proposed to classify irregular consumption as seen in studies by Dsir et al. [10], and another research by Ghori et al. [12]; the concepts, however, are hugely hinged upon classifier ensemble randomization fundamentals as described in a study conducted by Liu, Ting & Zhou [13]. A hybridized detection architecture involving a support vector machine (SVM) model, Naive Bayes algorithm and decision tree has been proposed by Harshit Saxenal [14] – the KDD99 dataset was used to evaluate the model effectiveness. Firstly, the SVM model was trained to classify data into normal or abnormal samples. The decision tree algorithm was then employed to provide details on the specificity of (known) attacks as per the abnormal samples. Where the decision tree model falls short in identifying unknown attacks, the Naive Bayes algorithm provides expedient benefit. In effect, Saxena's hybrid model can best be described as one that thrives on complementarity, and the analysis did show that it was able to reach an accuracy level of 99.62%. A convolutional auto-encoder was adopted for the extraction of payload features by Yu et al. [15], and the model was tested on the CTU-UNB dataset. The packets were initially converted to images to ensure optimization, after which, the training of the auto-encoder was actualized for the purpose of feature extraction. Classification of packets was based on learned features, and the analyses showed a precision rate of 98.44% while the recall rate was 98.40%. Min et al. [16] used a text-based CNN method for the detection of payload attacks with the model tested on the ISCX 2012 dataset. Statistical features for this work were from packet headers and, while the content features were sourced from payloads that were subsequently made subjected to concatenation. The concatenated payloads were thereafter encoded using a skip-gram word embedding method and extraction of content features followed suit. The result showed that this intrusion detection framework had a 99.13% accuracy rate (Fig. 1).

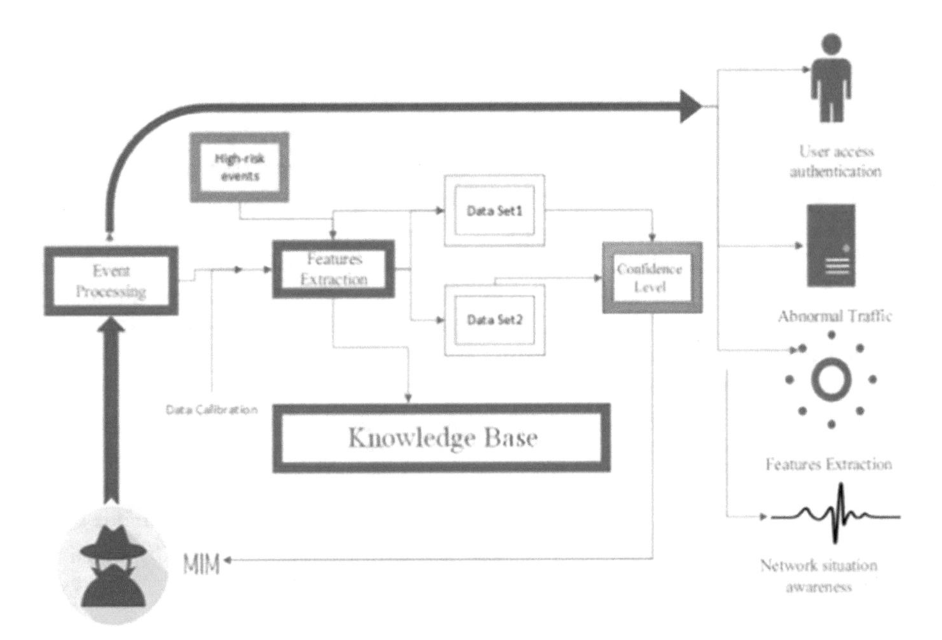

Fig. 1. AI Flow diagram for network security anomaly detection

In a bid to boost the speed of detection, Kuttranont et al. [17] designed a modified K-nearest neighbour (KNN) algorithm with parallel computing methods being used to carry out accelerated calculations on graphics processing units. In designing this model, the authors factored in the uneven nature of data distribution, and it was apparent from the result that the model performed exceptionally on sparse data. As the model was tested on the KDD99 dataset, an accuracy level of 99.30% was realized even as comparative analysis revealed that the incorporation/use of a graphics processing unit made the model 30 times faster. In another development, Meng et al. [18] created a KNN-based model to specifically address the issue of false alarms. Five different threat levels were involved in this work, and the model was trained to rank the alerts. It was discovered that the model succeeded in reducing the volume of false alerts to the tune of 89%. Going further into the possible algorithms and structures/models that could be applied in anomaly detection using AI network security models, we have probabilistic models which have over time grown into an effective idiom which is used in explaining real-world problems of anomaly detection in energy consumption using randomly generated variables, such as building models defined by probabilistic relationships; a practical illustration of such could be found in [13]. A framework using artificial neural network (ANN) and fuzzy clustering was implemented by Gaikwad et al. [19]. The main goal was to create a model that is capable of the challenges of poor precision and weak stability that are some times associated with intrusion detection systems. They utilized fuzzy clustering for dimensionality reduction and then trained the subsets with ANN. The authors utilized the restore point for project database, registry keys, installed programs and system files roll back in this work, and the results were found to be promising. An anomaly network intrusion detection system based on a neuro-fuzzy algorithm was described by Mohajerani et al. [20]. This algorithm features a three-tier architecture that is designed for efficiency. The first tier has intrusion detection agents (IDAs) that scrutinize network activities and act to report any unusual behavioral pattern, compared to the second tier, which has agents that detect the status of the local area network, acting on observable network traffic and Tier-1 agents' reports. The third-tier agents then generate higher-level reports and execute data correlation before alerting about a possible intrusion. It is worth noting that the agents in this model are of four different categories – ICMPAgent, PortAgent, TCPAgent and UDPAgent – with each of them monitoring traffic within its domain. In order to generate data, Bayesian maximum likelihood models are used to identify anomaly profiles of time-series patterns. Bayesian network models are used to detect anomalies in categorical and mixed/oscillated quantities of data transmitted in such networks in Jakkula & Cook [21]. In the subsequent paragraph, a particular instance of thedevelopment of smart AI to detect anomalies in power consumption. Statistical algorithms are used by Liu et al. [13] to classify anomalies by identifying extremes based on the standard deviation, while the authors in Coma-Puig et al. [22] use both statistical models and clustering schemes to detect power consumption anomalies. The use of Naive Bayes algorithms to detect anomalies caused by electricity theft attacks is suggested in Janakiram et al. [25]. Janakiram et al. [25], for example, use a Bayesian network to capture conditional dependencies between data and then classify anomalies. Zimek et al. [26] introduces a predictive prediction method based on a generalized additive model for detecting irregular energy consumption activity in real time. With little or no correspondence to classification problems, measuring the efficiency of anomaly detection algorithms is more difficult. Liu et al. [13] introduced an unsupervised neural net model that was able to detect new attacks, as well as known ones, in real time. The challenges emanating from single-level structures were addressed with a hierarchical intrusion detection framework that was created through neural networks built upon principal components analysis (PCA). Labib & Vemuri [23] also introduced some sort of intelligent intrusion detection model, the Network Self-Organizing Maps (NSOM) that identifies or classifies network behaviours as usual or irregular. The model constantly pre-processes data obtained from network ports and then moves on to carry out feature selection in preparation for the classification stage, which is done by sequentially examining packets. The traffic reflects usual network behaviour clustered around certain cluster centre(s) while the traffic shows irregularities clustered separately. The State Transition Analysis Tool – a knowledge-based model – was proposed by Ilgun et al. [23]. This model works by analysing the series of changes that are reflected in traffic data, with the objective of differentiating between secure state and target compromised state. The model entails three primary components, namely an inference engine, a knowledge base and a decision engine. The inference engine is responsible for monitoring the changes in the state of pre-processed audit data and then contrasts these with the states present in the knowledge base. Finally, the decision engine is in place to evaluate the matching precision of the inference engine and also determines the actions to be executed. On the one hand, since true anomalies are uncommon in nature, the ground reality of anomalies remains unknown. Anomaly detection algorithms, on the other hand, often generate an anomalous score for each item. If the anomalous scores of objects are greater than a certain threshold, they are called anomalies. It's difficult to set a proper threshold for each submission ahead of time. Real anomalies would be ignored if the threshold was set too high; otherwise, certain items that were not true anomalies would be incorrectly identified as possible anomalies. It's difficult to set a proper threshold for each submission ahead of time. Real anomalies would be ignored if the threshold was set too high; otherwise, certain items that were not true anomalies would be incorrectly identified as possible anomalies.

Further, our analytic results inform us about the effect of AI-based techniques to increase the performance of anomaly detection instead of classic methods (Table 1).

Table 1. Showing the summary of some selected experiments on AI-based Intrusion

Authors	Datasets	Source	Techniques	AI Algorithms	Result (Accuracy rate)
Meng et al	Private	Log	Rule-base	KNN	89.00%
Goes chal	KDD99	Traffic flow	Flow statistic feature	SVM, Decision tree Naïve Bayes	99.60%
Zeng et al	ISCX2012	Packet	Payload analysis	CNN, LSTM, auto-enco	99.98%
Min et al	ISCX2012	Packet	Payload analysis	CNN	81.96%
Wang et al	DARPA 1998; ISCX2012	Session	Session sequence feature	CNN	99.13%
Varto uhi et al	CSIC2010	Log	Text analysis	Isolate forest	97.27%
Teng et al Goes chal	KDD99	Traffic flow	Traffic grouping	SVM	99.51%
Hu et al	DARPA 2000	Packet	Packet parsing	Fuzzy C-means	88.32%
Rigak i et al Meng et al	Private	Packet	Payload analysis	GAN	89.82%
Zhan g et al	KDD99	Traffic flow	Flow deep learning	Flow deep learning	98.30%

4 Evaluation Methods for Grading Efficiency of Anomaly Detection Models

4.1 Reflects the Proportion of Anomalies

R-precision [27]. The value of R-precision will be very small since the number of true anomalies is small in comparison to the size of the dataset. As a result, it contains less data. In this case (of correlation coefficients), metrics employed in conducting assessment include but are not limited to correlation coefficients such as Spearman's rank similarity and Pearson correlation.

The tests/evaluation methods listed beneath are commonly used to test the efficiency of anomaly detection methods in general. To begin with, in the test of Precision at t (P@t), provided that there exists a dataset D with N properties, P@t is defined as the proportion of true anomalies, AD, to the top t possible anomalies found by the detection method; that is,

$$P@t = \frac{|a \in A|rank(a) \leqslant t|}{t} \tag{1}$$

It is fairly common knowledge that it is quite in hurdle to set the value of t for each application. Setting t equal to the number of anomalies in the ground truth is a standard strategy. Another criterion used in evaluation is the receiver operating characteristic, commonly known as ROC. The true (false, as the case may be) positive rate reflects the proportion of anomalies (inliers) ranked among the top t possible anomalies, and the receiver operating characteristic (ROC) curve is a graphical plot of the true positive rate against the false positive rate. The ROC curve for a random model appears to be diagonal, while the ROC curve for a good ranking model will produce true anomalies first, resulting in the region under the corresponding curve (AUC) covering all available space, as given by Schubert et al. [26]. As a result, the AUC is often used to test the efficiency of anomaly detection algorithms numerically. The Rank power (RP). Both the precision and AUC parameters ignore anomaly rating characteristics. Intuitively, an anomaly ranking algorithm would be considered more accurate if it places true anomalies at the top of the list of anomaly candidates and usual observations at the bottom. Such a metric is rank power, which assesses the exhaustive ranking of true anomalies. In a formal manner of mathematical/statistical description, its definition is given as:

$$\text{RankPower} = \frac{n(n+1)}{2\sum_{i=1}^{n} R_i} \tag{2}$$

The number of anomalies at the top of potential artifacts is given from the above expression n, and the rank of the i-th true anomaly is Ri. A greater value of t means better results for a fixed value of t. The rank power equals one when all of the anomaly candidates are real anomalies. Again, there is the yardstick of R-precision. The proportion of true anomalies within the top of possible anomalies found, where the number of ground truth anomalies is measured by calculation, puts a greater focus on the possible deviations at the top. More information on co-efficient measurements can be found in Schubert et al. [26] and the references therein. Instead of measuring precision on an individual basis, average precision (AP) refers to the mean of precision scores across all anomaly object ranks. It is mathematically expressed as:

$$\mathrm{AP} = \frac{1}{|\mathrm{a}|}\sum_{\mathrm{t}=1}^{|\mathrm{a}|} \mathrm{P@t} \tag{3}$$

As expressed from the equation above, P@t, in this case is given as the precision at t as well.

5 Conclusions and Future Work

The perception and development of networks that will be in existence in the near or distant future from a global view, with centralized regulatory administration, helps to facilitate the control of traffic in networks of the big data environment, effectuating convenience and effectiveness. It should, however, be said that a large number of anomalies traffic detection are required, more often than not, to be detected by way of a vast collection of data samples; merging this, together with the abnormal spike of growth in traffic, a corresponding dip in the efficiency of detection is noticed. The need for introducing artificial intelligence to the sphere of anomaly detection is majorly informed by the dynamism of networks of late. This implies thus, that with such dynamism comes bending the rules as to what is normal and what is not, reshaping the general conception of normality in networks. Be that as it may, to create a structure where it would be easier to detect and thwart intrusions into systems, or networks, the concept of anomaly detection would continue to prove instrumental to the cause. Furthermore, anomaly detection is synonymous with the prediction of anomalies, changes observed in functionality, and tracking unprecedented patterns in the distribution of data in networks. Anomaly detection could be a potential, crucial tool in various sectors, ranging from financial security, medicine, natural sciences, manufacturing firms to electricity firms; the array is infinite. Incorporating, or embedding the concept of Artificial Intelligence into the detection of anomalies is justified; because of the fact that it possesses special skill sets in the mold of accuracy/precision, automation, performing evaluations in real time, and self-learning (with streams of training data, of course) to bolster the efficacy with which anomaly detection is implemented.

References

1. Abdiyeva-Aliyeva, G., Hematyar, M., Bakan, S.: Development of system for detection and prevention of cyber attacks using artificial intelligence methods. In: 2021 2nd Global Conference for Advancement in Technology (GCAT), pp. 1–5 (2021). https://doi.org/10.1109/GCAT52182.2021.9587584
2. Hu, W., Liao, Y., Vemuri, V.R.: Robust support vector machines for anomaly detection in computer security. In: Proceedings of the International Conference on Machine Learning and Applications—ICMLA 2003, pp. 168–174. Los Angeles, CA, USA, 23–24 June 2003
3. Feng, B., Li, Q., Pan, X., Zhang, J., Guo, D.: Groupfound: an effective approach to detect suspicious accounts in online social networks. Int. J. Distrib. Sens. Netw. **13**(7), 1550147717722499 (2017)

4. Pastore, V., Zimmerman, T., Biswas, S., Bianco, S.: Annotation-free learning of Plankton for classification and anomaly detection. Sci Rep **10**, 1–15 (2020)
5. Shi, Z., Li, P., Sun, Y.: An outlier generation approach for one-class random forests: an example in one-class classification of remote sensing imagery. In: 2016 IEEE International geoscience and remote sensing symposium (IGARSS), pp. 5107–5110 (2016)
6. Ruff, L., Vandermeulen, R., Goernitz, N., Deecke, L., Siddiqui, SA., Binder, A., et al.: Deep one-class classification. In: Dy, J., Krause, A. (eds.) In: Proceedings of Machine Learning Research, vol. 80, pp. 4393–402. PMLR, Stockholmsmässan, Stockholm Sweden (2018)
7. Chalapathy, R., Menon, A.K., Chawla, S.: Anomaly detection using one-class neural Networks. arXiv:1802.06360 (2018)
8. Oza, P., Patel, V.M.: One-class convolutional neural network. IEEE Signal Process Lett. **26**(2), 277–281 (2019)
9. Zheng, Z., Yang, Y., Niu, X., Dai, H., Zhou, Y.: Wide and deep convolutional neural Networks for electricity-theft detection to secure smart grids. IEEE Trans. Ind. Inf. **14**(4), 1606–1615 (2018)
10. Dsir, C., Bernard, S., Petitjean, C., Heutte, L.: One class of random forests. Pattern Recognit. **46**(12), 3490–3506 (2013)
11. Wagner, C., François, J., State, R., Engel, T.: Machine learning approach for ip-flow record anomaly detection. In: Domingo-Pascual, J., Manzoni, P., Palazzo, S., Pont, A., Scoglio, C. (eds.) NETWORKING 2011. LNCS, vol. 6640, pp. 28–39. Springer, Heidelberg (2011). https://doi.org/10.1007/978-3-642-20757-0_3
12. Ghori, K., Imran, M., Nawaz, A., Abbasi, R., Ullah, A., Szathmary, L.: Performance Analysis of machine learning classifiers for non-technical loss detection. J. Ambient Intell. Human. Comput. 1–16 (2020). https://doi.org/10.1007/s12652-019-01649-9
13. Liu, X., Iftikhar, N., Nielsen, P.S., Heller, A.: Online anomaly energy consumption detection using lambda architecture. In: Madria, S., Hara, T. (eds.) Big data Analytics and knowledge discovery, pp. 193–209. Springer International Publishing, Cham (2016)
14. Saxena, H., Richariya, V.: Intrusion detection in KDD99 dataset using SVM-PSO and feature reduction with information gain. Int. J. Comput. Appl. **98**(6), 25–29 (2014). https://doi.org/10.5120/17188-7369
15. Yu, Y., Long, J., Cai, Z.: Network intrusion detection through stacking dilated convolutional autoencoders. Secur. Commun. Netw. **2017**, 4184196 (2017)
16. Min, E., Long, J., Liu, Q., Cui, J., Chen, W.: TR-IDS: anomaly-based intrusion detection through text-convolutional neural network and random forest. Secur. Commun. Netw. **2018**, 1–9 (2018). https://doi.org/10.1155/2018/4943509
17. Kuttranont, P., et al.: Parallel KNN and neighborhood classification implementations on GPU for network intrusion detection. J. Telecommun. Electron. Comput. Eng. (JTEC) **9**, 29–33 (2017)
18. Meng, W., Li, W., Kwok, L.F.: Design of intelligent KNN-based alarm filter using knowledge-based alert verification in intrusion detection. Secur. Commun. Netw. **2015**(8), 3883–3895 (2015)
19. Gaikwad, D.P., Jagtap, S., Thakare, K., Budhawant, V.: Anomaly based intrusion detection system using artificial neural network and fuzzy clustering. Int. J. Eng. Res. Technol. **1**(9), 1–6 (2012)
20. Mohajerani, M., Moeini, A., Kianie, M.: NFIDS: a neuro-fuzzy intrusion detection system. In: Proceedings 10th IEEE International Conference on Electronics, Circuits and Systems, vol. 1, pp. 348–351 (2003)
21. Jakkula, V., Cook, D.: Outlier detection in smart environment structured power datasets. In: 2010 sixth international conference on intelligent environments, pp. 29–33 (2010)

22. Coma-Puig, B., Carmona, J., Gavaldà, R., Alcoverro, S., Martin, V.: Fraud detection in energy consumption: a supervised approach. In: 2016 IEEE international Conference on data science and advanced analytics (DSAA), pp. 120–129 (2016)
23. Labib, K., Vemuri, R.: NSOM: A tool to detect denial of service attacks using self-organizing maps. Department of Applied Science University of California, Davis Davis, California, U.S.A., Tech. Rep. (2002)
24. Ilgun, K., Kemmerer, R.A., Porras, P.A.: State transition analysis: a rule-based intrusion detection approach. IEEE Trans. Software Eng. **21**(3), 181–199 (1995)
25. Janakiram, D., Kumar, A.V.U.P., Reddy, V.A.M.: Outlier detection in wireless sensor Networks using Bayesian belief networks. In: 2006 1st international conference on communication systems software middleware, pp. 1–6 (2006)
26. Zimek, A., Gaudet, M., Campello, R.J.G.B., Sander, J.: Subsampling for efficient and effective unsupervised outlier detection ensembles. In: Proceedings of the 19th ACM SIGKDD International Conference on Knowledge Discovery and Data Mining, KDD 2013, pp. 428–436. USA (2013)
27. Craswell, N.: R-precision. In: Liu, L., Ozsu, M. (eds.) Encyclopedia of Database Systems, p. 2453. Springer, Berlin, Germany (2009)

Mathematical Modeling of the Antenna Devices of the Microwave Range

Islam J. Islamov(✉), Mehman H. Hasanov, and Elmar Z. Hunbataliyev

Department of Radioengineering and Telecommunication, Azerbaijan Technical University, H. Javid Ave. 25, AZ 1073 Baku, Azerbaijan
icislamov@mail.ru

Abstract. In the work, mathematical modeling of antenna devices in the microwave range was carried out. A method is proposed for calculating the fields of aperture antennas in the time domain, in which the antiderivative impulse response of the antenna as a function of the observation point can in many cases be expressed in elementary functions in the entire half-space in front of the aperture. On its basis, the characteristic features of pulsed fields of aperture antennas of various shapes for different observation points are shown. Flyby diagrams of a circular aperture are obtained at various flyby radius, as well as the amplitude profile of the electromagnetic field of a circular aperture along the electric field strength vector at various distances, the amplitude profile of the electromagnetic field of circular apertures corresponding to the successive addition of Fresnel zones along the electric field strength vector, longitudinal amplitude profiles of small and large circular apertures and the amplitude profile of the electromagnetic field of the annular aperture corresponding to two adjacent Fresnel zones along the electric field strength vector. The proposed calculation method is generalized for field distributions decreasing towards the edges of the aperture. The proposed modeling method is generalized for rectangular, circular and annular apertures. It is shown that for a monochromatic signal when measuring the radiation pattern with a receiving aperture antenna at a finite distance, the measurement error is minimized when the probe size is about half the size of the antenna.

Keywords: Electromagnetic Field · Waveguide · Mathematical Modeling · Antenna · Microwave Devices

1 Introduction

The relevance of mathematical modeling of multibeam antenna arrays is associated with the development of radar systems, communications and means of controlling the electronic environment. Multibeam antennas provide increased radio network capacity with improved spectral efficiency and better user experience.

The multibeam phased antenna array has a multi-lobe radiation pattern. Usually it has several independent inputs and outputs, each of which has its own radiation pattern – its own beam. A multi-beam phased antenna array provides a parallel view of space, i.e. in space, a plurality of rays are simultaneously formed, located discretely in the

D. J. Hemanth et al. (Eds.): ICAIAME 2022, ECPSCI 7, pp. 160–175, 2023.
https://doi.org/10.1007/978-3-031-31956-3_14

scanning sector. Signals arriving at a phased array antenna from different directions can be separated and transmitted to different antenna ports, i.e. separated in space. The subject of research work is a diagram-forming system. The beam forming system is the main and one of the most expensive links in a multibeam antenna array. Its task is to form the amplitude-phase distribution at the inputs of the radiators of the antenna array. To increase the directivity, and hence the gain of the antenna system, it is necessary to minimize mainly the phase errors in the phase distribution. Therefore, mathematical modeling of multibeam antenna devices is an urgent task.

Until recently, the structure and characteristics of the electromagnetic field of the emitting aperture were most often considered in the far zone, which is associated with the practical need to analyze the characteristics of antennas [1–5]. However, in recent years, the problems of energy transfer have been intensively solved using electromagnetic waves [6–10].

Non-contact energy transfer is a priority task of electrodynamics, microwave technology and antenna technology. A necessary property of an electromagnetic field excited for contactless energy transfer is its localization in a limited volume of space. The indicated property is possessed by the electromagnetic field of the radiating aperture under certain conditions, which determines the actual the nature of this work.

The purpose of the article is to generalize and systematize the experience of developing software for modeling the characteristics of the electromagnetic field of radiating apertures in the intermediate and far zones.

It should be noted that the work within the framework of the declared subject already exists. Modeling of the electromagnetic field of a rectangular aperture was carried out in [11–15] to ensure the study of material for the analysis of Fresnel diffraction. As applied to practical issues, similar results for a rectangular aperture are developed in [16–19]. However, the works [20–23] are limited only to the presentation of simulation results, leaving aside the issues of implementation and the specifics of the approach to the simulation itself, which excludes the possibility of developing ideas, improving computational procedures, and developing similar programs for studying non-planar apertures with different amplitude and phase distributions. At the same time, works [24–26] pay due attention to the formalization of modeling algorithms, but proceed from the need to analyze the field in the far zone or find the equivalent opening impedance. The authors suggest that this article will fill the indicated gap within the stated subject. The presence of a large number of examples in the indicated sources for rectangular apertures allows us not to duplicate these results, giving preference to modeling circular and annular apertures.

2 Formulation of the Problem

By a flat radiating aperture, we mean a fragment S of the plane P, on which the distribution of the tangential components of the electric and magnetic field strength vectors is given. For an example in Fig. 1, the plane xOy is considered as the surface P. The tangential components of the field characteristics on the surface P are given as follows

$$\dot{\vec{E}}_m\Big|_S = \vec{x}_0 \dot{E}_0(x, y),\ \dot{\vec{H}}_m\Big|_S = \vec{y}_0 \frac{1}{W} \dot{E}_0(x, y), \tag{1}$$

where

$$\dot{E}_0(x, y) = \begin{cases} E_0(x, y)e^{j\phi E_0(x,y)}, (x, y) \in S \\ 0, (x, y) \notin S. \end{cases}$$

In accordance with the principle of equivalence equivalent currents can be written as:

$$\dot{\vec{\chi}}_m^M = -\left[\vec{z}_0, \dot{\vec{E}}_m\right]\Big|_S = -\left[\vec{z}_0, \vec{x}_0\right]\big|_S \dot{E}_0 = -\vec{y}_0 \dot{E}_0,$$

$$\dot{\vec{\chi}}_m^E = -\left[\vec{z}_0, \dot{\vec{H}}_m\right]\Big|_S = \left[\vec{z}_0, \vec{y}_0\right]\big|_S \frac{1}{W} \dot{E}_0 = -\vec{x}_0 \frac{1}{W} \dot{E}_0.$$

Since there are crossed electric and magnetic surface currents at each point of the aperture, the aperture itself can be represented by a set of Huygens elements of size $l_e \times l_h$, where l_e is the size along the electric field strength vector, l_h is the size along the magnetic field strength vector. The position of each Huygens element is characterized by the radius vector $\vec{r}'$ and the surface normal vector $\vec{n}_0$ (orientation vector).

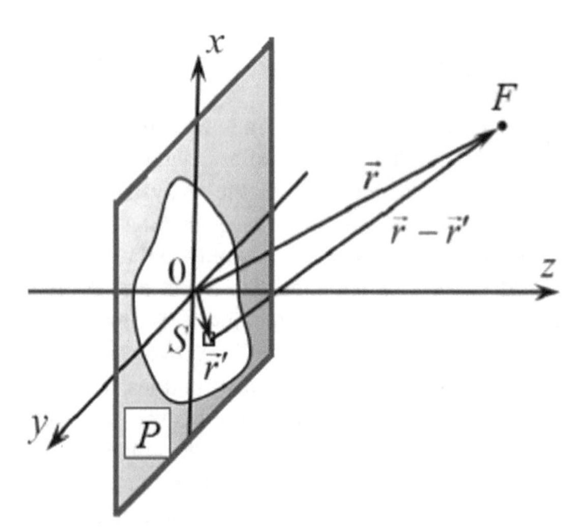

Fig. 1. Radiating aperture

In a rigorous formulation of the problem, the division of the aperture into Huygens elements is assumed to be continual, and the elements themselves to be infinitesimal. However, in the framework of a rigorous approach, obtaining solutions to the problem under consideration in the general case is difficult, therefore, in what follows, finite element partitions of the aperture are considered, with $l_{e,h} \leqslant \lambda$, where λ is the wavelength.

Thus, at the location of each Huygens element, there is formally an elementary electric emitter with a length l_e and a current of $\dot{I}_m^E = \frac{\dot{E}_0}{W} l_h$, as well as an elementary magnetic emitter with a length l_h and a current of $\dot{I}_m^M = \dot{E}_0 l_e$.

The characteristics of the electromagnetic field at a certain point F, characterized by the radius vector $\vec{r}$, based on the superposition principle, can be defined as the vector sums of the corresponding characteristics of the fields created by each elementary emitter:

$$\dot{\vec{E}}_m = \left\{ \left(\dot{\vec{E}}^{E}_{m,n_e,n_h} + \dot{\vec{E}}^{M}_{m,n_e,n_h} \right)_{n_e=0,n_h=0}^{N_e-1,N_h-1}, \right.$$
$$\dot{\vec{H}}_m = \left\{ \left(\dot{\vec{H}}^{E}_{m,n_e,n_h} + \dot{\vec{H}}^{M}_{m,n_e,n_h} \right)_{n_e=0,n_h=0}^{N_e-1,N_h-1}, \right. \tag{2}$$

where N_e is the number of Huygens elements that fit on S along the direction of the electric field strength vector; N_h is the number of Huygens elements that fit on S along the direction of the magnetic field vector; (n_e, n_h) is the total number of the Huygens element on S; $\dot{\vec{E}}_{m,n_e,n_h}$, $\dot{\vec{H}}_{m,n_e,n_h}$- characteristics of the electromagnetic field created by an elementary electric emitter oriented along the electric field strength vector on the aperture, length l_e and current $\dot{I}^E_m = \frac{\dot{E}_0}{W} l_h$, the position of which is given by the position vector of the Huygens element corresponding to number (n_e, n_h); $\dot{\vec{E}}^{M}_{m,n_e,n_h}$, $\dot{\vec{H}}^{M}_{m,n_e,n_h}$- electromagnetic field characteristics created by an elementary magnetic emitter oriented along the vector magnetic field strength at the aperture, length l_h and current $\dot{I}^M_m = \dot{E}_0 l_e$, the position of which is given by the position vector of the Huygens element with the number (n_e, n_h). The task requires, based on the given conditions (1), to find the characteristics of the electromagnetic field (2).

3 Approach to Mathematical Modeling of the Electromagnetic Field of Radiative Apertures

Electrodynamic calculations are characterized by increased complexity; calculations are complex and vector; different coordinate systems are used to describe the processes, most often Cartesian and spherical in the framework of our work. Providing these calculations is in itself a time-consuming task: the development of a program for electrodynamic calculations is a complex creative process, which necessitates its division into several stages, each of which solves one or another subtask, and the program itself is developed as a set of modules.

To provide vector calculations, a separate module has been developed, a feature of which is the implementation of operations with complex vectors. The calculation functions of the vector calculation module are used by the simulation module an elementary electric emitter, an elementary magnetic emitter simulation module, a Huygens element simulation module, and, finally, the aperture itself. The aperture modeling module is the computing module of the upper hierarchy and is directly used by the control and interface module.

The control and interface module provides direct interaction with the user, interprets user commands for configuring the aperture, and presents calculation results.

The characteristics of the electromagnetic field of an elementary electric emitter are determined by the known expressions:

$$\begin{cases} \dot{\vec{E}}_m^E = \dot{p}_m^{est} j\dot{k}\dot{W}\left(\vec{r}_0\frac{2}{\dot{k}r}\left(j+\frac{1}{\dot{k}r}\right)(\vec{r}_0,\vec{p}_0)+(\vec{r}_0(\vec{r}_0,\vec{p}_0)-\vec{p}_0)\right) \\ \times\left(1-j\frac{1}{\dot{k}r}-\frac{1}{(\dot{k}r)^2}\right)G(r), \\ \dot{\vec{H}}_m^E = (\vec{p}_0,\vec{r}_0)\dot{p}_m^{est}\dot{k}\left(j+\frac{1}{\dot{k}r}\right)G(r), \end{cases} \quad (3)$$

where $\dot{p}_m^{est} = \dot{I}_m^E l_e = \frac{\dot{E}_0}{\dot{W}} l_h l_e$ is the current moment of an elementary electric radiator; $\dot{W}$ – wave resistance; $\dot{k}$ is the wave number; $\vec{p}_0$ – orientation vector; $G(r) = \frac{1}{4\pi}\frac{e^{-jkr}}{r}$ – Green's function; $\vec{r'}$ is the position vector.

The written expressions give the following program model: an elementary electric radiator is characterized by a complex current amplitude $\dot{I}_m$, length l_e, orientation vector $\vec{p}_0$, position radius vector $\vec{r'}$. These vectors are specified in the Cartesian basis of the global coordinate system. To calculate the field of an elementary electric radiator from the known coordinates of the investigated point in space in the global system, its radius vector is found, then its radius vector in its own (local) coordinate system of the elementary electric radiator $\vec{r} - \vec{r'}$, and then the vectors of the electric and magnetic fields are calculated in accordance with (3).

The characteristics of the electromagnetic field of an elementary magnetic emitter are found as

$$\begin{cases} \dot{\vec{H}}_m^M = \dot{p}_m^{mst}\frac{j\dot{k}}{\dot{W}}\left(\vec{r}_0\frac{2}{\dot{k}r}\left(j+\frac{1}{\dot{k}r}\right)(\vec{r}_0,\vec{p}_0)+(\vec{r}_0(\vec{r}_0,\vec{p}_0)-\vec{p}_0)\right) \\ \times\left(1-j\frac{1}{\dot{k}r}-\frac{1}{(\dot{k}r)^2}\right)G(r), \\ \dot{\vec{E}}_m^E = -(\vec{p}_0,\vec{r}_0)\dot{p}_m^{mst}\dot{k}\left(j+\frac{1}{\dot{k}r}\right)G(r), \end{cases} \quad (4)$$

where $\dot{p}_m^{mst} = \dot{I}_m^M l_h = \dot{E}_0 l_h l_e$- moment of an elementary magnetic emitter.

The electromagnetic field of the Huygens element is calculated as the vector sum of the fields of crossed elementary electric and elementary magnetic emitters.

When initializing the Huygens element, its length is specified along the electric field vector l_e and along the magnetic field vector l_h, the electric field vector $\dot{\vec{E}}_0$ and the normal vector to the plane of the Huygens element $\vec{n}_0$ (orientation vector), as well as the position vector $\vec{r}_0$. Further, the position vector is specified in as position vectors for elementary electric and elementary magnetic emitters.

At the location of the Huygens element, there is an elementary electric emitter with a length l_e and a current, as well as an elementary magnetic emitter with a length l_h and a current $\dot{I}_m^E = \frac{\dot{E}_0}{\dot{W}} l_h$. The orientation of the elementary electric emitter is determined by the direction of the specified electric field strength vector $\dot{I}_m^M = \dot{E}_0 l_e$, the orientation of the elementary magnetic emitter is determined direction of the vector product $\left[\vec{n}_0, \dot{\vec{E}}_0\right]$.

A rectangular aperture is represented as a set of Huygens elements, and its electromagnetic field as a superposition of the fields of these elements.

When initializing the aperture, it is assumed that the electric field strength vector $\dot{\vec{E}}_0$ is oriented along the x axis, the aperture length along this vector L_x, the aperture length along the magnetic field strength vector L_y, and the number of Huygens elements along the coordinate axes N_x and N_y are specified. The aperture is considered to be located so that its center, determined by the center of the middle Huygens element, corresponds to the origin of the coordinate system, and it itself lies in the xOy plane. To fulfill this condition, $N_{x,y}$ must be odd numbers. In a C++ program, it is convenient to represent a rectangular aperture as a two-dimensional array of objects of the "Huygens element" class.

To obtain an annular aperture, first a square one is created, into which the outer contour of the ring can be inscribed, then the Huygens elements are turned off, the centers of which fall inside the shading region or are located outside the outer contour of the ring. When initializing the annular aperture, it is considered that the electric field strength vector $\dot{\vec{E}}_0$ is oriented along the x axis, the outer and inner diameters of the annular aperture D and d, as well as the number of Huygens elements along the outer diameter N_D are specified. The aperture is considered to be located so that its center, determined by the center of the middle Huygens element, corresponds to the origin of the coordinate system, and it itself lies in the xOy plane. To fulfill this condition, N_D must be odd. Setting $d = 0$ corresponds to a circular aperture (Fig. 2).

In the program developed by the authors, the user sets the frequency or wavelength, the size of the aperture along the coordinate axes, selects the sampling scheme for the given aperture, and the dimensions of the aperture elements are immediately calculated and displayed, which makes it possible to make sure that the dimensions of the elements are much smaller than the wavelength, that is, provided correctness of calculations.

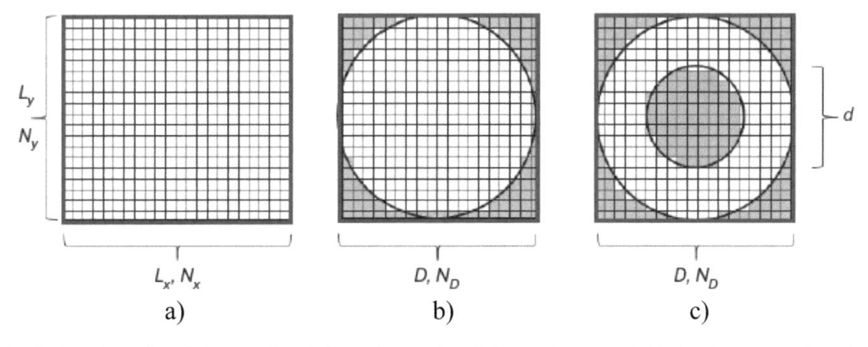

Fig. 2. Rectangular (a), circular (b), and annular (c) apertures and their decomposition into Huygens elements.

As a result of the work, the program presents fly-by diagrams in the *xOz* and *yOz* planes. The flyby diagram is the dependence of the normalized amplitude of the characteristic of the electromagnetic field on the position of a point in space on a circle of a selected radius (circle radius), characterized by an angle measured from the aperture plane, in a selected plane (Fig. 3). Flight diagrams are built in polar coordinates, but it is possible to observe them in Cartesian coordinates as well. Flying diagrams show 4 curves corresponding to the amplitudes of the radial, azimuthal, meridional components, as well as the resulting amplitude of the electric field strength vector. Normalization is carried out to the maximum of the resulting amplitude.

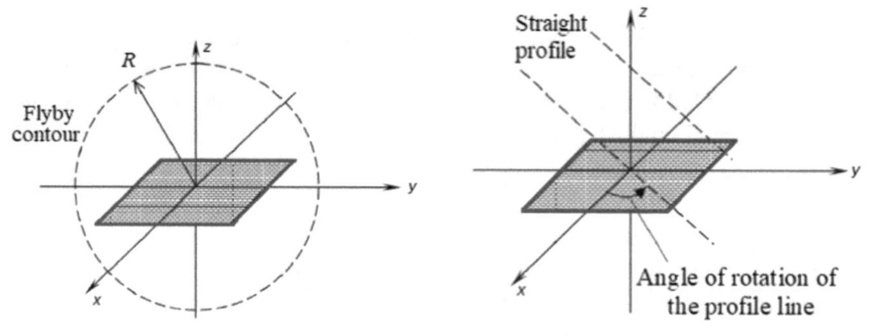

Fig. 3. Flyby contour in the *yOz* plan

Fig. 4. Reference straight line of the amplitude profile

Depending on the user-defined fly-by radius, the fly-by pattern can correspond to the intermediate or far zone of the aperture.

The fly-by pattern in the far field is a radiation pattern. The second result presented by the program is the amplitude profile of the electric field strength vector – this is the dependence of the amplitude of the component or the electric field strength vector itself at a point in space on its position on a straight line parallel to the aperture plane and oriented in the direction characterized by a given angle measured from the abscissa axis (Fig. 4). The angle of rotation of the profile line is set by the user. The amplitude profile of the field is a non-normalized characteristic.

4 Some Results of Modeling the Electromagnetic Field of a Radiant Aperture and Their Analysis

Examples of flyby diagrams of a circular radiating aperture are shown in Fig. 5. The aperture itself lies in the xOy plane, the electric field strength vector is directed along the abscissa axis. The left figures correspond to the diagrams in the xOz plane, the right ones correspond to the diagrams in the yOz plane. The diagram for the meridional component is shown in blue, the diagram in red for the radial component, and in green for the azimuthal component of the electric field strength vector. The black color shows the diagram for the electric field strength vector itself. The diagrams correspond to the principle of radiation at infinity: as the distance from the emitter increases, the radial component decays, and the wave becomes spherical, which corresponds to the transition to the far zone. In this case, the flyby pattern transforms into a radiation pattern.

Figure 6 contains the results of modeling a circular aperture, the diameter of which is much larger than the wavelength $D = 1$ m; $\lambda = 1$ cm; $D/\lambda = 100$. The ratio of the aperture diameter to the wavelength has the meaning of the relative aperture diameter. The figure shows the amplitude profiles of the characteristics of the electromagnetic field along the abscissa axis at various distances from the aperture plane. In the intermediate zone, when a red profile is observed corresponding to the radial component, one can approximately assume that the electromagnetic field is localized in a cylindrical volume ("wave tube"), the diameter of which is equal to the diameter of the aperture. As the distance from the aperture to the reference line increases, the amplitude profile begins to "smear", and the radial component of the electric field strength vector becomes negligibly small compared to the meridional or azimuthal components. The smearing of the profile thus already takes place in the intermediate zone.

Figure 7 shows the wavelength $\lambda = 10$ cm and the distance to the profile reference line $z = 10$ m, and the graphs correspond to a gradual increase in the aperture diameter by a value corresponding to the addition of one Fresnel zone. The sequence of figures shows that the "spotlight" effect is more pronounced the more Fresnel zones are open. Approximately starting from the tenth Fresnel zone, the profile width at a level of 0,5 becomes equal to the aperture diameter (at a unit amplitude of the electric field strength at the aperture). In a more general case, you can set the number of the Fresnel zone n_0, starting from which the field localization takes place, then the aperture diameter should coincide with the diameter of this zone, that is, we can write

$$D = 2\sqrt{n_0 L \lambda}, \tag{5}$$

whence the length of the "wave tube"

$$L = \frac{D^2}{4 n_0 \lambda}. \tag{6}$$

Or, dividing the left and right sides of (6) by the wavelength, we write

$$\frac{L}{\lambda} = \frac{1}{4 n_0} \left(\frac{D}{\lambda} \right)^2. \tag{7}$$

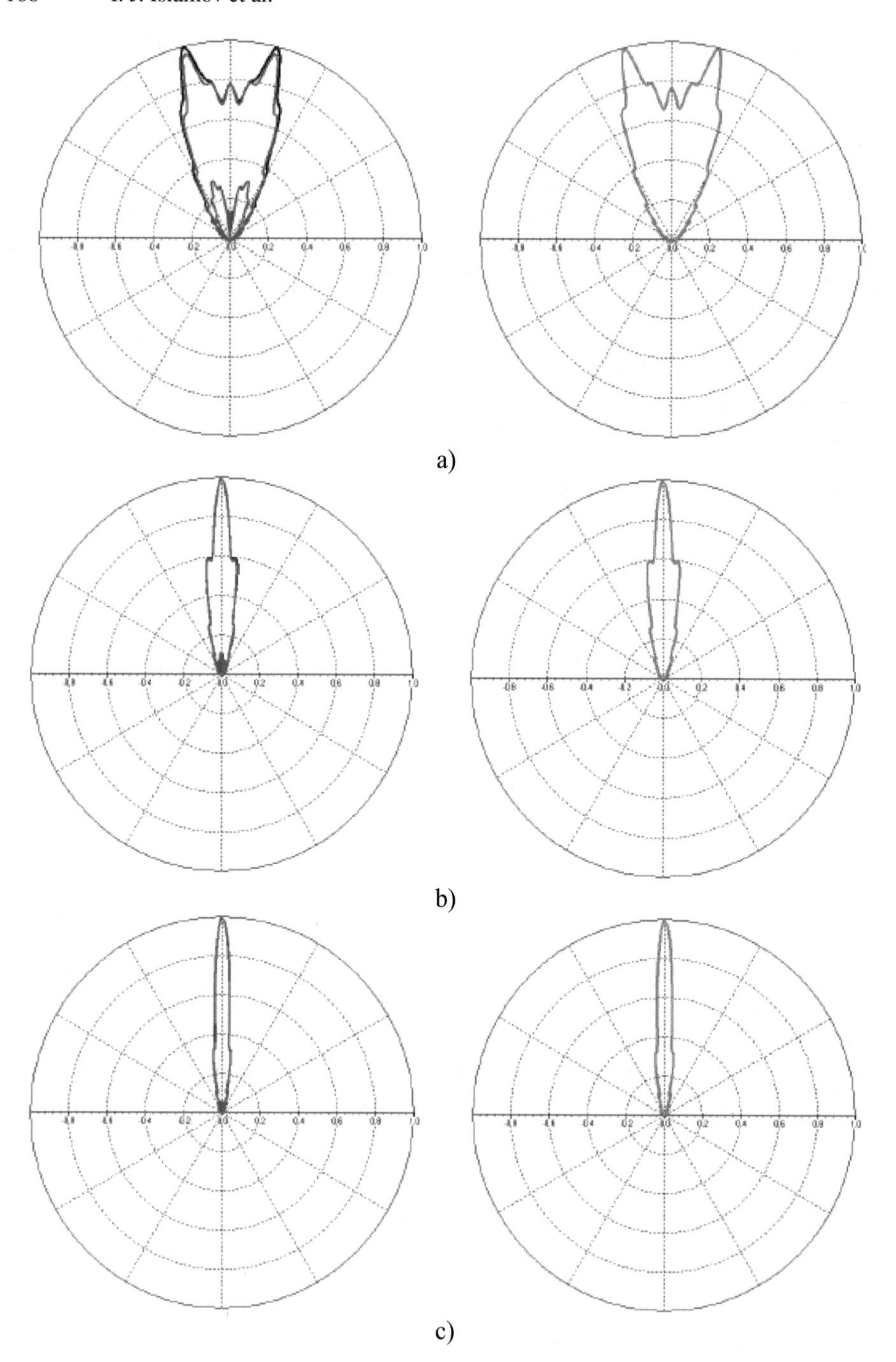

Fig. 5. Fly-by diagrams of a circular aperture (diameter $D = 1$ m, wavelength $\lambda = 10$ cm) for various fly-by radius R: a) $R = 100$ cm; b) $R = 200$ cm; c) $R = 300$ cm; d) $R = 500$ cm

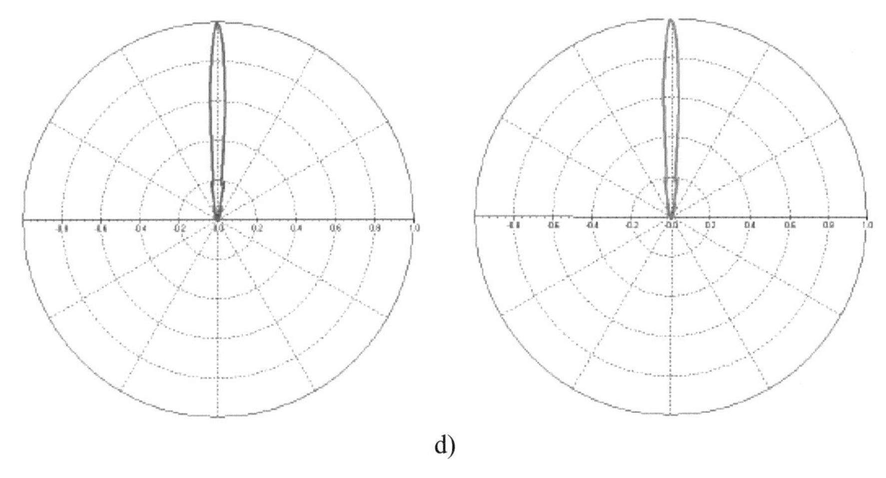

d)

Fig. 5. (*continued*)

The written expression shows that the relative length of the wave tube L/λ is directly proportional to the square of the relative diameter of the aperture, and the proportionality coefficient does not exceed 0,25 and depends on the choice of the number of the Fresnel zone, which must fit on the aperture when it is observed from a distance $z < L$.

The choice of n_0 can also be made on the basis of an analysis of the longitudinal amplitude profile of the electromagnetic field above the center of the aperture. As an example, in fig. Figure 8 shows the results of calculating the longitudinal profile for small (D/λ = 10) and large (D/λ = 100) circular apertures. The coordinates of the local extrema of the longitudinal amplitude profile (Fig. 8) approximately correspond to those applicates, when observing the aperture from which it is represented as a set of a certain number of complete Fresnel zones.

In general, the profiles have characteristic fluctuations as the longitudinal coordinate increases until the moment when the aperture observed from the current point above the center does not correspond to the first Fresnel zone, after which the fluctuations stop and the amplitude profile line monotonically decreases. When the length of the wave tube is limited by the last extremum, one should choose $n_0 = 1$ in formula (7).

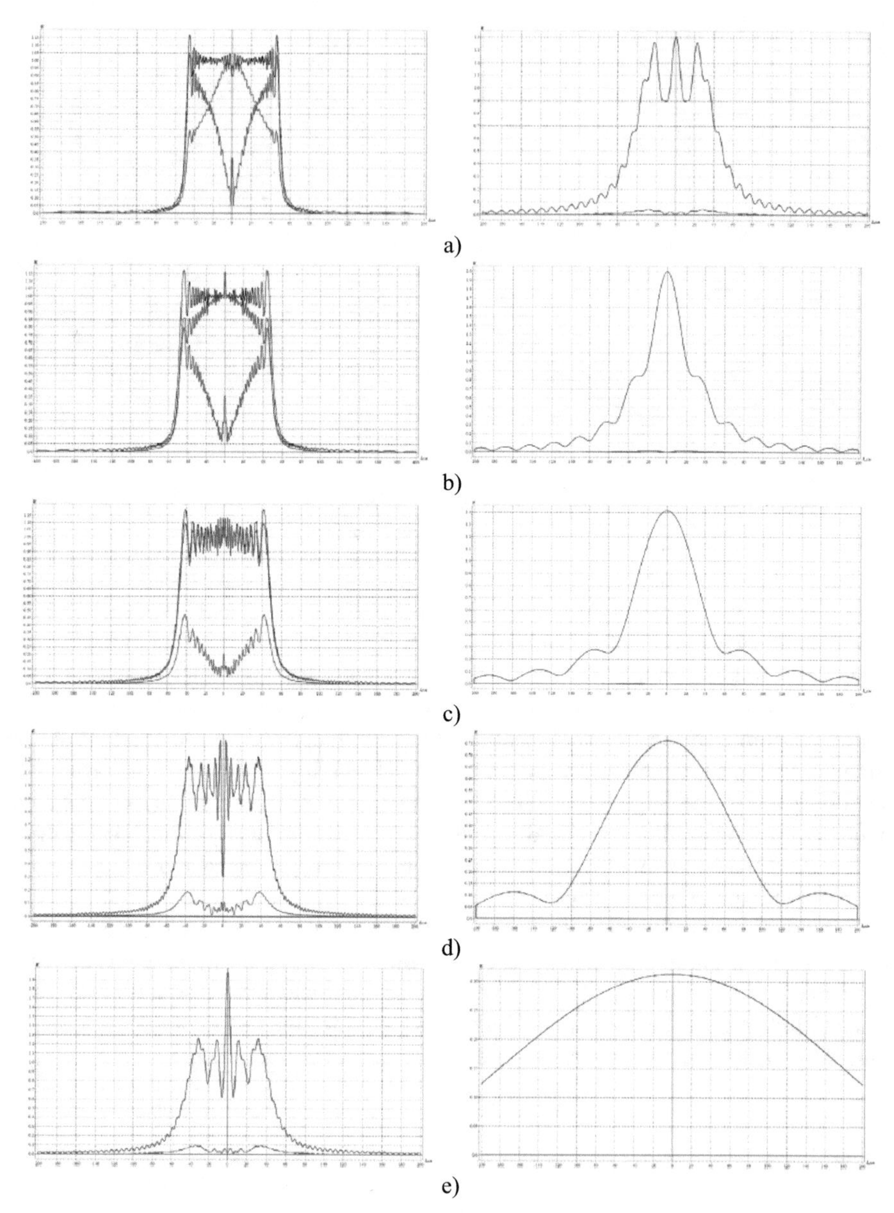

Fig. 6. Amplitude profile of the electromagnetic field of a circular aperture (diameter $D = 1$ m, wavelength $\lambda = 1$ cm) along the electric field strength vector at different distances z: a) $z = 25$ cm, $z = 10$ m; b) $z = 50$ cm, $z = 25$ m; c) $z = 1$, $z = 50$ m; d) $z = 2{,}5$ m, $z = 100$ m; e) $z = 5$ m, $z = 250$ m

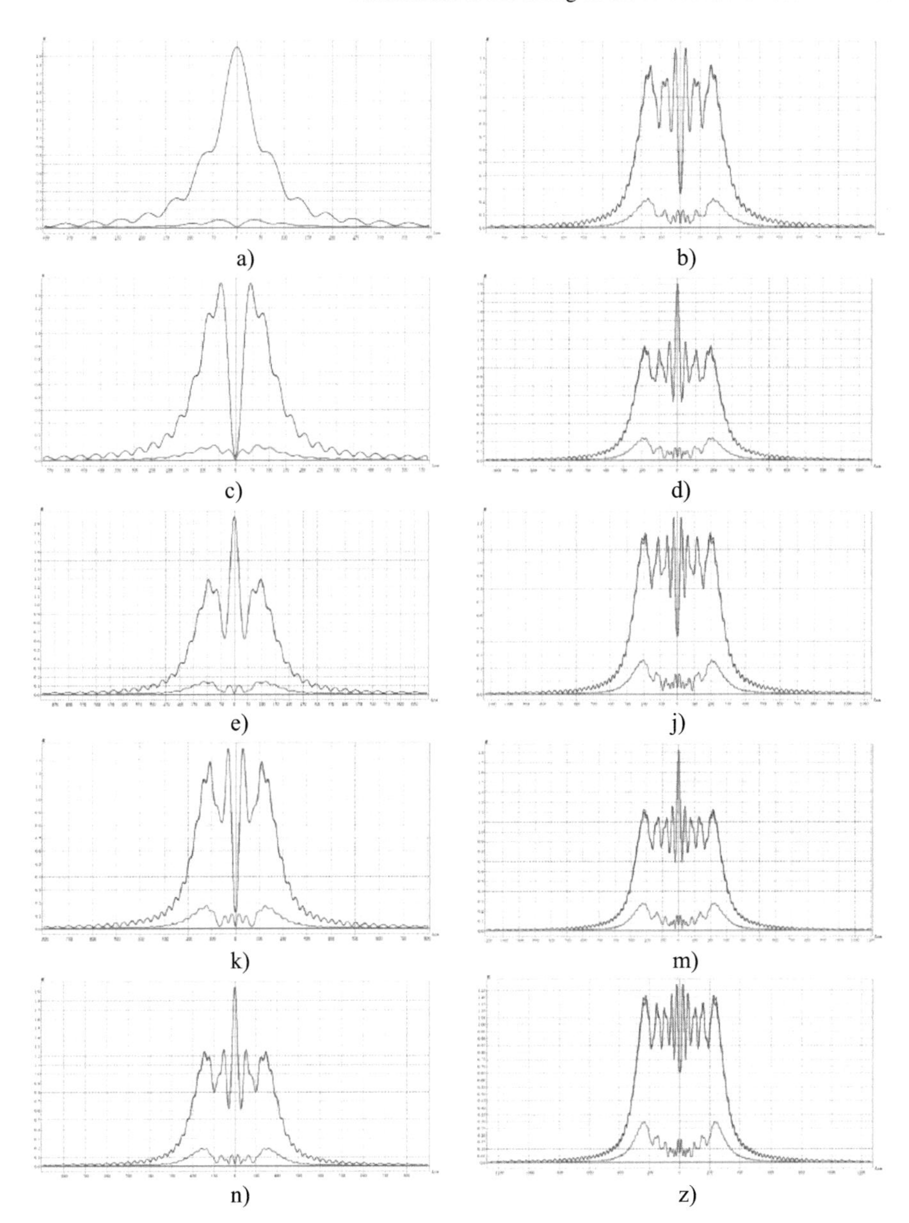

Fig. 7. Amplitude profile of the electromagnetic field of round apertures corresponding to the successive addition of Fresnel zones along the electric field strength vector (wavelength $\lambda = 10$ cm, $z = 10$ m): a) 1st Fresnel zone; b) 1–6 Fresnel zone; c) 1–2 Fresnel zone; d) 1–7 Fresnel zone; e) 1–3 Fresnel zone; j) 1–8 Fresnel zone; k) 1–4 Fresnel zone; m) 1–9 Fresnel zone; n) 1–5 Fresnel zone; z) 1–10 Fresnel zone

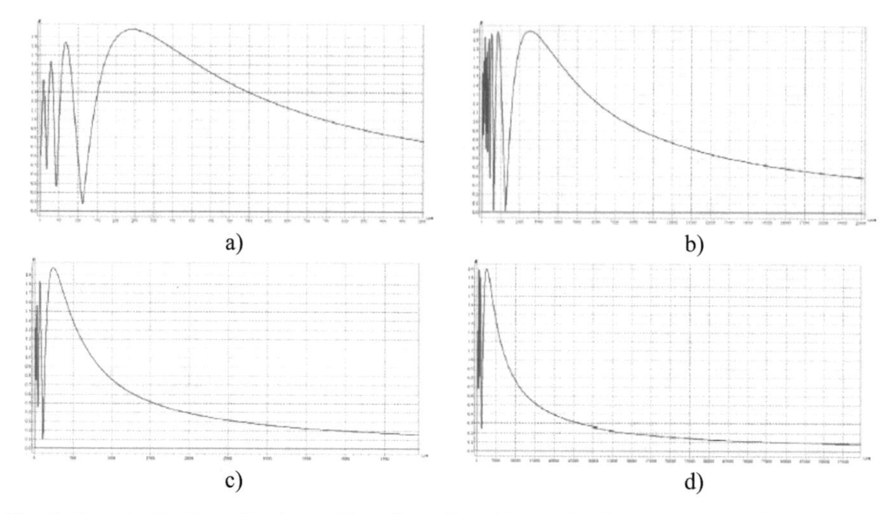

Fig. 8. Longitudinal amplitude profiles of small and large circular apertures: a) $D = 1$ m, $\lambda = 10$ cm, $0 < z < 10$ m; b) $D = 1$ m, $\lambda = 1$ cm, $0 < z < 200$ m; c) $D = 1$ m, $\lambda = 10$ cm, $0 < z < 50$ m; d) $D = 1$ m, $\lambda = 1$ cm, $0 < z < 1000$ m

Examples of numerical simulation show that the last local extremum of the longitudinal profile falls on the intermediate zone of the aperture. The known conditional boundary of the far zone falls at relative distances $2(D/\lambda)^2$ and corresponds to a decrease in the longitudinal profile below 20% of its maximum value at the local extremum of the intermediate zone.

On Fig. 9 shows the amplitude profiles of the electromagnetic field of annular apertures corresponding to two adjacent Fresnel zones. It is usually assumed that two neighboring Fresnel zones with large numbers cancel each other out. However, such compensation is best observed on the longitudinal amplitude profile of the field inside the wave tube, and the neighboring zones separated into the annular emitter themselves have a transverse profile quite different from zero.

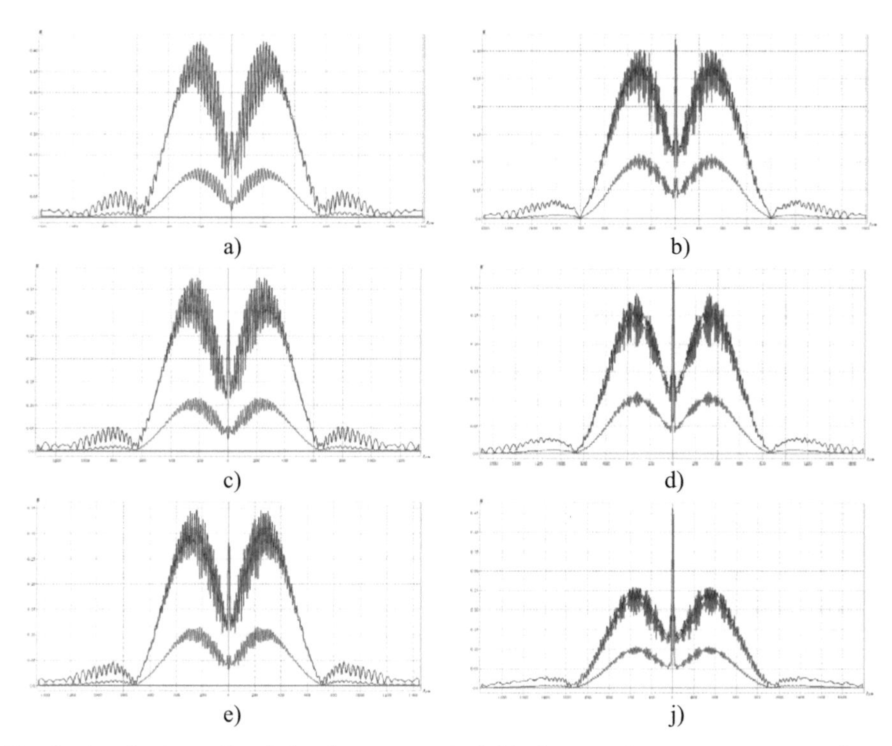

Fig. 9. Amplitude profile of the electromagnetic field of an annular aperture corresponding to two neighboring Fresnel zones along the electric field strength vector (wavelength $\lambda = 10$ cm, $z = 10$ m, zone numbers are indicated below the graph): a) 7–9; b) 14–16; c) 9–11; d) 16–18; e) 11–13; j) 18–20.

5 Conclusion and Recommendation

The paper proposes a method for calculating the fields of aperture antennas in the time domain, in which the primitive impulse response of the antenna as a function of the observation point can in many cases be expressed in elementary functions in the entire half-space in front of the hole. Diagrams of the passage of a circular aperture at various radius of passage, as well as the amplitude profile of the electromagnetic field of a circular aperture along the vector of the electric field strength at various distances, have been obtained; the amplitude profile of the electromagnetic field of a circular aperture, corresponding to the sequential addition of the Fresnel zones along the vector of the electric field strength; longitudinal profiles of the amplitudes of the small and large round holes and the profile of the amplitudes of the electromagnetic field of the annular hole, corresponding to two adjacent Fresnel zones along the vector of the electric field strength. The proposed method of calculation is generalized for field distributions decreasing towards the edges of the aperture. The proposed modeling method is generalized to rectangular, round and annular holes. It is shown that for a monochromatic signal when measuring the radiation pattern with a receiving aperture antenna at a finite distance, the measurement error is minimized when the probe size is about half the size of the antenna.

References

1. Islamov, I.J., Ismibayli, E.G., Gaziyev, Y.G., Ahmadova, S.R., Abdullayev, R.: Modeling of the electromagnetic feld of a rectangular waveguide with side holes. Prog. Electromagn. Res. **81**, 127–132 (2019)
2. Islamov, I.J., Shukurov, N.M., Abdullayev, R.Sh., Hashimov, Kh.Kh., Khalilov, A.I.: Diffraction of electromagnetic waves of rectangular waveguides with a longitudinal. In: IEEE Conference 2020 Wave Electronics and its Application in Information and Telecommunication Systems (WECONF). INSPEC Accession Number: 19806145 (2020)
3. Khalilov, A.I., Islamov, I.J., Hunbataliyev, E.Z., Shukurov, N.M., Abdullayev, R.Sh.: Modeling microwave signals transmitted through a rectangular waveguide. In: IEEE Conference 2020 Wave Electronics and its Application in Information and Telecommunication Systems (WECONF). INSPEC Accession Number: 19806152 (2020)
4. Islamov, I.J., Ismibayli, E.G.: Experimental study of characteristics of microwave devices transition from rectangular waveguide to the megaphone. IFAC-PapersOnLine. **51**(30), 477–479 (2018)
5. Ismibayli, E.G., Islamov, I.J.: New approach to definition of potential of the electric field created by set distribution in space of electric charges. IFAC-PapersOnLine. **51**(30), 410–414 (2018)
6. Islamov, I.J., Ismibayli, E.G., Hasanov, M.H., Gaziyev, Y.G., Abdullayev, R.: Electrodynamics characteristics of the no resonant system of transverse slits located in the wide wall of a rectangular waveguide. Prog. Electromagnet. Res. Lett. **80**, 23–29 (2018)
7. Islamov, I.J., Ismibayli, E.G., Hasanov, M.H., Gaziyev, Y.G., Ahmadova, S.R., Abdullayev, R.: Calculation of the electromagnetic field of a rectangular waveguide with chiral medium. Prog. Electromagnet. Res. **84**, 97–114 (2019)
8. Islamov, I.J., Hasanov, M.H., Abbasov, M.H.: Simulation of electrodynamic processes in the cylindrical-rectangular microwave waveguide systems transmitting information. In: Aliev, R.A., Kacprzyk, J., Pedrycz, W., Jamshidi, M., Babanli, M., Sadikoglu, F.M. (eds.) ICSCCW 2021. LNNS, vol. 362, pp. 246–253. Springer, Cham (2022). https://doi.org/10.1007/978-3-030-92127-9_35
9. Islamov, I.J., Hunbataliyev, E.Z., Zulfugarli, A.E.: Numerical simulation of characteristics of propagation of symmetric waves in microwave circular shielded waveguide with a radially inhomogeneous dielectric filling. Cambridge Univ. Press: Int. J. Microw. Wireless Technol. **14**(6), 761–767 (2021)
10. Islamov, I.J., Hunbataliyev, E.Z., Abdullayev, R.Sh., Shukurov, N.M., Hashimov, Kh.Kh.: Modelling of a microwave rectangular waveguide with a dielectric layer and longitudinal slots. In: International Symposium for Production Research, Antalya, pp. 550–558 (2021)
11. Abdulrahman, S.M.A., Anthony, E.S., Konstanty, S.B., Amin, A.: Flexible meander-line antenna array for wearable electromagnetic head imaging. IEEE Trans. Antennas Propag. **69**(7), 4206–4211 (2021)
12. Mohammad, M.F.: A wideband antenna using high gain fractal planar monopole antenna array for RF energy scavenging. Int. J. Antennas Propag. **2020**, 3489323 (2020)
13. Pinuela, M., Mitcheson, P.D., Lucyszyn, S.: Ambient RF energy harvesting in urban and semi-urban environments. IEEE Trans. Microw. Theory Tech. **61**(7), 2715–2726 (2013)
14. Zhou, M., Shojaei Baghini, M., Kumar, G.: Broadband bent triangular omnidirectional antenna for RF energy harvesting. IEEE Antennas Wirel. Propag. Lett. **15**, 36–39 (2016)

15. Yi-Ming, Z., Shuai, Z., Guangwei, Y., Gert, F.P.: A wideband filtering antenna array with harmonic suppression. IEEE Trans. Microw. Theory Tech. **68**(10), 4327–4339 (2020)
16. Zhao, Y., Guangjun, W., Wei, H., Daniele, I., Yongjun, H., Jian, L.: Microwave airy beam generation with microstrip patch antenna array. IEEE Trans. Antennas Propag. **69**(4), 2290–2301 (2020)
17. Ao, L., Kwai-Man, L.: Single-layer wideband end-fire dual-polarized antenna array for device-to-device communication in 5G wireless systems. IEEE Trans. Veh. Technol. **69**(5), 5142–5150 (2020)
18. Botao, F., Liangying, L., Kwok, L.C., Yansheng, L.: Wideband widebeam dual circularly polarized magnetoelectric dipole antenna/array with meta-columns loading for 5G and beyond. IEEE Trans. Antennas Propag. **69**(1), 219–228 (2020)
19. Muhammad, M.H., Muzhair, H., Adnan, A.K., Imran, R., Farooq, A.B.: Dual-band B-shaped antenna array for satellite applications. Int. J. Microw. Wirel. Technol. **13**(8), 851–858 (2021)
20. Yuchen, M., Junhong, W., Zheng, L., Yujian, L., Meie, C., Zhan, Z.: Planar annular leaky-wave antenna array with conical beam. IEEE Trans. Antennas Propag. **68**(7), 5405–5414 (2020)
21. Avishek, D., Durbadal, M., Rajib, K.: An optimal circular antenna array design considering the mutual coupling employing ant lion optimization. Int. J. Microw. Wirel. Technol. **13**(2), 164–172 (2021)
22. Kai, G., Xiaoxiang, D., Li, G., Yanwen, Z., Zaiping, N.: A broadband dual circularly polarized shared-aperture antenna array using characteristic mode analysis for 5G applications. Int. J. RF Microwave Comput. Aided Eng. **31**(3), 234–243 (2021)
23. Pooja, P., Aneesh, R. K., Gopika, R., Chinmoy, S.: Quad antenna array design for microwave energy harvesting. In: International Conference on Wireless and Optical Communications (WOCC) (2021)
24. Daniel, C., Stavros, V., Jeffrey, A.N.: Imageless shape detection using a millimeter-wave dynamic antenna array and noise illumination. IEEE Trans. Microw. Theory Tech. **70**(1), 758–765 (2022)
25. Yusifbayli, N., Guliyev, H., Aliyev, A.: Voltage Control System for Electrical Networks Based on Fuzzy Sets. In: Aliev, R.A., Yusupbekov, N.R., Kacprzyk, J., Pedrycz, W., Sadikoglu, F.M. (eds.) WCIS 2020. AISC, vol. 1323, pp. 55–63. Springer, Cham (2021). https://doi.org/10.1007/978-3-030-68004-6_8
26. Ibrahimov, B.G.: Research and estimation characteristics of terminal equipiment a link multiservice communication networks. Autom. Control. Comput. Sci. **46**(6), 54–59 (2010)

Design and Simulation of the Miniaturized Dual-Band Monopole Antenna for RFID Applications

Kayhan Çelik(✉)

Department of Electrical and Electronics Engineering, Faculty of Technology, Gazi University, Ankara, Turkey
kayhancelik1923@gmail.com, kayhancelik@gazi.edu.tr

Abstract. In this paper, design and simulation of the modified classical patch antenna shaped, miniaturized dual-band monopole antenna is presented for RFID applications. The novel antenna geometry has a dimension of 25 x 30 x 1.6 mm^3 which uses the FR4 dielectric material with the relative permittivity of 4.4 and loss tangent of 0.02. The antenna has a conventional rectangular radiating patch and defected ground structure which has the additional stubs for improving the impedance bandwidth of the antenna at the desired frequencies. The antenna resonates at the 2.4 GHz / 5.8 GHz with the bandwidths of 170 MHz and 500 MHz respectively. In the text, the main performance quality values of the antenna as *S(1,1)*, radiation patterns and gain are investigated and results are given systematically. The presented antenna has a low profile and basic structure which makes it suitable for the RFID systems.

Keywords: Microstrip monopole antenna · RFID · Multiband antenna · Wireless communication

1 Introduction

Radio frequency identification (RFID) is a recent emerging technology that operates with the help of the radio frequency (RF) signals for the aim of transferring the data between the objects [1]. RFID is a wireless system which is made up of basically from the components called antenna, a transceiver and a transponder. The reader (interrogator) is formed by the combination of the antenna and the transceiver which has a network connection for data exchanging and processing and communicates with the tag [2]. Tags or transponders are the tiny devices and they have ability of receiving and using the low power RF waves for storing and transmitting data to the readers. In addition to this, tags can be classified as an active and passive depending on whether they have batteries or not.

When the academic papers are examined that, it can be inferred that RFID has different applications in many various sectors. For instance, RFID-based e-healthcare system and cloud-based RFID healthcare systems [3, 4], manufacturing control [5, 6],

D. J. Hemanth et al. (Eds.): ICAIAME 2022, ECPSCI 7, pp. 176–185, 2023.
https://doi.org/10.1007/978-3-031-31956-3_15

pet animal or livestock tracking [7], cargo tracking [8], vehicle tracking [9], food supply chain [10], bus detection device for the blind people [11] can be cited as prime examples. The electromagnetic spectrum has a lot of frequency bands such as 130 kHz, 13.5 MHz, 900 MHz, 2.4 GHz, 5.8 GHz and 24 GHz which are assigned and reserved for RFID applications [12] and many antenna designs have been carried out depending on these frequency bands for the diverse applications. For instance, the dual-band (2.2–2.6 GHz and 5.3–6.8 GHz) microstrip-fed modified F-shaped antenna is proposed by Parchin et al. [2]. In another work, a bunched dual-band antenna presented by Panda et al. [13]. The dual-band mechanism of the system is accomplished by the 9-shaped folded antenna. The design of the folded strip dual band antenna which has a stub in the ground plane for the using in applications of the WLAN and RFID is also submitted by Panda et al. [14]. The dual band mechanism of the antenna is accomplished by the folded radiating element and stub which is added to the ground plane. A novel dual-band (0.92 GHz and 2.45 GHz) single-layer, diamond-shaped antenna is presented by Sabran et al. [15]. The dual band mechanism of the antenna is utilized by the size and form of the radiating element and it has a high gain and efficiency values. The alternative two setup which has the E-shaped slot for the operation of the antenna at the various RFID frequencies (2.45 GHz, 5.8 GHz and 0.92 GHz, 2.45 GHz) are designed with the high gains values by Abu et al. [16]. The tree-like fractal shaped RFID antenna which one works on the bands of 842 and 922 MHz with the bandwidths of 4.4% and 3.1% respectively is presented by Liu et al. [17]. A simple dual-band (2.45 and 5.8 GHz) monopole antenna for the RFID and WLAN applications is presented by Karli and Ammor [18]. The proposed antenna has a characteristic of the simple structure and compact size with the omnidirectional radiation patterns. A compact dual-band (2.4 GHz and 5.2 GHz) monopole antenna for the portable devices covering WLAN and RFID technologies is presented by Pandey and Mishra which has a consistent omnidirectional radiation patterns and has a gain (>2.5 dBi) [19]. The dual-band microstrip antenna which has a double inverted L-shaped slots combined in the radiating patch is presented by Hamraoui et al. [20] for the 0.915 GHz and 2.45 GHz RFID frequencies bands. A low-cost, inkjet-printed RFID tag antenna consisting of nested-slot configuration and parallel strips for the healthcare application at the 900 MHz is implemented by Sharif et al. [21]. In this paper, design and analyses of the classical patch antenna shaped dual-band antenna is presented for the RFID applications. In the section II, concept of the antenna is presented. In section III, obtained results of the antenna are presented given. Lastly, the paper is finished with the conclusions.

2 Concept and Design

In this part of the paper, design and analyses of the antenna that operates at the 2.45 GHz and 5.8 GHz RFID frequency bands are presented. The general form of the intended antenna with the parameters is given in Fig. 1. The antenna has a classical rectangular patch shape with the defected ground structure which makes it a monopole antenna. The monopole antennas are the common antenna type which are the designed in many shapes [22]. The copper parts as patch and ground plane of the antenna are etched on the two separate surfaces (top and bottom) of the dielectric material which is the FR-4 with the relative permittivity of 4.4 and loss tangent of 0.02. The optimized antenna has an overall size of 25 x 30 x 1.6 mm^3 and all the parameter values that make up the antenna are tabulated in Table 1.

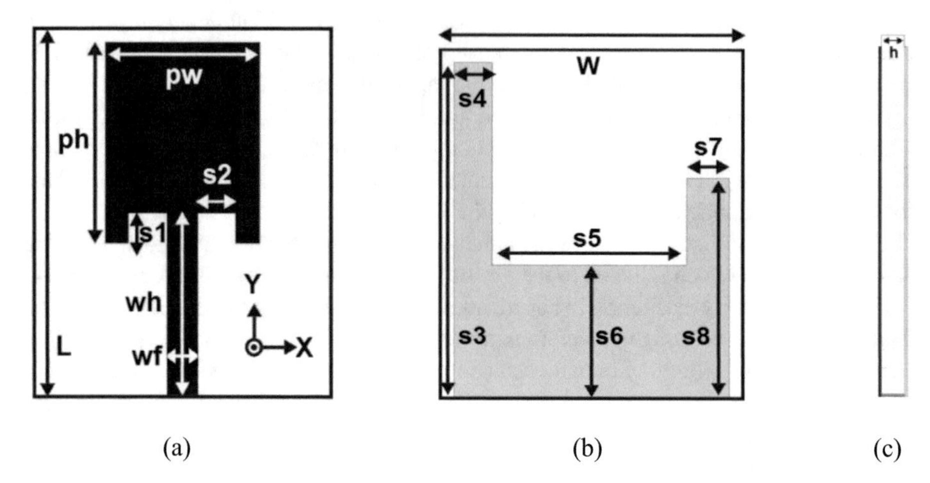

Fig. 1. The general shape of the proposed antenna.

Equation 1 gives the fundamental design formula of the rectangular monopole antenna which is found by equating its area to the equivalent cylindrical monopole antenna [23]. In this formula, *W x L* shows the antenna width and length respectively and *p* is *gap* between the ground plane and the radiating patch.

$$f_L = \frac{7.2}{(l + r + p)} = \frac{7.2}{(L + \frac{W}{2\pi} + p)} GHz \quad (1)$$

The design steps of the recommended antenna can be expressed as follows, respectively. Firstly, as seen Fig. 2(a) the rectangular monopole antenna with the resonance frequency of the 3.4 GHz is implemented. The length of the ground plane is nearly equal to the λ/4 at the 3.4 GHz. At the second step as seen in Fig. 2(a), the first stub is added to left side of the ground plane for the enhancing impedance matching at the lower frequency values. According to Fig. 3, this process decreases the antenna resonance frequency from 3.4 GHz to 2.7 GHz.

At the third step, the another stub (Fig. 2(c)) is added to right side of the ground plane. Consequently, it is clear that the lower cut-off point of the suggested structure decreases further and reaches approximately to the 2.5 GHz. In addition to this, the impedance matching increases at the interested upper cut-off frequency value which means that the antenna radiates at the frequency of nearly 6.5 GHz as seen from Fig. 3.

Table 1. The parameter values of the recommended antenna.

Parameter	Value (mm)
L	30
W	25
Pw	13.3
Ph	16
Wh	15
Wf	2.13
H	1.6
s1	2.5
s2	3.3
s3	29
s4	3.16
s5	16.34
s6	11.5
s7	3.54
s8	19

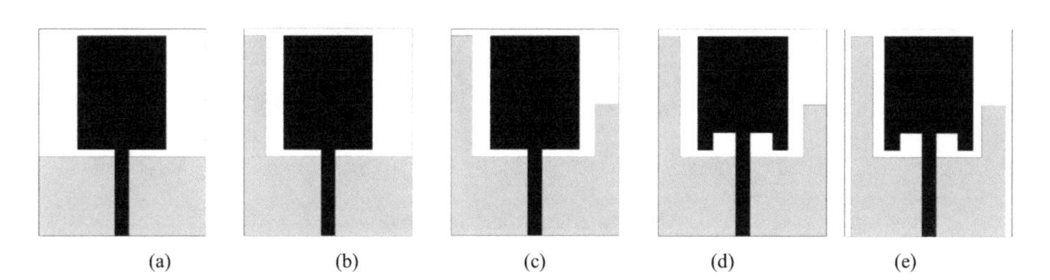

Fig. 2. Design steps of the proposed antenna.

In the fourth step (Fig. 2(d)), two inset slots are added to the junction point of the feeding line and the patch in order to shift the radiation bandwidth that occurs at high frequencies values. It is easily understood from the figure that there is a slight change in the lower cutoff frequency as a result of this process, while the upper working frequency decreases to the desired value, that is, from 6.5 GHz to 5.8 GHz. In the last step (Fig. 2(e)), the width of the ground plane is narrowed to bring the antenna operating frequencies exactly to the desired band ranges. At the end of these processes, the antenna operates from 2.35 GHz to 2.52 GHz and operates from 5.5 GHz to 6 GHz with the bandwidths of 170 MHz and 500 MHz respectively. Figure 4 illustrates the fabricated antenna.

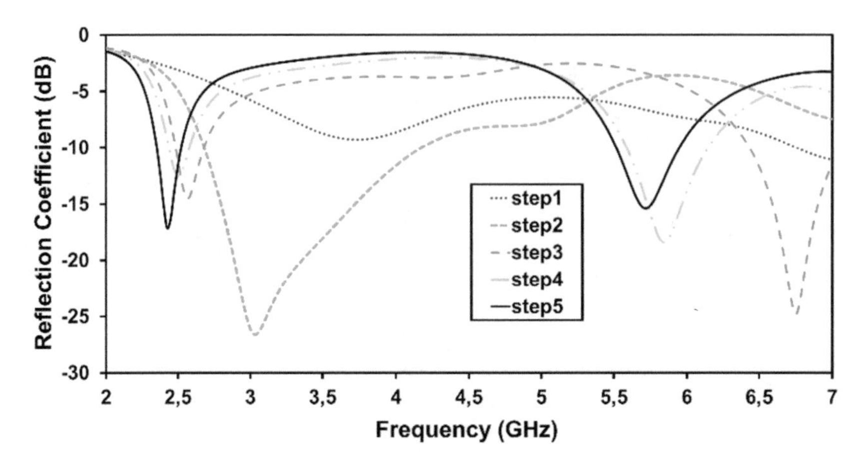

Fig. 3. S(1,1) values of the antenna design steps.

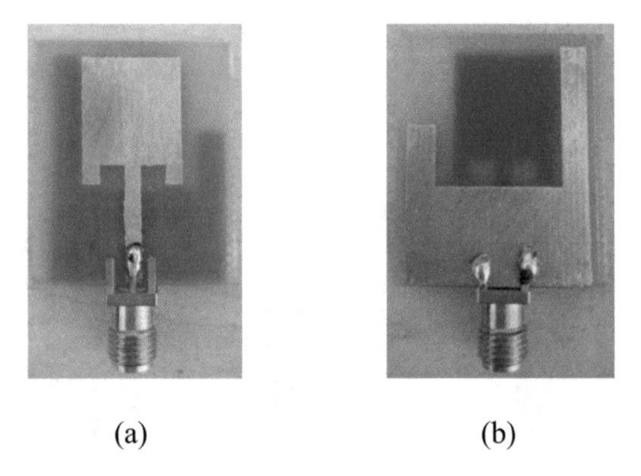

(a) (b)

Fig. 4. Fabricated antenna.

3 Results and Discussion

In this part of the paper, classical antenna performance parameters of the designed antenna are described. The crosscheck of *S(1,1)* obtained from measurement and simulation are given in Fig. 5. It could be easily deduced from the graph that, the values obtained from the measurement and simulation are compatible with each other. However, slight differences between them are thought to be caused by manufacturing and tolerance errors.

The radiation pattern of the proposed antenna at the operating frequencies are illustrated in Fig. 6. According to the figure, at the 2.4 GHz, the suggested antenna has a quasi-non-directional radiation pattern at the Y-Z plane (phi = 90) and it can be said that it has a quasi- directional pattern at the X-Z plane (phi = 0). In addition to this, at the X-Y plane (theta = 90), it has a quasi-bidirectional radiation pattern. At the 5.8 GHz band, it is easily understood that the radiation direction does not change much at the

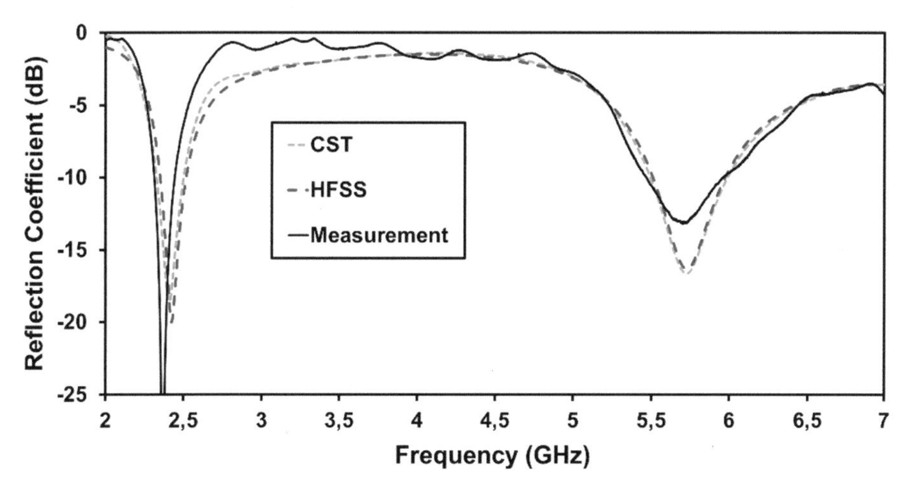

Fig. 5. S(1,1) values of the antenna.

phi = 0, while the radiation pattern changes in other planes and 3 main lobes occur especially for the value of theta 90.

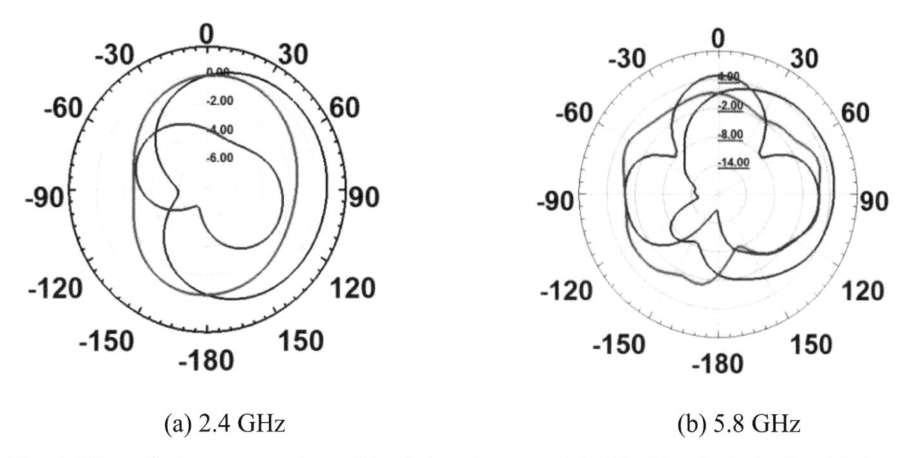

Fig. 6. The radiation pattern values of the designed antenna (phi (black) = 0, phi (red) = 90, theta (blue) = 90)

The change graph of the gain according to the frequency is given in Fig. 7. According to the figure the proposed antenna at the operating frequencies have gain values of 2.25 dB and 5.35 dB respectively. The effects of antenna parameter values on the reflection coefficient are also examined and presented below. For example, the variation of the reflection coefficient depending on the variation of the patch width is given in Fig. 8. According to this graph, increasing the patch width from 8 mm to 16 mm decreases the lower operating frequency from 2.66 GHz to 2.15 GHz. It also makes the similar

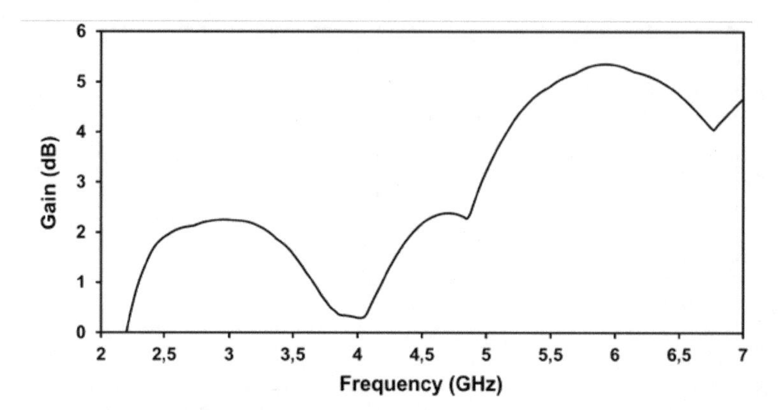

Fig. 7. The changing of gain according to the frequency.

variation on the upper frequency band such as decreasing it from 6.5 GHz to 5 GHz. So the patch width is taken as 13.3 mm.

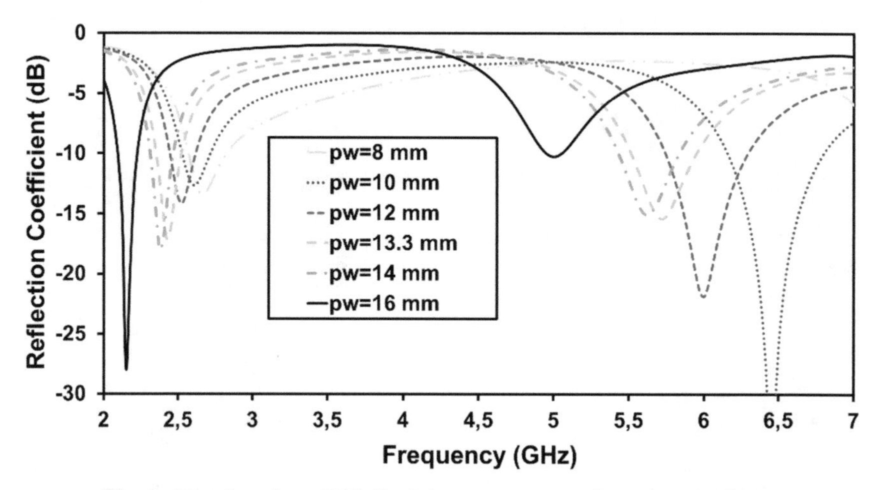

Fig. 8. The changing of S(1,1) of the antenna according to patch width.

Table. Keyword Abstract The variation of the reflection coefficient depending on the variation of the patch length is given in Fig. 9. According to this graph, increasing the patch length from 8 mm to 16.5 mm does not change the lower operating frequency, however, only increases the level of impedance matching at this frequency band. At the upper operating frequency, increasing the patch length from 8 mm to 14 mm improves the overall impedance matching, while an additional increase from 14 mm to 16.5 mm reduces the operating frequency from 6.1 to 5.8 GHz. The optimum patch height is taken as 16 mm.

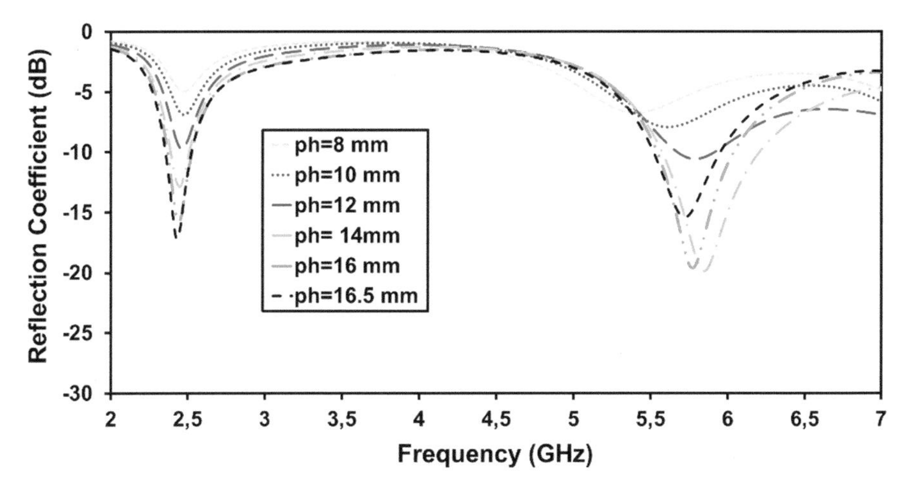

Fig. 9. The changing of S(1,1) of the antenna according to patch height.

The variation of the reflection coefficient depending on the variation of the ground length is given in Fig. 10. According to this graph, increasing the ground height from 6 mm to 11.5 mm increases the lower operating frequency from nearly 2 GHz to 2.4 GHz. At the upper operating frequency, increasing the ground length from 6 mm to 9 mm increases the operating frequency towards to 7 GHz and the changing it from 9 mm to 11.5 mm increases the impedance matching in the operating frequency range.

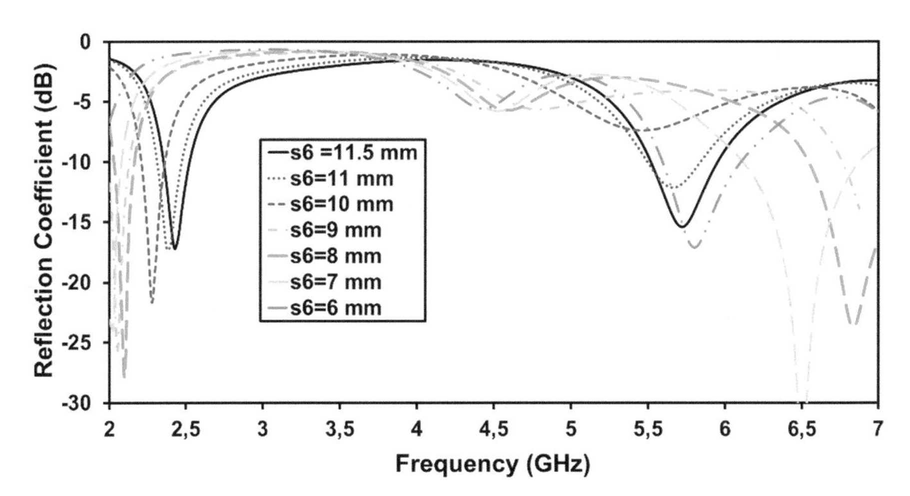

Fig. 10. The changing of S(1,1) of the antenna according to ground length.

The comparison of the proposed antenna with the similar ones in the literature is given in the Table 2. According to the table, the operation frequency bands of the antennas are almost same. When the antennas are compared with each one, the proposed antenna has the smallest size. In terms of the gain, the antenna has the greatest gain value at the higher frequency band, in addition to this, at the lower band it is also has a significant gain value.

Table 2. The comparison chart of the proposed antenna.

Ref.	Dual Band	Resonance Frequency (GHz)	Bandwidth (GHz) S11 < -10dB	Size W x L x h (mm^3)	Maximum gain (dBi)
[2]	Yes	2.4 5.8	2.2–2.6 5.3–6.8	38 x 45 x 1.6	2 4
[13]	Yes	2.46 5.22	2.31 - 2.70 4.03- 6.19	30 x 38 x 1.6	> 2.5
[14]	Yes	2.4 5.8	2.05 GHz - 2.86 5.55 GHz-6.14	35 x 50 x 1.6	3.7 3.57
[19]	Yes	2.4 5.2	0.6 2.52	38 x 30 x 1.6	> 2.5
[20]	Yes	0.915 2.4	0.900–0.928 2.4–2.5	48 x 47 x 1.6	N. A
Proposed antenna	Yes	2.4 5.8	2.35- 2.52 5.5 – 6	25 x 30 x 1.6	2.25 5.35

4 Conclusion

In this communication, design and analyses of the classical patch antenna shaped antenna with the defected ground structure which one has a double stub for the improving impedance matching at the desired frequency bands was presented. The proposed antenna has an easy structure for implementing and it also has a compact size. In addition to these features, it has a considerable gain values at the operating frequency bands. As a result of these features, it can be inferred that the implemented antenna could be suitable candidate for the dual band RFID applications.

Conflicts of Interest. No conflict of interest was declared by the authors.

References

1. Yüksel, M.E., Fidan, H.: Performance analysis of tree-based tag anti-collision protocols for RFID systems. Electrica **19**(2), 182–192 (2019)
2. Ojaroudi Parchin, N., Jahanbakhsh Basherlou, H., Abd-Alhameed, R.A., Noras, J.M.: Dual-band monopole antenna for RFID applications. Future Internet **11**(2), 31 (2019)
3. Chen, M., Gonzalez, S., Leung, V., Zhang, Q., Li, M.: A 2G-RFID-based e-healthcare system. IEEE Wirel. Commun. **17**(1), 37–43 (2010)

4. Fan, K., Zhu, S., Zhang, K., Li, H., Yang, Y.: A lightweight authentication scheme for cloud-based RFID healthcare systems. IEEE Network **33**(2), 44–49 (2019)
5. Chen, R.S., Tu, M.A.: Development of an agent-based system for manufacturing control and coordination with ontology and RFID technology. Expert Syst. Appl. **36**(4), 7581–7593 (2009)
6. Huang, G.Q., Zhang, Y.F., Chen, X., Newman, S.T.: RFID-enabled real-time wireless manufacturing for adaptive assembly planning and control. J. Intell. Manuf. **19**(6), 701–713 (2008)
7. Floyd, R.E.: RFID in animal-tracking applications. IEEE Potentials **34**(5), 32–33 (2015)
8. He, W., Tan, E.L., Lee, E.W., Li, T.Y.: A solution for integrated track and trace in supply chain based on RFID & GPS. In: 2009 IEEE Conference on Emerging Technologies & Factory Automation, pp. 1–6. IEEE (2009)
9. Bajaj, D., Gupta, N.: GPS based automatic vehicle tracking using RFID. Int. J. Eng. Innova. Technol. (IJEIT) **1**(1), 31–35 (2012)
10. Hong, I.H., et al.: A RFID application in the food supply chain: A case study of convenience stores in Taiwan. J. Food Eng. **106**(2), 119–126 (2011)
11. Noor, M.Z.H., Ismail, I., Saaid, M.F.: Bus detection device for the blind using RFID application. In: 2009 5th International Colloquium on Signal Processing & Its Applications, pp. 247–249. IEEE (2009)
12. Siakavara, K., Goudos, S., Theopoulos, A., Sahalos, J.: Passive UHF RFID Tags with Specific Printed Antennas for Dielectric and Metallic Objects Applications. Radioengineering 26(3) (2017)
13. Panda, J.R., Saladi, A.S.R., Kshetrimayum, R.S.: A compact printed monopole antenna for dual-band RFID and WLAN applications. Radioengineering **20**(2), 464–467 (2011)
14. Panda, J.R., Kshetrimayum, R.S.: A printed 2.4 GHz/5.8 GHz dual-band monopole antenna with a protruding stub in the ground plane for WLAN and RFID applications. Progress In Electromagnetics Research **117**, 425–434 (2011)
15. Sabran, M.I., Rahim, S.K.A., Rahman, A.Y.A., Rahman, T.A., Nor, M.Z.M.: A dual-band diamond-shaped antenna for RFID application. IEEE Antennas Wirel. Propag. Lett. **10**, 979–982 (2011)
16. Abu, M., Hussin, E.E., Amin, M.A., Raus, T.M.: Dual-Band e-Shaped Antenna for RFID Reader. J. Telecommu. Electro. Comp. Eng. (JTEC) **5**(2), 27–31 (2013)
17. Liu, G., Xu, L., Wu, Z.: Dual-band microstrip RFID antenna with tree-like fractal structure. IEEE Antennas Wirel. Propag. Lett. **12**, 976–978 (2013)
18. Karli, R., Ammor, H.: (2015) Rectangular patch antenna for dual-band RFID and WLAN applications. Wireless Pers. Commun. **83**(2), 995–1007 (2015)
19. Pandey, A., Mishra, R.: Compact dual band monopole antenna for RFID and WLAN applications. Materials Today: Proceedings **5**(1), 403–407 (2018)
20. El Hamraoui, A., El Hassan Abdelmounim, J.Z., Errkik, A., Bennis, H., Latrach, M.: A Dual-band Microstrip Slotted Antenna for UHF and Microwave RFID Readers. TELKOMNIKA (Telecommunication, Computing, Electronics and Control) **16**(1), 94–101 (2018)
21. Sharif, A., Ouyang, J., Yan, Y., Raza, A., Imran, M.A., Abbasi, Q.H.: Low-cost inkjet-printed RFID tag antenna design for remote healthcare applications. IEEE Journal of Electromagnetics, RF and Microwaves in Medicine and Biology **3**(4), 261–268 (2019)
22. Paga, P.G., Nagaraj, H.C., Shashidhara, K.S., Dakulagi, V., Yeap, K.H.: Design and Analysis of Printed Monopole Antenna with and without CSRR in the Ground Plane for GSM 900 and Wi-Fi. Electrica **22**(1), 92–100 (2022)
23. Ray, K.P.: Design aspects of printed monopole antennas for ultra-wide band applications. Int. J. Antennas and Propagation (2008)

Exploring the Driven Service Quality Dimensions for Higher Education Based on MCDM Analysis

Aleyna Sahin(✉), Mirac Murat, Gul Imamoglu, Kadir Buyukozkan, and Ertugrul Ayyildiz

Department of Industrial Engineering, Karadeniz Technical University, Trabzon 61080, Turkey
{aleynasahin,miracmurat,gulimamoglu,kbuyukozkan,
ertugrulayyildiz}@ktu.edu.tr

Abstract. Increasing competition in higher education forces universities to take steps to improve the quality of service. Pre-determining crucial factors in the provision of educational services is significant to ensure student satisfaction, and thus to strengthen the bond of students with the university. The perception of students and employers about the quality of the education service offered is a key factor in positioning students, who are future employees, in the labor market. The agility of university services plays a key role in the continuation of education and training activities, especially in challenging situations such as pandemics. At this point, different perceptions about the quality of the service emerge for different universities. In this study, the problem of evaluating the service quality of universities is discussed. The aim is to find out the factors that involve all stakeholders in the evaluation process and prioritize the factors based on stakeholders' belief in overall service quality. In this context, the traditional dimensions of service quality models were examined, and the model was extended to address comprehensive service quality dimensions for higher educations that are especially to be used in extraordinary processes such as pandemics. Then, the weights of the determined service quality evaluation dimensions were determined by a multi-criteria decision making (MCDM) method, considering the expert opinions. At this point, Pythagorean fuzzy sets were preferred to effectively reflect experts uncertainty that the consulted may experience in the decision-making process. With this study, the first service quality model is introduced in the literature for higher education institutions, supported by a fuzzy MCDM method.

Keywords: MCDM · Service Quality · Higher Education · Fuzzy Logic

1 Introduction

Successfully operated higher education and its internationally recognized performance are vital pillars of stable economic growth and, thus, societal improvement. Societies equipped with a conveniently qualified workforce drive the development, advancement, and innovation of the world [1]. Societies want to be the determinants of this global

D. J. Hemanth et al. (Eds.): ICAIAME 2022, ECPSCI 7, pp. 186–196, 2023.
https://doi.org/10.1007/978-3-031-31956-3_16

progress and to make the most of its blessings. Therefore, today, the importance of nationwide and international competitiveness of higher education institutions (universities) has gradually become more appreciated in global improvement [2].

The increasing awareness of competition in higher education forces universities to pay more attention to the quality of educational services [3]. Efforts to understand factors related to satisfaction levels of stakeholders, especially of students, regarding educational services are further supported because student satisfaction is highly effective in university image and sustainable education. Therefore, when considering the provision of educational services, universities can be the leading party in determining the current satisfaction level and improving it only with their ability to predetermine the factors considered or will be considered important.

The importance of the quality of educational services is better understood in compelling situations [4]. Measures taken and new practices implemented in order to maintain educational services in compelling situations such as the pandemic, where the 'normal' is no longer valid, greatly affect the quality of educational services [5, 6]. In such periods, the quality perception in the 'normal' can rapidly deteriorate and evolve to levels requiring much more effort to restore. For this reason, even in the 'new normal', the agility of educational services to be maintained healthy way has a key importance to satisfying a realistic quality perception.

Agile service systems, which are required to be affected by changes at a minimum level and to maintain quality at least at its current level, must have correctly prioritized factors considering the expectations of students [7]. Thus, the university can shift its focus to the factors that need to be addressed as a priority in order to maintain and improve the satisfaction level that is essentially the overall students' expectation.

In line with increasing competition, there is an accelerating research trend to identify factors affecting student satisfaction and the dimensions of service quality for universities, especially in the last two decades.

To identify the factors that influence undergraduate student happiness in higher education, DeShields et al. [8] administered a questionnaire to 160 students at a state university in Pennsylvania. Likewise, to determine the main factors that affect student satisfaction in higher education, Zineldin et al. [9] conducted a questionnaire with 39 characteristics to 1641 students from Turkey and obtained seven factors from factor analysis. Douglas et al. [10] conducted a survey on 350 undergraduate students in the UK to determine critical quality factors in higher education. The authors utilized the qualitative critical incident technique to analyze surveys and concluded that access, attentiveness, availability, and communication have an influence on loyalty. Goyal et al. [11] applied an Analytic Hierarchy Process (AHP) to prioritize the factors that make up the perception of quality in higher education institutions from the viewpoint of students. The results reveal that employment opportunities, exchange programs, and practical exposure of the courses were determined as vital factors. Abdullah [12] proposed a method to analyze the quality of higher education institutions' services, namely HEdPERF, consisting of six dimensions and 41 attributes; on the other hand, Teeroovengadum et al. [13] aimed to determine the relevant factors to propose a hierarchical evaluation model for the quality of higher education services. Bozbay et al. [14] focused on the evaluation of perception of service quality in Turkish universities. For this purpose, a modified SERVQUAL questionnaire was organized and carried out in 168 students. Sohail and Hasan [15] aimed to understand how service quality affects student satisfaction. To do so, a SERVPERF questionnaire was applied to 279 students from universities in Saudi Arabia. To assess whether independent variables significantly affect dependent variables, the authors utilized structural equation modeling. The findings demonstrated a significant relationship between four factors and student satisfaction: tangibility, reliability, responsiveness, and assurance.

This paper aims to determine the factors that influence students' perceptions of the value of educational services offered by universities and the variables that should be given priority in order to preserve this perception. In addition to the five classical dimensions of the SERVQUAL model, which is frequently used in the evaluation of service systems, the main factors that can measure students' expectations even in situations shaped by developing technology, changing needs and unexpected circumstances have been determined by literature research. Then, these factors have been weighted with SWARA (Stepwise Weight Assessment Ratio Analysis) by consulting expert opinions to properly reflect the perceived quality of today's higher educational services from the student's perspective. While conducting SWARA, the experts were asked to make an evaluation using linguistic terms corresponding to Pythagorean fuzzy numbers to consider the vagueness that may be experienced during the evaluation in the decision-making process.

The remainder of the study is organized as follows. The next section introduces the method of SWARA and its fuzzy extension Pythagorean Fuzzy SWARA. The relevant literature on SWARA and the use of Pythagorean fuzzy sets is also given in this section. In the third section, Pythagorean fuzzy SWARA is practically applied to determine the importance of the factors regarding educational services. The findings are discussed in the last section.

2 Pythagorean Fuzzy SWARA

SWARA, developed by Keršulienė et al. [16], is one of the expert-oriented multi-criteria decision-making (MCDM) methods used to determine the weights of criteria. In this method, the evaluations are subjectively determined on the basis of expert opinions. The selection of experts who will make the evaluation is essential, as experts' opinions are a key element for this method. In addition, each expert may have different levels of knowledge, experience, and expertise in the subject to be evaluated. For this reason, it would not be a realistic approach to give equal weight to all experts in aggregating expert opinions, unlike most studies in the literature. Giving weight to each expert according to their level of expertise will make the evaluation results more meaningful and valid.

To address uncertainty, SWARA is extended using various fuzzy sets. Raad et al. use SWARA with Multiple Objective Optimization based on the Ratio Analysis plus Full Multiplicative (MULTIMOORA) with triangular fuzzy sets for dry port site selection [17]. Dahooie et al. utilize SWARA under an intuitionistic fuzzy environment for selection advertising media for online games [18]. Yazdi et al. propose SWARA for determining weights about banking performance indicators using hesitant fuzzy sets

[19]. In order to examine crowdfunding platforms for microgrid project investors, Wu et al. introduce a multi-SWARA (M-SWARA) using q-rung orthopair fuzzy numbers [20].

In the literature, few studies extended SWARA with Pythagorean fuzzy sets. Rani et al. propose Pythagorean fuzzy SWARA VIseKriterijumska Optimizcija I Kaompromisno Resenje (VIKOR) to handle the solar panel selection problem [21]. He et al. employ the SWARA-MULTIMOORA combination under an interval-valued Pythagorean fuzzy environment to model the growing sustainable community-based tourism. [22]. Cui et al. propose using the Pythagorean fuzzy SWARA with Combined Compromise Solution (CoCoSo) technique to pinpoint the major obstacles preventing the manufacturing sector from adopting the IoT (Internet of Things) [23]. Ramya et al. use the hesitant Pythagorean fuzzy SWARA-WASPAS (Weighted Aggregated Sum Product Assessment) methodology to solve a thermal energy storage technique selection problem [24]. Alipour et al. present an extended Entropy-SWARA-Complex Proportional Assessment (COPRAS) approach under the Pythagorean fuzzy environment to solve the hydrogen components and the fuel supplier selection problem [25]. Similarly, Alrasheedi et al. use an extended Pythagorean fuzzy Entropy-SWARA-WASPAS method to evaluate sustainable suppliers in manufacturing companies [26]. Lastly, Kamali Saraji et al. propose a Pythagorean fuzzy-SWARA-Technique of Order Preference Similarity to the Ideal Solution (TOPSIS) hybrid approach to evaluate the European Union (EU) progress towards sustainable energy development [27].

In this study, we employ Pythagorean fuzzy SWARA (PF-SWARA) to determine the importance weight of the service quality dimensions for universities. Yager [28] developed Pythagorean fuzzy sets (PFSs) based on intuitionistic fuzzy sets (IFSs) to deal with fuzziness in decision-making problems. PFSs give experts a wider range than IFSs to express their opinions on uncertainty in linguistic terms. The steps of PF-SWARA, including expert evaluation, are as follows:

Step 1: Evaluate experts' skills according to the linguistic terms given in Table 1.

Table 1. Linguistic terms to evaluate experts

	Linguistic Term	μ	v
AS	Absolutely Skilled	0.95	0.1
VS	Very Skilled	0.8	0.25
MS	More Skilled	0.7	0.35
S	Skilled	0.5	0.55
LS	Less Skilled	0.45	0.6
VLS	Very Less Skilled	0.15	0.9

Then, calculate the weight of each expert using Eq. 2. Let d be the number of experts, and $E_k = (\mu_k, v_k)$ be the Pythagorean fuzzy number for the evaluation of expert k.

Let X is a fixed set and $\mu_{\tilde{p}}(x) : X \mapsto [0, 1]$ represents the degree of membership of the element $x \in X$ to $\tilde{P}$. $v_{\tilde{p}}(x) : X \mapsto [0, 1]$ [0, 1] represents the degree of non-membership of the element $x \in X$ to $\tilde{P}$. $\pi_{\tilde{p}}(x)$ is the hesitancy degree and it can be calculated by Eq. 1 [29].

$$\pi_{\tilde{p}}(x) = \sqrt{1 - \mu_{\tilde{p}}(x)^2 - v_{\tilde{p}}(x)^2} \tag{1}$$

$$w_k = \frac{\left(\mu_k^2 + \pi_k^2\left(\frac{\mu_k^2}{\mu_k^2+v_k^2}\right)\right)}{\sum_{k=1}^{d}\left(\mu_k^2 + \pi_k^2\left(\frac{\mu_k^2}{\mu_k^2+v_k^2}\right)\right)} \tag{2}$$

Step 2: Create a decision matrix by gathering expert opinion on each criterion using the language terms listed in Table 2. The evaluation of criterion i by expert k is shown by $A_{ik} = (\mu_{ik}, v_{ik})$.

Table 2. Linguistic terms for evaluating criteria

	Linguistic Term	μ	v
EHI	Extremely High Important	0.85	0.25
VHI	Very High Important	0.8	0.3
HI	High Important	0.7	0.35
MHI	Medium High Important	0.6	0.5
MI	Medium Important	0.45	0.65
MLI	Medium Low Important	0.35	0.75
LI	Low Important	0.3	0.85
VLI	Very Low Important	0.25	0.9
ELI	Extremely Low Important	0.15	0.95

Step 3: To create a Pythagorean fuzzy decision matrix, aggregate expert opinions on the criteria.

$$z_i = Y(\mu_i, v_i) = Y\left(\sqrt{1 - \prod_{k=1}^{d}\left(1 - \mu_{ik}^2\right)^{w_k}}, \prod_{k=1}^{d}\left(v_{ik}\right)^{w_k}\right) \tag{3}$$

where $Y(\mu_i, v_i)$ represents the aggregated weight for criterion i.

Step 4: For each criterion, determine the normalized crisp score values $S^*(i)$.

$$S^*(i) = \frac{(S(i) + 1)}{2} \tag{4}$$

$$S(i) = (\mu_i)^2 - (v_i)^2 \quad (5)$$

where S(i) is a score function for criterion i.

Step 5: Sort the criteria based on the crisp score values in descending order S*(i).

Step 6: By comparing normalized score values, determine the relative importance of each criterion's score value to determine the second preferred criterion.

Step 7: Calculate the comparative coefficient (k_i) for each criterion

$$k_i = \begin{cases} 1, & i = 1 \\ s_i + 1, & i > 1 \end{cases} \quad (6)$$

Step 8: Estimate recalculated weights (q_i)

$$q_i = \begin{cases} 1, & i = 1 \\ \frac{q_{(i-1)}}{k_i}, & i > 1 \end{cases} \quad (7)$$

Step 9: Determine the criteria weights according to Eq. 8.

$$w_i = \frac{q_i}{\sum_{i=1}^{n} q_i} \quad (8)$$

Let n be the number of criteria.

3 Weighting the Service Dimensions

Three anonymous experts' opinions are used to calculate the weights of the criteria. First, experts are evaluated using the linguistic terms shown in Table 2 with respect to their expertise in related fields. The first, second, and lastly third experts are evaluated as MS, S, and VS, respectively. Thus, the weights of Expert-1, Expert-2, and Expert-3 are determined as 0.370, 0.209, and 0.421, respectively, based on Eq. 2.

After determining the expert weights, each service dimension is evaluated by each expert, and thus the decision matrix is constructed. Experts use the linguistic terms given in Table 2 to evaluate the service dimensions. Table 3 presents the decision matrix.

Then the opinions are converted to Pythagorean fuzzy numbers. Equation 3 is used to aggregate expert opinions. Table 4 presents the aggregated opinions.

The crisp value of each dimension is determined by Eq. 4. Then, the dimensions are ordered in descending order. Steps 6, 7, and 8 are performed based on the score values. The crisp score values s_i, k_i, and p_i values are given in Table 5.

Lastly, Eq. 8 is used to calculate the final weight of each dimension as presented in Fig. 1.

Table 3. Decision matrix

	Expert-1	Expert-2	Expert-3
Tangibles	EHI	HI	HI
Reliability	EHI	EHI	EHI
Responsiveness	HI	VHI	HI
Reputation	VHI	MI	MHI
Accessibility	HI	HI	MHI
Resilience	MHI	VHI	HI
Digital technology	EHI	VHI	VHI
Safety	VHI	VHI	HI
Social life	MHI	HI	HI
Environmental impact	LI	MI	MHI

Table 4. Aggregated decision matrix

	μ	v	π
Tangibles	0.770	0.309	0.558
Reliability	0.850	0.250	0.464
Responsiveness	0.725	0.339	0.599
Reputation	0.677	0.437	0.592
Accessibility	0.662	0.407	0.629
Resilience	0.696	0.387	0.605
Digital technology	0.820	0.280	0.498
Safety	0.764	0.320	0.561
Social life	0.667	0.399	0.629
Environmental impact	0.487	0.643	0.592

Reliability is determined as the most important dimension among the ten dimensions. That is, students care about universities' full compliance with their commitment to service delivery even in challenging situations. Digital technology is another significant dimension as might be expected in line with its simplifying effect. Considering the current age of technology, university decision makers should follow technological developments and take steps in this direction.

Table 5. Application results of PF-SWARA

	Crisp	s_i^*	k_i	p_i
Reliability	0.830	-	1.000	1.000
Digital Technology	0.797	0.033	1.033	0.968
Tangibles	0.749	0.049	1.049	0.923
Safety	0.740	0.008	1.008	0.916
Responsiveness	0.705	0.035	1.035	0.885
Resilience	0.667	0.038	1.038	0.852
Social life	0.643	0.024	1.024	0.832
Accessibility	0.637	0.006	1.006	0.827
Reputation	0.634	0.003	1.003	0.824
Environmental impact	0.412	0.222	1.222	0.675

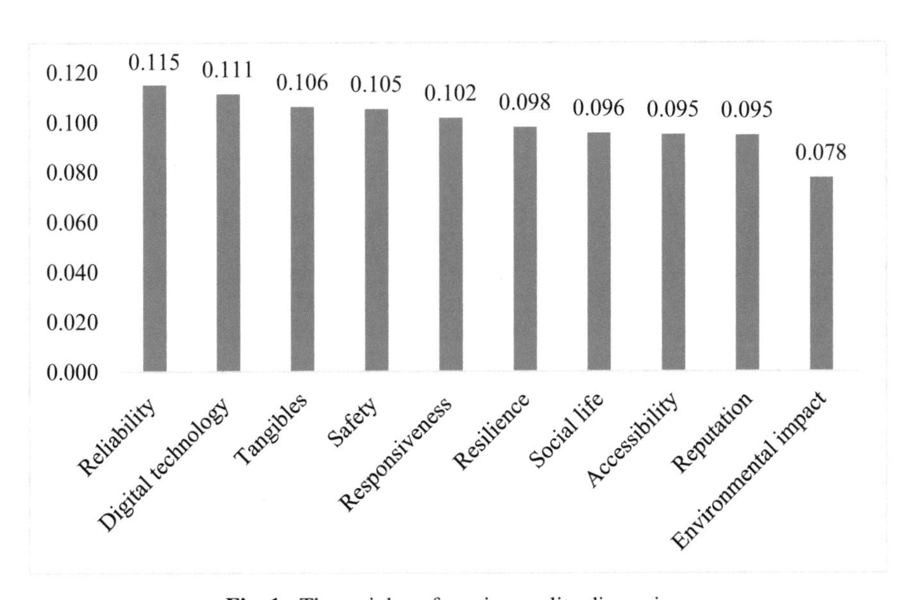

Fig. 1. The weights of service quality dimensions

4 Conclusion

In this study, the importance levels of service quality dimensions related to higher education are aimed to determine. To do this, the current literature on service quality is reviewed to determine the dimensions that affect overall service quality from the student's perspective. The service quality dimensions are collected under ten titles. In other words, ten service dimensions, reliability, digital technology, tangibles, safety, responsiveness, resilience, social life, accessibility, reputation, and finally environmental impact are specified, which should be considered to evaluate the quality of service of higher education institutions. Then, opinions are taken from multiple experts to evaluate the dimensions. Three different experts evaluate the dimensions in linguistic terms. Pythagorean fuzzy numbers are utilized to convert linguistic terms into mathematical expressions. One of the most effective MCDM methodologies, SWARA, is used to determine the weights of each dimension in a Pythagorean fuzzy environment. Reliability is determined as the most significant service quality dimension for higher education institutions according to opinions taken from three different experts. Therefore, it may be claimed that institutional decision-makers should give reliability issues more consideration in order to improve service quality. It is now crucial to deliver the promised services to students without delay in order to increase the level of service.

As future suggestions, the number of service dimensions can be increased, or the dimensions can be handled with sub-dimensions sets to make more detailed analysis. Different MCDM methods can be used, and the results of this study can be compared. In addition, different higher education institutions can be compared using the proposed Pythagorean fuzzy SWARA methodology.

Acknowledgement. This work is supported by Karadeniz Technical University Scientific Research Projects Coordination Unit. Project Number: FAY-2022–10123.

References

1. Labas, I., Darabos, E., Nagy, T.O.: Competitiveness - higher education. Stud. Univ. Arad. "Vasile Goldis" Arad – Econ. Ser. **26**(1), 11–25 (2016)
2. Bileviciute, E., Draksas, R., Nevera, A., Vainiūte, M.: Competitiveness in higher education: the case of university management. J. Compet. **11**(4), 5–21 (2019)
3. de Jager, J., Gbadamosi, G.: Predicting students' satisfaction through service quality inhigher education. Int. J. Manag. Educ. **11**(3), 107–118 (2013)
4. Beltman, S., Mansfield, C.F.: Resilience in education: an introduction. In: Wosnitza, M., Peixoto, F., Beltman, S., Mansfield, C.F. (eds.) Resilience in Education, pp. 3–9. Springer, Cham (2018)
5. Yang, R.: China's higher education during the covid-19 pandemic: some preliminary observations. High. Educ. Res. Dev. **39**(7), 1317–1321 (2020)
6. Jena, P.K.: Impact of pandemic covid-19 on education in India. Int. J. Curr. Res. **12**(07), 12582–12586 (2020)
7. Davey, B., Parker, K.R.: Technology in education: an agile systems approach. In: Proceedings of the 2010 InSITE Conference, pp. 297–306 (2010)
8. DeShields, O.W., Kara, A., Kaynak, E.: Determinants of business student satisfaction and retention in higher education: applying Herzberg's two-factor theory. Int. J. Educ. Manag. **19**(2), 128–139 (2005)
9. Zineldin, M., Akdag, H.C., Vasicheva, V.: Assessing quality in higher education: new criteria for evaluating students' satisfaction. Qual. High. Educ. **17**(2), 231–243 (2011)
10. Douglas, J.A., Douglas, A., McClelland, R.J., Davies, J.: Understanding student satisfaction and dissatisfaction: an interpretive study in the UK higher education context. Stud. High. Educ. **40**(2), 329–349 (2015)
11. Goyal, A., Gupta, S., Chauhan, A.K.: Prioritizing the factors determining the quality in higher educational institutions—an application of fuzzy analytic hierarchy process. J. Public Aff. **22**(4), e2647 (2021)

12. Abdullah, F.: The development of HEdPERF: a new measuring instrument of service quality for the higher education sector. Int. J. Consum. Stud. **30**(6), 569–581 (2006)
13. Teeroovengadum, V., Kamalanabhan, T.J., Seebaluck, A.K.: Measuring service quality in higher education: development of a hierarchical model (HESQUAL). Qual. Assur. Educ. **24**(2), 244–258 (2016)
14. Bozbay, Z., Baghirov, F., Zhang, Y., Rasli, A., Karakasoglu, M.: International students' service quality evaluations towards Turkish universities. Qual. Assur. Educ. **28**(3), 151–164 (2020)
15. Sohail, M.S., Hasan, M.: Students' perceptions of service quality in Saudi universities: the SERVPERF model. Learn. Teach. High. Educ. Gulf Perspect. **17**(1), 54–66 (2021)
16. Keršulienė, V., Zavadskas, E.K., Turskis, Z.: Selection of rational dispute resolution method by applying new step-wise weight assessment ratio analysis (SWARA). J. Bus. Econ. Manag. **11**(2), 243–258 (2010)
17. Raad, N.G., Rajendran, S., Salimi, S.: A novel three-stage fuzzy GIS-MCDA approach to the dry port site selection problem: a case study of Shahid Rajaei Port in Iran. Comput. Ind. Eng. **168**, 108112 (2022)
18. Dahooie, J.H., Estiri, M., Janmohammadi, M., Zavadskas, E.K., Turskis, Z.: A novel advertising media selection framework for online games in an intuitionistic fuzzy environment. Oeconomia Copernicana **13**(1), 109–150 (2022)
19. Karbassi Yazdi, A., Spulbar, C., Hanne, T., Birau, R.: Ranking performance indicators related to banking by using hybrid multicriteria methods in an uncertain environment: a case study for Iran under covid-19 conditions. Syst. Sci. Control Eng. **10**(1), 166–180 (2022)
20. Wu, X., Dinçer, H., Yüksel, S.: Analysis of crowdfunding platforms for microgrid project investors via a q-rung orthopair fuzzy hybrid decision-making approach. Financ. Innov. **8**(1) (2022)
21. Rani, P., Mishra, A.R., Mardani, A., Cavallaro, F., Štreimikiene, D., Khan, S.A.R.: Pythagorean fuzzy SWARA-VIKOR framework for performance evaluation of solar panel selection. Sustain. **12**(10) (2020)
22. He, J., Huang, Z., Mishra, A.R., Alrasheedi, M.: Developing a new framework for conceptualizing the emerging sustainable community-based tourism using an extended interval-valued Pythagorean fuzzy SWARA-MULTIMOORA. Technol. Forecast. Soc. Change **171**, 120955 (2021)
23. Cui, Y., Liu, W., Rani, P., Alrasheedi, M.: Internet of Things (IoT) adoption barriers for the circular economy using Pythagorean fuzzy SWARA-CoCoSo decision-making approach in the manufacturing sector. Technol. Forecast. Soc. Change **171**, 120951 (2021)
24. Ramya, L., Narayanamoorthy, S., Kalaiselvan, S., Kureethara, J.V., Annapoorani, V., Kang, D.: A congruent approach to normal wiggly interval-valued hesitant Pythagorean fuzzy set for thermal energy storage technique selection applications. Int. J. Fuzzy Syst. **23**(6), 1581–1599 (2021)
25. Alipour, M., Hafezi, R., Rani, P., Hafezi, M., Mardani, A.: A new Pythagorean fuzzy-based decision-making method through entropy measure for fuel cell and hydrogen components supplier selection. Energy **234**, 121208 (2021)
26. Alrasheedi, M., Mardani, A., Mishra, A.R., Rani, P., Loganathan, N.: An extended framework to evaluate sustainable suppliers in manufacturing companies using a new Pythagorean fuzzy entropy-SWARA-WASPAS decision-making approach. J. Enterp. Inf. Manag. **35**(2), 333–357 (2022)
27. Kamali Saraji, M., Streimikiene, D., Ciegis, R.: A novel Pythagorean fuzzy-SWARA-TOPSIS framework for evaluating the EU progress towards sustainable energy development. Environ. Monit. Assess. **194**(1), 1–19 (2021)

28. Yager, R.R.: Pythagorean fuzzy subsets. In: Proceedings of the 2013 Joint IFSA World Congress and NAFIPS Annual Meeting, IFSA/NAFIPS 2013, pp. 57–61 (2013)
29. Ayyildiz, E., Taskin Gumus, A.: Pythagorean fuzzy AHP based risk assessment methodology for hazardous material transportation: an application in Istanbul. Environ. Sci. Pollut. Res. **28**(27), 35798–35810 (2021)

Deep Learning-Based Traffic Light Classification with Model Parameter Selection

Gülcan Yıldız[1(✉)], Bekir Dizdaroğlu[2], and Doğan Yıldız[3]

[1] Department of Computer Engineering, Ondokuz Mayis University, Samsun, Turkey
gulcan.ozer@omu.edu.tr

[2] Department of Computer Engineering, Karadeniz Technical University, Trabzon, Turkey
bekir@ktu.edu.tr

[3] Department of Electrical-Electronics Engineering, Ondokuz Mayis University, Samsun, Turkey
dogan.yildiz@omu.edu.tr

Abstract. Considering the existence of autonomous vehicles, it is seen that many studies have been done on the traffic light classification recently. Automatic determination of traffic lights can significantly prevent traffic accidents. As the number of vehicles on the road increases daily, such a classification process becomes crucial. The classification process appears to result in higher accuracy using deep learning approaches. In this study, a deep learning-based classification process is performed for traffic lights. A convolutional neural network model with efficient parameters is proposed. Additively, hyperparameter adjustment is made. In addition to this, the effects of color spaces and input image sizes on the classification results are investigated. There are four classes of images with red, yellow, green, and off tags in the database used. When the results are examined, it is seen that the classification accuracy of over 96% is achieved.

Keywords: autonomous driving · convolutional neural network · deep learning · hyperparameter selection · traffic light classification

1 Introduction

Since transportation is an essential part of human life, it is significantly related to human health, well-being, and safety. Many innovations and developments that are likely to occur in transportation systems are expected from communication, sensing, and processing technologies [1, 2]. In the 21st century, there have been significant increases in road transport due to different needs and reasons, and the road networks of countries have become unable to handle this traffic load day by day. For this reason, Traffic Control Devices (TCDs) have been developed to direct and warn drivers and regulate traffic rules. TCDs are traffic infrastructure units that communicate with drivers, such as pavement signs, traffic lights, and signs, providing efficient and safe transportation for drivers. These units are practical systems that control the traffic rules set for everyone who uses the highways in common. However, TCDs become dysfunctional when rules are not followed, which can have dangerous consequences. For example, if a driver does

D. J. Hemanth et al. (Eds.): ICAIAME 2022, ECPSCI 7, pp. 197–217, 2023.
https://doi.org/10.1007/978-3-031-31956-3_17

not stop at a red light at an intersection in an area where the stop at red light rule applies, it could result in an accident that could result in loss of life and property. Driver errors can be intentional or unintentional due to distraction, stress, faulty TCDs, and misunderstandings [3]. Traffic accidents can be prevented by monitoring the environment with Driver Assistance Systems (DAS) and warning the driver (intervention if necessary) in critical situations. In recent years, different automotive companies have developed many Advanced Driver Assistance Systems (ADAS) [4]. Modern ADASs are mostly vision-based, but ADASs also use light sensing, range (LIDAR), radio sensing, range (radar) and other sensing technologies [5].

One of the most important working areas of ADASs is traffic lights (TLs). TLs regulate traffic flow by informing traffic members about the right of way, such as pedestrians and vehicles using common roads for transportation [3].

The structure of traffic lights is similar in many countries. Generally, they consist of three vertically arranged circular indicators in green, yellow, and red [6]. In autonomous vehicles, traffic light recognition technology provides information about the status, color, and number of TLs and the lanes controlled by each TL [7–9]. While designing a traffic light recognition system that makes accurate recognition, it may be necessary to cope with different difficulties due to the variety of outdoor conditions and the ego movement of the vehicle. Some of these difficulties can be listed as follows [10]:

1. The ever-changing environment,
2. Interference from light sources such as streetlamps and billboards,
3. The effect of different weather and lighting conditions,
4. The size and viewing angle of the traffic lights change due to the ego movement of the vehicle,
5. Various appearances of traffic lights; for example, some traffic lights have a countdown timer, some do not,
6. The existence of traffic signs indicating different meanings with different shapes,
7. Auto white balance and autofocus function of smartphones and fixed cameras, which may cause blur or color distortion,
8. The lack of real-time traffic light recognition algorithm [11–13].

Although the formal design of traffic lights is easy using different methods, identifying traffic lights is not an easy process. The essential features that this technology must have to cope with difficult traffic conditions are:

1. The recognition system should be robust [14].
2. Algorithm work should be real-time [15].

Many studies have been carried out in the literature to recognize traffic lights. For this purpose, feature-based approaches have been widely adopted in the first phase [16–21]. For example, in [22], shape, color, structure, and geographic location information were used to develop an algorithm that scores the possible positions of TLs. In [10], a threshold model was created in ellipsoid geometry in HSV color space to extract color regions. In addition, a kernel function is proposed to identify candidate regions of the traffic light and combine the two heterogeneous features used. With the traffic light detected in [11], the prediction was made by applying the Viterbi algorithm to find the best existing situation sequence. Using the color distribution in [23], TL candidates are divided into three vertical regions. In [16], arrow-type TL candidates were classified by 2D independent component analysis and nearest neighbor classification of Gabor image features. [24] considered support vector machine (SVM) for classification based on HSV histograms. In [25], a 21-element feature vector was classified using a Joint Boost classifier.

Feature-based approaches for recognizing TLs generally do not perform well in various weather conditions and brightness. Therefore, deep learning models that mimic neural decision-making have been proposed in recent years to solve the problem of classification and recognition of TLs [3]. The use of neural networks dramatically increases the accuracy of TL recognition in dynamic scenarios [26–29]. In these methods, the primary goal is to create a region of interest (ROI), and the neural network determines the status and color of TLs [30–32]. John et al. [33] used image-processing techniques to extract texture, color, and shape features of ROI. Then, Multi-Layer Perceptron (MLP) and an artificial neural network defined the TL state. In these studies, pre-processing reduces the amount of computation [34]. The adaptive template matching system proposed in [35] is compared with the learning-based AdaBoost graded Haar feature classifier. A neural network was used to determine the status of TLs found in [36] and [37].

Some studies for the classification of TLs are based on cascade classifiers. For example, De Charette and Nashashibi analyzed TLs from different countries with a customized Adaptive Pattern Matching method. First, the method was tested on urban roads [7] and then was applied to TLs from China, France, and the USA [35]. The accuracy rate was obtained in these studies is 95%. In another study, many TL image sets were used to train the proposed ADAS, and only red and green lights were considered [8]. The histogram of oriented gradients (HOG) features and SVM are also recognizers used in this field [6]. Cascading classifiers based on Haar properties were also used in [38] and [39].

In this study, we propose a new convolutional neural network model with a low number of parameters for classifying traffic lights. Since hyperparameter tuning is a challenging process in deep learning networks, we conduct training trials and compare the test accuracy results to make this tuning. In addition, we increase the number of training set samples with data augmentation and enable the training to be done more efficiently. Moreover, we make changes to the input image given to the model, except for the model parameters. These changes are the input image size and image color space. As a result, we conduct training trials on all these parameters and have produced the best-performing model for the network.

The contributions of this study are as follows:

- A convolutional neural network model is proposed that has efficient model parameters and thus produces more accurate results.
- For the input image, training of the model is carried out by considering the size and color changes.
- Ten hyperparameter tuning are made by training and testing the model with different values. These hyperparameters are the number of filters, filter kernel size, activation function, and merge layer, the number of neurons in the fully connected hidden layer,

dropout rate, batch size, optimization algorithm, learning rate, and the number of epochs.

The reminder of this paper is organized as follows. Section 2 presents the proposed architecture. There are pre-processing stages and explanations of the images used for model training. It also gives detailed information about the model. Section 3 includes the experiments and results of the study. Information about the data set used is given. Training and testing processes for hyperparameter selection and their results are submitted. Final section concludes this study.

2 Proposal Architecture

2.1 Color Space

Color spaces contain essential information. It has a distinctive feature, especially in the field of computer vision. It can be used to extract meaningful knowledge from the given image. Many studies in the literature show that color spaces are used in different subjects, precisely recognition [40–43].

Many color spaces are used in TL studies. One of the most preferred color spaces in studies is RGB [44–48]. Grayscale color space is included in studies [49, 50]. [28] highlights that the CIELab color space contributes to extracting the region of interest for detection. In [51], CIELab is preferred as a color space as it was less affected by differences in lighting. HSV color space has been preferred quite a lot in the studies in the literature. The detection process of [52] uses LUV color space. Spotlights are found via the intensity channel L. The YCrCB color space is used to access the candidate ROI in the detector phase of [38]. The region is hereby applying thresholds on the Cr and CB color channels.

There are also studies in which more than one color space is used. In [11], HSV and grayscale; in [53], grayscale, RGB, HSV, and LUV color spaces are used. During the recognition phase, [4] tested the model using six different color spaces, namely RGB, normalized RGB, RYG, YCbCr, HSV, and CIE Lab, and achieved the best performance with RGB color space.

2.2 Image Resizing

CNN architecture can train images of the same size due to its network structure. In order to meet this condition, the image resizing process is usually performed [54]. In addition, it has been seen that the image sizing process affects the performance in classification studies [43, 55]. One of the situations that cause this result may be that the datasets mainly consist of images with different image sizes. Considering all these situations, training and testing processes have been carried out with different image sizes to achieve the highest performance in this study. The resize choosing and process details used in our study are given in Sect. 3.

2.3 Data Augmentation

Data augmentation is the operation of producing new labelled data by making changes to the existing training set. Various data augmentation methods are available. One of them is generating new data by performing some conversions on the actual data. Another method is accomplished by adding new objects to the actual data [56]. When the performance obtained as a result of training the model by performing data augmentation in the studies on the classification problem and the results of the training performed on the unprocessed training data were compared, it was observed that higher classification accuracy was achieved with augmentation [55, 57]. Thus, it is ensured that training can be done better by increasing the amount of data in case the training data set is small. Better learning is provided by increasing the data diversity, and this situation also better reflects the differences in the real world.

This study performed the data augmentation phase by applying the width shifting, height shifting, zooming, shearing, and rotation operations to the training set. The applied operations and range values are given in Table 1.

Table 1. Data Augmentation Arguments

Operation	Range value
Width shifting	0.1
Height shifting	0.1
Zooming	0.15
Shearing	0.15
Rotation	10°

2.4 Layer Information

In this study, a deep network model based on CNN architecture is proposed for traffic light classification. This architecture includes four convolution blocks. The blocks include the convolution layer, batch normalization, and ReLU (Rectified Linear Unit) activation layer respectively. In our approach, the padding parameter is "same," and the stride value is set to 1. There is also the max-pooling layer after the first and third convolution blocks. Then the add layer, which will provide a skip connection, is implemented. In this layer, the outputs of the third and fourth convolution blocks are collected. Then, the output of the network is vectorized and sent to the hidden, fully connected layer. This layer has 32 neurons. Dropout is added to prevent overfitting. Finally, to classify the 4-class dataset, the model architecture is completed with the last fully connected layer with four neurons and Softmax. The architectural details used in this study are given in Table 2 and visually Fig. 1. Here, the input image size is 32×32, and color space is RGB.

Table 2. Details and layer information of the proposed architecture in this study

Layer	Kernel	Stride	Channels	Output size	Parameters
Convolutional layer	5 × 5	1	64	32 × 32	4864
Batch Normalization ReLU					256
Max-pooling layer	2 × 2	1	64	16 × 16	
Convolutional layer	3 × 3	1	64	16 × 16	36928
Batch Normalization ReLU					256
Convolutional layer	5 × 5	1	96	16 × 16	153696
Batch Normalization ReLU					384
Max-pooling layer	2 × 2	1	96	8 × 8	
Convolutional layer	3 × 3	1	96	8 × 8	83040
Batch Normalization ReLU					384
Add Flatten			96	8 × 8	
Fully-Connected			32		196640
BatchNormalization					128
Dropout (0.5)					
Fully-Connected			**4**		132
Activation (Softmax)					

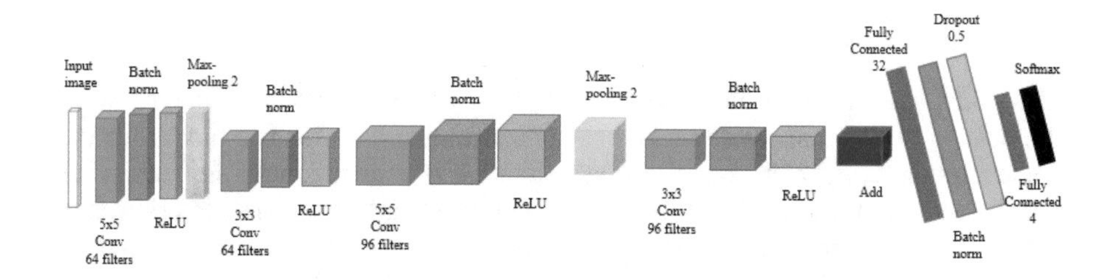

Fig. 1. The details of proposed CNN architecture in this study

3 Experiments and Results

3.1 Dataset

Bosch dataset, Udacity simular, and Udacity car dataset [58] were used to train and test the proposed architecture. There are 9259 TL images in the training set and 1030 TL images in the test set. The pixel dimensions of the images are 32 × 32. The dataset consists of four classes with labels red, green, yellow, and off.

3.2 Hyperparameter Tuning

There are many hyperparameters in CNN networks. Making hyperparameter tuning on these is a very laborious task. The change in the value of each parameter can lead to significant performance change in the network. It is almost impossible to adjust the parameters to get the best performance value. Therefore, in this study, we compared the performance results by making hyperparameter changes to produce a model with high accuracy. We conducted training trials to determine the parameter values of the model. Parameter values have been selected from the high accuracy values obtained from our experiments. In this section, comparison tables for all trials will be given.

In our study, training processes were carried out to select the parameters of the model. The parameter with the highest accuracy in the relevant unit was selected by choosing the initial values and keeping the other parameters constant. By updating, other units were examined, respectively. Five pieces of training were carried out for the parameter values in each unit, and the value providing the maximum accuracy in the test results was selected as the parameter. The highest performing value for one parameter was kept constant, and that value was used in the other parameter. Thus, by sequentially updating, the model with the highest accuracy was achieved in our experiments. In the following, the selection of parameter values according to the accuracy of the test results is shown and the comparison tables are given. Before starting the trainings, initially selected model parameters are given in Table 3.

Table 3. Initially selected model parameters

The number of filters	32–32-64–64	Batch size	32
Filter kernel size	5–3-5–3	Optimization algorithm	SGD
Activation function	ReLU	Learning rate	0.001
Merging layer	Add	The number of Epochs	50
Number of Neurons in The Fully Connected Hidden Layer	32	Input image size	32 × 32
Dropout rate	0.5	Color Space	RGB

Operations are performed on Google Colab using Tensorflow and Keras libraries. Categorical cross-entropy is used as the loss function. 10% (ten percent) of the training set is reserved as the validation set.

In this section, experiments were done for the parameters, respectively, and they are given together with the training and test results.

a) *The number of filters selection:* When building the Convolution layer, it is necessary to give a few parameters that will significantly affect the performance, one of them is the number of filter sizes. Setting this number is also crucial for the number of parameters. The number of filters selected high will decrease performance in the training phase. In this study, we tried different numbers (16, 32, 64, 96, 128, and 256) of combinations to determine the number of filter sizes suitable for our model. As a result of the training, we achieved the highest test accuracy with 64-64–96-96 filters for four convolution layers. The test accuracy results of the model according to the number of the convolutional filter is given in Fig. 2.

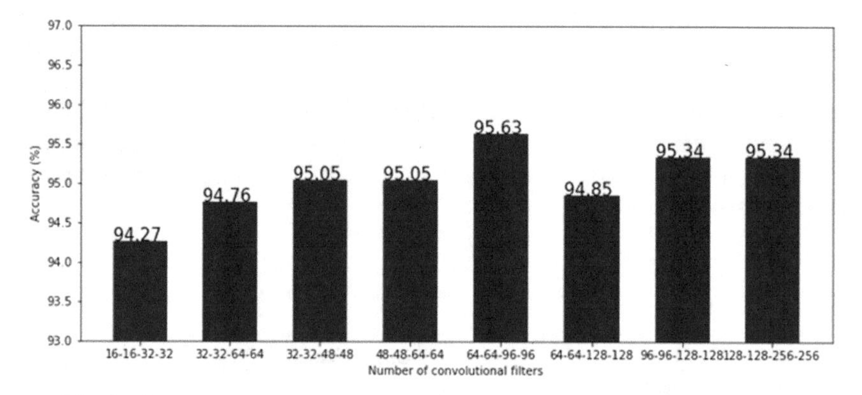

Fig. 2. The test accuracy results of the model according to the number of filters

b) *Filter kernel size:* Another critical parameter of the Convolution layer is the filter kernel size. Filter size significantly affects the number of parameters and performance. Therefore, low values are preferred in studies. Our study used 3×3, 5×5, and 7×7 filter kernel size values. As in the number of filters, pieces of training were made with different values for the four convolution layers. Figure 3 shows the test accuracy results. In the 5-3-5-3 nomenclature given here, the value in 4 layers and layers is, e.g., 3, 3×3 kernel size; 5 represents a kernel size of 5×5. When the results are examined, it is seen that the highest accuracy is reached with 5-3-5-3 kernel size values.

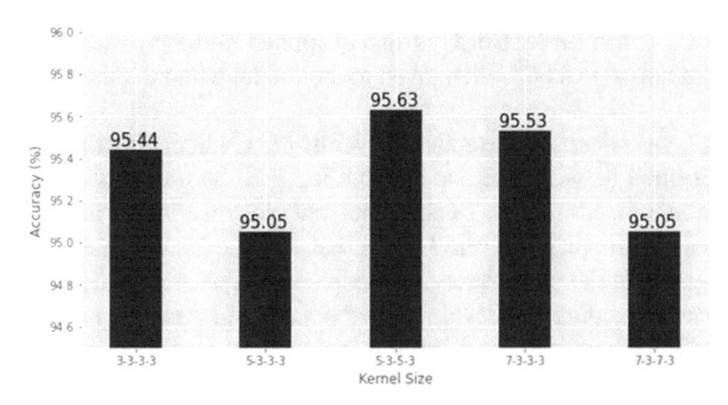

Fig. 3. Test accuracy results using different number of filter kernel sizes

iii) *Activation function:* One of the essential hyperparameters in deep networks is the activation function. This function produces the nonlinearity necessary for learning. The activation functions used in our study are ReLU, Leaky ReLU, Parametric ReLU (PReLU), and Exponential Linear Unit (ELU). ReLU is the most used activation function in the literature. In ReLU, all negative values return 0, and positive values return their own value. In gradient-based methods, negative values cannot be trained with RELU. To prevent this, the Leaky ReLU activation function, which can handle negative values with a low value, has been introduced.

Parametric ReLU is similar in structure to Leaky ReLU, and negative input values are multiplied by a small value [59]. But the situation here is different from Leaky ReLU. The value of it can be learned with an appropriate function.

ELU was suggested by Clevert et al. [60] as an alternative to ReLU. The two functions produce the same results for positive values, while the ELU includes an exponential multiplication factor for negative values [61].

In this study, trials have been made for the activation function. Although the results of the two functions are high (Fig. 4), the widely used RELU is chosen. In addition,

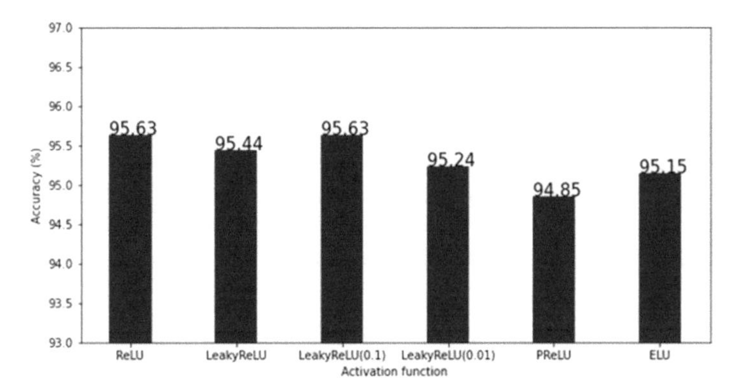

Fig. 4. Test accuracy results of the model according to different activation functions

the training result with Leaky ReLU is also examined in the last stage, but the highest accuracy could not be reached when given to the model with all values.

iv) *Merging layer selection:* To reduce the vanishing gradient effect, a merge operation was performed between the last convolution and the previous convolution layer. Add, Maximum, Minimum, Concatenate, and Average layers in the Keras library were used for the merge layer. The highest accuracy was achieved with the Add layer (Fig. 5). In the present case, the skip connection was made. Skip connection is achieved by adding low-level features obtained in previous layers to high-level features [29]. Thus, it is considered to increase the classification performance by obtaining strong features and to reduce the vanishing gradient effect [62].

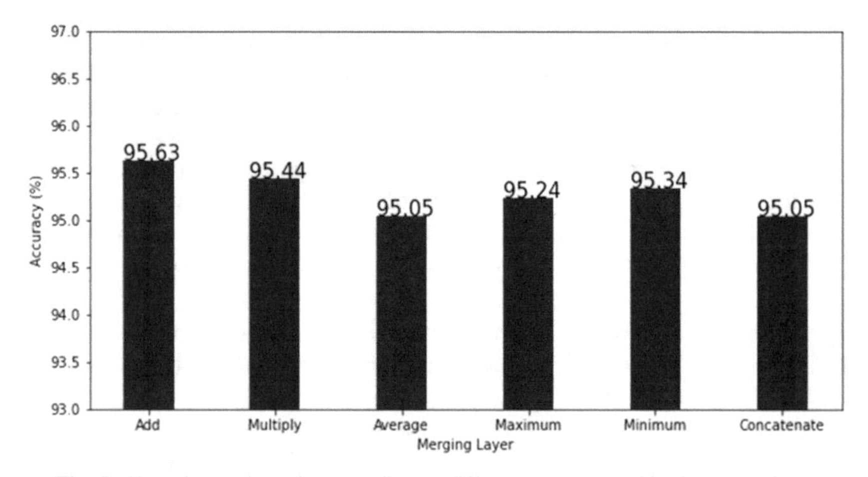

Fig. 5. Experimental results according to different classes used in the merge layer

e) *Number of Neurons in The Fully Connected Hidden Layer:* Another important hyper-parameter is the number of neurons in the fully connected hidden layer. The different number of fully connected layers can give better results on deep and shallow CNN architecture. According to the data set and architecture, the change in the number of fully connected layers affects the performance [63]. This study conducted training experiments to determine the Number of Neurons in The Fully Connected Hidden Layer. Test accuracy results obtained according to different number of neurons in the fully connected hidden layer are given in Fig. 6. Thus, the best performance was obtained by giving 32 neurons to the network.

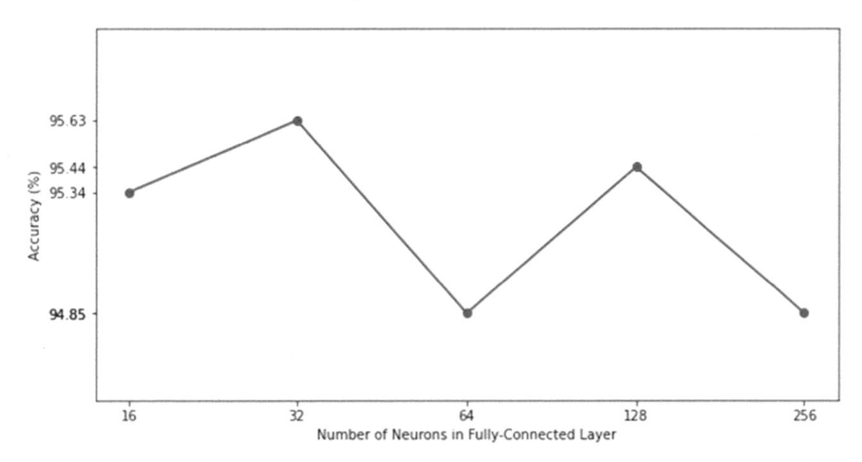

Fig. 6. Model test results according to the number of neurons in the fully connected hidden layer

f) *Dropout rate:* Dropout [64] is a layer frequently used in the literature to avoid overfitting. A certain percentage of randomly selected neurons in the layer is removed from the network. All connections to and from these neurons are annihilated. According to the dropout percentage rates, the highest test accuracy value was obtained at 0.5 (Fig. 7).

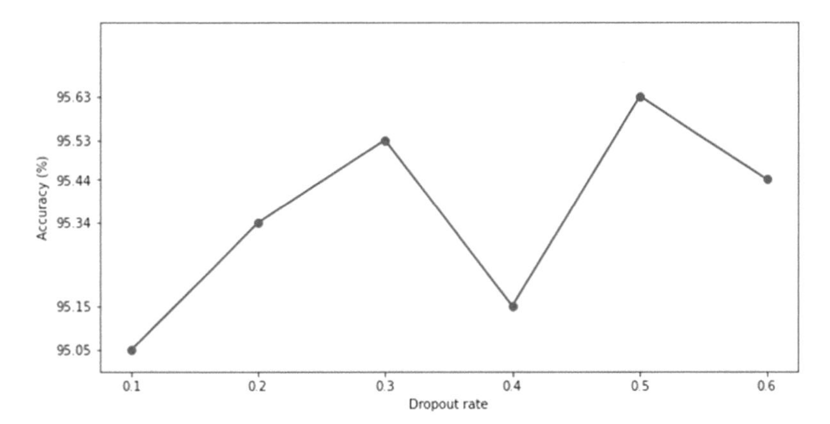

Fig. 7. Dropout rate selection

g) *Batch size:* Batch refers to the number of images fed to the network for training at one time. In [65], it was stated that instead of high batch size, small batch size and learning rate would provide better performance. When the results of our study are examined in Fig. 8, it is seen that the highest test accuracy result was achieved with 32 batch size compared to the others.

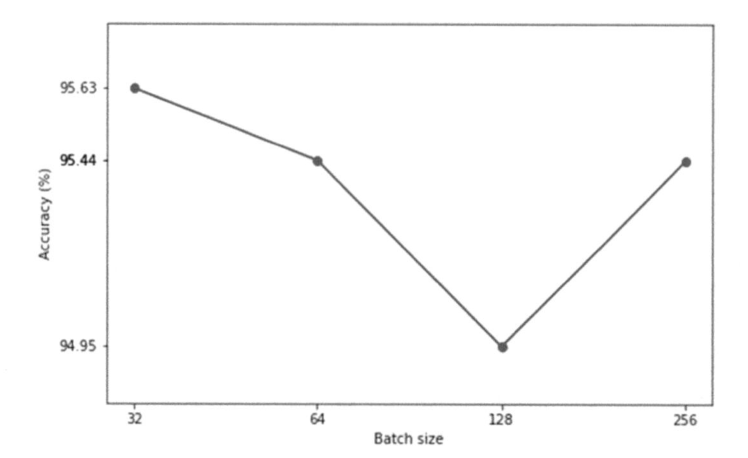

Fig. 8. Accuracy output based on batch size value

h) *Optimization algorithm:* The effect of the difference in optimization algorithms on the results is also investigated. The learning rate is initially 0.001. Accordingly, when the training is done, the highest accuracy is achieved with the SGD algorithm. The test results are given in Fig. 9.

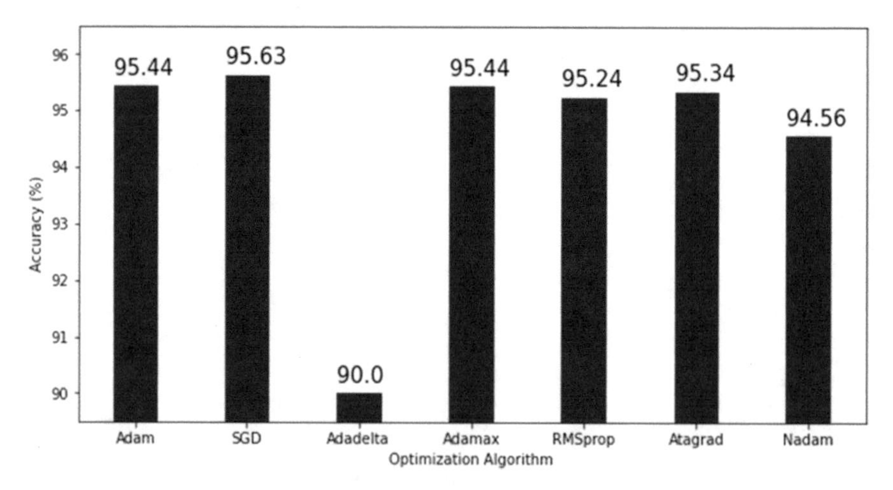

Fig. 9. Accuracy results for different optimization algorithms used in the model

i) *Learning rate selection*: Learning rate is one of the most significant hyperparameters. Fine-tuning is very important for training the network. In the case that the learning rate is chosen too small, overfitting may take place. If it is selected too large, the minimum error cannot be reached, and the training becomes different, deviation may occur [66].

In our study, first, training experiments were carried out by choosing a constant learning rate value. Accordingly, the same highest performance was achieved with 0.5, 0.01, and 0.001 (Fig. 10-a). In the next step, these values are selected as starting learning rate, and the network is retrained with the cosine decay class [67] and the decay value of 100000. According to the test results, the highest accuracy value of 96.21% was obtained with a 0.5 learning rate. Test results are shown in Fig. 10-b.

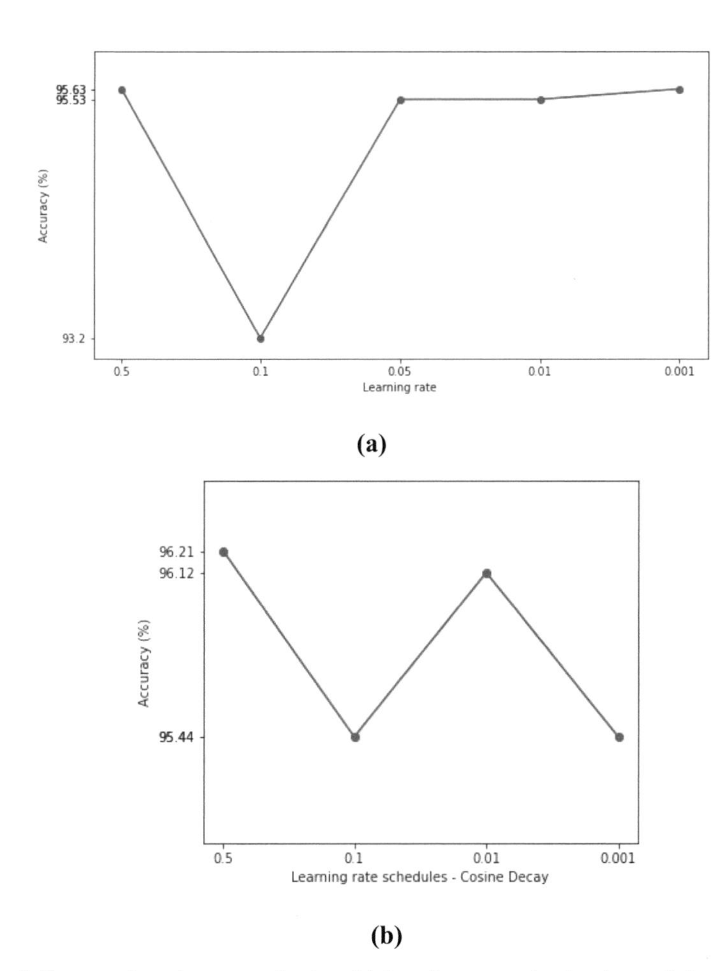

Fig. 10. **a)** Constant learning rate selection, **b)** learning rate selection by applying the cosine decay class.

j) *Number of epochs*: While determining the number of epochs, the effect of different initial learning rate values used with the cosine decay class on test accuracy was also investigated. Model test results according to number of epochs and learning rate combinations are given in Table 4. In the model, the highest test accuracy is reached with a learning rate of 0.01 and the number of epochs 100. Thus, in the next stages, 0.01 was used as the learning rate and 100 as the number of epochs.

Table 4. Model test results according to number of epochs and learning rate combinations (LR: learning rate)

LR / Number of Epochs	50	100	200
0.5	96.21%	95.73%	95.63%
0.01	96.12%	**96.41**%	95.73%

k) *Input image size:* The model was trained and tested to investigate the effect of input image size on performance. 32 × 32, 45 × 45, 60 × 60, and 75 × 75 values were chosen as the image size. According to the test results, the highest accuracy was reached with the 32 × 32 input image size taken at the beginning (Fig. 11). Even more, there is quite a distance between it and other results; by far, it is superior to other results.

The accuracy and loss graphs in the training phase are given with Fig. 12. When the graphics are examined, it can be seen that 32-input image size produces higher accuracy and lower loss value.

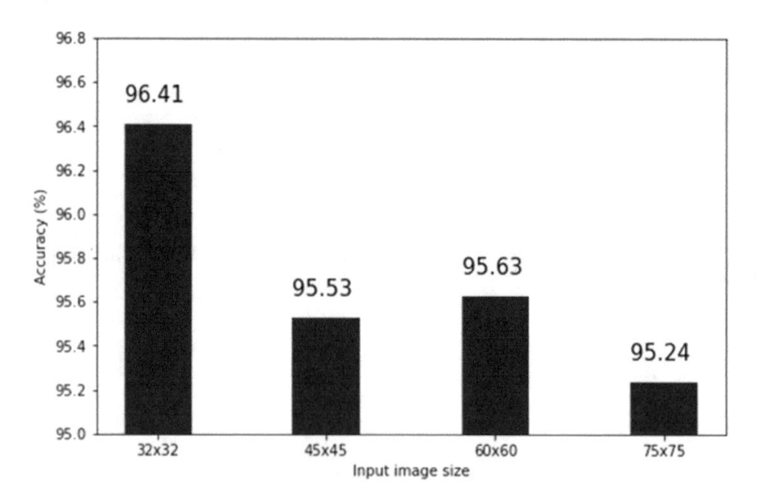

Fig. 11. Model testing with different input image size

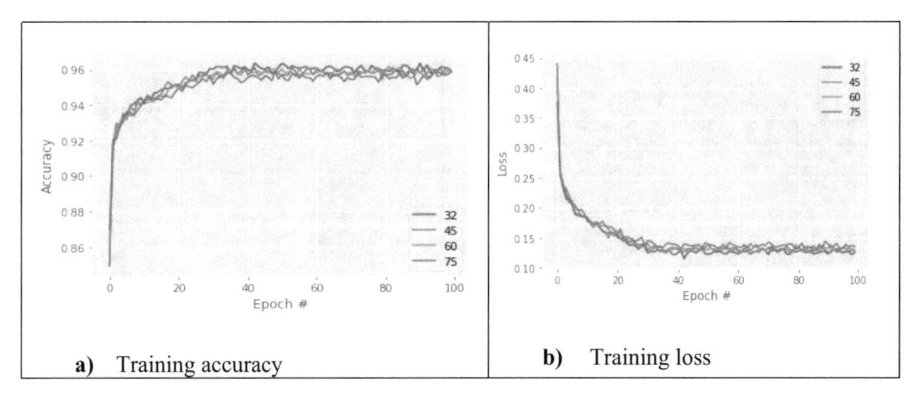

Fig. 12. a) The training accuracy, b) training loss graph for input image size.

xii) *Color spaces:* The final section, the model with all obtained parameter values, is trained on different color spaces. When the results were examined, it was observed that higher accuracy was obtained with the RGB color space we selected at the beginning. Test accuracy results are given in Fig. 13.

In addition, the training accuracy and training loss graph is given in Fig. 14. When the graph is examined, it is seen that the RGB color space produces significantly better results than the others.

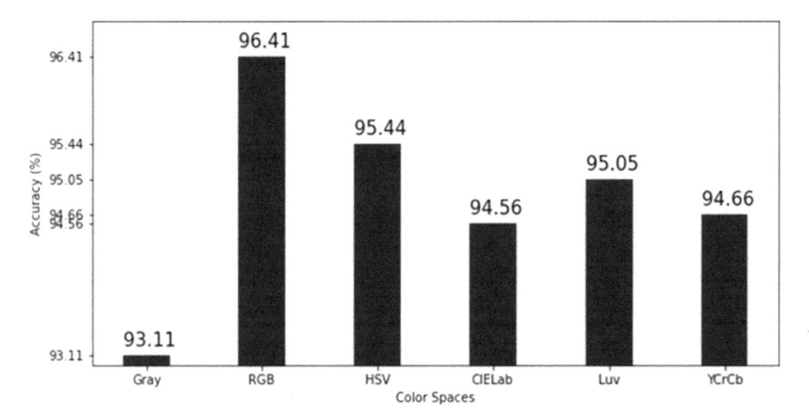

Fig. 13. Test accuracy according to color space

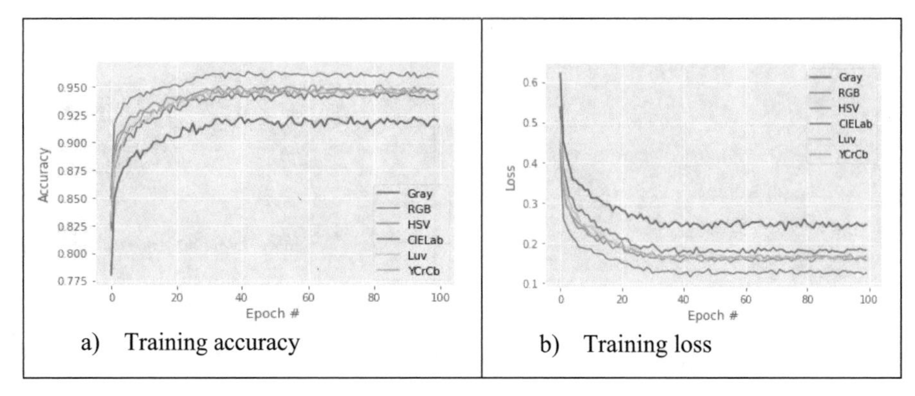

Fig. 14. a) The training accuracy, b) training loss graph for color spaces.

After all parameter selections were made, the total number of parameters of the network was calculated as 476,708. Apart from accuracy, other metrics were applied to the model obtained because of all parameter selections on the test dataset. All metric results are given in Table 5. In addition, the confusion matrix of the model is presented in Fig. 15. When it is examined, it is seen that images with red and green tags are labeled more accurately than yellow and off tags. Additionally, it is important to note that the actual red light is never labeled green.

Table 5. Metric results for the model

	Precision	Recall	F1-score	Support
Red	0.95	0.98	0.97	384
Yellow	0.89	0.73	0.80	56
Green	0.98	0.99	0.99	511
Off	0.92	0.85	0.88	79
Accuracy			0.96	1030
Macro avg	0.94	0.89	0.91	1030
Weighted avg	0.96	0.96	0.96	1030

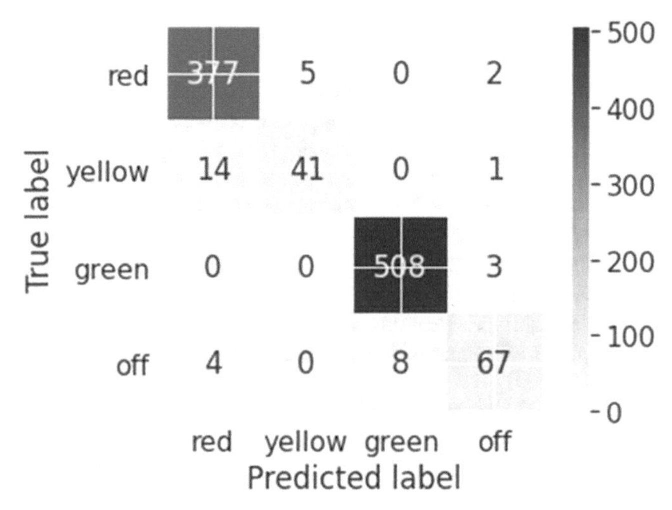

Fig. 15. Confusion matrix of traffic light classification

4 Conclusions

This study proposes a new deep neural network for traffic light classification. We applied the data augmentation process, which increases the amount of data in the model by producing synthetic data and affects the model performance. We conducted training trials for model parameters. The parameters used are the number of filters, filter kernel size, activation function, merge layer, the number of neurons in the fully connected hidden layer, dropout rate, batch size, optimization algorithm, learning rate, and the number of epochs. The effect of each parameter and its values on performance has been examined. In addition, different values were used for the size and color selection of the image given to the input at the beginning, and the results were compared. The highest test accuracy is 96.41%, which is achieved with a model with RGB color space and 32 × 32 input size.

In addition, when the confusion matrix results were examined, it was seen that the red light was not labelled as green in any of the images. This situation prevents the most common traffic violation of crossing a red light and does not cause traffic accidents.

References

1. Sussman, J.S.: Perspectives on intelligent transportation systems (ITS). Springer Science & Business Media (2008)
2. Papadimitratos, P., De La Fortelle, A., Evenssen, K., Brignolo, R., Cosenza, S.: Vehicular communication systems: enabling technologies, applications, and future outlook on intelligent transportation. IEEE Commun. Mag. **47**(11), 84–95 (2009)
3. Jensen, M.B., Philipsen, M.P., Møgelmose, A., Moeslund, T.B., Trivedi, M.M.: Vision for looking at traffic lights: issues, survey, and perspectives. IEEE Trans. Intell. Transp. Syst. **17**(7), 1800–1815 (2016)

4. Kim, H.-K., Park, J.H., Jung, H.-Y.: An efficient color space for deep-learning based traffic light recognition. J. Adv. Transp. **2018**, 2365414 (2018)
5. Kukkala, V.K., Tunnell, J., Pasricha, S., Bradley, T.: Advanced driver-assistance systems: a path toward autonomous vehicles. IEEE Consumer Electron. Mag. **7**(5), 18–25 (2018)
6. Diaz, M., Cerri, P., Pirlo, G., Ferrer, M.A., Impedovo, D.: A survey on traffic light detection. In: Murino, V., Puppo, E., Sona, D., Cristani, M., Sansone, C. (eds.) New Trends in Image Analysis and Processing – ICIAP 2015 Workshops: ICIAP 2015 International Workshops, BioFor, CTMR, RHEUMA, ISCA, MADiMa, SBMI, and QoEM, Genoa, Italy, September 7–8, 2015, Proceedings, pp. 201–208. Springer International Publishing, Cham (2015). https://doi.org/10.1007/978-3-319-23222-5_25
7. De Charette, R. Nashashibi, F.: Real time visual traffic lights recognition based on spot light detection and adaptive traffic lights templates. In: 2009 IEEE Intelligent Vehicles Symposium, pp 358–363. IEEE (2009)
8. Fairfield, N., Urmson, C.: Traffic light mapping and detection. In: 2011 IEEE International Conference on Robotics and Automation, pp 5421–5426. IEEE (2011)
9. Possatti, L.C., et al.: Traffic light recognition using deep learning and prior maps for autonomous cars. In: 2019 International Joint Conference on Neural Networks (IJCNN), pp. 1–8. IEEE (2019)
10. Liu, W., et al.: Real-time traffic light recognition based on smartphone platforms. IEEE Trans. Circuits Syst. Video Technol. **27**(5), 1118–1131 (2016)
11. Gomez, A.E., Alencar, F.A., Prado, P.V., Osorio, F.S., Wolf, D.F.: Traffic lights detection and state estimation using hidden markov models. In: 2014 IEEE Intelligent Vehicles Symposium Proceedings, pp. 750–755. IEEE (2014)
12. Omachi, M., Omachi, S.: Traffic light detection with color and edge information. In: 2009 2nd IEEE International Conference on Computer Science and Information Technology, pp. 284–287. IEEE (2009)
13. Roters, J., Jiang, X., Rothaus, K.: Recognition of traffic lights in live video streams on mobile devices. IEEE Trans. Circuits Syst. Video Technol. **21**(10), 1497–1511 (2011)
14. Chiang, C.-C., Ho, M.-C., Liao, H.-S., Pratama, A., Syu, W.-C.: Detecting and recognizing traffic lights by genetic approximate ellipse detection and spatial texture layouts. Int. J. Innovative Comput., Inform. Control **7**(12), 6919–6934 (2011)
15. Greenhalgh, J., Mirmehdi, M.: Real-time detection and recognition of road traffic signs. IEEE Trans. Intell. Transp. Syst. **13**(4), 1498–1506 (2012)
16. Cai, Z., Li, Y., Gu, M.: Real-time recognition system of traffic light in urban environment. In: 2012 IEEE Symposium on Computational Intelligence for Security and Defence Applications, pp. 1–6. IEEE (2012)
17. Diaz-Cabrera, M., Cerri, P., Medici, P.: Robust real-time traffic light detection and distance estimation using a single camera. Expert Syst. Appl. **42**(8), 3911–3923 (2015)
18. Hosseinyalamdary, S., Yilmaz, A.: A Bayesian approach to traffic light detection and mapping. ISPRS J. Photogramm. Remote. Sens. **125**, 184–192 (2017)
19. Lee, S.-H., Kim, J.-H., Lim, Y.-J., Lim, J.: Traffic light detection and recognition based on Haar-like features. In: 2018 International Conference on Electronics, Information, and Communication (ICEIC), pp. 1–4. IEEE (2018)
20. Saini, S., Nikhil, S., Konda, K.R., Bharadwaj, H.S., Ganeshan, N.: An efficient vision-based traffic light detection and state recognition for autonomous vehicles. In: 2017 IEEE Intelligent Vehicles Symposium (IV), pp. 606–611. IEEE (2017)
21. Wang, K., Xiong, Z.: VISUAL ENhancement method for intelligent vehicle's safety based on brightness guide filtering algorithm thinking of the high tribological and attenuation effects. J. Balkan Tribol. Assoc. **22**(2A), 2021–2031 (2016)

22. Zhang, Y., Xue, J., Zhang, G., Zhang, Y., Zheng, N.: A multi-feature fusion based traffic light recognition algorithm for intelligent vehicles. In: Proceedings of the 33rd Chinese control conference, pp. 4924–4929. IEEE (2014)
23. Barnes, D., Maddern, W., Posner, I.: Exploiting 3D semantic scene priors for online traffic light interpretation. In: 2015 IEEE Intelligent Vehicles Symposium (IV), pp. 573–578. IEEE (2015)
24. Nienhüser, D., Drescher, M., Zöllner, J.M.: Visual state estimation of traffic lights using hidden Markov models. In: 13th International IEEE Conference on Intelligent Transportation Systems, pp. 1705–1710. IEEE (2010)
25. Haltakov, V., Mayr, J., Unger, C., Ilic, S.: Semantic segmentation based traffic light detection at day and at night. In: Gall, J., Gehler, P., Leibe, B. (eds.) Pattern Recognition: 37th German Conference, GCPR 2015, Aachen, Germany, October 7–10, 2015, Proceedings, pp. 446–457. Springer International Publishing, Cham (2015). https://doi.org/10.1007/978-3-319-24947-6_37
26. Bach, M., Stumper, D., Dietmayer, K.: Deep convolutional traffic light recognition for automated driving. In: 2018 21st International Conference on Intelligent Transportation Systems (ITSC), pp. 851–858. IEEE (2018)
27. Chen, Z., Huang, X.: Accurate and reliable detection of traffic lights using multiclass learning and multiobject tracking. IEEE Intell. Transp. Syst. Mag. **8**(4), 28–42 (2016)
28. John, V., Yoneda, K., Liu, Z., Mita, S.: Saliency map generation by the convolutional neural network for real-time traffic light detection using template matching. IEEE Trans. Comput. Imaging **1**(3), 159–173 (2015)
29. Lee, E., Kim, D.: Accurate traffic light detection using deep neural network with focal regression loss. Image Vis. Comput. **87**, 24–36 (2019)
30. Hirabayashi, M., Sujiwo, A., Monrroy, A., Kato, S., Edahiro, M.: Traffic light recognition using high-definition map features. Robot. Auton. Syst. **111**, 62–72 (2019)
31. Wang, J.-G., Zhou, L.-B.: Traffic light recognition with high dynamic range imaging and deep learning. IEEE Trans. Intell. Transp. Syst. **20**(4), 1341–1352 (2018)
32. Wang, K., Huang, X., Chen, J., Cao, C., Xiong, Z., Chen, L.: Forward and backward visual fusion approach to motion estimation with high robustness and low cost. Remote Sens. **11**(18), 2139 (2019)
33. John, V., Yoneda, K., Qi, B., Liu, Z., Mita, S.: Traffic light recognition in varying illumination using deep learning and saliency map. In: 17th International IEEE Conference on Intelligent Transportation Systems (ITSC), pp. 2286–2291. IEEE (2014)
34. Wang, K., Tang, X., Zhao, S., Zhou, Y.: Simultaneous detection and tracking using deep learning and integrated channel feature for ambint traffic light recognition. J. Ambient. Intell. Humaniz. Comput. **13**, 271–281 (2021)
35. De Charette, R., Nashashibi, F.: Traffic light recognition using image processing compared to learning processes. In: 2009 IEEE/RSJ International Conference on Intelligent Robots and Systems, pp. 333–338. IEEE (2009)
36. Franke, U., Pfeiffer, D., Rabe, C., Knoeppel, C., Enzweiler, M., Stein, F., Herrtwich, R.: Making bertha see. In: Proceedings of the IEEE International Conference on Computer Vision Workshops, pp. 214–221 (2013)
37. Lindner, F., Kressel, U., Kaelberer, S.: Robust recognition of traffic signals. In: IEEE Intelligent Vehicles Symposium, 2004, pp. 49–53. IEEE (2004)
38. Kim, H.-K., Park, J.H., Jung, H.-Y.: Effective traffic lights recognition method for real time driving assistance systemin the daytime. Int. J. Electr. Comput. Eng. **5**(11), 1429–1432 (2011)
39. Shen, Y., Ozguner, U., Redmill, K., Liu, J.: A robust video based traffic light detection algorithm for intelligent vehicles. In: 2009 IEEE Intelligent Vehicles Symposium, pp. 521–526. IEEE (2009)

40. Kasaei, S.H., Ghorbani, M., Schilperoort, J., van der Rest, W.: Investigating the importance of shape features, color constancy, color spaces, and similarity measures in open-ended 3D object recognition. Intel. Serv. Robot. **14**(3), 329–344 (2021)
41. Wang, S.-J., et al.: Micro-expression recognition using color spaces. IEEE Trans. Image Process. **24**(12), 6034–6047 (2015)
42. Yang, J., Liu, C., Zhang, L.: Color space normalization: enhancing the discriminating power of color spaces for face recognition. Pattern Recogn. **43**(4), 1454–1466 (2010)
43. Yildiz, G., Dizdaroğlu, B.: Convolutional neural network for traffic sign recognition based on color space. In: 2021 2nd International Informatics and Software Engineering Conference (IISEC), pp. 1–5. IEEE (2021)
44. Aneesh, A., Shine, L., Pradeep, R., Sajith, V.: Real-time traffic light detection and recognition based on deep retinanet for self driving cars. In: 2019 2nd International Conference on Intelligent Computing, Instrumentation and Control Technologies (ICICICT), pp 1554–1557. IEEE (2019)
45. Behrendt, K., Novak, L., Botros, R.A.: deep learning approach to traffic lights: Detection, tracking, and classification. In: 2017 IEEE International Conference on Robotics and Automation (ICRA), pp. 1370–1377. IEEE (2017)
46. Chen, X., Luo, C.: Fast traffic light recognition using a lightweight attention-enhanced model. In: 2021 The 4th International Conference on Machine Learning and Machine Intelligence, pp. 15–20 (2021)
47. Golovnin, O.K., Yarmov, R.V.: Universal convolutional neural network for recognition of traffic lights and road signs in video frames. In: Solovev, D.B., Kyriakopoulos, G.L., Venelin, T. (eds.) SMART Automatics and Energy: Proceedings of SMART-ICAE 2021, pp. 459–468. Springer Nature Singapore, Singapore (2022). https://doi.org/10.1007/978-981-16-8759-4_48
48. Kim, H.-K., Shin, Y.-N., Kuk, S., Park, J.H., Jung, H.-Y.: Nightime traffic light detection based on svm with geometric moment features. Int. J. Comput. Inform. Eng. **7**(4), 472–475 (2013)
49. Ji, Y., Yang, M., Lu, Z., Wang, C.: Integrating visual selective attention model with HOG features for traffic light detection and recognition. In: 2015 IEEE Intelligent Vehicles Symposium (IV), pp. 280–285. IEEE (2015)
50. Trehard, G., Pollard, E., Bradai, B., Nashashibi, F.: Tracking both pose and status of a traffic light via an interacting multiple model filter. In: 17th International Conference on Information fusion (FUSION), pp. 1–7. IEEE (2014)
51. Sooksatra, S., Kondo, T.: Red traffic light detection using fast radial symmetry transform. In: 2014 11th international conference on electrical engineering/electronics, computer, telecommunications and information technology (ECTI-CON), pp. 1–6. IEEE (2014)
52. Philipsen, M.P., Jensen, M.B., Møgelmose, A., Moeslund, T.B., Trivedi, M.M.: Traffic light detection: A learning algorithm and evaluations on challenging dataset. In: 2015 IEEE 18th International Conference on Intelligent Transportation Systems, pp. 2341–2345. IEEE (2015)
53. Du, X., Li, Y., Guo, Y., Xiong, H.: Vision-based traffic light detection for intelligent vehicles. In: 2017 4th International Conference on Information Science and Control Engineering (ICISCE), pp. 1323–1326. IEEE (2017)
54. Boltaevich, M.B.: Estimation affects of formats and resizing process to the accuracy of convolutional neural network. In: 2019 International Conference on Information Science and Communications Technologies (ICISCT), pp. 1–5. IEEE (2019)
55. Haque, W.A., Arefin, S., Shihavuddin, A., Hasan, M.A.: DeepThin: a novel lightweight CNN architecture for traffic sign recognition without GPU requirements. Expert Syst. Appl. **168**, 114481 (2021)

56. Hassan, E.T., Li, N., Ren, L.: Semantic consistency: the key to improve traffic light detection with data augmentation. In: 2020 IEEE Intelligent Vehicles Symposium (IV), pp. 1734–1739. IEEE (2020)
57. Cubuk, E.D., Zoph, B., Mane, D., Vasudevan, V., Le, Q.V.: Autoaugment: Learning augmentation policies from data. arXiv preprint arXiv:180509501 (2018)
58. Traffic light classifer: https://github.com/JunshengFu/traffic-light-classifier (2017). Accessed May 2022
59. He, K., Zhang, X., Ren, S., Sun, J.: Delving deep into rectifiers: Surpassing human-level performance on imagenet classification. In: Proceedings of the IEEE international conference on computer vision, pp. 1026–1034 (2015)
60. Clevert, D.-A., Unterthiner, T., Hochreiter, S.: Fast and accurate deep network learning by exponential linear units (elus). arXiv preprint arXiv:151107289 (2015)
61. Rasamoelina, A.D., Adjailia, F., Sinčák, P.: A review of activation function for artificial neural network. In: 2020 IEEE 18th World Symposium on Applied Machine Intelligence and Informatics (SAMI), pp. 281–286. IEEE (2020)
62. Taghanaki, S.A., et al.: Select, attend, and transfer: Light, learnable skip connections. In: Suk, H.-I., Liu, M., Yan, P., Lian, C. (eds.) Machine Learning in Medical Imaging: 10th International Workshop, MLMI 2019, Held in Conjunction with MICCAI 2019, Shenzhen, China, October 13, 2019, Proceedings, pp. 417–425. Springer International Publishing, Cham (2019). https://doi.org/10.1007/978-3-030-32692-0_48
63. Basha, S.S., Dubey, S.R., Pulabaigari, V., Mukherjee, S.: Impact of fully connected layers on performance of convolutional neural networks for image classification. Neurocomputing **378**, 112–119 (2020)
64. Srivastava, N., Hinton, G., Krizhevsky, A., Sutskever, I., Salakhutdinov, R.: Dropout: a simple way to prevent neural networks from overfitting. The J. Mach. Learn. Res. **15**(1), 1929–1958 (2014)
65. Kandel, I., Castelli, M.: The effect of batch size on the generalizability of the convolutional neural networks on a histopathology dataset. ICT Express **6**(4), 312–315 (2020)
66. Smith, L.N.: A disciplined approach to neural network hyper-parameters: Part 1--learning rate, batch size, momentum, and weight decay. arXiv preprint arXiv:180309820 (2018)
67. Loshchilov, I., Hutter, F.: Sgdr: Stochastic gradient descent with warm restarts. arXiv preprint arXiv:160803983 (2016)

Investigation of Biomedical Named Entity Recognition Methods

Azer Çelikten[1] (✉), Aytuğ Onan[2], and Hasan Bulut[3]

[1] Manisa Celal Bayar University, Software Engineering, Manisa, Turkey
azer.celikten@cbu.edu.tr
[2] İzmir Katip Çelebi University, Computer Engineering, İzmir, Turkey
aytug.onan@ikcu.edu.tr
[3] Ege University, Computer Engineering, İzmir, Turkey
hasan.bulut@ege.edu.tr

Abstract. Biomedical named-entity recognition is the process of identifying entity names such as disease, symptom, drug, protein, and chemical in biomedical texts. It plays an important role in natural language processing, such as relationship extraction, question-answer systems, keyword extraction, machine translation, and text summarization. Biomedical domain information extraction can be used for early diagnosis of diseases, detection of missing relationships between biomedical entities such as diseases and chemicals, and determination of drug interactions and side effects. Since biomedical texts contain domain-specific words, complicated phrases, and abbreviations, named entity recognition in this domain is still a challenging task. In this study, we first investigated methods for named entity recognition in the biomedical domain. These methods are classified into four categories: dictionary-based, rule-based, machine learning, and deep learning methods. Recent advances such as deep learning and transformer-based biomedical language models have helped to achieve successful results in the named entity recognition task. Second, we conduct an experimental study on an annotated dataset called MedMention which is available to researchers. Finally, we present our experimental results and discuss the challenges and opportunities of the existing methods. The experimental study shows that the most successful method for extracting diseases and symptoms from biomedical texts is BioBERT, with an F1 score of 0.72.

Keywords: Biomedical Information Extraction · Biomedical Named Entity Recognition · Deep Learning · Machine Learning · Deep Contextualized Embeddings

1 Introduction

Information extraction is the process of extracting important, valuable, and useful information from large amounts of unstructured or semi-structured data. Information extraction in biomedical texts includes methods developed to obtain entity names such as diseases, symptoms, drugs, drug side effects, proteins, genes, and other biological species

D. J. Hemanth et al. (Eds.): ICAIAME 2022, ECPSCI 7, pp. 218–229, 2023.
https://doi.org/10.1007/978-3-031-31956-3_18

from biomedical texts and to determine useful relationships among these entities. Information extraction consists of two basic steps: Named Entity Recognition (NER) and Relationship Extraction. NER is a subset of information extraction and refers to the process of assigning entity names in the text to a particular category. Relationship extraction involves determining the semantic relationship between two or more entity names. In this study, we investigated NER methods for the biomedical domain, which is the first step of information extraction. Electronic patient records, physician notes, patient reports, and medical articles can be examples of biomedical texts. With the increase of digitization, the number of biomedical texts in the digital environment has greatly increased. For many natural language processing applications, including question-and-answer systems, text summarization, and machine translation, extracting entity names from biomedical literature is crucial [1]. Chatbots using question-answering systems must be able to parse the entity names (disease, symptom, etc.) in the question to provide the appropriate answer [4]. Moreover, NER is an important step in summarizing texts [5] and extracting keywords from texts [6].

Analyzing the sources of important information about diseases, drugs, proteins, and genes and extracting information using appropriate methods are helpful in solving many problems. Some of these problems are prediction of diseases, determination of risk factors in diseases, early diagnosis of diseases, detection of unknown relationships between diseases, determination of drug-drug interactions and side effects of drugs. Thanks to advances in natural language processing and achievements in deep learning and transfer learning, entity name extraction from biomedical texts is gaining popularity among researchers, and effective biomedical text mining models are being developed. In the biomedical domain, various natural language processing methods are required due to the complex expressions typical of this domain. Recently, thanks to biomedical language models trained on large biomedical texts, very successful results have been obtained in NER as well as in many other natural language processing tasks [2, 3].

In this study, we investigate the methods for biomedical NER, related works and data sets. We also propose a method by fine-tuning transformer-based models for the task of biomedical NER. Although there are several methods for disease recognition, there are few studies on identifying symptoms from medical texts. Moreover, this is a preliminary step for extracting relations between disease and symptoms. For this purpose, we rearranged the MedMention [25] corpus, that is annotated by the experts and designed originally for entity linking. We use four Bert-based deep contextualized embedding models on this corpus to identify disease and symptom names.

In Section 2, related works, recent biomedical NER methods, and datasets are introduced and compared. Section 3 explains an experimental study for identifying disease and symptoms. Section 4 contains the results of the experiments and discussion part. In the last section, conclusions and future works are presented.

2 Related Works, Methods, and Datasets

For NER tasks in the biomedical domain, several approaches have been suggested. They are divided into four groups as dictionary-based, rule-based, machine learning, and deep learning methods.

2.1 Dictionary Based Methods

The first methods for identifying biomedical names rely on dictionary-based approaches. These methods require predefined dictionaries consisting of extensive lists of names for entity types. The entity names in the texts are determined based on their occurrence in these dictionaries. Yang et al., developed a dictionary based biomedical NER method. The steps of their method were dictionary creation and expansion, word matching, and post-processing. The proposed method was tested on the JNLPBA dataset and achieved an F-score of 68.80% [7]. MetaMap [8], is another dictionary based biomedical NER method, developed by the National Library of Medicine. In this method, biomedical entity names are matched with the Unified Medical Language System (UMLS) concepts such as *disease and syndrome* and *sign and symptom.* In some studies, biomedical texts were annotated using MetaMap tool [10, 11]. Due to the possibility of additional entities, dictionary-based algorithms are distinguished by better accuracy but lower recognition rates. The dictionaries are also insufficient or incomplete for many other biological things, even if they exist for names of common biological entities like genes and diseases.

2.2 Rule Based Methods

In systems that use the rule-based approach, entities are identified by manually defined rules based on text patterns. In other words, rule-based approaches use structured rule patterns to discover probable entities [9]. Fukuda et al. proposed a rule-based entity name recognition system called PROPER [10]. They discovered that in this method, fundamental phrases like 'p53' and concatenated terms like 'receptor' represented protein names. As a result, entity characteristics such protein name patterns, protein name nomenclature, variety in entity styles, and different prefixes and suffixes were used. They tested the proposed method on a small dataset. As a result of their work, a precision and recognition score of over 90% was obtained.

Rule-based methods rely on the predefined patterns that depend on certain textual attributes of an entity class. In other words, these techniques need rules and patterns that are distinct to each entity type. This is a laborious procedure, and the knowledge of domain experts is required to design rules. Most traditional approaches based on dictionary and rule-based methods have shown significant improvements in robustness but rely heavily on well-defined dictionaries and a set of words in hand-crafted rules.

2.3 Machine Learning Methods

In machine learning-based methods, models are trained using training data consisting of features and labels, and the performance of the models is measured using test data. In machine learning-based systems, the NER task is formulated as a sequence labelling problem. In this method, the tokens in a sequence are annotated by one of the corresponding labels B, I or O. This process is also referred to as IOB tagging. The example sequence labelled using this method is shown in Fig. 1. When a disease name consists of more than one word such as lung cancer, it is tagged as 'B-disease' or 'I-disease'. B-disease represents the beginning of the disease name and I-disease denotes the words in the disease name. If the related word does not contain any word in the disease name such as *patients, with*, it is labelled as O. In machine learning approaches, the main factors that affect the performance of NER are the hand-generated features and the algorithms used. Feature engineering and expert knowledge are important to achieve successful results. Machine learning-based approaches achieve good results on task-specific datasets. Several machine learning models have been developed for biomedical NER, such as Hidden Markov Models (HMM), Support Vector Machines (SVM), and Conditional Random Fields (CRF). Kazama et al. created an SVM model with a set of attributes such as "word feature", "word type", "prefix", "suffix", "subword", and "antecedent class" [13]. Performance evaluation using the GENIA corpus yielded an F1 score of 56.5% in identifying protein names and an F1 score of 51% in identifying all entity names. In contrast, Zhou et al. obtained an F1 score of 75.8% and 66.6% in protein type and all type recognition, respectively, in the GENIA corpus with the Bio-Entity Recognizer using the HMM. The Viterbi method is used by CRF, which has been shown to be successful in a variety of entity name recognition tasks [14]. This method computes a probability distribution for the possible tags and determines the most likely tag sequence. Although CRF is widely used in NER tasks, it is used as a layer in some deep learning architectures, such as recurrent neural networks, to determine the optimal tag sequence.

A CRF-based model was proposed by McDonald and Pereira (2005) to determine the most useful features to find gene and protein names in texts. After training with 7500 MEDLINE sentences, F1 score of 82.4% was obtained using the proposed BioCreative I Gene Mention Dataset system [15].

Machine learning-based methods are of limited use for NER systems because they require expert work to identify features, and feature development and selection of effective features strongly affect system performance.

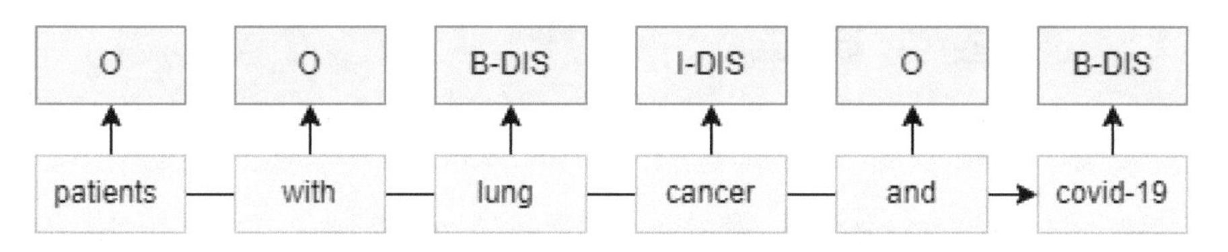

Fig. 1. Sequence Labelling Approach

2.4 Deep Learning Methods

Deep Learning methods use various neural network architectures to model NER problems and many other NLP tasks. Recently, Convolutional Neural Networks (CNN) and Long-Short-Term Memory (LSTM) models have been widely used in NER. In Deep Learning methods, NER task is formulated as a sequence-to-sequence architecture where the input is a sequence of words represented as embedding vectors. Thanks to Deep Learning methods and available hardware systems, not only the huge amount of data can be processed, but also the features are extracted automatically. As input, the words are represented as embeddings in the vector space using the traditional word vector models such as word bag, word2vec, glove, and fast text or the pre-trained word embedding models developed specifically for the biomedical domain, such as BioWordVec [3] and BioBERT [2].

Zhu et al.(2018) develop a method called GRAM-CNN for biomedical NER. In this method the input word initially consists of part of speech (POS) tag and character vectors. The word vector and POS tag are then merged with the character vectors of each letter in the word. Then CNN feature vectors are created by extracting the attributes of each word using GRAM-CNN architecture. For IOB tagging, the sequence labelling method was performed for IOB tagging and CRF is applied to determine the optimal tag based on the results of GRAM-CNN. The architecture of GRAM-CNN is shown in Fig. 2. On three biomedical datasets for various entities, this method was evaluated. For the Biocreative II dataset, the NCBI dataset, and the JNLPBA dataset, it received an F1 score of 87.26%, 87.26%, and 72.57%, respectively [12].

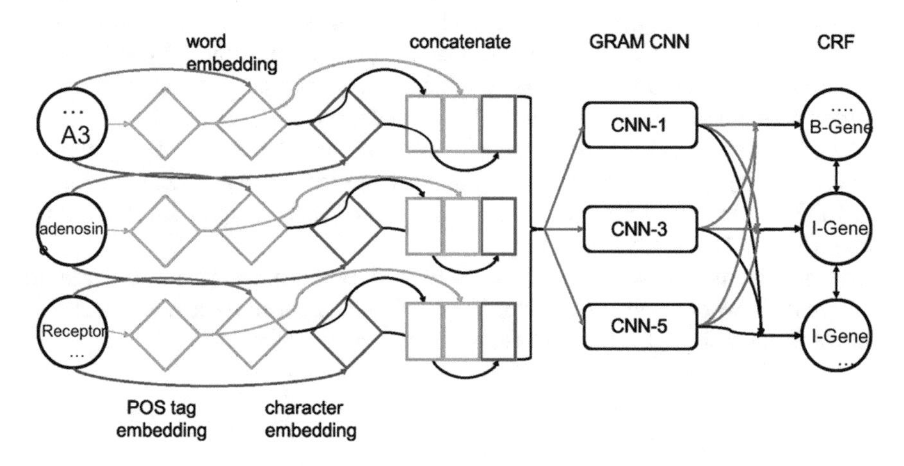

Fig. 2. Architecture of GRAM-CNN Method [12]

Luo et al. (2017) proposed a chemical-NER system using a BiLSTM architecture with CRF layer. As shown in Fig. 3, in this method, the input tokens are tagged with B, I, and O tags and the model predicts the corresponding label. In the embedding layer, the word embeddings are given as input to the model. In the BiLSTM layer, the LSTM computes a context representation of the sequence in the left and right directions. The final CRF layer finds the optimal tags among all possible labels by computing confidence scores. They also add an attention layer to the basic BiLSTM-CRF architecture to capture document-level attention to similar entities. They achieved 91.14% and 92.57% F1-scores on BC5CDR and CHEMDNER datasets, respectively [16].

Dictionary-based, rule-based, and machine learning approaches often rely on extensive dictionaries, rules, or well-structured dataset. Deep Learning-based algorithms that are independent of manually created features and find the features automatically have recently supplanted conventional methods.

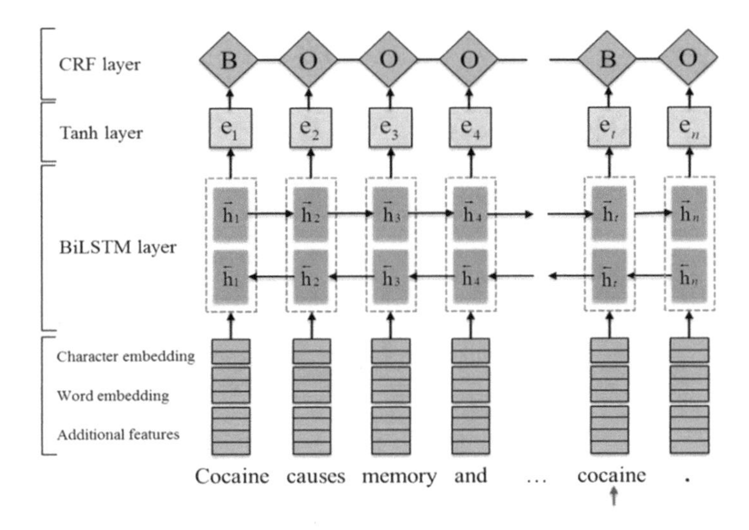

Fig. 3. Basic Structure of BiLSTM-CRF architecture [16]

2.5 Deep Contextualized Embeddings

Significant success has been achieved in many NLP tasks through transformer-based deep contextualized embeddings. BERT and BERT driven models have been re-trained and fine-tuned according to task requirements. Some BERT-based models can be retrained on texts from different domains to create language models that can be used for domain specific NLP tasks. After retraining, A new layer is added, and the weights are optimized to be able to use it for a specific task. This is fine-tuning of a pre-trained embedding model. While BioBERT [2] was trained on biomedical Pubmed and PMC articles (see Fig. 4), ClinicalBioBERT [17] is initialized from BioBERT and trained on electronic health records. Scientific publications from variety domains were used to train SciBERT [18]. RoBERTa [19] modifies key hyperparameters in BERT and trains with much larger mini-batches and learning rates.

In this study, we propose to examine the efficacy of deep contextualized embeddings for disease and symptom identification using a sequence labelling technique and four transformer-based models named BioBERT, Clinical Bert, RoBERTa, and SciBERT.

2.6 Annotated Datasets for Biomedical NER

In this section, we provide an overview of the characteristics, usage, and content of common annotated datasets created for biomedical NER. In supervised learning, the datasets used for NER tasks should be labelled. Labelling entities in texts in the biomedical domain is a difficult and tedious process that requires the assistance of expert physicians. Even if the entities are annotated by the experts, it is difficult to open the datasets, including electronic medical records, radiology reports, and pathology reports, to researchers because of the ethical issues related to patient privacy. Therefore, public scientific articles are used to create the annotated datasets. NCBI [20], CHEMDNER [21], BC5CDR [22],

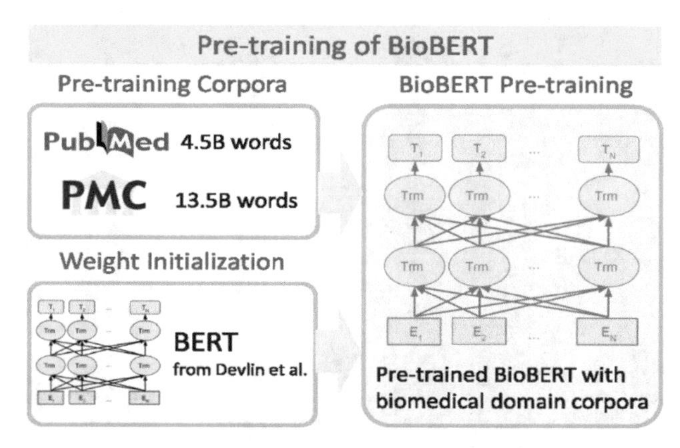

Fig. 4. Pre-training process of BioBERT [2]

JNLPBA [23], and Species-800 [24] are the datasets studied in this context. Information about the datasets can be found in Table 1.

NCBI: The NCBI [20] disease corpus is a tagged resource created by a team of 14 disease name discovery experts. It comprises of 6892 disease names matched to MeSH and OMIM descriptions and 793 PubMed papers.

CHEMDNER: The CHEMDNER [21] is a corpus of BioCreative IV for drug and chemical name identification and contains a total of 84,355 manual names for chemical substances. There are 19,805 unique chemical names extracted from 10,000 thoroughly reviewed abstracts. Although most abstracts mention at least one chemical (8301 abstracts in total), a number of abstracts that were also used for true-negative verification do not mention the substance.

BC5CDR: The BC5CDR [22] corpus was created as part of the BioCreative V organization for the task of disease and chemical name recognition. BioCreative V was developed for tasks based on disease name recognition and chemical-disease relationship extraction. It consists of 1500 items with 4409 labelled chemical names, 5818 disease names, and 3116 associations between diseases and chemicals.

JNLPBA: JNLPBA [23] is a biomedical dataset from the GENIA corpus version 3.02. As a result of a controlled search in MEDLINE, 2000 abstracts were annotated with DNA, RNA, protein, cell type, and cell culture labels.

Species-800: [24] is a dataset of 800 Pubmed abstracts for taxonomy entity name recognition. 100 abstracts were selected from 8 categories (bacteriology, botany, entomology, medicine, mycology, protistology, virology, and zoology).

Although there are datasets for most biomedical entity types, there are few studies on symptom identification. Since extracted entities are needed to extract relationships between diseases and symptoms, an annotated dataset is required for this purpose. Therefore, we rearranged the MedMention [25] dataset, which contains many UMLS concepts including diseases and symptoms, to develop a NER model for identifying diseases and symptoms.

Table 1. Comparison of Biomedical NER Datasets

Dataset	Entity	Content
NCBI	Disease	793 Pubmed Abstracts
CHEMDNER	Chemical, Drug	8301 Pubmed Abstracts
BC5CDR	Disease, Chemical	1500 Pubmed Abstracts
JNLPBA	DNA, RNA, Protein, Cell type, Cell line	2000 Pubmed Abstracts
Species-800	Taxonomies	800 Pubmed Abstracts

3 Experimental Study

In this paper, we conduct an experimental study to identify disease and symptom names from texts. We have redesigned the Medmention [25] dataset, originally developed for entity linking, for this purpose. We develop a biomedical NER approach for disease and symptom recognition by fine tuning four BERT-based deep contextualized embeddings.

3.1 Dataset

We use the Medmention corpus [25], which contains 4392 randomly selected articles (titles and abstracts) from the Pubmed database. In this corpus, many UMLS concepts, including disease and symptom names, were labelled by experts. After annotation of the articles, the dataset is presented in Pubtutor format. To measure the quality of MedMentions annotation, eight articles with a total of 469 concepts were randomly selected from the annotated corpus, and two biologists who were not involved in the annotation task reviewed four articles each. The agreement between the reviewers and the annotators on the labelling accuracy was 97.3%.

In this corpus, the names of diseases and symptoms are labelled as T047 and T184, respectively. Since the names other than disease and symptom names are outside the scope of this work, the corpus needs to be restructured according to the purpose of the work. An example of an excerpt from the corpus is shown in Fig. 5.

3.2 Pre-Processing

Extraction of Abstracts with Disease and Symptom Labels: In this section, article abstracts with symptoms and diseases were extracted from 4392 abstracts. In this process, symptoms must be considered in the summary. At the end of this process, the number of abstracts decreased to 649.

Removal of Tags other than Disease and Symptom Tags: In this section, all tags except T184 (symptom) and T047 (disease) have been removed.

Re-labelling the Abstracts with B,I,O Tags: In the final step, the abstracts of the articles are tokenized and labelled with the B, I and O tags. An example of a sample sentence in IOB format can be found in Fig. 6.

```
27274524|t|Eosinophilic Gastroenteritis as a Rare Cause of Recurrent Epigastric Pain
27274524|a|Eosinophilic gastroenteritis (EGE) is a rare inflammatory disorder of gastrointestinal tract characterized by
eosinophilic infiltration of the bowel wall. It can mimic many gastrointestinal disorders due to its wide spectrum of
presentations. Diagnose is mostly based on excluding other disorders and a high suspicion. Here we report a case of 26 year
old man with a history of sever epigastric pain followed by nausea, vomiting since a few days before admission with final
diagnosis of EGE.
27274524    0     28    Eosinophilic Gastroenteritis     T047    C1262481   ==> Disease
27274524    34    38    Rare    T080    C0522498
27274524    39    44    Cause   T169    C0015127
27274524    48    73    Recurrent Epigastric Pain        UnknownType    C0743544
27274524    74    102   Eosinophilic gastroenteritis     T047    C1262481
27274524    104   107   EGE     T047    C1262481
27274524    119   140   inflammatory disorder    T047    C1290884
27274524    144   166   gastrointestinal tract   T022    C0017189
27274524    167   180   characterized   T052    C1880022
27274524    184   209   eosinophilic infiltration        T046    C0333390
27274524    217   227   bowel wall      T023    C0021853
27274524    247   273   gastrointestinal disorders       T047    C0017178
27274524    285   315   wide spectrum of presentations   T033    C0243095
27274524    317   325   Diagnose        T033    C0011900
27274524    345   354   excluding       T169    C0332196
27274524    355   360   other   T080    C0205394
27274524    361   370   disorders       T047    C0012634
27274524    377   381   high    T080    C0205250
27274524    382   391   suspicion       T041    C0242114
27274524    401   414   report a case   T170    C0007320
27274524    421   425   year    T079    C0439234
27274524    426   429   old     T079    C0580836
27274524    430   433   man     T098    C0025266
27274524    441   448   history T033    C0262926
27274524    452   473   sever epigastric pain    T184    C0232493   ==> Symptom
27274524    486   492   nausea  T184    C0027497
```

Fig. 5. A sample abstract from MedMention dataset. (27274524 refers to the PubMed ID, t refers to title, a refers to abstract)

Token	Tag
Eosinophilic	B-DIS
gastroenteritis	I-DIS
(	O
EGE	B-DIS
)	O
is	O
a	O
rare	O
inflammatory	B-DIS
disorder	I-DIS
of	O
gastrointestinal	O
tract	O
characterized	O
by	O
eosinophilic	O
infiltration	O
of	O
the	O
bowel	O

Token	Tag
man	O
with	O
a	O
history	O
of	O
sever	B-SYM
epigastric	I-SYM
pain	I-SYM
followed	O
by	O
nausea	B-SYM
,	O
vomiting	B-SYM
since	O
a	O
few	O
days	O
before	O
admission	O
with	O

Fig. 6. A sample of tokens after IOB Tagging (B-DIS refers the disease, B-SYM refers the symptoms)

3.3 Experimental Settings

We employed four distinct pre-trained contextual embeddings to test the efficacy of contextual embeddings for biomedical NER tasks: BioBERT, Clinical BioBERT, SciBERT (scivob-cased), and RoBERTa (base). We conducted our experiments using Python, Google Colab platform and GPU runtime. NER model from Simple Transformers library was used for fine tuning of the contextual embedding models.

For training, 80% of the dataset was used, and for testing, 20%. The parameters of the NER models are presented in Table 2. To evaluate the embedding models we use precision, recall, and F1-score classification metrics because we adressed NER problem as token classification.

Table 2. Parameters of NER models

Parameter	Value
num_train_epochs	4
learning_rate	2e-4
train_batch_size	32
eval_batch_size	32
max_seq_length	128

4 Discussion and Results

According to experimental results presented in Table 3, the BioBERT model performs better than the other embedding models on disease and symptom extraction since it was trained on Pubmed literature. Clinical BioBERT is the second successful model because its architecture is based on BioBERT. The SciBERT model performs less than the BioBERT and clinical BioBERT models, which is understandable given that it was trained on a corpus that consists of not only biomedical but also publications from a variety of fields. RoBERTa is a language model that uses the BERT architecture pre-trained on general text. The results show that the domain-specific embeddings are more effective than other contextualized embeddings in biomedical NER.

Table 3. Average Results of NER Methods for disease-symptom extraction

Model	Precision	Recall	F1-Score
RoBERTa	0.72	0.59	0.65
SciBERT	0.75	0.66	0.70
Clinical BioBERT	0.77	0.66	0.71
BioBERT	**0.77**	**0.68**	**0.72**

5 Conclusion and Future Work

In this paper, biomedical NER methods and dataset in the literature are reviewed and a sequence labelling approach is proposed to address the biomedical NER problem. Deep contextualized embeddings for both the general and the medical domains are performed for identifying disease and symptoms. As a result of the experimental study on a dataset that consisting of biomedical texts, we obtained an F1 score of 0.72 for BioBERT, which is a much better performance than three contextual embeddings. Since NER is the preliminary step for relation extraction, the developed models can be used to extract relationships between diseases and symptoms. We are planning to apply relation extraction methods to determine hidden relations between diseases and symptoms. In addition, we will employ contextual embeddings with neural network architectures for NER tasks as future work.

References

1. Li, J., Sun, A., Han, J., Li, C.: A survey on deep learning for named entity recognition. IEEE Trans. Knowl. Data Eng. **34**(1), 50–70 (2020)
2. Lee, J., et al.: BioBERT: a pre-trained biomedical language representation model for biomedical text mining. Bioinformatics **36**(4), 1234–1240 (2020)
3. Zhang, Y., Chen, Q., Yang, Z., Lin, H., Lu, Z.: BioWordVec, improving biomedical word embeddings with subword information and MeSH. Sci. Data **6**(1), 1–9 (2019)
4. Kaddari, Z., Mellah, Y., Berrich, J., Bouchentouf, T., Belkasmi, M.G.: Biomedical question answering: a survey of methods and datasets. In: 2020 Fourth International Conference On Intelligent Computing in Data Sciences (ICDS), pp. 1–8. IEEE (2020)
5. Aramaki, E., Miura, Y., Tonoike, M., Ohkuma, T., Masuichi, H., Ohe, K.: Text2table: Medical text summarization system based on named entity recognition and modality identification. In: Proceedings of the BioNLP 2009 Workshop, pp. 185–192 (2009)
6. Çelikten, A., Uğur, A., Bulut, H.: Keyword extraction from biomedical documents using deep contextualized embeddings. In: 2021 International Conference on INnovations in Intelligent SysTems and Applications (INISTA), pp. 1–5 (2021). https://doi.org/10.1109/INISTA52262.2021.9548470
7. Yang, Z., Lin, H., Li, Y.: Exploiting the performance of dictionary-based bio-entity name recognition in biomedical literature. Comput Biol Chem **32**(4), 287–291 (2008)
8. Aronson, A.R.: Effective mapping of biomedical text to the UMLS metathesaurus: the metamap program. In: Proceedings of the AMIA Symposium, p. 17. American Medical Informatics Association (2001)
9. Kang, N., Singh, B., Afzal, Z., et al.: Using rule-based natural language processing to improve disease normalization in biomedical text. J. Am. Med. Inform. Assoc. **20**(5), 876–881 (2013)
10. Fukuda, K.I., Tsunoda, T., Tamura, A., Takagi, T.: Toward information extraction: identifying protein names from biological papers. In Pac. Symp. Biocomput. **707**(18), 707–718 (1998)
11. Khordad, M., Mercer, R.E., Rogan, P.: A machine learning approach for phenotype name recognition. In: Proceedings of COLING 2012, pp. 1425–1440 (2012)
12. Zhu, Q., Li, X., Conesa, A., Pereira, C.: GRAM-CNN: a deep learning approach with local context for named entity recognition in biomedical text. Bioinformatics **34**(9), 1547–1554 (2018)
13. Kazama, J., Makino, T., Ohta, Y., et al.: Tuning support vector machines for biomedical named entity recognition. In: Proceedings of the ACL-02 Workshop on Natural Language Processing in the Biomedical Domain-vol. 3, pp. 1–8. Association for Computational Linguistics (2002)
14. Kazkılınç, S., Adalı, E.: Koşullu Rastgele Alanlar ile Türkçe Haber Metinlerinin Etiketlenmesi. Türkiye Bilişim Vakfı Bilgisayar Bilimleri ve Mühendisliği Dergisi, **5**(2) (2012)
15. McDonald, R., Pereira, F.: Identifying gene and protein mentions in text using conditional random fields. BMC Bioinform. **6**(1), 1–7 (2005)
16. Luo, L., et al.: An attention-based BiLSTM-CRF approach to document-level chemical named entity recognition. Bioinformatics **34**(8), 1381–1388 (2018)
17. Alsentzer, E., Murphy, J.R., Boag, W., Weng, W.H., Jin, D., Naumann, T., McDermott, M.: Publicly available clinical BERT embeddings. arXiv preprint arXiv:1904.03323 (2019)
18. Beltagy, I., Lo, K., Cohan, A.: SciBERT: A pretrained language model for scientific text. arXiv preprint arXiv:1903.10676 (2019)

19. Liu, Y., et al.: Roberta: A robustly optimized bert pretraining approach. arXiv preprint arXiv:1907.11692 (2019). Doğan, R.I., Leaman, R., Lu, Z.: NCBI disease corpus: a resource for disease name recognition and concept normalization. J. Biomed. Inform. **47**, 1–10 (2014)
20. Doğan, R.I., Leaman, R., Lu, Z.: NCBI disease corpus: a resource for disease name recognition and concept normalization. J. Biomed. Inform. **47**, 1–10 (2014)
21. Krallinger, M., et al.: The CHEMDNER corpus of chemicals and drugs and its annotation principles. J. Cheminform. **7**(1), 1–17 (2015)
22. Li, J., et al.: BioCreative V CDR task corpus: a resource for chemical disease relation extraction. Database **2016**, baw068 (2016). https://doi.org/10.1093/database/baw068
23. Kim, J.D., Ohta, T., Tsuruoka, Y., Tateisi, Y., Collier, N.: Introduction to the bio-entity recognition task at JNLPBA. In: Proceedings of the international joint workshop on natural language processing in biomedicine and its applications, pp. 70–75 (2004)
24. Pafilis, E., et al.: The species and organisms resources for fast and accurate identification of taxonomic names in text. PLoS ONE **8**(6), e65390 (2013)
25. Mohan, S., Li, D.: Medmentions: A large biomedical corpus annotated with umls concepts. arXiv preprint arXiv:1902.09476 (2019)

Numerical Solutions of Hantavirus Infection Model by Means of the Bernstein Polynomials

Şuayip Yüzbaşı[1(✉)] and Gamze Yıldırım[1,2]

[1] Department of Mathematics, Faculty of Science, Akdeniz University, Antalya, Turkey
syuzbasi@akdeniz.edu.tr, suayipyuzbasi@gmail.com, gamzeyildirim@akdeniz.edu.tr
[2] Department of Mathematics, Faculty of Basic Science, Gebze Technical University, Kocaeli, Turkey

Abstract. A collocation method is presented for solving the Hantavirus infection model in this study. The method is based on the Bernstein polynomials. Firstly, the Bernstein polynomials are written in matrix form. Next, the assumed solutions for Hantavirus infection model are written in matrix forms. Also, the nonlinear terms in the model, the derivatives of the solution forms and initial conditions are written in matrix forms. Then, the model problem is transformed into a system of nonlinear algebraic equations by using the collocation points and these matrix forms. Moreover, an error estimation technique is offered and then the residual improvement technique is given. The proposed method is applied for different values of N by selecting the parameters $\alpha = 1, \beta = 0.5, \gamma = 20, \delta = 0.1, \eta = 10, \theta = 10$ in the model. Application results are presented in table and graphs. Moreover, a comparison is made with the result of another method in the literature.

1 Introduction

Hantavirus is an infectious diseases of animal origin. This disease firstly was reported in Canada in 1994. In recent years, for the numerical solutions of the Hantavirus infection model, an exponential matrix method [23], the differential transformation method [10], the variational iteration method [9] have been studied by many researchers. Along the leading ideas of extended thermodynamics, Barbera, Curro and Valenti studied a hyperbolic reaction-diffusion model for describing the hantavirus infection in mice population [6]. Abramson and Kenkre analyzed the traveling waves in a model of the Hantavirus infection in deer mice [2]. Abramson and Kenkre studied a model of the infection of Hantavirus in deer mouse, Peromyscus maniculatus, based on biological observations of the system in the North American Southwest [1]. Allen, Langlais and Phillips developed, analyzed, and numerically simulated an SI epidemic model for a host with two viral infections circulating within the population [3]. Allen, McCormack and

D. J. Hemanth et al. (Eds.): ICAIAME 2022, ECPSCI 7, pp. 230–243, 2023.
https://doi.org/10.1007/978-3-031-31956-3_19

Jonsson formulated and studied two new mathematical models for hantavirus infection in rodents [4].

In recent years, two-step almost collocation method [8], spectral collocation method [11], rational Chebyshev collocation method [17], Legendre-Gauss-Radau collocation method [19], an exponential matrix method [24], block hybrid collocation method [21], Taylor collocation and Adomian decomposition method [7], RKN-type Fourier collocation method [18], Laguerre collocation method [26,27] exponential Fourier collocation method [20] have been studied for solving system of differential equations.

Recently, various methods by using Bernstein polynomials have been studied for solutions of pantograph equations [12], linear integro-differential equations [15], linear second-order partial differential equations with mixed conditions [16], Lane-Emden type equations [13], high order initial and boundary values problems [14], linear third-order differential equations [5], fractional Riccati type differential equations [25], nonlinear Fredholm-Volterra integro-differential equations [22]. The aim of this paper is to develop a collocation method by using the Bernstein polynomials for numerical solutions of the Hantavirus infection model given by

$$\begin{cases} S'(d) = \alpha(S(d)+I(d)) - \beta S(d) - \frac{S(d)(S(d)+I(d))}{\gamma} - \delta S(d)I(d), \\ I'(d) = -\beta I(d) - \frac{I(d)(S(d)+I(d))}{\gamma} + \delta S(d)I(d) \end{cases} \quad 0 \leq d < b < \infty \tag{1}$$

with the initial conditions $S(0) = \eta, I(0) = \theta$. Here, the populations of the susceptible mice and the populations of the infected mice, respectively, are represented by $S(d)$ and $I(d)$. $\alpha, \beta, \gamma, \delta, \eta$ and θ show the constant parameters. δ, α and β represent, respectively, infection rates, birth rates and death rates. γ represents the environmental parameter. η denotes the initial number of the susceptible mice and θ represents the initial number of the infected mice.

The approximate solutions of (1) are examined as

$$\begin{cases} S_N(d) = \sum_{i=0}^{N} a_{1,i} B_{i,N}(d) \\ I_N(d) = \sum_{i=0}^{N} a_{2,i} B_{i,N}(d) \end{cases} \tag{2}$$

Here, $a_{1,i}, a_{2,i}$ and $B_{i,N}(d)$ represent, respectively, the unknown coefficients of $S_N(d)$, the unknown coefficients of $I_N(d)$, and the Bernstein polynomials. The Bernstein polynomials are defined by [12,25]

$$B_{n,N}(d) = \sum_{k=0}^{N-n} (-1)^k \binom{N}{n} \binom{N-n}{k} \frac{d^{n+k}}{b^{n-k}}. \tag{3}$$

There are many studies based on the Bernstein polynomials in the literature. The results of these studies have been quite effective and successful. For this reason, the Bernstein polynomials are used in this study.

Another aim of this study is to offer an error estimation method and then to make the residual improvement technique. Another aim of this study is to prepare a code of the suggested method in MATLAB and to apply the presented methods. Obtaining effective results as a result of the application and presenting the results in table and graphs is another aim of the study.

2 Method of Solution

Firstly, the Bernstein polynomials are expressed in matrix form [5,12–16,22,25] as follows

$$\mathbf{B}(d) = \mathbf{D}(d)\mathbf{M}^T \tag{4}$$

where $\mathbf{B}(d) = \left[B_{0,N}(d)\ B_{1,N}(d) \cdots B_{N,N}(d) \right]$, $\mathbf{D}(d) = \left[1\ d \cdots d^N \right]$ and

$$\mathbf{M} = \begin{bmatrix} m_{00} & m_{01} & \cdots & m_{0N} \\ m_{10} & m_{11} & \cdots & m_{1N} \\ \vdots & \vdots & \ddots & \vdots \\ m_{N0} & m_{N1} & \cdots & m_{NN} \end{bmatrix}, \quad m_{ij} = \begin{cases} \frac{(-1)^{j-i}}{b^j} \binom{N}{i} \binom{N-i}{j-i}, & i \le j \\ 0, & i > j. \end{cases}$$

Now, the solutions (2) can be written in forms

$$\begin{cases} S_N(d) = \mathbf{B}(d)\mathbf{A}_1 \\ I_N(d) = \mathbf{B}(d)\mathbf{A}_2 \end{cases} \tag{5}$$

where $\mathbf{A}_1 = \left[a_{1,0}\ a_{1,1} \cdots a_{1,N} \right]^T$ and $\mathbf{A}_2 = \left[a_{2,0}\ a_{2,1} \cdots a_{2,N} \right]^T$.

Then, the matrix relation (4) is substituted in (5) and so the Bernstein polynomial solutions of (1) become as follows:

$$\begin{cases} S_N(d) = \mathbf{D}(d)\mathbf{M}^T\mathbf{A}_1 \\ I_N(d) = \mathbf{D}(d)\mathbf{M}^T\mathbf{A}_2. \end{cases} \tag{6}$$

Secondly, taking the first derivative of (6), we have

$$\begin{cases} S_N^{'}(d) = \mathbf{D}(d)\mathbf{B}^T\mathbf{M}^T\mathbf{A}_1 \\ I_N^{'}(d) = \mathbf{D}(d)\mathbf{B}^T\mathbf{M}^T\mathbf{A}_2 \end{cases} \tag{7}$$

where

$$\mathbf{B} = \begin{bmatrix} 0 & 0 & 0 & \cdots & 0 & 0 \\ 1 & 0 & 0 & \cdots & 0 & 0 \\ 0 & 2 & 0 & \cdots & 0 & 0 \\ \vdots & \vdots & \vdots & \ddots & \vdots & \vdots \\ 0 & 0 & 0 & \cdots & N & 0 \end{bmatrix}.$$

On the other hand, using (6), nonlinear terms in the Hantavirus infection model (1) are written in the matrix forms

$$\begin{cases} S_N(d)S_N(d) = \mathbf{D}(d)\mathbf{M}^T\mathbf{A}_1\mathbf{D}(d)\mathbf{M}^T\mathbf{A}_1, \\ S_N(d)I_N(d) = \mathbf{D}(d)\mathbf{M}^T\mathbf{A}_1\mathbf{D}(d)\mathbf{M}^T\mathbf{A}_2, \\ I_N(d)S_N(d) = \mathbf{D}(d)\mathbf{M}^T\mathbf{A}_2\mathbf{D}(d)\mathbf{M}^T\mathbf{A}_1, \\ I_N(d)I_N(d) = \mathbf{D}(d)\mathbf{M}^T\mathbf{A}_2\mathbf{D}(d)\mathbf{M}^T\mathbf{A}_2. \end{cases} \tag{8}$$

As the third, using (6), the matrix relations of the initial conditions $S(0) = \eta$ and $I(0) = \theta$ are written as, respectively,

$$\begin{cases} S_N(0) = \eta \rightarrow \mathbf{D}(0)\mathbf{M}^T\mathbf{A}_1 = \eta \\ I_N(0) = \theta \rightarrow \mathbf{D}(0)\mathbf{M}^T\mathbf{A}_2 = \theta \end{cases} \tag{9}$$

or

$$\begin{cases} \mathbf{U}\mathbf{A}_1 = \eta,\, \mathbf{U} = \mathbf{D}(0)\mathbf{M}^T, \\ \mathbf{V}\mathbf{A}_2 = \theta,\ \mathbf{V} = \mathbf{D}(0)\mathbf{M}^T. \end{cases} \tag{10}$$

As the next step, the matrix relations (6), (7) and (8) are substituted in the Hantavirus infection model (1) and thus we get the matrix relation

$$\begin{cases} \mathbf{D}(d)\mathbf{B}^T\mathbf{M}^T\mathbf{A}_1 - (\alpha-\beta)\mathbf{D}(d)\mathbf{M}^T\mathbf{A}_1 - \alpha\mathbf{D}(d)\mathbf{M}^T\mathbf{A}_2 + \frac{1}{\gamma}\left(\mathbf{D}(d)\mathbf{M}^T\mathbf{A}_1\mathbf{D}(d)\mathbf{M}^T\mathbf{A}_1\right) \\ \qquad + \left(\frac{1}{\gamma}+\delta\right)\mathbf{D}(d)\mathbf{M}^T\mathbf{A}_1\mathbf{D}(d)\mathbf{M}^T\mathbf{A}_2 = 0, \end{cases}$$

and

$$\begin{cases} \mathbf{D}(d)\mathbf{B}^T\mathbf{M}^T\mathbf{A}_2 + \beta\mathbf{D}(d)\mathbf{M}^T\mathbf{A}_2 + \frac{1}{\gamma}\left(\mathbf{D}(d)\mathbf{M}^T\mathbf{A}_2\mathbf{D}(d)\mathbf{M}^T\mathbf{A}_1\right) - \frac{1}{\gamma}\left(\mathbf{D}(d)\mathbf{M}^T\mathbf{A}_2\mathbf{D}(d)\mathbf{M}^T\mathbf{A}_2\right) \\ \qquad +\delta\mathbf{D}(d)\mathbf{M}^T\mathbf{A}_1\mathbf{D}(d)\mathbf{M}^T\mathbf{A}_2 = 0. \end{cases} \tag{11}$$

(12)

Now, we consider the collocation points which are described by

$$d_i = \frac{b}{N}i, \quad i = 0, 1, ..., N. \tag{13}$$

The collocation points d_i are written instead of d in (11)–(12) and so, we have

$$\begin{bmatrix} \mathbf{W}_0\mathbf{A}_1 + \mathbf{X}_0\mathbf{A}_2 \\ \mathbf{Y}_0\mathbf{A}_1 + \mathbf{T}_0\mathbf{A}_2 \\ \vdots \\ \mathbf{W}_N\mathbf{A}_1 + \mathbf{X}_N\mathbf{A}_2 \\ \mathbf{Y}_N\mathbf{A}_1 + \mathbf{T}_N\mathbf{A}_2 \end{bmatrix} = \begin{bmatrix} 0 \\ 0 \\ \vdots \\ 0 \\ 0 \end{bmatrix} \tag{14}$$

where

$$\begin{aligned} &\mathbf{W}_i = \mathbf{D}(d_i)\mathbf{B}^T\mathbf{M}^T - (\alpha-\beta)\mathbf{D}(d_i)\mathbf{M}^T + \frac{1}{\gamma}\left(\mathbf{D}(d_i)\mathbf{M}^T\mathbf{A}_1\mathbf{D}(d_i)\mathbf{M}^T\right), \\ &\mathbf{X}_i = -\alpha\mathbf{D}(d_i)\mathbf{M}^T + \left(\frac{1}{\gamma}+\delta\right)\mathbf{D}(d_i)\mathbf{M}^T\mathbf{A}_1\mathbf{D}(d_i)\mathbf{M}^T, \\ &\mathbf{Y}_i = \frac{1}{\gamma}\left(\mathbf{D}(d_i)\mathbf{M}^T\mathbf{A}_2\mathbf{D}(d_i)\mathbf{M}^T\right), \\ &\mathbf{T}_i = \mathbf{D}(d_i)\mathbf{B}^T\mathbf{M}^T + \beta\mathbf{D}(d_i)\mathbf{M}^T - \frac{1}{\gamma}\left(\mathbf{D}(d_i)\mathbf{M}^T\mathbf{A}_2\mathbf{D}(d_i)\mathbf{M}^T\right) + \delta\mathbf{D}(d_i)\mathbf{M}^T \\ &\mathbf{A}_1\mathbf{D}(d_i)\mathbf{M}^T. \end{aligned}$$

Finally, systems (10) and (14) are expressed as a single system and the obtained last system is solved in MATLAB. By solving this new system we get the Bernstein coefficient matrices $\mathbf{A}_1$ and $\mathbf{A}_2$ in (6). Substituting the calculated coefficient matrices $\mathbf{A}_1$ and $\mathbf{A}_2$ in (6), the Bernstein polynomial solutions of (1) are obtained. Note here that the dimensions of the matrices in 14 are $2(N \times 1)$.

3 Error Estimation Method

Let's define the error functions for the Hantavirus infection model (1) as

$$\begin{cases} e_{1,N}(d) = S(d) - S_N(d), \\ e_{2,N}(d) = I(d) - I_N(d). \end{cases} \tag{15}$$

By substituting the Bernstein polynomial solutions, which obtained in previous section, in (1), we have the residual functions

$$\begin{cases} R_{1,N} = S_N^{'} - \alpha(S_N + I_N) + \beta S_N + \frac{S_N(S_N+I_N)}{\gamma} + \delta S_N I_N, \\ R_{2,N} = I_N^{'} + \beta I_N + \frac{I_N(S_N+I_N)}{\gamma} - \delta S_N I_N. \end{cases} \tag{16}$$

Here, $S(d)$ and $I(d)$ represent the exact solutions of (1). $S_N(d)$ and $I_N(d)$ represent the Bernstein polynomial solutions of (1). $e_{1,N}(d)$ and $e_{2,N}(d)$ represent the actual error functions of (1). $R_{1,N}(d)$ and $R_{2,N}(d)$ represent the residual functions of (1).

Since the Bernstein polynomial solutions of (1) yield the initial conditions of the Hantavirus infection model (1), we can write

$$S_N(0) = \eta, \quad \text{and} \quad I_N(0) = \theta. \tag{17}$$

Now, subtracting (16)–(17) from Hantavirus infection model (1) and the initial conditions $S(0) = \eta$ and $I(0) = \theta$, the error problem is obtained as

$$\begin{cases} e_{1,N}^{'} = (\alpha - \beta - \frac{2S_N+I_N}{\gamma} - \delta I_N)e_{1,N} - \frac{1}{\gamma}e_{1,N}^2 + (\alpha - \frac{S_N}{\gamma} - \delta S_N)e_{2,N} + (-\frac{1}{\gamma} - \delta)e_{1,N}e_{2,N} + R_{1,N}, \\ e_{2,N}^{'} = (-\beta - \frac{2I_N+S_N}{\gamma} + \delta S_N)e_{2,N} - \frac{1}{\gamma}e_{2,N}^2 + (-\frac{I_N}{\gamma} + \delta I_N)e_{1,N} + (-\frac{1}{\gamma} + \delta)e_{1,N}e_{2,N} + R_{2,N}, \\ e_{1,N}(0) = 0, \quad e_{2,N}(0) = 0. \end{cases} \tag{18}$$

Note that here, using (15), $e_{1,N}(d) + S_N(d)$ is written instead of $S(d)$. Similarly, $e_{2,N}(d) + I_N(d)$ is written instead of $I(d)$.

Consequently, by solving this error problem (18) according to the Bernstein collocation method in Sect. 2, we have the estimated error functions

$$\begin{cases} e_{1,N,M}(d) = \sum_{i=0}^{M} a_{1,i}^{*} B_{i,N}(d) \\ e_{2,N,M}(d) = \sum_{i=0}^{M} a_{2,i}^{*} B_{i,N}(d). \end{cases} \tag{19}$$

Thus, we can get the improvement polynomial solutions by

$$\begin{cases} S_{N,M}(d) = S_N(d) + e_{1,N,M}(d) \\ I_{N,M}(d) = I_N(d) + e_{2,N,M}(d), \end{cases} \tag{20}$$

and the errors of them become as follows:

$$\begin{cases} E_{1,N,M}(d) = S(d) - S_{N,M}(d) \\ E_{2,N,M}(d) = I(d) - I_{N,M}(d). \end{cases} \tag{21}$$

4 Numerical Results and Discussion

In this section, the parameters of the Hantavirus infection model (1) and the initial conditions $S(0) = \eta$ and $I(0) = \theta$, are selected as $\alpha = 1, \beta = 0.5, \gamma = 20, \delta = 0.1, \eta = 10, \theta = 10$. The results obtained according to the selected parameters are shown in tables and graphics.

According to the selected parameters, the Hantavirus infection model (1) becomes

$$\begin{cases} S'(d) = \alpha(S(d) + I(d)) - 0.5\,S(d) - \frac{S(d)(S(d)+I(d))}{20} - 0.1S(d)I(d), \\ I'(d) = -0.5I(d) - \frac{I(d)(S(d)+I(d))}{20} + 0.1S(d)I(d), \\ S(0) = 10, \quad I(0) = 10. \end{cases} \tag{22}$$

According to the Bernstein collocation method in Sect. 2, the Bernstein polynomial solutions for $N = 3$ are investigated as

$$\begin{cases} S_3(d) = \sum_{i=0}^{3} a_{1,i} B_{i,N}(d) \\ I_3(d) = \sum_{i=0}^{3} a_{2,i} B_{i,N}(d). \end{cases} \tag{23}$$

Using (6), the Bernstein polynomial solutions of (1) are sought in forms

$$\begin{cases} S_3(d) = \mathbf{D}(d)\mathbf{M}^T\mathbf{A}_1 \\ I_3(d) = \mathbf{D}(d)\mathbf{M}^T\mathbf{A}_2 \end{cases} \tag{24}$$

or

$$\begin{cases} S_3(d) = \begin{bmatrix} 1 \; d \; d^2 \; d^N \end{bmatrix} \begin{bmatrix} 1 & 0 & 0 & 0 \\ -3 & 3 & 0 & 0 \\ 3 & -6 & 3 & 0 \\ -1 & 3 & -3 & 1 \end{bmatrix} \begin{bmatrix} a_{1,0} \\ a_{1,1} \\ a_{1,2} \\ a_{1,3} \end{bmatrix} \\ I_3(d) = \begin{bmatrix} 1 \; d \; d^2 \; d^N \end{bmatrix} \begin{bmatrix} 1 & 0 & 0 & 0 \\ -3 & 3 & 0 & 0 \\ 3 & -6 & 3 & 0 \\ -1 & 3 & -3 & 1 \end{bmatrix} \begin{bmatrix} a_{2,0} \\ a_{2,1} \\ a_{2,2} \\ a_{2,3} \end{bmatrix} \end{cases}. \tag{25}$$

The collocation points for the range $[0, 5]$ are determined as $d_0 = 0, d_1 = 5/3, d_2 = 10/3, d_3 = 5$. According to system (14), we can write

$$\begin{bmatrix} \mathbf{W}_0\mathbf{A}_1 + \mathbf{X}_0\mathbf{A}_2 \\ \mathbf{Y}_0\mathbf{A}_1 + \mathbf{T}_0\mathbf{A}_2 \\ \mathbf{W}_1\mathbf{A}_1 + \mathbf{X}_1\mathbf{A}_2 \\ \mathbf{Y}_1\mathbf{A}_1 + \mathbf{T}_1\mathbf{A}_2 \\ \mathbf{W}_2\mathbf{A}_1 + \mathbf{X}_2\mathbf{A}_2 \\ \mathbf{Y}_2\mathbf{A}_1 + \mathbf{T}_2\mathbf{A}_2 \\ \mathbf{W}_3\mathbf{A}_1 + \mathbf{X}_3\mathbf{A}_2 \\ \mathbf{Y}_3\mathbf{A}_1 + \mathbf{T}_3\mathbf{A}_2 \end{bmatrix} = \begin{bmatrix} 0 \\ 0 \\ 0 \\ 0 \\ 0 \\ 0 \\ 0 \\ 0 \end{bmatrix} \tag{26}$$

where

$$\begin{aligned}
&\mathbf{W}_i = \mathbf{D}(d_i)\mathbf{B}^T\mathbf{M}^T - 0.5\mathbf{D}(d_i)\mathbf{M}^T + \tfrac{1}{20}\left(\mathbf{D}(d_i)\mathbf{M}^T\mathbf{A}_1\mathbf{D}(d_i)\mathbf{M}^T\right),\\
&\mathbf{X}_i = -\mathbf{D}(d_i)\mathbf{M}^T + \left(\tfrac{1}{20}+0.1\right)\mathbf{D}(d_i)\mathbf{M}^T\mathbf{A}_1\mathbf{D}(d_i)\mathbf{M}^T,\\
&\mathbf{Y}_i = \tfrac{1}{20}\left(\mathbf{D}(d_i)\mathbf{M}^T\mathbf{A}_2\mathbf{D}(d_i)\mathbf{M}^T\right),\\
&\mathbf{T}_i = \mathbf{D}(d_i)\mathbf{B}^T\mathbf{M}^T + 0.5\mathbf{D}(d_i)\mathbf{M}^T - \tfrac{1}{20}\left(\mathbf{D}(d_i)\mathbf{M}^T\mathbf{A}_2\mathbf{D}(d_i)\mathbf{M}^T\right) + 0.1\mathbf{D}(d_i)\mathbf{M}^T\\
&\mathbf{A}_1\mathbf{D}(d_i)\mathbf{M}^T,\\
&\mathbf{A}_1 = \left[a_{1,0}\; a_{1,1}\; a_{1,2}\; a_{1,3}\right]^T,\\
&\mathbf{A}_2 = \left[a_{2,0}\; a_{2,1}\; a_{2,2}\; a_{2,3}\right]^T.
\end{aligned}$$

The matrix relations of the initial conditions are

$$\begin{cases}\mathbf{D}(0)\mathbf{M}^T\mathbf{A}_1 = 10\\ \mathbf{D}(0)\mathbf{M}^T\mathbf{A}_2 = 10\end{cases} \tag{27}$$

or

$$\begin{cases}\left[1\,0\,0\,0\right]\begin{bmatrix}1 & 0 & 0 & 0\\ -3 & 3 & 0 & 0\\ 3 & -6 & 3 & 0\\ -1 & 3 & -3 & 1\end{bmatrix}\begin{bmatrix}a_{1,0}\\ a_{1,1}\\ a_{1,2}\\ a_{1,3}\end{bmatrix} = 10,\\ \left[1\,0\,0\,0\right]\begin{bmatrix}1 & 0 & 0 & 0\\ -3 & 3 & 0 & 0\\ 3 & -6 & 3 & 0\\ -1 & 3 & -3 & 1\end{bmatrix}\begin{bmatrix}a_{2,0}\\ a_{2,1}\\ a_{2,2}\\ a_{2,3}\end{bmatrix} = 10.\end{cases} \tag{28}$$

From here, the obtained system by combining (26) and (28) is solved with the help of MATLAB. So, by calculating the coefficient matrices $\mathbf{A}_1$ and $\mathbf{A}_1$, they are written in (24). As a result, the Bernstein polynomial solutions become as below:

$$\begin{cases}S_3(d) = 9.3875 - 3.146d + 1.2633d^2 - 0.13133d^3\\ I_3(d) = 9.7095 - 4.8528d + 1.1148d^2 - 0.089204d^3.\end{cases} \tag{29}$$

The graphs of populations of the susceptible mice $S_N(d)$ for $N = 5$ with present method (the Bernstein polynomial solutions), the graphs of populations of the susceptible mice $S_{N,M}(d)$ for $(N, M) = (5, 6)$ with present method (the improvement Bernstein polynomial solutions) and the graphs of populations of the susceptible mice $S_N(d)$ for $N = 5$ with exponential matrix method (EMM) [23] are shown in Fig. 1. The graphs of populations of the infected mice $I_N(d)$ for $N = 5$ with present method (the Bernstein polynomial solutions), the graphs of populations of the infected mice $I_{N,M}(d)$ for $(N, M) = (5, 6)$ with present method (the improvement Bernstein polynomial solutions) and the graphs of populations of the infected mice $I_N(d)$ for $N = 5$ with exponential matrix method (EMM) [23] are displayed in Fig. 2.

We show the absolute residual errors of the solution $S_N(d)$ for $N = 5, N = 8$ and $N = 10$ in Fig. 3. We display the absolute residual errors of the solution $I_N(d)$ for $N = 5, N = 8$ and $N = 10$ in Fig. 4. Similarly, these residual errors are also given for various points of d_i in Table 1. The estimated error function of the solution $S_N(d)$ is presented for $(N, M) = (3, 4)$ in Fig. 5. The estimated error function of the solution $I_N(d)$ is displayed for $(N, M) = (3, 4)$ in Fig. 6.

In Fig. 7, the absolute residual errors of the solution $S_N(d)$ for $N = 8$ are compared with EMM [23]. In Fig. 8, the absolute residual errors of the solution $I_N(d)$ for $N = 8$ are compared with EMM [23].

It is interpreted from Fig. 3, Fig. 4 and Table 1 that the errors decrease while the value of N in the method increases. It is said from Fig. 5 and Fig. 6 that the results of the estimation errors for $(N, M) = (3, 4)$ are quite satisfactory. Thanks to this error estimation method, in case the exact solution is not known, it is possible to comment on the error made in the method. According to Fig. 1 and Fig. 2, it can be said that there was a decrease in the populations of susceptible mice and infected mice at 3 days. It is also seen that the populations of the susceptible mice are more resistant. On the other hand, in Fig. 1 and Fig. 2, the present method is also compared with EMM [23]. According to this comparison, it is seen that the presented method for $N = 5$ gives similar results with EMM [23] for $N = 5$. However, when Fig. 7 and Fig. 8 are compared for $N = 8$, it is observed that the presented method yields more successful results than EMM [23] for $N = 8$.

Table 1. The residual errors for $N = 5, N = 8, N = 10$

d_i	$\lvert R_{1,5}(d)\rvert$	$\lvert R_{1,8}(d)\rvert$	$\lvert R_{1,10}(d)\rvert$	$\lvert R_{2,5}(d)\rvert$	$\lvert R_{2,8}(d)\rvert$	$\lvert R_{2,10}(d)\rvert$
0	5.0826e-02	9.4103e-05	1.1497e-09	1.0854e-02	7.4129e-05	9.4962e-10
0.2	4.6743e-01	1.9294e-05	1.5016e-10	5.4908e-02	1.6077e-05	1.2854e-10
0.4	5.6875e-01	2.4660e-06	1.1187e-11	6.3661e-02	2.2701e-06	9.9637e-12
0.6	4.6227e-01	1.7734e-07	5.8617e-13	4.5978e-02	2.2169e-07	7.8107e-13
0.8	2.7702e-01	2.6901e-07	3.4087e-13	2.0301e-02	3.5244e-07	3.0151e-13
1	8.8123e-02	1.3546e-06	2.4607e-13	3.1990e-03	1.1529e-06	8.2710e-13

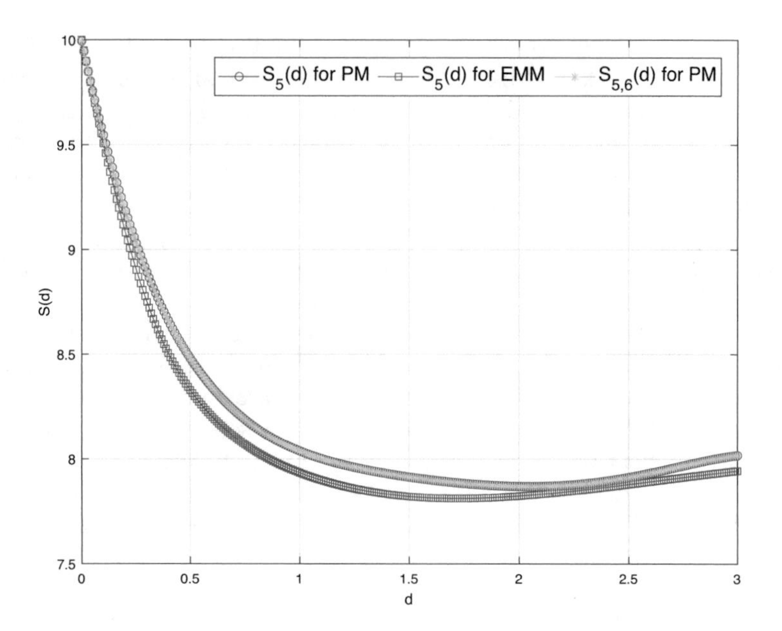

Fig. 1. Comparison of solutions $S(d)$ with EMM [23]

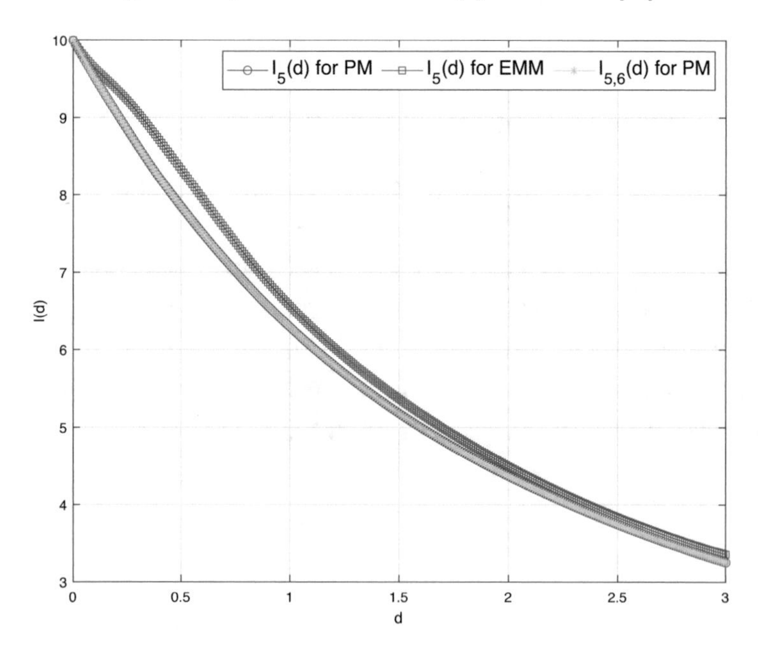

Fig. 2. Comparison of solutions $I(d)$ with EMM [23]

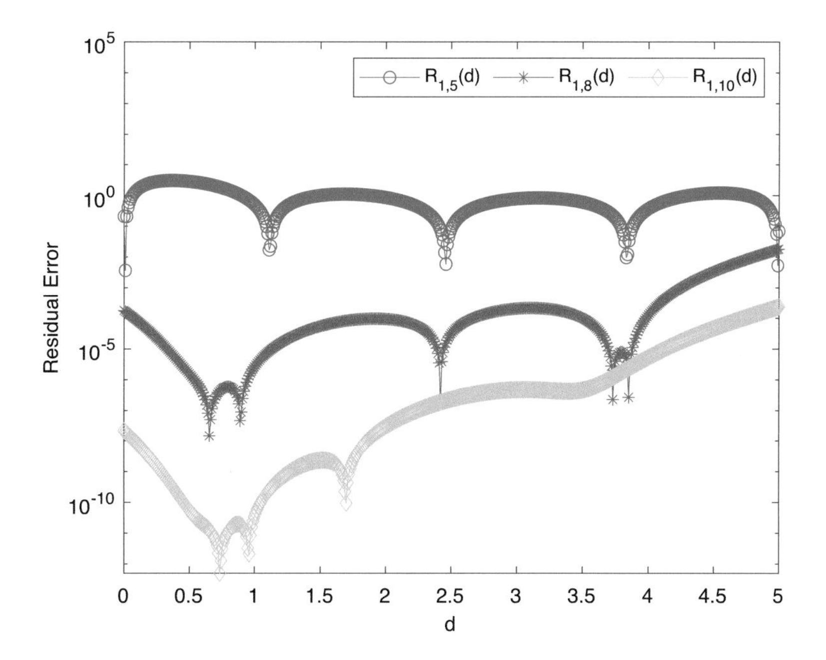

Fig. 3. The residual errors of $S_N(d)$ for $N = 5, N = 8, N = 10$

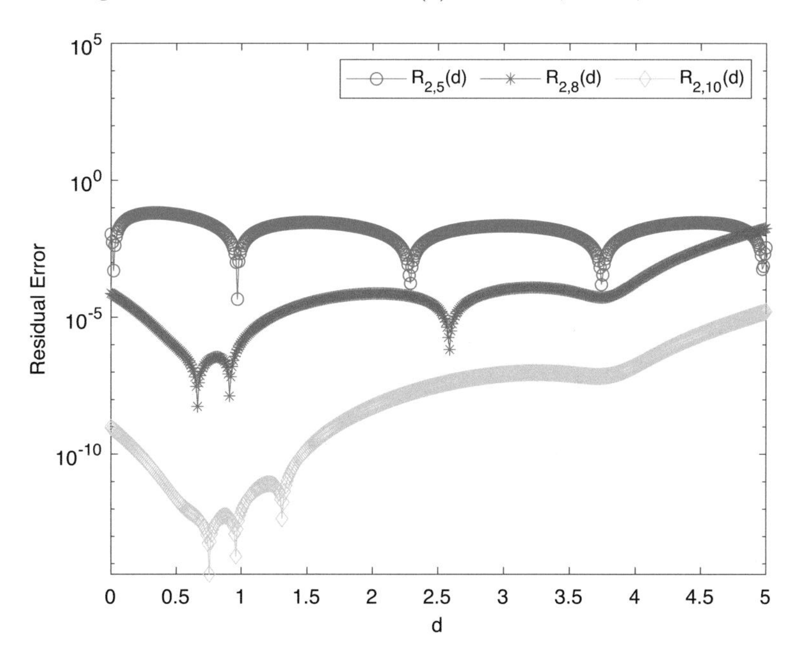

Fig. 4. The residual errors of $I_N(d)$ for $N = 5, N = 8, N = 10$

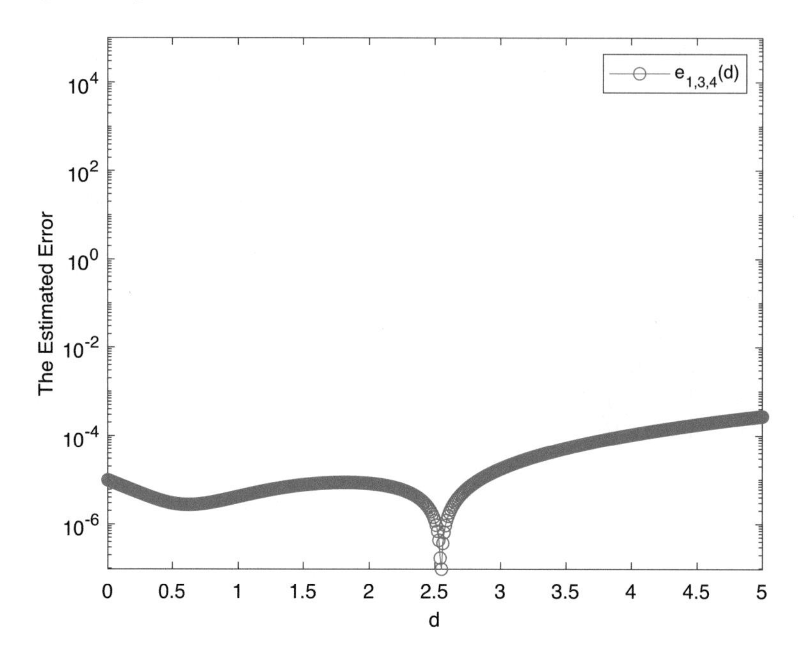

Fig. 5. The estimated error of $S_N(d)$ for $(N, M) = (3, 4)$

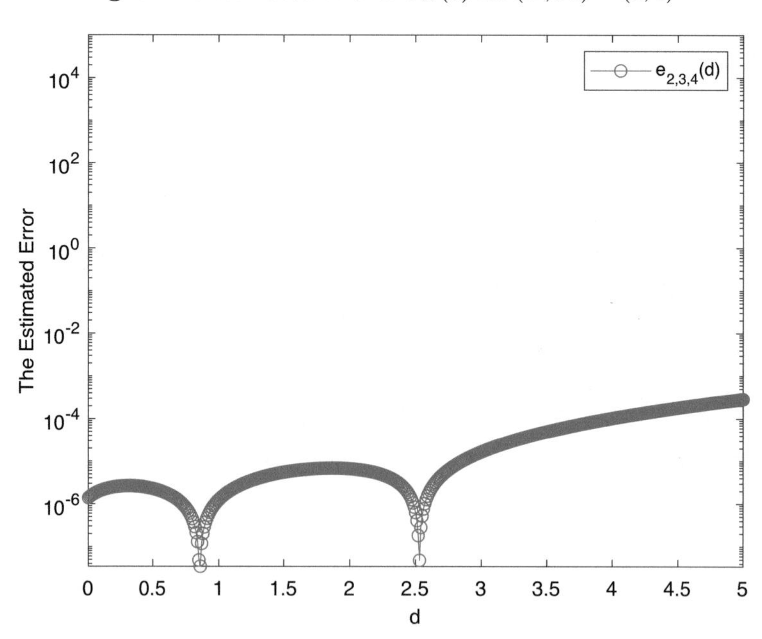

Fig. 6. The estimated error of $I_N(d)$ for $(N, M) = (3, 4)$

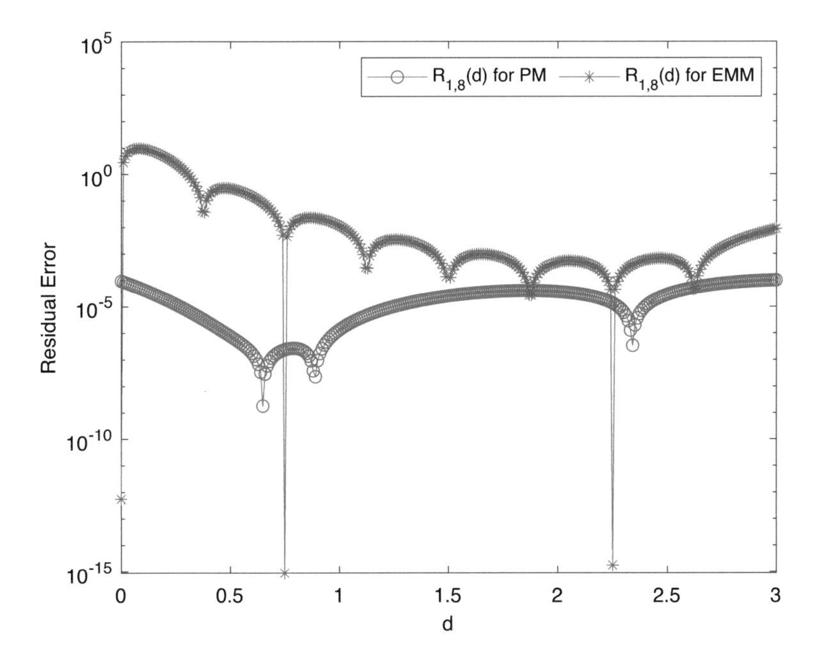

Fig. 7. Comparison of residual errors for $S_N(d)$ with EMM [23]

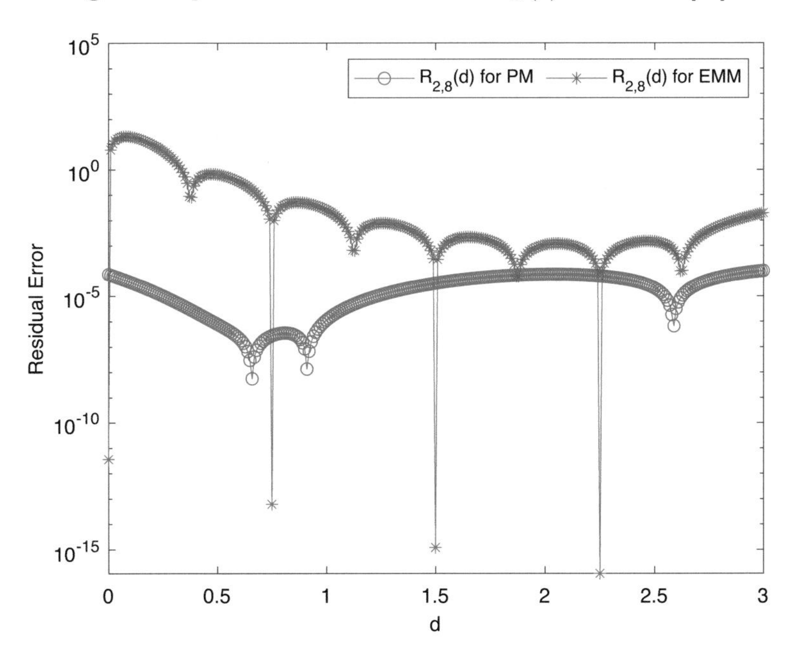

Fig. 8. Comparison of residual errors for $I_N(d)$ with EMM [23]

5 Conclusion

We present a collocation method based on the Bernstein polynomials for the Hantavirus infection model in this study. Also, we give an error estimation method by using the residual function and we offer the residual improvement technique. Then, we make numerical applications of the method for the parameters $\alpha = 1, \beta = 0.5, \gamma = 20, \delta = 0.1, \eta = 10, \theta = 10$. According to the selected parameters, we generate a code of the method in MATLAB. Then, we show the obtained numerical results in table and graphs. Also, we make a comparison with exponential matrix method (EMM) [23] in the literature. According to the comparisons, it is observed that the presented method is more effective than EMM [23]. According to the table and graphics, it is seen that the method is successful and effective.

References

1. Abramson, G., Kenkre, V.M.: Spatiotemporal patterns in the Hantavirus infection. Phys. Rev. E **66**, 011912 (2002)
2. Abramson, G., Kenkre, V.M., Yates, T.L., Parmenter, R.R.: Traveling waves of infection in the hantavirus epidemics. Bull. Math. Biol. **65**, 519–534 (2003)
3. Allen, L.J., Langlais, M., Phillips, C.J.: The dynamics of two viral infections in a single host population with applications to hantavirus. Math. Biosci. **186**, 191–217 (2003)
4. Allen, L.J., McCormack, R.K., Jonsson, C.B.: Mathematical models for hantavirus infection in rodents. Bull. Math. Biol. **68**, 511–524 (2006)
5. Aydin, T.A., Sezer, M., Kocayigit, H.: Bernsteinn polynomials approach to determine timelike curves of constant breadth in Minkowski 3-space. Commun. Math. Model. Appl. **3**, 9–22 (2018)
6. Barbera, E., Curro, C., Valenti, G.: A hyperbolic reaction-diffusion model for the hantavirus infection. Math. Methods Appl. Sci. **31**, 481–499 (2008)
7. Bildik, N., Deniz, S.: Implementation of Taylor collocation and Adomian decomposition method for systems of ordinary differential equations. In: AIP Conference Proceedings, vol. 1648, p. 370002 (2015)
8. D'Ambrosio, R., Ferro, M., Jackiewicz, Z., Paternoster, B.: Two-step almost collocation methods for ordinary differential equations. Numer. Algorithms **53**, 195–217 (2010)
9. Goh, S.M., Ismail, A.I.M., Noorani, M.S.M., Hashim, I.: Dynamics of the Hantavirus infection through variational iteration method. Nonlinear Anal. Real World Appl. **10**, 2171–2176 (2009)
10. Gökdoğan, A., Merdan, M., Yildirim, A.: A multistage differential transformation method for approximate solution of Hantavirus infection model. Commun. Nonlinear Sci. Numer. Simul. **17**, 1–8 (2012)
11. Guo, B.Y.,Wang, Z.Q.: A spectral collocation method for solving initial value problems of first order ordinary differential equations. Discrete Contin. Dyn. Syst.-B **14**, 1029 (2010)
12. Işik, O.R., Güney, Z., Sezer, M.: Bernstein series solutions of pantograph equations using polynomial interpolation. J. Differ. Equ. Appl. **18**, 357–374 (2012)

13. Işik O. R., Sezer, M.: Bernstein series solution of a class of Lane-Emden type equations. Math. Probl. Eng. **2013** (2013)
14. Işik, O.R., Sezer, M., Güney, Z.: A rational approximation based on Bernstein polynomials for high order initial and boundary values problems. Appl. Math. Comput. **217**, 9438–9450 (2011)
15. Işik, O.R., Sezer, M., Güney, Z.: Bernstein series solution of a class of linear integro-differential equations with weakly singular kernel. Appl. Math. Comput. **217**, 7009–7020 (2011)
16. Işik, O.R., Sezer, M., Güney, Z.: Bernstein series solution of linear second-order partial differential equations with mixed conditions. Math. Methods Appl. Sci. **37**, 609–619 (2014)
17. Sezer, M., Gülsu, M., Tanay, B.: Rational Chebyshev collocation method for solving higher-order linear ordinary differential equations. Numer. Methods Partial Differ. Equ. **27**, 1130–1142 (2011)
18. Wang, B., Meng, F., Fang, Y.: Efficient implementation of RKN-type Fourier collocation methods for second-order differential equations. Appl. Numer. Math. **119**, 164–178 (2017)
19. Wang, Z.Q., Guo, B.Y.: Legendre-Gauss-Radau collocation method for solving initial value problems of first order ordinary differential equations. J. Sci. Comput. **52**, 226–255 (2012)
20. Wu, X., Wang, B.: Exponential Fourier collocation methods for first-order differential equations. In: Recent Developments in Structure-Preserving Algorithms for Oscillatory Differential Equations, pp. 55–84. Springer, Singapore (2018). https://doi.org/10.1007/978-981-10-9004-2_3
21. Yap, L. K., Ismail, F., Senu, N.: An accurate block hybrid collocation method for third order ordinary differential equations. J. Appl. Math. **2014** (2014)
22. Yüzbaşı, Ş: A collocation method based on Bernstein polynomials to solve nonlinear Fredholm-Volterra integro-differential equations. Appl. Math. Comput. **273**, 142–154 (2016)
23. Yüzbaşi, Ş, Sezer, M.: An exponential matrix method for numerical solutions of Hantavirus infection model. Appl. Appl. Math. Int. J. (AAM) **8**, 98–115 (2013)
24. Yüzbaşı, Ş, Sezer, M.: An exponential matrix method for solving systems of linear differential equations. Math. Methods Appl. Sci. **36**, 336–348 (2013)
25. Yüzbaşı, Ş: Numerical solutions of fractional Riccati type differential equations by means of the Bernstein polynomials. Appl. Math. Comput. **219**, 6328–6343 (2013)
26. Yüzbaşı, Ş, Yıldırım, G.: A Laguerre approach for solving of the systems of linear differential equations and residual improvement. Comput. Methods Differ. Equ. **9**, 553–576 (2021)
27. Yüzbaşı, Ş, Yıldırım, G.: Laguerre collocation method for solutions of systems of first order linear differential equations. Turk. J. Math. Comput. Sci. **10**, 222–241 (2018)

Instance Segmentation of Handwritten Text on Historical Document Images Using Deep Learning Approaches

Umid Suleymanov(✉), Vildan Huseynov, Ilaha Manafova, Asgar Mammadli, and Toghrul Jafarov

E-GOV Development Center, Baku , Azerbaijan
{u.suleymanov,v.huseynov,i.manafova,a.mammadli,t.jafarov}@asan.gov.az

Abstract. Handwritten text segmentation is one of the essential initial steps for higher-level document processing tasks such as text recognition. The topic of handwriting segmentation and recognition of archive documents, especially with the presence of printed characters, has been encountered rarely in academic research. In this paper, we tried to address the task of segmentation and at later stage recognition of handwritten archive documents to purify and extend information databases from the past years. In our case, we defined the problem as an instance-based image segmentation task, for which we propose two different methods; one-stage and two-stage architectures. In the one-stage approach, we introduced the application of DeepLab3 and U-Net-based architectures with Watershed transformation that is utilized for fixing handwritten instances overlapping problems at the pixel level. Whereas, two-stage architecture encapsulates both detection and segmentation. Our approach to the mentioned task includes the application of the Mask R-CNN, one of the pre-trained COCO Instance Segmentation models using Detectron2. We have chosen Mean Average Precision at the different intersection over union(mAPIoU) for a binary class metric to evaluate the performance of the aforementioned models. Experimental results indicate that U-Net based architecture outperforms two-stage model such as Mask R-CNN on this task.

Keywords: Deep Learning, Handwritten Segmentation · Instance segmentation · Mask-R-CNN · U-Net · DeepLab3

1 Introduction

In document analysis, one of the crucial parts is the detection or segmentation of the text areas. This step is important for further handwritten words or characters recognition. In other words, failure in segmentation leads to the failure in correct recognition. In the case of handwritten texts, there exist fewer computer vision solutions because of the lack of available data compared to the digital ones. In general, research has been rarely done on datasets containing both printed and

D. J. Hemanth et al. (Eds.): ICAIAME 2022, ECPSCI 7, pp. 244–253, 2023.
https://doi.org/10.1007/978-3-031-31956-3_20

handwritten scripts. In the section Dataset, we present and describe the dataset that includes dense text and sophisticated structure. Additionally, Section post-process provides information about how we deal with the overlaps using certain conventional post-processing methods. In general, when it comes to text segmentation, the research has two different types of approaches, which are classic computer vision methods and deep learning based solutions. The classical techniques, possess several disadvantages, for example, with them generalization and in some cases, extraction of the correct information is impractical. Before diving into the deep learning solutions, we attempted to solve our task in traditional ways. Considering that, most of the templates in our images have horizontal black lines, to approximate the coordinates of the rectangle around the text, we retrieved those horizontal features firstly utilizing Sobel [1] filters followed by erosion. Nevertheless, we did not obtain the desired results both due to the imperfections of the lines, for example, some had been faded and not all the document images were appropriate for the application of this method. In section Models, we describe our deep learning based approaches, namely DeepLab3 and U-Net-based architectures with Watershed [2] transformation and Mask R-CNN provided by Detectron2 and later comparing various models in the Discussions section.

2 Related Works

Many traditional methods have handled this problem as follows: removal of associated components (CCs) and classification of removed CCs into two classes. Lemaitre [1] proposes extracting line segments from the image at low resolution and words employing a distance threshold between connected components. Their approach does not utilize any deep learning based techniques. While manual feature extraction techniques are effective approaches, it has also been shown that deep learning approaches [2] often dominate handmade algorithms in the domain of representing complex shape features which is very hard to express using manual approaches. There are several deep learning based approaches utilized for handwritten text recognition such as Moysset and al. [3–5], which propose a combination of Multi-Dimensional Long short Term Memory (MDLSTM) neural network combined with convolutional layers to predict a bounding box around the line. Those methods give superb results but are limited to horizontal lines.

Junho et. al. [6] offers a method of separating handwritten and machine-printed components which is mixed and overlapped in documents. Junho et al. [6] propose a new architecture that performs pixel-level classification with a convolutional neural network. For neural network optimization, they suggest a cross-entropy-based loss function to mitigate the class imbalance problem. They demonstrate that the experimental results on synthetic and real images are satisfactory. Although the recommended network is trained only with synthetic images, it also increases the OCR rate of real documents. Specifically, they managed to increase the OCR rate for machine-printed text is increased from 0.8087 to 0.9442.

In their paper, for Handwritten Text Line Segmentation (HTLS), Li et al. [7] propose a Line Counting formulation. With the method, the number of text lines is counted at every pixel location. Based on the formulation, the DNN model is also proposed to perform HTLS. Their contributions [6] include Line Counting formulation; a specifically designed network for that formulation; Update the cumsum activation function to make sure that outputs are monotonic. Arun et al. [8] use a bounding box approach for non-touching characters and a pixel-based approach otherwise. It proposes a solution for character segmentation. Machine learning (KNN) is used only for the recognition of characters.

3 Dataset Description

3.1 Data Collection

The dataset includes images collected from online sources each containing 20–25 pairs of handwritten and printed words belonging to different time intervals. The documents have been written in the Azerbaijani language and we have labeled the handwritten text with the help of a Computer Vision Annotation Tool (CVAT) [9] to have the polygons in the COCO format. The characteristics of the problem itself present several challenging and unique issues, solving which requires special approaches. As an example, we can show the existence of numerous distortions in old documents, the alphabet changes from Latin to Cyrillic and then back to Latin during different periods in addition to the challenges caused due to the unconstrained nature of the handwriting styles.

3.2 Challenges

The essence of the handwritten text segmentation is a complicated process considering the flexibility of people's handwriting styles which can even vary for a person. In addition, the existence of the both printed and handwritten text, sometimes with those characters overlapping one another, as well as noises present in the documents in the form of added notes and numerous lines corresponding to the templates depending on the document category introduces more challenges to the already complex task. Since the majority of the documents are quite old, the background mostly is not clear and stained, and some textual data have faded away. Another an important point to account is related to the changing of the alphabet, alternating between Cyrillic and Latin during diverse periods in Azerbaijan.

3.3 Preprocessing

Considering that we are dealing with a segmentation task, we needed to segment and label our data on a word-level. For this purpose, we have utilized the CVAT and segmented the data manually. Polygons were drawn around the words and given just one label, followed by annotations being extracted in the COCO format. Those annotations include the coordinates of the points making up the masks and bounding box information for each mask. The images had approximately *1200* × *1600* resolution (H × W), and they were resized to 768 × 512 and Gaussian Smoothing was added. The final dataset consists of document images and their corresponding annotations. The number of words was around 11000 and the dataset was split into training and test sets with a 0.8/0.2 ratio. Additionally, images have been converted to gray scale and augmentation of them have been carried out with the help of Albumentations [10] library. Horizontal and vertical flip, rotation and cutout also have been applied to get better results. Due to the fact that our images were significantly bright, we had to apply Contrast Limited Adaptive Histogram [11] equalization with the help of OpenCV [12] library.

4 Models

4.1 U-Net

U-net [13] architecture is symmetric one with encoder, bottleneck, and decoder parts and is proposed for solving the semantic segmentation tasks in computer vision. The encoder block consists of some convolutional blocks where each of them has 2 convolutional layers with a batch norm [14] and ReLU activation function. After each convolutional block max pooling operation is applied for 2 times dimension reduction. In the decoder (growth) path the number of filters is reduced by half in each previous block, the input is up-sampled 2 times using transposed convolution or interpolation methods and later is concatenated with the symmetrical layer in the encoder block. For each decoder layer batch, norm and ReLU are also applied. For the first model, we choose ResNet50 [15] with pretrained on Imagenet [16] dataset as an encoder block where many layers could be compressed as one block as shown in Fig. 1, where R1 denotes first three blocks of ResNet, R2 fourth block, R3 fifth block, R4 denotes sixth, and R5 denotes seventh block.

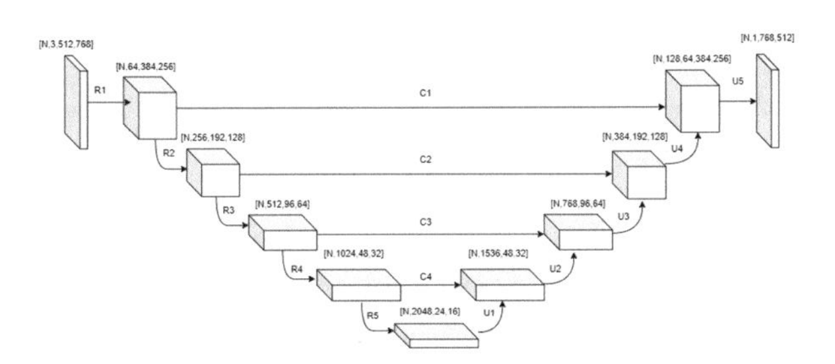

Fig. 1. Unet architecture with resnet encoder. *R* is ResNet's block, *U* up-sampling block, *C* concatenate.

In the case of the second model, we are inspired by LineCounter [7] method and modified U-Net by using counter block as the bottleneck as shown in Fig. 3.

Vertical and horizontal GRU [17] propagate global information along those directions. The proposed architecture consists of the encoder, a counter block that contains three convolutions, each convolution has batch norm and ReLu activation, two recurrent layers, and decoder block being symmetric to the encoder (Fig. 2).

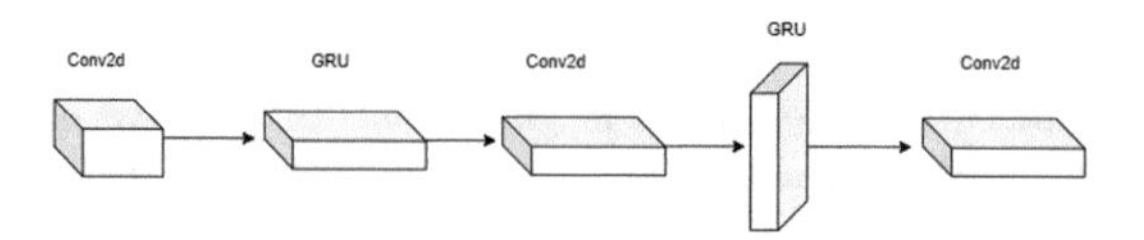

Fig. 2. Bottleneck with Conv2d*(3 × 3)* and bi-directional *GRU*

4.2 Deeplab3

Deeplab [18] is a deep convolutional network with an encoder and Atrous Spatial Pyramid Pooling has convolutional layers with different scales and pooling. The encoder of this model is ResNet-50 as in the original paper, but we use first four blocks instead of seven and we have to change ASPP dilation rates to (5, 7, 9). Before going into ASPP, atrous convolution with stride two is applied to the fourth block of ResNet50. Atrous convolutional block reduces feature map into 512 before applying ASPP layer. All layers in ASPP are concatenated and then up-sampled.

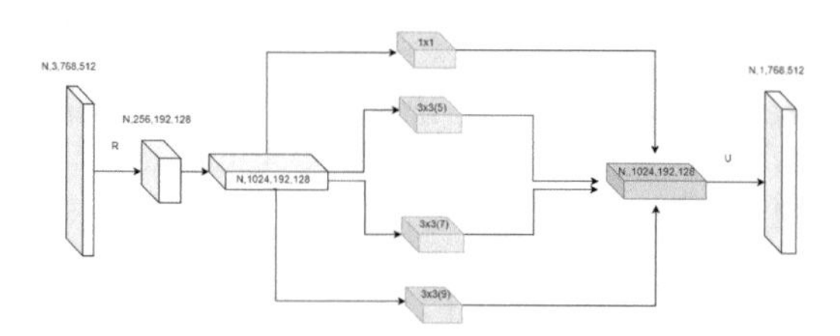

Fig. 3. *R* is ResNet blocks, *green* are ASPP blocks, *red* is concatenating block, *U* is up-sampling.

4.3 Mask-R-CNN

In our project, pre-trained Mask R-CNN [19] from Detectron2 [20], which is a PyTorch-based library developed by Facebook AI Research (FAIR), has been utilized as a base model. The framework provides state-of-the-art neural network architectures for both object detection and segmentation tasks. Out of three various backbone combinations provided, Feature Pyramid Network (FPN) and ResNet mixture has been chosen since it was claimed that the finest results are achieved when it comes to the speed and accuracy tradeoff. Mask R-CNN is a region-based and state-of-art Convolutional Neural Network (CNN) when it comes to instance segmentation. It is an extended version of the Faster R-CNN [21]. While Faster R-CNN outputs the class label and bounding box coordinates for the given object on the images, Mask R-CNN in addition to them generates the object mask.

5 Training Protocol

5.1 Loss Function

Our labeled images are binary masks and as a loss function, we choose pixel-wise cross-entropy.

$$L(y,x) = \frac{1}{HW}\sum_{i=1}^{H}\sum_{j=1}^{W} -(y_{ij}log(x_{ij}) + (1-y_{ij})log(1-x_{ij})))$$

For optimizing loss function we used Adam [22] optimizer ($b1 = 0.9$, $b2 = 0.999$ with *learning rate* = 2e−4 and *weight decay* = 1e−5) with lambda scheduler (*step = 0.65*)

5.2 Postprocessing

We apply classic computer vision algorithms such as erosion and dilation for removing the small noises in our predicted mask. One of the essential problems in our documents is the existence of overlapping characters and as a solution, we use the watershed algorithm. The watershed transformation is based on the notion that any grayscale image can be used as a topographic surface. Because of the overlapping areas before the watershed step we use Morphological Opening [23] with kernel *ratio* of **1:2** which horizontally divides the objects and improves mAPIoU. To partition the image into various sections, the surface is exposed to the overflow of the water starting from its minima, however, simultaneously, avoiding the water running from separate sources from being mixed by creating barriers. Those regions are considered the catchment basins, and these barriers, also called watershed lines, mark the borders between the basins. In general, the watershed transform operates on the image gradient, whereas the catchment basins represent the uniform gray level components of the image. The last step of our post-processing is an approximation of the connected area with rectangle [24]. Figure 4 shows an example of watershed transformation.

6 Experimental Results

6.1 Metrics

We apply *F1*, *IOU* and *mAPIoU* metrics to validate our methods. The mAP IoU is calculated using IoU, threshold values ranging from 0.5 to 0.95 with a step

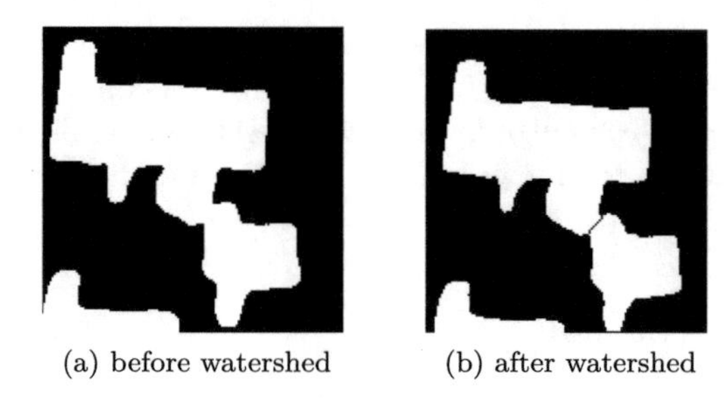

(a) before watershed (b) after watershed

Fig. 4. An example of watershed transformation on mask where words overlapped.

size of 0.05, and average precision values evaluated in each threshold. When the predicted object and the ground truth correspond to each other over the set IoU threshold, it is considered as a true positive. If no ground truth exists for the relevant predicted object, then it is a false positive while the other way around means the false negative. Later, the average is calculated using the precision values for each IoU threshold respectively, resulting in the average precision score for a single image. Figure 5 presents comparison of performance of the models on validation dataset.

$$mAPIoU = \frac{1}{|thresholds|} \sum_{t=0.5}^{0.95} \frac{TP(t)}{TP(t) + FP(t) + FN(t)}.$$

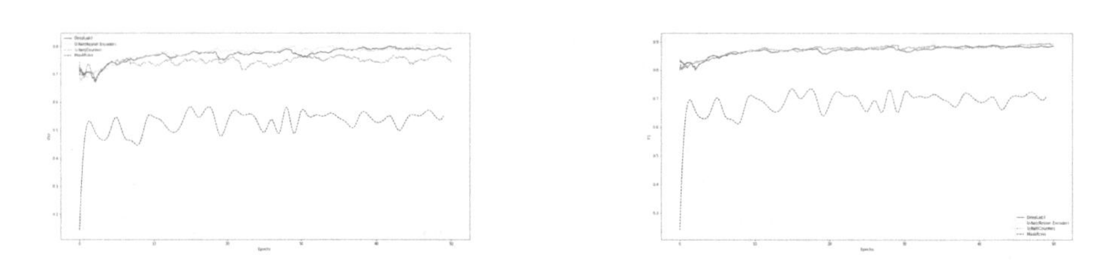

Fig. 5. IoU(left) and **F1**(right) scores on validation dataset.

6.2 Models Comparison

If we take into account that the main metric is very sensitive, the 2 stage architecture as Mask R-CNN model shows rather worse results on this task than proposed semantic segmentation approaches with watershed algorithm. We assume that the two-stage architectures have difficulties in detecting bounding boxes at the first stage, therefore, they perform worse in the gray images. Furthermore, their performance suffers due to the partitioning of each world region, therefore, lowering the IoU score. Our experimental method with Counter bottleneck in the U-Net model was not more effective for this task than for line counting in documents as described in the original article. Figure 6 demonstrates the output masks on a single image (Table 1).

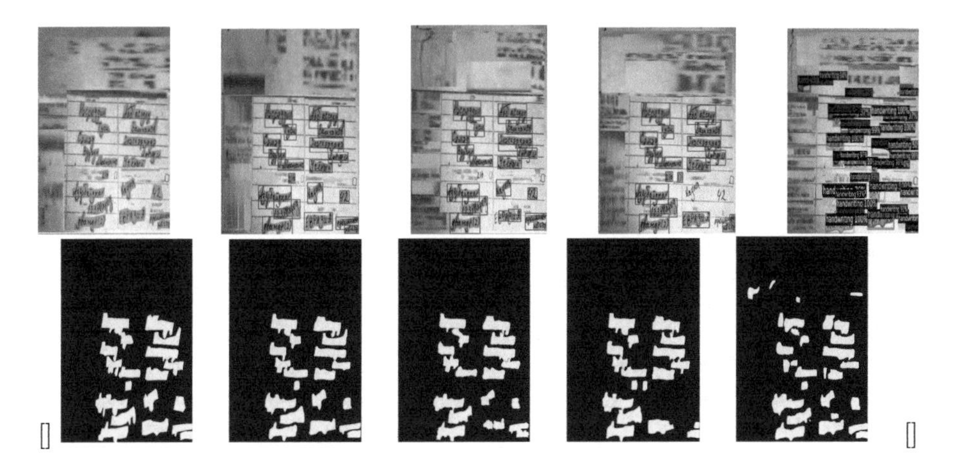

Fig. 6. Output comparison of the models with ground truth. Left to right: **GT**, **U-Net**(ResNet), **DeepLab3**, **U-Net**(counter), **Mask R-CNN**

Table 1. Experimental results of architectures on test dataset

Model	F1	IOU	mAPIoU(05–0.95)	mAPIoU(0.5)
U-Net(Resnet encoder)	**0.85**	**0.75**	**0.32**	**0.61**
U-Net(Counter)	0.83	0.72	0.27	0.55
DeepLab3	0.84	0.73	0.28	0.57
Mask R-CNN	0.71	0.55	0.28	0.43

7 Conclusion

Considering the importance of handwriting segmentation for further recognition tasks, in this paper, we address the former problem and for this, several models and architectures are proposed and compared based on the chosen metrics. The mentioned task is challenging due to several factors, such as the dataset containing different alphabets, having considerable noise as well as overlapping and touching characters. Initially, we tried the traditional methods such as identifying the horizontal lines using the Sobel operator, however, later switching the focus to the deep learning based ones, as a result of which we have achieved much better outcomes. While testing the solution, failures happened due to the numerous overlaps of the characters. In future work, we intend to improve our current post-processing techniques to better and effectively separate the segmentation masks of the words.

References

1. Lemaitre, A., Camillerapp, J., Coüasnon, B.: A perceptive method for handwritten text segmentation. In: Document Recognition and Retrieval XVIII, vol. 7874, p. 78740C, International Society for Optics and Photonics (2011)

2. LeCun, Y., Bengio, Y., Hinton, G.: Deep learning. Nature **521**(7553), 436–444 (2015)
3. Moysset, B., Louradour, J., Kermorvant, C., Wolf, C.: Learning text-line localization with shared and local regression neural networks. In: 2016 15th International Conference on Frontiers in Handwriting Recognition (ICFHR), pp. 1–6, IEEE (2016)
4. Moysset, B. , Adam, P., Wolf, C., Louradour, J.: Space displacement localization neural networks to locate origin points of handwritten text lines in historical documents. In: Proceedings of the 3rd International Workshop on Historical Document Imaging and Processing, pp. 1–8 (2015)
5. Moysset, B., Kermorvant, C.. Wolf, C.: Full-page text recognition: Learning where to start and when to stop. In: 2017 14th IAPR International Conference on Document Analysis and Recognition (ICDAR), vol. 1, pp. 871–876, IEEE (2017)
6. Jo, J., Koo, H.I., Soh, J.W., Cho, N.I.: Handwritten text segmentation via end-to-end learning of convolutional neural networks. Multimedia Tools Appl. **79**(43), 32137–32150 (2020)
7. Li, D., Wu, Y., Zhou, Y.: Linecounter: Learning handwritten text line segmentation by counting. In: 2021 IEEE International Conference on Image Processing (ICIP), pp. 929–933, IEEE (2021)
8. Arun, M., Arivazhagan, S., Rathina, D.: Handwritten text segmentation using pixel based approach. In: 2019 3rd International Conference on Trends in Electronics and Informatics (ICOEI), pp. 791–796, IEEE (2019)
9. Sekachev, B., et al.: "opencv/cvat: v1.1.0" (2020)
10. Buslaev, A., Iglovikov, V.I., Khvedchenya, E., Parinov, A., Druzhinin, M., Kalinin, A.A.: Albumentations: fast and flexible image augmentations. Information **11**(2), 125 (2020)
11. Reza, A.M.: Realization of the contrast limited adaptive histogram equalization (CLAHE) for real-time image enhancement. J. VLSI Sign. Process. Syst. Sign. Image Video Technol. **38**(1), 35–44 (2004)
12. Bradski, G.: "The OpenCV Library." Dr. Dobb's J. Soft. Tools (2000)
13. Ronneberger, O., Fischer, P., Brox, T.: U-Net: convolutional networks for biomedical image segmentation. In: Navab, N., Hornegger, J., Wells, W.M., Frangi, A.F. (eds.) MICCAI 2015. LNCS, vol. 9351, pp. 234–241. Springer, Cham (2015). https://doi.org/10.1007/978-3-319-24574-4_28
14. Ioffe, S., Szegedy, C.: Batch normalization: accelerating deep network training by reducing internal covariate shift. In: International Conference on Machine Learning, pp. 448–456, PMLR (2015)
15. He, K., Zhang, X., Ren, S., Sun, J.: Deep residual learning for image recognition. In: Proceedings of the IEEE Conference on Computer Vision and Pattern Recognition, pp. 770–778 (2016)
16. Deng, J., Dong, W., Socher, R., Li, L.-J., Li, K., Fei-Fei, L.: Imagenet: a large-scale hierarchical image database. In: 2009 IEEE Conference on Computer Vision and Pattern Recognition, pp. 248–255. IEEE (2009)
17. Chung, J., Gulcehre, C., Cho, K., Bengio, Y.: Empirical evaluation of gated recurrent neural networks on sequence modeling. arXiv preprint arXiv:1412.3555 (2014)
18. Chen, L.-C., Papandreou, G., Schroff, F., Adam, H.: Rethinking atrous convolution for semantic image segmentation. arXiv preprint arXiv:1706.05587 (2017)
19. He, K., Gkioxari, G., Dollár, P., Girshick, R.: "Mask R-CNN.;; In: Proceedings of the IEEE International Conference on Computer Vision, pp. 2961–2969 (2017)
20. Wu, Y., Kirillov, A., Massa, F., Lo, W.-Y., Girshick, R.: "Detectron2" (2019). https://github.com/facebookresearch/detectron2

21. Ren, S., He, K., Girshick, R., Sun, J.: Faster R-CNN: towards real-time object detection with region proposal networks. In: Advances in Neural Information Processing Systems, vol. 28 (2015)
22. Kingma, D.P., Ba, J.: Adam: a method for stochastic optimization. arXiv preprint arXiv:1412.6980 (2014)
23. Goutsias, J., Heijmans, H.J., Sivakumar, K.: Morphological operators for image sequences. Comput. Vis. Image Underst. **62**(3), 326–346 (1995)
24. Freeman, H., Shapira, R.: Determining the minimum-area encasing rectangle for an arbitrary closed curve. Commun. ACM **18**(7), 409–413 (1975)

Synthetic Signal Generation Using Time Series Clustering and Conditional Generative Adversarial Network

Nurullah Ozturk(✉) and Melih Günay

Akdeniz University, Antalya, Turkey
ozturknurullah97@gmail.com, mgunay@akdeniz.edu.tr

Abstract. In recent years, artificial data became inevitable need for almost any of the deep learning applications. Most of the published works on data generation uses deep networks for signal generation without preprocessing and adequate characterization. Therefore, not properly summarizing data leads to dramatic loss of key aspects of the signal. In this study, we propose a generic way for creating signal that includes important features of the authentic data. Our approach involves time series clustering and Generative Adversarial Networks for grouping and simulating signals. Even with the very small amount of data, the model can effectively split data set into meaningful clusters and generates signals that have high monotonic associations to corresponding cluster. We finally report on experimental results of different time series clustering techniques used for preprocessing and the outcomes of different approaches are compared statistically for both synthetic and real data.

Keywords: Synthetic Data · Dynamic Time Warping · Deep Neural Networks · Unsupervised Learning · Clustering

1 Introduction

The way to convey information between electronic devices held by a function called the signal [1]. That phenomenon could be used for representation of different data types, including medical data. The process of extracting information from medical records for better diagnosis or tracking health condition are struggles with the fact that the enormous number of non observable features in high dimensional data. However, in the last decade, increasing research interest and works on deep learning models helped us to overcome from the limitation of traditional algorithms and hand-designed features [2] Unfortunately such significant improvements towards processing high dimensional data brings different types of problems to deal with. Deprivation of data is still the one of the most encountered problem for achieving more accurate results. The cost of manually labeling is so high that researchers eventually started to focus on generating and using synthetic data [3–5]. After first introduced to the literature in 2014

D. J. Hemanth et al. (Eds.): ICAIAME 2022, ECPSCI 7, pp. 254–263, 2023.
https://doi.org/10.1007/978-3-031-31956-3_21

[6], generative adversarial network also drawn attention for generating synthetic data with multiple features and that framework also used in this paper.

The purpose of this study is to establish a generic method for processing high-quality synthetic signals. Because signals are the fundamental phenomenon in computer science and other subjects, understanding and processing signals can help in the development of other fields. Signals may behave in an unexpected way due to their sensitivity to environmental conditions. So that, before training a machine learning or deep learning model, signals should be carefully analyzed to avoid from any misinterpretation. Therefore, the preprocessing step is necessary to boost performance of the predictor model. We utilized multiple time series clustering on the signal data set to promote preprocessing and demonstrated that it can improve feature extraction, hence the overall performance of the trained model. As previously stated, we used a deep learning technique to generate signals because the number of features in the signal data set could be too large and advanced mathematical structures are more likely the best option for doing such complex feature extraction operations. We modified the original generative adversarial network [6] to be able to perform generating signals. For each clustering algorithm used, we analyzed the performance of generated signals and found that our generic model is capable of generating signals that are similar to the training set.

Generative Adversarial Network

The study of using deep learning as a generative model started with the generative adversarial network (GAN) [6]. Goodfellow *et al.* proposed the GAN which is a deep learning framework that consist of two models that compete each other in a form of zero-sum game. The generator model aims on generating inputs that can confuse the discriminator model. On the other hand, discriminator aims to successfully distinguish between real and fake inputs. Input of the generator model in GAN is an array of random numbers. Therefore, there is no control over generator model to generate certain type of features. In the same year GAN was released, the paper named Conditional Generative Adversarial Nets [7] proposed the way of splitting clusters in high dimensional space by using labels as an encoded input.

The zero-sum game between discriminator and generator not always halts on the saddle point. Usually, one dominates the other and that leads to collapse of the network model and observing poor quality outputs. The Wasserstein GAN [8], introduced by Arjovsky *et al.* in 2017, shared an alternative designing way for minimizing mode collapse for a GAN model.

Signal Generation Models

The majority of the GAN related works for generating realistic signals have been based around using recurrent or conditional networks as a GAN model and application specific mathematical equations for providing evaluation metric to the corresponding generative network. In this study, we preferably aim on constructing generic model for better understanding and overview of the GAN based signal generation approaches. Yet, evaluation of the GAN models are not inclusive enough to be generic [9]. So that, this section includes more comprehensive analysis towards previous works rather than diving deep inside their implementations.

To generate biosignals like electrocardiogram (ECG) or electroencephalogram (EEG), Harada *et al.* [10] proposed a GAN model that has been constructed with LSTM layers. In that work, artificial biosignals that are similar to train set but having low frequencies were generated. In LSTM-GAN architecture, the generative model can show similar collective behaviours of data set but it has no local sensitivity. The reason behind that phenomenon is, as Harada *et al.* also concluded in the paper, LSTM behaves like low-pass filter. In contrast to that, Brophy *et al.* [11] proposed WGAN model for generating time series data, where it takes 2d grayscale image as input and produces 2d output. Brophy *et al.* used a customized low-pass filter due to high frequency of generated signals.

Another concern for medical signal generation works is, if there is any leak of private information of the patients caused by generated signal. Esteban *et al.* [4] proposed a novel evaluation method for generating medical data which they called "Train on Synthetic, Test on Real" (TSTR). In TSTR approach, generated signals were used to train GAN model and then its tested on real samples. So that, applicability of synthetic data for other platforms can be assured beforehand. In the later pages of the work, Esteban *et al.* also shared latent space analysis of generated signals to see if the GAN model is actually memorising the training set or not. The benefit of using that approach is that it helps us to develop an intuition about the trained GAN model.

2 Method

The clustering aims to split unlabeled data set into the homogeneous groups that are sharing similarities within the group and dissimilarities to the other groups as well. [12] The clustering have been used for various type of fields [13], including medicine [14,15], to better understand characteristics of unlabeled data set. The retrieved features can be used to optimize predictive models so that they know which pathway to take in the majority of situations. This is especially important in unsupervised learning. In signal generation, like other fields, that should be considered as well. The fact that types of signals vary, it is unclear that what kind of operations should be held for all types of signals as a prior to extract information from them. As a result, it is dependent on specific type or the application. Although there may not be a perfect clustering approach for the type of signal used, it should be decided and implemented somehow.

Learning a similar set of functions is simple for a deep learning model. As a result, as the number of clusters decreases, a deep learning model requires more time and data to learn and represent each type of signal. However, as the

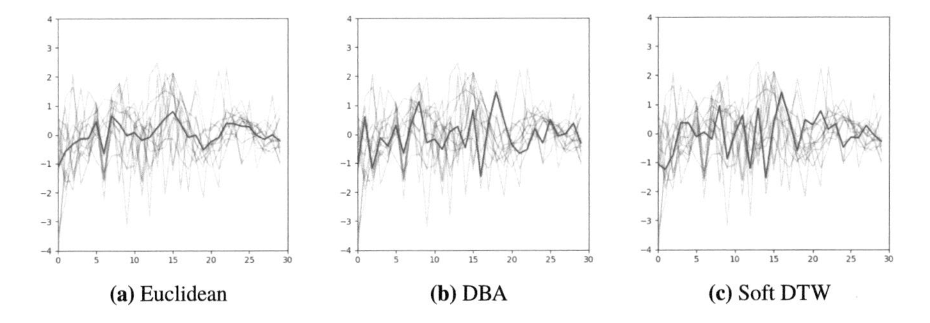

(a) Euclidean (b) DBA (c) Soft DTW

Fig. 1. Result of the different k-means clustering methods applied on a small portion of training set

number of clusters increases, it becomes easier for a deep learning model to learn each individual subset. The reason for this phenomenon is that if we consider the number of clusters to be the maximum value, which is equal to the number of data in the data set, each cluster represents one data. Because deep learning models typically have thousands of parameters, assigning each node to a single data solves the optimization problem. In contrast to that notion, if no clustering is applied to a data set, it is a single blob. So the deep learning model must find the most of the key features from the data set and assign each its node to one or multiple descriptions. It may appear to be the better approach, but such a perfect structure is not guaranteed and is costly. As a result, clustering has a significant impact on the entire learning process.

In this study, we used standard Euclidean, The DTW Barycenter Averaging (DBA) [16] and Soft-DTW [17] k-means clustering methods for preprocess the signal data to demonstrate that within class correlation of the clusters can be preservable in generative models for any method. The key distinction between the clustering algorithms used in this study is whether they are time domain sensitive or not. As shown in the Fig. 1, DBA and Soft-DTW perform finding optimum alignment over time dependent sequences [18] hence they are less affected by the fact that most of the points are far away from each other at any particular time. However, because standard euclidean implementation does not do time dependent analysis and instead takes the average of the input, the outcome becomes a smooth signal.

The goal of discriminator in the GAN model is to achieve high classification accuracy on binary output. In other words, discriminator is simply a deep learning solution to classification problem with an additional feature. For that reason, how well it distinguishes between training labels are actually strongly correlated with its overall performance. So far, 1d convolutional networks performed state of art results for classification of signals. It was reported in literature that 1d convolutional network outperformed 2d convolutional network and other machine learning algorithms on ECG signal classification task [19]. Therefore, we built our proposed GAN model, which illustrated in Fig. 2, up on 1d convolutional network where it downsamples the input signal in discriminator model and

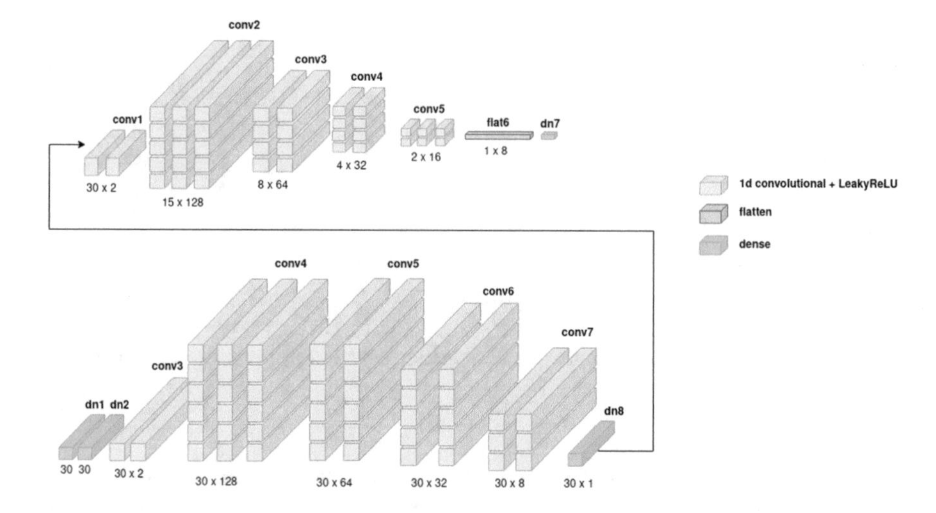

Fig. 2. Overview of proposed generic model. Output layers of generator (bottom) and discriminator (top) models are activated with the hyperbolic tangent activation function and the sigmoid activation function, respectively. The stack of convolutional layers used multiple times to increase depth of the network

upsamples the input array of random numbers and encoded label in generator model.

The discriminator model takes two 1d data as input; the signal, and the associated label. Since deep networks only work with numerical variables, labels are converted to numerical format beforehand. Both input layers are transformed into initial convolutional layers and multiple convolutional layers are then stacked on top of each other according to the specified shapes in Fig. 2. The output of discriminator, where it predicts whether input signal is fake or not, is a binary output. Therefore it is activated with sigmoid function [20] and binary cross-entropy used as loss function of model. On the other hand, the generator model takes two 1d data as input; the random noise and the label. As in discriminator model, both input layers are transformed into initial convolutional layers and multiple convolutional layers are then stacked on top of each other. Figure 2 also shows the shapes of the layers in the generator model and the connection between output layer of generator with the input layer of discriminator. Since the real signal comes from the training set and the fake signal comes from the generator model, the output shape of the generator is identical to the input shape of the discriminator. As a result, generated signals can be properly fed into the discriminator throughout the training process. Even though the relu function is the most preferred activation function for convolutional layers [21], for each of the 1d convolutional layers, we used LeakyRelu [22] as an activation function, which is similar to the relu function but it has small slope for negative values.

3 Results and Discussion

Although signals can be classified under various subjects and evaluation of generative models mostly depends on their targeted application [9], we provide more naive and inclusive approach to evaluate performance of generated signals which is monotonic associations. The term monotonic association is a statistical approach that is used to measure similarity between two variables in which the value of one variable increases as the other variables also increases, or where the value of one variable increases as the other variable also decreases. The key distinction between monotonic and other associations is that it can measure similarity even though variables increase or decrease at different rates which is called non linear association. Signals are also obtained in a non-linear form in the majority of circumstances. As a result, while measuring signal similarity, more meaningful results can be obtained by using non-linear approaches. In this experiment, the Spearman method used for measuring correlations between signals since it is the most appropriate choice to determine monotonic associations [23]

Table 1. Within class correlation scores for each clustering approach. Class names are symbolic and only related to corresponding rows

	Class 1	Class 2	Class 3	Class 4	Class 5
Original clusters	0.03	0.09	**0.88**	0.10	0.24
Euclidean	0.10	**0.90**	**0.40**	**0.62**	**0.43**
DBA	**0.90**	0.22	**0.65**	**0.36**	0.14
SoftDTW	0.15	**0.53**	**0.90**	0.24	0.23

Dataset

The data set used in this experiment consist of 480 samples and length of each sample is 30. Out of 480 samples, labels were distributed over 5 classes equally (each class has 96 samples). A total of 50 samples are taken from the set to be used as test samples. As shown in the Table 1, except for the class 3, there is no significant within class correlations in the data set. Signals in class 3 are mostly sigmoid-like functions.

Experimental results

To simulate effectiveness of clustering techniques as preprocessing method to signal data, we used time series clustering algorithms on training set to extract more information and construct related classes. Table 1 demonstrates that how well clustering algorithms performed on extracting correlated features. For each clustering approach, we separately trained our proposed GAN model, which illustrated in Fig. 2, on training set. The number of 1d convolutional layers in both discriminator and generator model was 5. The weights were updated using the Adam optimizer [24] with a learning rate of 0.0002 and a decay rate of 0.5. The number of training epochs was set to 500 and signal values are in data set normalized between 0 and 1.

Table 2. Correlation scores between generated signals and original set with the default label

	Test Class 1	Test Class 2	Test Class 3	Test Class 4	Test Class 5
Label 1	0.008	0.01	0.02	0.02	0.02
Label 2	0.003	0.03	0.02	0.03	0.04
Label 3	−0.02	0.17	**0.87**	0.08	0.26
Label 4	0.01	0.009	−0.01	0.02	−0.02
Label 5	−0.04	−0.03	−0.008	−0.06	0.12

In the first experiment, original training set trained on the proposed GAN model. Table 2 presents result of correlation scores between generated and real signal without using any time series clustering technique. Initially training set had only one strong correlation score, third column of first row in the Table 1, and that correlation is preserved while generating synthetic signals. In other words, the GAN model trained on a non-clustered training set, generated signals that are highly correlated with test class 3.

Table 3. Correlation scores between generated signals and test data where Euclidean clustering algorithm used as preprocessing method

	Test Cluster 1	Test Cluster 2	Test Cluster 3	Test Cluster 4	Test Cluster 5
Label 1	0.05	0.04	−0.02	−0.12	0.016
Label 2	0.03	**0.90**	−0.03	0.005	0.11
Label 3	−0.01	−0.03	**0.33**	−0.38	−0.39
Label 4	−0.15	0.11	−0.006	**0.53**	0.05
Label 5	0.04	0.02	−0.29	0.19	**0.34**

In the second experiment, training set has clustered by using standard Euclidean method. As a result, out of 5 classes, 4 of them had high within class correlation scores, which can be observable in the second row of Table 1. Similar to previous experiment, the proposed GAN model in Fig. 2 trained on the data set with the same hyperparameters. As shown in Table 3, high correlations between generated signals and 4 test classes are obtained.

Table 4. Correlation scores between generated signals and test data where DBA clustering algorithm used as preprocessing method

	Test Cluster 1	Test Cluster 2	Test Cluster 3	Test Cluster 4	Test Cluster 5
Label 1	**0.86**	−0.006	−0.02	0.03	0.03
Label 2	−0.03	0.11	−0.15	0.002	0.01
Label 3	0.20	−0.13	**0.41**	−0.14	0.06
Label 4	−0.08	0.07	−0.34	**0.29**	−0.14
Label 5	0.14	0.03	0.18	−0.20	0.11

In the third experiment, training set has clustered by using DBA method. The number of iterations for the barycenter was set to 10. Even though it uses DTW algorithm to perform clustering, we obtained less correlated classes in Table 1 compared to Euclidean method. Similar to previous experiments, the proposed GAN model in Fig. 2 trained on the data set with the same hyperparameters. As shown in Table 4, high correlations between generated signals and 3 test classes are obtained.

Table 5. Correlation scores between generated signals and test data where Soft-DTW clustering algorithm used as preprocessing method

	Test Cluster 1	Test Cluster 2	Test Cluster 3	Test Cluster 4	Test Cluster 5
Label 1	0.01	0.12	−0.04	−0.03	−0.01
Label 2	−0.03	**0.41**	0.17	0.11	−0.19
Label 3	0.02	0.06	**0.89**	0.08	−0.04
Label 4	0.04	0.20	0.05	0.09	−0.16
Label 5	0.03	−0.24	−0.25	−0.07	0.18

In the last experiment, training set has clustered by using Soft-DTW method. Parameter value for the Soft-DTW metric was set to 0.01. However, only two classes exhibited higher within class correlation scores. Similar to previous experiments, the proposed GAN model in Fig. 2 trained on the data set with the same hyperparameters. As shown in Table 5, high correlations between generated signals and 2 test classes are obtained.

4 Conclusion

In this work, we proposed generic GAN model for generating synthetic signals. We also performed different time series clustering algorithms as preprocessing method to extract more related features and demonstrated that within class correlations can be preservable in deep networks with the robustness of CGAN approach. We concluded that 1d convolutional layers is effective enough to be use in deep networks as signal processing unit. Beside that, we also confirmed that WGAN implementation significantly reduces mode collapse due to training time.

We are not recommending to use of either recurrent or 2d convolutional networks for generating signals without customized implementation. That is because recurrent networks behave like low-pass filter and to perform 2d convolutional operations, 1d signals must be transformed into 2d form. That means continuous signals has to be splitted or reorganized. As long as there is no unique algorithm specified, it contaminates monotonic associations of the signal.

References

1. Priemer, R.: Introductory signal processing. In: Advanced Series in Electrical and Computer Engineering. World Scientific Publishing Company (1990). https://books.google.com.tr/books?id=5AM8DQAAQBAJ
2. Liu, W., Wang, Z., Liu, X., Zeng, N., Liu, Y., Alsaadi, F.E.: A survey of deep neural network architectures and their applications. Neurocomputing **234**, 11–26 (2017). https://doi.org/10.1016/j.neucom.2016.12.038
3. Nikolenko, S.I.: Introduction: the data problem. In: Synthetic Data for Deep Learning. SOIA, vol. 174, pp. 1–17. Springer, Cham (2021). https://doi.org/10.1007/978-3-030-75178-4_1
4. Esteban, C., Hyland, S.L., Rätsch, G.: Real-valued (medical) time series generation with recurrent conditional GANs (2017). https://doi.org/10.48550/ARXIV.1706.02633
5. Smith, K.E., Smith, A.O.: Conditional GAN for timeseries generation (2020). https://doi.org/10.48550/ARXIV.2006.16477
6. Goodfellow, I., et al.: Generative adversarial nets. Adv. Neural Inf. Process. Syst. **27**
7. Mirza, M., Osindero, S.: Conditional generative adversarial nets (2014). https://doi.org/10.48550/ARXIV.1411.1784
8. Arjovsky, M., Chintala, S., Bottou, L.: Wasserstein GAN (2017). https://doi.org/10.48550/ARXIV.1701.07875
9. Theis, L., Oord, A.V.D., Bethge, M.: A note on the evaluation of generative models (2015). https://doi.org/10.48550/ARXIV.1511.01844
10. Harada, S., Hayashi, H., Uchida, S.: Biosignal generation and latent variable analysis with recurrent generative adversarial networks (2019). https://doi.org/10.48550/ARXIV.1905.07136
11. Brophy, E., Wang, Z., Ward, T.E.: Quick and easy time series generation with established image-based GANs (2019). https://doi.org/10.48550/ARXIV.1902.05624
12. Warren Liao, T.: Clustering of time series data-a survey. Pattern Recognit. **38**(11), 1857–1874 (2005). https://doi.org/10.1016/j.patcog.2005.01.025
13. Kavitha, V., Punithavalli, M.: Clustering time series data stream - a literature survey (2010). https://doi.org/10.48550/ARXIV.1005.4270
14. Toft, P., et al.: On clustering of fMRI time series. NeuroImage **5**
15. Goutte, C., Hansen, L., Liptrot, M., Rostrup, E.: Feature-space clustering for fMRI meta-analysis. Hum. Brain Mapp. **13**, 165–83 (2001). https://doi.org/10.1002/hbm.1031

16. Petitjean, F., Ketterlin, A., Gançarski, P.: A global averaging method for dynamic time warping, with applications to clustering. Pattern Recognit. **44**(3), 678–693 (2011). https://doi.org/10.1016/j.patcog.2010.09.013
17. Cuturi, M., Blondel, M.: Soft-DTW: a differentiable loss function for time-series (2017). https://doi.org/10.48550/ARXIV.1703.01541
18. Berndt, D.J., Clifford, J.: Using dynamic time warping to find patterns in time series. In: KDD Workshop, vol. 10, No. 16, pp. 359–370 (1994)
19. Li, D., Zhang, J., Zhang, Q., Wei, X.: Classification of ECG signals based on 1D convolution neural network. In: 2017 IEEE 19th International Conference on e-Health Networking, Applications and Services (Healthcom), pp. 1–6. IEEE (2017). https://doi.org/10.1109/HealthCom.2017.8210784
20. Goodfellow, I., Bengio, Y., Courville, A.: Sigmoid Units for Bernouilli Output Distributions. The MIT Press, Cambridge (2017)
21. Nair, V., Hinton, G.E.: Rectified linear units improve restricted Boltzmann machines. In: ICML, pp. 807–814 (2010). https://icml.cc/Conferences/2010/papers/432.pdf
22. Xu, B., Wang, N., Chen, T., Li, M.: Empirical evaluation of rectified activations in convolutional network (2015). https://doi.org/10.48550/ARXIV.1505.00853
23. Schober, P., Boer, C., Schwarte, L.: Correlation coefficients: appropriate use and interpretation. Anesth. Analg. **126**, 1 (2018). https://doi.org/10.1213/ANE.0000000000002864
24. Kingma, D.P., Ba, J.: Adam: a method for stochastic optimization. arXiv preprint arXiv:1412.6980 (2014)

Transfer Learning Based Flat Tire Detection by Using RGB Images

Oktay Ozturk[1] and Batuhan Hangun[2(✉)]

[1] School of Computing, Wichita State University, Wichita, KS 67260, USA
oxozturk1@shockers.wichita.edu
[2] Istanbul, Turkey
batuhanhangun@gmail.com

Abstract. Traffic accidents have caused many casualties through the years. Most of the time those accidents are driver-related but there are also many non-negligible vehicle-related ones. A report indicates that 35% of vehicle-related accidents are directly related to tires. This emphasizes the importance of inspecting tires carefully to avoid undesired situations. Many manufacturers have been using computers for automated inspection for a long time. Computer-based inspections of tires are also common practice. The majority of tire fault detection research in the literature focuses on manufacturing line inspection and employs X-ray imaging. Tire inspection must be ongoing till the tire is changed due to a necessity at some point. In most countries, however, it is part of the mandatory periodic vehicle inspection. An automated tire inspection system, which uses RGB images to detect flaws on tires and is mounted at inspection stations, has the potential to reduce inspection time. To begin, we updated the dataset using various image manipulation techniques in order to improve generalization during the training phase. Second, we utilized conventional CNN, DenseNet201, InceptionResNetV2, and Xception in this work to identify flat tires in an RGB image data set that included flat tires, normal tires, and non-tire situations. As a result we obtain 80%, 98%, 97%, and 100% F1-Score from CNN, DenseNet201, InceptionResNetV2, and Xception respectively.

Keywords: Transfer learning · deep learning · flat tire detection · automated inspection · driver safety

1 Introduction

Each year too many people lose their lives in traffic accidents. Most of the time, those accidents are driver related and some studies focused on improving the driver safety [11,16,23]. Besides driver-related accidents, vehicle-related accidents have taken many lives. A report from 2015 indicates that 2% of all accidents were vehicle-related. 35% of these accidents were related to a problem with tires

B. Hangun—Independent Researcher

D. J. Hemanth et al. (Eds.): ICAIAME 2022, ECPSCI 7, pp. 264–273, 2023.
https://doi.org/10.1007/978-3-031-31956-3_22

[22]. The National Highway Traffic Safety Administration's report states that 612 motor vehicle accidents that resulted in fatalities that occurred in 2019 were related to tires [4]. There were 1.4 billion motor vehicles in 2019, and more than 70 million motor vehicles were sold in 2020 [1,2]. Depending on the increasing number of motor vehicles in traffic, risk of the fatal accidents due to tires will increase. The first measure to be taken to prevent tire-related accidents is to inspect tires just right after they are out of the production.

Computer-based inspection of tires are used frequently in production. One of the earliest studies on computer-based tire inspection was published by Chen et al. in 1993. Their study provided an algorithm that consists of image acquisition and image processing steps, which is used to inspect tire treads. Due to quality control requirements, the tire inspection process consists of two separate parts which are the Groove Inspection (GI) and the Surface Inspection (SI). In the GI, groove cracks that may appear in the tire tread materials were inspected. For identifying the groove cracks, image processing, and pattern recognition techniques were used. Secondly, in SI, surface defects on the tread were examined. As opposed to the groove cracks, successful imaging of the tread surface is easier. To detect surface defects, pattern recognition techniques were used. In the study, it was reported that the suggested technique could successfully detect various defects on the tire treads [8]. In 2003, Abou-Ali and Khamis suggested an intelligent system named TIREDDX which can be used in the tire production process and service. By developing TIREDDX, it was aimed to develop an integrated diagnostic procedure that significantly reduces the defect detection time, and is a training instrument to guide less-experienced personnel [6]. Even though there are some steps that are completed by electronic systems, in the tire defect detection, some steps are still completed by human operators who use their self-experience. Although human experts can successfully detect the defects in tires, using individuals in an automated production system slows the process down. For this reason, the human intervention must be as small as possible. In their study, Wang et al. proposed a system that can automatically inspect tire images to detect inner bubble defects in a tire. Experiments showed that their system's accuracy rate was about 85% and the deviation between the system's result and reference was about to 5mm [28].

As a result of the improvements in the field of Artificial Intelligence (AI), Artificial Neural Networks (ANN) have also become one of the fundamental tools in tire defect detection. In the study of Tada et al., a Convolutional Neural Network (CNN) based method was used to classify defects that appear on the inner surface of a tire. They classified tires as "good", "quasi-good", and "defective". The study's successful results showed that using a CNN was effective if there were objects which have parts that have varying shapes [26]. Since it is important to detect bubble defects, Chang and Wang provided a CNN-based method to detect bubble defects in tires. The proposed method had accuracy up to 89.16%. Besides their high accuracy rate, the provided method also was successful at reducing the false positive rate [7]. Fang et al. suggest a backpropagation (BP) neural network-based system to detect defects in the tires. They designed a MATLAB program that has a GUI interface for interaction. As a first step, the user selects the tire image, then the image was preprocessed by using threshold transformation, edge detection, and seed filling. The output of this step was sent into BP neural network and output analysis results were displayed on the screen. Their results showed that the designed system was successful at detecting defects and reporting them [12]. The study of Cui et al. aimed to improve the accuracy rating in the tire defect classification process. They tried to design a method that is based on deep learning and can operate flawlessly with limited training samples that are acquired under varying scene illumination. The proposed method has achieved a 98.47% success rate and has had a satisfactory classification performance [10]. Zhang et al. provided a deep learning-based image recognition system to identify the tire damage. The proposed system consists of image pre-processing, feature extraction, and classification steps, and can automatically classify the type of damage in tire images. Experiments have shown that the suggested method has an accuracy of about 67%. Even though the accuracy rate is not high, it still highlights the feasibility and effectiveness of deep learning on tire damage recognition systems [29]. In a recent study by Li et al., a defect detection method that is based on Faster Regional CNN (Faster-R CNN) was used to spot the small size defects in the belt layer of radial tires. Their method was successful in reducing the false negative and false positive rates of small-size defects [18].

The literature research conducted on tire defect detection generally focuses on production line inspection and operates on X-ray images. Tire inspection must be a continuous process until the tire is replaced at some time as a result of a requirement. On the other hand, it is the part of the periodic vehicle inspection that is officially required in most countries. An automated tire inspection system installed at inspection stations, which operates on RGB images, and can detect defects on tires, has the potential to decrease the time spent on the inspection. In one of our previous study, we demonstrated the success rates of various CNNs to solve a classification problem about the solar panel condition. InceptionV3, VGG-19, and ResNet-50 provide 100%, 99%, and 91% F1-Score, respectively [21]. As a result of previous satisfactory results, we used conventional CNN [20], DenseNet201 [15], InceptionResNetV2 [25], and Xception [9] to detect flat tires in an RGB image data set which contains images of flat tires, and normal tires. Experiments show that we achieved 80% to 100% F1-Score by using the mentioned neural networks.

The rest of this study is organized as follows. In Sect. 2, we briefly explained the data set we used. Details of the method we used, and success rate evaluation is given in Sect. 3 and Sect. 4. Finally, in chapter Sect. 5, we discuss the results and possible future studies.

2 Dataset

This study uses the "Full vs Flat Tire Images" dataset [14] for detecting flat tires. This dataset contains 900 images with three classes that are based on the amount of air in the tires. These classes are labeled as "flat tire", "no tire", and "full tire". We only use "flat tire" and "full tire" classes for this study that are a total of 600 images. We use Tensorflow to alter these pictures with random rotation, Gaussian distortion, random color, and random brightness to boost neural network generalization for identifying flat and non-flat tires via transfer learning [5]. This data was split into three parts training, testing, and validation process. 80% were used for training, 10% used for testing, and 10% used for validation. Figure 1 shows an example of enhanced pictures.

Fig. 1. The example augmented images for training data.

3 Method

3.1 Deep Learning

Deep learning is a subtype of machine learning that outperforms machine learning by requiring more training data. Deep learning models are utilized in image classification, object identification, and semantic segmentation research. Their deep structures and hierarchical nature enable them to learn high-level features; when the image is processed via the learning model's pipeline, more abstract features are derived from the high-level features. All of these processes, however, have extremely high computing costs. Due to the development of modern

AI, there are different approaches to deep learning. Convolutional Neural Networks (CNN), Recurrent Neural Networks (RNN), and Generative Adversarial Networks (GAN) are the most popular ones in today's research environment (GANs) [13]. An example of a fully-connected neural network is given in Fig. 2

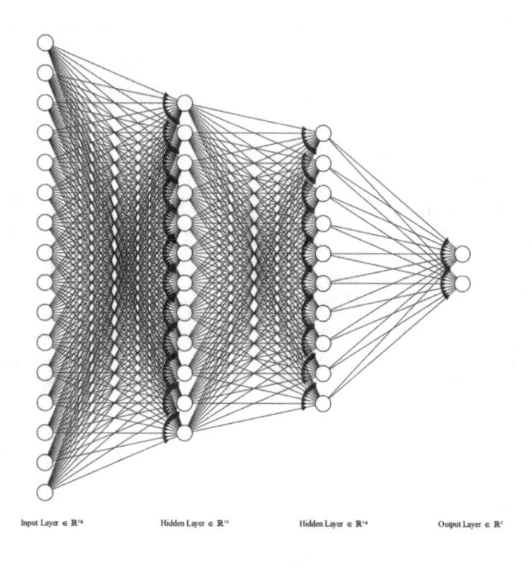

Fig. 2. General representation of ANN architecture (Generated by using [3])

CNN models are often regarded as the most accurate image classification methods. A CNN is mainly composed of many convolutional layers, each of which has one or more input and output layers. Following the convolution process, a batch normalization procedure normalizes the input batch by recentering and rescaling in a standard CNN model. After normalization, the ReLU (Rectified Linear Unit) activation function is used to add non-linearity into the data, resulting in fewer vanishing gradient concerns than other activation functions. Following the activation function, a dimension reduction operation, usually a max-pooling or average-pooling layer, reduces the model's spatial size and computation cost. When all of the steps are done, the features are flattened into a thick fully-connected layer.

3.2 Convolutional Neural Networks (CNNs)

Convolutional networks [17], commonly known as convolutional neural networks or CNNs, are a type of neural network that is used to process data using a grid-like structure. Time-series data, which can be imagined as a 1D grid capturing samples at regular intervals, and picture data, which can be imagined as a 2D grid of pixels, are two examples. Convolutional networks have had a lot of success in real-world applications. The name "convolutional neural network" refers to the network's use of convolution operations. Convolutional networks are basically neural networks with at least one layer that uses convolution instead of standard matrix multiplication. Example CNN architecture can be seen in Fig. 3.

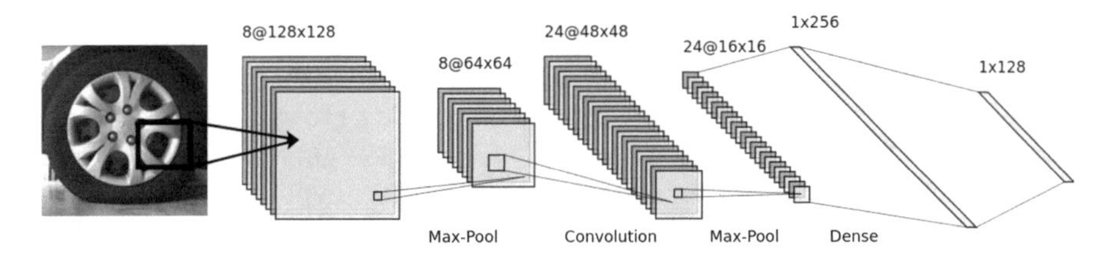

Fig. 3. The representation of CNN architecture for our study (Created by using the graphics generated by [3])

3.3 Transfer Learning

Transfer learning is a machine learning approach in which a model developed for one task is used as the foundation for a model developed for another activity. Given the massive processing and time resources necessary to create neural network models on these challenges, as well as the significant jumps in skill that they give on related problems, using pre-trained models as the starting point on computer vision tasks is a frequent method in deep learning.Transfer learning and domain adaptation are terms used to describe the process of using what has been learned in one environment to increase generalization in another. This extends the concept introduced in the preceding part, in which we moved representations from an unsupervised to a supervised learning task [13]. Depiction of transfer learning for our study can be seen in Fig. 4.

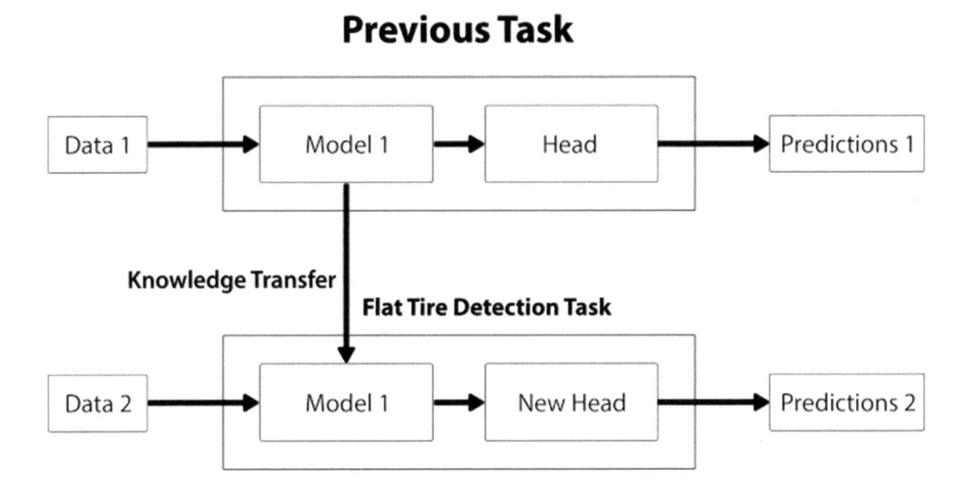

Fig. 4. The representation of transfer learning for our study (Adapted from [24]).

Hyperparamaters. In machine learning, hyperparameters are used to fine-tune the methods used in the learning process. The values of other parameters, such as node weights, are determined by the initial training process. Model hyperparameters, which refer to the model selection task and cannot be inferred while fitting the machine to the training set, and algorithm hyperparameters, which have no effect on the model's performance but affect the speed and quality of the learning process, are two types of hyperparameters. The topology and size of a neural network are examples of model hyperparameters. Learning rate and mini-batch size are two examples of algorithm hyperparameters. The hyperparameters used in this study are given in Table 1.

Table 1. Hyperparameters used in the implementations

Hyperparameter	Value
Number of Epochs(for CNN)	20
Number of Epochs(for pre-trained models)	5
Number of Classes	2
Learning Rate	0.0001
Batch Size Rate	8
Minimum Delta	0.006

3.4 Evaluation Metrics

In order to evaluate the detection of a flat tire on the fine-tuned deep learning models, accuracy, recall, and F-Score metrics have been evaluated for each method.

Accuracy. Accuracy is one statistical criterion for evaluating classification models. The accuracy is calculated by dividing the number of correct predictions by the total number of input samples [19].

$$Accuracy = \frac{TP + TN}{TP + TN + FP + FN} \tag{1}$$

Precision. By dividing the number of positive samples classified by the total number of positive samples classified, the precision is determined. The precision metric assesses the model's ability to correctly classify a positive sample [19].

$$Precision = \frac{TP}{TP + FP} \tag{2}$$

Recall. The recall is determined by dividing the total number of positive samples by the proportion of positive samples that were accurately identified as positive. The model's ability to detect positive samples is measured by the recall. The more positive samples identified, the higher the recall [19].

$$Recall = \frac{TP}{TF + FN} \tag{3}$$

F1 Score. The F-score, sometimes called the F1-Score, is a model accuracy statistic that determines how accurate a model is on a given dataset. It's used to assess classification algorithms, which divide data into "positive" and "negative" or much more classes. The F-score is a technique for combining the precision and recall of a model. It is defined as the harmonic mean of the model's accuracy and recall [27].

$$F_1 = 2 \times \frac{Precision \times Recall}{Precision + Recall} \tag{4}$$

4 Experimental Results

To classify flat and non-flat tires, we use three neural network architectures: Conventional CNN, DenseNet201, InceptionResNetV2, and Xception. The quantitative findings of each pre-trained network for a flat tire dataset are shown in Table 2. Scores for accuracy, precision, recall, and F1 measure have been provided for each approach.

Table 2. Classification results in different neural networks

Algorithm	Accuracy	Precision	Recall	F1 Score
Conventional CNN	80%	81%	80%	80%
DenseNet201	98%	98%	98%	98%
InceptionResNetV2	97%	97%	97%	97%
Xception	**100%**	**100%**	**100%**	**100%**

The results suggest that each approach is quite effective. When compared to Conventional CNN, InceptionResNetV2, and DenseNet201, Xception performs the best. Conventional CNN, InceptionResNetV2, and DenseNet201, on the other hand, do not perform poorly. As a consequence, we can see that each neural network used to identify flat tires has a high F1-Score rate.

5 Conclusion

This article proposed a smart tire classification application based on different CNN approaches using a tire dataset. The quantitative comparison presented gives accuracy, recall, precision, and F1 scores. We had over 80% of the success rate for each neural network. Additionally, a custom dataset created by images taken from various scenes which might or might not contain tires can be used to increase the system's robustness and could reveal a broader perspective to train and evaluate these neural networks. Finally, we point out the untapped potential of deep learning models for detecting different types of tire conditions. For future work, other types of neural networks could be used to compare results with DenseNet201, InceptionResNetV2, and Xception to decide which neural network is most suitable.

Our study shows that CNN models are suitable for tire classification purposes. Even though we used a personal computer to train and classify the tire images, pre-trained models might be used on embedded systems and single-board computers (SBC) with low-end specs. Another aspect to emphasize is, classification in this study was not done in real-time. For prospective studies, it will be proper to test this study's approach using custom-built image acquisition and classification system because a custom-built and portable system would be more practical.

References

1. Global car sales by key markets, 2005–2020 – charts – data & statistics – iea. https://www.iea.org/data-and-statistics/charts/global-car-sales-by-key-markets-2005-2020. Accessed 10 Feb 2021
2. How many cars are there in the world? https://drivetribe.com/p/how-many-cars-are-there-in-the-dqbpAzrATLOOSgDfRrgkjQ?iid=H-GqU0ALSdyjiEmSlJIb3g. Accessed 10 Feb 2021
3. NN-SVG Publication-ready NN-architecture schematics. https://alexlenail.me/NN-SVG/index.html. Accessed 17 May 2022
4. Tires. https://www.nhtsa.gov/equipment/tires
5. Abadi, M., et al.: TensorFlow: Large-scale machine learning on heterogeneous systems (2015). https://www.tensorflow.org/, software available from tensorflow.org
6. Abou-Ali, M.G., Khamis, M.: TIREDDX: an integrated intelligent defects diagnostic system for tire production and service. Expert Syst. Appl. **24**(3), 247–259 (2003)
7. Chang, C.-Y., Wang, W.-C.: Integration of CNN and faster R-CNN for tire bubble defects detection. In: Barolli, L., Leu, F.-Y., Enokido, T., Chen, H.-C. (eds.) BWCCA 2018. LNDECT, vol. 25, pp. 285–294. Springer, Cham (2019). https://doi.org/10.1007/978-3-030-02613-4_25
8. Chen, P., Shubinsky, G.D., Jan, K.H., Chen, C.A., Sidla, O., Poelzleitner, W.: Inspection of tire tread defects using image processing and pattern recognition techniques. In: Vision, Sensors, and Control for Automated Manufacturing Systems, vol. 2063, pp. 14–21. International Society for Optics and Photonics (1993)
9. Chollet, F.: Xception: Deep learning with depthwise separable convolutions (2016). https://doi.org/10.48550/ARXIV.1610.02357, https://arxiv.org/abs/1610.02357
10. Cui, X., Liu, Y., Zhang, Y., Wang, C.: Tire defects classification with multi-contrast convolutional neural networks. Int. J. Pattern Recognit. Artif. Intell. **32**(04), 1850011 (2018)

11. Şener, A.Ş, Ince, I.F., Baydargil, H.B., Garip, I., Ozturk, O.: Deep learning based automatic vertical height adjustment of incorrectly fastened seat belts for driver and passenger safety in fleet vehicles. Proc. Ins. Mech. Eng. Part D: J. Autom. Eng. **236**(4), 639–654 (2022). https://doi.org/10.1177/09544070211025338
12. Fang, L., Yanxue, L., Kai, Y.: Tire x-ray detection method based on BP neural network. In: Proceedings of the 2018 International Conference on Image and Graphics Processing, pp. 64–67 (2018)
13. Goodfellow, I., Bengio, Y., Courville, A.: Deep Learning. MIT Press, Cambridge (2016). http://www.deeplearningbook.org
14. Hammell, B.: Full vs flat tire images. https://www.kaggle.com/rhammell/full-vs-flat-tire-images (2021). Accessed 10 Feb 2021
15. Huang, G., Liu, Z., van der Maaten, L., Weinberger, K.Q.: Densely connected convolutional networks (2016). https://doi.org/10.48550/ARXIV.1608.06993, https://arxiv.org/abs/1608.06993
16. Khana, I.A., Ahmedb, S.Z., Iqbalc, M.: Driver safety system for drowsiness, heart attack, object detection, and internal temperature control of car with real-time wireless communication
17. LeCun, Y., et al.: Backpropagation applied to handwritten zip code recognition. Neural Comput. **1**(4), 541–551 (1989)
18. Li, P., Dong, Z., Shi, J., Pang, Z., Li, J.: Detection of small size defects in belt layer of radial tire based on improved faster R-CNN. In: 2021 11th International Conference on Information Science and Technology (ICIST), pp. 531–538 (2021). https://doi.org/10.1109/ICIST52614.2021.9440580
19. Olson, D.L., Delen, D.: Advanced Data Mining Techniques. Springer, Berlin (2008)
20. O'Shea, K., Nash, R.: An introduction to convolutional neural networks (2015). https://doi.org/10.48550/ARXIV.1511.08458, https://arxiv.org/abs/1511.08458
21. Ozturk, O., Hangun, B., Eyecioglu, O.: Detecting snow layer on solar panels using deep learning. In: 2021 10th International Conference on Renewable Energy Research and Application (ICRERA), pp. 434–438 (2021). https://doi.org/10.1109/ICRERA52334.2021.9598700
22. Singh, S.: Critical reasons for crashes investigated in the national motor vehicle crash causation survey. Technical report (2015)
23. Sprajcer, M., et al.: New parents and driver safety: What's sleep got to do with it? a systematic review. A Systematic Review
24. Sufian, A., Ghosh, A., Sadiq, A.S., Smarandache, F.: A survey on deep transfer learning to edge computing for mitigating the COVID-19 pandemic. J. Syst. Architect. **108**, 101830 (2020)
25. Szegedy, C., Ioffe, S., Vanhoucke, V., Alemi, A.: Inception-v4, inception-ResNet and the impact of residual connections on learning (2016). https://doi.org/10.48550/ARXIV.1602.07261, https://arxiv.org/abs/1602.07261
26. Tada, H., Sugiura, A.: Defect classification on automobile tire inner surfaces using convolutional neural networks. In: 2017 International Conference on Computing, Communication, Control and Automation (ICCUBEA), pp. 1–6 (2017). https://doi.org/10.1109/ICCUBEA.2017.8463768
27. Van Rijsbergen, C.J.: Information retrieval, 2nd Ed. Newton, MA (1979)
28. Wang, Y.C., Lin, J.C., Yang, H.Y.: Quantitative post-processing module of online automatic image inspection for inner bubble defects in a tire. DEStech Transactions on Engineering and Technology Research (imeia) (2016)
29. Zhang, S., Wu, Y., Chang, J.: Design of tire damage image recognition system based on deep learning. In: Journal of Physics: Conference Series, vol. 1631, p. 012015. IOP Publishing (2020)

Cyber Threats and Critical Infrastructures in the Era of Cyber Terrorism

Zeynep Gürkaş-Aydin and Uğur Gürtürk(✉)

Faculty of Engineering, Computer Engineering, AVCILAR, Istanbul University-Cerrahpaşa, Istanbul 34320, Turkey
zeynepg@iuc.edu.tr, 23ugurgurturk23@gmail.com

Abstract. Critical infrastructures are characterized as organizational and physical structures of vital importance to a nation. Critical infrastructure problems are significant because they have the potential to significantly affect values such as state and society security, economy, health and welfare, and the relationship of critical infrastructure to all other infrastructures. Damage to these structures has a significant negative impact country's social, psychological, and economic structure. When the literature is examined, it is seen that the attacks on critical infrastructures globally are primarily aimed at energy, health, disaster relief and emergency management systems. In this study, based on all these reasons, critical infrastructure types and the most severe attacks against critical infrastructures are included. Then, in order to present a perspective in this context, information about the solutions of these problems is given. Additionally, we sought to ascertain global trends in critical infrastructure protection and what needs to be done in Türkiye regarding internal security in this area.

Keywords: Critical Infrastructure · Cyber Attack · Cyber Security · Cyber Terrorism · Information Technology

1 Introduction

Although the scope of critical infrastructure systems is wide, these systems generally include agriculture, water, electricity grid, transportation, communication systems, technology systems, various public and private sector structures. These sectors are highly essential to ensure the continuation of daily life. Failure of just one part of the infrastructure will cause other dependent systems to fail, affecting the entire infrastructure where the failure occurred. In such an environment, the flexibility and security of infrastructure become a national priority because any damage or disruption to the services or operations provided by the critical infrastructure can have potentially dangerous consequences for the security and economy of the states and the well-being of the citizens of the state. Cyber threats are no longer limited to infiltrating computer systems, stealing information from systems and inserting false information into systems. It appears as an asymmetrical war type to the extent that it damages the communication systems, computer systems, energy and transportation networks, and military command and control

D. J. Hemanth et al. (Eds.): ICAIAME 2022, ECPSCI 7, pp. 274–287, 2023.
https://doi.org/10.1007/978-3-031-31956-3_23

systems of a country that can be considered critical. The increase in crimes committed in cyberspace reveals that states should increase their security measures in this area. It is seen that the vast majority of cyber attacks target corporate/government systems. The development of cyber security strategies has slowly gained momentum after 2008, as the trend of simple cyber attacks has shifted to large-scale state-sponsored attacks. The Turkish Penal Code and the European Cybercrime Convention have distinct practices in this area. The protection of cyberspace's critical infrastructures is a top priority for NATO member countries. Both the legal field and technical practices related to the protection of critical infrastructures have made significant progress in member countries working on this issue. In the first section of this study, the definition of critical infrastructure is described in depth, along with the various types of such infrastructure. The second section investigates the most significant attacks against critical infrastructures in recent years and their corresponding countermeasures. In the final section, an attempt is made to determine global trends in critical infrastructure protection and what Türkiye must do to improve its internal security.

2 Related Works

2.1 General Evaluation of the Studies on the Protection of Critical Infrastructures from Cyber Attacks in Türkiye

In the study of Demirci (2021), it is aimed to emphasize the duties and responsibilities of the Disaster and Emergency Management Presidency (AFAD), which works in coordination with the institutions and is one of the leading responsible institutions for the protection of critical infrastructures in Türkiye [1]. Choraś et al. (2016) presented methods and evaluations in modeling cyber security issues on critical infrastructure. In addition, they examined the attacks in this area to draw attention to the importance of cyber attacks in critical infrastructures [2]. In the study of Kınık (2016), a descriptive analysis was made and a conceptual framework was drawn about cyber security and critical infrastructures. Determinations were made regarding a rising theme of international security. After that, the legal status of the Caspian Basin was evaluated and an international security strategy was presented [3]. Karabacak (2011) defines critical infrastructure and evaluations about the dependence of critical infrastructures on cyber infrastructure. In addition, cyber security incidents for critical infrastructures are mentioned. After the current situation of our country is explained, the studies that need to be carried out for critical infrastructure security are included. The article is concluded with a future prediction about cyber threats [4]. Ak (2019) mentions in his study about carried out for the protection of critical infrastructures in the world and his point of view in this sense. In addition, a perspective on the work that needs to be done in Türkiye is provided. This theoretical study has been carried out in the form of a literature review [5]. Genco (2021), in his study, primarily included the definition of critical infrastructure. First, he touched upon the types of critical infrastructure and the work done on Critical Infrastructure in Türkiye, the USA and the EU. Then, the threats against these structures were analyzed, and an evaluation of countermeasures against attacks against these structures was presented [6].

In his study, Karabacak (2011) has included evaluations of the construction of critical infrastructure and the probability of critical infrastructures to cyber infrastructure. After presenting the current situation in our country, the studies that need to be carried out for critical infrastructure security are included. The article is concluded with a future perspective on cyber threats [8]. In this context, the studies of Ünver (2011) and others have been prepared to draw attention to the issue of critical infrastructures, which have become the most crucial target of threats in the cyber environment. The report primarily includes the definition and scope of critical infrastructures. Subsequently, the regulatory framework and studies on critical infrastructures in the United States (USA), European Union (EU) and Japan are mentioned. Finally, the current situation in Türkiye was mentioned and recommendations regarding in Türkiye have been made [11]. Karaca (2019) has focused on crimes that can be committed using information technologies in his study. In addition, in the study, studies on the protection of critical infrastructures in Türkiye and in the world against cyber attacks and the measures to be taken against attacks on these infrastructures are mentioned [12]. Erkal (2018) has concentrated on energy and assessed energy security risks in light of critical infrastructure security concerns. In addition, the energy policies in the USA, EU, China, Russia and Türkiye were mentioned, and a projection is made by looking at the current situation and for the future studies [14]. In the study of Çotak (2019), cyber risk insurance, which offers protection against cyber threats and is revealed by insurance companies, is formed in the world and in Türkiye, its operation, application methods, developments, the guarantees provided, in which cases the protection mechanism comes into play, and the driving force in the use of cyber risk insurance examines the factors [22]. In their study, Kutlu (2018) et al., based on the relationship between e-government and security, included the main reasons for the transition to e-government applications in the world and Türkiye. After examining the different dimensions of the security infrastructure of public services, various examples, experiences and studies in the world are revealed [27]. In his study, Karaman (2020) has evaluated the responsibilities and rights of citizens in the e-government system within the framework of the modern security concept. This evaluation has been examined about the Turkish Presidency model, which entered into force in 2018 [34]. In his thesis, Altunyuva (2020) approached the concept of energy security in a holistic way and examined the issue from the individual to the state, from the state to the interstate and supranational levels. In addition, it also scanned the current energy policies, which are prepared by taking into account Türkiye's energy outlook and targets [35]. In his study, Köylü (2017) reveals what kind of a Security Management System should be established according to internationally accepted standards, procedures, and practices to prevent threats to civil aviation in Türkiye. The main body of the research consists of aviation enterprises operating in the Turkish Civil Aviation System, which is responsible for implementing SMS [40]. In the study of Dilek (2021) and others they have aimed to examine the effects of production and supply chain areas that emerged with industry 4.0 on technological developments in terms of the logistics sector [42].In the study of Akçakanat (2021) and others, cyber security threats for businesses and information on managing these risks are discussed. They provided reviews of the top ten banks' cyber security and information technology activities by size and the data obtained from the reports in this area [44].

2.2 Cyber Attacks by Critical Infrastructure Types

In their studies, Daş (2019) et al. mentioned attacks on critical infrastructures. In addition, they included an evaluation of the Internet of Things. They discussed the types of attacks in this area as well as an Internet of Things perspective. In addition to mentioning intrusion detection systems, they provided a comprehensive literature review of academic studies in this field. [7]. Avcı (2022) et al. have briefly mentioned smart transportation vehicles in their article are briefly mentioned; in the second part, the structure of smart transportation and smart transportation vehicles are explained and examples of the types of cyber attacks are given. Again, the multi-layered security model designed against cyber attacks is introduced in the second part. The last part presents the results of cyber attacks on smart transportation vehicles and recommendations [15]. Çetin's (2017) study is about cyber security, policies and plans, other country practices, the International Atomic Energy Agency (IAEA) recommendations on this issue, the shortcomings of cyber security in Turkish legislation, and how to improve it. In addition, it covers related activities and responsibilities of institutions and individuals [16]. Birol's (2019) study questions how accurately physical and cyber risk analyses can be made in pipelines and whether the necessary precautions are adequately implemented to ensure Critical Energy Infrastructure Security within the Union. Two main hypotheses were tried to be evaluated in the study. First, the existing infrastructures of the supplier and transit countries that ensure the EU's energy security are insufficient and the EU cannot make sufficient progress in this regard. The second is that the EU has taken essential steps in this regard and has cooperated with transit and supplier countries sufficiently [18]. Torun (2022) et al. have conducted a research on consumers on the Internet. The detection of malware in the field of e-commerce on the personal computers of the users and the threats that will cause the sharing of personal data have been examined [21]. Aksoy et al. firstly have discussed cyber, cyber space, cyber crime conceptually and summarized the research in general terms. In the second rt, under the concept of cyber crime, the factors that lead to the emergence of cyber crimes, the costs of cyber attacks to companies, the types of cyber attacks, and the consequences of cyber attacks on companies and countries are discussed. [23]. In his thesis, Keleştemur (2018) explains the cyber attack methods and weapons necessary to carry out an effective cyber intelligence activity and the defense methods and systems that are important for countering intelligence activities [24]. In the study of Eliacik (2018), it is seen that the historical development of 'Energy', which constitutes the most important building block, has been examined in our daily life. The place of nuclear energy within the scope of renewable and sustainable energy has been determined and its importance and usage areas have been explained in detail [26]. Gündüz (2020) et al. in their study, critical cyber security objectives and requirements in smart grid data transmission networks are presented. In addition, the purpose of the study is to assess the attacks that violate the basic security requirements for ensuring secure communication in smart networks, evaluate the existing cyber security solutions, and develop a cyber security-centered perspective for the design of the secure network protocol architecture [28].

In their study, Tekke (2021) et al. evaluated the mass media and social media within the scope of cyberspace in the context of manipulative contents and digital ethical principles. At the same time, it is among the study's aims to detail the cyber security strategies carried out against the threats in cyberspace in the USA, China and Russia [33]. Topçu (2021), in his study, first deals with the formation of the information society and then define information security and cyber security. Later, the cyber security discourses of Myriam Dunn Cavelty and Laura Fichtner, respectively, were examined. After that, it focuses on cyber threats that are expected to increase in the future. Finally, what kind of measures to be taken against cyber security threats are explained [37]. The objective of the study by Tütüncü (2021) is to identify the themes around which scientific publications are concentrated in order to identify the areas in which the Internet of Things is used in the management of the COVID-19 epidemic, and to determine the areas that can contribute to the aforementioned literature by utilizing science maps [41]. The study of Meral (2021) et al. focuses on the detection of unplanned outages in web services used in data sharing between public institutions, and Windows services are used to shorten the detection times of unplanned outages. Then, the working details of Windows services are summarized. In addition, the findings related to the applied technique are included. Finally, the results of the study are given [43].

2.3 Policy Determination for the Protection of Critical Infrastructures

Ercan (2015) first mentioned the introduction and usage areas of ICTs within the scope of his thesis. Cyber security and types of cyber attacks, cyber defense activities, the current situation in our country on cyber security, critical infrastructures, and penetration tests applied to critical infrastructures are explained in the study and opinions on the subject is made at the end [9]. Söğüt (2019) et al. are aimed at attacking Industrial Control Systems (SCADA), one of the most important critical infrastructures. [10]. In this context, Kıran (2021] et al. have put forward a method by blending the above-mentioned layered architecture and scenario-based approach with additional asset-based accepted risk analysis and management methods and international standards/procedures/frameworks on the subject. This method aims to apply the existing accepted literature on critical infrastructures easily and effectively instead of the new risk analysis and management approach [13]. Akkutay's work (2017) approaches the subject from a public perspective within the framework of international law. In this work, in addition to drawing attention to the threats posed by cyber attacks in terms of international flights and airports, international legal rules that states must comply with within the framework of current international law are emphasized [17].

In his study, Bıçakçı (2014) has focused on the emergence of the Internet and focused on increasing cyber problems in this sense. In addition, information is given on cyberspace by providing an overview of the Developing Hacker Culture. Cyber threats to NATO are considered and NATO's perspective on critical infrastructures is presented [19]. The primary objective of Ozkaynak's (2016) study is to develop a model that can be proven secure in accordance with the provisions regarding the privacy of private life, which is protected by legal regulations, and the protection of data in the form of trade secrets, during the use of biometric and personal health data in identity verification processes [20]. In his thesis, Güzeloğlu (2017) has tried to discuss what ' security' means in the literature by drawing a roadmap from general to specific. The aim here is to show that the level of analysis, which is one of the most important discussion topics in the discipline of international relations, also causes controversy in security studies and to draw attention to how even the definition of security changes when the level of analysis changes [25]. In their study Memiş (2020) et al. have focused on data privacy in smart city applications. They have included the threats through big data and the open state of big data in smart cities [29]. Efe (2021), while focusing on the relationship between industry 4.0 (4IR) and AI, draws attention to the cyber security risks brought by this fusion and analyzes and evaluates the precautions that must be taken, the studies that must be conducted, and the controls that must be developed [30]. In his study, Başkan (2020) sought to devise a method for effectively protecting critical facilities against terrorist threats [31]. Şanlısoy et al. (2021) analyzed the development of cryptocurrencies and the status of cryptocurrencies in terms of their monetary functions. Then, first of all, currency wars are discussed. Then it is tried to evaluate how cryptocurrencies, currency wars and international monetary relations will interact in the future by creating different scenarios [32]. Karaarslan (2017) et al., have discussed how P2P and blockchain-based architecture and its elements work. Considering the system's reliability, they specified which security services this architecture provides. Examples from academic studies on the use of this structure for cyber security are given, problems in blockchain systems are discussed, and new approaches for solving them are mentioned [36]. In his study, Aydaner (2019) aimed to prevent consumers from being victims of their online shopping and make them feel safe. It is aimed to measure the effect of social engineering and cyber security awareness of young consumers on their online shopping intentions; tactical and precautionary information is conveyed [38]. Erdem's (2020) work focuses on ' cyber space' one of the space fields, but deals with ' outer space' in the parts deemed necessary and considers ' cyber space' as a new power competition field in 21st century International Relations aims to demonstrate [39].

3 Cyber Terrorism and Cyber Attacks on Critical Infrastructure

The use of the cyber world to carry out terrorist activities has led to the emergence of the concept of cyber terrorism. In general, cyber terrorism manifests itself in attacks aimed at damaging critical national infrastructures by using computer networks or making these infrastructures completely unusable. The realization purposes of cyber terrorism are generally to obtain, change or use information in structures such as state-protected telecommunications and national security networks, or to use it in terrorist acts. In addition, cyber terrorism aims to harm or inflict pain on people for a political purpose. Unlike traditional terrorism, cyber terrorism has some features. These can be listed as the use of cyber terrorism methods, the reasons for the cyber terrorist attack, and the deliberate use of information technologies by the attacker. Cyber terrorist attacks can carry more deadly dangers than traditional terrorist attacks [10]. Today, the protection of critical facilities is one of the most critical problems. Especially with on September 11, 2001, and the terrorist attacks in EU countries, many studies have started on the security of critical facilities in the USA, EU, and other developed countries. Critical facilities are of high importance for a country for any reason, like war, terrorist attacks, natural disasters, etc. These facilities can significantly affect the country if any damage is intended to them. Weaknesses in these facilities will cause security gaps, affect social life and cause economic losses [31]. In Fig. 1, the main critical infrastructure facilities

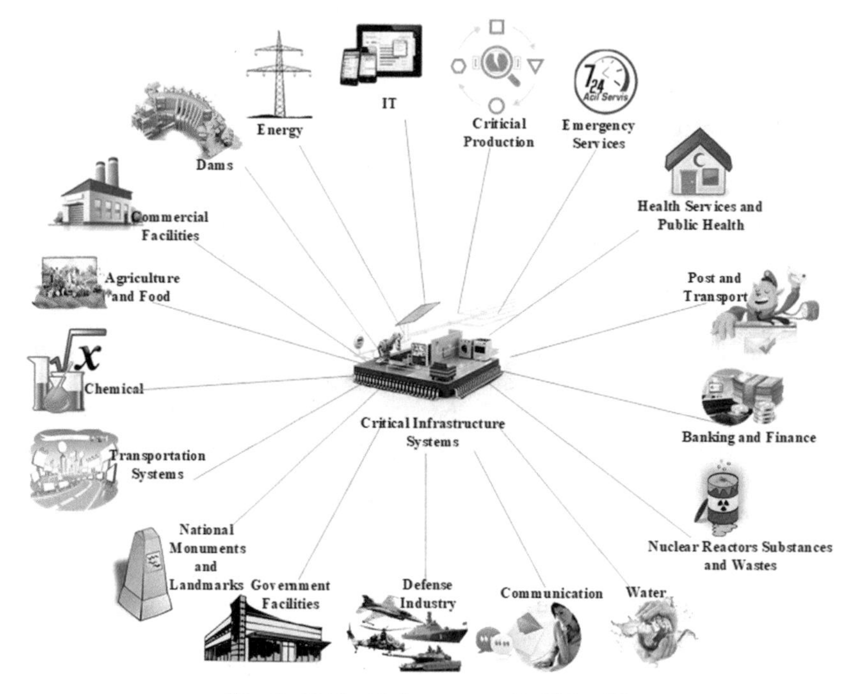

Fig. 1. Critical Infrastructure Facilities [7]

are presented. This section covers the significant major attacks on critical infrastructure facilities.

3.1 Logical Bomb

During the Cold War in 1982, computer technology; had become an increasingly important tool of attack for the CIA. The US exploded the Siberian gas pipeline without using a missile, bomb, or other weapons. In the 'logic bomb' method, which is accepted as one of the first known offensive or offensive cyber attacks, a code was added to the computer system controlling the Siberian gas pipeline, and thanks to this code, the computer that manages the system was confused. Siberia was extensively damaged without any physical intervention. A logic bomb is an unwanted code inserted into a software program that is only triggered after a specific event has occurred. Logic bombs are harmless and invisible to the system until the code is awakened to remove malicious instructions. Logic bomb viruses can be created to inflict any type of damage to the system they run on. It can rely on one or more triggers before hitting the road. Time-based logic bombs, also called time bombs, are a common form where the actions the code performs only take place on a specific date and time [45].

3.2 Sony Online Entertainment

In 2011, it was revealed that the account information of 100 million people was stolen in the attacks on PlayStation Network and Sony Online Entertainment. It has been said that these details include essential information such as personal information and credit card numbers. Examples of seized information include name, address, date of birth, e-mail addresses, passwords, and security questions. The damage caused by this attack is estimated to be between $1 billion and $2 billion [46].

3.3 Sven JASCHAn's Virus

The virus, which Sven Jaschan, an 18-year-old student, published on the Internet in 2004, instantly infected millions of computers worldwide and disabled computer systems. This cyber incident, which even caused the cancellation of Delta Air Lines' intercontinental flights, can be an example of attacks on critical infrastructures. Sven, whom Microsoft promised 250 thousand dollars to the finder, was caught three months later and proudly claimed this attack. Sven's damage is estimated to be close to $500 million [47].

3.4 The Oldsmar Attack

The Oldsmar Attack starting on February 5, 2021 was conducted using TeamViewer to log into a facility operator's console. The attacker, who accessed the session a few more times in the same day, did not attack in his first access. However, later on the day of the attack, the attacker re-entered the operator console and, using the facility's human machine interface (HMI), temporarily increased the sodium hydroxide content of the Water to toxic levels. First, one of the employees noticed an intrusion into the computer system that he could access remotely but thought his superiors did it. Then, that same afternoon, the same employee noticed that the mouse cursor on the computer screen was moving independently of them. This time, the amount of sodium hydroxide in the water was tried to be increased from 100 parts per million 11,100 parts. The operator observed this abnormal behavior and ordered the levels to return to typical values. Thereupon, the employee, who intervened in the situation, reversed the hacker's action, which was active for 3–5 min in the system, and restored the sodium hydroxide level in the water. Sodium hydroxide, also known as caustic soda, is used to regulate the amount of acid in water. Although this substance is also present in some hygiene products such as soap, exposure to high amounts can lead to burns, hair loss and other complications. In addition, if swallowed, it causes great harm to the mouth, throat and stomach. It was not known whether the hacker made this attempt from the USA or from outside the country regarding the incident, which was investigated. This attack can be described as one of the most significant attacks on critical infrastructures in recent history [48].

3.5 Titan Rain Attacks

Titan Rain is the name given to a series of cyber attacks on American computer systems that occurred on computer networks at the Pentagon in 1998. Attacks on major DoD structures have taken place, including the Redstone Arsenal, NASA, and Lockheed Martin. It is also estimated that between 10 and 20TB of confidential files were stolen. Cyber attacks were in the form of cyber espionage, in which attackers could obtain sensitive information from computer systems. Unofficial investigations to determine the cause of the attacks have not been fully proven, although it has been shown that the Chinese government did [49].

3.6 Hannaford Bros Attacks

Food retailer Hannaford Bros., which was exposed to one of the attacks against many American Retail chains between 2005–2008, has been subjected to a series of attacks starting in 2007. These attacks, which are thought to be made by Alberto Gonzalez and his team using the SQL Injection method, are one of the important attacks on the food industry. It is estimated that 4.2 million credit card numbers and sensitive information were seized during this period. It is estimated that this attack cost the company $252 million [50].

3.7 2013 Singapore Cyber Attacks

The Anonymous hacking group carried out a series of cyberattacks against the Singaporean government in 2013. According to the hackers, the attacks responded to the establishment of web censorship regulations by the government. The cyberattacks lasted several days and focused on influential people's social media accounts alongside influential websites [51].

3.8 OpIsrael Attacks

OpIsrael is the name given to a series of cyberattacks that spanned websites considered Israeli. It has been reported that the objective of a group attacking Anonymous is to erase the word 'Israel' from the Internet. The cyberattacks began on the eve of Holocaust Remembrance Day on April 7, 2013 and included database leaks, database hijacking and falsifications. Websites, including websites owned by schools, Israeli newspapers, small businesses, non-profits, and banks [52].

3.9 Estonia Cyber Attacks

Cyber terrorism is carried out by using Internet, computer and telecommunication technologies for terrorist purposes. Cyberterrorism aims to damage critical infrastructure facilities that threaten people's property, life or health via the Internet. The 2007 cyber attacks on Estonia are among the most famous examples of cyber terrorism. The attacks that started on April 27, 2007, the parliament, ministries, banks, social media and other similar structures also aimed to damage the websites of Estonian organizations. In addition, these attacks against Estonia caused the deterioration of relations between Estonia and Russia. The basis of these attacks is the decision of the Estonian government to carry the Bronze. As a result, Estonia's institutions and organizations, websites could not be accessed due to cyber attacks. In addition, cyber attackers shared fake news on Estonia's social media [53].

3.10 Shady RAT Attacks

Shady RAT (Remote Access Tool) attacks, which were first included in a report prepared by McAfee in 2011, are the most significant cyber attack ever, and the identity of the perpetrators could not be identified. In the report published by McAfee, it was stated that the first traces of Shady RAT date back to 2009 and that it was the attacks suffered by a defense company, and it was stated that even if the attacks were noticed and necessary precautions were taken, the infiltration continued for about a month from the first time it started. It is believed that the attacks on government institutions and businesses in 14 countries, and even international organizations such as the United Nations, began before the Chinese government supported the 2008 Summer Olympic Games. The main targets of the hackers, who seem to attack in an organized way, have been states, organizations, large companies, defense companies and even international Olympic committees. The USA, Japan, Taiwan, the United Kingdom, India, South Korea, Vietnam and Canada are among the states that suffered the most. According to the report, it is seen that such attacks, called Advanced Persistent Threat, started in 2006 and slowed down in 2010. The attacked areas include energy, industry, electronics and communications, defense areas and energy research laboratory, as well as various Korean steel and construction companies. The report has been evaluated that an attack of this scale cannot be carried out by a non-state actor alone [54].

3.11 Shamoon Virus

Shamoon is a type of virus linked to the infection of computer systems and cyber espionage on computers in the energy sector. Shamoon, also known as Disttrack, was carried out on August 15, 2012 by a group of hackers known as ' Swords of Interruption of Justice' for the firm's computer systems called Saudi Amarco Company. It also can enter the company's computers and delete and change many files. The group claimed responsibility for the attacks affecting operations on the company's 30,000 workstations. The primary purpose of this attack was to stop the oil and gas production of Saudi Arabia, the biggest oil exporter. Although these attacks continued in the following years, it was seen that some precautions could be taken [55].

3.12 Attack on the Ukrainian Power Grid

In 2015, a cyberattack was carried out on an electricity distribution network in Ukraine's Ivano-Frankivsk region. As a result, people in this region, where approximately 225 thousand people live, were left without electricity for hours. Energy companies stated that the attack took place through the BlackEnergy virus. In the attacks on December 23, 2015, Ukrainian Kyivoblenergo, a regional electricity distribution company, reported service interruptions to customers; It was stated that the connection of 23 35 kV substations was cut off for three hours. It has been observed that approximately 225,000 customers were affected by these attacks. After the attack, the Ukrainian government stated that a cyber attack caused interruptions by the Russian security services. Following these allegations, the US government and Ukraine conducted investigations and offered assistance to determine the outage's root cause. The attackers targeted phishing e-mails and manipulated Microsoft Office documents containing BlackEnergy malware and other malicious software to keep pace with power companies' Information Technology (IT) networks [56].

3.13 Stuxnet Attack

Stuxnet virus was noticed as a result of the research conducted by Ralph Langner and his team in 2010, upon the complaints of some customers that their computers shut down spontaneously. In the report published on the Stuxnet attack, which went down in history as an attack that has the power to affect the future of the cyber world, it was announced that the viruses called Trojan-Spy.0485 and MalwareCryptor.Win32.Inject.gen.2 was added to the antivirus base. However, when Langner tried to test the worm in his lab, he realized it was not working, and thought it might have been specially crafted to attack a specific location. Although it infects any Windows computer, it can find through research; it primarily targets Siemens brand controllers. After locating one, it employs a complex fingerprinting process that includes checking model numbers, configuration details, and even downloading the program code from the controller to ensure that it's the correct program before determining that the target has been successfully identified. The virus, which spread through Windows operating systems, has caused critical damage to information systems in dozens of countries in a short time. The virus, which specifically targeted the systems used in the nuclear energy sector, has impeded the occurrence of extraordinary situations in the reactors and paved the way for nuclear catastrophes. Edward Snowden, a former NSA (United States National Security Agency) employee, stated that the United States and Israel collaborated to create the Stuxnet computer virus to disrupt Iran's nuclear program. Iran is the country most affected by the virus, accounting for 58.85% damage. Iran has lost millions of dollars. In addition, the centrifuge machines in nuclear power plants spun at extraordinary speeds [57–60].

4 Conclusions

Cyber attack is defined as a concept that expresses planned and coordinated attacks on the critical infrastructures of states or institutions, such as stealing, changing, corrupting data, damaging with malware, deactivating, and revealing confidential information. It has been observed that cyber attacks have taken place intensively in Türkiye recently. Protection of critical infrastructures is vital in order to be prepared against a possible cyber attack in the future. For countries to fight against cyber attacks, it is of great importance to determine these infrastructures and establish a defense mechanism for these infrastructures. In addition to the concept of war,even though traditional armies are still the most influential power, it becomes necessary to add cyber armies to these armies simultaneously. States, institutions, organizations and individuals also have a presence on the Internet and it is seen that they need to be protected physically. The majority of these attacks on critical infrastructures are conducted by the military, according to the research conducted. Attacks on vital infrastructures are one of the most pressing concerns that states must address. Examining the cyber attacks that have occurred in the past is essential in raising awareness in terms of cyber attacks that may occur in the future. Examining the relevant literature reveals that cyber security vulnerabilities are typically the result of human actions. Even though these actions occurred in virtual environments, they have severe consequences for individuals and institutions. Ensuring cross-border cooperation, in which all countries collaborate, not only at the country level, is an essential factor in setting a deterrence against cybercrime as a common goal. Considering cyber security as a whole and the active participation of all stakeholders requires national and international cooperation. In this context, it is considered that in future studies, protecting critical infrastructures against cyber threats and conducting risk analysis in this sense will mean risk analysis in critical infrastructures. Critical infrastructures are getting more and more complex day by day. In order to protect these structures, institutions and organizations must be ready at the point of attacks on the critical infrastructures of countries. For this reason, a clear definition of the risk analysis will make it easier to apply to similar systems. It will be efficient to measure the country's level by increasing the valuation criteria of future studies and by including the institutional plans in the examination.

References

1. Demirci, K.: Kritik altyapilarda siber güvenlik ve afad üzerinden bir değerlendirme. Nazilli İktisadi ve İdari Bilimler Fakültesi Dergisi **2**(2), 54–64 (2021)
2. Choraś, M., Kozik, R., Flizikowski, A., Hołubowicz, W., Renk, R.: Cyber threats impacting critical infrastructures. In: Managing the Complexity of Critical Infrastructures, pp. 139–161. Springer, Cham (2016)
3. Kınık, H., Güntay, V.: Siber güvenlik temelinde kritik altyapılar ve hazar havzası. J. Int. Soc. Res **9**(47), 252–252 (2016). https://doi.org/10.17719/jisr.2016.1373
4. Karabacak, B.: Kritik altyapılara yönelik siber tehditler ve Türkiye için siber güvenlik önerileri. Siber Güvenlik Çalıştayı, Ankara **29**, 1–11 (2011)
5. Tarik, A.K.: İç güvenlik yönetimi açisindan kritik altyapilarini korunmasi. In: ASSAM Uluslararası Hakemli Dergi, pp. 42–51 (2019)
6. Genco, A.: Türkiye'de kritik altyapi ve kritik altyapiya yönelik tehditler. Kamu Yönetimi ve Teknoloji Dergisi **2**(2), 38–46 (2020)
7. Daş, R., Gündüz, M.Z.: Analysis of cyber attacks in IoT-based critical infrastructures. Int. J. Inform. Secur. Sci. **8**(4), 122–133 (2020)
8. Karabacak, B.: Kritik altyapılara yönelik siber tehditler ve Türkiye için siber güvenlik önerileri. Siber Güvenlik Çalıştayı, Bilgi Güvenliği Derneği, Ankara **29**, 1–11 (2011)
9. Ercan, M.: Kritik altyapıların korunmasına ilişkin belirlenen siber güvenlik stratejileri (Master's thesis, Sosyal Bilimler Enstitüsü) (2015)
10. Söğüt, E., Erdem, O.A.: Endüstriyel kontrol sistemlerine (scada) yönelik siber terör saldırı analizi. Politeknik Dergisi **23**(2), 557–566 (2020)
11. Ünver, M., Canbay, C., Özkan, H.B.: Kritik altyapıların korunması. Bilgi TeNnolojileri ve Koordinasyon Dairesi Başkanlığı, Mayıs (2011)
12. Karaca, M.: Kritik altyapılara yönelik bilişim suçları; Türkiye ve AB uygulamaları (Master's thesis, Fen Bilimleri Enstitüsü) (2019)
13. Kıran, E., Soğukpınar, İ: Kritik Altyapılarda Siber Risk Analizi ve Yönetimi. Türkiye Bilişim Vakfı Bilgisayar Bilimleri ve Mühendisliği Dergisi **14**(1), 1–19 (2021)
14. Erkal, H.Y.: Enerji Güvenliğine Yönelik Tehditler Ve Enerji Güvenliği Politikalarindaki Değişim. Ahi Evran Üniversitesi İktisadi ve İdari Bilimler Fakültesi Dergisi **2**(2), 63–78 (2018)
15. Avcı, I., Özarpa, C., Özdemir, M., Kınacı, B.F., Kara, S.A.: Akıllı ulaşım araçlarında siber güvenlik ve çok katmanlı güvenlik önlemi. Akıllı Ulaşım Sistemleri ve Uygulamaları Dergisi **5**(1), 22–35 (2022)
16. Çetin, M.: Nükleer tesislerde siber emniyet, siber saldırı senaryoları, sonuçları ve savunma sistemleri (2017)

17. Akkutay, A.İ: Sivil havaciliğa yönelik gerçekleştirilen siber saldirilar: uygulanacak uluslararasi hukuk kurallari, yetki ve sorumluluk. Türkiye Adalet Akademisi Dergisi **32**, 151–196 (2017)
18. Birol, S.: Boru Hatları Özelinde Avrupa Birliği Kritik Enerji Altyapı Güvenliği, PhD Thesis. Marmara Universitesi (Türkiye) (2019)
19. Bicakci, S.: NATO'nun gelişen tehdit algısı: 21. yüzyılda siber güvenlik. Uluslararası İlişkiler Dergisi **10**(40), 100–130 (2014)
20. Özkaynak, F.: Sosyal Güvenlik Kurumu Biyometrik Kimlik Doğrulama Sisteminin Problemleri ve Olası Çözüm Önerileri. Fırat Üniversitesi Mühendislik Bilimleri Dergisi **28**(2), 185–188 (2016)
21. Torun, N.K., Torun, T.: Kötü amaçli yazilimlarin e-ticaret içerisinde siber güvenlik açisindan incelenmesi. Sakarya İktisat Dergisi **11**(1), 1–16 (2022)
22. Çotak, A.: Sigortacılık sektöründe Siber güvenliği, dünyada ve Türkiye'Deki gelişmelerin Incelenmesi, PhD Thesis. Marmara Universitesi (Türkiye) (2019)
23. Aksoy, B., Erilli, N.A.: Siber suçların siber saldırılara maruz kalan şirketlerin hisse senedi fiyatları üzerindeki etkileri. İşletme Bilimi Dergisi **9**(2), 3–4 (2021)
24. Keleştemur, S.A.: Siber istihbaratın kamu güvenliği için rolü ve önemi (Master's thesis, Sosyal Bilimler Enstitüsü) (2018)
25. Güzeloğlu, M.: Enerji güvenliği bağlamında kritik enerji altyapı güvenliği, NATO konsepti ve Bakü-Tiflis-Ceyhan (BTC) Petrol Boru Hattı örneği (Master's thesis, TOBB ETÜ Sosyal Bilimler Enstitüsü) (2017)
26. Eliaçik, C.F.: Uluslararası hukuk bağlamında nükleer santrallerin siber güvenliği (Master's thesis, Sosyal Bilimler Enstitüsü) (2018)
27. Kutlu, Ö., Sevinç, İ, Kahraman, S.: Türkiye'de e-devlet uygulamalarında güvenlik risklerinin analizi. Electron. Turk. Stud. **13**(21), 129–156 (2018). https://doi.org/10.7827/TurkishStudies.13874
28. Gündüz, M.Z., Daş, R.: Akıllı şebekelerde iletişim altyapısı ve siber güvenlik. J. Inst. Sci. Technol. **10**(2), 970–984 (2020)
29. Memiş, L., Melikali, G.Ü.Ç.: Akıllı kentlerde verinin gizliliği ve güvenliği: ilkeler ve yaklaşimlar. Güvenlik Bilimleri Dergisi, (International Security Congress Special Issue), pp. 95–112 (2020)
30. Efe, A.: Yapay Zekâ ve Endüstri 4.0 İlişkisinin Siber Güvenlik Perspektifinden Analizi. Adnan Menderes Üniversitesi Sosyal Bilimler Enstitüsü Dergisi **8**(1), 123–143 (2021)
31. Başkan, O.: Kritik bölge savunma planlaması için koruma modelleri temelinde bütünleşik bir çözüm yaklaşımı (2020)
32. Şanlısoy, S., Çiloğlu, T.: Uluslararası Parasal İlişkiler ve Kur Savaşları Bağlamında Kripto Para Birimlerinin Geleceği. Nobel Yayınevi (2021)
33. Tekke, A., Aybala, L.: Sosyal Medyada Etik, Bilgi Manipülasyonu ve Siber Güvenlik. Akademik İncelemeler Dergisi **16**(2), 44–62 (2021)
34. Karaman, Z.T.: E-Devletin güvenlik bağlantılı sorumlulukları ve e-vatandaşın hakları. Bitlis Eren Üniversitesi İktisadi ve İdari Bilimler Fakültesi Akademik İzdüşüm Dergisi **5**(1), 1–20 (2020)
35. Altunyuva, F.B.: Çok boyutlu bir kavram olarak enerji güvenliği: Türkiye'de enerji güvenliği alanındaki politika yapıcı aktörlerin rolü (2021)
36. Karaarslan, E., Akbaş, M.F.: Blokzinciri tabanli siber güvenlik sistemleri. Uluslararası Bilgi Güvenliği Mühendisliği Dergisi **3**(2), 16–21 (2017)
37. Topcu, N.: Siber güvenlik: tehditler ve çözüm yollari. Cyberpolitik J. **6**(12), 155–181 (2021)
38. Aydaner, G.:Genç tüketicilerin sosyal mühendislik ile siber güvenlik farkındalıklarının online alışveriş niyetleri üzerindeki etkisinin ölçülmesi (Master's thesis, Sosyal Bilimler Enstitüsü) (2019)

39. Erdem, T., Turan, S.: 21. Yüzyılda uluslararası ilişkilerde yeni güç rekabet sahası: Siber Uzay (2020)
40. Köylü, S.M.: Sivil havacılık güvenlik yönetim sistemi (Master's thesis, Sosyal Bilimler Enstitüsü) (2022)
41. Tütüncü, D.: Salgın hastalıkların yönetiminde nesnelerin interneti kullanımı: COVID-19 örneği. Sağlık Akademisyenleri Dergisi **8**(2), 169–177 (2021)
42. Dilek, Ş, İncaz, S.: Küreselleşme sürecinde teknolojik dönüşümün lojistik sektörüne etkileri. Beykoz Akademi Dergisi **9**(2), 30–49 (2021)
43. Meral, E., Acar, S.: Web Servislerde Meydana Gelen Plansız Kesintilerin Tespiti. Uluslararası Yönetim Bilişim Sistemleri ve Bilgisayar Bilimleri Dergisi **5**(2), 212–225 (2021)
44. Akçakanat, Ö., Özdemir, O., Mazak, M.: İşletmelerde siber güvenlik riskleri ve bilgi teknolojileri denetimi: bankaların siber güvenlik uygulamalarının incelenmesi. Mehmet Akif Ersoy Üniversitesi Uygulamalı Bilimler Dergisi **5**(2), 246–270 (2021)
45. Cardenas, A., Amin, S., Sinopoli, B., Giani, A., Perrig, A., Sastry, S.: Challenges for securing cyber physical systems. In: Workshop on future directions in cyber physical systems security, vol. 5, issue 1 (2009)
46. Law, U.: Kişisel verilerin yurt dişi sunuculara aktarilmasinin bilişim sistemleri çerçevesinde türk ve avrupa hukuku açisindan incelenmesi. 3. Uluslararası Yönetim Bilişim Sistemleri Konferansı, vol. 63 (2016)
47. Georgescu, C., Tudor, M.: Cyber terrorism threats to critical infrastructures Nato's role in cyber defense. Knowledge Horizons. Economics **7**(2), 115 (2015)
48. Cervini, J., Rubin, A., Watkins, L.: Don't drink the cyber: extrapolating the possibilities of oldsmar's water treatment cyberattack. Int. Conf. Cyber Warfare Secur. **17**(1), 19–25 (2022). https://doi.org/10.34190/iccws.17.1.29
49. Lewis, J.A.: Computer espionage, titan rain and china. Center for Strategic and International Studies-Technology and Public Policy Program, 1 (2005)
50. Sherstobitoff, R.: Anatomy of a data breach. Inform. Secur. J.: A Global Perspect. **17**(5–6), 247–252 (2008)
51. Wei, L.H.: The challenges of cyber deterrence. J. Singap. Armed Forces **41**, 12–22 (2015)
52. Beraldo, D.: Movements as multiplicities and contentious branding: lessons from the digital exploration of# Occupy and# Anonymous. Inform. Commun. Soc **25**, 1098–1114 (2020)
53. Kardelya, O.V.: Cyber terrorism as a new type of terrorism and the 2007 cyber attacks on Estonia as an example of cyberterrorism (Doctoral dissertation, National Aviation University) (2021)
54. Alperovitch, D.: Revealed: operation shady rat. White Paper, Threat Research, McAfee, pp. 1–14 (2011)
55. Bronk C, Tikk-Ringas E.: Hack or attack? Shamoon and the Evolution of Cyber Conflict (2013)
56. Case, D.U.: Analysis of the cyber attack on the Ukrainian power grid. Electricity Information Sharing and Analysis Center (E-ISAC), vol. 388, pp. 1–29 (2016)
57. Orak, M.: Siber ordular ve siber savaşlar. Kamu Yönetimi ve Teknoloji Dergisi **3**(2), 214–226 (2022)
58. Kesik, M.: Siber güvenliği yeniden düşünmek: Stuxnet örnek olayı (Master's thesis, Sosyal Bilimler Enstitüsü) (2021)
59. Peychev, A.: What is the measured response to a cyber attack on critical infrastructures? Institute for Security and International Studies (ISIS), Sofia (2022)
60. Kumar, A., Choi, B.J.: Benchmarking machine learning based detection of cyber attacks for critical infrastructure. 2022 International Conference on Information Networking (ICOIN), pp. 24–29 (2022)

Secure Data Dissemination in Ad-Hoc Networks by Means of Blockchain

Cansin Turguner(✉), Engin Seven, and Muhammed Ali Aydin

Engineering Faculty, Computer Engineering, Istanbul University Cerrahpasa, Istanbul, Turkey
{cansin.turguner,engin.seven}@ogr.iuc.edu.tr,
aydinali@iuc.edu.tr

Abstract. The study of Ad-Hoc networks is one of the promising areas of computer science. Unmanned aerial vehicles (UAV), fully and semi-autonomous smart vehicles or combination of the both items with fix ground stations are the main application area of the ad-hoc networks. Due to the nature of these networks, they have heterogeneous structure, mobility, large geographic operation areas and energy demand. Besides these challenging demands of ad-hoc networks, to ensure security of communication between mobile nodes and fix station in ad-hoc networks is an essential issue. Some approaches related to this problem area are not sufficient as the generally the ad-hoc networks have de-centralize systems. So, to prevent the network from cyber-attacks, de-centralize precautions will be convenient such as Block-chain. In this study, the path of ensuring secure communication in complex ad-hoc networks using block-chain mechanism is proposed by using different geographic distribution algorithms such as Greedy Perimeter Stateless Routing (GPSR), Dynamic Source Routing protocol (DSR), Gauss Markov security performance of the proposed method is analyzed regarding attack resilience and power consumption of nodes for future studies.

Keywords: ad-hoc networks · secure communication · cyber security · block-chain · geographical algorithms

1 Introduction

Ad-hoc networks are wireless networks which nodes are power seeker, de-centralized and connected each other without any conventional network devices [1]. They have distinguished missions and functions associated where they are used. In many areas such as industrial, military and even daily life, they are designed and implemented by their service environment [2]. Vast majority of ad-hoc network implementations are in mobile ad-hoc network area. There are generally two main problem areas for mobile ad hoc networks which are vehicular ad-hoc network (VANET) and flying ad-hoc network (FANET) [3]. In this study, firstly the two main problem area are introduced. Then the main related researches which are best fitted in the study of FANETs are shown and explained. The detail block-chain schema and functions of the layer approach are clarified. The propose model secure data dissemination in ad-hoc networks by means

D. J. Hemanth et al. (Eds.): ICAIAME 2022, ECPSCI 7, pp. 288–297, 2023.
https://doi.org/10.1007/978-3-031-31956-3_24

of block-chain technology (SDDB) comes up with a solution. The implantation of this model and additionally Markov Chain distribution and communication algorithm are shown up in this study. Moreover, future vision of this model is discussed in the final part.

1.1 VANETs

Autonomous and semi- autonomous vehicles constitute the nodes of VANET. These nodes can communicate with each other directly and/or roadside units (RSUs). RSUs are the infrastructures which are located in the path of VANET [4]. The purpose of these units is to support huge amount of data transmission. Especially some VANETs have special mission which is mainly collecting data. Besides communication data, these cumulative data have deep impact for VANETs. Radar data, camera views, GPS location information are the examples of collecting data. Vehicles can either collect the data for analyzing or share them with other units [5]. By the reason of limited resources, collecting and sharing become more challenging issue. Even implementing these fundamental functions of VANETs are mighty, to sustain integrity, availability and privacy of the data is more complex and critical.

1.2 FANETs

UAV networks are one of the significant practice are of ad-hoc networks. Autonomous decision making, 3D direction movement and rapid motion abilities make the network wide and geographically independent as well as frequent changes can occur in the network [6]. Therefore some simple tasks such as data transmission can be interrupted. The reason of this interruption is sometimes limited resources or occasionally cyber security issues. To understand how to achieve security countermeasure for ad-hoc networks, firstly well developed network topology shall be implemented. Huge size of UAV networks need clustering structure. Despite clustering term refers hierarchical and centralized views of ad-hoc networks, by means of block-chain, the ad-hoc network topology will be decentralize and more resistant against cyber threats [7].Communication security is essential for both military and industry UAVs [20]. Cryptography including encryption can be exercised in data transmission.In Fig. 1, the topology of the Ad-hoc System is shown with security threats. In the related studies of this paper, some approaches are analyzed regarding this problem area. In addition, more rational way is came up with the proposed method using block-chain.

2 Related Works

2.1 Block-Chain Approaches

Vast majority of block-chain studies are discussed on UAV network security. Main inclusive work on Block-chain application is [1]. This study gives researchers some important paths to achieve well developed application of block-chain in FANETs [8]. Moreover, UAV types, regulations, architecture, some important roles of block-chain in the network, security side of block-chain, weakness of block-chain and more are included in the review. Deeply, how to support secure data transmission in UAV networks is discussed in [7]. Privacy, integrity, availability of service are fundamental issues which the system shall have resistance against internal and external attacks. To achieve secure data dissemination in UAV networks, types, ability, specialties of nodes and network environment have also key roles. Block-chain approach are useful for heterogeneous and de-centralized systems which is extensively focused by researchers like in [8, 9], proposed key management for FANETs are based on block-chain.

2.2 Agent Based Communication

In agent based communication [10], agent programming which is de-centralize, autonomous and flexible is discussed and directed to the problem of security platform. In this study, cartographic algorithms such as RSA and key distribution methods are implemented by means of agent programming. As known, wide majority of cartographic approaches in ad-hoc networks are challenging issue which the algorithms demand more memory and power [20].

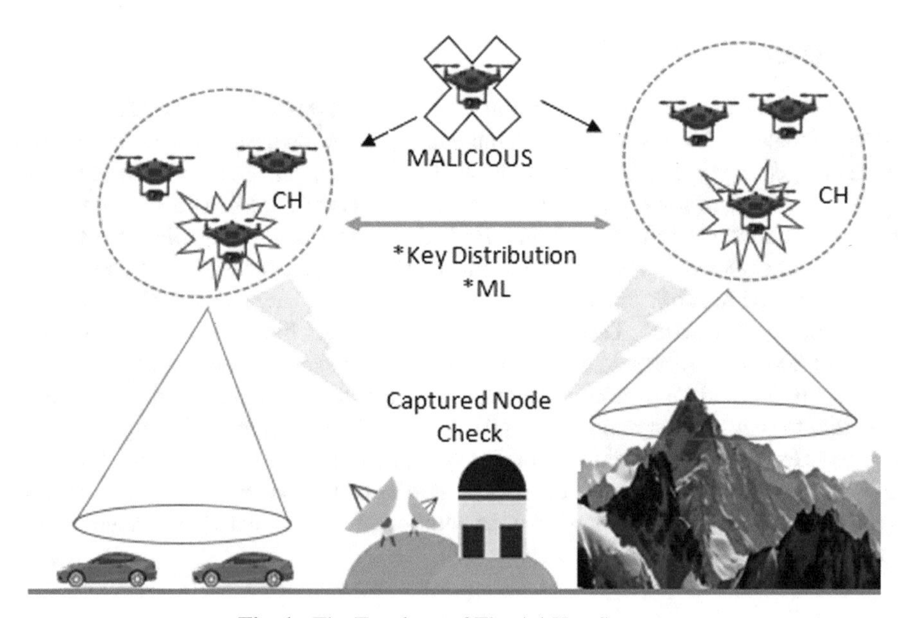

Fig. 1. The Topology of The Ad-Hoc System

Agent based fault tolerance mechanism works as an upper layer security precaution. This medium layer is claimed for especially availability of network services [20]. Flexibility and self decision making functions are the key roles of agent programming which the the nodes are able to make replicas in undesirable state of network. Therefore, the system operations are not effected even proposed security mechanism in progress.

2.3 Agent Based Communication

As a physical layer protection, jamming aid nodes method is proposed in paper [11]. Apart from conventional usage of ground jamming stations, the UAVs are charged for interrupting unwanted signal activities [11]. The purpose of this study is to keep the nodes in a distance which is suitable to establish secure and fine communication. On the other hand, this type of systems is semi centralized that is unwieldy. Moreover, regarding security target the system may have more vulnerabilities than the de-centralize structures. Also, there will be some optimization problem which researcher shall take care [12]. Some nodes have jamming activities so they can be easily noticeable by malicious attacks. So, the availability of service can be affected deeply by captured nodes.

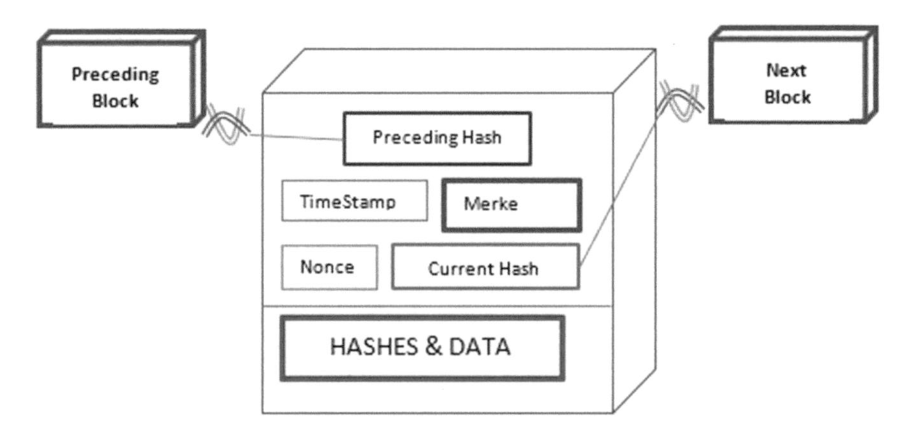

Fig. 2. The Block-Chain Connection Schema

3 Proposed Model

3.1 Block-Chain Structure

Layers generates block-chain structure [9]. These are data layer, network layer, consensus layer, incentive layer, contract layer and application layer.

3.1.1 Data Layer

The data layer has the timestamped blocks of data. This layer is divided into two parts which are body and header. The header covers previous blocks hash. In the block schema, all blocks contain preceding and next one's hash. The process of chain establishes by this link between blocks. The chain connection system is shown in Fig. 2. In this figure header contains Timestamp, Merkle root, Nonce and Hashes [13]. Timestamp is time of the block creation. Merkle Root is the tree model for storing the transactions which is time manner. Nonce is the specific number which is randomly created by miners.

3.1.2 Network Layer

The second layer is the network layer. The functions of this layer are to control the transactions by means of authentication and transmission. Generally, this connection establishes as Peer to peer (P2P). Encryption algorithm for the signs is used for authentication. Key matching and verification are of block-chain is practiced in this layer.

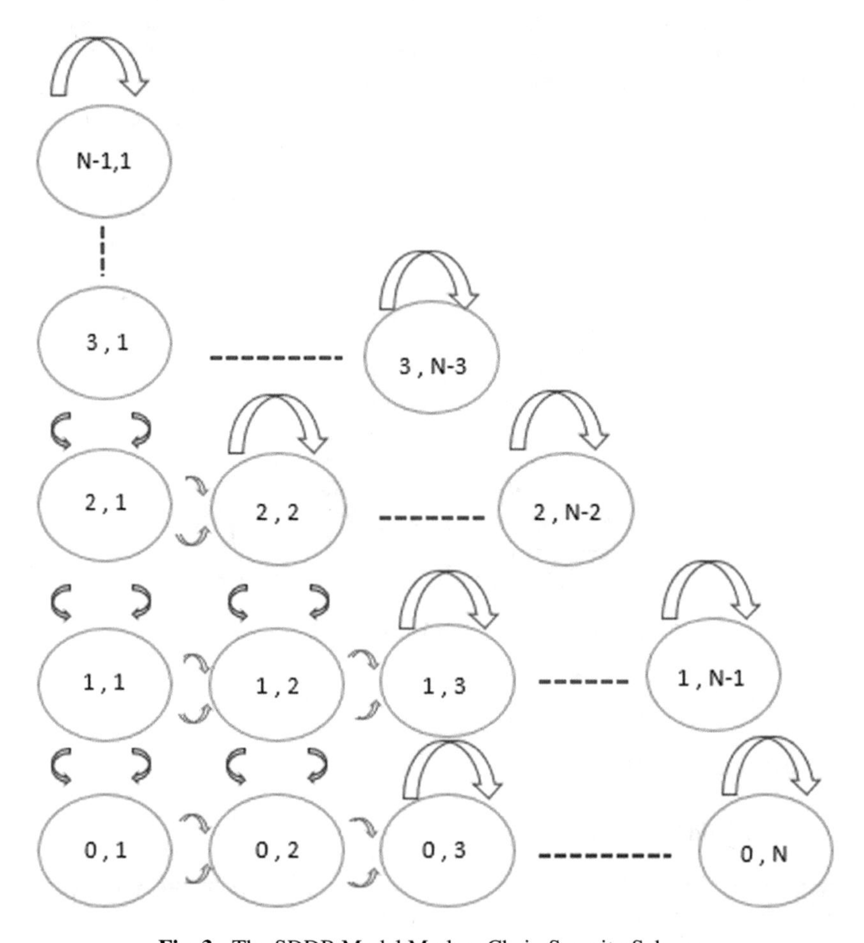

Fig. 3. The SDDB Model Markov Chain Security Schema

3.1.3 Consensus Layer

In Block-chain technology, consensus of nodes is very important. So, the invalidated party is not allowed to join communication. Some algorithms exist to obtain consensus [14]. They are, Proof of Stake (PoS), Proof of Work (PoW), Practical Tolerance (PBFT), and, Delegated Proof of Stake (DPoS).

3.1.4 Incentive Layer

This layer implies one of the popular usages of block-chain which effects economic benefits. Therefore, miners mine more power to gain more digital coin as well as constitution of power.

3.1.5 Contract Layer

The agreement between two units is approved in this layer. Security function of block-chain technology is performed only while the nodes deal with each other in specific terms. Moreover, detail communication, algorithms, coding area, encryption for validation are practiced in the contract layer [15].

3.1.6 Application Layer

As the FANET network problem area is discussed in this paper, mainly the application layer side of block-chain technology is used and mentioned. Financial protocols, internet of things applications [16], codes for machine learning algorithms etc. many experiment and study fields are studied in this layer. Energy control mechanisms [16], program implementations, some cryptographic approaches which are the interests of this paper.

3.2 Block-Chain Structure

In this study, besides the security for FANETs is provided by block-chain technology which provides trusted key exchange mechanism, additionally Markov Chain algorithm is implemented to detect any malicious nodes in the network [17]. According to researches even the block-chain layer schema usage solves the vast majority of security issue,the network unfortunately is still vulnerable and also it is not possible to fill this intrusion hole completely. So Markov Chain Security layer is added to the system to check periodically transmission for any malicious nodes existence. This part of SDDB model is shown in Fig. 3. Markov Chain process is a moment which many transmissions take place regarding and taking account of preceding situation of the network. So this sophisticated decision making algorithm provides performance of the network. This decision making based on prediction which after cumulative transition is made up and the condition of the nodes at a certain time is in sight. In this time, decision making is done to reach the better option. Rather than probabilistic algorithms such as Ant Colony Optimization [19], this stage by stage approach of Markov chain works to show probability of the state in limited opportunities. This probability is not always change regardless of the preceding states in the network [18]. If the probability of communication between specific node gets higher abnormally, then there is sink node which identifies as the friend node which is captured.

3.3 The SDDB Model

The environment that is considered consists of clusters of UAVs and ground station (GS). Apart from FANETs, VANETs have road side units(RSU) additionally. But in our system naturally usage of units like RSU is meaningless. So, there is a ground station for the system start-up and periodically check. Like in Fig. 1, clustering approach is used for redundancy. If one of the clusters has a problem such as security attacks, continues of operation, physical measurement by sensors or loss of communication data, the next cluster shall take the mission and prioritize the missions which it has. Therefore, the network has clusters that the number of clusters is dependent to amount of mission, environment, the UAVs capability. In this study, key distribution and update by done by block-chain process besides the Markov chain approach is for mainly intrusion detection. The process for the SDDB model is shown Fig. 4.

Algorithm 1 The SDDB Algorithm

```
FANET start-up;
UAVs initialize;
GS initialize;
Chronometer;
N = numberofUAVs;
UAV=create UAVs;
CH=create UAVs(master)
for each UAV do
    GS send keys to UAV(n) from chain;
end for
while mission continue do
    if chronometer is up then
        for each CH do
            GS check malicious node;
            if malicious node suspected (Markov Security) then
                if malicious node is CH then
                    Nearest CH takes the duty;
                    Delete malicious CH;
                    Update ML;
                    Update keys from GS;
                else
                    Delete UAV;
                end if
            else
                CH updates other UAVs in cluster;
            end if
        end for
```

Fig. 4. The SDDB Algorithm

3.3.1 Start-up for the SDDB Model

The start-up takes place with defining of nodes (UAVs). The registration numbers of all UAVs are listed in member list (ML). The rank of UAVs such as cluster head (CH) is assigned from ML. GS generates public and private keys and spread them to the nodes. So, all the nodes have public and private keys and naturally keep the private key in secret. Then there is another list for CH which only nodes in this list have authorization for communicate with other clusters. In the cluster, other nodes have only different authority when assigned for its duty. Otherwise, the rest of UAVs have the same power. All of the nodes have the list of CH and ML. These lists are updated after if any malicious node is detected which is discussed in Markov process security schema. All nodes have cluster index matching with their registration info in ML, address, Merkle tree, previous and current hashes, generator and data.

3.3.2 The Process for SDDB Model

The GS generates first block and starts up the UAVs by sending specific blocks. This transmission and the transmission by GS in the future takes place in secure area. CH nodes spread CH lists and public keys to each other. Therefore, the secure communication is established. The key change and distribution can take place only if there is malicious node suspected in the network or ant cluster node is not on duty anymore because of any reasons. Markov Process Chain security process is done by CHs for each cluster. If at any time, there is unusual communication probability then the system considers there is an attack. Moreover, GS calls for a new initialization for this cluster. In a specific range, the CHs are checked by GS if any CH detects any unpredictable probability of transmission. So, the keys in CHs for encryption are updated by GS.

4 Conclusion

In this paper, the application of secure data dissemination by means of block-chain with Markov Chain supported was explained. The problem ad-hoc network fields VANETs and FANETs were exposed. The limitations and vulnerabilities of the network was shown. Even if it seems that vehicular and flying ad-hoc networks have the same security issues, to establish well developed security infrastructure and intrusion detection in FANETs is more challenging than VANETs. FANETs have more mobility including three-dimension movement and the signal coverage which is fundamental for transmission. The signal is prone to fraction, reflection and absorption problems. The study area of VANETs is in the limit of roads but FANETs have no road so the control of the communication is more tough. Even the solving security issue seems too far, latest technology of the block-chain suggests encryption approach which is less weak comparing with traditional ones. Also, additionally usage of security and communication process of Markov Chain with the Block-chain makes the network more robust.

5 Future Studies

In the future study, implementation and simulation of this proposed model is planned. Then according to results of SDDB, optimization algorithms for the model will be used to achieve more secure, reliable, power saver network systems.

References

1. Turguner, C.: Secure fault tolerance mechanism of wireless Ad-Hoc networks with mobile agents. In: 2014 22nd Signal Processing and Communications Applications Conference (SIU), pp. 1620–1623 (2014). https://doi.org/10.1109/SIU.2014.6830555
2. Bujari, A., et al.: Flying ad-hoc network application scenarios and mobility models. Int. J. Distrib. Sensor Netw. **13**(10), 1–17 (2017)
3. Shukla, A., Xiaoqian, H., Karki, H.: Autonomous tracking and navigation con-troller for an unmanned aerial vehicle based on visual data for inspection of oil and gas pipelines. In: 2016 16th International Conference on Control, Automation and Systems, ICCAS, pp.194–200. IEEE (2016)
4. Xu, W., et al.: Internet of vehicles in big data era. IEEE/CAA J. Automatica Sinica **5**(1), 19–35 (2018)
5. Kang, J., et al.: Blockchain for secure and efficient data sharing in vehicular edge computing and networks. IEEE Internet Things J. **6**(3), 4660–4670 (2019). https://doi.org/10.1109/JIOT.2018.2875542
6. Lakew, D.S., et al.: Routing in flying ad hoc networks: a comprehensive survey. IEEE Commun. Surv. Tut. **22**(2), 1071–1120 (2020)
7. Jensen, I.J., Selvaraj, D.F., Ranganathan, P.: Blockchain technology for networked swarms of unmanned aerial vehicles (UAVs). In: 2019 IEEE 20th International Symposium on "A World of Wireless, Mobile and Multimedia Networks" (WoWMoM), pp. 1–7 (2019). https://doi.org/10.1109/WoWMoM.2019.8793027
8. Tan, Y., Liu, J., Kato, N.: Blockchain-based key management for heterogeneous flying ad hoc network. IEEE Trans. Industr. Inf. **17**(11), 7629–7638 (2021). https://doi.org/10.1109/TII.2020.3048398
9. Alladi, T., et al.: Applications of blockchain in unmanned aerial vehicles: a review. Veh. Commun. **23**, 100249 (2020). https://doi.org/10.1016/j.vehcom.2020.100249
10. Turguner, C.: Secure data dissemination in MANETs by means of mobile agents. In: 2014 10th International Conference on Communications (COMM), pp. 1–4 (2014). https://doi.org/10.1109/IC-Comm.2014.6866761
11. Li, A., Zhang, W.: Mobile jammer-aided secure UAV communications via trajectory design and power control. China Commun. **15**(8), 141–151 (2018). https://doi.org/10.1109/CC.2018.8438280
12. Cumanan, K., Xing, H., Xu, P., et al.: Physical layer security jamming: theoretical limits and practical designs in wireless networks. IEEE Access **5**, 3603–3611 (2017)
13. Cui, G., Shi, K., Qin, Y., Liu, L., Qi, B., Li, B.: Application of block chain in multi-level demand response reliable mechanism. In: 2017 3rd International Conference on Information Management, ICIM, pp. 337–341. IEEE (2017)
14. Xie, J., et al.: A survey of blockchain technology applied to smart cities: research issues and challenges. IEEE Com-mun. Surv. Tutor. **21**(3), 2794–2830 (2019)
15. Kosba, A., Miller, A., Shi, E., Wen, Z., Papamanthou, C.: Hawk: the blockchain model of cryptography and privacy-preserving smart contracts. In: 2016 IEEE (2016)

16. Elloini, N., Pahl, C., Helmer, S.: A decision framework for blockchain platforms for iot and edge computing. In: SCITEPRESS (2018)
17. Alnaghes, M.S., Gebali, F.:A Markov chain model for securing link layer in mobile ad hoc networks. In: 2015 SAI Intelligent Systems Conference (IntelliSys), pp. 971–975 (2015). https://doi.org/10.1109/IntelliSys.2015.7361260
18. Estahbanati, M.M., Rasti, M., Hamami, S.M.S.: A mobile ad hoc network routing based on energy and Markov chain trust. In: 7'th International Symposium on Telecommunications (IST'2014), pp. 596–601 (2014). https://doi.org/10.1109/ISTEL.2014.7000775
19. Turguner, C., Sahingoz, O.K.: Solving job shop scheduling problem with Ant Colony Optimization. In: 2014 IEEE 15th International Symposium on Computational Intelligence and Informatics (CINTI), pp. 385–389 (2014). https://doi.org/10.1109/CINTI.2014.7028706
20. Turguner, C., Sahingoz, O.K.: The study of experimental data transmission in wireless sensor networks. In: 2015 23nd Signal Processing and Communications Applications Conference (SIU), pp. 2222–2225 (2015). https://doi.org/10.1109/SIU.2015.7130317

Battery Charge and Health Evaluation for Defective UPS Batteries via Machine Learning Methods

Mehmetcan Çelik[1](✉), İbrahim Tanağardıgil[2], Mehmet Uğur Soydemir[1], and Savaş Şahin[1]

[1] Faculty of Engineering and Architecture, Department of Electrical and Electronics Engineering, İzmir Katip Çelebi University, İzmir, Türkiye
mehmetcancelik360@gmail.com
[2] Tescom Sanayi ve Ticaret A.Ş. Ar-Ge Bölümü, İzmir, Turkey

Abstract. In this study, the Dual ARM Cortex-M0+ 133 MHz RP2040 microcontroller-based data acquisition card was developed, which enables the measurement of voltage, current, and temperature values of the batteries connected in series in the Uninterruptible Power Supply (UPS). Owing to the developed card, UPS battery variables are observed continuously, so the dataset of the battery was prepared and used in machine learning regression methods. As a result of the regression methods, estimation processes were made for the State of Charge, State of Health, and the remaining discharge time. Root mean square error (RMSE), mean absolute error (MAE), and R^2 score values were used to compare the regression models' performances. Considering the comparison of RMSE, MAE, and R^2 score values showed that the eXtreme Gradient Boosting Regression method gives better results according to the obtained dataset.

Keywords: State of Charge · State of Health · Remaining Discharge Time · Machine Learning · RP2040

1 Introduction

The development of technology has caused data to occupy an important place today. According to some studies, data is produced as 2.5 exabytes per day [1]. This incremental data storage inspires new projects with artificial intelligence. Machine learning, a sub-branch of artificial intelligence, develops the concept of data and can perform operations in different areas such as classification, prediction, and clustering [2]. Data and machine learning concepts serve different areas, leading to various innovations. These two interconnected concepts have formed the idea for the realization of a study of Uninterruptible Power Supply (UPSs) that has a significant impact in the field of electricity. UPS is an electronic device that contains battery groups used in health, military, and industry, which prevents the main supply of any electrical-electronic device from being cut off [3]. Thanks to UPSs, energy is provided in critical places where electricity should never be cut off. Systems have been developed for the regular operation of UPSs, which

D. J. Hemanth et al. (Eds.): ICAIAME 2022, ECPSCI 7, pp. 298–308, 2023.
https://doi.org/10.1007/978-3-031-31956-3_25

are essential in maintaining electrical energy. The Battery Monitoring System (BMS), which is actively used in vehicles today, can easily monitor information about the battery [4]. This system used in vehicles might be integrated into the UPS, and the status of the batteries can be continuously monitored. UPS, critically important in electrical energy, can be made more up-to-date by predicting problems that may occur due to integrating BMS into a constantly monitored system.

In this study, a 12 V–9 A volt lead-acid battery in the UPS was used [5]. To develop the BMS system, a data acquisition card based on RP2040, which provides the battery's voltage, current, and temperature values, was designed [6]. With the designed card, the battery was charged 14 V–1 A, 14 V–1.5 A, 14 V–2 A, and discharged 9.5 V–2.5 A. The dataset of the battery was created by performing the charge-discharge process continuously. The collected data was transferred to the computer using the UART TTL communication protocol, and the resulting dataset was applied to machine learning regression methods. By using regression methods, estimations of State of Charge (SOC), State of Health (SOH), and the remaining time for discharge of the battery were made. The battery that may have a problem was observed with the values found as a result of the SOH estimation. To compare the accuracy of different regression algorithms used in machine learning, Root Mean Square Error (RMSE), Mean Absolute Error (MAE), and R^2 score values were calculated.

In the remaining part of the paper, the studies related to the project are examined in the second section, the data acquisition card and machine learning regression methods used in the formation of the project are explained in the third section, and the results of the project are evaluated in the fourth section. In the final section, it is mentioned the work that can be done in the future.

2 Related Works

Chandran et al. (2021) studied the SOC estimation of The Panasonic 18650FP battery cell. The Panasonic 18650FP battery cell dataset was compiled at McMaster University, Ontario, Canada, by the Department of Mechanical Engineering. Machine learning regression algorithms were used for SOC estimation with the Panasonic 18650FP battery cell dataset. These are artificial neural network (ANN), support vector machine (SVM), linear regression (LR), Gaussian process regression (GPR), ensemble bagging (EBa), and ensemble boosting (EBo). The regression methods' accuracy was tested by comparing Mean Square Error (MSE) and RMSE performance metrics [7].

Yusuf et al. (2019) compared SOC estimation with machine learning regression methods for the battery. The machine learning methods used in SOC estimation are Artificial Neural Network, Support Vector Machine regression, Linear regression, Ridge regression, and Lasso regression. They used data collected from the battery management system (BMS) database of the College of Engineering - Center for Environmental Research & Technology, University of California, Riverside, for machine learning regression methods. They obtained the results of machine learning regression algorithms using the dataset. MAE, RMSE, and Mean Absolute Percentage Error (MAPE) performance metrics were used to compare the accuracy of machine learning regression algorithms. When the MAE, RMSE, and MAPE values were examined, the most accurate result was seen in the SVM regression method [8].

Ronanki et al. (2019) conducted a study that estimated SOC for lithium-ion batteries. When the study was examined, it was seen that machine learning regression methods were used for SOC estimation. These algorithms are random forest (RF) regression, support vector regression, and neural network (NN) regression. They used the batteries' data for the estimation process, namely the dynamic stress test (DST) and the US06 highway drive cycle (US06). It was seen that the experiments were carried out at different temperatures for the estimation processes. MAE and COD performance metrics were used to compare the accuracy of the regression methods. When the regression methods were compared for the US06 dataset at 0 °C, 25 °C, and 45 °C, the MAE values of the temperature values were observed as 2.59, 2.88, and 2.88, respectively, and it was seen that the most accurate result was given in the RF regression algorithm. As for the COD performance metric, it was observed that the worst result occurred in the NN algorithm as the temperature increased toward 45 °C [9].

Li et al. (2021) studied SOC estimation for lithium-ion batteries. They used the Gaussian process regression with a gated recurrent unit kernel method for SOC estimation. It was seen that the Panasonic 18650PF battery cell under a series of electric vehicle drive cycles and the 18-Ah Li-ion battery cell under high-rate pulse discharge test (HRPDT) collected the dataset they used for the estimation process. They used MAE and MAX performance metrics to measure the accuracy of the regression method used [10].

Zhang et al. (2018) studied SOH estimation for lithium-ion batteries. In the study, they used GPR for SOH estimation. It was seen that the data for GPR estimation was taken from the data repository of the NASA Ames Prognostics Center of Excellence. Lithium-ion 18650 battery cells were used to collect the dataset. They used four batteries labeled 5, 6, 7, and 33 to demonstrate the SOH behavior in different conditions. The tests of these batteries were carried out at a room temperature of 24 °C. They used the RMSE performance metric to measure the accuracy of labeled batteries [11].

Hu et al. (2020) studied battery health. NASA, CALCE, and A123 datasets were used for health status estimation. Artificial neural networks, support vector machine, relevance vector machine, and Gaussian process regression from machine learning regression methods were applied and compared. Comparing the RMSE and MAE values found that the Gaussian process regression method was the most accurate result [12].

3 Method and Material

3.1 Developed Data Acquisition Card

As in Fig. 1, techniques were developed to measure voltage, current, and temperature values for developing the data acquisition card. A voltage divider circuit was used for voltage measurement. The value taken over the voltage divider circuit was read through the analog-digital converter (ADC) input of the RP2040 microcontroller used on the card. ACS712 current sensor is used for current measurement, while the DS18B20 temperature sensor was used for temperature measurement [13]. The values read over RP2040 were transferred to the computer with the UART TTL communication protocol. A dataset containing voltage, current, temperature, and time values was created with the data received continuously during the charge and discharge process of the UPS battery.

Fig. 1. Data Acquisition Card

3.2 Machine Learning Regression Methods

In the developed project, the codes of the machine learning regression methods were implemented in the Spyder environment with the Python programming language. The dataset was prepared for use in regression processes. After data preprocessing, estimation processes were made with each regression method, as in Fig. 2.

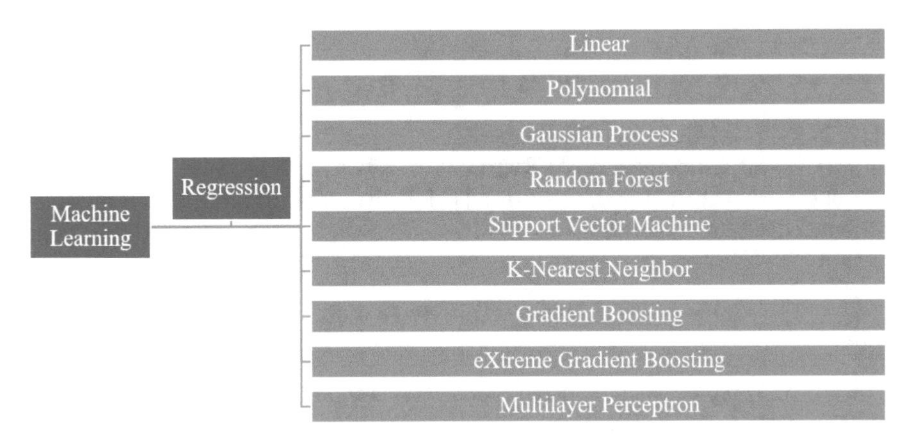

Fig. 2. Machine Learning Regression Methods

Linear Regression: According to the resulting dataset, it is the linear approach used to find the most relevant result. It is one of the estimation methods frequently used in supervised learning [8]. A general linear regression equation is given below.

$$y = \beta_0 + \beta_1 \times x \tag{1}$$

In Eq. 1, y is the dependent variable, x is the independent variable, β_0 is the regression constant, and β_1 is the slope.

Polynomial Regression: It is an estimation method used in the dataset that does not show linear distribution. A general polynomial regression equation is given below.

$$y = \theta_0 + \theta_1 x_1 + \theta_2 x_2^2 \tag{2}$$

In Eq. 2, y is the dependent variable, θ_0 is the regression constant, x_1 is the first independent variable, θ_1 is the coefficient for the first independent variable, x_2 is the second independent variable, θ_2 is the coefficient for the second independent variable.

Gaussian Process Regression: Another algorithm used for estimation is Gaussian Process Regression. It is used in regression operations for multivariate randomly distributed data. It is a kernel-based regression algorithm [12].

$$P((y_i|f(x_i), x_i) \sim N(y_i|h(x_i)^T\beta + f(x_i).\sigma^2 \tag{3}$$

In Eq. 3, σ^2 is noise variance, β is a coefficient vector, $f(x_i)$ is observation x_i.

Random Forest Regression: Random Forest regression generates many decision trees. It estimates by taking the average numerical values obtained from the decision trees. RF Regression generates regression trees using X training data, expressed as $X = x_1, x_2, x_3, \ldots, x_n$. In this method, each tree produces corresponding k outputs and can be expressed as $T_1(x), T_2(x), \ldots\ldots T_k(x)$ [9]. To calculate the result, the average of the estimation values of all trees is found by the equation given in Eq. 4.

$$RF(X) = \frac{1}{k}\sum_{k=1}^{k} T_k(x) \tag{4}$$

Support Vector Machine (SVM) Regression: It is a supervised learning regression algorithm to estimate non-parametric data [14]. It gives more accurate results on widely distributed datasets.

$$Y_i = \sum_{i}^{N} W.K(x_i, x) + B \tag{5}$$

In Eq. 5, Y_i represents predicted output, W is weight, K is kernel trick, (x_i, x) are support vectors, and B is bias.

k-Nearest Neighbor (KNN) Regression: KNN is a type of supervised learning mainly used in machine learning classification problems. Based on a fundamental constant k, it classifies according to its nearest neighbor. When the KNN algorithm is used in regression methods, it calculates as a numeric value by taking the average of the k training points closest to a test point (x_t). The distance of each training point to the x_t a point can be found by the Euclidian distance [15],

$$d(x_t, x_i) = \sqrt{\sum_{n=1}^{N} w_n(x_{t,n} - x_{i,n})^2} \tag{6}$$

N is the number of features, $x_{t,n}$ and $x_{i,n}$ are n th feature values of the testing point x_t and training point x_i, respectively. w_n is the weight assigned to the n th feature.

Gradient Boosting (GB) Regression and eXtreme Gradient Boosting (XGB) Regression: The gradient boosting algorithm is one of the boosting algorithms that help to reduce bias error in machine learning [16].

$$F(x;\ P) = F\left(x;\ \{\beta_m,\ \alpha_m\}_1^M\right) = \sum\nolimits_{m=1}^{M} \beta_m\, h(x;\ \alpha_m) \tag{7}$$

where, $F(x; P)$ denotes x function with P parameters, β denotes the weight of each node, α the parameter in the model. P represents the parameter of the model. $\varphi(P)$ is the likelihood function of the loss function $F(x; P)$.

$$\varphi(P) = E_{y,x}L(y,\ F(x;\ P)) \tag{8}$$

If the $m - 1$ number is derived from the number of models, the derivative is used to determine the direction of the fastest fall of the lost function. i.e. g_m.

$$g_m = \{g_{jm}\} = \left\{\left[\frac{\partial_\varphi(P)}{\partial_{pj}}\right]_{p=p_{m-1}}\right\} \tag{9}$$

The next step is to determine the gradient direction for the likelihood function. It can be expressed with ρ_m.

$$\rho_m = arg\ min\ \varphi(\rho_{m-1} - \rho_m g_m) \tag{10}$$

Finally, $f_m(x)$ for the mth model can be derived according to the following equation.

$$f_m(x) = -\rho_m g_m(x) \tag{11}$$

The extreme gradient boosting algorithm is the optimized version of the gradient boosting algorithm. It reduces model variances and normalizes the loss function. XGBoost improves loss function with Taylor expansion. It prevents over-fitness while increasing the complexity for trees to learn. It is a powerful computing tool for solving machine learning problems [16].

Multilayer Perceptron (MLP) Regression: Artificial neural networks are systems that provide mathematical modeling of the human brain. MLP, one of the types of ANN, was used as the regression method in the study. A multilayer perceptron (MLP) is a feedforward artificial neural network that produces a set of outputs from a set of inputs. It is a deep learning method that uses a backpropagation algorithm to train the network [17].

$$h_i = f^{(1)}\left(b_i^{(1)} + \sum_{j=1}^{n_h} w_{ij}x_j\right) \tag{12}$$

h_i is hidden neuron output, w_{ij} is the weight of each node, x_j are the inputs and b_i is the bias value

$$z = f^{(2)}\left(b^{(2)} + \sum_{j=1}^{n_h} v_j h_j\right) \tag{13}$$

In Eq. 13, n_h is the hidden neuron, v_j is the weight vector, which represents the weight connecting hidden unit j to the output neuron.

4 Discussion and Results

4.1 SOC Estimation

The relationship between the inputs was examined according to the collected dataset. When the dataset was examined, it was seen that the voltage value changed depending on the time. Voltage values were estimated according to time in both charge and discharge states. The estimation process was done with machine learning regression methods. RMSE, MAE, and R^2 score performance metrics were used to measure the accuracy of the estimation algorithms. RMSE is the square root of the mean of the square of all errors. It is always greater than MAE because it penalizes larger values than MAE [17]. MAE is a performance metric that takes the absolute square of the difference between the observed and predicted values. The fact that the MAE value is close to zero indicates the accuracy of the applied regression method [17]. R^2 is a statistical method that measures how close the data are to the fitted regression line. It has a percentage value. It has values ranging from 0 to 1. A value close to 1 indicates high training accuracy [17]. The equation is expressed as the sum of squares.

$$RMSE = \sqrt{\frac{\sum_{i=1}^{n}(y_i - \widehat{y_i})^2}{n}} \tag{14}$$

$$MAE = \frac{1}{n}\sum\nolimits_{i=1}^{n} y_i - \widehat{y_i} \tag{15}$$

$$R^2 = 1 - \frac{\sum_{i=1}^{n}(y_i - \hat{y}_i)^2}{\sum_{i=1}^{n}(y_i - \overline{y}_i)^2} \tag{16}$$

In Eqs. 14 and 15, n represents the number of observations, y_i actual value, $\widehat{y_i}$ is predicted value. In Eq. 16, $\overline{y}_i$ represents the mean of the response variables.

The performance metric values of machine learning regression algorithms for SOC estimation in charge-discharge states were calculated. Values and graphs of performance metrics were shown in Fig. 3 and Fig. 4 for charge states and Fig. 5 and Fig. 6 for discharge states, respectively. As a result, the comparison was made according to the performance metric values. It was seen that the most accurate estimation was done with the XGB regression algorithm. The fact that the XGB regression algorithm gives the most accurate result can be explained as follows. There are some factors that affect the accuracy of the system in machine learning methods. Examples of these factors are overfitting and bias-variance trade-offs. The XGB algorithm provides these two factors correctly by combining gradient boosting, bagging-bootstrap, and features randomness [16]. As a result, providing these two factors increases the accuracy of the system.

4.2 SOH Estimation

The batteries were charged with 1 A, 1.5 A, and 2 A, respectively, and discharged with 2.5 A. The time completed for charge-discharge is known as the cycle. To be able to decide on the battery health status, the cycle completion time was based. The cycle time of defective batteries differs from healthy batteries. At the end of 12 cycles in Fig. 7, the completion time of the cycle charged with the same current value is close. As a result, when the cycle times were examined, it was seen that the battery was healthy.

Algorithm	RMSE	MAE	R^2
Linear Regression	1.9465	1.0448	0.9884
Polynomial Regression	1.2719	0.4806	0.9951
Gaussian Process Regression	4.1824×10^{-9}	3.6014×10^{-9}	1.0
Random Forest Regression	0.1763	0.0322	0.9999
SVM Regression	1.5535	0.2916	0.9926
KNN Regression	0.2995	0.0664	0.9997
Gradient Boosting Regression	0.2515	0.1237	0.9998
XGB regression	0.0488	0.0426	0.9999
MLP Regression	3.6489	3.3641	0.9594

Fig. 3. Charge State Performance Metrics

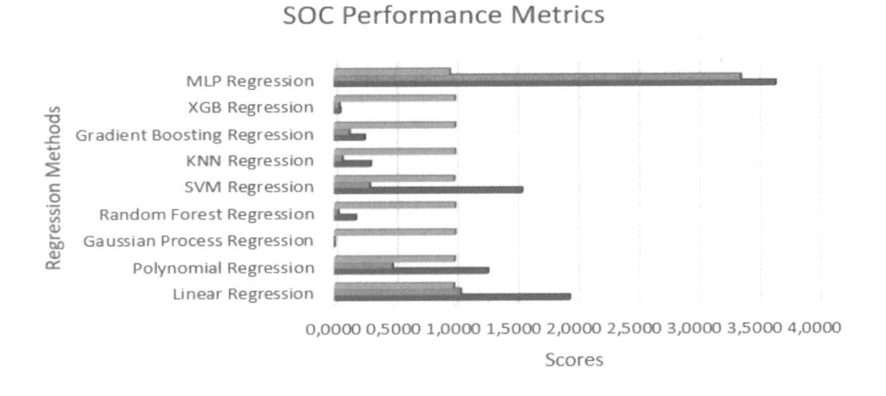

Fig. 4. Charge State Performance Metrics Graph

Algorithm	RMSE	MAE	R^2
Linear Regression	2.7833	2.0141	0.9725
Polynomial Regression	0.9409	0.4914	0.9969
Gaussian Process Regression	2.8377×10^{-9}	2.3664×10^{-9}	1.0
Random Forest Regression	0.0719	0.0258	0.9999
SVM Regression	1.1029	0.2918	0.9956
k-NN Regression	0.1632	0.0522	0.9999
Gradient Boosting Regression	0.1502	0.1125	0.9999
XGB regression	0.0283	0.0227	0.9999
MLP Regression	2.8608	2.0580	0.9709

Fig. 5. Discharge State Performance Metrics

4.3 Remaining Discharge Time Estimation

The voltage value and time of the battery at any given moment can be estimated using the XGB regression method (Fig. 8), which gives the most accurate result among the regression methods. The remaining time for discharge can be found using the voltage

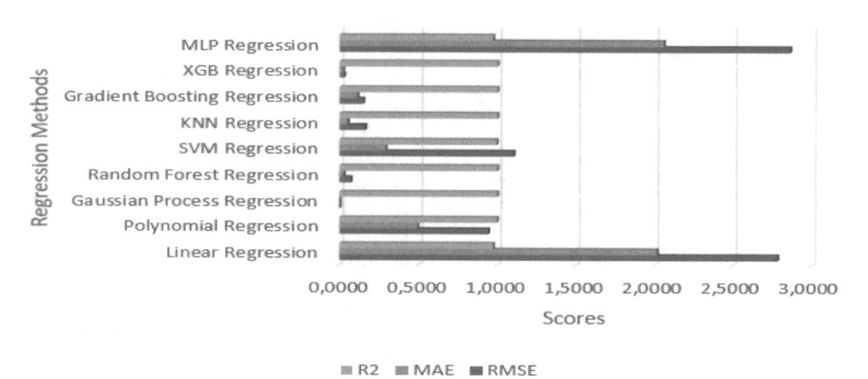

Fig. 6. Discharge State Performance Metrics Graph

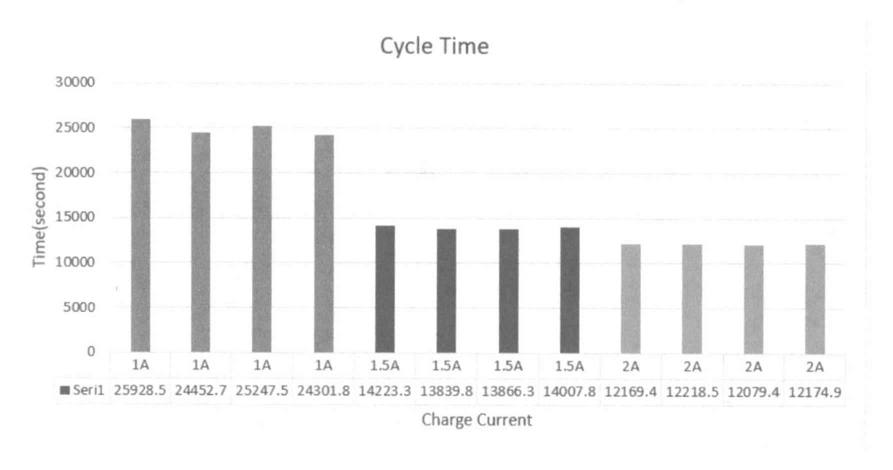

Fig. 7. SOH Cycle Time Graph

value and the time value corresponding to the voltage value. In the study, it was seen that the battery was discharged in 6752 s with 2.5 A. Moreover, the remaining time for discharge was determined using the total time known for discharge and the time value corresponding to the voltage value. The Remaining Discharge Time equation is given below.

$$Remaining\ Time = Total\ Time - Predicted\ Time \tag{17}$$

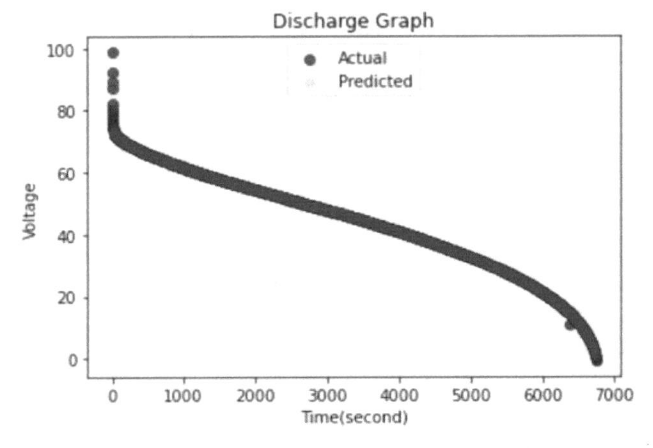

Fig. 8. XGB Regression Discharge Estimation Graph

5 Conclusions and Future Work

In this study, a dataset was collected for SOC, SOH, and remaining discharge time estimation of UPS batteries, and the regression models of the collected dataset were analyzed. RMSE, MAE, and R^2 score values were used to compare the regression models used. According to the performance metrics used, it was seen that the most accurate result was given in XGB regression algorithm. When the collected dataset and regression models were examined, it was realized that new studies could be carried out with the collected datasets of batteries in different environments or parameters. By increasing the size of the dataset, it was seen that deep learning algorithms could be used for SOC, SOH, and discharge remaining time estimations.

References

1. Aktan, E.: Big data: application areas, analytics and security dimension. J. Inf. Manag. **1**(1), 1–22 (2018)
2. The MathWorks, Inc. (n.d.) How machine learning works. https://www.mathworks.com/discovery/machine-learning.html
3. Chandrokar, M.C., Divan, D.M., Banerjee, B.: Control of distributed UPS systems. In: Proceedings of 1994 Power Electronics Specialist Conference-PESC 1994, vol. 1, pp. 197–204. IEEE (June 1994)
4. Li, R., Liu, C., Luo, F.: A design for automotive CAN bus monitoring system. In: 2008 IEEE Vehicle Power and Propulsion Conference, pp. 1–5. IEEE (September 2008)
5. TESCOM (2022). https://www.tescom-ups.com/tr/urunler/akuler-34
6. Raspberry Pi (2022). https://www.raspberrypi.com/products/rp2040/
7. Chandran, V., Patil, C.K., Karthick, A., Ganeshaperumal, D., Rahim, R., Ghosh, A.: State of charge estimation of lithium-ion battery for electric vehicles using machine learning algorithms. World Electr. Veh. J. **12**(1), 38 (2021)
8. Hasan, A.J., Yusuf, J., Faruque, R.B.: Performance comparison of machine learning methods with distinct features to estimate battery SOC. In: 2019 IEEE Green Energy and Smart Systems Conference (IGESSC), pp. 1–5. IEEE (November 2019)

9. Sidhu, M.S., Ronanki, D., Williamson, S.: State of charge estimation of lithium-ion batteries using hybrid machine learning technique. In: IECON 2019-45th Annual Conference of the IEEE Industrial Electronics Society, vol. 1, pp. 2732–2737. IEEE (October 2019)
10. Xiao, F., Li, C., Fan, Y., Yang, G., Tang, X.: State of charge estimation for lithium-ion battery based on Gaussian process regression with deep recurrent kernel. Int. J. Electr. Power Energy Syst. **124**, 106369 (2021)
11. Yang, D., Zhang, X., Pan, R., Wang, Y., Chen, Z.: A novel Gaussian process regression model for state-of-health estimation of lithium-ion battery using charging curve. J. Power Sources **384**, 387–395 (2018)
12. Hu, X., Che, Y., Lin, X., Onori, S.: Battery health prediction using fusion-based feature selection and machine learning. IEEE Trans. Transp. Electr. **7**(2), 382–398 (2020)
13. Özdisan (2022). https://ozdisan.com/optolar-vesensorler/sensorler/sicaklik-sensorleri/DS18B20
14. The Mathworks, Inc. (n. d.). Understanding Support Vector Machine Regression. https://www.mathworks.com/help/stats/understanding-support-vector-machine-regression.html
15. Hu, C., Jain, G., Zhang, P., Schmidt, C., Gomadam, P., Gorka, T.: Data-driven method based on particle swarm optimization and k-nearest neighbor regression for estimating capacity of lithium-ion battery. Appl. Energy **129**, 49–55 (2014)
16. Chang, Y.C., Chang, K.H., Wu, G.J.: Application of eXtreme gradient boosting trees in the construction of credit risk assessment models for financial institutions. Appl. Soft Comput. **73**, 914–920 (2018)
17. Berecibar, M., et al.: Online state of health estimation on NMC cells based on predictive analytics. J. Power Sources **320**, 239–250 (2016)

Covid-19: Automatic Detection from X-Ray Images Using Attention Mechanisms

Cemil Zalluhoğlu(✉) and Cemre Şenokur

Computer Engineering, Hacettepe University, Ankara, Turkey
cemil@cs.hacettepe.edu.tr

Abstract. The Covid-2019 pandemic, which started spreading in Wuhan, China (2019), has made people's daily lives hard to maintain. Hospitals work at maximum capacity because of the rapidly growing cases and the patients coming in for the Covid tests. Testing kits are hard to acquire. Therefore not everyone could be tested except they have obvious symptoms. Therefore it is critical to detect positive cases as soon as possible with less effort and less human work. Doctors can detect Covid from a patient's chest X-Ray images. This work tries to obtain an algorithm that can detect the positive cases as accurately as possible without needing a human eye. In this work, two different methods have been used, both learning the patterns of positive and negative Covid cases; one is a baseline method and the other uses attention with merged X-Ray datasets. The proposed method resulted in efficient accuracies; the best is 0.86.

Keywords: Covid-19 · Deep Learning · Computer Vision · X-Ray · Attention

1 Introduction

COVID-2019 is a new type of coronavirus which is a pneumonia disease first seen in Wuhan, China, in December 2019 and it has spread to the rest of the world [1]. The current data shows that more than 400 million cases are seen and more than six million of those cases resulted in death [2].

Coronavirus is a common, mostly not dangerous kind of virus that causes an infection in the nose and upper throat. Covid-2019 is caused by the SARS-CoV-2 type of coronavirus that triggers respiratory tracts upper or lower. It can be dangerous and deadly when it affects lower tracts.

The most common symptoms of Covid-2019 are fever, fatigue, and cough. However, because these symptoms are similar to the flu, it is hard to determine without testing or screening.

In detecting Covid, Covid testing kits have been widely used, but there is not enough supply of these kits in some countries and sometimes the tests give false-positive results. That pushed the doctors to use chest images to detect

D. J. Hemanth et al. (Eds.): ICAIAME 2022, ECPSCI 7, pp. 309–319, 2023.
https://doi.org/10.1007/978-3-031-31956-3_26

the disease. Radiographic images (X-Ray) are helpful in the early detection of Covid-2019 disease, but because of the X-Ray's lower density resolution, it can be read as no findings when it is a Covid positive case in the very early stages of the disease. In this case, CT imaging is more effective in detection [3].

At the time of the outbreak, many researchers have used deep learning techniques to detect coronavirus from X-Ray and CT images. Many different datasets have been collected and used. Most of the works got good results in their detection. However, according to some studies in the literature in X-Ray images, the algorithms learn not just the disease but the features of the used dataset [4]. That is why a recent data collection has started by researchers, containing X-Ray images from different sources to minimize dataset feature learning. This is an open dataset, still in the works, gaining more images in time [5].

In this study, we use a merged X-Ray dataset collected from different open-source datasets to prevent the algorithm from learning dataset features. Training is done with two different methods; one is a base deep learning algorithm and the other uses an attention mechanism. The baseline method is trained with four deep learning architectures, namely VGG-16, Densenet-161, ResNet-34, and ResNet-50, while the attention method is trained with ResNet-34 and ResNet50.

The rest of the paper is organized as follows. Section 2 gives a brief overview of related works. Section 3 introduces the merged X-Ray dataset in this work. Section 4 presents the methodology of our work. Evaluation metrics and experimental results are reported in Sect. 5 and finally Sect. 6 presents the conclusions.

2 Related Work

Kassan et al. [5] used Cohen's dataset containing 117 X-Ray and 20 CT images for positive cases and added negative cases the same amount from the Kaggle dataset. They tried 16 deep learning models with six different machine learning algorithms.

Maguolo et al. [4] worked on the generalization of dataset features and showed that deep learning algorithms learn dataset features on X-Ray images. They extracted the chest cavities from the images and trained their program with these new images.

Kedia et al. [6] proposed a novel deep neural network model called CoVNet-19 to find features of covid positive X-Ray images. They got an accuracy of 98.28% from a 3-class classification from a dataset containing 1628 covid, 2341 normal, and 2345 pneumonia.

Apostolopoulos et al. [7] used two different datasets composed from public datasets, one containing three classes Covid, Pneumonia, No Finding, other containing only two. They used State-of-Art CNN models, which have been proposed in recent years, using transfer learning strategy. Their work was about seeing how different deep learning algorithms work on datasets and comparing the results. They got the best results with MobileNetV2 and VGG19, respectively 98.75%, 97.40% accuracy on two-class classification and 93.48%, 92.85% accuracy on 3-class classification.

Hussain et al. [8] proposed a new model called CoroDet. CoroDet is a 22-layered CNN model. Nine of them are 2D convolution layers, the other nine 2D Max Pooling layers for reducing the size, one Flatten layer, two Dense layers, and one activation layer which uses Leaky ReLu. CoroDet works on two, three and four class classifications. For three-class classification, they used Covid, Normal, and Pneumonia samples, and for four-class classification, they used Covid, Normal, Pneumonia Viral, Pneumonia Bacterial samples. They compared their model with ten other existing models. As a result, they got 99.1%, 94.2%, 91.2% for two, three, and four class classifications which are better results than existing methods.

Jain et al. [9] used 1832 images for their work. They proposed a new method using State-of-art CNNs, containing image preprocessing, data augmentation, training of deep learning ResNet50 network, training ResNet-101. Their work determines bacterial and virus-induced pneumonia and no finding cases on X-Ray images. First ResNet50 training is to determine viral-induced pneumonia, bacterial-induced pneumonia, and normal cases, and second ResNet101 training is to detect the presence of COVID-19 from positive viral-induced pneumonia cases. They got a result of 97.77% accuracy.

3 Dataset

Four different open-source datasets are merged into one compact dataset in this work. The first one is Cohen's dataset [10] containing 478 Covid positive and 18 No Finding images at the time of project creation. As mentioned in Sect. 1, Cohen's is a dataset that is still in the collection phase. It holds many samples from many different hospitals and the patients are in different age ranges and disease stages. It is a very diverse dataset.

The second dataset is the COVID-19 Chest X-Ray Dataset Initiative's dataset [11]. This dataset holds 55 Covid Positive images. After this merge, in order to get more accurate results, the project needed more No Finding images. Therefore as a third dataset, we used the RYDLS-20 dataset [12] which holds samples from COVID-19, MERS, SARS, No Findings, and a few more viruses. However, this works only looks at the detection between Covid-19 and No Finding samples, so only those merged to the works dataset.

Merging the three would be enough because the RYDLS-20 dataset had enough No Finding images, but in order to increase the diversity in the No Findings samples, a fourth dataset has been added to the works dataset, which is NIH dataset [13]. The NIH dataset is composed of many samples with different lung diseases and no findings. For this work, only no finding samples are extracted.

As a result, a new dataset has been established in this work. There is a need for a more extensive and diverse dataset to train CNNs. Even though Cohen's dataset has big data for Covid positives, it had enough no finding samples and as opposite site, the RYDLS-20 dataset does not have enough Covid positive samples.

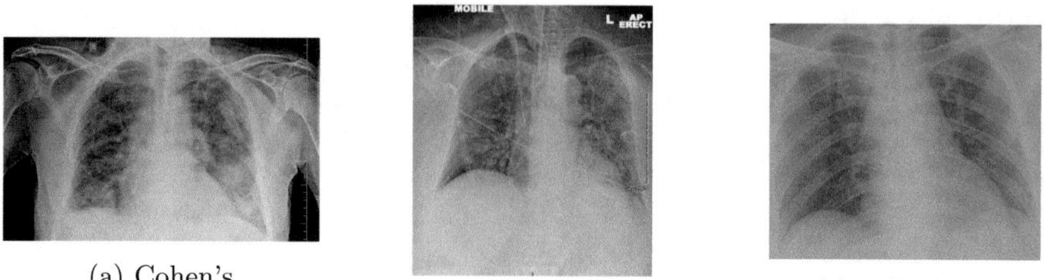

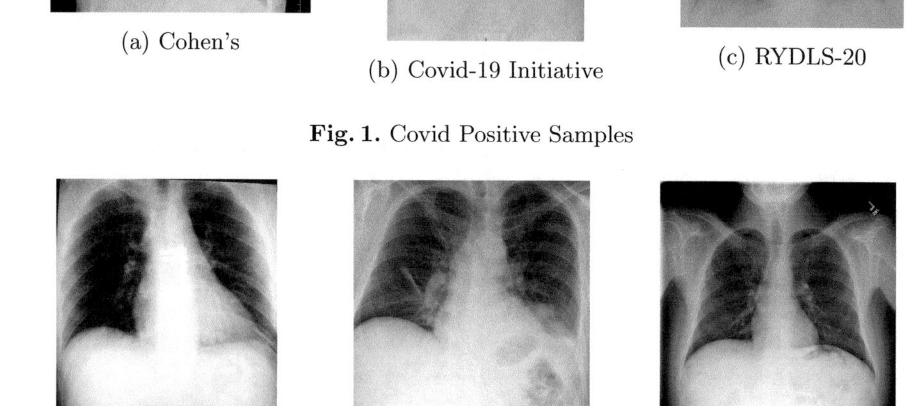

(a) Cohen's (b) Covid-19 Initiative (c) RYDLS-20

Fig. 1. Covid Positive Samples

(a) Cohen's (b) RYDLS-20 (c) NIH

Fig. 2. No Finding Samples

We demonstrate the Covid positive samples in Fig. 1, and we show No Finding samples in Fig. 2 from all dataset used can be seen.

This works new dataset contains 623 Covid Positive and 623 No Finding samples. The whole ratio can be seen in Table 1. The merged dataset will be published together with the article.

Table 1. Dataset Merge Ratio

Data/Dataset	Cohen's	COVID-19 Dataset Initiative	RYDLS-20	NIH	Total
Covid	478	55	90	-	623
No Finding	18	-	301	304	623

We create a setup for evaluation. In this setup, 70% of the images are used for training and 30% for testing in each class. The division has made manually. Table 2 shows the training, test and total numbers of our collected dataset.

Table 2. Train/Test Splits of Merged Dataset

Dataset	Covid-2019	No Finding
Train	436	436
Test	187	187
Total	623	623

4 Methodology

Convolutional Neural Network is a deep learning algorithm which consists of a set of layers that learns the features of the input data. These layers are convolution, pooling and fully connected layers. Convolution layer is the layer that maps the input's features with the set of kernels inside it. A kernel is a window (matrix) which holds weights, at first assigned randomly then with each training epoch the weights are tuned to the features. During feature mapping the kernel slides through the input, takes dot product and creates a feature map from resulting scalar values. Pooling layer shrinks the feature maps to a smaller size. During this operation most dominant features are preserved. Fully connected layer is the output layer. These layers flattens the feature map and generates the classification [14].

This work uses three different CNN models for image classification which are VGG, ResNet and DenseNet. VGG is a pretrained CNN model proposed by K.Simonyan and A.Zisserman [15]. The input goes through multiple convolutions with small receive size (3×3). ResNet [16] was proposed with the purpose of solving the vanishing gradient problem. It uses residual mapping which means the previous layer's state passed to another. DenseNet [17] utilizes dense connections between layers through Dense Blocks where all layers are connected and each layer receives the knowledge from all preceding layers.

4.1 Baseline

This work uses a baseline CNN model that loads pretrained models and trains with these models as base. Figure 3 shows a basic pipeline of this method architecture. After taking the inputs, the model starts preprocessing them. As remarked in Sect. 3, this work uses a merged dataset and the image sizes are not

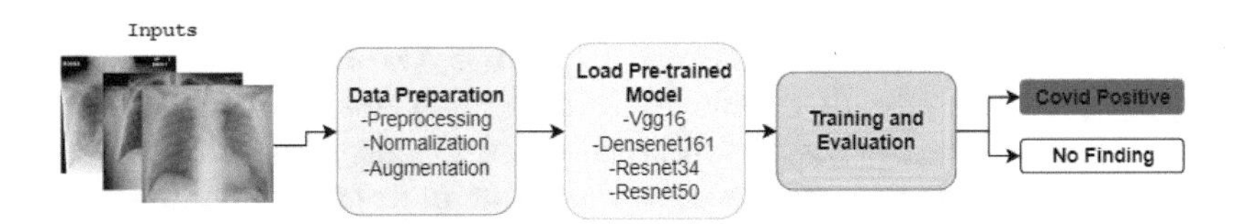

Fig. 3. Baseline Architecture Pipeline

equal. In order to get better results the images are all resized to 224×224 and normalized. As seen from Table 3 this works dataset is not very large but since to train a model a large dataset is better augmentation has been added to the train set and images are randomly rotated and flipped.

In training part four models are used. DenseNet161, VGG16, ResNet34 and ResNet50. The desired model is loaded after the preprocessing and the fully connected (FC) layers of the training part is tuned according the chosen models architecture. All the models used in this work are trained with 100 epochs, a learning rate of 0.00001, a dropout of 0.5 and the batch size for train set was 128 [18].

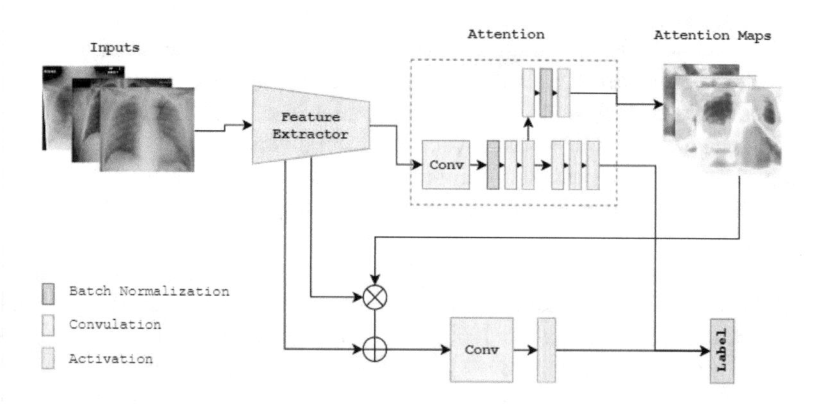

Fig. 4. Attention Structure

4.2 Attention Mechanism

Attention is widely used in computer vision, sequential models with recurrent neural networks. This work used soft attention approach and [19] 's framework with ResNet34 and ResNet50 base, which is a framework that acquired successful results. The structure of the mechanism could be seen in Fig. 4. Feature extractor contains multiple convolutional layers and extracts feature maps of the given inputs. In Attention mechanism feature maps passes a Convolutional layer, then goes through two batch normalizations, two convolutions, two activations and extracts attention maps. After attention maps extracted these maps are applied to feature maps with soft attention mechanism [20]. Attention maps are multiplied with feature maps and the resulting matrices are summed. Lastly perception takes the resulting matrices (maps) and convolutes them with the used ResNet models. [21] All the models used in attention are trained with 100 epochs, a learning rate of 0.0001 and the batch size for train set was 128.

5 Experiments

5.1 Evaluation Metrics

In this study as evaluation metrics True Negative, False Negative, True Positive, False Positive, Accuracy, Precision, Recall, Specificity and F1 score are used. True Negative (TN) is when the actual and the predicted data is false. False Negative (FN) is when the actual data is positive but predicted false. True Positive (TP) is when both the actual and predicted classes are true. False Positive (FP) is when the actual data is false but predicted true.

Accuracy is the correct predictions over all predictions made.

$$Accuracy = \frac{TP + TN}{TP + TN + FP + FN} \tag{1}$$

Precision is the proportion of the patients that the model predicted as Covid positive or are actually Covid positive.

$$Precision = \frac{TP}{TP + FP} \tag{2}$$

Recall is the proportion of patients that actually are Covid positive and predicted as Covid positive by the model.

$$Recall = \frac{TP}{TP + FN} \tag{3}$$

Specificity is the proportion of patients that are not Covid Positive and predicted as not Covid.

$$Specificity = \frac{TN}{TN + FP} \tag{4}$$

F1-score is the harmonic average of precision and sensitivity. It shows the similarity rate between predicted and true classes. [15]

$$F1 - score = \frac{2 * Precision * Recall}{Precision + Recall} \tag{5}$$

5.2 Experiments

In this section, the experimental results are reported and discussed. A comparison is made between the baseline CNN models and models with attention mechanism. A detailed result for baseline is reported in Table 3. In this study, for baseline CNN ResNet50, ResNet34, Densenet161 and VGG16 are used as pre-trained models. Among these models from looking accuracies, ResNet50 shows the best result with an 83% accuracy. If we look at precisions, underlined in Table 3, VGG16 is showing a better performance than others at predicting true positive cases among all positive predicted cases.

A detailed result for attention model is reported in Table 4. In this study ResNet50 and ResNet34 used as pre-trained models for attention model. Among

Table 3. Baseline Results

Model/Results	Precision	Recall	Specificity	F1 Score	Accuracy
Vgg16	0.673797	0.933333	0.744770	0.7826	0.8130
ResNet34	0.491979	0.920000	0.653285	0.6411	0.7995
ResNet50	0.604278	0.918699	0.705179	0.7290	0.8394
DenseNet161	0.470588	0.907216	0.642599	0.6197	0.8177

Table 4. Attention Results

Model/Results	Precision	Recall	Specificity	F1 Score	Accuracy
ResNet34	0.875000	0.875000	0.882353	0.8750	86.6310
ResNet50	0.851190	0.851190	0.866310	0.8512	84.4919

these models from looking accuracies and precision ResNet34 shows the best result with an 86% accuracy.

From comparing Table 3 and 4 it could be seen that the attention mechanism predicts more accurately and performs better. By using the attention mechanism, not only accuracies are increased, but there was a 7% of increase in precision which means with attention, the ratio of true predicted positive among all predicted positive cases is bigger. The attention model is also better at the ratio of true negative cases among all negative cases, which could be seen from the specificity increase.

In Fig. 5, confusion matrix of ResNet34 in attention model is reported. From this matrix, we can see that with the attention mechanism, the models could

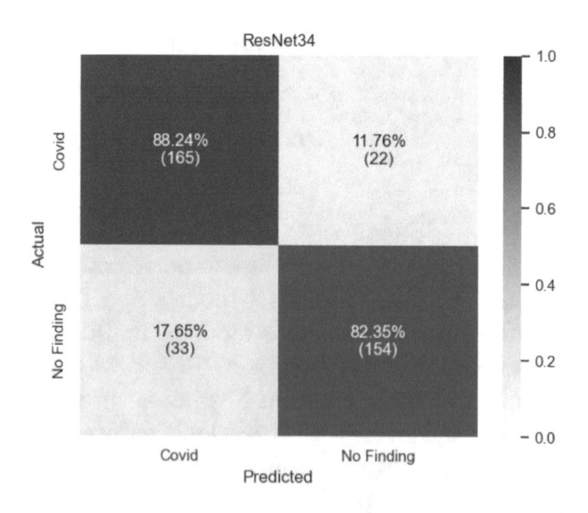

Fig. 5. Confusion Matrix of Attention Model

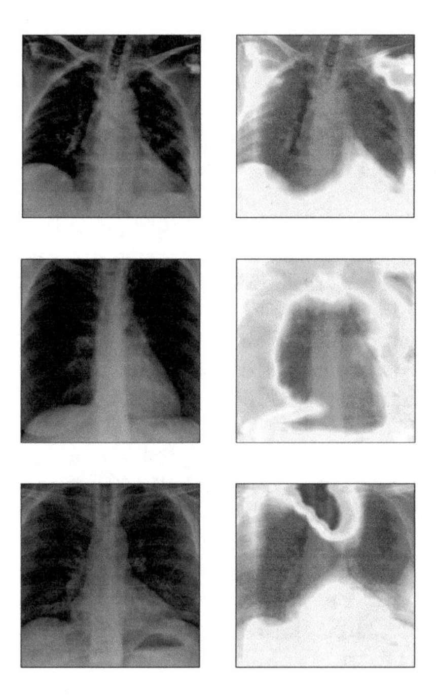

Fig. 6. Sample attention maps from dataset. Left column shows the original X-Ray images, and the right column shows the affected regions

predict not only true no findings but also the covid cases. That is the reason the precisions are higher than baseline CNN.

In Fig. 6 attention maps from this works trained model is shown. In these maps the concentration is at ground-glass opacity areas showing Covid symptoms of the X-Ray Scans.

6 Conclusion

This study looks into the Covid detection from X-Rays scans using a baseline CNN and attention using the CNN model. In this work, a new dataset is collected from several open-source databases in order to train and test these models. As in other studies in the literature, we have obtained our results with the base CNN architectures on the new dataset. In addition to studies in the literature, we have added attention mechanisms in our study. This work compares the evaluation results from baseline and attention methods and reports better performance on attention methods with the ResNet34 model.

Conflict of Interest. The authors declared no potential conflict of interest statements with respect to the research, authorship, and/or publication of this article.

References

1. Wang, C., Horby, P.W., Hayden, F.G., Gao, G.F.: A novel coronavirus outbreak of global health concern. Lancet **395**(10223), 470–473 (2020)
2. COVID-19 Coronavirus Pandemic. https://www.worldometers.info/coronavirus/. Accessed 01 Mar 2022
3. Zu, Z.Y., et al.: Coronavirus disease 2019 (COVID-19): a perspective from china. Radiology **296**(2), 15–25 (2020)
4. Maguolo, G., Nanni, L.: A critic evaluation of methods for COVID-19 automatic detection from X-ray images. Inf. Fusion **76**, 1–7 (2021)
5. Kassania, S.H., Kassanib, P.H., Wesolowskic, M.J., Schneidera, K.A., Detersa, R.: Automatic detection of coronavirus disease (COVID-19) in X-ray and CT images: a machine learning based approach. Biocybern. Biomed. Eng. **41**(3), 867–879 (2021)
6. Kedia, P., Katarya, R., et al.: CoVNet-19: A deep learning model for the detection and analysis of COVID-19 patients. Appl. Soft Comput. **104**, 107184 (2021)
7. Apostolopoulos, I.D., Mpesiana, T.A.: COVID-19: automatic detection from X-ray images utilizing transfer learning with convolutional neural networks. Phys. Eng. Sci. Med. **43**(2), 635–640 (2020)
8. Hussain, E., Hasan, M., Rahman, M.A., Lee, I., Tamanna, T., Parvez, M.Z.: Corodet: a deep learning based classification for COVID-19 detection using chest X-ray images. Chaos Solitons Fractals **142**, 110495 (2021)
9. Jain, G., Mittal, D., Thakur, D., Mittal, M.K.: A deep learning approach to detect COVID-19 coronavirus with X-ray images. Biocybern. Biomed. Eng. **40**(4), 1391–1405 (2020)
10. Cohen, J.P., Morrison, P., Dao, L., Roth, K., Duong, T.Q., Ghassemi, M.: COVID-19 image data collection: prospective predictions are the future. arXiv preprint arXiv:2006.11988 (2020)
11. Chung, A.: COVID-19 Chest X-ray Dataset Initiative (2022). https://github.com/agchung/Figure1-COVID-chestxray-dataset
12. Pereira, R.M., Bertolini, D., Teixeira, L.O., Silla, C.N., Jr., Costa, Y.M.: COVID-19 identification in chest X-ray images on flat and hierarchical classification scenarios. Comput. Methods Programs Biomed. **194**, 105532 (2020)
13. Wang, X., Peng, Y., Lu, L., Lu, Z., Bagheri, M., Summers, R.M.: Chestx-ray8: hospital-scale chest X-ray database and benchmarks on weakly-supervised classification and localization of common thorax diseases. In: Proceedings of the IEEE Conference on Computer Vision and Pattern Recognition, pp. 2097–2106 (2017)
14. Ghosh, A., Sufian, A., Sultana, F., Chakrabarti, A., De, D.: Fundamental concepts of convolutional neural network. In: Balas, V.E., Kumar, R., Srivastava, R. (eds.) Recent Trends and Advances in Artificial Intelligence and Internet of Things. ISRL, vol. 172, pp. 519–567. Springer, Cham (2020). https://doi.org/10.1007/978-3-030-32644-9_36
15. Simonyan, K., Zisserman, A.: Very Deep Convolutional Networks for Large-Scale Image Recognition (2015)
16. He, K., Zhang, X., Ren, S., Sun, J.: Deep Residual Learning for Image Recognition (2015)
17. Huang, G., Liu, Z., van der Maaten, L., Weinberger, K.Q.: Densely Connected Convolutional Networks (2018)
18. Berrimi, M., Hamdi, S., Cherif, R.Y., Moussaoui, A., Oussalah, M., Chabane, M.: COVID-19 detection from XRAY and CT scans using transfer learning. In: 2021 International Conference of Women in Data Science at Taif University (WiDSTaif), pp. 1–6. IEEE (2021)

19. Fukui, H., Hirakawa, T., Yamashita, T., Fujiyoshi, H.: Attention branch network: learning of attention mechanism for visual explanation. In: Proceedings of the IEEE/CVF Conference on Computer Vision and Pattern Recognition, pp. 10705–10714 (2019)
20. Zalluhoglu, C., Ikizler-Cinbis, N.: Comparison of 2D and 3D attention mechanisms for human (collective) activity recognition. Signal Image Video Process. 1–8 (2021)
21. Hamed, G., Marey, M.A.E.-R., Amin, S.E.-S., Tolba, M.F.: Deep learning in breast cancer detection and classification. In: Hassanien, A.-E., Azar, A.T., Gaber, T., Oliva, D., Tolba, F.M. (eds.) AICV 2020. AISC, vol. 1153, pp. 322–333. Springer, Cham (2020). https://doi.org/10.1007/978-3-030-44289-7_30

Lexicon Construction for Fake News Detection

Uğur Mertoğlu(✉) and Burkay Genç

Department of Computer Engineering, Hacettepe University, Ankara, Turkey
ugurmertoglu@itu.edu.tr, bgenc@cs.hacettepe.edu.tr

Abstract. With the digitization of media, an immense amount of news data has been generated by online sources, including mainstream media outlets as well as social networks. However, the ease of production and distribution resulted in the circulation of fake news as well as credible, authentic news in other words. The pervasive dissemination of fake news has extreme negative impacts on individuals and society. Therefore, fake news detection has recently become an emerging topic as an interdisciplinary research field that is attracting significant attention from many research disciplines, including social sciences and linguistics. In this study, we propose a method for detecting fake news in Turkish, which can be used in other agglutinative languages as well. Our method is mainly based on a lexicon approach, including a scoring system to facilitate the detection of the fake news. We contribute to the literature by collecting a novel, large scale, and credible dataset of Turkish news, and by constructing the first fake news detection lexicon for Turkish.

Keywords: Turkish · Fake News · Lexicon · Agglutinative Languages · Gdelt · Lexicon-based Detection

1 Introduction

Even though the preference to obtain news from traditional media sources like TV, press etc. is still not negligible, the consumption of digital media sources is rapidly increasing. As a result of computer-mediated communication (CMC), people began to follow the news via the internet rather than traditional methods, notably through social media. In 2019, about 2.82 billion users were reported[1] to use social media worldwide and as of January 2022, there were 4.95 billion internet users worldwide at that time which constitutes 62.5% of the global population. The research[2] conducted in 2019 in USA points to an observation of "closing of the gap" between the number of consumers who prefer to get their news from TV and those who do so online. When it comes to potential influence on society, there seems to be a domino-effect of online news through platforms such as Facebook, Twitter and Whatsapp, providing global connectivity. This effect is

[1] Social Media statistics and facts [online]. Website https://www.statista.com/statistics/ [accessed 26 April 2022].

[2] Pew Research Center [online]. Website https://www.pewresearch.org/fact-tank/ [accessed 03 August 2019].

D. J. Hemanth et al. (Eds.): ICAIAME 2022, ECPSCI 7, pp. 320–336, 2023.
https://doi.org/10.1007/978-3-031-31956-3_27

far more influential on society in both positive and negative ways than the conventional news platforms do.

Fake news has become a serious problem in international society for the past decade. With the widespread use of Internet in this technology-driven era, lately there has been a lot of discussions on "fake news", because nowadays it is being used to serve as a political, economic, and even strategic means of interest. Being exposed to "fake news" at an unprecedented pace in various forms (propaganda, misinformation, disinformation, manipulation etc.), it appears most readers are involuntary to question the credibility of these news and to distinguish whether they are fake or not.

Having a huge potential to manipulate people's perception of reality, fake news causes major problems in politics, media, advertising, tourism, national security, and healthcare systems, society polarization (conflicts and violence among ethnic groups, refugees, and immigrant) and negatively affects people's view of the world through growing mistrust among people. For instance, the causes behind the outbreak of the Arab Spring, which first emerged in Tunisia and spread across the Arab World in quite a short time, are numerous such as imbalances in income, corruption, widespread poverty etc. However, it was some fake news distributed in social media that ignited the spread of the events by making the situation even more chaotic.

Due to the rapid changes in communication patterns and technologies, not only people are vulnerable to the bombardment of information and news but also the governments and organizations are. Moreover, identifying the truth of the news is a demanding task in all respects. The issue has come to such a point that social media platforms and technology companies like Facebook, Twitter and Google etc. have started to work on finding solutions to the issue to preserve their reputation. They have been committed to fighting fake news through a combination of technology and human review, including removing fake accounts, partnering with fact-checkers, and promoting news literacy. In the last 4–5 years, in accordance with the growing academic studies about the subject, international panels and conferences and many activities have begun to be organized. For instance, in 2017 the term "fake news" was chosen by the Collins Dictionary as the "Word of the Year 2017". Again in 2017, The Fake News Challenge (FNC-1), a machine learning competition/task between AI community, journalists, and fact-checkers, was organized to develop tools towards fake news detection. However, it seems that the state of art systems of these efforts needs to be enhanced. Because, now the struggle mostly depends on crowdsourcing, fact-checking organizations, specific efforts, and third-party tools etc. With more than a billion pieces of content posted every day, we know that fact-checkers cannot review every story manually. Therefore, looking into automated ways to identify fake news and act on a bigger scale using computer scientific methods has a reasonably strong motivation.

Throughout this study, we will use the term "Valid" to represent "Non-fake" news. Our aim is not to validate the content of news from a sociological or technological perspective in an era we live, with ongoing discussions on concepts such as post-truth. From our standing point, a news story on a politician claiming that "the moon is green" is valid news if the politician indeed made the claim. Similarly, we do not try to detect falsehood in a news story about "how climate change does not exist". Hence, we deliberately do not use the terms "True News" or "Correct News" as that implies the correctness of the content. Rather, we use "Valid News" to show that the story is credible, or worth "considering". Our contributions in this paper are as follows:

- We present a novel lexicon-based approach to fake news detection in Turkish. We developed the first Fake News Lexicon for Turkish, named FaNLexTR containing 4 different categorizations derived from a comprehensive corpus.
- We present a workflow to construct FaNLexTR, which can also be used for other languages, especially other agglutinative ones, if the structural diversity of the languages are properly addressed.

The rest of the paper is structured as follows: Sect. 2 presents a review of the relevant literature. The data and methodology we use is discussed in Sect. 3. In Sect. 4, the results and evaluations are presented. Subsequently, we provide conclusions and propose possible future study topics in the Final Section.

2 Literature Review

When we consider the studies in the literature, especially the natural language processing (NLP) and text mining studies, it can be said that the research on textual deception vary on many different sub-domains and constitutes a different but closely associated discipline with the other text classification domains. We visualized it in Fig. 1, in order to make this comprehensive literature survey more understandable. In terms of fake news detection, several methods like NLP Analysis (Syntactic, Rhetoric, Statistical, and Semantic etc.), Network/Graph Analysis, and Source Analysis and hybrid models combining different methods of each have been used.

Verbal Communication	Computer Mediated Communication	Law, Security and Science	Journalism
• Psychological/Clinical Cases	• Emails (Spam, phishing, bot etc.)	• Court Judgements	• Politic, Economic Interests
• Questionnaires, Interviews	• Forum, Blogs, Online Services	• Forensic Science (Police interrogations)	• Regional, Global Rivalry/Competition
• Case Scenarios	• Social Media (Fake accounts, tweets, posts, links etc.)	• Intelligence Reports	• Social Events (Revolts, Protests, Disturbance etc.)
• Customer Services Talks	• Reviews (Product, hotel, services etc.)	• Forgery on Documents	• Company/Brand Competition
• Empirical Data (Polygraph Tests)	• Advertising	• Suspect/Witness Reports	• Exploitation of Social Media/Internet

Fig. 1. Basic variations of textual deception in the literature

There are studies under the topic of machine learning methods aimed at deception detection. In this study, we focused on textual deception detection. In the literature, there are large body of studies applying NLP-Machine Learning-AI which aimed at detecting, predicting, and classifying according to the text which have been focused on several domains. For example, some of the domains used in some studies are categorizing social media posts [1], detecting spam posting [2], gender deception in online communication [3], and deceptive opinions in online reviews [4]. Although the chronology may also be linked to earlier, following the 2016 US presidential election many have expressed concerns about the effects of fake news [5] and the topic has gained popularity. Thus, the number of academic studies has increased since then.

In one of the pioneering works on fake news detection, Conroy et al. (2015) make a summarization and explains the existing methodologies related to the problem [6]. The listed methodologies mainly focus on linguistic approaches, network approaches, source credibility approaches, semantic approaches, and hybrid approaches. The range of studies include relatively simple linguistic approaches, such as "bag of words" as well as complicated deep learning methods. Hancock and Markowitz (2014) tried intensively to find clues of deception using n-grams, part of speech taggers (POS) and other syntactic analysis methods [7]. Others tried to improve their results by following a linguistic approach as a complementary tool [8, 9].

There are also studies which use rhetorical-base detection. In one of the studies, Rubin and Lukoianova (2015) used rhetorical structure theory (RST) as the analytic framework to identify systematic differences between deceptive and truthful stories in terms of their coherence and structure [10]. Rhetorical structure along with the syntactic patterns and discourse constituent parts especially when used with word2vec models can also be used as a remarkable detection marker for revealing informal structure of news, which is a good indicator for fake news. In a study focused on discourse level, rhetorical structures are used as vector space modeling applicants for predicting whether a report is truthful or deceptive for English news [11]. When considered from this point of view, the methods used in these studies are parallel to the methods used for solving author identification problems.

Sarcasm detection which can also be attributed to a sub-domain of fake news is closely related with rhetorical structure. In one of the works, the authors have developed models based on a pre-trained convolutional neural network (CNN) for extracting sentiment, emotion, and personality features for sarcasm detection [12]. While some researchers use traditional machine learning approaches pointing to drawbacks of deep learning techniques about stance detection [13], some researchers have applied deep learning techniques reporting shortcomings of more traditional machine learning techniques to figure out spamming behavioral challenges [14].

In our study, we present a new analysis method that can be used alone or together with the existing ones. Our method is based on fake news lexicons which are derived from a large corpus of fake and valid news texts. Lexicon-based approaches are used in many types of NLP analyses. For example, in sentiment analysis, a sentiment lexicon is frequently used for sentiment classification. Promising performance is reported for sentiment analysis in the study of Kang and Park [15]. Especially when the goal of the study is to detect polarity, lexicon-based approaches are widely used as in the studies [16–18]. However, to the best of our knowledge, there is no study using lexicons which have fake-valid polarities of language components. One of the most important reasons for this absence is the non-existence of fake lexicons belonging to the language. To lead the field in this respect, we developed lexicons for each of "raw word", "raw word and its part of speech tagger (POS)", "root/stem of word" and "morphemes" in Turkish language with a Fake/Valid value. We generate these lexicons utilizing a large, labeled corpus of news texts produced by our team using the GDELT (Global Database of

Events, Language and Tone) Project[3] datasets, verified data taken from a fact-checking organization "teyit.org" and online news data manually verified by our research team.

We build our framework in an adaptive manner for future studies. For instance, our general-purpose lexicon may also be adapted to the specific domains in a similar way as done in one of the studies in sentiment analysis in Turkish [19]. Lexicon-based approaches can also be studied for to be compared with machine learning based approaches in fake news detection as done again in a sentiment analysis study for Turkish [20].

3 Data and Methodology

A considerable number of studies which are found in the literature use publicly available datasets in English, given the fact that collecting and labeling such data from various online resources is a time-consuming process. Moreover, these studies generally focus on fake news detection via machine learning techniques. It is known that relatively more successful results can be obtained by studying a specific domain in machine learning problems. Therefore, these studies mostly focus on a specific domain, such as politics, sports, or satirical news. Aiming to construct a robust, general purpose fake news lexicon in Turkish, we had to gather our data in a similar vein as the studies on developing general purpose sentiment lexicons. Our data encompasses many types of fake news. To this end, the verification and homogeneity of the news is our primary objective. We used Zemberek[4], the Turkish NLP engine, to obtain roots of words, Part of Speech taggers (POS) and suffixes. We propose a novel and almost fully automated methodology to construct a general purpose Turkish fake news lexicon.

3.1 Data Collection and Preparation

In agglutinative languages such as Turkish, words are extended by suffixes to create new words. Therefore, a single root word can be extended in tens of different ways to obtain many different words. One must sample as many combinations of these extensions as possible to build a qualified lexicon. Hence, we concentrated on gathering a large body of news texts which we can use as a training set for FaNLexTR and other possible future research studies. While creating our initial database of news texts, the validation and accuracy of data labeling was one of our highest priorities. To this end, we used the archives of the GDELT project to obtain the URLs of approximately 100k news published by 3 major, authentic news agencies in Turkey. Hence, our news texts database does not include any news from local or hard to validate news sources. For the fake news part of the database is constructed by articles from "teyit.org", a fact-checking organization in Turkey, which tags the news as Fake or True, and publishes them on its web page. Furthermore, we include a hand curated collection of fake and valid news obtained from various online sources manually verified by our research team. We refer to this collection as Manually Verified News, MVN. These three methods that we prefer while collecting

[3] GDELT Project [online]. Website https://www.gdeltproject.org [accessed 03 August 2019].

[4] https://code.google.com/p/zemberek.

data are significant in terms of representing methods for verification of news in real world. These methods consist of artificial intelligence projects and tools, fact-checking organizations, and human effort, respectively.

Our dataset collection steps, and main phases of the process are outlined in Fig. 2. In the first phase, we scraped the data from the web, cleaned HTML tags and other extra materials, obtained lean news content, and stored them in files. The final dataset consists of 84734 unique news articles, which belong to 11 different news domains, uniformly distributed within a time mostly[5] between years 2017 and 2019. The corpus statistics including type, source, class off the dataset, number of news texts within the dataset, average word count and sentence count per document are shown in Table 1. In the second phase, we ran all news texts through the Zemberek pipeline to obtain the unique words in each text, as well as the associated information for each word, such as root of the word, part of speech information and the ordered set of suffixes.

Table 1. The collected news data statistics

Dataset Type	Source	Class	Count	Avg. W.C.	Avg. S.C.
Train	GDELT	VALID	82691	243	15
	MVN		855	169	11
	Teyit.org	FAKE	902	119	9
	MVN		286	102	8
Test	GDELT	VALID	22	188	13
	MVN		188	170	10
	Teyit.org	FAKE	89	114	10
	MVN		121	106	9

As seen from this table, the fake news count is quite small compared to valid ones. It may seem like the imbalanced nature of the dataset can cause a handicap; however, this is not an unexpected issue, as real-world datasets are mostly composed of "normal" observations with only a small percentage of "abnormal" or "interesting" examples. It should also be noted that falsehood diffuses significantly farther, faster, deeper, and wider than the truth in all categories [21]. Moreover, we do not use any classical machine learning models, some of which are highly sensitive to the imbalance of training sets, in our study. Furthermore, one can easily note that the average word count and sentence count statistics are much smaller in fake news when compared with valid news. This is an expected phenomenon as fake news creators mostly use a much simpler language whereas valid news content creators prefer to use more sophisticated expressions. These statistics are strong candidates of being useful features in a machine learning based study for detecting fake news.

[5] There are also some news belongs to the years 2015 and 2016 (mainly composed of fake ones recurring every year). But they are small in number compared with the majority (2017–2019).

From now on, we refer to the valid news training dataset as D_{tr}^{V}, valid news test dataset as D_{ts}^{V}, fake news training dataset as D_{tr}^{F}, and fake news test dataset as D_{ts}^{F}.

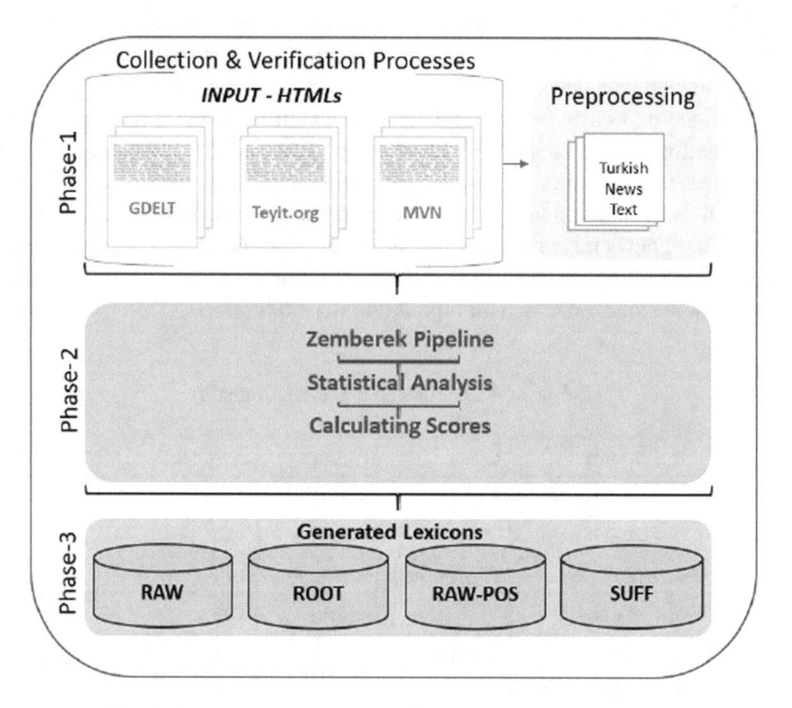

Fig. 2. Main phases of FaNLexTR Development Framework

3.2 Methodology

Turkish is an agglutinative language where suffixes are added to the end of a word to change its meaning and use in a sentence. Hence, we have the following structure in a Turkish word:

$$W = R + S_1 + S_2 + ... + S_k$$

Here, W is the raw form used in text, R is the root of the word, and S_i's are the suffixes. Although, the root word, R, generally determines a major part of the meaning of W, certain suffixes can have a significant effect on the meaning. Indeed, even the part of speech information of the word may be determined by suffixes. Hence, we decided to split our analysis in four different classes of information: RAW words, ROOTs, RAW+POS (part of speech), and SUFFixes. For each of these classes, we aim to generate a separate lexicon and compare them at the end to see which class lexicon provides the best classifier for validity of Turkish news. The main idea behind our approach is as follows: we first compute the frequency score of each term (raw, root, raw+pos or suffix) in D_{tr}^{V}, we then compute the corresponding frequency scores for D_{tr}^{F}, and finally we compute a novel metric named Fake/Valid score for each document to be tested by summing up the Fake/Valid frequencies of the terms contained within the document.

3.2.1 Lexicon Generation

The first thing we need to do to construct a lexicon based on the above four classes is to extract the corresponding terms of that class from the news datasets. To this end, we process the D^V_{tr} and D^F_{tr} datasets and extract the full lists of RAW, ROOT, RAW+POS and SUFF terms within the documents. For each term in each list, we compute two scores: one based on its frequency in D^V_{tr}, and another based on its frequency in D^F_{tr}. The scores $S^V_{t,C}$ and $S^F_{t,C}$ for term t and lexicon class C are formally represented in Eq. 1, 2;

$$S^V_{t,C} = \frac{\sum_{d \in D^V_{tr}} f_d(t)}{\sum_{d \in D^V_{tr}} \sum_{x \in T_C} f_d(x)} \tag{1}$$

$$S^F_{t,C} = \frac{\sum_{d \in D^F_{tr}} f_d(t)}{\sum_{d \in D^F_{tr}} \sum_{x \in T_C} f_d(x)} \tag{2}$$

In these equations, T_C represents the set of all terms within the lexicon of class C, where C is one of RAW, ROOT, RAW+POS or SUFF. Also, $f_d(x)$ represents the frequency of term x in document d.

After computing the fake and valid term scores for all four lexicons, we obtained the resulting statistics shown in Table 2. Note that each lexicon now contains terms from both D^V_{tr} and D^V_{ts} datasets. We now explain these results briefly.

Table 2. The lexicons developed

Lexicon	Unique Term	Common Terms	Only in Fake	Only in Valid
LEX-RAW	443174	10166	456	432552
LEX-ROOT	63237	3830	77	59330
LEX-RAW+POS	457187	10301	466	446420
LEX-SUFF	9930	898	12	9020

The LEX-RAW lexicon is the most intuitive lexicon where we only consider the default form of the words as they occur within the texts. Hence, there is a high number of unique words in this lexicon. However, when we consider words common to both valid and fake news texts, the number falls to 10166 unique words. This is the result of two mechanisms in action: first, D^F_{tr} is much smaller than D^V_{tr} and hence contains much less words; second, as we have mentioned earlier the language used in fake news is much simpler, resulting in less unique words. On the other hand, there are 456 unique words which only exist in the fake news texts. When we examine these words, we see that these are mostly informal words and exclamations which cannot be used in formal, valid news texts. The existence of these words in a text is also a very strong indicator of

Table 3. LEX-RAW lexicon data examples[6]

Raw Word	Fake Score	Valid Score	Raw Word	Fake Score	Valid Score
Ardından	14.80	21.80	Zaten	60.62	18.60
Bulunan	7.85	17.09	Bulunmuş	0.6	0.1
Terör	1.21	14.96	Terörist	3.32	2.11
İlk	10.29	20.12	Son	17.21	15.59
Tutuklanarak	0	1.19	Serbest	3.62	2.95
Yok	5.74	6.22	Değil	11.79	7.07
Doğrultusunda	0	1.59	Gibi	32.32	17.60
Korktu	0	0.01	Tırstı	0.3	0
Güvenilmez	0	0.05	Kaypak	0.6	0
Gündem	0	0.01	Çalkalanırken	0.6	0

the fakeness of the text. Some interesting term examples and their associated fake and valid scores in LEX-RAW are provided in Table 3.

When we only consider the roots of the words, the number of unique terms drops down to 63237. The number of common terms in fake and valid texts are again much less than the total number of terms. We have detected 77 root terms that are unique to fake news. These are mostly slang or made-up words that are not part of the formal language. Again, these are strong indicators of fake news. Selected term examples and their associated fake and valid scores in LEX-ROOT are provided in Table 4.

Table 4. LEX-ROOT lexicon data examples

Root	Fake Score	Valid Score	Root	Fake Score	Valid Score
Süs	0	0.29	Janjan	1.21	0
Ara	28.70	38.90	Bul	8.15	7.26
Değer	6.64	13.37	Haber	17.82	7.49
Sahtekâr	0	0.03	Sahte	1.81	1.56
Ehil	0	0.12	Tıs	0.6	0
Hilekâr	0	0.004	Fırıldak	2.41	0.001
Net	0.91	1.55	Gibi	32.31	17.86
Düzgün	0	0.29	Paçoz	0.3	0
Asker	4.83	8.43	Coni	1.21	0
Yüzde	1.21	4.07	Tahmini	1.21	0.03

[6] In Tables 3, 4, 5 and 6, all scores are multiplied by 10000 for readability.

The RAW+POS lexicon uses the raw forms of the words paired with their part of speech tags. Therefore, the same word can exist multiple times in this lexicon, each time paired with a different POS tag. We can see from Table 2 that there are 457187 unique word-POS pairs, of which 466 is unique to fake news and 10301 are common. Selected term examples and their associated fake and valid scores in LEX-RAW+POS are provided in Table 5.

Table 5. LEX-RAW+POS lexicon data examples

Raw (POS)	Fake Score	Valid Score	Raw (POS)	Fake Score	Valid Score
Teröristlerce (Noun)	0	0.27	Terörist (Noun)	3.32	2.11
Apaçık (Adjective)	0	0.02	Alenileşen (Adjective)	1.21	2.33
Bitmemiş (Verb)	0	0.01	Bitmiyormuş (Verb)	0.3	0
Sımsıkı (Adverb)	0	0.001	Yapışırcasına (Adverb)	0,3	0
Uyanık (Noun)	0	0.07	Uyanık (Adjective)	1.21	0.1
Olay (Noun)	6.04	15.59	Olaylara (Noun)	0.6	0.3
Biz (Pronoun)	11.17	12.33	Siz (Pronoun)	1.81	0.1
Sınır (Noun)	1.21	1.38	Hadleri (Noun)	0.6	0

The last lexicon is based on the suffixes. This is probably the most interesting lexicon we have constructed, considering that no suffix lexicon exists in the literature, and this can only be done for agglutinative languages. In the suffix lexicon, we considered all possible sub-sequences of the suffixes of a word. For example, if $R + S_1 + S_2 + S_3$ is a word that exists in LEX-RAW, then in LEX-SUFF we consider S_1, S_2, S_3, $S_1 + S_2$, $S_2 + S_3$ and $S_1 + S_2 + S_3$ as terms. Overall, there were 9930 detected suffix sequences, 898 were common in both news types and only 12 belonged exclusively to fake news. Selected term examples and their associated fake and valid scores in LEX-SUFF are provided in Table 6.

So far, the construction of four lexicons has been explained. Next, we show how to use these lexicons to evaluate the validity/fakeness of a document.

3.2.2 Document Evaluation

In this part, we explain and demonstrate how to use the generated lexicons to evaluate the validity/fakeness of a given document. First, we need to note that considering we have generated 4 different lexicons based on different terms, we will be conducting 4 different analyses. The text we will use for demonstration purposes is provided in

Table 6. LEX-SUFF lexicon data examples

Suffixes	Fake Score	Valid Score	Example Word
Caus-Caus-Neg-FutPart-A3pl-Acc	4.11	0	ARA **-t-tır-ma-yacak-lar-ı**
Inf2-P3pl-Narr	1.54	0	AT **-ma-ları-ymış**
A3pl-Loc-Rel-P2sg-Abl	3.86	0	ZAMAN **-lar-da-ki-n-den**
PresPart-P3sg-Narr	1.54	0	GİD **-en-i-ymiş**
A3pl-Dat	111.91	9.49	İNSAN **-lar-a**
With-A3pl-P2sg	0	5.8	TALİH **-li-ler-in**
Able-Aor-A1pl	0	5.00	KAÇ **-abil-ir-iz**

Table 7. In summary, the text boasts about Cuba as a holiday destination and reports a few statistics about the country.

Table 7. A fake news text from our test dataset

Title	İNANILMAZ AMA DOĞRU
Text	Ta Küba! Kim gidecek demeyin? Heralde bu yaz tatil listenizdeki yer Küba olmalı. 47 yıldır cinayet işlenmedi. 58 yıldır tecavüz ve istismar suçu işlenmedi. Hatta 5 yıldır hırsızlık bile olmadı. Herkese eşit maaş. Vergi yok. Hemen herşey ücretsiz ya da sudan ucuz. Gezin görün!

To analyze and evaluate a document we execute the following steps: first, we parse the text and extract the related terms for each lexicon, then we sum up the corresponding fake and valid scores of terms over all text. The resulting sum of fake scores of terms is called the document fake score, $S^F_{D,C}$, and the sum of valid scores of terms is called the document valid score, $S^V_{D,C}$. The formal definitions of both are provided below;

$$S^V_{D,C} = \sum\nolimits_{t \in T^D_C} S^V_{t,C} \tag{3}$$

$$S^F_{D,C} = \sum\nolimits_{t \in T^D_C} S^F_{t,C} \tag{4}$$

In Eq. 3, T^D_C represent the set of all terms in document D with respect to lexicon class C. Once these two scores are computed for each lexicon class, we compare to see which one is greater. If $S^V_{D,C}$ is greater than $S^F_{D,C}$, then we label the document as VALID with respect to lexicon C, otherwise we label it FAKE;

$$L_D = \begin{cases} \text{VALID, if } S^V_{D,C} > S^F_{D,C} \\ \text{FAKE, otherwise} \end{cases} \tag{5}$$

Let us now look at the example news text in Table 7. This example demonstrates a click-bait, a text that is generated to provoke clicks. Most of the time, click-baits are hidden between valid news texts and contain false information. This text is one of the shortest in our test dataset. The reason we have chosen it is to be able to fit the complete analysis into these pages. An analysis conducted on a larger text follows the same steps as this one.

We will start by the LEX-RAW lexicon. With respect to this lexicon, the text contains 39 different terms. Of these terms we want to mention a few that have striking differences between $S^V_{t,C}$ and $S^F_{t,C}$. Let us start with "heralde", which means "in any case". However, there is an important issue here: the correct spelling of this term is "herhalde". However, the middle 'h' in this word is a very weak 'h' and in daily speech mostly it is not pronounced. Still, a valid and respected news source should use the correct form of the word. We see from the LEX-RAW lexicon that this word (***"heralde"***) is existing in both valid and fake news; however, it is almost 600 times more frequently used in fake news. The difference between $S^V_{t,C}$ and $S^F_{t,C}$ in this case is 0.603. Another interesting word is "bile" which means "even" in English, as used in the sentence "There wasn't even a theft in 5 years.". We can see from the lexicon scores that "bile" is a very frequently used word, both in valid and fake news. However, the difference between $S^V_{t,C}$ and $S^F_{t,C}$ is very large, making use of "bile" a strong indicator of fake news. A few more interesting terms are provided in Table 8 (All scores are multiplied by 10000 and rounded down for a nicer presentation).

Overall, the $S^F_{D,C}$ and $S^V_{D,C}$ scores are computed as 70.37 and 50.20 (scores multiplied by 10000 for better readability). This is a clear win for $S^F_{D,C}$ and hence the LEX-RAW lexicon labels the text as FAKE.

Table 8. LEX-RAW terms from the news text in Table 7

Term	$S^F_{t,C}$	$S^V_{t,C}$
heralde	0.604	0.001
listenizdeki	0.3020	0.0003
herşey	0.604	0.0286
görün	0.906	0.0418
tecavüz	2.7181	0.2810
bile	8.1544	2.5641
doğru	7.5504	4.8678

When we consider the root lexicon, LEX-ROOT, we again detect 39 root terms. However, there is a critical issue here. Extracting roots from Turkish words is not a simple feat, and Zemberek, the Turkish NLP engine we use is not perfect. Therefore, we notice some incorrect terms when we manually examine the results. However, to keep things as automated as possible, we do not fix these mistakes. Once again ***"heralde"*** tops the list of interesting terms as it is also detected as a root term. The overall $S^F_{D,C}$ and $S^V_{D,C}$ scores are 504.06 and 493.46, respectively. It is a closer call, but still a win for $S^F_{D,C}$.

Next, we look at the LEX-RAW+POS lexicon which pairs raw words with their POS tags. Although this creates some variety, the overall scores are similar to the LEX-RAW lexicon: 62.517 for $S^F_{D,C}$ and 44.634 for $S^V_{D,C}$. Once again, a decisive win for $S^F_{D,C}$.

The last lexicon is the suffix lexicon, LEX-SUFF. As we have mentioned earlier, this lexicon only uses the suffix morphemes of the words, completely ignoring the root. In this news text we observed 25 different terms (suffix sets). One of the interesting suffixes to note is the A2pl (second person plural) suffix. This suffix has a $S^F_{t,C}$ of 16.980 and a $S^V_{t,C}$ of 4.077. This shows that, despite being used in valid news, the use of A2pl is much more frequent in fake news. Naturally, one does not expect a news text to be written with direct references to the reader. However, in this example the word "**demeyin**", which literally means "do not say", is directed at the reader as if the author of the text is speaking to the reader. The use of this suffix is significantly penalized by our lexicon. As a result, $S^F_{D,C}$ and $S^V_{D,C}$ become 981.9913 and 954.4907, respectively. Once again, the document is classified as FAKE.

In this section, we concisely explained our methodology for classifying a document as fake or valid. In the next section, we provide batch testing results and evaluate the findings.

4 Results and Evaluation

In the previous section, we have explained how we constructed 4 different lexicons and how these lexicons are used for fake news detection in Turkish. In this section, we outline the results of our experiments to demonstrate the success ratio of each lexicon and present a comparative analysis. We would like to remind that our task is to label potential fake news for facilitating further examination. Hence, we focus on minimizing the amount of fake news which were accidentally labeled as valid news. In the rest of this section, we provide several statistics computed from the labeling of documents residing in the test datasets. Note that, these documents have not been introduced to the training phase where we constructed the lexicons. And we meticulously examined the test set consisting of unique news articles. Hence, it is possible that there exist terms in these documents which do not exist in the lexicons. If we observe such a word in a document, we assume a term score of 0 for both fake and valid scores of that term.

The tests are done on all documents in D^V_{ts} and D^F_{ts} for each lexicon type. Hence, 210 valid and 210 fake documents were tested. We provide confusion matrices as well as error statistics for each lexicon separately in Tables 9, 10, 11 and 12.

Table 9 shows the confusion matrix for LEX-RAW lexicon. Out of the 210 FAKE test cases, LEX-RAW was able to label 195 of them as FAKE, and 15 was erroneously labeled as VALID news. This results in a recall value of 0.929. The precision is lower than recall, however still acceptably high at a value of 0.802. Overall accuracy is 0.85. Successfully identifying 93% of fake news without generating an unacceptable number of false positives, the LEX-RAW lexicon becomes the most promising among the four.

Presented in Table 10, LEX-ROOT lexicon includes only roots of words and achieves a recall of 0.919, only slightly below LEX-RAW, failing to identify two additional fake news texts. However, a larger number of false positives are also generated, resulting in a lower precision than LEX-RAW. Overall LEX-ROOT becomes the second most

Table 9. Confusion matrix and error statistics for the LEX-RAW lexicon

Prediction	Actual	
	FAKE	VALID
FAKE	195	48
VALID	15	162

Precision	Recall	Accuracy	F1
0.802	0.929	0.850	0.861

successful lexicon and shows that root terms of words carry a lot of information regarding the validity of the text.

Table 10. Confusion matrix and error statistics for the LEX-ROOT lexicon

Prediction	Actual	
	FAKE	VALID
FAKE	193	64
VALID	17	146

Precision	Recall	Accuracy	F1
0.751	0.919	0.807	0.827

Next, we turn to LEX-RAW+POS, whose results are presented in Table 11. Although the LEX-RAW+POS lexicon terms carry more information than LEX-RAW terms, both the recall and precision values are worse. Compared to LEX-ROOT, LEX-RAW+POS has a worse recall, a slightly better precision, and the same overall accuracy. However, considering recall is our primary goal, this lexicon ranks third among the four.

Table 11. Confusion matrix and error statistics for the LEX-RAW+POS lexicon

Prediction	Actual	
	FAKE	VALID
FAKE	188	59
VALID	22	151

Precision	Recall	Accuracy	F1
0.761	0.895	0.807	0.823

Finally, we present the results for the LEX-SUFF lexicon in Table 12. Although, we achieve the worst scores in every statistic in this lexicon, we still believe that the results are impressive and worth discussing. The most impressive feat here is the fact that LEX-SUFF does not contain any information about the actual words used in the text. It simply uses the suffix information to evaluate the text as fake or valid. Considering only the suffix groups used within the text, LEX-SUFF achieves an impressive 0.814 recall ratio, corresponding to 171 hits in 210 fakes. Although, this is lower than the other lexicons' recall ratios, LEX-SUFF achieves this completely unaware of the content of the text. This is a feature that can only be obtained from agglutinative languages and a clear sign that similar studies including sentiment analysis can benefit from this finding.

Table 12. Confusion matrix and error statistics for the LEX-SUFF lexicon

Prediction	Actual	
	FAKE	VALID
FAKE	171	60
VALID	39	150

Precision	Recall	Accuracy	F1
0.740	0.814	0.764	0.776

5 Conclusions

In this paper, we presented the first known scholarly study for detecting fake news in Turkish news media based on lexicons. Our study includes collecting a large data set of labeled (fake/valid) Turkish news texts, generating four different lexicons based on these datasets, and providing a highly successful model for evaluating news texts using these lexicons. The lexicons we constructed differs by the terms used to generate them. Using the powerful agglutinative structure of the Turkish language we generate a raw words lexicon, a root words lexicon, a raw word with part of speech tags lexicon, and a suffix lexicon. Although similar studies with respect to raw words, root words and part of speech tags have been conducted in the literature for different purposes, a lexicon generated using only the suffixes of words is a unique contribution of our paper to the literature.

Our results show that the lexicon generated using the raw forms of the words (as they are used in the text) is the most successful in detecting fake news. In our experiments, this lexicon achieved a recall ratio of 0.929, without generating an unacceptable level of false positives. The lexicon based on root words, and the lexicon based on raw words paired with part of speech tags are also quite successful in detecting fake news. The suffix lexicon came last; however, it achieved a significantly high recall of 0.814. Although, this is lower than the other lexicons, considering that there is no content awareness in this lexicon, this is an impressive ratio. This is a clear indication that suffixes carry a lot of information in agglutinative languages and should be directly considered in similar studies, such as sentiment analysis.

In the future, we are expecting to enlarge our training datasets with more labeled fake news, further increasing our recall and precision values. We are planning to turn this study into a web service that can be used to validate news texts automatically. We are planning to experiment with reduced lexicons containing only a very small percentage of the lexicons presented in this study to understand whether a core lexicon can be extracted without compromising the recall and precision values. We also consider it useful to state that to avoid some common terms to dominate the Fake/Valid metric computation of a document, we did not multiply the term scores with their frequencies within the document. We want to emphasize that alternative document metrics can be computed based on our lexicons, however in our experiments we could not achieve a better result by the alternatives we tried. Hence, here we presented our most successful approach to this day.

Finally, we are expecting to merge all four lexicons to come up with an ensemble model that has a higher recall and precision than all four individual lexicons.

In this study, we focused on lexicon based fake news detection and showed that it can be very effective in Turkish fake news detection. However, we note that there are many other alternative methods for fake news detection, including stylometry analysis, lexical diversity analysis, punctuation analysis, n-gram based lexicons, etc. It is also possible to merge all these studies under a machine learning model, each individual analysis providing a feature value for the learning algorithm. Through this kind of approach, we conjecture that very high recall and precision values can be achieved.

Acknowledgement. The authors would like to thank fact-checker organization "teyit.org"[7] for sharing the data they own. We would also like to thank the students in our research team who helped us in the verification process of the collected news.

References

1. Rubin, V.L.: Deception detection and rumor debunking for social media. In: The SAGE Handbook of Social Media Research Methods, p. 342. Sage (2017)
2. Inuwa-Dutse, I., Liptrott, M., Korkontzelos, I.: Detection of spam-posting accounts on Twitter. Neurocomputing **315**, 496–511 (2018). https://doi.org/10.1016/j.neucom.2018.07.044
3. Ho, S.M., Lowry, P.B., Warkentin, M., Yanyun, Y., Hollister, J.M.: Gender deception in asynchronous online communication: a path analysis. Inf. Process. Manag. **53**(1), 21–41 (2016). https://doi.org/10.1016/j.ipm.2016.06.004
4. Fusilier, D.H., Montes-y-Gómez, M., Rosso, P., Cabrera, G.: Detecting positive and negative deceptive opinions using PU-learning. Inf. Process. Manag. **51**(4), 433–443 (2015). https://doi.org/10.1016/j.ipm.2014.11.001
5. Allcott, H., Matthew, G.: Social media and fake news in the 2016 election. J. Econ. Perspect. **31**(2), 211–236 (2017). https://doi.org/10.1257/jep.31.2.211
6. Conroy, N.J., Rubin, V.L., Chen, Y.: Automatic deception detection: methods for finding fake news. In: Proceedings of the Association for Information Science and Technology (ASIST), pp. 1–4 (2015)
7. Markowitz, D.M., Hancock, J.T.: Linguistic traces of a scientific fraud: the case of Diederik Stapel. PLoS ONE **9**(8), e105937 (2014). https://doi.org/10.1371/journal.pone.0105937
8. Zhang, H., Fan, Z., Zeng, J., Liu, Q.: An improving deception detection method in computer-mediated communication. J. Netw. **7**(11), 1811–1816 (2012). https://doi.org/10.4304/jnw.7.11.1811-1816
9. Ott, M., Cardie, C., Hancock, J.T.: Negative deceptive opinion spam. In: Proceedings of the 2013 Conference of the North American Chapter of the Association for Computational Linguistics: Human Language Technologies, pp. 497–501 (2013)
10. Rubin, V.L., Lukoianova, T.: Truth and deception at the rhetorical structure level. J. Am. Soc. Inform. Sci. Technol. **66**(5), 905–917 (2015). https://doi.org/10.1002/asi.23216
11. Rubin, V.L., Conroy, N.J., Chen, Y.: Towards news verification: deception detection methods for news discourse. In: Proceedings of the Hawaii International Conference on System Sciences, pp. 5–8 (2015)
12. Poria, S., Cambria, E., Hazarika, D., Vij, P.: A deeper look into sarcastic tweets using deep convolutional neural networks. In: Proceedings of the International Conference on Computational Linguistics, pp. 1601–1612 (2016)

[7] https://www.teyit.org/

13. Masood, R., Aker, A.: The fake news challenge: stance detection using traditional machine learning approaches. In: KMIS, pp. 126–133 (2018)
14. Mikolov, T., Chen, K., Corrado, G., Dean, J.: Efficient estimation of word representations in vector space. In: Proceedings of the International Conference on Learning Representations (ICLR) (2013)
15. Kang, D., Park, Y.: Review-based measurement of customer satisfaction in mobile service: sentiment analysis and VIKOR approach. Expert Syst. Appl. **41**(4), 1041–1050 (2014). https://doi.org/10.1016/j.eswa.2013.07.101
16. Deng, Z., Luo, K., Yu, H.: A study of supervised term weighting scheme for sentiment analysis. Expert Syst. Appl. **41**(7), 3506–3513 (2014). https://doi.org/10.1016/j.eswa.2013.10.056
17. Eirinaki, M., Pisal, S., Singh, J.: Feature-based opinion mining and ranking. J. Comput. Syst. Sci. **78**(4), 1175–1184 (2012). https://doi.org/10.1016/j.jcss.2011.10.007
18. Sağlam, F., Genç, B., Sever, H.: Extending a sentiment lexicon with synonym–antonym data sets: SWNetTR++. Turk. J. Electr. Eng. Comput. Sci. **27**(3), 1806–1820 (2019). https://doi.org/10.3906/elk-1809-120
19. Demiröz, G., Yanıkoğlu, B., Tapucu, D., Saygın, Y.: Learning domain-specific polarity lexicons. In: IEEE 12th International Conference on Data Mining Workshops, Brussels, Belgium, pp. 674–679 (2012)
20. Türkmenoğlu, C., Tantuğ, A.C.: Sentiment analysis in Turkish media. In: International Conference on Machine Learning, Beijing, China, pp. 1–11 (2014)
21. Vosoughi, S., Roy, D., Aral, S.: The spread of true and false news online. Science **359**(6380), 1146–1151 (2019). https://doi.org/10.1126/science.aap9559

A Transfer Learning Approach for Skin Cancer Subtype Detection

Burak Kolukısa[1], Yasin Görmez[2(✉)], and Zafer Aydın[1]

[1] Department of Computer Engineering, Abdullah Gül University, Kayseri, Turkey
{burak.kolukisa,zafer.aydin}@agu.edu.tr

[2] Faculty of Economics and Administrative Sciences, Management Information Systems, Sivas Cumhuriyet University, Sivas, Turkey
yasingormez@cumhuriyet.edu.tr

Abstract. The second most fatal disease in the world is cancer. Skin cancer is one of the most common types of cancer and has been increasing rapidly in recent years. The early diagnosis of this disease increases the chance of treatment dramatically. In this study, deep learning models are developed for skin cancer subtype detection including a standard convolutional neural network (CNN), VGG16, Resnet50, MobileNet, and Xception. The parameters of the standard CNN model are regularized using batch normalization, dropout, and L2-norm regularization. The hyper-parameters of this model are optimized using grid search, in which early stopping is used to optimize the number of epochs. For the rest of the models, transfer learning strategies are employed with and without fine-tuning as well as re-training from scratch. Data augmentation is performed for increasing the number of samples in the training set further. The performances of the models are evaluated on a Kaggle dataset that is developed for binary classification of skin images as malignant or benign. The best prediction accuracy of 87.88% is achieved using ResNet50 as the convolutional neural network model, which is re-trained from scratch and with data augmentation applied.

Keywords: Skin Cancer Classification · Cancer Detection · Deep Learning · Transfer Learning · Computer Vision

1 Introduction

Skin is the largest organ of our body; it regulates the body's heat and protects it against external influences. Sunlight is the most important factor for and primary cause of skin cancers. Skin cancer occurs due to uncontrolled proliferation of the DNA structures of cells on the skin. It is one of the leading cancer types, especially in Australia, United States, and Europe [1]. In the United States, 20% of Americans get skin cancer, 76,000 malignant melanomas are diagnosed each year, and 10,000 deaths occur due to this disease. Fortunately, in the early diagnosis of this skin cancer, curability is over 92% [2].

The main diagnostic and prognostic parameters of melanoma are the vertical thickness of the lesion, three dimensional shape, irregularities at the border of the lesion,

D. J. Hemanth et al. (Eds.): ICAIAME 2022, ECPSCI 7, pp. 337–347, 2023.
https://doi.org/10.1007/978-3-031-31956-3_28

and non-uniform pigmentation of various colors. Examination of dermoscopic images is a time-consuming and challenging task and requires a dermatologist's experience. Dermatologists believe that the diagnosis can be automatically made with the variety of lesions and confounding factors encountered in practice. For this reason, a computer-aided system has become important for the analysis of dermoscopic images and has become a research area [3–5]. Many researchers have been working in skin cancer classification and various skin cancer classification systems have been proposed for the last two decades using different techniques including image processing, computer vision, machine learning, and deep learning. As a result of their satisfactory performance on image classification tasks, convolutional neural networks (CNNs) have gained increasing popularity, especially for medical analysis, including skin cancer classification [6]. In the literature, studies compare the computer-aided systems with the results of dermatologists or increase the performance results by applying some preprocessing steps and different deep learning architectures. The study by Brinker collaborated with 157 dermatologists from 13 German hospitals [7]. A CNN network was trained with a total of 12,378 dermoscopic images, which outperformed 136 of 157 dermatologists on a binary classification task for a set of 100 dermoscopic test images. In the study by Maron, it has been shown that the transfer learning method ResNet50 trained using 11,444 dermoscopic images outperformed dermatologists in binary and multi-class classification tasks [8]. In that study, a total of 112 dermatologists have been cooperated with from 13 German hospitals. Jain et al. extracted 5 geometric features (i.e. Area, Perimeter, Greatest Diameter, Circularity Index, Irregularity Index), from segmented skin lesion to detect melanoma skin cancer [9]. Alquran et al. used gray level co-occurrence matrix to extract features and applied principal component analysis as a dimension reduction technique. They obtained 92.1% accuracy using a support vector machine classifier [10]. Linsangan et al. used geometric features of the skin lesion and achieved 90.0% accuracy with a k-nearest neighbors classifier [11]. Daghrir et al. proposed an ensemble model which uses a k-nearest neighbors, a support vector machine and a convolutional neural network and achieved 88.4% accuracy [12]. Jinnai et al. proposed faster, region-based convolutional neural network (RCNN) and obtained 91.5%, 83.3% and 94.5% accuracy, sensitivity and specificity, respectively for binary classification of skin lesions [13]. Nawaz et al. proposed RCNN along with fuzzy k-means clustering and achieved 95.40%, 93.1%, and 95.6% on the ISIC-2016, ISIC-2017, and PH2 datasets, respectively [14].

In this study, convolutional neural networks are trained for classification of skin cancer as benign or malignant. The following tasks are performed:

- Implementing a standard convolutional neural network with hyper-parameter optimization
- Applying transfer learning by freezing the convolutional base layers and retraining the top layers (e.g. fully connected multi-layer perceptron (MLP) layers).
- Applying fine-tuning by unfreezing a few last convolutional layers and retraining these and the top layers.
- Re-train the transfer learning model from scratch.

The rest of the paper is organized as follows. Section 2 introduces the data sets and elaborately describes the CNN and pre-training methodologies. Section 3 shows the performance evaluations of different pre-training algorithms and discusses the outcomes. Finally, the last section concludes the paper.

2 Materials and Methods

2.1 Dataset

We worked on a publicly available skin cancer data set, which is downloaded from Kaggle [15]. Figure 1 contains example images from this data set, which contains a total of 3297 image samples (1800 malignant and 1497 benign). The resolution of the images is 224×224 pixels. This data set is already partitioned into a train set and a test set in Kaggle. In addition, we obtained a validation set for hyper-parameter optimization by selecting 20% of the training set images via stratified random sampling and storing the remaining samples as the updated train set. As a result, the train set contains 2349 images (1152 malignant and 1197 benign), the validation set includes 528 images (288 malignant and 240 benign), and the test set has 660 images (360 malignant and 300 benign).

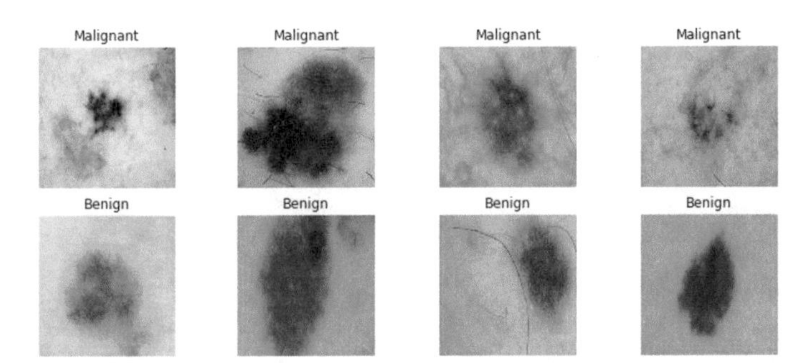

Fig. 1. Skin cancer images of malignant and benign classes.

2.2 Performance Metrics

Accuracy is an important measure especially when we have a symmetric dataset. For the skin cancer diagnosis problem accuracy may not be enough to determine the performance of a classifier. For example, it is costly to diagnose a sick patient as healthy. For this reason, we also employ other performance metrics, such as precision, recall, and F-measure, in addition to accuracy.

The traditional confusion matrix shown in Table 1 helps to assess the performance of the model in several aspects. In this table, positive corresponds to the malignant class and negative to the benign class.

Accuracy: The overall accuracy is the ratio of the correctly predicted labels to the total number of predictions made as shown in Eq. (1).

$$Accuracy = \frac{TP + TN}{TP + FN + FP + TN} \tag{1}$$

Table 1. Traditional Confusion Matrix

	Predicted Positive	Predicted Negative
Actual Positive	TP	FN
Actual Negative	FP	TN

Precision: is the ratio of correctly predicted sick patients to all sick predictions as shown in Eq. (2).

$$Precision = \frac{TP}{TP + FP} \tag{2}$$

Recall: is the ratio of correctly predicted sick patients to all sick labeled patients as shown in Eq. (3).

$$Recall = \frac{TP}{TP + FN} \tag{3}$$

F-measure: is the harmonic mean of precision and recall metrics as shown in Eq. (4).

$$F - measure = \frac{2}{\frac{1}{Precision} + \frac{1}{Recall}} \tag{4}$$

2.3 Classification Models

In this study, our goal is to classify skin images as benign or malignant. For this purpose, we implemented various convolutional neural networks (CNNs) including a standard CNN (similar to the one in Fig. 2), VGG16 [16], ResNet50 [17], MobileNet [18], and Xception [19]. We trained these models from scratch using the train set introduced in Sect. 2.1. We also performed transfer learning as an alternative strategy to train VGG16, Resnet50, MobileNet, and Xception networks.

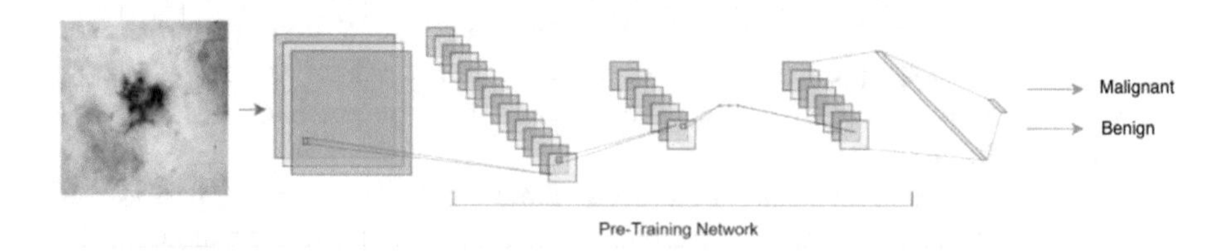

Fig. 2. The general structure of a convolutional neural network model

Transfer learning can be defined as storing information obtained while solving one problem (task 1) and using that information to solve another but related problem (task

2). After the design step, a model is first trained using the dataset for task 1. The purpose of transfer learning is to build a new model for task 2 by benefiting from the knowledge of the model developed for task 1. This can be achieved by transferring the model architecture and the weight parameters of a subsection of the first model, which learned a certain concept well, to the second one. The remaining parts of the second model are then trained using the dataset for task 2. The process of transfer learning is summarized in Fig. 3.

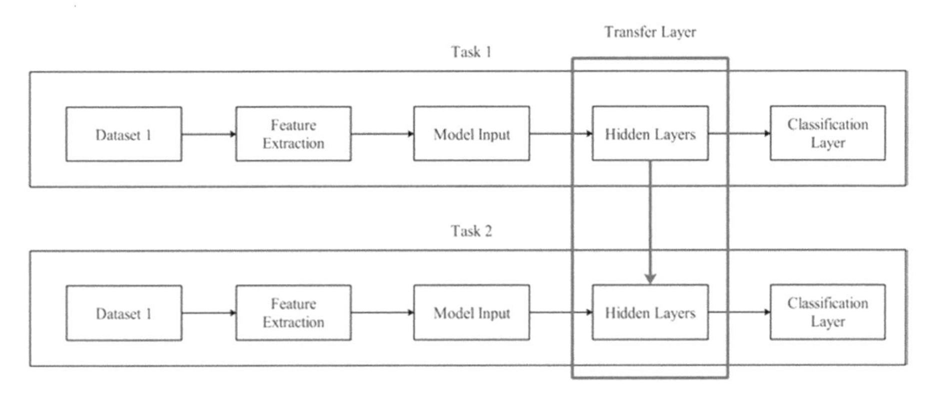

Fig. 3. Summarize of transfer learning process

To achieve the best prediction performance, we implemented the following four approaches. The first one implements a standard convolution neural network for which the hyper-parameters are optimized by grid search. The structure of this CNN with the hyper-parameters on each layer is presented in Table 2. In addition, the hyper-parameter grid of this model is shown in Table 3. Due to random weight initialization, model training and validation steps are repeated five times for each hyper-parameter combination and the best performing configuration is retrieved. Once the optimums are found the model is trained on train set using the optimum hyper-parameters and is tested on the test set. Note that in this work, the hyper-parameter optimization is performed for the first approach only. The second up to the fourth approaches train VGG16, ResNet50, MobileNet, and Xception models. The second approach applies transfer learning by freezing the convolutional base layers and re-training the top layers (i.e. fully connected MLP layers). The third approach performs fine-tuning while doing transfer learning by unfreezing the last few layers of the convolutional base and retraining these and the top layers. As a result of fine-tuning, we unfreeze the last four layers (starting with block5_conv1), the last 10 layers (starting with conv5_block3_1_con), the last 26 layers (starting with conv_dw_9), and the last 6 layers (starting with block14_sepconv1) on VGG16, ResNet50, MobileNet, and Xception models, respectively. The last approach re-trains VGG16, ResNet50, MobileNet, and Xception models from scratch (i.e. all layers are re-trained). In the second and third approaches, the weights of the frozen layers are taken from the models trained on ImageNet dataset.

In each approach, after the flatten layer of the convolutional base, we applied an MLP network, which has a single hidden layer. Due to the binary-class classification problem,

Table 2. Layer Configurations of the Standard CNN Model

Layer Name	CNN Models
L1	Conv2D(32 × (3,3))
L2	MaxPoolling2D (2,2)
L3	Conv2D(64 × (3,3))
L4	MaxPoolling2D (2,2)
L5	Conv2D(128 × (3,3))
L6	MaxPoolling2D (2,2)
L7	Conv2D(128 × (3,3))
L8	MaxPoolling2D (2,2)
L9	Flatten()
L10	Dropout(0.5)
L11	Dense(512)
L12	Dense(1)

the activation function of the last layer is sigmoid, optimizer is RMSprop (only for the first approach it is Adam) with a binary cross-entropy loss function. The mini-batch size is selected as 32, the maximum number of epochs is set to 30 and early stopping is used based on the validation loss. If the validation loss did not improve for 10 epochs the models stopped training. Apart from the above information, during the experiments, the learning rate of transfer learning models is set to 2e−5. Data augmentation [20] has been applied to the best model obtained so far due to the limited number of samples available in train set. For this purpose, we have increased the variety and number of training images by applying operations such as rotating, flipping, cutting, and zooming on the data. All the approaches are implemented using Keras library of Python programming language [21].

Table 3. Hyper-Parameter Grid of the Standard CNN Model

Parameters	Values
Dropout Rate	0.2, 0.3, 0.4, 0.5
L2 Norm Regularization	0.001, 0.0001
Learning rate	0.01, 0.001, 0.0001
Epochs	30

3 Results and Discussion

In this study, a standard CNN, VGG [16], Resnet50 [17], MobileNet [18], and Xception [19] models are implemented for skin cancer classification. Firstly, a standard CNN model is trained for which the hyper-parameters are optimized with grid search as explained in Sect. 2.3. The accuracy of this model is obtained as 73.64% with the following optimum hyper-parameters: dropout = 0.2, learning rate = 0.001, L2 norm regularization = 0.0001, and optimizer set to Adam. In the next step, the second, third, and fourth approaches are implemented for VGG16, MobilNet, ResNet50, and Xception models. The ResNet50 model trained from scratch obtained the best results with an accuracy of 86.67%. When data augmentation (DA) is also applied to the ResNet50 model, the accuracy further increased to 87.88%. Table 4 shows the performance results of the standard CNN and transfer learning methods on skin cancer classification dataset obtained from Kaggle.

Table 4. Performance Results of Skin Cancer Classification

Approach	Model	Accuracy	F-Measure	Precision	Recall
First	Standard CNN	73.64%	72.20%	68.54%	79.27%
Second	VGG16	82.28%	79.60%	83.57%	77.36%
Second	Resnet50	45.45%	61.80%	45.45%	100%
Second	MobileNet	65.45%	37.42%	86.18%	25.35%
Second	Xception	66.21%	42.52%	82.60%	30.04%
Third	VGG16	82.60%	77.03%	77.54%	80.41%
Third	Resnet50	44.80%	60.61%	44.79%	100%
Third	MobileNet	72.20%	60.03%	82.74%	51.01%
Third	Xception	67.40%	54.67%	78.23%	45.62%
Fourth	VGG16	85.61%	81.26%	85.93%	79.51%
Fourth	Resnet50	86.67%	84.31%	84.66%	88.08%
Fourth	MobileNet	83.64%	80.28%	83.16%	80.58%
Fourth	Xception	83.94%	79.88%	78.98%	83.72%
Fourth DA	**Resnet50**	**87.88%**	**85.95%**	**84.59%**	**90.47%**

From these results, we can observe that some of the transfer learning models, which did not have any hyper-parameter optimization, performed better than our base model, which is implemented with hyper-parameter optimization. The best performance is obtained when ResNet50 model is retrained from scratch with data augmentation. This shows that the image data for skin cancer classification problem has its own characteristic features that are not fully captured by models pre-trained using the ImageNet dataset.

The statistical significance of the improvements depends on the number of samples in a dataset. Sometimes a small improvement can be a significant because the number of samples in test dataset is high. On the contrary, a high improvement in accuracy may not be significant because the number of samples in the test set is low. To assess the statistical significance of the improvements made in prediction accuracy, p-values are computed using two-tailed Z-tests, which compare the accuracy of standard CNN to the accuracy of VGG16, ResNet50, MobileNet, and Xception models trained using the second, third, and fourth approaches (see Table 5). For the fourth approach, the results without data augmentation are used.

Table 5. P-values for comparing the accuracy of standard CNN with VGG16, ResNet50, MobileNet, and Xception trained using second, third, and fourth approaches

	VGG16	ResNet50	MobileNet	Xception
Second	**0.0003**	0.00001	0.00208	0.00496
Third	**0.00018**	**0.00001**	0.57548	0.01778
Fourth	**0.00001**	**0.00001**	**0.00001**	**0.00001**

In this table, bold cells correspond to cases where VGG16, ResNet50, MobileNet or Xception obtained better accuracy than standard CNN. According to these results, if we use a p-value threshold of 0.05, it can be seen that all transfer learning models (i.e. VGG16, ResNet50, MobileNet, and Xception), which use the fourth approach (i.e. when these models are trained from scratch) obtains a statistically significant improvement over the standard CNN model for skin cancer classification. For the second approach (i.e. when only the top layers are re-trained for transfer learning), only VGG16 obtained significantly better accuracy than the standard CNN and other models performed worse. For the third approach (i.e. when fine-tuning is performed for transfer learning), VGG16 and ResNet50 performed significantly better than the standard CNN, which performed significantly better than Xception. The standard CNN model also obtained better results than MobileNet, which is trained using the third approach, however this difference is not significant.

In addition to the above analysis, the performance improvement of each transfer learning model is analyzed when second and third, second and fourth, and third and fourth training approaches are compared. For this purpose, p-values are computed using two-tailed Z-tests, which are shown in Table 6. The numbers in boldface show statistically significant improvements when the three training approaches (second up to the fourth) are compared.

According to these results, if we use a p-value threshold of 0.05, it can be seen that only MobileNet obtained a significant improvement when second and third approaches are compared. The third approach performed better than the second one for MobileNet, VGG16 and Xception and it performed worse for ResNet50. However, the accuracy improvements for VGG16 and Xception and the decrease of accuracy for ResNet50 are not statistically significant. When the second and the fourth approaches are compared, the fourth approach performed better than the second one for all the models but this improvement is statistically significant for ResNet50, MobileNet, and Xception. The same behavior is also observed when the third and the fourth approaches are compared.

Table 6. P-values for comparing the accuracy of transfer learning models with respect to training approaches

Model Name	Second – Third	Second – Fourth	Third - Fourth
VGG16	0.88076	0.11642	0.15272
ResNet50	0.8181	**0.00001**	**0.00001**
MobileNet	**0.01174**	**0.00001**	**0.00001**
Xception	0.65994	**0.00001**	**0.00001**

To summarize the results obtained in Tables 5 and 6, the best performances are obtained when the fourth training approach is used, which re-trains the transfer learning models from scratch and the improvements over the other training approaches are statistically significant except for the VGG16 model, which still performs the best with the fourth approach.

Finally, the accuracy of the ResNet50 model trained using the fourth approach with data augmentation is compared to the performance of ResNet50 trained using the fourth approach without data augmentation. For this purpose, a two-tailed Z-test is performed and a p-value of 0.5287 is obtained, which shows that the performance improvement of the data augmentation approach is not statistically significant for ResNet50 when the p-value threshold is set to 0.05.

4 Conclusion

In this study, deep learning models are developed for skin cancer subtype classification. Two main types of model training strategies are employed including training from scratch and transfer learning (with and without fine-tuning). The best prediction accuracy is achieved when the models are trained from scratch. The accuracy further improved slightly when the number of training set images is increased using data augmentation.

As a future work, the hyper-parameters can be optimized for the second, third, and fourth approaches, which may improve the accuracy further. In addition, other model architectures can also be implemented for transfer learning such as DenseNet, EfficientNet, Inception, and NasNet. Finally, the performance of deep learning models can be compared with standard machine learning models that use feature extraction methods for images.

References

1. Marks, R.: Epidemiology of melanoma. Clin. Exp. Dermatol. **25**, 459–463 (2000). https://doi.org/10.1046/j.1365-2230.2000.00693.x
2. Leiter, U., Buettner, P., Eigentler, T., Garbe, C.: Prognostic factors of thin cutaneous melanoma: an analysis of the central malignant melanoma registry of the German dermatological society. J. Clin. Oncol.: Off. J. Am. Soc. Clin. Oncol. **22**, 3660–3667 (2004). https://doi.org/10.1200/JCO.2004.03.074

3. Ariel, I.M.: Malignant melanoma of the female genital system: a report of 48 patients and review of the literature. J. Surg. Oncol. **16**, 371–383 (1981). https://doi.org/10.1002/jso.2930160411
4. Sober, A.J.: Diagnosis and management of skin cancer. Cancer **51**, 2448–2452 (1983). https://doi.org/10.1002/1097-0142(19830615)51:12+%3c2448::AID-CNCR2820511311%3e3.0.CO;2-L
5. Mahbod, A., Schaefer, G., Wang, C., Ecker, R., Ellinge, I.: Skin lesion classification using hybrid deep neural networks. In: ICASSP 2019 - 2019 IEEE International Conference on Acoustics, Speech and Signal Processing (ICASSP), pp. 1229–1233 (2019). https://doi.org/10.1109/ICASSP.2019.8683352
6. Romero Lopez, A., Giro-i-Nieto, X., Burdick, J., Marques, O.: Skin lesion classification from dermoscopic images using deep learning techniques. In: 2017 13th IASTED International Conference on Biomedical Engineering (BioMed), pp. 49–54 (2017). https://doi.org/10.2316/P.2017.852-053
7. Brinker, T.J., et al.: Deep learning outperformed 136 of 157 dermatologists in a head-to-head dermoscopic melanoma image classification task. Eur. J. Cancer **113**, 47–54 (2019). https://doi.org/10.1016/j.ejca.2019.04.001
8. Maron, R.C., et al.: Systematic outperformance of 112 dermatologists in multiclass skin cancer image classification by convolutional neural networks. Eur. J. Cancer **119**, 57–65 (2019). https://doi.org/10.1016/j.ejca.2019.06.013
9. Jain, S., Jagtap, V., Pise, N.: Computer aided melanoma skin cancer detection using image processing. Procedia Comput. Sci. **48**, 735–740 (2015). https://doi.org/10.1016/j.procs.2015.04.209
10. Alquran, H., et al.: The melanoma skin cancer detection and classification using support vector machine. In: 2017 IEEE Jordan Conference on Applied Electrical Engineering and Computing Technologies (AEECT), pp. 1–5 (2017). https://doi.org/10.1109/AEECT.2017.8257738
11. Linsangan, N.B., Adtoon, J.J., Torres, J.L.: Geometric analysis of skin lesion for skin cancer using image processing. In: 2018 IEEE 10th International Conference on Humanoid, Nanotechnology, Information Technology, Communication and Control, Environment and Management (HNICEM), pp. 1–5 (2018). https://doi.org/10.1109/HNICEM.2018.8666296
12. Daghrir, J., Tlig, L., Bouchouicha, M., Sayadi, M.: Melanoma skin cancer detection using deep learning and classical machine learning techniques: a hybrid approach. In: 2020 5th International Conference on Advanced Technologies for Signal and Image Processing (ATSIP), pp. 1–5 (2020). https://doi.org/10.1109/ATSIP49331.2020.9231544
13. Jinnai, S., Yamazaki, N., Hirano, Y., Sugawara, Y., Ohe, Y., Hamamoto, R.: The development of a skin cancer classification system for pigmented skin lesions using deep learning. Biomolecules **10**, 1123 (2020). https://doi.org/10.3390/biom10081123
14. Nawaz, M., et al.: Skin cancer detection from dermoscopic images using deep learning and fuzzy k-means clustering. Microsc. Res. Tech. **85**, 339–351 (2022). https://doi.org/10.1002/jemt.23908
15. Skin Cancer: Malignant vs. Benign. https://www.kaggle.com/datasets/fanconic/skin-cancer-malignant-vs-benign. Accessed 15 Aug 2022
16. Simonyan, K., Zisserman, A.: Very Deep Convolutional Networks for Large-Scale Image Recognition (2015). http://arxiv.org/abs/1409.1556, https://doi.org/10.48550/arXiv.1409.1556
17. He, K., Zhang, X., Ren, S., Sun, J.: Deep Residual Learning for Image Recognition. Presented at the (2016)
18. Howard, A.G., et al.: MobileNets: Efficient Convolutional Neural Networks for Mobile Vision Applications (2017). http://arxiv.org/abs/1704.04861, https://doi.org/10.48550/arXiv.1704.04861

19. Chollet, F.: Xception: Deep Learning With Depthwise Separable Convolutions. Presented at the (2017)
20. Team, K.: Keras documentation: Image data preprocessing. https://keras.io/api/preprocessing/image/. Accessed 16 Aug 2022
21. Keras: Deep Learning for humans (2022). https://github.com/keras-team/keras

Competencies Intelligence Model for Managing Breakthrough Projects

Sergey Bushuyev(✉), Igbal Babayev, Natalia Bushuyeva, Victoria Bushuieva, Denis Bushuiev, and Jahid Babayev

Kyiv National University of Construction and Architecture, Povitroflotsky Avenue, 31, Kyiv 03680, Ukraine
sbushuyev@ukr.net

Abstract. An analysis of existing models for the formation, evaluation and development of breakthrough competencies in project management has shown the practical absence of research in this area of knowledge. A conceptual model of breakthrough competencies in the management of innovative projects and programs is presented. The model is based on four interconnected areas of breakthrough competencies - Emotional Intelligence, Social Intelligence, Cognitive Intelligence and Managerial Intelligence. Within each area, sets of competencies that support the implementation of breakthrough projects are defined. A case of applying this model to the implementation of the double degree project of the Kyiv National University of Civil Engineering and Architecture and the Dortmund University of Applied Sciences within the framework of the implementation of European Union DAAD VIMACS and ERASMUS+ WORK4CE projects is presented.

Keywords: Modelling · Conceptual model · Competencies · Innovation program · Breakthrough projects

1 Introduction

Modern approaches to the development of competencies in the context of innovative approaches and the implementation of breakthrough projects and programs do not demonstrate effectiveness, form several gaps in the required competencies and do not have intellectual support for models. It follows from this that a new approach is needed to model the systems of competencies of innovative projects and programs that implement breakthrough technologies. Breakthrough projects have significant specifics and require special competencies aimed at emotional, social, cognitive and managerial areas in the form of a holistic model. Such a model should ensure the success of breakthrough projects based on the balance of these competencies. At the same time, the created values of breakthrough projects should actively migrate among the stakeholders. The acquired new knowledge creates the foundation for breakthrough projects. The creative activity of man, which transforms nature, as a consequence, hinders and inhibits the creative activity of the cause, i.e. nature, which seeks to improve man. A hypothetical way to

D. J. Hemanth et al. (Eds.): ICAIAME 2022, ECPSCI 7, pp. 348–359, 2023.
https://doi.org/10.1007/978-3-031-31956-3_29

solve this problem is to clarify the fundamental difference between the level of innovative technology used by nature and which so far man has been able to master. The cognitive process, evolving and improving in itself, aimed at simply expanding needs, may need to be adjusted concerning the unknown motives of nature's behaviour. The emergence of information systems promises to provide a means of flexible expansion of innovative resources in the construction industry, which can point the way to the transition to such innovative technologies that do not conflict with existing regulations and standards. These conclusions should be considered at the level of hypotheses. The time has come for the development of society when it is necessary to flexibly adapt their innovative technologies to natural ones to prevent and avoid global troubles. The vast majority of man-made technologies are based on imitation and copying of various natural processes and phenomena. Agile technologies are no exception, they try to model the creative behaviour of the manager and are based on deep historical traditions of different cultures. Previously, the main object of various innovative technologies in construction was an individual or group, the task was to accumulate knowledge, and organize new behaviour in adverse external conditions. The traditions of these schools cover various aspects of activity: philosophy, preaching, commerce, intelligence, diplomacy, and politics. Now, in connection with the rapid development of information technology, a new association has emerged, consisting of the deeper use of computer systems and networks in innovation: artificial intelligence systems, and expert systems. The trend of such penetration is significantly growing and expanding, so there is a need for a new organization of construction activities with the wide involvement of information technology.

2 Analysis of Recent Research and Publication

Management of the formation of breakthrough competencies is systematic processes, creation, preservation, and distribution, which are used as elements of intellectual capital necessary for the success of the organization in competition. At the same time, special importance is attached to the strategy of effective use of intellectual assets to increase productivity, and efficiency and create new values. As a basic model of competencies, the authors propose to use the standards, (IPMA OCB 2013) and (IPMA ICB 4.0 2015), (P2M 2015), and Agile (2017).

The intelligence models and methods for managing breakthrough projects are the tools for creating an effective system of competencies for project management teams (NCIS 2000). Intellectual competencies form the core of project management in general. In the process of managing breakthrough projects, such a core plays a decisive role in their success. In Pherson R. H., Heuer R. J. (2020) the following products are defined: analytical, search, knowledge products and systems products Obradović V., et al. (2018). Each type of product of intellectual activity makes it possible to form potential breakthrough competencies in the management of innovative projects (Bondar et al. 2021), (Forsberg et al. 2005).

Analytic products include results analysis, research area structure analysis, breakthrough technology market profile, network analysis, risk and opportunity analysis, goal profile, and rapid assessment of the state and prospects of projects (Pherson et al. 2020), (Slivitsky 2006).

The search products are associated with the uncertainty, variety and quality of information (Jae-Yoon 2011), (Belack 2019).

Knowledge products serve as a basis for further development of the model itself and quality maintenance and balancing of socio-cognitive space (Drouin et al. 2021).

The system products are designed to ensure that the appropriate breakthrough models are available for its efficient operation and to minimize inefficient practices such as the use of multiple information systems platforms (Bushuyev et al. 2017), (Bushuyev et al. 2020).

The proposed model identifies the following groups of competencies:

1. Leadership and commitment
2. Setting goals and coordinating teams
3. Standardization and creativity
4. Intelligent analytics

The intelligence assessment of the competencies of breakthrough projects would allow sharing between stakeholders to be done more easily, benefiting all parties.

Consider the key principles of Agile transformation in the organization:

1. Ignoring immunity to change

Transformation in an organization occurs only when the people in it change.

But people do not change, even if they want to. Remember your New Year's promises. Many do not even make plans, because they remember how bitter it was to realize that for a year and did not activate the season ticket to the gym or not take up their English. People do not change, even if they are threatened with death. It turned out that when cardiologists warn patients that they will die if they do not change their lifestyle (do not go on a diet, do not exercise, do not quit smoking), only one in seven patients changes their lives.

We have immunity to change: we reject the new, the unknown and cling to habitual beliefs. Immunity protects against fears.

2. Fear of becoming unnecessary to the company. Fear of losing authority and status. Fear of losing yourself.

When we try to change the thinking, behaviour, and culture in the organization, we are faced with this immunity, and therefore any change is difficult, painful and long.

We do the simple, we don't do the important

Where will cross-functional teams come from if we have 1–2 independent professionals and an army of assistants in each region?

Where will self-organization come from, if in fact, we have strict subordination?

Where will teamwork come from if motivation and reward are individual?

How will employees suddenly become happy if we do not give them resources, do not remove bureaucratic obstacles in their work, but only add more control, rallies and Agile - coaches?

We put aside complex and key things, those that meet a lot of resistance and make it interesting and enjoyable. It is much more interesting to organize pieces of training, stick stickers, and hold demo days and team sessions, rather than rebuild motivation and hiring systems.

Cognitive analysis and modelling are used in studies of organizational systems development as one tool. The purpose of cognitive modelling is to generate and test a model of the observed situation in the system to obtain a system model that can explain its behaviour in the observed situation and development. A poorly structured system means any dynamic, i.e. functioning in a time system, in the structure and functioning of which an important role is played by the human factor. It is the presence of the human factor, for various manifestations of which it is almost impossible to build accurate mathematical models, that allows us to consider the following system poorly structured system.

Dissatisfaction with the level of trust in the organization is often perceived by management, but ideas about the causes and possible ways to change the situation in the system are vague and contradictory. If we manage to formalize these ideas, it is possible to develop models and methods of decision-making in poorly structured situations of trust formation.

It is sometimes said that management is a successful experience. But the experience of management is well accumulated under two conditions:

- the problem situation has a verbal description, a formalized idea;
- there are many cases of confirmation of experience to deduce patterns.

Cognitive maps, allowing you to display subjective perceptions of the studied situation, are ways to formalize the idea of the studied situation of trust in the system. Instead of spatial relations in it, as a rule, the relations of influence, causality, and passing of events are allocated.

Research and reflection on changes in the situation over time (compiling a sequence of cognitive models allows you to gain experience in analyzing and managing the situation).

Most often, the cognitive map is presented in the form of a weighted graph, in which the vertices are compared factors, and the edges - are the weight of a school.

$$G = [V, E],$$

where V is the set of vertices (concepts);

the vertices Vi belong to the set V;

$i = 1, 2, ..., k$. are elements of the system under study;

E - many arcs, arcs reflect the relationship between the vertices. Arcs are weighted by indicators of trust and communications.

Vertices V most often indicate a qualitative representation of the system element. Among the identified five types of cognitive maps by the type of relationships used to study the problems of trust formation and study the causes of conflicts, it is advisable to use cognitive maps that represent the impact, causality and system dynamics (causal cognitive maps).

The study of the interaction of factors allows us to assess the spread of influence on the cognitive map, and changes in their state (value). Analysis of cognitive maps allows us to identify the structures of the problem (system), find the most significant factors influencing it, to assess the influence of factors on each other. Given the specifics of modern organizations, it should be borne in mind that, due to increasing globalization, small groups can be virtual teams, i.e. representatives of small groups can be removed territorially and may exist in different national cultures. The possible territorial remoteness of team members requires increased attention to the mechanisms of coordination of actions of members of territorially distributed teams.

Representatives of different centres of influence can be distributed in the company territorially (i.e. belong to different elements of the OBS). That is, a team working together is not necessarily a team whose members work together. That is, the interaction of members of such teams can often be built only verbally. Modern information technology allows you to create virtual teams for employees of geographically distributed offices to jointly participate in the implementation of organizational development programs (Wolff 2014).

The task of building trust in virtual teams is quite new for many modern companies. In such conditions, when building trust, much attention is paid to creating a single database for decision-making, knowledge bases necessary for building community and understanding the value of diversity. Virtual cooperation largely depends on the level of trust and the shared vision and the provision of publicly available principles of business. One of the approaches to building trust can be, in particular, additional efforts to adapt methods of doing business depending on the uniqueness of national cultures, ethical principles, the legal framework of specific countries, differences in levels of economic development and, consequently, methods of remuneration.

Lack of informal communication, language barriers, and differences in national cultural traditions, including differences in ethical norms and laws of countries can often serve as barriers to establishing trust. Methods of building trust for virtual teams have their specifics. Arrangements for joint work are formalized first, there should be no situations where the rules of interaction are determined intuitively. Proposing, discussing and accepting a set of rules and regulations by virtual team members that allow and limit the actions of team members can provide a higher level of trust.

For the elements of the system identified in the model - small groups of stakeholders, homogeneity is formed, formed on a professional basis or is related to the specifics of the tasks. Therefore, it is necessary to take into account the differences in the subcultures of groups related to their professional activities. It makes sense to measure the professional homogeneity of the group, the index of professional cohesion, and the sociometric professional status of group members. At least one feature must be recognized and significant for team members, it was homogeneous.

The professional homogeneity index of the group is defined as the ratio of positive, negative and neutral (zero) interpersonal assessments obtained during group testing. The group can be characterized by a low, medium or high degree of professionalism.

The mathematical record of the formula for calculating the index (I_{hom}) looks like this:

$$Ihom = [(AFP - Obo)/[Nrp * (Nrp - 1))] * 100\%,$$

where AFP is the number of mutually positive evaluations;

O_{bo} the number of mutually negative evaluations;

N_{rp} - the number of members of the test group.

The sociogram constructed based on results of testing in the group will allow visualizing a condition of the group from the point of view of its homogeneity. Since the team must be homogeneous on at least one basis, recognized and important to team members, it is possible to study its homogeneity not only on a professional basis but also on any other basis, for example, it may be an indicator of creative homogeneity.

Similarly, the degree of trust of group members in each other can be examined.

Experts have identified some common mistakes made by managers when creating a team:

1. Selection of the team on the principle of "psychological compatibility". It is much more important to unite the team based on a common goal and joint activities.
2. Reformation of formed groups on the principle of potential "psychological compatibility". "It destroys the foundations of the team - the experience of interaction and the experience of strengthening each other." In the new team, the resistance of the conscious and the unconscious will be great. It is much more effective to use the experience of interaction in new conditions.
3. Underestimation of the value of diversity. Groups made up of dissimilar individuals are more effective than groups with similar views. Understanding the value of diversity helps to improve relationships between team members and, consequently, strengthens trust in a small group.

Group cohesion is a measure of the mutual attraction of group members to each other and the group. Cohesion is expressed in the desire to stay in the group, in the desire to cooperate in solving common problems and to preserve the group. The more cohesive the group, the tighter group control over the views and actions of its members. An atmosphere of attentiveness and mutual support is created in a close-knit group. The downside of excessive cohesion is the reluctance of its members to think critically and make serious decisions due to the development of the group's unanimity. This tendency arises due to the tendency to conformism. (Optional: On conformism: a person with a higher intellectual level is less conformist than a person with a low intellectual level. An educated person is also usually confident in the accuracy of his statement and does not feel the need for support from the group).

When a group becomes too cohesive and has common expectations, the following flaws arise in the decision-making process:

1. Group decisions have a small number of options, opportunities outside this range are rejected or not considered at all
2. Initially set goals are not reviewed or challenged
3. Newly identified risks shall not be taken into account so as not to call into question the course of action initially chosen.
4. The courses of action rejected by the group from the beginning are not considered again in light of new information.
5. Experience and knowledge of external experts are not involved.
6. When new information is discovered, the group gives priority to information that supports its initial hypotheses and ignores conflicting information.
7. The group does not think about how bureaucratic inertia or resistance of the organization can hinder the implementation of the chosen political line.

3 Competencies Intelligent Model for Managing Breakthrough Projects

In real practice project managers apply a conceptual model with four domains of competencies:

1. Emotional Intelligence (EI) competencies of Result Orientation, Initiative, Flexibility, and Self-Confidence;
2. Social Intelligence (SI) competencies of Empathy, Influence, Networking, and Team Leadership. They also showed significantly more cognitive competencies in Systems Thinking and Pattern Recognition. The behaviour pattern combines EI, SI;
3. Cognitive Intelligence (CI) competencies being key to effectiveness in Acquisitions of knowledge, Creativity and Innovation, Artificial Intelligence and Modeling in an organization;
4. Managerial intelligence (MI) competencies like Strategy, Culture and Values, Planning and Control, Opportunity and Risk Management.

The four-domain conceptual model of competencies for Breakthrough Projects is shown in Fig. 1.

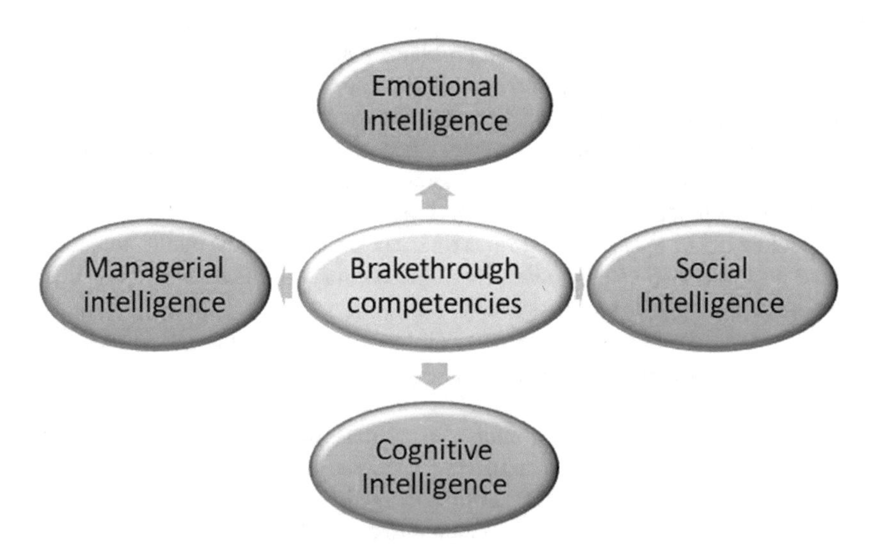

Fig. 1. The four-domain conceptual model of competencies for Breakthrough Projects

Let's define the formula for the success of projects based on breakthrough competencies and the proposed concept.

Use the concept of "Emotional energy" based on EI of the project manager and the project team.

Let us single out two emotional states of the project manager and the team "As is" before emotional infection and "As will be" after emotional infection. At the same time, the team develops such competencies as leadership, self-confidence, and result orientation.

The second component of the model is "Entrepreneurial potential" as one of the resources influencing the SI.

The third component of the model is the level of breakthrough (innovative) maturity of the project team based on Cognitive readiness.

The fourth component defines Managerial competence (MI) and plays the role of integrator of all grope of competencies in breakthrough projects (A Systems Approach 2009).

Let's look at the case study according to the application of the proposed conceptual model. It had been developed by an assessment Double degree Master's program in preparation for Project Managers at Kyiv National University of Construction and Architecture. At the end of this program grope of 20 students had been assessed according to the four domain conceptual model of breakthrough competencies.

The project team's competence was assessed using the IPMA OCB (2013) and IPMA ICB 4 (2015) models (Table 1).

Table 1. Results of assessment competencies level according to Benchmark of project success

Competence Name	Benchmark	Assessment level
Emotional Intelligence		
Result Orientation	7	8
Initiative	7	6
Flexibility	7	9
Self-Confidence	7	7
Social Intelligence		
Empathy	7	8
Influence	7	9
Networking	7	7
Team Leadership	7	6
Cognitive Intelligence		
Acquisitions of knowledge	7	8
Creativity and Innovation	7	9
Artificial Intelligence	7	7
Modeling by vision	7	8
Managerial intelligence		
Strategy	7	7
Culture and Values	7	8
Planning and Control	7	8
Opportunity and Risk	7	9

As a result of analyses, there are two competencies, where the assessment level low than Benchmark. As the result of analyses, there are 2 competencies, where the assessment level is low the Benchmark. These are Initiative and Team Leadership competencies. To be successful project team need to improve these two competencies (Fig. 2).

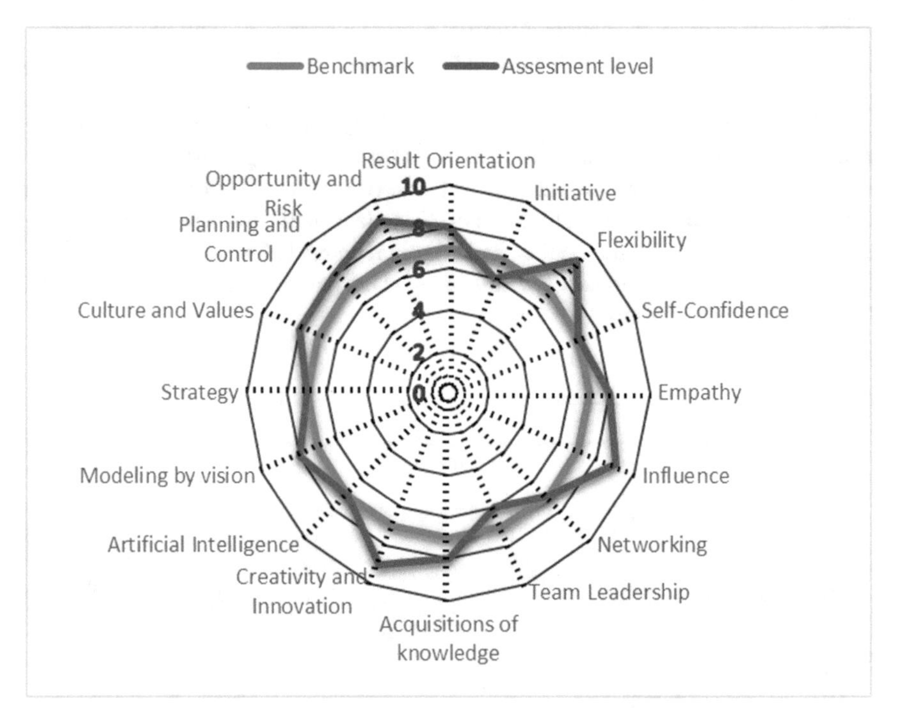

Fig. 2. Results of the case study of assessment by the four-domain conceptual model of competencies for Breakthrough Projects success

The project manager decided on the initial step to organize 2 pieces of training for the project team. The first training was devoted to the development of the initiative of the breakthrough project team. The second concerned leadership. As a result, the assessment of the team's competence has changed significantly and is given in Table 2.

The results of the evaluation of Breakthrough competencies compared to the benchmark are shown in Fig. 3.

As a result of the pieces of training of the project team, the assessments of breakthrough competencies in almost all cases exceeded the level of the benchmark. This indicates the readiness of the project team for its successful implementation.

4 General Process of Application Four-Domain Conceptual Model of Competencies for Breakthrough Projects

Let's discuss step by step process for applying the proposed conceptual model of competencies for Breakthrough Projects' success.

Table 2. Results of assessment competencies level according to Benchmark of project success after pieces of training

Competence Name	Benchmark	Assessment level
Emotional Intelligence		
Result Orientation	7	8
Initiative	7	8
Flexibility	7	9
Self-Confidence	7	9
Social Intelligence		
Empathy	7	8
Influence	7	9
Networking	7	8
Team Leadership	7	8
Cognitive Intelligence		
Acquisitions of knowledge	7	8
Creativity and Innovation	7	9
Artificial Intelligence	7	7
Modelling by vision	7	8
Managerial intelligence		
Strategy	7	8
Culture and Values	7	8
Planning and Control	7	8
Opportunity and Risk	7	9

Work break down the structure of the Breakthrough Projects

Step 1. Initiation

Definition of the Mission, vision and expected result of a breakthrough project. Profiling project goals

Installing a benchmark by directions

Generation of alternative directions for breakthrough projects

Entropy estimation for each direction

Breakout Direction Choice

Assessment of the competence of each member of the project team and the team as a whole.

Step 2. Implementation

Competence gap analysis of the team and the benchmark

Formation of an operational training program for a breakthrough project team

Using the conceptual model of breakthrough competencies for the successful implementation of the project

If the goals are not achieved, the next alternative is selected and Step 1 is repeated.

Fig. 3. Results of the case study of assessment by the four-domain conceptual model of competencies for Breakthrough Projects success

Step 3. Closing down

Results assessment
Lesson learns
Good practice

After achieving the goal of a breakthrough project, a report is generated and an analysis of the level of competence of the project team is carried out.

5 Conclusion

The proposed conceptual model of competencies for Breakthrough Projects' success covers four interrelated competencies. These areas are guided by the emotional state of the project team, social status, and cognitive and managerial capabilities. To analyze the success of a breakthrough project, a benchmark assessment is used, which allows you to identify problematic competencies and, at the stage of project initiation, plan the necessary corrective actions to develop insufficient project competencies. The given example of the implementation of the program for the preparation of masters with double diplomas confirmed the effectiveness of the proposed model. The typical work structure given in the article allows you to successfully carry out breakthrough projects.

The following areas should be highlighted as areas for future research:

- generation of alternative directions for breakthrough projects;
- building an entropy model for managing the uncertainty of a breakthrough project;
- substantiation of the level of benchmarks in different areas of competence.

References

IPMA Organisational Competence Baseline (IPMA OCB), 67 p. IPMA (2013)

Individual Competence Baseline for Project, Programme & Portfolio Management. Version 4, 415 p. International Project Management Association (2015)

Obradović, V., Todorović, M., Bushuyev, S.: Sustainability and agility in project management: contradictory or complementary? In: IEEE 13th International Scientific and Technical Conference on Computer Sciences and Information Technologies, CSIT 2018 (2018)

Pherson, R.H., Heuer, R.J.: Structured Analytic Techniques for Intelligence Analysis, 384 p. SAGE Publications Inc. (2020)

Jae-Yoon, J., Chang-Ho, C., Jorge, C.: An entropy-based uncertainty measure of process models. Inf. Process. Lett. **111**(3), 135–141 (2011)

Belack, C., Di Filippo, D., Di Filippo, I.: Cognitive Readiness in Project Teams. Reducing Project Complexity and Increasing Success in Project Management, 252 p. Routledge/Productivity Press, New York (2019)

Drouin, N., Müller, R., Sankaran, S., Vaagaasar, A.-L.: Balancing leadership in projects: role of the socio-cognitive space. Proj. Leadersh. Soc. **2**, 100031 (2021)

Bushuyev, S., Murzabekova, A., Murzabekova, S., Khusainova, M.: Develop breakthrough competence of project managers based on entrepreneurship energy. In: Proceedings of the 12th International Scientific and Technical Conference on Computer Sciences and Information Technologies, CSIT 2017 (2017)

P2M: A guidebook of Program & Project Management for Enterprise Innovation. 3rd edn., 366 p. (2015)

The Availability of Information on the National Intelligence Mode, 42p. NCIS (2000)

Bondar, A., Bushuyev, S., Bushuieva, V., Onyshchenko, S.: Complementary strategic model for managing entropy of the organization. In: CEUR Workshop Proceedings, vol. 2851, pp. 293–302 (2021). http://ceur-ws.org/Vol-2851/paper27.pdf12

Forsberg, K., Mooz, H., Cotterman, H.: Visualizing Project Management, 3rd edn. John Wiley and Sons, New York, pp. 108–116, 242–248, 341–360 (2005)

Slivitsky, A.: Value migration. Mann, Ivanov & Ferber, 432 p. (2006)

Bushuyev, S., Bushuiev, D., Zaprivoda, A., Babayev, J., Elmas, Ç.: Emotional in-fection of management infrastructure projects based on the agile transformation. In: CEUR Workshop Proceedings, vol. 2565, pp. 1–12 (2020)

Agile Practice Guide: Paperback. USA, Project Management Institute, 210 p. (2017)

A Systems Approach to Planning, Scheduling, and Controlling, 10th edn. Wiley, New Jersey, 1120 p. (2009)

Wolff, C.: A structured process for transferring for academic research into innovation projects – Pimes case study. Int. J. Comput. **13**(4), 227–239 (2014)

Secure Mutual Authentication Scheme for IoT Smart Home Environment Using Biometric and Group Signature Verification Methods

Hisham Raad Jafer Merzeh[1(✉)], Mustafa Kara[2], Muhammed Ali Aydın[3], and Hasan Hüseyin Balık[1]

[1] Graduate School of Science and Engineering, Faculty of Electrical and Electronics Engineering, Yildiz Technical University, Istanbul 34220, Turkey
hisham.merzeh@mtu.edu.iq, balik@yildiz.edu.tr

[2] Hezârfen Aeronautics and Space Technologies Institute, Computer Engineering Department, National Defense University, Istanbul 34000, Turkey
mkara@hho.msu.edu.tr

[3] Faculty of Engineering, Istanbul University-Cerrahpaşa, Istanbul 34320, Turkey
aydinali@istanbul.edu.tr

Abstract. Smart equipment in smart home environments is ubiquitous and widely separated nowadays. Thus, smart things are connected through the network to make it accessible everywhere and remotely controlled. Nevertheless, connection to the internet made smart things vulnerable to hacking, exploitation, and compromised incorrectly, and guaranteeing its security has been a broad concern. Many research studies discuss these security problems and propose different solutions to solve these problems. However, the nature of smart devices and the lack of resources makes securing their significant challenges. Most of the existing solutions for these issues are based on single server architecture with low concern for privacy and anonymity. This paper proposes a new authentication approach for authenticating users and devices in a smart home environment. This approach is an improvement on the existing methods. This study combines the group signature scheme with the biometric signature to propose an authentication mechanism based on decentralized architecture. First, we utilize the biometric data using a fuzzy extractor algorithm to authentic the legitimate user of the device. Then, the user device is authenticated as a group member of the smart home using a group signature. The group signature is a Blockchain-based scheme that allows a group member to sign their request for remote access or control. The proposal is received as a group request instead of a specific group member. Every group member has their group private key used for signing the request and the group public key used for request verification. Compared with existing approaches, this mutual authentication technique can provide high security and reliability in addition to privacy and anonymity. This research integrates biometric user data, group signature, message authentication code, elliptic curve integrated encryption, and Blockchain to increase the security and reliability of the authentication mechanism. We analyze the security features of our proposed scheme by comparing it with the existing scheme.

Keywords: IoT · authentication · blockchain · group signature

D. J. Hemanth et al. (Eds.): ICAIAME 2022, ECPSCI 7, pp. 360–373, 2023.
https://doi.org/10.1007/978-3-031-31956-3_30

1 Introduction

Smart devices, represented by the Internet of Things devices (IoT), provide remote access and control in the smart home environment. This capability is convenient but carries out risky. The vulnerable devices are exposed to the risk of an exploit like conducting surveillance or performing other suspicious activities on the users that cause privacy disclosure and other serious problems. This highlights the need to design a robust authentication solution in term of security and efficiency, to authenticate remote users, ensure the highest level of security, and protect user privacy by complying with the General Data Protection Regulation (GDPR) [1].

In this study, entity verification is realized by biometric verification between the device and the person. Using biometrics security to authenticate users to access the user device provides some advantages [2]. These advantages prevent security vulnerabilities such as password sharing, forgetting, guessing or stealing, loss, damage, and sharing smart cards by users before blockchain authentication of endpoint phone devices. Thus, the first step is mutual authentication of the users with the device. Person authentication in accessing the user device in the smart home application was carried out through biometric data authentication. It provides Integrity, Authorization, and Non-Repudiation in the smart home system over the proposed authentication method.

A group signature based authentication scheme proposed by Chao Lin *et al.* [3] called HomeChain. This scheme based on decentralized architecture using Blockchain. This model can provide user privacy, as the request is issued in the name of the group instead of a specific member. All transactions signed with group signature. This scheme used asynchronous communication between users and home gateway by sending all transactions to the Blockchain to be stored, then the concerned party will read this transaction from the Blockchain. Even if the asynchronous communication do not provide a notification mechanism to notify the concerned party with the new request which need for extra manipulation for constantly check for any new request. But this approach can provide high anonymity and privacy consideration. In spite of all the security features that can this scheme provide but still the mutual authentication between the user (group member) and the user device(mobile, tablet, etc.) is not considered. These devices are vulnerable to theft or used by unauthorized persons, even if a password or secrete key is used. These vulnerability still exist. To solve these issues we proposed an authentication system for smart home applications. Privacy is considered and complies with GDPR using some new development for the existing authentication schemes proposed in [3] to enhance their security and performance level, using different authentication models. GDPR enhanced the individual's data privacy and rights and built on the obligations and responsibilities of data controllers [4, 5]. Personal data is any information relating to an identified or identifiable living person. Those personal data can be categorized into different types, for example, Racial or Ethnic Origin, Political opinions, People's health records, Personal sex life behaviors, etc. [6, 7].

In this study, our main objective provides secure authentication for the smart home device, user device, and the users. The proposed method is used a group signature to anonymize the device request for remote access or control. In addition, we used biometric verification to authenticate the user while using the user's device to access the system.

The user's biometric data can provide a high level of authentication information and reliably obtain cryptographic keys using a fuzzy extractor as a cryptographic method.

Our key research contributions are summarized as follows.

- Propose a secure authentication scheme for smart home environment.
- Improve the existing scheme by increasing the security features and authentication capability
- Utilize the user's biometric data with Blockchain and group signature to propose a new authentication approach between group members and home IoT devices and securely verify user's access requests.
- The proposed scheme complies with GDPR and privacy considerations.

The rest of this article was organized as follows. Studies in the literature to establish a secure authentication model are introduced in Sect. 2. Section 3 describes the proposed architecture of our authentication model in smart home applications and classifies the threat model categorically. Section 4 discusses the security analysis and formal validation of this plan. Finally, we conclude the article by writing the conclusion part with Sect. 5.

2 Related Works

The literature has proposed that concerned with providing security and privacy. Zhihua Cui *et al.* [8] proposed an authentication scheme that can be used for IoT systems consisting of multi clusters, clusters heads, and base stations. Based on the local Blockchain for each base station and combined all base stations with the public blockchain. Ma Zhaofeng *et al.* [9] proposed a decentralized authentication in an IoT environment called BlockAuth. This work proposed a model that can increase the fault tolerance using a decentralized Blockchain-based model, compared with centralized, traditional authentication models that are single side and poor with fault tolerance. A. Patwary *et al.* [10] proposed a Blockchain authentication model based on device location info to improve the authentication mechanism. They assume that all devices can be agreed upon and provide trusted location information. Chao Lin *et al.* [3] proposed a group signature-based decentralized authentication scheme using Blockchain in an IoT environment called HomeChain. Zhang *et al.* [11] proposed an authentication scheme for Blockchain-based mobile-edge computing. They used a group signature to validate the block created with Blockchain. The block validation is applied according to the signature. The block that belongs to the group that creates it. Jiang *et al.* [12] proposed an authentication approach used for vehicular ad-hoc networks. They adopted group signature for authentication in addition to the trusted region authority to achieve conditional privacy and anonymity. Zhang *et al.* [13] proposed an authentication scheme for Vehicular Ad Hoc Networks by combining Group Session Key with batch group signature to achieve computation efficiency. This work proposes a multiple group model, and each group creates a group signature that all group members can use to sign transactions and anonymously communicate with other remote devices. However, previous literature can mitigate security and privacy issues. Still, there are critical challenges that must be solved.

3 Proposed Authentication Scheme

This section briefly describes the basic background information needed to explain and analyze the proposed scheme, which is shown in Fig. 1. Topics are Fuzzy Extractor and Group Signature, respectively. Moreover, this part explains the proposed authentication approach for the smart homes.

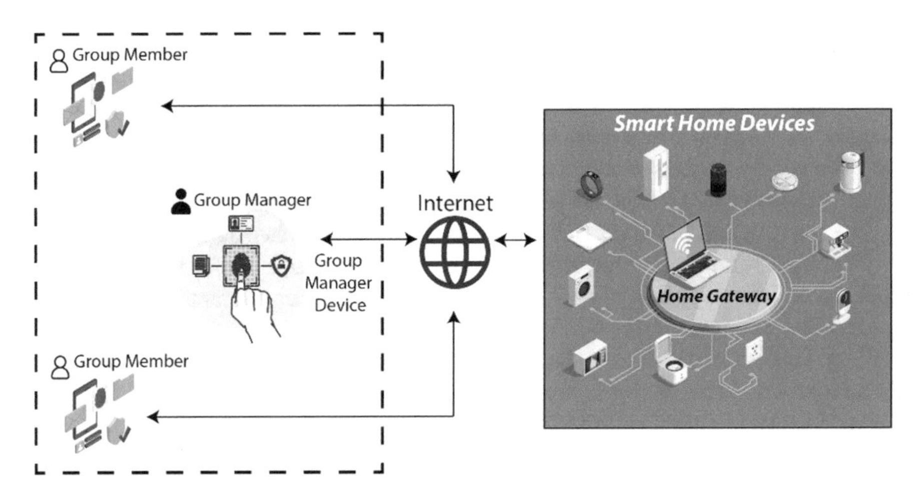

Fig. 1. Secure Communication Structure in the Smart Home System

The communication architecture of a smart home consists of group members and a group manager (as family members) connected to the smart home through the internet. The home gateway is used as a local server to manage the IoT smart devices that are connected wire or wireless to the local network. The home gateway manages the task requests from the group members and sends control or access requests to the home devices, then sends the response and device status to the request sender.

3.1 Group Signature

The group signature scheme in [3] which is the blockchain-based system, used an algorithm to sign the group members' requests for remote access or control. This can provide anonymously access to the system [14, 15]. The request is addressed as a group request instead of as an individual group member. In case of any suspicious behavior, the scheme can provide a tracking process that allows the group manager to identify group members with specific requests for investigation purposes. Moreover, the scheme with the short length of the group signature can reduce the communication overhead and provide a lightweight process. Therefore, it is suitable for IoT verification systems.

3.2 Fuzzy Extractor

Fuzz extractors are a biometric tool that allows user authentication using a template created from the user's biometric data as the key. It is a cryptographic method designed to reliably obtain cryptographic keys from a noisy environment over the corresponding device [16]. It uses a hashing algorithm and produces different results even with very small differences [17]. Reproduction of actual biometric data is very difficult in common practice and is affected by noise in the environment during data acquisition. To avoid this influence, a fuzzy extraction method is preferred, which can extract a random string and general information from the biometric template with a certain fault tolerance (t). In the data generation process of the device for biometric verification, the fuzzy extractor extracts the noise from the biometric data it has using general information and fault tolerance (t) and obtains the original biometric key data with the necessary methods.

M Suppose that $M = \{0, 1\}m$ biometric data points is a finite m-dimensional metric space, d: $M \times M \rightarrow Z+$ is a distance function used to calculate the distance between two points based on the selected metric; L is the number of bits of the output string and t is the error tolerance, where Z+ is the set of all positive integers [18].

The fuzzy extractor (M, L, t) is defined by the following two algorithms:

Gen: This is a probabilistic algorithm that takes a biometric information fi $\in$ M as input and outputs a key data bi $\in \{0, 1\}$ and a general replication parameter pari. In other words, Gen (fi) $= \{$bi, pari$\}$.

Rep: This is a deterministic algorithm that takes a noisy biometric information fi$'$ $\in$ M and a general parameter pari related to fi and then generates the biometric key data, that is, bi value. In other words, satisfying the condition d (fi, fi$'$) $\leq$ t gives the result Rep (fi$'$, pari) = bi.

3.3 The Proposed Scheme

- We assume that the group consists of family members. In addition, the group manager is chosen from one of the family members (exp: one of the parents).
- In the beginning all group members including the manager are registered their biometric information using fuzzy extractor authentication method. This method takes biometric information fi then generate key data bi and a general replication parameter pari. Gen(fi) $= \{$bi, pari$\}$.
- The group manager Initializes parameters in the home gateway by invoking the initialization Algorithm 1 and Enroll Algorithm 2 to get the group private keys gski and group public key gpk. The group private key gpki is allocated by the group manager for each member to be used for signing their transactions. The group public key is used to verify the transactions. This key is stored in the home gateway. The group manager also invokes public parameter generation to get the public parameter. That is used with the Elliptic Curve Integrated Encryption Scheme (ECIES) [19].

Algorithm 1 uses initialization parameters q, P1, P2 by the group manager as inputs and calculates cyclic groups G1, G2, G_T, where G_T is a bilinear pairing. Then generate public system parameters PP, private key of the group manager *sk*, tracing key *tk*, and group public key *gpk* as outputs.

Algorithm 1 initialization

Inputs : **q , P1, P2 ;**
Output : ***(*PP*, sk, tk, gpk)***
Initialization:
d, s, u = random()
calculate:
G1, G2, G_T = cyclic groups of order *q*
D=d . P1
S=s . P2
U=u . P1
sk = (d, s)
tk=u
gpk = *(D, S ,U)*
PP = (q,G1,G2,GT , e, P1, P2,H(·)),

Algorithm 2 is used for the enrollment process. The purpose of the algorithm gets the system public parameter, the group manager's private key as inputs, extract parameters from the public parameter PP, and *sk* private key of the group manager, to generate the group member's Tag_i, and *gski* private key of the group member. Finally, the group member's tag is added to the corresponding user ID_i and Group ID GU_i in the member list.

Algorithm 2 Enroll

Inputs: PP, ***sk , ID_i ,GU_i***
Output tag*i* , *gsk_i*
Initialize:
x_i= random()
(q,G1,G2,GT , e, P1, P2,H(·)) = PP
(d,s) = sk
Calculate:
$Z_i = (d - xi)(s.\ xi)^{-1}\ .\ P1$
$Tag_i = H(x_i \cdot Z_i)$
$gsk_i = (x_i, Z_i)$
MemberList ***(ID_i , GU_i)***= Tag_i

Figure 2 depicts the steps of the authentication procedure to authenticate the user via biometrical data, getting group signature key pair, uploading control or accesses request by the group member, and getting a request by the home gateway. Then, the home gateway uploads the response and task status to the request sender. Moreover, the manager can add a malicious user to the revocation list in case of suspicious behavior. All these steps are detailed and described below.

1. Biometrical User verification is essential to ensure data protection and users' privacy. Biometrical data with fuzzy extractor method verifies the user. Furthermore, fingerprint validation is preferred instead of other methodology because mobile

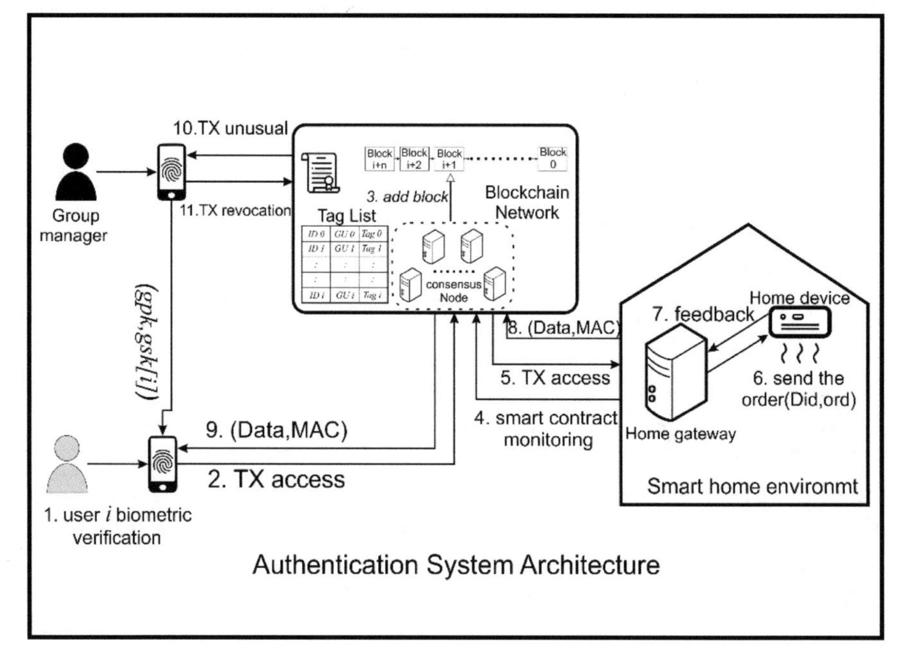

Fig. 2. Authentication System Architecture

devices commonly support fingerprint readers. Moreover, it is easy to use, fast and accurate. These reasons motivate this work to use fingerprints as biometric data in this scheme and satisfying the condition d (fi, fi′) ≤ t gives the result Rep (fi ′, pari) = bi.

2. A new public/private key pair (pk, sk) is generated using the key pair generation when group members request access or control with the home gateway. Then the transaction is created by the group member according to his/her requirement. Finally, this transaction is signed using the *Gsign* in Algorithm 3.

   $TX_{access} = GSign_{gski}\,(Enc_{pkhg}\,(ver||pk||Did||ord))$, where the *ver* is the is the current version, *pk* is the member public key, *Did* is the home device id, *ord* is the control order to change the device status, *Enc* encryption algorithm. pk_{hg} home gateway's public key, gsk_i member's private key.

   Algorithm 3 is used to sign a message with a group signature using public system parameters, group members' private key, group public key, and message. Extract parameters from PP, *gpk*, gsk_i and generate a message signature.

Algorithm 3 Group signing

Inputs: PP, *gski, gpk,* msg
Outputs: Signature $(C1, C2, c, w)$
Initialization:
$k = random\ ()$
$(q, G1, G2, GT, e, P1, P2, H(\cdot)) = PP$
$(D, S, U) = gpk$
$(x_i, Z_i) = gsk_i$
Calculate:
$C1 = k \cdot P1$
$C2 = xi \cdot Zi + k \cdot U$
$Q = e(U, S)k.$
$c = H(C1, C2, Q, H(\text{msg}))$
$w = kc + xi$
Signature = $(C1, C2, c, w)$

3. The member's request is uploaded using *Upload_request* in Algorithm 6 to the smart contract and creates a new block.
4. The home gateway keeps checking for new requests that are delivered to the smart contract.
5. Once the group member uploads a new request, the home gateway uses the *Get_request* function from Algorithm 6.
6. Using Group signature Verify in Algorithm 4 to check the transaction, if it pass and not found in revocation transaction, then the transaction is decrypted by the home gateway using its private key sk_{hg} to get the transaction information (*ver, pk, Did, ord*) and then sent the control order to the targeted home device.
7. After the home device receives the control order, the device sends the feedback to the home gateway. The feedback presents the result of the task execution or the current device status.
8. After receiving the feedback from the home device, the home gateway encrypt the feedback using member's public key *pk* and generate the corresponding MAC for the data using home gateway private key sk_{hg}
   Data $= \text{Enc}_{pk}(\text{info})$, MAC $= \text{MAC}_{key}(\text{Data})$, key $= pk^{sk}{}_{hg}$. Using *Upload_response* function in algorithm 6 to upload data to the smart contract.
9. The group member received response using Get_result function from Algorithm 6, then using his private key *sk* to re-compute $\text{MAC}^* = \text{MAC}_{key^*}$ (Data), where $key^* = pk^{sk}{}_{hg}$. If the result is verified and MAC* = MAC. Then decrypt the data by *Dec* algorithm using the group member private key *sk*, to get the response info corresponding to the sent request.
10. If any unusual transaction T_{xunusual} is detected, the group manager can trace the request back to the group member that sent the request. Group manager use the GTrace Algorithm 5 to obtain the member tag then seek the member list for corresponding tag.

11. After detecting a member with abnormal behavior and extracting its info, the group manager can selectively revoke the group member. Using *Add_revocation_list* function from Algo.6 to add the member info to the *revocation_list* as a transaction $TX_{revocation}$.

Algorithm 4 is used to verify the group signature by getting the candidate message with the signature pair and group public key as an input, then extracting parameters from the σ signature to verify the message. The output is either true or false.

Algorithm 4 Group signature Verify

Inputs: $(msg, \sigma), gpk$
Outputs: (True / false)
Initialization:
$(D, S, U) = gpk$
$(C1, C2, c, w) = \sigma$
Calculate:
$Q^{|} = [e(C2, S) \cdot e(w \cdot P1, P2)] / [e(c \cdot C1 + D, P2)]$
if $c = H(C1, C2, Q^{|}, H(msg))$ *then*
    return *True*
else
    return *False*

Algorithm 5 is used for Group member tracing by getting the signature and tracing key as inputs, extracting parameters from signature σ and calculating the *Tag*, searching the *Tag* in the *member list*, and returning the signer's identity as an output.

Algorithm 5 Group trace

Inputs: σ, tk
Outputs: IDi, GUi
Initialization:
$(C1, C2, c, w) = \sigma$
$u = tk$
Calculate:
$Tagi = H(C2 - u \cdot C1)$ // $Tagi = H(xi \cdot Zi)$
$IDi, GUi, = MemberList.find(Tagi)$

Algorithm 6 is used as a smart contract management function responsible for managing the requests, revoking operations, and the block creation. The algorithm consists of three groups of functions. The first group is to manage requests. The user uses the *upload request* function to upload its request task. The home getaway uses the *get request* function to get the request while waiting for the response. *Delete request* is used to clear the responded task from the request list. This function is used only by the group manager. The second group is to manage the responses. The home gateway uses the *upload response* function to send the response for the requested task. The user uses the *get result* function to get the result of the requested task from the home gateway. The third group is to manage the revocation list. *Add revocation list* function is used to add a malicious user to the revocation list by the group manager. The group manager uses the *delete revocation list* function to remove the specific user from the revocation list. Finally, the *get revocation list* function is used to *get the revocation* list and malicious users' information.

Algorithm 6 Smart Home Contract

Inputs: parameters of the required function
Outputs: the response of the revoked function
Initialization:
Request as a structure *Group_signature, result*
Group_Manager is the contract initiator. *request_list* is mapping *address* with *Request. revocation_list.*
Calculation:
Request management Functions:
Upload_request: Inputs are *address*, *Group_signature.* Store them in *request_list* to the specific *address.*
Get_request: Inputs is *address***.** output is *Group_signature.*from *request_list.*
Delete_request: Inputs is *address***.** Clear *result*, *Group_signature* from *request_list* for specific *address.*
Response management Functions:
Upload_responce: Input are *address*, *state***.** Store the *state* to *result* in the *request_list* for specific *address.*
Get_result: Input is *address***.** Output is *result* from the *request_list* for specific *address.*
Revocation list management Functions:
Add_revocation_list: input is *private_Info.* Search the *Revocation_list,* if not exist, push *private_Info to revocation_list.*
Delete_revocation_List: input is *private_Info.* Search the *Revocation_list,* if exist, remove *private_Info* from *revocation_list.*
Get_revocation_list: output is *revocation_list.*

In the proposed architecture, user devices not on the blockchain cannot authenticate. In this way, malicious devices will be blocked. However, in the first stage of the attack, if the malicious user who somehow gains access to a user device registered in the blockchain system wants to take control of this device, biometric verification will be performed on the related device. Then, the person whose biometric verification is approved by the user device will be able to make an access request from the relevant device. Thus, only the person who is confirmed to be the device owner will be able to initiate the device verification process.

The user provides the Password and Biometrics information, respectively, to the user device. The device generates biometric data (e.g., fingerprint, iris, face recognition) via the Gen function of the fuzzy extractor function. Biometric data is digested and stored by the user's phone device with the SHA-256 hashing algorithm. Also, a value is created for verification when more than one error is entered. After applying the XOR function to the biometric summary value with the person's password, the output value is encrypted and stored in the phone device (Key-Value). The purpose of person authentication is authenticating a user against the characteristics of a particular biometric data stream. The user is trying to use the phone from the interface of the relevant device and biometric authentication. Every time the entity unlocks his/her communicator, the device sensor recognizes the entity with the correct data (Biometric Hash = Biometric Hash'). The biometric data obtained with the sensor is matched with the stored mathematical model for authentication. The method is designed to protect against fraudulent unlock attempts by various techniques. With entity authentication, the first verification step of the proposed scheme, privacy-based accessibility is provided within the system before the access request. Suppose biometric verification cannot be achieved after more than one attempt. In that case, biometric data is entered initially, and the user's phone device requests the password entered during registration. A new key value is created and compared to the old generated value. If there are different values resulting from the comparison, the device will not turn on. In addition, since a password is added to the biometric authentication system, additional security is provided with a password system in certain cases, and the verification process is supported.

- If the user device, has not been used for more than 48h
- If the device has not been logged in using a password in the last 3 days and biometric data in the last 4h
- If the user phone device has received a remote lock command
- Password entry is mandatory after three unsuccessful biometric authentication attempts.

4 Security Analysis

We analyzed the security features of the proposed scheme by examining its robustness against different types of attacks and security risks. Thus, the affective aspects of our study are revealed. The security infrastructure for a smart home includes integrity, confidentiality, availability, and authentication mechanisms, respectively. Table 1 shows the most basic security requirements needed to prevent an attack by categorizing possible attack types according to security principles.

Resistance of Various Attacks:

1. *User Impersonation Attack:* Only legitimate group members can access the authorized home device using mutual authentication. So that our scheme can prevent any impersonation, in addition, any unusual behavior can be detected and revoked.

Table 1. Overview of Security Principles and Impacts on Proposed Mechanism

Security Concern	Confidentiality	Integrity	Availability	Authentication
Denial of Service			✓	✓
Replay Attack	✓	✓		✓
MITM	✓	✓		✓
Substitution Attack	✓	✓		✓
Spoofing Attack		✓		✓
Sybil Attack		✓	✓	✓

2. *DDoS Attack:* The decentralized architecture we used that is based on blockchain technique can resist the DDoS attack. Moreover, this solution is compelling and reliable with cryptocurrency, e.g., Bitcoin, to prevent this attack.
3. *Modification Attack:* Every transaction made by a group member or the home gateway must be signed with a group signature and message authentication code. So any modification of data in any transaction can be detected, and the transaction revoked.
4. *Replay Attack:* The legitimate group member generates a new key pair for every new request. This key pair is used to generate the message authentication code. In addition, the fresh key pair makes the home gateway detect any replay attack.
5. *Man-in-the-Middle Attack:* Mutual authentication used in our scheme and the other security features like a key exchange, data encryption, and message authentication code can prevent any MITM attack.
6. *Guessing password attack:* The password alone is not enough to authenticate the user. Both password and biometric information are needed. Moreover, the device is locked after a specific unsuccessful password entry.
7. *Stolen password attack:* The same reason with guessing a password, even if the attacker gains the password, it still needs the legitimate user biometric data.
8. *Stolen device attack:* Suppose that the attacker steals the user's device. The password and the biometric data are hashed and stored in an encrypted format. The attacker cannot gain this critical data even if the attacker has the password. Still not possible to authenticate without legitimate user biometric data. The scheme in [11] assumes that the device is secure and used only by its owner.

Table 2 compares the proposed and existing schemes in [11] in terms of resistance to different security attacks.

Table 2. Comparing Proposed Scheme with Existing Scheme

Resistance of Security Attacks	HomeChain [3]	Proposed Scheme
User Impersonation Attack	✓	✓
DDoS Attack	✓	✓
Modification Attack	✓	✓
Replay Attack	✓	✓
Man-in-the-Middle Attack	✓	✓
Guessing password attack	×	✓
Stolen password attack	×	✓
Stolen device attack	×	✓

5 Conclusions and Future Work

Smart home applications are quickly becoming a part of our daily lives. The main reason for the increasing demand for smart systems is to make life easy and more comfortable. To achieve these goals, more efforts have to spend to secure these systems and minimize the threats that can appear. In this paper, we proposed a secure authentication scheme that can be used to provide secure access to smart devices in the smart home environment. First, we based on biometric information and password to verify the legitimate owner of the user device. Secondly, the user device must be authenticated to allow creating access requests and remote control for the home devices. To achieve this approach, we base on decentralized architecture using a blockchain system to provide reliability and high fault tolerance. We utilized the smart contract to control the transaction and create new blocks for requests. In every smart home, there is a gateway that connects all sensors and smart devices in the home. This gateway is responsible for managing the user's access requests and forwarding feedback on the device status and the request execution. Before the user's request is forwarded to the smart device, the home gateway verifies the request by checking the group signature. If it is approved, the request is sent to the smart device. The main idea of using group signatures is to provide anonymously access to the system. Communication overhead can be reduced with the short length of the group signature and by providing a lightweight process. Furthermore, we increased the security features of the existing scheme, like device owner verification, stolen device risk, guessing, and stolen password attack, in addition to the existing security features. For these reasons, the proposed scheme is more reliable and suitable for smart home environments.

In future studies, the experimental environment of the proposed model will be created and performance evaluation will be made with traditional methods for key exchange mechanism.

References

1. Jafer Merzeh, H.R., Kara, M., Aydın, M.A., Balık, H.H.: GDPR compliance IoT authentication model for smart home environment. Intell. Autom. Soft Comput. **31**(3), 1953–1970 (2022)

2. Mahfouz, A., Mahmoud, T.M., Eldin, A.S.: A survey on behavioral biometric authentication on smartphones. J. Inf. Secur. Appl. **37**, 28–37 (2017)
3. Lin, C., et al.: HomeChain: a blockchain-based secure mutual authentication system for smart homes. IEEE Internet Things J. **7**(2), 818–829 (2019)
4. Hoofnagle, C.J., Sloot, B.V.D., Borgesius, F.Z.: The European union general data protection. Inf. Commun. Technol. Law **28**(1), 65–98 (2019)
5. Tikkinen-Piri, C., Rohunen, A., Markkula, J.: Eu general data protection regulation: changes and implications for personal data collecting companies. Comput. Law Secur. Rev. **34**(1), 134–153 (2018)
6. Loideain, N.N., Adams, R.: From alexa to siri and the GDPR: the gendering of virtual personal assistants and the role of data protection impact assessments. Comput. Law Secur. Rev. **36**, 105366 (2020)
7. Hussain, F., Hussain, R., Noye, B., Sharieh, S.: Enterprise API security and GDPR compliance: design and implementation perspective. IT Prof. **22**(5), 81–89 (2020)
8. Cui, Z., et al.: A hybrid blockchain-based identity authentication scheme for multi-WSN. IEEE Trans. Serv. Comput. **13**(2), 241–251 (2020)
9. Ma, Z., Meng, J., Wang, J., Shan, Z.: Blockchain-based decentralized authentication modeling scheme in edge and IoT environment. IEEE Internet Things J. **8**(4), 2116–2123 (2020)
10. Patwary, A.A.-N., et al.: FogAuthChain: a secure location-based authentication scheme in fog computing environments using blockchain. Comput. Commun. **162**, 212–224 (2020)
11. Zhang, S., Lee, J.-H.: A group signature and authentication scheme for blockchain-based mobile-edge computing. IEEE Internet Things J. **7**(5), 4557–4565 (2020)
12. Jiang, Y., Ge, S., Shen, X.: AAAS: an anonymous authentication scheme based on group signature in VANETs. IEEE Access **8**, 98986–98998 (2020)
13. Zhang, C., Xue, X., Feng, L., Zeng, X., Ma, J.: Group-signature and group session key combined safety message authentication protocol for VANETs. IEEE Access **7**, 178310–178320 (2019)
14. Ho, T., Yen, L., Tseng, C.: Simple-yet-efficient construction and revocation of group signatures. Int. J. Found. Comput. Sci. **26**(5), 611–624 (2015)
15. Aitzhan, N.Z., Svetinovic, D.: Security and privacy in decentralized energy trading through multi-signatures, blockchain and anonymous messaging streams. IEEE Trans. Depend. Secure Comput. **15**(5), 840–852 (2018)
16. Kara, M., Şanlıöz, Ş.G., Merzeh, H.R., Aydın, M.A., Balık, H.H.: Blockchain based mutual authentication for VoIP applications with biometric signatures. In: 2021 6th International Conference on Computer Science and Engineering (UBMK). IEEE (2021)
17. Das, A.K.: A secure and effective biometric-based user authentication scheme for wireless sensor networks using smart card and fuzzy extractor. Int. J. Commun. Syst. **30**(1), e2933 (2017)
18. Gowthami, J., Shanthi, N.: Secure Fuzzy Extractor based remote user validation scheme for Wearable devices. www.ijert.org. Accessed 31 May 2021
19. Hankerson, D., Vanstone, S., Menezes, A.: Guide to elliptic curve cryptography. Comput. Rev. **46**(1), 13 (2005)
20. Drev, M., Delak, B.: Conceptual model of privacy by design. J. Comput. Inf. Syst. **61**, 1–8 (2021)

Fuzzy Method of Creating of Thematic Catalogs of Information Resources of the Internet for Search Systems

Vagif Gasimov(✉)

Azerbaijan Technical University, Baku, Azerbaijan
vaqif.qasimov@aztu.edu.az

Abstract. The article is devoted to the research and development of methods for the automatic creation of thematic catalogs of Internet information resources for search engines. To this end, the existing approaches to solving this problem are considered and it is found out that the task still remains unresolved to the end. Based on this, the use of a fuzzy model of information search, developed earlier by the author of this article, is proposed here. The main sets and the relations between them included in the fuzzy model of information search are determined. The main sets of this model are a set of Internet information resources (documents) that need to be divided into categories by topic similar to the library and information system, a set of thematic catalogs, a set of system terms used to describe documents and thematic catalogs, a set of synonyms and associating terms. And as fuzzy relations are fuzzy relations between a set of documents (information resources) and a set of system terms, fuzzy relations between a set of thematic catalogs and a set of system terms, fuzzy relations between a set of terms and a set of synonyms, fuzzy relations between a set of queries and lots of terms. Next, the membership functions of these fuzzy relations are determined. On the basis of certain sets and fuzzy relations, methods have been developed for determining the thematic profiles of Internet information resources and distributing them into predefined thematic catalogs, as well as improving the quality of thematic catalogs.

Keywords: Search Engine · Search System · Thematic Catalog · Thematic Profiles · Fuzzy Model · Fuzzy Relations

1 Introduction

To build effective information retrieval systems, it is necessary to study the information search environment as a whole, to develop methods of indexing, thematic partitioning, search and presentation of information resources and search queries. Search engine models should include both the necessary sets (sets of documents, subject catalogs, terms, synonyms, etc.) and the relationships between these sets [1–5].

In practice, two main ways of creating search engines are used: indexes and subject directories. Research shows that search engines targeting subject directories have higher precision scores than those targeting automatic indexes. However, the completeness of

D. J. Hemanth et al. (Eds.): ICAIAME 2022, ECPSCI 7, pp. 374–382, 2023.
https://doi.org/10.1007/978-3-031-31956-3_31

automatic indexes usually far exceeds the completeness of subject directories. Automatic catalogs are more flexible and adaptable, ie. Indexer programs (robots, spiders, spiders, etc.) can periodically update the database without much difficulty.

Existing indexing methods are created on the basis of two mechanisms [6, 7, 10]:

- adding terms (key words) from the thematic rubricator according to semantic analysis;
- extraction of terms (according to the thematic rubricator) from the body of information resources.

The first approach requires the use of a powerful semantic apparatus or manual indexing by experts in the field of the indexed resource. Methods based on semantic analysis are difficult to implement and generally underdeveloped. Manual indexing is more accurate, however, creating a good thematic catalog depends on the professionalism of the staff and requires a lot of intellectual work [2, 6–10]. Traditionally, the work of creating and adapting subject catalogs is performed manually by an administrator or operator of a search engine.

Further in the article, indexing using of the second type methods is considered, i.e. when indexing, the terms are extracted from the content of information resources according to the thematic rubricator. However, the applied additional methods make it possible to assign terms from the thematic rubricator to the documents.

Indexing methods make it possible to extract the most important terms from the body of an information resource and calculate their importance coefficients. A good example of such a method is the statistical method, with the help of which a good result is achieved if at the same time a reference book of "stop words" exceptions is used, i.e. a reference book of service words, verbs, pronouns, etc. [2–5].

The weak point of this method is the following [3, 5, 8–11]:

- different authors may use different terms, the meaning of which is very close (possibly identical), for example, "information retrieval" and "information search";
- in an information resource, English, Latin or other equivalents may be used instead of terms;
- as a result of indexing, a keyword with a large weighting factor is allocated from the source, but it is not a common term in this area and is not included in the thematic rubricator, and other most significant terms in this area may turn out to be synonyms of this keyword;
- one term may associate other terms that are quite similar in subject matter.;
- the languages of sources on the same topic may be different.

Without taking into account the above mentioned circumstances, it is impossible to achieve a sufficient level of indexing, on which the search result directly depends. The way out of this situation is the use of reference books and dictionaries of synonyms, associating words, as well as translators.

These problems are much better solved in thematic catalogs. As mentioned above, thematic catalogs have disadvantages. In this sense, the study of the problem of creating automatic thematic catalogs is promising. The essence of this problem lies in automating the work of the "aggregator" of the information and library system, i.e. the distribution of documents into thematic catalogs by automatically determining the subject of the document and finding the most relevant thematic catalogs.

Further in the article the model of the information space constructed with the help of fuzzy mathematics is considered, the methods allowing to create automatic thematic catalogs without human intervention and to improve the quality of thematic catalogs with the use of synonyms and associating terms are described. For this purpose, the apparatus of fuzzy sets and relations is used, as well as the Bellman-Zadeh approach [5, 13, 14].

2 Fuzzy Model of Information Search

Here we propose a fuzzy model of information search, within which the method of creating thematic catalogs of Internet resources will be implemented. For this purpose, all the sets used in this model and the relationships between them are described below. Within the framework of the proposed model, such sets as sets of documents, sets of thematic catalogs, sets of terms and keywords, and sets of relationships between these sets are considered.

Let $D = \{d_i\}_I$ is the set of Internet information resources that need to be divided into categories by topic, similar to a library and information system. The thematic catalog is determined by a set of classification directions, i.e. thematic profiles. The thematic catalog is denoted by $C = \{C_l\}_L$ and is considered as the relation of a set of thematic profiles to a set of terms, which is a thematic rubricator of Internet information resources. Each thematic profile is defined by its own descriptors, keywords or other lexical units that are included in the thematic rubricator and are called terms. The set of terms of the thematic profile C_l, denoted by T_l^C. It should be noted that these sets may partially overlap, i.e. the same terms may be included in several sets.

The combination of the sets of terms of all the thematic catalogs of the search engine makes up the set of terms of the system $T = \{t_j\}_J$, which makes up the thematic rubricator for the search engine. As a thematic rubricator, you can use any universal bibliographic classifier, such as UDC, BBC, etc. Another approach is possible, which is based on many modern information search engines for the Internet. In this case, the thematic rubricator, i.e. a lot of terms are created by the information search engine itself or its administrator, as a database of indexes or metadata, and are supplemented in the process of the system functioning.

Let's introduce another set - a set of synonyms and associating words of terms $t_j \in T$, $j = \overline{1, J}$, which is represented as a fuzzy relation. For simplicity, we will assume that synonyms and associating terms make up one set $S = \{s_v\}_V$.

Considering the above, the information search can be represented as a five [1, 3, 4, 12]:

$$IR = \{C, D, T, Q, R\}, \quad (1)$$

where IR is the result of an information Retrieval and is given to the user in the form of a vector (list) of names, addresses and other details of information sources, Q are user requests. $R = \{R^D, R^C, R^S, R^Q\}$ - a set of relationships that defines relationships such as "information resource – term" (R^D), "thematic catalog - term" (R^C), "term - synonym"

(R^S) and "query - term" (R^Q). It should be noted that all the relationships defined here and considered further are fuzzy.

The thematic catalog is represented as a fuzzy relational matrix R^C of dimension LxJ, the rows of which correspond to thematic profiles, and the columns correspond to terms. Thus, each thematic profile is defined in the following form:

$$R^c = \{\varphi_{lj}\}_{L\times J}, \tag{2}$$

where ϕ_{lj}, $l = \overline{1, L}$, $j = \overline{1, J}$ it is a function of the relevance of the term t_j to the thematic profile C_l and receives values in the interval [0,1].

The values of the elements of the relationship matrices R^C and R^S (discussed below) are determined at the initial stage of creating an information retrieval system, for which the method of expert assessments is used, and in the process of functioning of the system, their values are adapted, i.e., trained.

Next, we denote the set of terms of the information resource d_i by T_i^D, which is a subset of the set T. The weighting coefficients of terms relative to each information resource are determined by the term's membership function to the set T_i^D, , which also receives values in the limit [0,1]. For simplicity, all terms T are included in the set T_i^D, but terms that are not important for d_i have zero weight coefficients.

Thus, the set of information resources D is represented as a fuzzy relational matrix R^D of dimension IxJ, the rows of which correspond to resources, and the columns correspond to terms:

$$R^D = \{\mu_{ij}\}_{I\times J}, \tag{3}$$

where μ_{ij} is the function of the term t_j belonging to the set of terms T_i^D, , the values of which are determined as a result of indexing an information resource in the limit [0,1].

3 Determination of Thematic Profiles of Information Resources

Now let's consider the problem of distributing Internet information resources by thematic profiles, i.e. dividing information resources into thematic catalogs. Based on the degree of relevance of terms to the information resource and the profiles of the thematic catalog, the thematic profile of the resource can be determined by the intersection of sets of relations R^C and R^D.

The task is set as follows: to find the most preferred (non-dominant) profile for the resource d_i among all the profiles of the thematic catalogs.

To solve the problem, first, an abstract (ideal) thematic profile is determined, which may be most suitable for this document. It is clear that such a profile should combine all the best aspects (relations) of the relevance of all possible profiles for this document in all terms. Next, the thematic profile among all profiles is determined, which is closest to the abstract thematic profile, taking into account all terms.

First, we will determine the degree of relevance of the information resource d_i for each profile of the thematic catalog C in relation to all terms. As noted above, in the search engine, the resource di and the thematic profile C_l are represented respectively by the relations $R_i^D = \{\mu_{i1}, \mu_{i2}, ..., \mu_{iJ}\}$, $i = \overline{1, I}$ and $R_i^C = \{\phi_{l1}, \phi_{l2}, ..., \phi_{lJ}\}$, $1 = \overline{1, L}$.

Then the relevance of the i-th information resource to the l-th thematic profile R_{il}^{DC} can be found as the intersection of fuzzy relations R_i^D and R_l^C.

As is known, the intersection of fuzzy sets is defined as the algebraic product of the corresponding elements of these sets [9], i.e. if $\eta_{ij}^l \in R_{il}^{DC}$, $1 = \overline{1, L}$, then $\eta_{ij}^l = \mu_{ij} \cdot \phi_{lj}$, $1 = \overline{1, L}$, $j = \overline{1, J}$, where η_{ij}^l is the degree of relevance of the subject of the i-th information resource to the l-th thematic profile C_l with respect to the j-th term.

Let C^{id} is an "ideal" thematic profile that combines all the best relevance ratios of all profiles to a given information resource for all its terms, then the thematic profile C^{id} can be determined by combining all fuzzy sets R_{il}^{DC}, i.e. finding maximum among all terms η_{ij}^l for all profiles:

$$\eta_{ij}^{id} = \max\left\{\eta_{ij}^1, \eta_{ij}^2, \ldots, \eta_{ij}^{id}\right\}, j = \overline{1, J}, \tag{4}$$

where η_{ij}^{id} is the degree of relevance of the i-th resource to the "ideal" thematic profile C^{id} in relation to the j-th term. Based on this, we can say that C^{id} is a relevant abstract profile for the resource d_i.

Next, from the set $\{C_l\}_L$ we find such C_l^*, the thematic profile of which is closest to the "ideal" profile of the C^{id} with respect to all terms of the information resource d_i. For this purpose, the total standard deviation λ_{il} of the relevance coefficients of all profiles C_l from the relevance coefficients of the "ideal" thematic profile of the C^{id} is calculated.

The thematic profile C_l^* with the minimum standard deviation $\lambda_i^* = \min\{\lambda_{i1}, \lambda_{i2}, \ldots, \lambda_{iL}\}$ is the best (most preferred) profile from the set $\{C_l\}_L$, i.e. his profile is the closest to the C^{id} profile and, accordingly, the most appropriate to the subject of the information resource. It follows that this resource must be included in the thematic catalog C according to the profile C_l^*.

Thus, it is possible to define one thematic profile for an information resource; in reality, it is required to determine not one, but all the closest thematic profiles, since resources can belong to several thematic profiles. So, for example, documents on medical partings can be included both in "Medical Instrumentation" and "Electronic Devices". For this purpose, δ_l, $l = \overline{1, L}$ threshold values are introduced for the degree of relevance of thematic profiles and it is required to find all thematic profiles, the standard deviation of which does not exceed a given threshold value, i.e. satisfy the condition: $\lambda_{il} \le \delta_l$, $l = \overline{1, L}$.

Let's look at a specific example below in order to visually demonstrate the proposed method and prove its adequacy. Suppose that the set of terms consists of ten terms ($J = 10$), and the thematic rubricator consists of four thematic catalogs ($L = 4$), that is:

$$T = \{t_1, t_2, t_3, t_4, t_5, t_6, t_7, t_8, t_9, t_{10}\}$$

and

$$C = \{C_1, C_2, C_3, C_4\}.$$

the relation R_l^C of thematic catalogs $C_l, l = \overline{1, 4}$ to terms $t_j, j = \overline{1, 10}$ is given by fuzzy relational table:

$$C_1(t_j/\phi_{1j}) = \{t_1/0.75;\ t_2/0;\ t_3/0.63;\ t_4/0;\ t_5/0.95;\ t_6/0.82;\ t_7/0;\ t_8/0;\ t_9/0.78;\ t_{10}/0\},$$

$$C_2(t_j/\phi_{2j}) = \{t_1/0;\ t_2/0.9;\ t_3/0.79;\ t_4/0.65;\ t_5/0.6;\ t_6/0;\ t_7/0.9;\ t_8/0;\ t_9/0.69;\ t_{10}/0.78\},$$

$$C_3(t_j/\phi_{3j}) = \{t_1/0.56;\ t_2/0.83;\ t_3/0;\ t_4/0;\ t_5/0.64;\ t_6/0;\ t_7/0.76;\ t_8/0.95;\ t_9/0;\ t_{10}/0.69\},$$

$$C_4(t_j/\phi_{4j}) = \{t_1/0,9;\ t_2/0.78;\ t_3/0.68;\ t_4/0.54;\ t_5/0.8;\ t_6/0.2;\ t_7/0.86;\ t_8/0;\ t_9/0.71;\ t_{10}/0.9\}.$$

Suppose that as a result of indexing the d_1 file by the search engine, the following degrees of importance of terms for this document were determined:

$$d_1(t_j/\omega_{1j}) = \{t_1/0;\ t_2/0.9;\ t_3/0.71;\ t_4/0.87;\ t_5/0.54;\ t_6/0;\ t_7/0.6;\ t_8/0;\ t_9/0.95;\ t_{10}/0.58\}.$$

Thus, the relationships between thematic catalogs and terms can be presented in the form of the following matrix.

$$R_l^C = \begin{pmatrix} 0.75 & 0.9 & 0.63 & 0 & 0.95 & 0.82 & 0 & 0 & 0.78 & 0 \\ 0 & 0.8 & 0.79 & 0.65 & 0.6 & 0 & 0.9 & 0 & 0.69 & 0.78 \\ 0.56 & 0.83 & 0 & 0 & 0.64 & 0 & 0.76 & 0.95 & 0 & 0.69 \\ 0 & 0.78 & 0.68 & 0.54 & 0.8 & 0.2 & 0.86 & 0 & 0.71 & 0.9 \end{pmatrix}.$$

And the description of documents in terms can be presented in the form of the following vectors.

$$R_1^D = \begin{pmatrix} 0 & 0.9 & 0.71 & 0.87 & 0.54 & 0 & 0.6 & 0 & 0.95 & 0.58 \end{pmatrix}.$$

Now let's calculate the prices of the coefficients η_{1j}^l:

$$\eta_{ij}^l = \omega_{ij} \cdot \phi_{lj},\ l = \overline{1,4},\ j = \overline{1,10}.$$

As a result, we will get the following table:

$$R_{jl}^{DC} = \begin{pmatrix} 0 & 0 & 0.45 & 0 & 0.51 & 0 & 0 & 0 & 0.74 & 0 \\ 0 & 0.81 & 0.56 & 0.57 & 0.32 & 0 & 0.54 & 0 & 0.66 & 0.45 \\ 0 & 0.75 & 0 & 0 & 0.35 & 0 & 0.46 & 0 & 0 & 0.4 \\ 0 & 0.68 & 0.48 & 0.47 & 0.43 & 0 & 0.52 & 0 & 0.67 & 0.52 \end{pmatrix}.$$

Let's calculate the prices of the weight coefficients of the abstract catalog C^{ab}:

$$\eta_{1j}^{ab} = \max_{l=\overline{1,4}}\left\{\eta_{1j}^l\right\},\ j = \overline{1,10}.$$

$$\left\{\eta_{1j}^{ab}\right\} = \left\{0\ 0.81\ 0.56\ 0.57\ 0.51\ 0\ 0.54\ 0\ 0.74\ 0.52\right\}.$$

Finally, let's calculate the standard deviation coefficients for all thematic catalogs:

$$\lambda_{11} = \frac{5}{10} \cdot \sum_{j=1}^{10}\left(\eta_{1j}^{ab} - \eta_{1j}^1\right)^2 = 0.78,$$

$$\lambda_{12} = \frac{6}{10} \cdot \sum_{j=1}^{10} \left(\eta_{1j}^{ab} - \eta_{1j}^{2} \right)^2 = 0.03,$$

$$\lambda_{13} = \frac{6}{10} \cdot \sum_{j=1}^{10} \left(\eta_{1j}^{ab} - \eta_{1j}^{3} \right)^2 = 0.74,$$

$$\lambda_{14} = \frac{7}{10} \cdot \sum_{j=1}^{10} \left(\eta_{1j}^{ab} - \eta_{1j}^{4} \right)^2 = 0.04\,.$$

Apparently, the thematic catalog, the profile of which is the least distanced from the profile of the abstract catalog, is C_2:

$$\lambda_1^* = \min_{l=\overline{1,4}} \{\lambda_{1l}\} = \min\{\lambda_{11}, \lambda_{12}, \lambda_{13}, \lambda_{14}\} = \lambda_{12} = 0.03.$$

This means that the thematic catalog C_2 is the thematic catalog closest to the thematic profile of the d_1 file, and therefore it is necessary to include the document d_1 in this thematic catalog.

4 Improving the Quality of the Thematic Catalog

In the previous section, it was noted that to improve the completeness and accuracy of the search, you can use a set of synonyms, associated words, thesauri and dictionaries, which are combined in one set $S = \{s_v\}_V = \{S_j\}_J$, where S_j is the set of synonyms and associated words of the term t_j, s_v - synonyms, associated word. For simplicity, in what follows, the set S will be called the set of synonyms. It should be noted that synonyms, in turn, are also terms, i.e. synonyms of terms of one information resource may be an important (frequently occurring) term of another resource. This means that the set of synonyms S is a subset of the set of terms T, i.e. $S \subset T$. Then, instead of the set S, you can use the set T. Therefore, instead of the relationship between the term and the synonym, you can consider the relationship between the terms, which is represented as a fuzzy relational matrix $R^S = \left\{ \nu_{jv} \right\}_{J \times J}$, where ν_{jv} is the membership function of the term t_v in the set of synonyms S_j, in other words, the degree of closeness terms t_j and t_v. If the term t_v is not synonymous with the term t_j, then $\nu_{jv} = 0$.

Synonyms and the degree of semantic proximity to their terms gives more knowledge about the subject of the information resource, which allows you to more accurately determine its thematic profile. If μ_{ij} is the coefficient of importance of the j-th term for the i-th information resource and ν_{jv} is the degree of proximity of the term t_j to the term t_v, we can calculate new coefficients of importance of the terms $\tilde{\mu}_{\text{ij}}$ for the information resource as follows:

$$\tilde{\mu}_{\text{ij}} = \max\left\{ \mu_{j1} \cdot \nu_{1j},\ \mu_{j2} \cdot \nu_{2j},\ \ldots,\ \mu_{j1} \cdot \nu_{1j}, \right\}, \text{j} = \overline{1{,}J}. \tag{5}$$

The new coefficients of the importance of terms are more improved (refined) and allow you to more accurately determine the subject of information resources. Using new values of $\tilde{\mu}_{\text{ij}}$ to solve the problem posed in the previous section allows you to achieve better results and create a more efficient thematic catalog.

Consider the following example in order to demonstrate the adequacy of the proposed method. Let us take the sets of T and C defined in the previous example, as well as and their relationships. Suppose that, in addition, a matrix of relations $R^S = \left\{ r_{jv}^T \right\}_{J \times J}$ is given:

$$R^S = \begin{pmatrix} S_1 \\ S_2 \\ S_3 \\ S_4 \\ S_5 \\ S_6 \\ S_7 \\ S_8 \\ S_9 \\ S_{10} \end{pmatrix} = \begin{pmatrix} 1 & 0.9 & 0 & 0 & 0 & 0 & 0 & 0 & 0 & 0 \\ 0.9 & 1 & 0 & 0 & 0 & 0 & 0 & 0 & 0 & 0 \\ 0 & 0 & 1 & 0.9 & 0 & 0 & 0 & 0 & 0 & 0 \\ 0 & 0 & 0.9 & 1 & 0 & 0 & 0 & 0 & 0 & 0 \\ 0 & 0 & 0 & 0 & 1 & 0 & 0 & 0 & 0 & 0 \\ 0 & 0 & 0 & 0 & 0 & 1 & 0 & 0 & 0.9 & 0 \\ 0 & 0 & 0 & 0 & 0 & 0 & 1 & 0 & 0 & 0 \\ 0 & 0 & 0 & 0 & 0 & 0 & 0 & 1 & 0 & 0 \\ 0 & 0 & 0 & 0 & 0 & 0.9 & 0 & 0 & 1 & 0.9 \\ 0 & 0 & 0 & 0 & 0 & 0 & 0 & 0 & 0.9 & 1 \end{pmatrix}$$

Then, according to the corresponding formula, we get:

$$R_1^D = \left\{ 0.81\ 0.9\ 0.78\ 0.87\ 0.54\ 0.86\ 0.6\ 0\ 0.95\ 0.86 \right\}.$$

By performing calculations using new values of the weight coefficients, we get the following values of the coefficient of the standard deviation of thematic profiles of catalogs from the thematic profile of the abstract catalog:

$$\lambda_{11} = 0.94,\ \lambda_{12} = 0.64,\ \lambda_{13} = 1.1,\ \lambda_{11},\ \lambda_{14} = 0.56.$$

It follows that, unlike the previous example, the thematic catalog C_4 is more relevant to the given document. So that:

$$\lambda_{14} = \min_{l=\overline{1,4}} \lambda_{1l} = 0.56.$$

Thus, through the multiplicity of synonyms and their relationship to terms, it is possible to more accurately determine the thematic profiles of documents and thematic catalogs. The profiles of the thematic catalaogs formed as a result will be more accurate in their content.

References

1. Gasimov, V.: Information search methods and systems. Textbook. Baku, p. 288 (2015)
2. Gasimov, V.: Search engines of library and information systems Internet. In: VI International Conference "Crimea-2000". Libraries and Associations in a Changing World, New Technologies and New Forms of Employment. Sudak. Autonomous Republic of Crimea. Ukraine. June 3–11 2000, pp. 240–244 (2000)

3. Gasimov, V.: Methods of information retrieval in Internet on the basis of fuzzy preference relations. Autom. Control Comput. Sci. Allerton Press, Inc. **37**(4), 62–67. New York (2003)
4. Gasimov, V.: Methods for construction of information retrieval systems based on a hierarchical model of the information space of the Internet. Autom. Control Comput. Sci. Allerton Press, Inc. **36**(1), 33–43. New York (2002)
5. Gasimov, V.: Simulation of information retrieval on the Internet on the basis of fuzzy knowledge. Autom. Control Comput. Sci. Allerton Press, Inc. **36**(2), 41–51. New York (2002)
6. Khramtsov, P.: Modeling and analysis of the work of information-search systems Internet. Open Systems. Moscow. № 6 (1996)
7. Gudivada, V.N.: Information search on World Wide Web. Comput. Weekly, Moscow, № 35, pp. 19–21, 26, 27 (1997)
8. Yanhong, Li.: Toward a qualitative search engine. IEEE Internet Comput. **2**(4), 24–29 July-August. 1998 (1998). (Internet: http://coputer.org/internet/
9. Okada, R., Lee, E.-S., Kinoshita, T., Shiratori, N.: A method for personalized web searching with hierarchical document clustering. Trans. Inf. Process. Soc. Japan **39**(4), 868–877 (1998)
10. Gasimov, V.: Methods of automatic creation of thematic catalogs of information resources of the internet for information and library systems. In: Proceedings of the VIII International Conference of Crimea-2001 "Libraries and Associations in a Changing World, New Technologies and New Forms of Cooperation" (Sudak, Crimea). Moscow, pp. 254–258 (2001)
11. Gasimov, V.: Development of full-text search system on information resources of Azerbaijan. In: VI International Scientific-Practical Conference "Problems of Creation, Integration and use of Scientific and Technical Information at the Present Stage". Kiev, pp. 15–16 December 16–17. 1999 (1999)
12. Gasimov, V.: Methods of information retrieval in computer networks with superfluous information resources. Monograph. Baku: Elm, p. 208 (2004)
13. Orlovskiy, S.A.: Problems of acceptance of decisions with inaccurate source information. M .: Nauka, p. 208 (1981)
14. Abbasov, A., Mamedova, M., Gasimov, V.: Fuzzy relational model for knowledge processing and decision making. Adv. Math. Res. **1**, 191–223. Nova Science Publishers Inc. New York (2002)

Scattered Destruction of a Cylindrical Isotropic Thick Pipe in an Aggressive Medium with a Complex Stress State

Sahib Piriev(✉)

Azerbaijan Technical University, H.Javid avenue 25, Baku AZ 1073, Azerbaijan
sahib.piriyev@aztu.edu.az

Abstract. The process of scattered destruction of a thick pipe with an aggressive filler, creating a uniform pressure at the inner bound of the pipe, was investigated. The results of some tests on the long-term strength of metals that are in an aggressive environment with a complex stress state show that many media lead to a several-fold decrease in the strength of a metal compared to vacuum, especially aggressive media (for example, high-sulfur fuels) lead to decrease by a factor of ten. Technical difficulties associated with conducting such tests lead to the need to elaboration of effective calculation methods to assess the impact of complex stress state and aggressive environment on the characteristics of long-term destruction of structural elements. The process of damageability is described by an integral operator of the hereditary type. The problem is solved taking into account the residual durability of the pipe material behind the destruction front.

Keywords: scattered destruction · aggressive medium · stress intensity · damageability

1 Introduction

The problem of determination of operating life of structures with defects of various nature and geometry appearing in the material in the process of loading appears to be one of the important tasks of scientific and technical progress. The problem of interaction of processes of deformation and destruction of materials with their structure and defects as primary ones arisen in the process of manufacturing as well as those appearing and developing in the process of loading is put forward as one of the main problems. Sometimes the external environment is of great importance because it exerts considerable influence on deformation and strength characteristics of the materials. Adequate description of deformation and fracture processes in such conditions for a complex stressed state of materials is the most important task of modern mechanics. Closely related to this range of problems is the problem of deformation and strength of polymeric and composite materials, which is important for rational design of structural elements, machines and aircraft. Many works are devoted to the problem of mathematical modelling of long-term strength of metals staying in aggressive environment, however the problem of analysis

D. J. Hemanth et al. (Eds.): ICAIAME 2022, ECPSCI 7, pp. 383–395, 2023.
https://doi.org/10.1007/978-3-031-31956-3_32

of influence of aggressive environment on time of structural elements failure is still very actual. The difficulty of studying influence of aggressive environment on the long-term strength of metals relates to the lack of experimental research. Tests on long-term strength of metal samples are mostly conducted in the air medium under uniaxial tension, because high-temperature tests in flat tension state are connected with considerable technical difficulties. Results of few long-term endurance tests of metals in aggressive environment under plane stress state show that many environments result in reduction of fracture time in comparison with vacuum by several times and especially aggressive environments (e.g., high sulphur fuels) - by tens of times. Technical difficulties of such tests lead to the necessity of development of effective calculation methods for estimating the influence of plane stressed state and aggressive medium on the characteristics of long-term failure of structural elements. One of the ways to investigate this issue is the structural-phenomenological approach. This approach has been demonstrated in [1–4].

In this paper, using the results of [1], the problem of determining the long-term durability of a damaged cylindrical pipe internally reinforced by a coaxial cylindrical layer of active material is solved. In the model used, the influence of the active medium is associated with the penetration of the components of the medium into the body due to the diffusion process.

A quantitative measure of the degree of the presence of a substance in the medium in a body is the concentration of the components of this substance in it.

2 Related Works

As an equation characterizing the distribution of the concentration of an aggressive medium in the body, the diffusion equation [5] is adopted:

$$\frac{\partial C}{\partial t} = div\,(D\,grad\;C) \tag{1}$$

with a zero initial value of the concentration C of the aggressive substance in the body and the boundary condition

$$C(\vec{r}, t)|_{\vec{r} \in S} = 1 \tag{2}$$

where $\vec{r}$ is the vector coordinate of the body point, S is the surface of the body, C is the concentration of components of the aggressive medium at a given point of the body, referred to its value on the surface of the body bound.

As in [1], we will assume that the properties of structural elements depend on the presence of environmental components in the body, which manifests itself in decreasing the short-term durability limit of the body's main material.

At a certain level of loading of the body, it begins to destruct gradually. In this regard, the external load is redistributed between the remaining undamaged parts of the body. The bound of the expanding destructed region of the body represents the destruction front, the propagation velocity of which determines the longevity or long-term durability.

The level of the stressed state of the structural element is characterized by an equivalent stress σ_E, which in this work is taken as the stress intensity σ_i. As a condition for the destruction of a structural element that is in contact with an aggressive medium, we will accept the condition for reaching the limit of short-term durability σ_T by this stress in the presence of a medium:

$$\sigma_E = \sigma_T \tag{3}$$

However, due to damage to the material of the body, the destruction will occur at a lower load. According to the hereditary theory of damageability [6], the criterion of destruction will look like this:

$$\left(1 + M^*\right)\sigma_E = \sigma_T \tag{4}$$

Where M^* is the integral operator of damageability. The limit of short-term durability σ_T is a function of the concentration of an aggressive substance in the body. In this paper, a linear approximation of this dependence is adopted:

$$\sigma_T(C) = \sigma_{T_0}(1 - \gamma\, C) \tag{5}$$

where $0 < \gamma < 1$ is the empirical constant.

The focal point of destruction occurs at a certain point in time t_0, , called the incubation period, at the point of the body where condition (4) is first carried out, as a result of which a destruction front arises and begins to propagate in the body. The body is destroyed if the velocity of the front of destruction turns to infinity, or when the destructive part covers the whole body.

A more precise approach to this problem is based on taking into account the presence of residual durability behind the destruction front, when the material of the body behind the destruction front retains to some extent the bearing capacity.

In this paper, this approach is implemented in the following way: it is assumed that if the condition (4) is satisfied, the material loses its ability to accumulate damage, it instantly qualitatively rearranges the structure, so that its behavior can be described by an ideal-elastic body model, but with sharply reduced values of the rigidity index of the Young's modulus of elasticity.

It is assumed that all the conditions ensuring the state of plane deformation are satisfied. Equations of state are described by physical relationships for an isotropic elastic-damaging medium [7]:

$$Э_{ij} = \frac{1}{2G}\left(1 + M^*\right)s_{ij}, \quad \varepsilon = \frac{1}{3K}\sigma \tag{6}$$

where ε, σ are the spherical parts, $Э_{ij}$ and s_{ij} are the deviators of the strain and stress tensors, M^* is integral heredity-type operator of damageability:

$$M^* s_{ij} = \sum_{i=1}^{n} \Phi\left(\sigma_i(t_k^+)\right) \int\limits_{t_k^-}^{t_k^+} M(t_k^+ - \tau)d\tau + \int\limits_{t_{n+1}^-}^{t} M(t - \tau)s_{ij}(\tau)d\tau \tag{7}$$

where $(t_k^-,\ t_k^+)$ are values of damage rank, $\Phi\left(\sigma_i(t_k^+)\right)$ is a function of healing defects, σ_i are stress intensity. For a monotonic, time-varying, stressed state, the operator of damageability (7) goes over into the usual hereditary elasticity operator:

$$M^* s_{ij} = \int_0^t M(t-\tau) s_{ij}(\tau) d\tau \tag{8}$$

and then the deformation relations (6) are identical to the corresponding physical relationships of the hereditary theory of elasticity, the stresses can be determined on the basis of the Volterra-Rabotnov correspondence principle and, in particular, when only the forces are specified on the bounds, the stresses are determined from the corresponding elasticity theory formulas [8].

Suppose that for the problem under consideration the conditions ensuring the state of plane deformation are satisfied. Then the construction is sufficiently represented by its transversal section, which is a concentric ring of outer radius b and inner radius a. At the bound $r = a$, a uniformly distributed pressure p is given (Fig. 1).

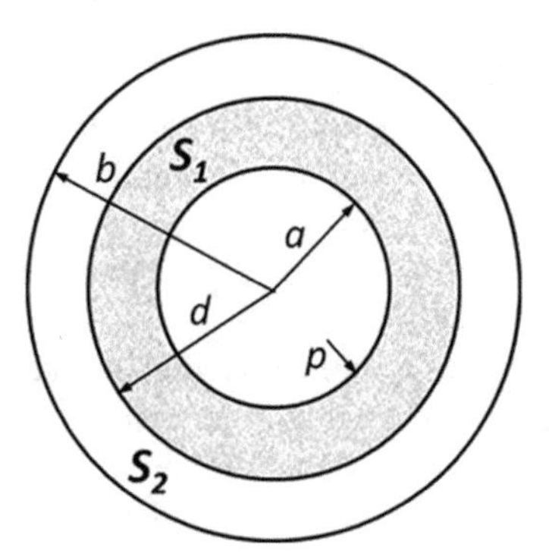

Fig. 1. Transversal section of the construction.

The work [9] is devoted to the study of this problem until the appearance of foci of destruction. In this paper, we investigate the further process of destruction associated with the motion of the destruction front. In this case, the outer annular region of the body is divided into two annular regions S_1 and S_2, where S_2 is the region of the not destroyed part of the body, S_1 is the region of the body behind the destruction front with power durability and stiffness characteristics.

Then the stresses in the not destroyed region have the form [10]:

$$\begin{aligned} \sigma_r &= \frac{qd^2}{b^2 - d^2}\left(1 - \frac{b^2}{r^2}\right); \\ \sigma_\theta &= \frac{qd^2}{b^2 - d^2}\left(1 + \frac{b^2}{r^2}\right); \end{aligned} \tag{9}$$

where q is the radial pressure at the bound between the destroyed and not destroyed zones. The axial normal stresses σ_z for planar deformation are determined as follows:

$$\sigma_z = \nu(\sigma_r + \sigma_\theta) \tag{10}$$

For simplicity, the pipe material is considered incompressible, i.e. Poisson's index is equal to $\nu = 0, 5$. Then formula (10) takes this form

$$\sigma_z = 0{,}5(\sigma_r + \sigma_\theta) \tag{11}$$

The intensity of stresses or the resulting stress is determined by the formula [11]:

$$\sigma_i = \frac{1}{\sqrt{2}}\sqrt{(\sigma_r - \sigma_\theta)^2 + (\sigma_r - \sigma_z)^2(\sigma_\theta - \sigma_z)^2} \tag{12}$$

Substituting the values of (11) into (12), we find

$$\sigma_i = \pm\frac{\sqrt{3}}{2}(\sigma_r - \sigma_\theta) \tag{13}$$

A simple analysis of the stress formulas (6) shows that the ring stress σ_θ is greatest. It reaches its maximum value on the inner contour of the pipe. Therefore, first the destruction will take place there, occupying in the future more and more new layers. Then

$$\sigma_{i\,\max} = \sigma_i|_{r=a} = \sqrt{3}\frac{pb^2}{b^2 - a^2} \tag{14}$$

Substituting this representation in the destruction criterion (4), we obtain an algebraic equation for determining the initial time of destruction of the inner boundary surface of the pipe [12, 13]:

$$(1 + M^*)p = \frac{\sigma_{T_0}}{\sqrt{3}}\frac{b^2 - a^2}{b^2} \tag{15}$$

or taking into account the constancy of the pressure p and the type of operator of damageability M*:

$$\int_0^{t_0} M(\tau)d\tau = \left(\frac{\sigma_{T_0}}{\sqrt{3}}\frac{b^2 - a^2}{b^2} - 1\right) \tag{16}$$

We give an explicit form for the initial destruction time for three kinds of kernels $M(t)$:

$$M(t) = m; \quad t_0 = \frac{1}{m}\left(\frac{\sigma_{T_0}}{\sqrt{3}}\frac{b^2 - a^2}{b^2} - 1\right) \tag{17}$$

$$M(t) = me^{-\alpha t};$$

$$t_0 = \frac{1}{\alpha}\ln\left\{1 + \frac{\alpha}{m}\left(1 - \frac{\alpha_{T_0}}{\sqrt{3}}\frac{b^2 - a^2}{b^2}\right)\right\}^{-1} \tag{18}$$

$$M(t) = me^{-\alpha}; 0 < \alpha < 1;$$

$$t_0 = \left\{\frac{1-\alpha}{m}\left(\frac{\alpha_{T_0}}{\sqrt{3}}\frac{b^2 - a^2}{b^2} - 1\right)\right\}^{\frac{1}{1-\alpha}} \tag{19}$$

According to (17)–(19), if the kernel is singular or constant, even the smallest pressure will still lead to destruction, although time t_0 increases with decreasing of p. In the case of a regular exponential kernel, there is always a lower bound:

$$p_* = \frac{\alpha}{\alpha + m} \frac{\sigma_{T_0}}{\sqrt{3}} \frac{b^2 - a^2}{b^2} \tag{20}$$

When $p < p_*$ destruction does not occur. This is explained by the upper limited possible volume of accumulated damageability for the exponential kernel of damageability.

However, the upper limit on the value of the internal pressure holds for all three kinds of kernel, i.e. there exists such p_{**}, that when $p \geq p_{**}$, the destruction occurs instantly on the application of the load. At this p_{**}, the same for all three types of kernels we have:

$$p_{**} = \frac{\sigma_{T_0}}{\sqrt{3}} \frac{b^2 - a^2}{b^2} \tag{21}$$

Substituting formula (13) into formula (4), we obtain:

$$\sigma_\theta - \sigma_r = \frac{2}{\sqrt{3}}(1 - M^*)\sigma_T \tag{22}$$

The equation of equilibrium can be written in the form [5]:

$$\frac{d\sigma_r}{dr} - \frac{\sigma_r - \sigma_\theta}{r} = 0 \tag{23}$$

Substituting formula (22) into formula (23), we obtain the stress component in the destroyed zone:

$$\begin{aligned} \sigma_r &= -p + \frac{2}{\sqrt{3}}(1 - M^*)\sigma_T(\tau)\ln\frac{r(\tau)}{a}; \\ \sigma_\theta &= -p + \frac{2}{\sqrt{3}}(1 - M^*)\sigma_T(\tau)\left(1 + \ln\frac{r(\tau)}{a}\right) \end{aligned} \tag{24}$$

From the condition of continuity of radial and tangential stresses at the destruction front $r = d$ using formulas (9)–(24), we obtain the following two equations:

$$\begin{cases} -\frac{qd^2(t)}{b^2 - d^2(t)}\left(\frac{b^2}{d^2(t)} - 1 =\right) - p + \frac{2}{\sqrt{3}}(1 - M^*)\sigma_T(\tau)\ln\frac{r(\tau)}{a}; \\ \frac{qd^2(t)}{b^2 - d^2(t)}\left(\frac{b^2}{d^2(t)} + 1\right) = -p + \frac{2}{\sqrt{3}}(1 - M^*)\sigma_T(\tau)\left(1 + \ln\frac{r(\tau)}{a}\right) \end{cases} \tag{25}$$

From the first equation of (25) we find

$$q = p - \frac{2}{\sqrt{3}}\left(\sigma_T(t)\ln\frac{d(\tau)}{a} - \int_0^t M(t - \tau)\sigma_T(\tau)\ln\frac{d(\tau)}{a}d\tau\right) \tag{26}$$

and substituting in the second Eq. (25), we obtain

$$\sqrt{3}p\left(\frac{b^2}{b^2-d^2(t)}+\int\limits_0^t M(t-\tau)\frac{b^2}{b^2-d^2(\tau)}d\tau\right)=\sigma_T(t)\left(2\ln\frac{d(\tau)}{a}\cdot\frac{b^2}{b^2-d^2(\tau)}+1\right) \tag{27}$$

It should be noted that in these equations the function $d(t)$ has the following structure:

$$d(t)=\begin{cases} a; & t\le t_0\\ d(t); & t>t_0\end{cases}$$

In the problem under consideration, for the concentration of the active substance in the pipe, the boundary conditions are taken in the form:

$$C(a;t)=1,\ \ C(b;t)=0 \tag{28}$$

We assume that the quasi-stationary distribution of the concentration of the active substance of the type takes place at the pipe thickness:

$$C=\frac{b-d(t)}{b-a} \tag{29}$$

The process of scattered destruction is investigated according to the scheme of L.M. Kachanov [14]. Destruction, starting at the inner bound of the pipe, where the intensity of stresses is maximum, develops to the outside. To determine the law of motion of the destruction front without considering the residual durability, we introduce the following dimensionless quantities

$$\frac{a}{b}=\beta_0\ ;\ \frac{r}{b}=\beta(t)\ ;0<\tau<t,\ \text{ and}$$

$$\frac{\sigma_{T_0}}{\sqrt{3}p}=g;\ M(t-\tau)=mK(t-\tau);\ \ m\tau=\varsigma;\ mt=s$$

Then, taking into account (29), from (27) we obtain the following nonlinear integral equation with respect to the dimensionless radius $\beta(t)$ of the destruction front:

$$\left(\frac{1}{1-\beta^2(s)}+\int\limits_0^t K(s-\varsigma)\frac{1}{1-\beta^2(\varsigma)}d\varsigma\right)=g\left(1-\gamma\cdot\frac{1-\beta(s)}{1-\beta_0}\right)\left(\frac{2\ln[\beta_0\cdot\beta(s)]+1}{1-\beta^2(s)}\right) \tag{30}$$

Then the solution of Eq. (30) with account of (26) is valid until the moment of detachment, that is, if the condition $q(t)>0$ is satisfied.

So, there are two possible ways of destruction the pipe: 1) because of detachment - the condition $q=0$ is carried out; 2) due to scattered destruction - when the front of destruction attains the outer bound $d=b$ and the condition $q(t)>0$ is always satisfied.

3 Method and Material

3.1 Diffusion in the Presence of a Chemical Reaction

In the presence of a chemical reaction of the first order (radioactive decay, constant confinement), the diffusion equation takes the form [15–17]:

$$\frac{\partial C}{\partial t} = D\frac{\partial^2 C}{\partial r^2} - kC \tag{31}$$

where k is the constant of an irreversible chemical reaction of the first order ($\sec^{-1}$).

In the stationary state, the condition $\partial C/\partial t = 0$ is carried out and equation

$$D\frac{\partial^2 C}{\partial r^2} - kC = 0 \tag{32}$$

The solution of Eq. (32) in the boundary conditions (28) is as follows:

$$C = \frac{e^{f(2+\beta_0-\beta)} - e^{f(\beta_0+\beta)}}{e^{2f} - e^{2f\beta_0}} \tag{33}$$

Here, introducing the following notation: $\zeta = \sqrt{\frac{k}{D}}, f = \zeta \cdot b$

and using (33) in (30), we obtain the following equation:

$$\left(\frac{1}{1-\beta^2(s)} + \int_0^t K(n-\varsigma)\frac{1}{1-\beta^2(\varsigma)}d\varsigma\right) = g\left(1 - \gamma \cdot \frac{e^{f(2+\beta_0-\beta)} - e^{f(\beta_0+\beta)}}{e^{2f} - e^{2f\beta_0}}\right)\left(\frac{2\ln[\beta_0 \cdot \beta(s)] + 1}{1-\beta^2(s)}\right). \tag{34}$$

generalization.

3.2 Numerical Method

For more complex types of kernels, a numerical method for solving the non-linear integral Eq. (30) is applied. We introduce the notation:

$$\begin{cases} A(s) = \left(1 - \gamma \cdot \frac{1-\beta(s)}{1-\beta_0}\right)\left(\frac{2\ln[\beta_0 \cdot \beta(s)]+1}{1-\beta^2(s)}\right) \\ B(s) = \frac{1}{1-\beta^2(s)};\ T(\varsigma) = \frac{1}{1-\beta^2(\varsigma)} \\ f(\beta(s)) = \frac{B(s)}{A(s)};\ \varphi(\beta(s), \beta(\varsigma)) = \frac{T(\varsigma)}{A(s)} \end{cases} \tag{35}$$

Then the integral Eq. (30) is replaced by:

$$f(\beta(s)) + \int_0^t M(s-\varsigma)\phi(\beta(s),\beta(\varsigma)) = g \tag{36}$$

The numerical solution (36) is based on replacing it with a discrete analog:

$$f(\beta_n) + h\sum_{i=1}^{n-1} M(s_n-\varsigma_i)\varphi(\beta_n,\beta_i) = G \tag{37}$$

where ς_i are the nodal points of the time grid: $\varsigma_i = ih$.

At each step, Eq. (37) is a nonlinear algebraic equation with respect to the coordinate of the destruction front β_n. The following iterative process is used for its solution:

$$\beta_n^{(k)} = \psi\left(\beta_n^{(k-1)}\right),$$

where.

$$\psi(\beta_n) = \beta_n + \mu\left[f(\beta_n) - h\sum_{i=1}^{n-1} M(s_n-\varsigma_i)\varphi(\beta_n,\beta_i) - g\right].$$

Here the parameter μ that ensures the convergence of the iterative process is selected in the course of a numerical experiment.

Numerical implementation was carried out for three types of kernels of the operator of damageability: singular $M(t) = mt^{-\alpha},\ \ 0 < \alpha < 1,\ M(t) = me^{-\alpha t}$ and constant $M(t) = m$ for the initial relative width of the pipe $\beta_0 = 0,5$.

Figures 2, 3, 4, 5, 6, 7, 8, 9 and 10 show the destruction front curves based on numerical calculation data.

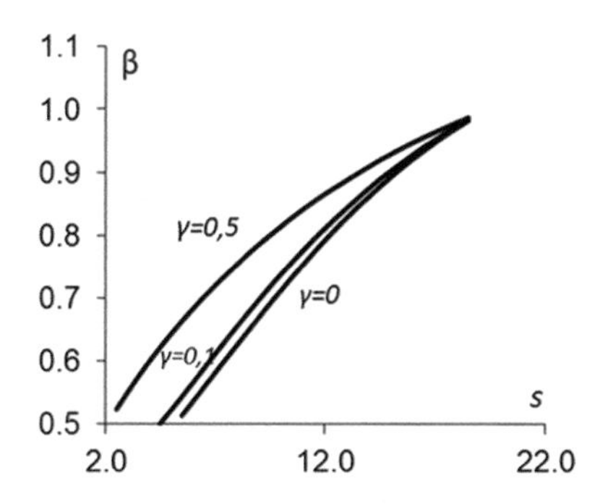

Fig. 2. Curves of the destruction front for the kernel $M = m = const$ (Eq. (30)).

As follows from the graphs, the motion of the destruction front occurs at a decreasing velocity. Calculations also showed that the presence of residual durability behind the destruction front has little effect on the character of the destruction front motion, but it strongly affects the time of onset of the detachment.

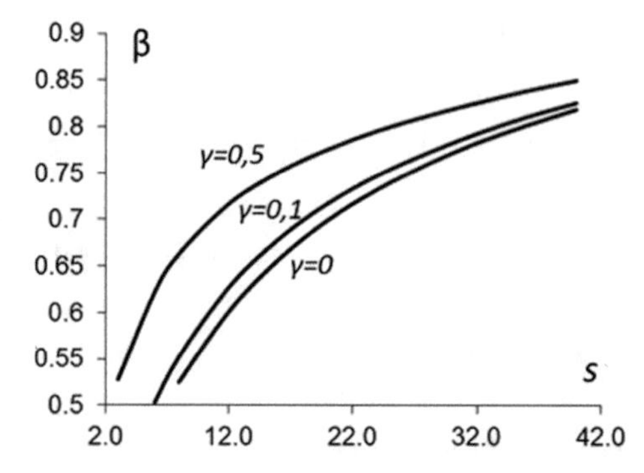

Fig. 3. Curves of the destruction front for the kernel $M = e^{-\alpha s}$ for $\alpha = 0.1$ (Eq. (30)).

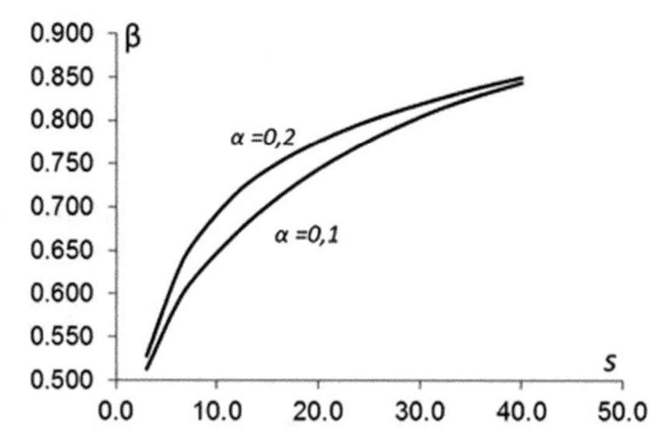

Fig. 4. Curves of the destruction front for the kernel $M = e^{-\alpha s}$ for $\gamma = 0.5$ (Eq. (30)).

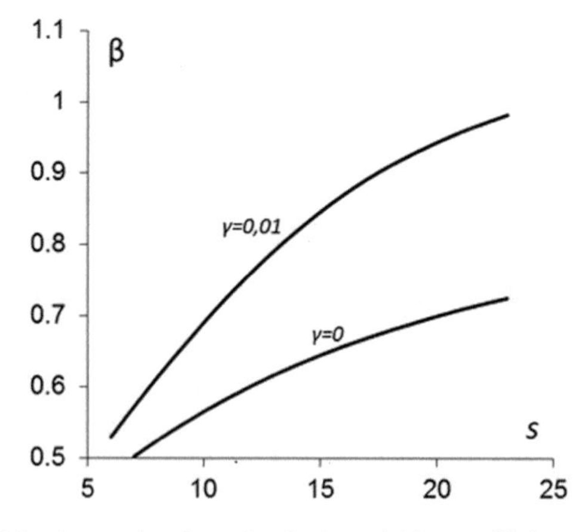

Fig. 5. Curves of the destruction front for the kernel $M = e^{-\alpha s}$ for $\alpha = 0.01$ (Eq. (34)).

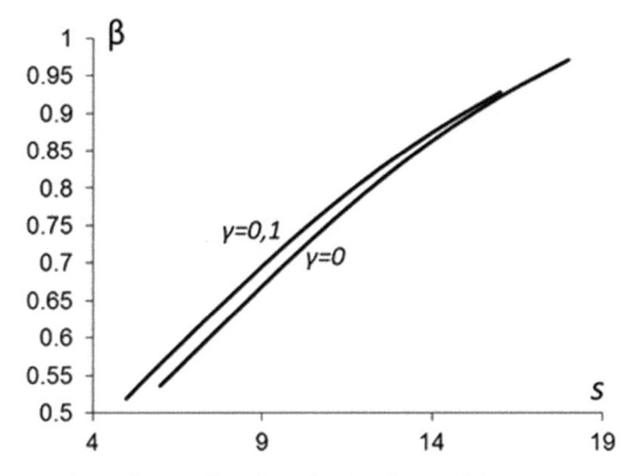

Fig. 6. Curves of the destruction front for the kernel $M = m = const$ (Eq. (34)).

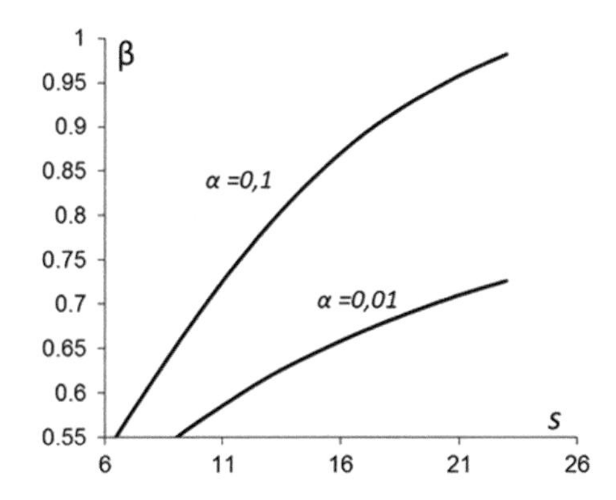

Fig. 7. Curves of the destruction front for the kernel $M = e^{-\alpha s}$ for $\gamma = 0.01$ (Eq. (34)).

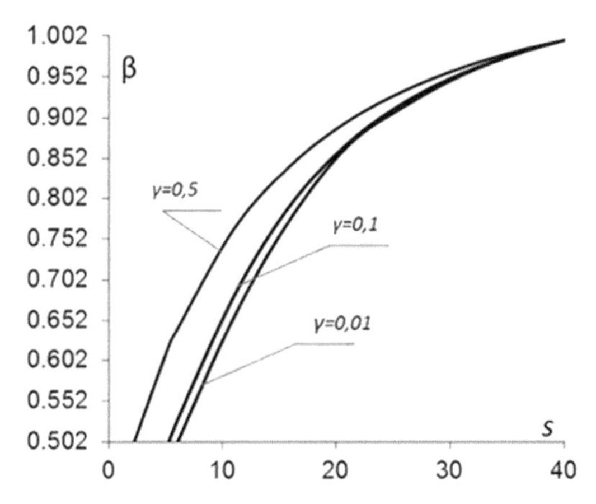

Fig. 8. Curves of the destruction front for the kernel $M = mt^{-\alpha}, s = m^{\frac{1}{1-\alpha}}t$ for $\alpha = 0.2$ (Eq. (30)).

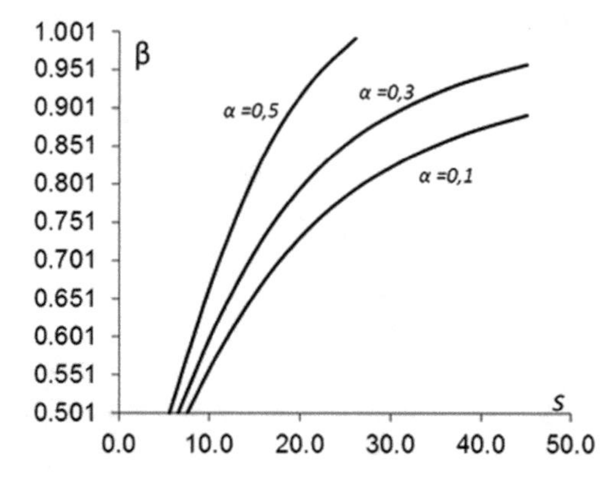

Fig. 9. Curves of the destruction front for the kernel $M = mt^{-\alpha}, s = m^{\frac{1}{1-\alpha}}t$ for $\gamma = 0.01$ (Eq. (30)).

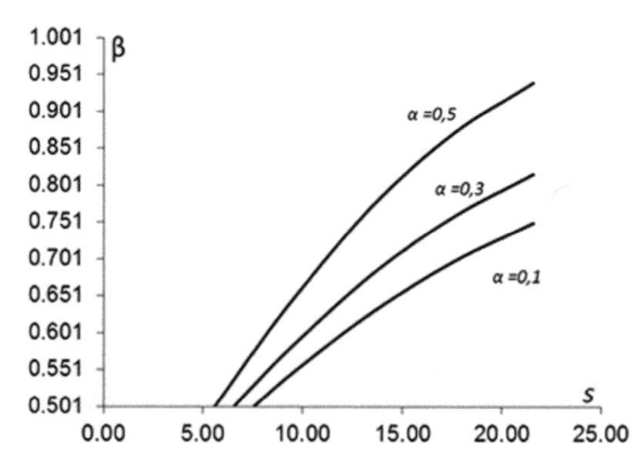

Fig. 10. Curves of the destruction front for the kernel $M = mt^{-\alpha}, s = m^{\frac{1}{1-\alpha}}t$ for $\gamma = 0.01$ (Eq. (34)).

4 Discussion and Results

The integral equation is derived with respect to the radial coordinate of the destruction front, taking into account the diffusion processes on the contact surface of the pipe with active filler, as well as the process of damageability the material of the pipe itself.

Explicit integral formulas are obtained for contact pressures at the destruction front of the surface of adhesion of the pipe to the active substance. The analysis of the relationship between the critical situations of detachment on the contact surface of a pipe with filler, as well as on the destruction front, and analysis of destruction due to the accumulation of a critical volume of damage is given.

References

1. Akhundov, M.B., Sadayev, A.: Dispersed failure of an anisotropic flywheel. Mech. Compos. Mater. **41**(1), 65–76 (2005). https://doi.org/10.1007/s11029-005-0008-x

2. Akhundov, M.B.: Diffuse destruction of a thick pipe under the action of alternating external pressure. In: Proceedings of the III All-Russian conference on the Theory of Elasticity with International Participation, Rostov, Russian Federation, pp. 53–55, (2004). (in Russian)
3. Akhundov, M.B., Sh, S.A.: Long destruction of structurally damaged composite pipe. Int. J. Cheminform. Chem. Eng. **3**(1), 37–47, (2013). https://doi.org/10.4018/ijcce.2013010104
4. Akhundov, M.B., Sadayev, A., Ayvazov, A.A.: An approach to solving the one-dimensional problem on compression of a viscoelastic layer dispersedly reinforced with elastic inclusions. Mech. Compos. Mater. **45**(4), 303–314 (2009). https://doi.org/10.1007/s11029-009-9090-9
5. Eremeev, V.S.: Diffusion and Stress, p. 182. Energoatomizdat, Moscow, (1984). (in Russian)
6. Kachanov, L.M.: Fundamentals of Fracture Mechanics, p. 312. Nauka, Moscow, (1974). (in Russian)
7. Kulagin, D.A., Lokoshenko, A.M.: Modeling the influence of an aggressive environment on the creep and long-term strength of metals in a complex stressed state. Mech. Solid, RAS, no. 1, pp. 188–199 (2004). (in Russian)
8. Levanov, A.V.: Macrokinetic problems in the general course of physical chemistry. In: Department of Operational Printing and Information, Faculty of Chemistry, p. 34. Moscow State University, Moscow (2016). (in Russian)
9. Lokoshenko, A.M.: Description of the long-term strength of metals using a probabilistic model. Bull. Engine Build. (Zaporozhye), no. 3, 102–105 (2008). (in Russian)
10. de Groot, A.C., Toonstra, J.: 10. In: Casuïstiek in de Dermatologie Deel I, pp. 35–36. Bohn Stafleu van Loghum, Houten (2009). https://doi.org/10.1007/978-90-313-7627-8_10
11. Lokoshenko, A.M.: Modeling of the Process of Creep and Long-Term Strength of Metals, p. 264. Publishing House of the Moscow State Industrial University, Moscow (2007). (in Russian)
12. Lokoshenko, A.M.: Statistical analysis of experimental data on the long-term strength of metals under a complex stress state. Aviat. Space Eng. Technol. 12, 122–126, (2009). (in Russian).
13. Piriev, S.A.: Long-term strength of a thick-walled pipe filled with an aggressive medium, with account for damage to the pipe material and residual strength. J. Appl. Mech. Tech. Phys. **59**(1), 163–167 (2018). https://doi.org/10.1134/s0021894418010200
14. Pisarenko, G.S., Mozharovsky, N.S.: Equations and boundary value problems of the theory of plasticity and creep, p. 492. Naukova Dumka, Kiev, (1981). (in Russian)
15. Rabotnov, V., Yu, N.: Elements of Hereditary Mechanics of Solids, p. 421. Nauka, Moscow (1977). (in Russian)
16. Vorobiev, A.K.: Diffusion Problems in Chemical Kinetics, Tutorial, p. 98. Moscow University Press, Moscow (2003). (in Russian)
17. Yu., V.: Suvorova nonlinear effects during deformation of hereditary media. Mech. Polymers 6, 976–980 (1977). (in Russian)
18. AlHajri, M.I., Alsindi, N., Ali, N.T., Shubair, R.M.: Classification of indoor environments based on spatial correlation of RF channel fingerprints. In: 2016 IEEE International Symposium on Antennas and Propagation (APSURSI), pp. 1447–1448. IEEE (2016)
19. AlHajri, M.I., Ali, N.T., Shubair, R.M.: Classification of indoor environments for IoT applications: a machine learning approach. IEEE Antennas Wirel. Propag. Lett. **17**(12), 2164–2168 (2018)
20. Kelleci, B.: VHDL ve Verilog ile Sayısal Tasarım, 1st edn. Seçkin Publishing, Turkey (2017)

Estimating the Resonance Frequency of Square Ring Frequency Selective Surfaces by Using ANN

Mehmet Yerlikaya and Hüseyin Duysak(✉)

Department of Electrical and Electronics Engineering, Karamanoglu Mehmetbey University, Karaman, Turkey
{myerlikaya,huseyinduysak}@kmu.edu.tr

Abstract. Frequency selective surfaces (FSSs) are periodic structures composed of symmetrical unit cells on dielectric substrates and with reflective or absorption properties. For this reason, FSSs are used in modern wireless communication elements such as antennas, filters and radomes at microwave and millimeter wave frequencies. Aperture (or ring) shaped FSSs act as a band-pass filter, while patch-shaped FSSs act as band-stop filters. In this paper, artificial neural network (ANN) analysis is performed to estimate the resonance frequency of square ring FSSs, which are especially used to increase the antenna gain. To start with, design parameters such as the dielectric constant of the substrate (ε_r), the substrate thickness (h_s), the substrate width (w_s), the inner width (w_i) and the outer width (w_o) of the square conductive ring are determined as ANN data inputs. Then, the resonant frequency (f_r) obtained by simulation for each combination of these input design parameters is set as the output data. The data set, which includes 108 simulation results in total, is divided into 81 training and 27 test data. ANN is trained with Levenberg Marquardt learning algorithm. According to the test results, the ANN can estimate the resonant frequency of the square ring FSS with high accuracy (mean absolute percentage error = 0.47 and mean squared error = 0.0072).

Keywords: Artificial neural network (ANN) · frequency selective surface (FSS) · resonant frequency · Levenberg-Marquardt learning algorithm · HFSS

1 Introduction

Frequency selective surfaces (FSS) are periodic structures consisting of conductive patches or cavities on an insulating substrate. FSSs are structures that are generally designed to reflect, transmit or absorb EM waves and whose characteristics change depending on the frequency of the incoming electromagnetic (EM) wave. The frequency response of the FSS varies depending on the material structure, shape and design of the unit element forming the FSS. When the frequency of the plane wave incident at any angle to the FSS surface and the resonance frequency of the FSS elements match, the incident wave will be completely or partially transmitted to the other side of the surface or reflected back from the surface [1]. FSSs can have band-stop, band-pass, low-pass and

D. J. Hemanth et al. (Eds.): ICAIAME 2022, ECPSCI 7, pp. 396–406, 2023.
https://doi.org/10.1007/978-3-031-31956-3_33

high-pass resonance characteristics depending on the shape of the performance. FSSs are widely used in the literature as radomes [2–4], absorbers [5–7], reflectors [8], and filters [9, 10]. Besides, FSS can also widely used in printed antennas as a metamaterial or superstrate to improve antenna gain or control the antenna beam [11–13].

In recent years, the use of artificial intelligence (AI) techniques has been mentioned in the solution or optimization of electromagnetic problems, as in many areas. Some of the AI techniques that are frequently used in engineering problems are as follows; deep learning, machine learning, fuzzy logic and neural networks [14].

Artificial neural network (ANN), which is one of the most commonly preferred techniques, has been used successfully in the field of EM modelling or inverse EM problems [15–20]. The study in [15] is one of the first applications of ANN in the field of EM. In the presented work, a method is presented that adapts the input data to search for traces of successive objects through a recognition step using a backpropagation ANN network. In [16] both the ANN and Extreme Learning Machines (ELM) models are used for to determining the operating frequency of the C-shaped notched antenna. In the study, ANN and ELM were compared and it was seen that the mean absolute error (MAE) value of the ANN model was lower than the ELM model. The analysis of experimental models and ANN model for path loss estimation is investigated in [17]. ANN based model gave better results compared to basic path loss models. In another study, an ANN-based synthesis method for nonuniform linear arrays with mutual coupling effects was presented to the literature [18]. In the proposed work, the positions and feeds of the array elements are optimized simultaneously.

In this paper, the ANN model and analysis has been presented to estimate the resonant frequency of the square ring FSS. The resonant frequencies of the FSSs have been derived by parametric analysis. According to parametric analysis, 108 data are obtained from simulated results made by Ansoft HFSS software. From 108 data sets, 86 datasets are used to train the ANN model and the remaining 22 are used for the testing process.

2 Square Ring FSS Structure

The proposed FSS unit cell consists of a square ring conductive patch placed on a one side of a dielectric substrate with dielectric constant of ε_r, thickness of h_s and width of w_s (Fig. 1). In Fig. 1, $\mathrm{w_p}$ and w_a are the outer width of the conductive patch and width of the annular ring, respectively. A total of 108 different FSS structures were simulated according to the input parameters (ε_r, h_s, w_s, w_p and w_a) given in Table 1.

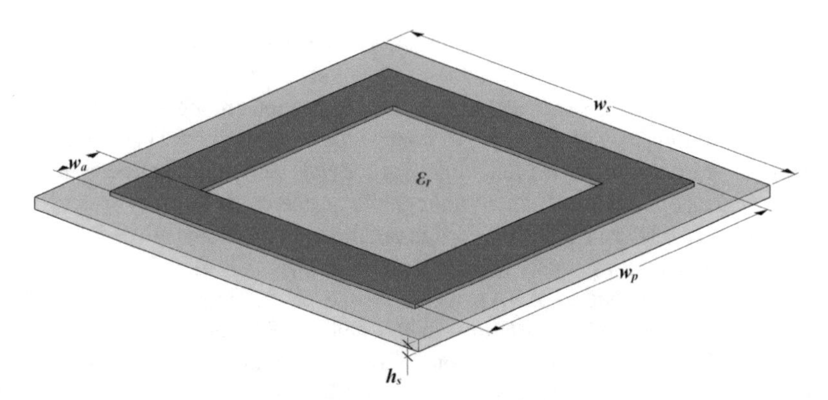

Fig. 1. 3D geometry of the square ring FSS

Table 1. Input parameters of the simulated FSSs

w_s	w_p	w_a	h_s	ε_r
8 mm	4 mm 6 mm 7.2 mm	0.8 mm 1.6 mm 3.2 mm	0.8 mm 1.6 mm	2.33 4.3 6.15
10 mm	5 mm 7.5 mm 9 mm	1 mm 2 mm 4 mm		

The simulations were performed by Ansoft HFSS software, which is the commercial electromagnetic design software of ANSYS. The values of 2.33, 4.3 and 6.15 were chosen as the dielectric substrate coefficients, which belong to the commercially available Rogers RT5870, FR-4 and Rogers RO3006 materials. Similarly, 0.8 mm and 1.6 mm values, which can be found common for all three substrates, were used as dielectric thickness. In the simulations made in the study, the transmission coefficient S_{21} as a function of frequency for both TE and TM polarized waves was examined and resonance points below −20 dB were determined. According to the simulations, the resonance frequency for a total of 108 FSS ranges from 4.9 GHz to 29.05 GHz.

3 Method and Material

3.1 Method

In this study, ANN algorithm is used to estimate FSS frequency. The results of the algorithm are evaluated with six evaluation criterias. The proposed method is summarized in Fig. 2.

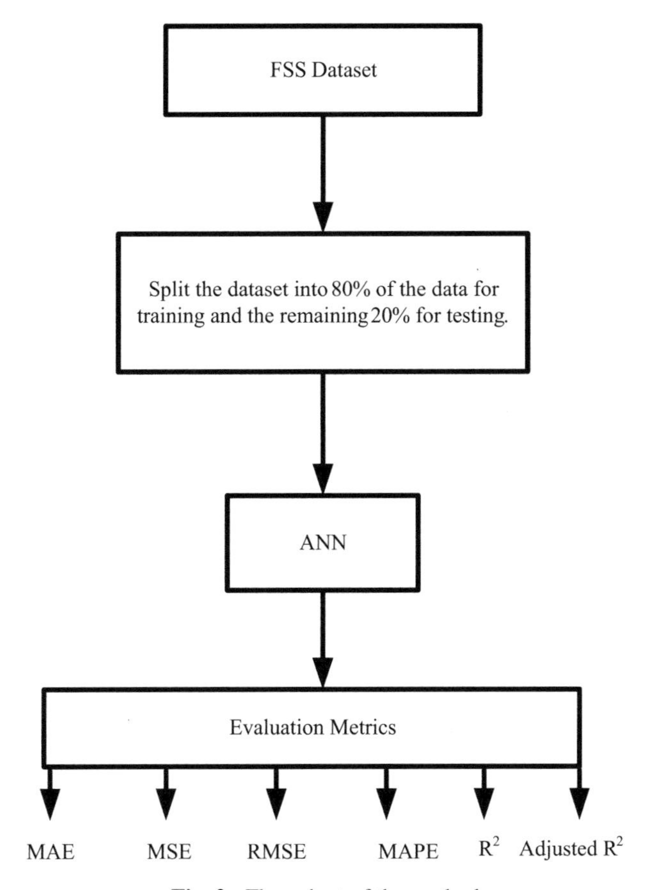

Fig. 2. Flow chart of the method

3.2 Artificial Neural Networks

The fundamental structure of ANN is given in Fig. 3. ANN model is basically composed of input, hidden and output layers. Each layer has neurons with weight values that are connected to each other. The weight values of the neurons are updated until the desired error rate is reached. Different learning algorithms are used in the training of these neurons. Some of backpropagation learning algorithms for ANN networks are gradient descent (GD), Powel-Beale conjugate gradient (PBCG), Fletcher-Powel conjugate gradient (FPCG), Polak-Ribiere conjugate gradient (PRCG), scaled con-jugate gradient (SCG), one step secant (OSS), Bayesian regularization (BR) and Levenberg-Marquardt (LM) [19, 20]. The classification or regression result is calculated by applying the activation function in the output layer. The parameters of the ANN model used in this study are given in the Table 2. The model includes one hidden layer with 7 neurons, while the learning algorithm is Levenberg–Marquardt. In addition, learning rate and epochs are tuned as 0.0001 and 1000, respectively.

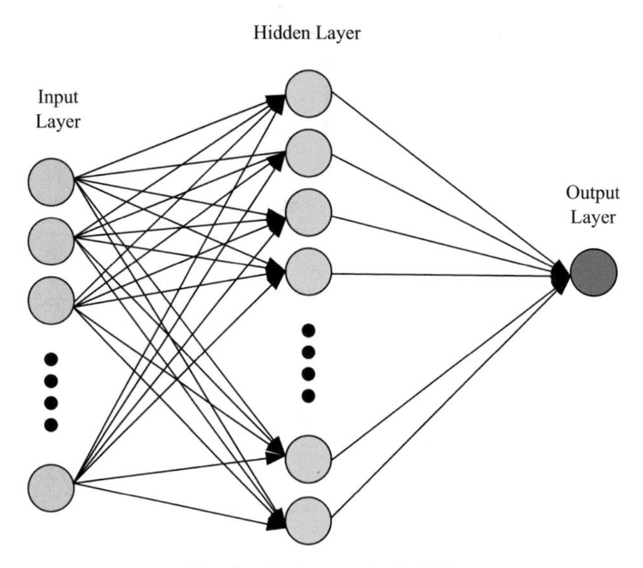

Fig. 3. Basic model of ANN

Table 2. Parameters of ANN

Hidden layer number	1
Neuron number	7
Learning algorithm	Levenberg–Marquardt
Learning rate	0.001
Epochs	1000

3.3 Evaluation Metrics

In order to determine the performance results of the ANN algorithm, six evaluation metrics are used. Equations of mean absolute error (MAE), mean absolute percentage error (MAPE), mean squared error (MSE), and root-mean-squared error (RMSE), R square (R^2) and adjusted R^2 (R^2_{adj}) are defined as follows,

$$MAE = \frac{1}{T}\sum_{t=1}^{T}\left|y_i - y_i'\right|$$

$$MAPE = \frac{1}{T}\sum_{t=1}^{T}\frac{\left|y_i - y_i'\right|}{Y} \times 100$$

$$MSE = \frac{1}{T}\sum_{t=1}^{T}\left(y_i - y_i'\right)^2$$

$$RMSE = \sqrt{\frac{1}{T}\sum_{t=1}^{T}\left(y_i - y_i'\right)^2}$$

$$R^2 = 1 - \frac{\sum\left(y_i - y_i'\right)^2}{\sum(y_i - \overline{y})^2}$$

$$R^2_{adj} = 1 - \frac{\left(1 - R^2\right)(T - 1)}{T - p - 1}$$

where y and y' are real and estimated values. i is data index. $\overline{y}$ is average value of y and T is the total number of data. p is the number of features.

4 Results

In this section, the estimation results of the FSS operating frequency with the ANN algorithm are given. Whereas 80% of the dataset is used for training, the rest of the dataset is used for testing.

4.1 Training Performance of ANN

The results for the training data are given in the Table 3. As can be seen from the results, good performance values are achieved for the training data set. The metric values corresponding to each frequency are given in Fig. 4. The MAPE value is in the range from 0 to 3, while other metrics are in the range from 0 to 0.5.

Table 3. Evaluation metrics for training data

MAPE	MAE	MSE	RMSE	R^2	R^2_{adj}
0.59	0.07	0.008	0.07	0.99	0.99

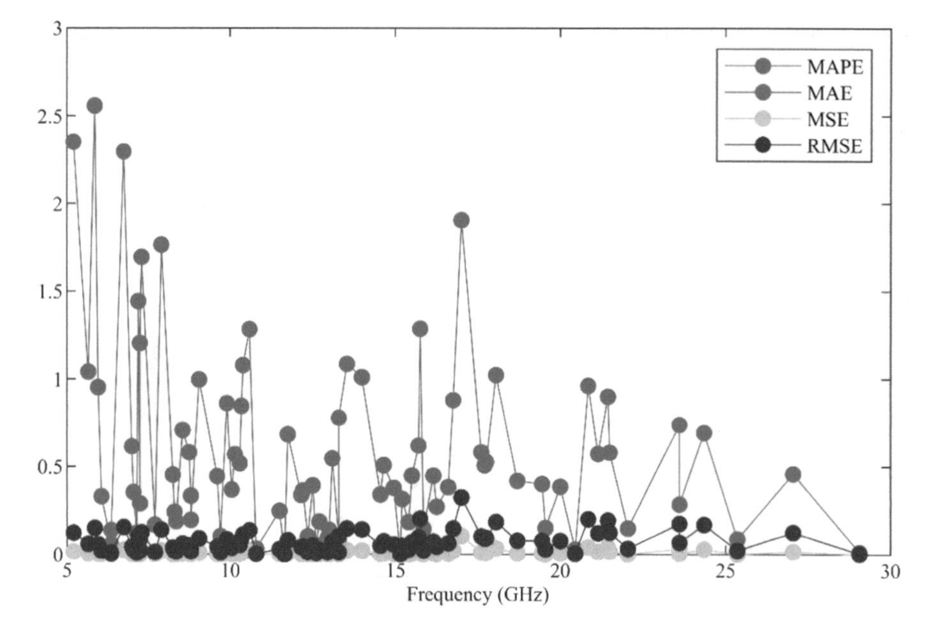

Fig. 4. Evaluation metric result corresponding to each frequency for training data

4.2 Testing Performance of ANN

Information for test dataset including FSS design parameters, the operating frequency of the FSS in the simulation environment (*fsim*) and the estimated frequency values (*fest*) with ANN are given in the Table 4. The metric results for the test dataset are given in Table 5. Metric results are achieved as MAPE = 0.47, MAE = 0.06, MSE = 0.0072, RMSE = 0.08, $R^2 = 0.99$ and $R^2_{adj} = 0.99$.

The metric results for each frequency are shown in Fig. 5. While MAPE results are between 0 and 1.5, MAE, MSE and RMSE results are between 0 and 0.5.

Table 4. Information for test dataset

FSS design parameters					FSS resonance frequency	
ε_r	w_s	h_s	w_p	w_a	f_{sim}	f_{est}
2.33	8	0.8	4	0.8	21.45	21.49
2.33	8	0.8	6	0.8	13.15	13.15
2.33	8	1.6	4	1.6	23.95	23.87
2.33	8	1.6	6	3.2	22	21.9
2.33	8	1.6	7.2	3.2	15.35	15.32
2.33	10	0.8	5	1	17.45	17.49

(*continued*)

Table 4. (*continued*)

FSS design parameters					FSS resonance frequency	
ε_r	w_s	h_s	w_p	w_a	f_{sim}	f_{est}
2.33	10	0.8	7.5	1	10.7	10.74
2.33	10	0.8	9	2	8.7	8.65
4.3	8	0.8	4	0.8	17.7	17.71
4.3	8	0.8	6	0.8	10.85	10.91
4.3	8	0.8	6	1.6	13.2	13.18
4.3	8	1.6	4	1.6	19	19.11
4.3	10	0.8	5	4	20.2	20.02
4.3	10	0.8	7.5	1	8.95	9
4.3	10	0.8	7.5	2	10.95	10.85
4.3	10	0.8	9	4	11.1	11.1
6.15	8	1.6	4	1.6	16.35	16.33
6.15	8	1.6	45	3.2	18.4	18.51
6.15	10	0.8	5	4	18.05	18.26
6.15	10	0.8	9	4	9.8	9.94
6.15	10	1.6	9	1	4.9	4.95
6.15	10	1.6	9	4	8.75	8.75

Table 5. Evaluation metrics for test dataset

MAPE	MAE	MSE	RMSE	R^2	R^2_{adj}
0.47	0.06	0.0072	0.08	0.99	0.99

In Fig. 6, the estimated values for all data are given. It is observed that there is a slight difference between the estimated and actual values.

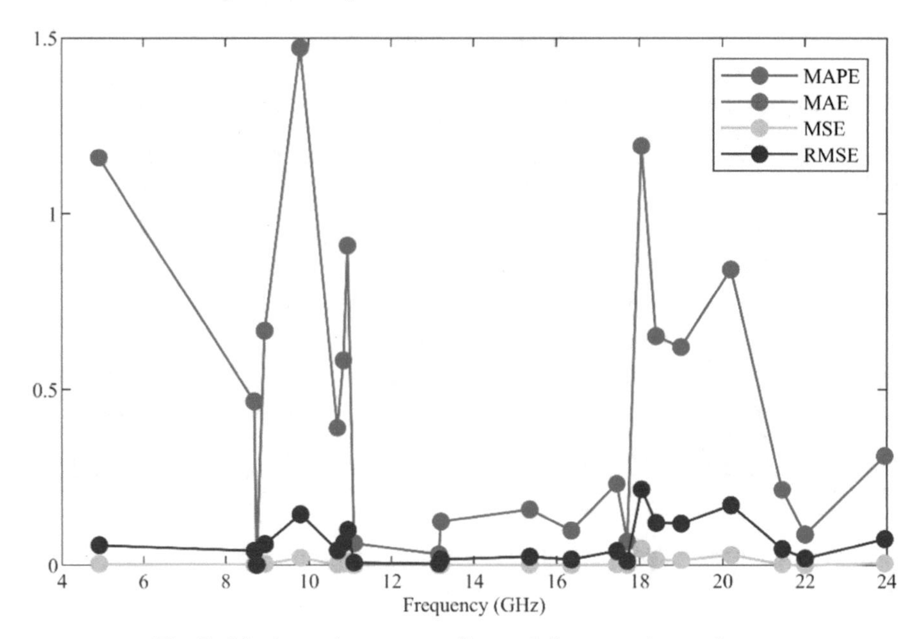

Fig. 5. Metric results corresponding each frequency in test dataset

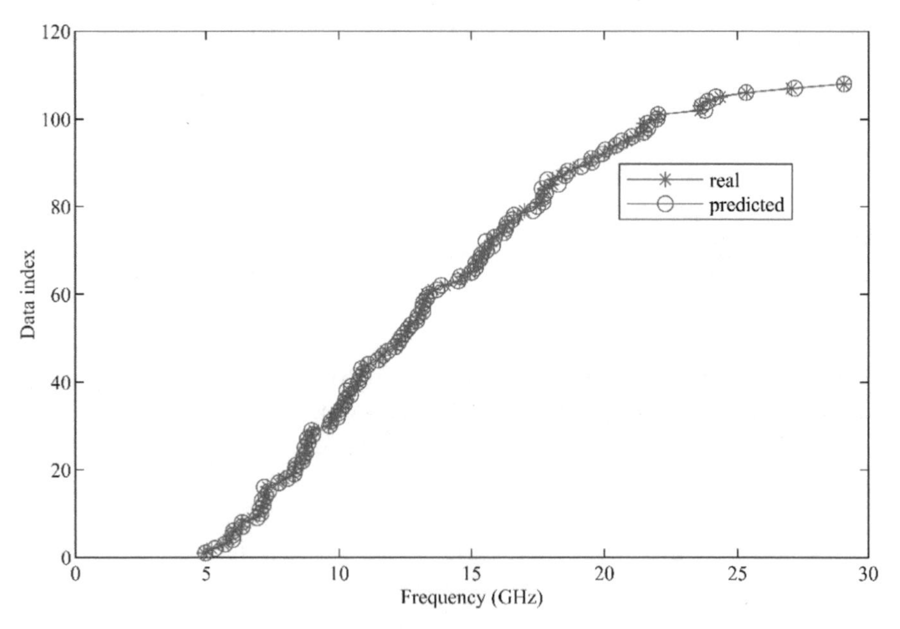

Fig. 6. Real and predicted values

5 Conclusions and Future Work

In this study, ANN is used to estimate the resonance frequency of FSS. The dataset is constituted using designed 108 FSS in the simulation environment. 80% of the data set is used for training and the rest was used for testing. The training performance of the algorithm has metric values of MAPE = 0.59, MAE = 0.07, MSE = 0.008, RMSE = 0.07, $R^2 = 0.99$, $R^2_{adj} = 0.99$. Test performance metrics is calculated as MAPE = 0.47, MAE = 0.06, MSE = 0.0072, RMSE = 0.08, $R^2 = 0.99$, $R^2_{adj} = 0.99$. The results demonstrate that modelled algorithm has a well regression-fitting model. The resonance frequency of the FSSs are estimated close to real values. In future studies, it is aimed to estimate the other parameters of FSS such as bandwidth and reflection loss by using different machine learning algorithms.

References

1. Munk, B.A.: Frequency Selective Surfaces: Theory and Design. Wiley, NY (2005)
2. Chen, H., Hou, X., Deng, L.: Design of frequency-selective surfaces radome for a planar slotted waveguide antenna. IEEE Antennas Wirel. Propag. Lett. **8**, 1231–1233 (2009)
3. Costa, F., Monorchio, A.: A frequency selective radome with wideband absorbing properties. IEEE Trans. Antennas Propag. **60**(6), 2740–2747 (2012)
4. Kim, J.H., Chun, H.J., Hong, I.P., Kim, Y.J., Park, Y.B.: Analysis of FSS radomes based on physical optics method and ray tracing technique. IEEE Antennas Wirel. Propag. Lett. **13**, 868–871 (2014)
5. Zabri, S.N., Cahill, R., Schuchinsky, A.: Compact FSS absorber design using resistively loaded quadruple hexagonal loops for bandwidth enhancement. Electron. Lett. **51**(2), 162–164 (2015)
6. Costa, F., Kazemzadeh, A., Genovesi, S., Monorchio, A.: Electromagnetic absorbers based on frequency selective surfaces. In: Forum Electromagn. Res. Methods Appl. Technol., vol. 37, No. 1, pp. 1–23 (2016)
7. Lin, C.W., Shen, C.K., Chiu, C.N., Wu, T.L.: Design and modeling of a compact partially transmissible resistor-free absorptive frequency selective surface for Wi-Fi applications. IEEE Trans. Antennas Propag. **67**(2), 1306–1311 (2018)
8. Wang, W.T., Gong, S.X., Wang, X., Yuan, H.W., Ling, J., Wan, T.T.: RCS reduction of array antenna by using bandstop FSS reflector. J. Electromagn. Waves Appl. **23**(11–12), 1505–1514 (2009)
9. Silva, P.H.D.F., Cruz, R.M.S., d'Assuncao, A.G.: Blending PSO and ANN for optimal design of FSS filters with Koch Island patch elements. IEEE Trans. Magn. **46**(8), 3010–3013 (2010)
10. Sushko, O., Pigeon, M., Donnan, R.S., Kreouzis, T., Parini, C.G., Dubrovka, R.: Comparative study of sub-THz FSS filters fabricated by inkjet printing, microprecision material printing, and photolithography. IEEE Trans. Terahertz Sci. Technol. **7**(2), 184–190 (2017)
11. Akbari, M., Gupta, S., Farahani, M., Sebak, A.R., Denidni, T.A.: Gain enhancement of circularly polarized dielectric resonator antenna based on FSS superstrate for MMW applications. IEEE Trans. Antennas Propag. **64**(12), 5542–5546 (2016)
12. Kundu, S.: A compact uniplanar ultra-wideband frequency selective surface for antenna gain improvement and ground penetrating radar application. Int. J. RF Microw. Comput.-Aided Eng. **30**(10), e22363 (2020)
13. Adibi, S., Honarvar, M.A., Lalbakhsh, A.: Gain enhancement of wideband circularly polarized UWB antenna using FSS. Radio Sci. **56**(1), 1–8 (2021)

14. Rajeev, S., Krishnamoorthy, C.S.: Artificial Intelligence and Expert Systems for Engineers. CRC Press, Boca Raton (1996)
15. Costamagna, E., Gamba, P., Lossani, S.: A neural network approach to the interpretation of ground penetrating radar data. In: Sensing and Managing the Environment. 1998 IEEE International Geoscience and Remote Sensing. Symposium Proceedingsb (IGARSS 1998), vol. 1, pp. 412–414 (1998). (Cat. No. 98CH36174)
16. Sabanci, K., Yiğit, E., Kayabasi, A., Toktas, A., Duysak, H., Aslan, M.F.: ANN and ELM based notch antenna operating frequency determination. In: 8th International Advanced Technologies Symposium (2017)
17. Eichie, J.O., Oyedum, O.D., Ajewole, M., Aibinu, A.M.: Comparative analysis of basic models and artificial neural network based model for path loss prediction. Prog. Electromagn. Res. M **61**, 133–146 (2017)
18. Gong, Y., Xiao, S., Wang, B.Z.: An ANN-based synthesis method for nonuniform linear arrays including mutual coupling effects. IEEE Access **8**, 144015–144026 (2020)
19. Kayabaşı, A., Sabancı, K., Yiğit, E., Toktaş, A., Yerlikaya, M., Yıldız, B.: Image processing based ANN with Bayesian regularization learning algorithm for classification of wheat grains. In: 2017 10th International Conference on Electrical and Electronics Engineering (ELECO), pp. 1166–1170 (2017)
20. Yigit, E., Duysak, H.: Determination of flowing grain moisture contents by machine learning algorithms using free space measurement data. In: IEEE Transactions on Instrumentation and Measurement, vol. 71, pp. 1–8 (2022). https://doi.org/10.1109/TIM.2022.3165740. Art no. 2507608

Application of Kashuri Fundo Transform to Population Growth and Mixing Problem

Haldun Alpaslan Peker[1(✉)] and Fatma Aybike Çuha[2]

[1] Department of Mathematics, Faculty of Science, Selçuk University, Konya, Turkey
hapeker@selcuk.edu.tr

[2] Graduate School of Natural and Applied Sciences, Selçuk University, Konya, Turkey

Abstract. Integral transforms are methods that provide great convenience in reaching the solutions of linear and nonlinear differential equations without making complex calculations. An integral transform can produce an algebraic result by integrating a given function with another function, called the kernel of the integral transform. The variety of integral transforms comes from the ability to transform the solution from complex mathematical calculations into simple algebraic methods. The Kashuri Fundo transform is the integral transform we consider in this study. The Kashuri Fundo integral transform is based on the Fourier integral and is of a very close connection with the Laplace transform. This transform was introduced by Kashuri and Fundo in order to find easier solutions to ordinary and partial differential equations. In this study, we show through a population growth problem and a mixing problem that the solutions of linear and nonlinear differential equations can be reached in a simple and understandable way using Kashuri Fundo integral transformation.

Keywords: Integral Transform · Kashuri Fundo Transform · Inverse Kashuri Fundo Transform · Population Growth · Mixing Problem

1 Introduction

It is possible to find differential equations in the background of many phenomena we encounter in every area of our lifes. Most laws in many different fields, especially in engineering, mathematics, physics, biology and biochemistry, are made more understandable by expressing them with differential equations, and thus it is seen that interpreting those laws becomes easier. The changes in the variables are associated with each other in the expressions in most scientific problems that have a wide place in the literature today. Differential equations are obtained in limit situations where the changes in these variables are infinitely small or differential changes [1]. Finding the solutions of these equations can sometimes be quite troublesome. For this reason, there are many various methods developed to reach the solution of these equations [2–4].

Integral transforms are one of these methods. Integral transforms are very practical methods used to find solutions to these equations. These transforms convert the original domain of the problems encountered in many different fields into another domain and

D. J. Hemanth et al. (Eds.): ICAIAME 2022, ECPSCI 7, pp. 407–414, 2023.
https://doi.org/10.1007/978-3-031-31956-3_34

enable the result to be reached in a very easy way. The basis of integral transforms is to create an equation that does not include derivatives of the dependent variable by multiplying each term in the equation with the kernel function and taking the integral of each term over the domain of the equation with respect to the independent variable. In addition, integral transforms are also used in the solution of initial and boundary value problems and integral equations in many different fields, especially in engineering and applied mathematics. Since it has various application areas, many integral transforms have been defined in the literature and still continues to be defined [5–20]. In this study, the integral transform we deal is Kashuri Fundo transform [21].

The Kashuri Fundo transform was introduced to the literature by Kashuri and Fundo in 2013 with the statement that it has a deep connection with the Laplace transform. Kashuri Fundo transform, like other integral transforms, converts differential equations to algebraic equations and produces a solution. In the literature, it is possible to come across many studies using Kashuri Fundo transform [22–31]. At the same time, it has been demonstrated by the studies that the Kashuri Fundo transformation is a suitable method to be used with other methods. Kashuri et al. combined Kashuri Fundo transform with homotopy perturbation method and introduced a new solution way to nonlinear differential equations [32]. Again, Kashuri et al. have studied some fractional differential equations by using Kashuri Fundo transform and expansion coefficients of binomial series [24]. Shah et al. applied the combination of this transform with the projected differential transform method to the time-fractional gas dynamics equations [33]. Sumiati et al., on the other hand, combined this transform with the Adomian decomposition method and defined the Kashuri Fundo decomposition method [34]. In this study, we have shown that effective results can be obtained by applying Kashuri Fundo transform to population growth and mixing problem.

The first model to which we will apply the Kashuri Fundo transform is population growth [35–39]. Population growth (a species, or an organ, or a cell, or a plant) was first described as a mathematical model by Thomas Malthus in 1798 in terms of a first order linear ordinary differential equation. This model was created based on the assumption that the total population expressed as $p(t)$ and the rate of growth of this population at time t is proportial. This assumption is expressed mathematically as

$$\frac{dp}{dt} = kp \tag{1.1}$$

$$p(0) = p_0 \tag{1.2}$$

where k is a proportionality constant and p_0 is the initial population at time $t = 0$.

Another model for which we will apply the Kashuri Fundo transform is the mixing problem [40]. The expression of this problem is as follows.

Let $a(t)$ be the quantity of sugar melted in the water in a receptacle at any one time. A solution that has a specific concentration C_{in} is pumped into the receptacle at a certain rate. The content of the receptacle is mixed steadily while the concentration C_{out} is pumped out at another certain rate. Let the water be pumped into and out of the receptacle at the same rate, and the quantity of water initially in the receptacle is a_0. We can express the mathematical model of this problem as follows

$$\frac{da}{dt} = C_{in} - C_{out}a(t) \tag{1.3}$$

$$a(0) = a_0 \tag{1.4}$$

2 Kashuri Fundo Transform

2.1 Definition of Kashuri Fundo Transform

Definition 1. We consider functions in the set F defined as [21],

$$F = \left\{ f(x) | \exists M,\ k_1,\ k_2 > 0 \ \text{ such that } \ |f(x)| \le Me^{\frac{|x|}{k_i^2}}, \ \text{ if } \ x \in (-1)^i \times [0, \infty) \right\} \tag{1.5}$$

For a function in the set defined above, M must be finite. $k_1,\ k_2$ may be finite or infinite.

Definition 2. Kashuri Fundo transform defined on the set F, denoted by $K(.)$, is defined as [21],

$$K[f(x)](v) = A(v) = \frac{1}{v} \int_0^\infty e^{\frac{-x}{v^2}} f(x)dx, \quad x \ge 0, \quad -k_1 < v < k_2 \tag{1.6}$$

Inverse Kashuri Fundo transform is expressed as $K^{-1}[A(v)] = f(x), \quad x \ge 0.$

Theorem 1 (Sufficient Conditions for Existence)
If $f(x)$ is piecewise continuous on $[0, \infty)$ and has exponential order $\frac{1}{k^2}$, then $K[f(x)](v)$ exists for $|v| < k$ [21].

2.2 Some Useful Properties of Kashuri Fundo Transform

Theorem 2 (Linearity)
Let the Kashuri Fundo transforms of $f(x)$ and $g(x)$ exist and c be a scalar. Then [21],

$$K[(f \pm g)(x)] = K[f(x)] \pm K[g(x)] \tag{1.7}$$

$$K[(cf)(x)] = cK[f(x)] \tag{1.8}$$

Theorem 3 (Kashuri Fundo Transform of the Derivatives)
Let's assume that the Kashuri Fundo transform of $f(x)$, denoted by $A(v)$, exists. Then [21],

$$K[f'(x)] = \frac{A(v)}{v^2} - \frac{f(0)}{v} \tag{1.9}$$

$$K[f''(x)] = \frac{A(v)}{v^4} - \frac{f(0)}{v^3} - \frac{f'(0)}{v} \tag{1.10}$$

$$K[f^{(n)}(x)] = \frac{A(v)}{v^{2n}} - \sum_{k=0}^{n-1} \frac{f^{(k)}(0)}{v^{2(n-k)-1}} \tag{1.11}$$

2.3 Kashuri Fundo Transform of Most Commonly Used Functions

Table 1. Kashuri Fundo Transform of Most Commonly Used Functions [21, 22]

$f(x)$	$A(v)$
1	v
x	v^3
x^n	$n!v^{2n+1}$
e^{cx}	$\frac{v}{1-cv^2}$
$\sin(cx)$	$\frac{cv^3}{1+c^2v^4}$
$\cos(cx)$	$\frac{v}{1+c^2v^4}$
$\sinh(cx)$	$\frac{cv^3}{1-c^2v^4}$
$\cosh(cx)$	$\frac{v}{1-c^2v^4}$
x^α	$\Gamma(1+\alpha)v^{2\alpha+1}$
$\sum_{k=0}^{n} c_k x^k$	$\sum_{k=0}^{n} k!c_k v^{2k+1}$

3 Applications

3.1 Application of Kashuri Fundo Transform to Population Growth

If we apply the Kashuri Fundo transform to the Eq. (1.1), we get

$$K\left[\frac{dp}{dt}\right] = K\left[kp\right] \tag{1.12}$$

Rearranging the left side of the Eq. (1.12) using the Eq. (1.9), we obtain

$$\frac{A(v)}{v^2} - \frac{p(0)}{v} = kA(v) \tag{1.13}$$

If we substitute the given initial condition into the Eq. (1.13), we get

$$A(v) = p_0\left(\frac{v}{1-kv^2}\right) \tag{1.14}$$

Having applied the inverse Kashuri Fundo transform bilaterally to the Eq. (1.14) and using Table 1, we get

$$p(t) = p_0 e^{kt} \tag{1.15}$$

which is exactly coincides with the existing results obtained by other methods [40–42].

3.2 Application of Kashuri Fundo Transform to Mixing Problem

Having applied the Kashuri Fundo transform bilaterally to the Eq. (1.3), we get

$$K\left[\frac{da}{dt}\right] = K[C_{in}] - K[C_{out}a(t)] \tag{1.16}$$

Rearranging the left side of the Eq. (1.16) using the Eq. (1.9), we obtain

$$\frac{A(v)}{v^2} - \frac{a(0)}{v} = C_{in}v - C_{out}A(v) \tag{1.17}$$

If we substitute the given initial condition into the Eq. (1.17), we get

$$A(v) = \frac{C_{in}v^3 + a_0 v}{1 + C_{out}v^2} \tag{1.18}$$

If we rearrange the Eq. (1.18), we get

$$A(v) = C_{in}\left[\frac{v}{C_{out}} - \frac{1}{C_{out}}\frac{v}{1 + C_{out}v^2}\right] + a_0\frac{v}{1 + C_{out}v^2} \tag{1.19}$$

Having applied the inverse Kashuri Fundo transform bilaterally to the Eq. (1.19) and using Table 1, we find

$$a(t) = \frac{C_{in}}{C_{out}} + \left(a_0 - \frac{C_{in}}{C_{out}}\right)e^{-tC_{out}} \tag{1.20}$$

which is in complete compliance with the existing result found by other method [40].

4 Conclusion and Future Work

Differential equations have a very important place in the mathematical modeling of phenomena in many fields, especially in applied mathematics, engineering and physics. They make these phenomena easier to understand and interpret. Therefore, their solutions are also very important. The solution of differential equations can be more complex than algebraic equations. For this reason, using integral transforms, which transform differential equations into algebraic equations, as a solution method helps us to reach the solution more clearly and easily. In this study, by using Kashuri Fundo transform, we transform the population growth model and mixing problem into algebraic equations and then reach their solutions without complex operations. As it can be understood from the steps we applied throughout the study, the Kashuri Fundo transform can be expressed as an effective, reliable and useful method in finding solutions to models expressed by ordinary differential equations such as population growth and mixing problem. By applying the Kashuri Fundo transform to different models in the applied fields such as physics, chemistry, biology, biochemistry, economics, etc., the application area can be further expanded.

References

1. Çengel, Y.A., Palm, W.J.: Mühendislik ve Temel Bilimler için Diferensiyel Denklemler. Translation ed. Tahsin Engin, İzmir Güven Kitabevi, İzmir, Turkey (2013). (in Turkish)
2. He, J.H.: Homotopy perturbation method: a new nonlinear analytical technique. Appl. Math. Comput. **135**(1), 73–79 (2003)
3. He, J.H.: Variational iteration method- a kind of nonlinear analytical technique: some examples. Int. J. Nonlinear Mech. **34**(4), 609–708 (1999)
4. Adomian, G.: Solving Frontier Problems of Physics: The Decomposition Method. Kluwer Academic Publishers, Boston, Mass, USA (1994)
5. Spiegel, M.R.: Schaum's Outlines of Laplace Transforms. McGraw Hill, New York (1965)
6. Lokenath, D., Bhatta, D.: Integral Transform and Their Applications. CRC Press, Boca Raton, Fla, USA (2014)
7. Bracewell, R.N.: The Fourier Transform and Its Applications. McGraw-Hill, Boston, Mass, USA (2000)
8. Beerends, R.J., Ter Morsche, H.G., Van Den Berg, J.C., Van De Vrie, E.M.: Fourier and Laplace Transforms. Cambridge University Press, Cambridge, UK (2003)
9. Elzaki, T.M.: The new integral transforms Elzaki transform. Glob. J. Pure Appl. Math. **7**(1), 57–64 (2011)
10. Deakin, M.A.B.: The 'Sumudu transform' and the Laplace transform. Lett. Editor (Received 24.03.1995) Int. J. Math. Educ. Sci. Technol. **28**(1), 159–160 (1997)
11. Khan, Z.H., Khan, W.A.: N-transform-properties and applications. NUST J. Eng. Sci. **1**(1), 127–133 (2008)
12. Maitama, S., Zhao, W.: New integral transform: Shehu transform a generalization of Sumudu and Laplace transform for solving differential equations. Int. J. Anal. Appl. **17**(2), 167–190 (2019)
13. Aboodh, K.S.: The new integral transform Aboodh transform. Glob. J. Pure Appl. Math. **9**(1), 35–43 (2013)
14. Zafar, Z.U.A.: ZZ transform method. Int. J. Adv. Eng. Glob. Technol. **4**(1), 1605–1611 (2016)
15. Mahgoub, M.M.A.: The new integral transform Mahgoub transform. Adv. Theor. Appl. Math. **11**(4), 391–398 (2016)
16. Sedeeg, A.K.H.: The new integral transform Kamal transform. Adv. Theor. Appl. Math. **11**(4), 451–458 (2016)
17. Mahgoub, M.M.A.: The new integral transform Mohand transform. Adv. Theor. Appl. Math. **12**(2), 113–120 (2017)
18. Elzaki, T.M., Elzaki, S.M.: On the new integral transform Tarig transform and systems of ordinary differential equations. Elixir Appl. Math. **36**, 3226–3229 (2011)

19. Shaikh, S.L.: Introducing a new integral transform: Sadik transform. Am. Int. J. Res. Sci. Technol. Eng. Math. (AIJRSTEM) **22**(1), 100–102 (2018)
20. Mahgoub, M.M.A.: The new integral transform Sawi transform. Adv. Theor. Appl. Math. **14**(1), 81–87 (2019)
21. Kashuri, A., Fundo, A.: A new integral transform. Adv. Theor. Appl. Math. **8**(1), 27–43 (2013)
22. Subartini, B., Sumiati, I., Sukono, R., Sulaiman, I.M.: Combined Adomian decomposition method with integral transform. Math. Stat. **9**(6), 976–983 (2021)
23. Kashuri, A., Fundo, A., Liko, R.: On double new integral transform and double Laplace transform. Eur. Sci. J. **9**(33), 82–90 (2013)
24. Kashuri, A., Fundo, A., Liko, R.: New integral transform for solving some fractional differential equations. Int. J. Pure Appl. Math. **103**(4), 675–682 (2015)
25. Shah, K., Singh, T.: A solution of the Burger's equation arising in the longitudinal dispersion phenomenon in fluid flow through porous media by mixture of new integral transform and homotopy perturbation method. J. Geosci. Environ. Prot. **3**(4), 24–30 (2015)
26. Shah, K., Singh, T.: The mixture of new integral transform and homotopy perturbation method for solving discontinued problems arising in nanotechnology. Open J. Appl. Sci. **5**(11), 688–695 (2015)
27. Helmi, N., Kiftiah, M., Prihandono, B.: Penyelesaian persamaan diferensial parsial linear dengan menggunakan metode transformasi Artion-Fundo. Buletin Ilmiah Matematika Statistika dan Terapannya **5**(3), 195–204 (2016)
28. Peker, H.A., Cuha, F.A.: Application of Kashuri Fundo transform and homotopy perturbation methods to fractional heat transfer and porous media equations. Therm. Sci. **26**(4A), 2877–2884 (2022)
29. Cuha, F.A., Peker, H.A.: Solution of Abel's integral equation by Kashuri Fundo transform. Therm. Sci. **26**(4A), 3003–3010 (2022)
30. Peker, H.A., Cuha, F.A., Peker, B.: Solving steady heat transfer problems via Kashuri Fundo transform. Therm. Sci. **26**(4A), 3011–3017 (2022)
31. Peker, H.A., Cuha, F.A.: Application of Kashuri Fundo transform to decay problem. SDU J. Nat. Appl. Sci. **26**(3), 546–551 (2022)
32. Kashuri, A., Fundo, A., Kreku, M.: Mixture of a new integral transform and homotopy perturbation method for solving nonlinear partial differential equations. Adv. Pure Math. **3**(3), 317–323 (2013)
33. Shah, K., Singh, T., Kılıçman, B.: Combination of integral and projected differential transform methods for time-fractional gas dynamics equations. Ain Shams Eng. J. **9**(4), 1683–1688 (2018)
34. Sumiati, I., Bon, S.A.: Adomian decomposition method and the new integral transform. In: IEOM Proceedings of the 2nd African International Conference on Industrial Engineering and Operations Management, pp. 1882–1887. IEOM, December 2020
35. Weigelhofer, W.S., Lindsay, K.A.: Ordinary Differential Equations and Applications: Mathematical Methods for Applied Mathematicians, Physicists, Engineers And Bioscientists. Woodhead Publishing, Delhi (1999)
36. Ahsan, Z.: Differential Equations and Their Applications, PHI, Delhi, India (2006)
37. Roberts, C.: Ordinary Differential Equations: Applications, Models and Computing. Chapman and Hall/CRC, Boca Raton, Fla, USA (2010)
38. Zill, D.G., Cullen, M.R.: Differential Equations with Boundary Value Problems. Thomson Brooks/Cole, USA (1996)
39. Bronson, R., Costa, G.B.: Schaum's Outlines of Differential Equations. McGraw-Hill, New York (2006)
40. Munganga, J.M.W., Mwambakana, J.N., Maritz, R., Batubenge, T.A., Moremedia, G.M.: Introduction of the differential transform method to solve differential equations at undergraduate level. Int. J. Math. Educ. Sci. Technol. **45**(5), 781–794 (2014)

41. Singh, G.P., Aggarwal, S.: Sawi transform for population growth and decay problems. Int. J. Latest Technol. Eng. Manag. Appl. Sci. **8**(8), 157–162 (2019)
42. Aggarwal, S., Gupta, A.R., Singh, D.P., Asthana, N., Kumar, N.: Application of Laplace transform for solving population growth and decay problems. Int. J. Latest Technol. Eng. Manag. Appl. Sci. **7**(9), 141–145 (2018)

Attack Detection on Testbed for Scada Security

Esra Söğüt(✉) and O. Ayhan Erdem(✉)

Faculty of Technology, Department of Computer Engineering, Gazi University, Ankara, Turkey
{esrasogut,ayerdem}@gazi.edu.tr

Abstract. Supervisory Control and Data Acquisition Systems (SCADA) performs inspection and monitoring tasks in critical infrastructures or facilities. Attackers target SCADA systems to damage these structures or facilities. Performance loss of SCADA systems can negatively affect the entire system or stop the operation of the entire system. Therefore, it has become necessary to provide cyber security of SCADA systems against attacks. In this study, the dataset obtained from the test bed containing the SCADA system was used. Different attack examples were applied to this test bed and the attack results were examined. Accordingly, the dataset includes DDoS attack data such as Modbus Query Flooding, ICMP Flooding and TCP SYN Flooding. Classification was made using machine learning algorithms to predict the attack type. In addition to machine learning, a method for feature selection is also used in the study. According to the results obtained, the highest success rates for both stages were obtained with Decision Tree, K-Nearest Neighbors Regressor and K-Nearest Neighbors Classifier.

Keywords: SCADA Systems · Critical Infrastructures · Modbus Protocol · DDoS Attacks · Feature Selection · Machine Learning

1 Introduction

Control and monitoring systems are used in different and comprehensive areas such as electrical networks, natural gas facilities, critical infrastructures such as nuclear power plants or sub-units of such structures. The most widely used of these are Supervisory Control and Data Acquisition Systems (SCADA). SCADA systems are taken part in a wide variety of processes such as delivering natural gas to homes, moving materials on belts in factories, and monitoring the movement of wind turbines. It ensures the monitoring and tracing of the transactions carried out within a system, intervening when necessary, and keeping the system operation under control [1]. It collects information about the system, makes analysis, helps to keep the system performance at the highest level and guides the operators who monitor the systems.

Continuity of operation of critical infrastructures, businesses, or factories depends on the smooth performance of the generation, transmission, distribution, or control operations. Ensuring this depends on the operation of SCADA systems without loss of performance. Therefore, SCADA systems become the target of cyber-attacks. It is vital for SCADA systems to minimize, detect and, if possible, prevent the negative effects of

D. J. Hemanth et al. (Eds.): ICAIAME 2022, ECPSCI 7, pp. 415–422, 2023.
https://doi.org/10.1007/978-3-031-31956-3_35

cyber-attacks. For this, many studies are carried out on the cyber security of SCADA systems.

Yılmaz et al. prepared a testbed in SCADA systems using two different Programmable Logic Controllers (PLCs). Man in the Middle and Denial of Service (DoS) attacks were conducted to examine PLCs in terms of cyber security vulnerability and analyzed security vulnerabilities [2]. Söğüt et al. used a dataset in which different attacks such as DoS, Reconnaissance, Command Injection and Response Injection were made for SCADA systems using Modbus/TCP protocol. They classified the attacks and achieved the highest success in attack detection with the Random Forest algorithm [3]. Altunay et al. discussed the anomaly-based attack detection studies in SCADA systems and examined the studies using Deep Learning techniques. Although Deep Learning techniques have high accuracy rates in anomaly detection, it has been stated that long training times are disadvantageous [4].

In addition to machine learning or deep learning methods, different methods are also used in attack detection. For example, reducing the attributes in the dataset and re-analyzing the analyzes accordingly are used in intrusion detection studies. Polat et al. showed the effects of feature selection processes on the detection of Distributed Denial of Service (DDoS) attacks [5]. Shitharth et al. proposed a new method for intrusion detection in SCADA systems and operationalized feature selection-based optimization processes [6].

2 Obtaining Test Bed and Data

The data set used in the study was created by Frazão et al. for research on the application of machine learning techniques to cyber security in industrial control systems [7, 8]. The test bed prepared for this simulates a cyber-physical system process controlled by the SCADA system using Modbus/TCP protocol. There is Variable Frequency Drive controlled by PLC. This driver runs the electric motor that simulates a liquid pump. The PLC communicates with the Arduino based Remote Terminal Unit and the Human-Machine Interface controlling the system. Various attacks have been performed on the test bed and system behavior has been monitored.

For this study, the section containing DDoS attacks such as Modbus Query Flooding, ICMP Flooding and TCP SYN Flooding was selected in the dataset prepared by Frazão et al. Each of these three attacks was made for one minute and half an hour of listening was made to catch the attacks.

3 Data Analysis Stages

This section provides information about data analysis. Operations are carried out under three headings.

3.1 Pre-processes

In order to make the dataset suitable for analysis, the datasets of these three attack sets were combined. In the obtained dataset, empty data is cleared. Meaningless data were transformed into meaningful ones. One of the same columns has been deleted. Repetitive data control was made, but no repetitive data was found. Labels have been made to group in the DestinationIP, Protocol and Class columns. Information on labeling is shown in Tables 1, 2 and 3.

Table 1. Labeling procedures for DestinationIp

No	Samples	DestinationIP
1	65248	172.27.224.250
2	14860	172.27.224.251
3	5756	172.27.224.70
4	410	255.255.255.255
5	62	172.27.224.255
6	17769	172.27.224.80
7	6	224.0.0.252
8	3	239.255.255.250
9	35	Ff02::1:2
10	6	Ff02::1:3
11	3	Ff02::c

Table 2. Labeling procedures for Protocol

No	Samples	Protocol
1	84849	TCP
2	18784	MODBUS/ TCP
3	369	DHCP
4	88	UDP
5	35	DHCPv6
6	21	BROWSER
7	12	LLMNR

Table 3. Labeling procedures for Class

No	Samples	Class
1	54415	Modbus Query Flooding
2	33026	ICMP Flooding
3	16717	TCP SYN Flooding

The data used in the DestinationIP feature is shown in Table 1. Grouping has been made for ease of analysis. Accordingly, tagging has been done for eleven different DestinationIP addresses. In Table 2, seven different labels are made for the Protocol feature. Accordingly, TCP and Modbus/TCP protocols are frequently used in the dataset. Labeling was made according to three attack types for the Class column, which will determine the classification. Looking at the packet numbers of the attacks, it was seen that the Modbus Query Flooding attack was the most common. TCP SYN Flooding attack has the least number of packets.

The No and SourceIP columns were deleted as they were not effective in data analysis. After these operations, the number of data decreased from 122959 rows to 104158 rows. The number of columns also decreased from 11 to 8. The dataset was made suitable for analysis by preprocessing. Examples of the dataset before (Table 4) and after (Table 5) preprocessing are shown below.

Table 4. The first 5 data representations of the state of the dataset before preprocessing

No1	No	Time	SourceIP	SourcePort	DestinationIP	DestinationPort	Protocol	Length	DeltaTime	Class
0	1	0.000000	172.27.224.251	54200	172.27.224.250	502	TCP	60	0.000000	1
1	2	0.009696	172.27.224.250	502	172.27.224.251	54200	TCP	60	0.009696	1
2	3	0.010556	172.27.224.250	502	172.27.224.251	54200	TCP	60	0.000860	1
3	4	0.010574	172.27.224.251	54200	172.27.224.250	502	TCP	60	0.000018	1
4	5	0.010752	172.27.224.250	502	172.27.224.251	54200	TCP	60	0.000178	1

Table 5. The first 5 data displays of the state of the dataset after preprocessing

Time	SourcePort	DestinationIP	DestinationPort	Protocol	Length	DeltaTime	Class
0.000000	54200	1	502	1	60	0.000000	1
0.000000	49174	1	502	1	60	0.000000	2
0.009696	502	2	54200	1	60	0.009696	1
0.064853	52972	1	502	1	60	0.064853	2
0.010556	502	2	54200	1	60	0.000860	1

3.2 Algorithms

The attack type is estimated by studying the three attacks mentioned. In case of a new attack, classification is made to predict which of these three attacks. For this,

Logistic Regression, Linear SVM, Random Forest, Decision Tree, K-Nearest Neighbors Classifier, K-Nearest Neighbors Regressor and Naive Bayes algorithms are used.

In all algorithm analysis processes, 66% of the dataset is used for training and the remaining 34% for testing. Relevant encodings are performed on the Phyton Pandas environment. The realization times of the data classification processes for each algorithm were also taken into account in the analysis processes.

3.3 Feature Selection

For data analysis, firstly, analyzes are carried out without feature selection. Then, using an algorithm, the importance rankings and usage rates of the features are obtained. By using this information, two separate operations are performed in all algorithm usages. These are: Performing the analyzes without Attribute Selection and by Performing the Attribute Selection Process.

Without Feature Selection Process: In this section, algorithms are applied without feature selection.

Performing Feature Selection: In this section, the features that are considered important in the analysis and that are used more frequently are selected. Random Forest algorithm is used for this. The order of importance of the features (Fig. 1) and their usage rates (Fig. 2) are given below.

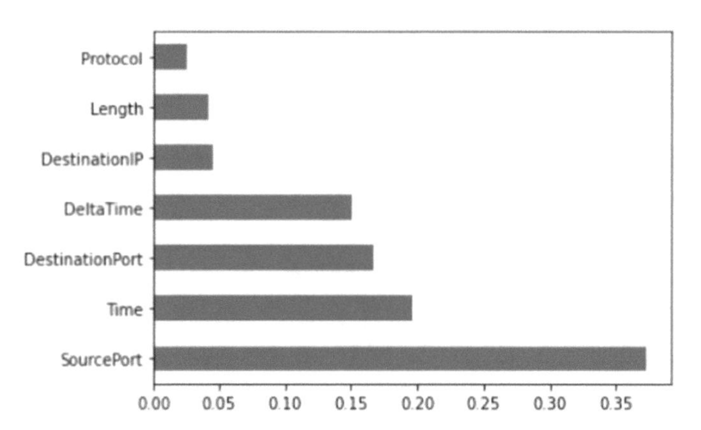

Fig. 1. The order of importance of the features for the Feature Selection Process

By using this information, insignificant features were removed from the dataset or not used in the analysis. These attributes are the Protocol, Length, and DestinationIP columns.

SourcePort:	0.373181
Time:	0.196722
DestinationPort:	0.167477
DeltaTime:	0.150663
DestinationIP:	0.044853
Length:	0.041535
Protocol:	0.025568

Fig. 2. Usage rates of features for the Feature Selection Process

Table 6. Results of classification processes

Algorithms No	Without Feature Selection		Performing Feature Selection	
	Accuracy Rate	Processing Time (sec)	Accuracy Rate	Processing Time (sec)
1	0.59	24.922	0.60	12.188
2	0.42	21.250	0.16	13.719
3	0.99	86.422	1.0	94.484
4	0.99	0.297	1.0	0.516
5	0.99	0.859	1.0	0.531
6	0.99	0.953	1.0	0.438
7	0.65	0.047	0.61	0.281

Table 7. Ranking of algorithms

No	Algorithms
1	Logistic Regression
2	Linear SVM
3	Random Forest
4	Decision Tree
5	K-Nearest Neighbors Classifier
6	K-Nearest Neighbors Regressor
7	Naive Bayes

4 Analysis Results

The attack type is determined by making classifications with the determined algorithms. Accordingly, the results obtained are given in Table 6 and the relevant algorithm rankings are given in Table 7.

According to the results achieved in this study, without the feature selection, best results are accomplished by Random Forest, Decision Tree, K-Nearest Neighbors Classifier and K-Nearest Neighbors Regressor algorithms. For the Processing Time Decision Tree achieved best result by being the fastest in the above-given algorithms. After the Feature Selection process, it is observed that the accuracy metrics of the given algorithms are slightly changed. According to that best accuracted rates are achieved by Random Forest, Decision Tree, K-Nearest Neighbors Classifier, K-Nearest Neighbor Regressor respectively. When these algorithms are ranked according to the Processing Time, K-Nearest Neigbors Regressor is achieved the best result. In addition to that, it also has been seen that there are also some algorithms that work fast but does not achieve good performance.

Feature Selection is caused the performance of some of the above-given algorithms to be higher. According to the Table-6, with the Feature Selection, the performance metrics of Logistic Regression, K-Nearest Neighbors Classifier and K-Nearest Neighbors Regressor algorithms has gone up and the Processing Times has gone down.

5 Conclusions and Comments

In this study, machine learning algorithms are used in order to detect attacks on Cyber Physical Systems and SCADA systems which use Modbus/TCP protocol. Classifications are made in data analysis to detect attacks and, also when there is a recent attack, the type of attack has been predicted using the given model. The achievement scores of the classification also show the detection rates of attack as well. In order to achieve better performance, the feature selection method has been used. According to that, the best classification performance has been achieved by Decision Tree, K-Nearest Neighbors Regressor and K-Nearest Neighbors Classifier algorithms.

For further studies, it is aimed to use different Feature Selection methods and in addition to that, achieve better performance results. Furthermore, it also is aimed by the researchers to use Feature Selection for different datasets and compare the results.

References

1. Aydın, H., Barışkan, M.A., Çetinkaya, A.: Siber Güvenlik Kapsamında Enerji Sistemleri Güvenliğinin Değerlendirilmesi. Güvenlik Bilimleri Dergisi **10**(1), 151–174 (2021)
2. Yılmaz, E.N., Sayan, H.H., Üstünsoy, F., Serkan, G., Karacayılmaz, G.: Cyber security analysis of DoS and MitM attacks against PLCs used in smart grids. In: Proceedings of the 7th International Istanbul Smart Grids and Cities Congress and Fair, pp. 36–40, IEEE, İstanbul (2019)
3. Söğüt, E., Erdem, O.A.: Endüstriyel Kontrol Sistemlerine (SCADA) Yönelik Siber Terör Saldırı Analizi. Politeknik Dergisi **23**(2), 557–566 (2020)

4. Altunay, H.C., Albayrak, Z., Özalp, A.N., Çakmak, M.: Analysis of anomaly detection approaches performed through deep learning methods in SCADA systems. In: Proceedings of the 3rd International Congress on Human-Computer Interaction, Optimization and Robotic Applications, pp. 1–6, IEEE, Ankara, June 2021
5. Polat, H., Polat, O., Çetin, A.: Detecting DDoS attacks in software-defined networks through feature selection methods and machine learning models. Sustainability **12**(3), 1035 (2020)
6. Shitharth, S., Prasad, K.M., Sangeetha, K., Kshirsagar, P.R., Babu, T.S., Alhelou, H.H.: An enriched RPCO-BCNN mechanisms for attack detection and classification in SCADA systems. IEEE Access **9**, 156297–156312 (2021)
7. Frazão, I., Abreu, P.H., Cruz, T., Araújo, H., Simões, P.: Denial of service attacks: detecting the frailties of machine learning algorithms in the classification process. In: Luiijf, E., Žutautaitė, I., Hämmerli, B. (eds.) Critical Information Infrastructures Security. CRITIS 2018. LNCS, vol. 11260, pp. 230–235. Springer, Cham (2019). https://doi.org/10.1007/978-3-030-05849-4_19
8. Internet: https://github.com/tjcruz-dei/ICS_PCAPS/releases/tag/MODBUSTCP%231. Cruz, T. Modbus TCP SCADA. Access Date: 10.02.2022

Conveyor Belt Speed Control with PID Controller Using Two PLCs and LabVIEW Based on OPC and MODBUS TCP/IP Protocol

Arslan Tirsi(✉), Mehmet Uğur Soydemir, and Savaş Şahin

Faculty of Engineering and Architecture, Department of Electrical and Electronics Engineering, İzmir Kâtip Çelebi University, İzmir, Turkey
arslan.tirsi@hotmail.com

Abstract. In this study, it is aimed to control the speed of a conveyor belt system based on the principles of observation, control, data storage and data transfer between systems, which became widespread with Industry 4.0. Speed control of the conveyor belt is achieved with the help of an Open Platform Communications Unified Architecture (OPC UA) server using multiple Programmable Logic Controllers (PLC) and Proportional-Integral-Derivative (PID) controller algorithm. With these PLCs, the simultaneous interoperability of different PLCs in industrial systems has been demonstrated. Therefore, PLCs were communicated via MODBUS TCP/IP communication protocol and configured as master-slave, and the OPC server was installed for PLCs, and PLCs were communicated with the LabVIEW program. With the OPC server, reading, writing, and data storage operations can be performed on LabVIEW, and the desired speed value for the PID controller can be adjusted in real-time via LabVIEW. Also, the conveyor belt system can be monitored in real-time via LabVIEW, the stability of the system can be observed, and fault prediction can be made by processing the speed and encoder data stored via LabVIEW.

Keywords: PLC · PID Controller · OPC · MODBUS TCP/IP · LabVIEW

1 Introduction

With industrial automation systems, it is possible to increase production quality, reduce costs, save time, and increase flexibility in production. With the developments in the industry, the need for the communication of industrial systems with each other, the monitoring of the systems, and the remote intervention of these systems have emerged. These requirements have led to the emergence of the concept of Industry 4.0 which has the purpose that is to be bringing together information technologies and industry. To ensure technology and industry collaboration, it needs communication protocols and communication standards. Two of these protocols and standards are Open Platform Communications (OPC) and MODBUS. With MODBUS TCP/IP and OPC, it is possible to ensure that hardware and software commonly used in the industry can communicate with each other, be programmed, observed, and data stored. Integrating an OPC-based

D. J. Hemanth et al. (Eds.): ICAIAME 2022, ECPSCI 7, pp. 423–433, 2023.
https://doi.org/10.1007/978-3-031-31956-3_36

system into the industry is easy with Programmable Logic Controllers (PLCs) already used in the industry. PLCs are extremely durable, easy to program, and safe control devices. Many different types of sensors, switches, and motors can be connected to PLCs. With the help of the OPC Server, it is possible to turn automation systems in factories where PLCs are used into intelligent systems that communicate with each other, make decisions according to a program, and analyze and store data. In these smart systems, sensors, encoders, and many other sensing elements are actively used and they keep the system work properly thanks to the feedback they give from the system they work. When it is necessary to monitor, control and analyze these smart systems installed with OPC, additional software is needed in the computer environment, LabVIEW is an extremely useful and multi-featured graphical programming engineering program that can perform these operations.

In this study, a conveyor belt speed control application was implemented on the LabVIEW platform via an OPC server. Control of the conveyor belt system was carried out over LabVIEW via OPC using Programmable Logic Controllers (PLCs), Proportional-Integral-Derivative (PID) Controller, and Pulse Width Modulation (PWM) technique. The PLC program containing the PID controller, and the control program created in LabVIEW work simultaneously to provide speed control of the motors that move the conveyor belt. The PWM signal that rotates the motors was applied by the PLC and the feedback signal from the encoder adjusted the PID controller output signal at the desired speed. In this way, motor speed control was provided.

The remaining content related to the subject of this article and the work done is proceeded as follows: In the Sect. 2, OPC, Industry 4.0, LabVIEW, and PID controller works in the literature are declared. In the Sect. 3, the methods, materials, and software used in this study are mentioned. The rest of the article is completed with the results obtained, some future studies, and conclusions sections.

2 Related Works

Aytaç (2018) shows that with the completion of the industry 4.0 transformation in factories, smart factories will become widespread, and cheaper, better quality, faster, and more efficient production will be made in these factories. Attention was drawn to the important points in these transformations in the factories. The reasons for the importance and necessity of the systems that communicate with each other, detect the environment with sensors, and perform data analysis in the production in the industry 4.0 revolution have been revealed [1].

Tuncay and Mahir (2016) showed that with the technological concept called the Internet of Things (IoT), the communication of devices with each other, real-time data flow, and cloud storage in the internet environment positively affect the production in the industrial environment. They have shown the benefits of these systems based on intelligent sensors to minimize human error and evaluate information in real-time by decision support systems [2].

Hudedmani et al. (2017) presented the advantages of using a Programmable Logic Controller (PLC) in automation systems. The most important of these advantages are the effective operation of the process, safety, and ease of programming. In addition, with the widespread use of PLCs, it has been shown that systems such as Supervisory control and data acquisition (SCADA) and Distributed control systems (DCS) have begun to be used effectively with PLCs [3].

Faroqi et al. (2018) presented the design of a controller to control the speed of a DC motor using the PWM method. To perform the control, they have realized the controller design by utilizing the PID controller. The PID tuning method was used in the Matrix Laboratory software to calculate the coefficients of the PID controller. They developed a 2-wheel mechanism to adapt the controller they designed to real life. The motors connected to the wheels were driven with the help of a motor drive and the distance traveled by the prepared mechanism and the elapsed time were measured. With the measurements made, they observed that the PID method works efficiently [4].

Ramazan et al. (2013) designed a computer-controlled training tool in which they can adjust the DC motor speed using PLC. They used the PWM signal generated by the PLC to control the motor and measured the speed of the motor with the tachogenerator. Using OPC, they provided the communication between both the ProfiLab-based interface and the PLC. In addition, the through OPC server was able to perform real-time monitoring on the training set they designed [5].

Chvostek et al. (2006) aimed to realize today's complex control systems without being dependent on the equipment required by the manufacturers. A MATLAB/Simulink and OPC-based study has been carried out to adapt the complex control systems made in the simulation environment to reality. A control system was designed using OPC Toolbox, which is its library in MATLAB/Simulink, and PLC communication with the OPC server was realized. It has also been observed over Simulink while the designed control system is working with PLC. They obtained the results graphically in the observations made on Simulink [6].

3 Method and Material

3.1 Overview of PID Controller and Pulse Width Modulation

In this study, the PID controller algorithm and Pulse Width Modulation techniques were used to controlling conveyor belt speed. With the help of these techniques, conveyor belt speed control is provided.

3.1.1 PID Controller

PID controller is a form of controller frequently used in industrial control systems. The fact that it is used quite frequently in applications such as motor speed and position control is effective in choosing this controller for the project. PID controller is an algorithm that calculates the error value by comparing the input and output quantities of the system in a closed system and generates a new control signal by using the proportional, derivative, and integral gain coefficients [7]. A Parallel PID controller diagram and its connection to the system are given in Fig. 1. To generate a new control signal, it is necessary to know the signal at the output of the system. The output signal can be measured with the encoder in the motor, the signal measured with the encoder can be called a feedback signal. To summarize the function of the PID controller in the project, by processing the feedback signal obtained from the encoder, it detects possible differences in conveyor belt speed and tries to equalize it to the reference value applied at the input. PID controller algorithm and coefficients are shown in Eq. 1.

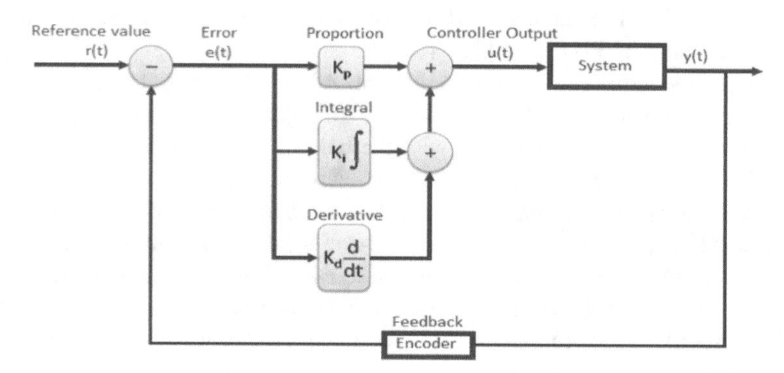

Fig. 1. PID Controller Diagram [8]

$$u(t) = K_p e(t) + K_i \int_0^t e(\tau)d\tau + K_d \frac{d}{dt} e(t) \tag{1}$$

In the above equation, K_p, K_d, K_i refers to the proportional, derivative, and integral gain coefficients, respectively. Also, $e(t)$ denotes the error signal and $u(t)$ denotes the controller output. The effects to be obtained after coefficient increase while setting the PID coefficients are shown in Table 1.

Table 1. Coefficient Increment Table [9]

Coefficient	Rise Time	Overshoot	Settling Time	Steady-State Error	Stability
K_p	Decrease	Increase	Small Change	Decrease	Degrade
K_i	Decrease	Increase	Increase	Eliminate	Degrade
K_d	Minor Change	Decrease	Decrease	No Effect	Improve if K_d small

3.1.2 Pulse Width Modulation

The Pulse Width Modulation (PWM) technique is used to drive the motor in the study. It generates a square wave based on the PWM technique and takes the frequency as a reference. The reason why it is preferred in the project is that it has less power loss and generally provides a quieter control. PWM frequency and duty cycle representation and calculation are shown in Fig. 2 and Eq. 2.

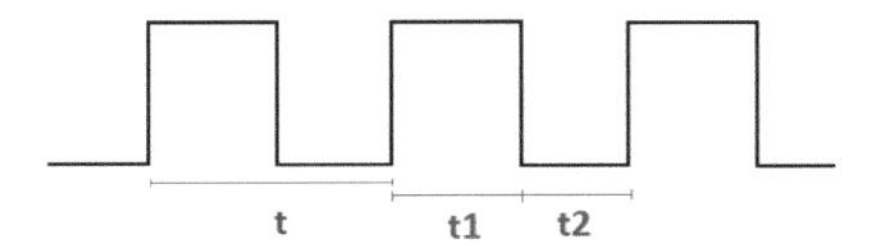

Fig. 2. Frequency and Duty Cycle

$$\text{Duty} - \text{Cycle} = \frac{\text{t1}}{t1 + t2} \tag{2}$$

3.2 Communication Standard and Protocol

Many communication protocols and standards have been developed to communicate different types of hardware and software from different manufacturers. In this study, OPC was used as the communication standard and MODBUS TCP/IP was used as the protocol. Many different variations of OPC have been introduced and OPC UA is used in this study. Systems such as OPC and MODBUS can cooperate with many other communication protocols. Figure 3 shows a connection diagram of an OPC UA-based system with other communication protocols and tools.

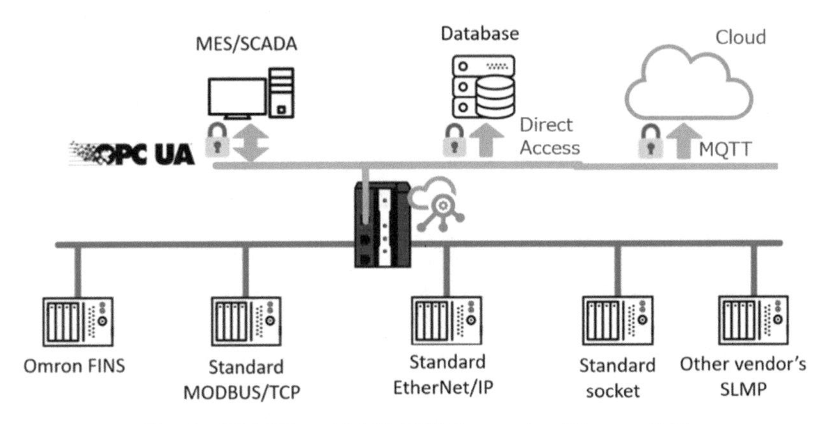

Fig. 3. OPC UA and MODBUS Connection Diagram [10]

3.2.1 OPC

OPC, one of the communication standards, enables the communication of different platforms. With the OPC server, the LabVIEW platform and PLCs were communicated, and real-time data transfer was provided. While there are OPC platforms developed by many different manufacturers, NI OPC was used in this study.

3.2.2 MODBUS TCP/IP

This communication protocol was developed by Modicon in 1979. It uses ethernet to move data between TCP/IP compatible devices. This communication protocol uses the Master/Slave method to establish communication between devices. During the project, while one master was used, another PLC set as a slave was used. With this method, data can be transferred at a speed of 100 Mbps [11]. It is a simple protocol to use and can be easily communicated with classical internet network tools.

3.3 Overview of Materials

The features and purposes of use of the equipment used in this project have been selected in accordance with the project. PLC is used to provide control and a DC motor is preferred to move the conveyor belt system. After these two main parts were selected, other parts that should be suitable were selected.

3.3.1 Programmable Logic Controller (PLC)

It is an extremely durable and programmable controller, which has input and output (I/O) units, a central processing unit (CPU), and a memory unit of its own, working in line with the program written in it and generally used in industrial areas. In this project, FX5U-32M model PLCs produced by Mitsubishi company were used and a DC motor was driven by the PWM method using the digital outputs on it and controlled with a PID controller. In the conveyor belt system to which DC motors are connected, there are more motors than the output number of a PLC. Therefore, two PLCs are used, and they are assigned as Master and Slave with MODBUS TCP/IP configurations. In this way, the number of inputs and outputs was increased and the communication of PLCs with each other was ensured.

CPU - It is the unit that regulates the arithmetic and logical operations of the PLC. In other words, it is the part that gives intelligence to the PLC.

Memory - The data stored in memory is used by the CPU. System data is permanently stored in ROM (Read Only Memory). Temporary information such as the status of inputs and outputs, timing and counters information are kept in RAM (Random Access Memory). In addition, memory units such as EPROM and PROM can be used in PLCs.

Input - It allows the input of the information received from the elements of the system to be controlled, such as sensors and buttons, to the PLC. This input information can be used according to the program programmed in the PLC.

Output - The output unit represents the output points of the PLC. Different types of outputs are available. Outputs may consist of circuits such as relays, transistors, and triacs. These outputs are active according to the situation in the program processed in the PLC. PLCs mostly contain digital outputs, but some models can also generate analog output signals (Fig. 4).

One of the features of the PLC used is that it has Transistor/Source output, this output is shown in Fig. 5. In this way, the PWM method can be used. While using the PWM method in conventional relay output PLCs may damage the PLC output, this problem is not encountered in PLCs with transistor outputs. The reason for this is that the PWM

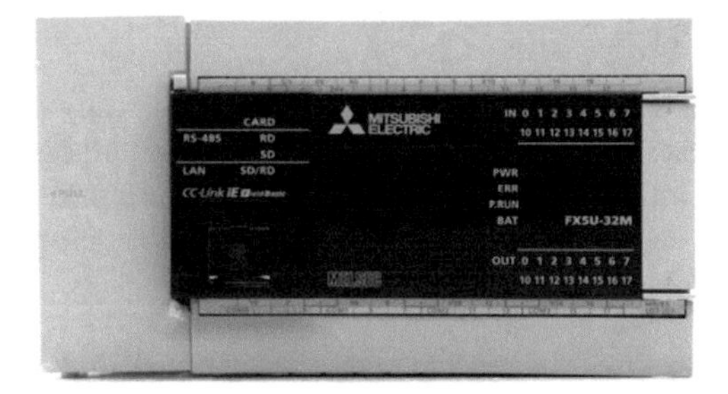

Fig. 4. Mitsubishi FX5U-32M Programmable Logic Controller [14]

method cannot reach the switching speed or cannot withstand this fast on-off operation because the relay output is a mechanical component.

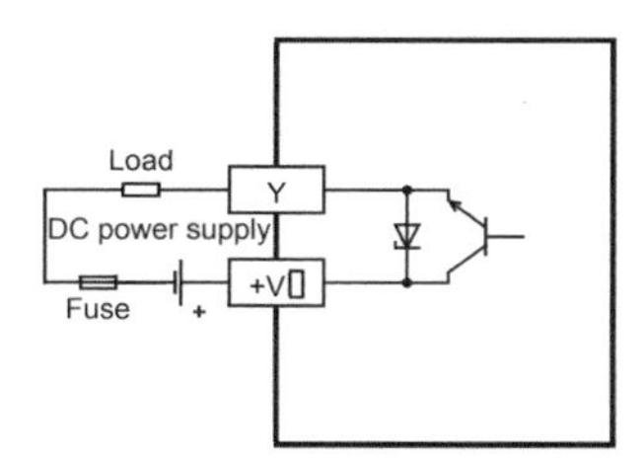

Fig. 5. Source Output Wiring [12]

3.3.2 DC Motor with Encoder

They are machines that work with direct current and convert electrical energy into mechanical energy. In the project, the speed of the conveyor belt was controlled by controlling the speed of the motor. The reason why it is selected with encoder is that PID controller is used. With the PID controller, the speed of the DC Motor was tried to be kept at the desired level. Data such as the speed and position of the motor are obtained by the encoder. 24V Pololu 37D Metal Gearmotor used in the project is shown in Fig. 6.

Fig. 6. Pololu 37D Metal Gearmotor [13]

3.3.3 Other Materials

Ethernet Switch: It is a network switching device with many ports on it. It allows devices to communicate with Ethernet to be connected. The purpose of use in the project is to connect PLCs with Ethernet connection for configuring MODBUS TCP/IP and communicating with each other. In addition, it is used to load the program written with GX Works3 software to PLCs by connecting a computer.

Motor Holder: Motors to be connected to the conveyor belt system need a holder to be connected to the system. The part designed via Autodesk Inventor software was printed by a 3D printer. The printed parts were mounted on the conveyor system together with the motors. The designed part and its assembled state are shown in Fig. 7.

Fig. 7. Designed Part and Mounted Motor Holder

3.4 Software

The two software used to perform the study are GX Works 3 and LabVIEW. While there is a program providing control in GX Works3, a visual control interface has been created in LabVIEW.

3.4.1 GX Works 3

Each PLC manufacturer offers its programming software to program the PLCs they produce. GX Works3 is the software offered by Mitsubishi for programming PLCs. The purpose of using this software in the project is to program the PLCs to be used. Speed control is provided by using the PID controller block on this software. With the program prepared over GX Works 3, LabVIEW works in coordination and the operation of the system can be observed.

3.4.2 LabVIEW

LabVIEW is a graphical programming environment where production test, validation, and control programs that engineers usually use can be created. LabVIEW blocks such as indicator and control were used. Thanks to the OPC connection, communication with the program prepared on GX Works3 can be provided, and control can be made via LabVIEW. As can be seen in Fig. 8, the programming made from the LabVIEW interface and the Tags created with the OPC Server is seen. With the adjustments to be made on the front panel, values such as proportional gain, integral time, and derivative gain, which are required for the PID controller, can be adjusted. In addition, the SET VALUE (DUTY CYCLE) control block is used to adjust the speed of the engine. During the project, Proportional gain was set to "200", Integral Time to "10" and Derivative Gain to "5". By comparing the Measured Value (Encoder) with the Set Value, the speed changes of the motor can be observed, as can be seen in the OUTPUT VALUE block.

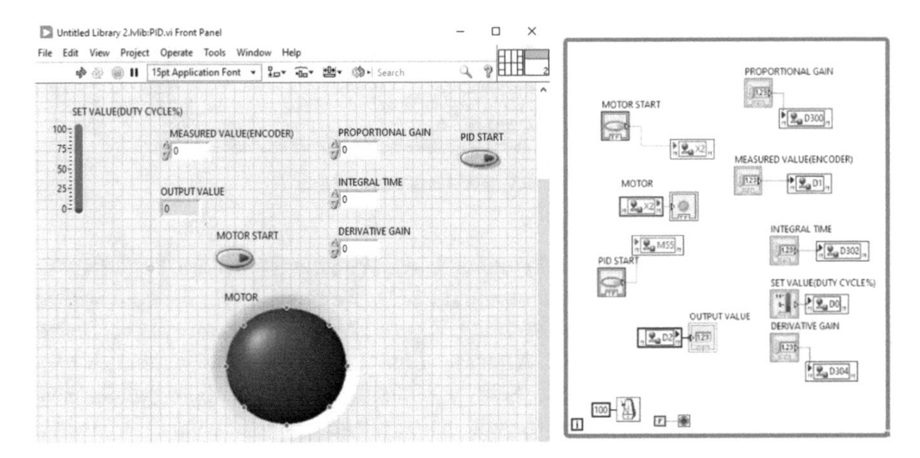

Fig. 8. LabVIEW Front and Program Panel

4 Discussion and Results

In this study, speed control of an OPC standard and MODBUS communication protocol-based conveyor belt was performed via LabVIEW. As a result of this study, the programs prepared over PLC and LabVIEW communicated and worked simultaneously. This work was carried out thanks to the OPC server, and any changes made via LabVIEW were also realized in real-time on the PLC. Increasing the control block's value in LabVIEW to increase the speed of the motor also increased the frequency of the PWM signal output by the PLC. In this way, the speed control of the motor was done in real-time by writing LabVIEW on the computer. Encoder data can also be viewed and recorded via LabVIEW. The encoder data is the actual speed data of the motor. These data are collected with the program shown in Fig. 9.

The key point in this study is the communication of hardware and software with each other. This communication was done using OPC and MODBUS, but this is a choice, not a necessity. For example, while the OPC server used in this study was NI OPC, many OPCs could be preferred as an alternative. There are small company-based differences

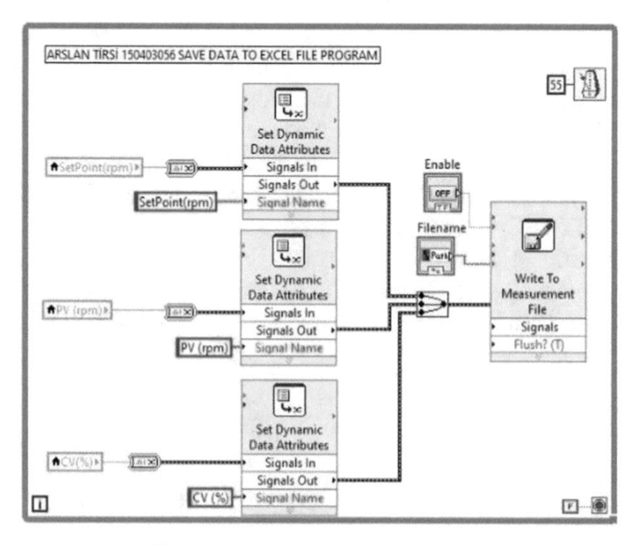

Fig. 9. LabVIEW Data Acquisition

between these OPC servers, which do the same job. With the OPC setup, LabVIEW and PLC were communicated, and the motor and PLC were mounted on the conveyor belt system where the system would be installed. Figure 10 shows the conveyor belt system to which the DC motor is connected. After real-time tests on the system, some problems encountered were resolved and the system was stabilized. One of these problems was achieving a stable operating range when adjusting the gain values in the PID. Since the desired operating range was achieved in a short time, auto-tuning was not tried. If the desired results were not obtained in the study, PID gain values would be tried to be found with the auto-tuning method.

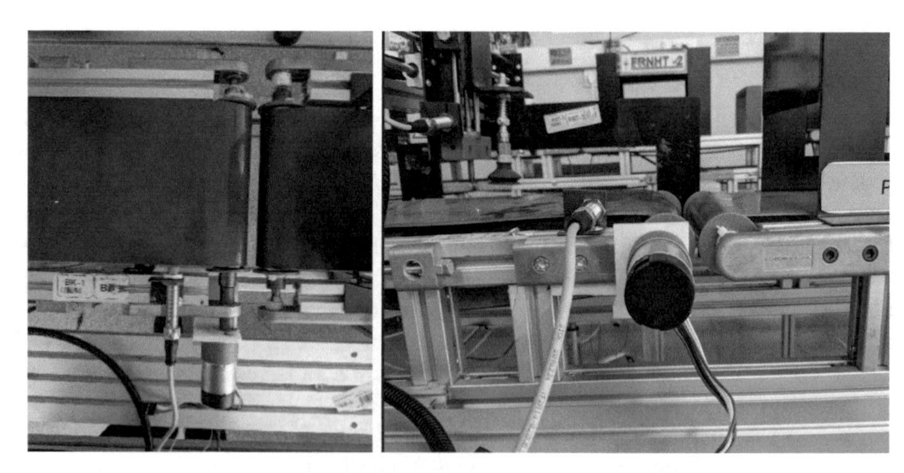

Fig. 10. Conveyor Belt System

5 Conclusions and Future Work

As a result of the study, it has been shown that the speed of a conveyor belt system can be controlled in real-time via a computer. Thanks to OPC, the LabVIEW platform on the computer and the PLC were successfully communicated, and the data sent via LabVIEW were processed in real-time by the PLC. In addition, with the communication of two PLCs over MODBUS TCP/IP, the applicability of this study to real systems has been proven by communicating with many PLCs in the industry. As for the control methods, it has been observed that the PID controller and PWM technique used in the study can be used efficiently in an OPC, PLC, and LabVIEW-based system. The future goal of the study, it is aimed to establish a wider and more advanced control network by integrating this system with a SCADA system.

References

1. Aytac, Y.: Endüstri 4.0 ve akıllı fabrikalar. Sakarya Üniversitesi Fen Bilimleri Enstitüsü Dergisi **22**, 546–556 (2018)
2. Kutay, M., Ercan, T.: Endüstride Nesnelerin Interneti (IoT) Uygulamaları. Afyon Kocatepe Üniversitesi Fen ve Mühendislik Bilimleri Dergisi **16**, 599–607 (2016)
3. Hudedmani, M.G., Umayal, R., Kabberalli, S.K., Hittalamani, R.: Programmable logic controller (PLC) in automation. Adv. J. Graduate Res. **2**, 37–45 (2017)
4. Faroqi, A., Ramdhani, M.A., Frasetyio, F., Fadhil, A.: Dc motor speed controller design using pulse width modulation. In: IOP Conference Series: Materials Science and Engineering, vol. 434, no. 1. IOP Publishing (2018)
5. Bayindir, R., Vadi, S., Goksucukur, F.: Implementation of a PLC and OPC-based DC motor control laboratory. In: 4th International Conference on Power Engineering, Energy and Electrical Drives, pp. 1151–1155 (2013). https://doi.org/10.1109/PowerEng.2013.6635773
6. Chvostek, T., Foltin, M., Farakas, L.: The adaptive PID controller using OPC toolbox. In: Proceedings of MATLAB Conference (2006)
7. Ogata, K.: Modern Control Engineering. Prentice-Hall, Hoboken (2010)
8. Smuts, J.F.: Process Control for Practitioners (2011)
9. Yu, B.: PID Control. The Chinese University of Hong Kong (2019)
10. Connecting PLCs to Various Software Packages with OPC UA. https://opcconnect.opcfoundation.org/2022/03/connecting-plcs-to-various-software-packages-with-opc-ua/
11. MODBUS Protokolü ve tüm özellikleri nelerdir? https://www.hubbox.io/tr/blog/veri-toplama/modbus-protokolu-ve-tum-ozellikleri-nedir
12. MELSEC iQ-F FX5U User's Manual (Hardware). https://dl.mitsubishielectric.com/dl/fa/document/manual/plcf/jy997d55301/jy997d55301u.pdf
13. Pololu Robotics & Electronics. https://www.pololu.com/category/116/37d-metal-gearmotors
14. MITSUBISHI CPU MODULE FX5U-32M. https://emea.mitsubishielectric.com/fa/products/cnt/plc/plcf/cpu-module/fx5u-32mt-ess.html

AI-Powered Cyber Attacks Threats and Measures

Remzi Gürfidan[1(✉)], Mevlüt Ersoy[2], and Oğuzhan Kilim[3]

[1] Isparta University of Applied Science, Yalvac Technical Sciences Vocational School, Isparta, Turkey
remzigurfidan@isparta.edu.tr
[2] Computer Engineering, Süleyman Demirel University, Isparta, Turkey
mevlutersoy@sdu.edu.tr
[3] Isparta University of Applied Science, Keçiborlu Vocational School, Isparta, Turkey
oguzhankilim@isparta.edu.tr

Abstract. We conducted a study examining the use of artificial intelligence (AI) and cybersecurity technology, which are two popular topics today. In the content of the study, you can find summary and vital information about artificial intelligence technology. In addition, brief information and some technical explanations about popular cyber threats and attacks are given. You can learn how artificial intelligence-supported cyber threats and attacks are carried out and what is done from current studies. You can also examine how the cyber-attack prevention and disposal systems developed with artificial intelligence were developed and the models implemented. It is predicted that this study will be an inspiration for researchers who want to work on new attack methodology or new cyber defense and attack avoidance techniques by combining artificial intelligence technology and cyber security field.

Keywords: Cyber Security · Cyber Attacks · Cyber Threads · Machine Learning · Artificial Intelligence

1 Introduction

After the Industry 4.0 industrial revolution report presented in 2013, the development of technological infrastructures and devices has accelerated. Developing technology areas accelerate the development of new technologies and these developments are closely followed by the universal society. Many platforms such as devices, communication and workstations used in daily life have been established with technology support. There is almost no platform left without the internet and informatics. The industry 4.0 industrial revolution includes autonomous robots, simulation technologies, big data, augmented reality, additive manufacturing, cloud technologies, blockchain, internet of things, artificial intelligence, and cyber security [1]. Among these topics, artificial intelligence technologies and the field of cyber security stand out in a remarkable way. Artificial intelligence technologies are widely used in various fields such as suggestion systems, language translations, healthcare services, voice assistants, and assistive robots.

D. J. Hemanth et al. (Eds.): ICAIAME 2022, ECPSCI 7, pp. 434–444, 2023.
https://doi.org/10.1007/978-3-031-31956-3_37

Cyber security is a field of study that has become very popular in recent years. Financial frauds, theft of accounts on electronic platforms, attacks on commercial sectors and platforms have been a hot topic in recent years. Another point that can be said more seriously is that international conflicts and wars are carried out with cyber-attack and defense methods. It would be simple and insufficient to characterize the information-based disruption and destruction efforts of power plants, production plants, official government accounts, attack, and defense systems as an ordinary technological field. The field of cyber security is so strong and comprehensive that due to the effectiveness of this field, it causes the fields of protection and security such as blockchain and cryptology to be constantly renewed.

2 Artificial Intelligence Technologies

Among the topics in Industry 4.0, artificial intelligence and cyber security have become popular topics today. Renewal of many systems currently in use by combining them with artificial intelligence technology and obtaining more successful and performance results increases the interest in this field. Many studies and algorithms related to the field of artificial intelligence continue to be developed. In Fig. 1, the sub-branches of artificial intelligence and the developed algorithms are shown in a diagram. The most vital requirement for the application of algorithms used in artificial intelligence technology is the data set required for the model to be trained. Many data that will serve the purpose you want to achieve are brought together and used in the model training to be developed. The increase in the number of data suitable for the concept and its heterogeneity are important factors that increase the success of the developed model.

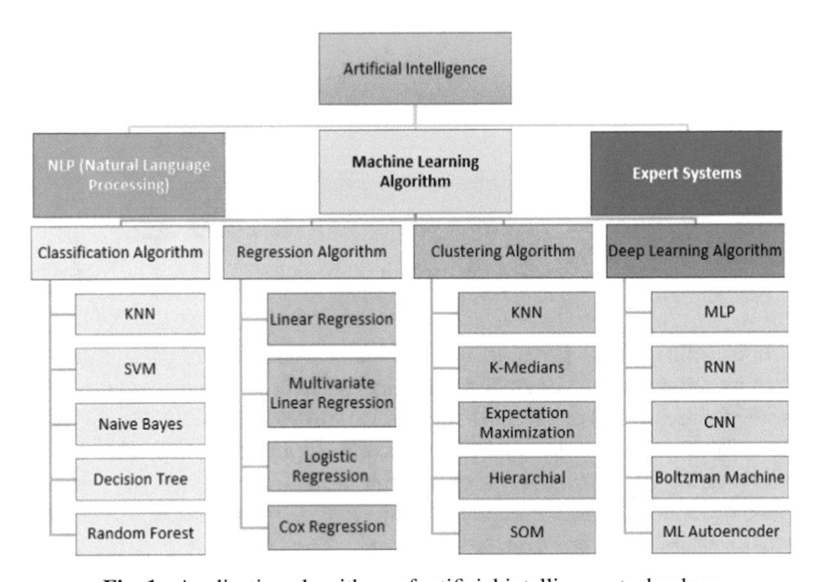

Fig. 1. Application algorithms of artificial intelligence technology

Machine learning algorithms are the most preferred algorithms in the applications of artificial intelligence. It has a rich variety as shown in Fig. 1. While labeled data sets are needed for machine learning algorithms, which are used extensively in classification, prediction and clustering processes, deep learning algorithms can give very successful results on unlabeled data.

3 Cyber Security and Cyber Threats

Information technology security is one of the most sensitive issues of digital transformation processes that have been increasing in recent years. All activities and processes carried out to protect the workflows and data of individuals and companies from external threats and unauthorized access can be grouped under the heading of cyber security. The concept of cyber security can be defined as all the efforts to defend and protect against unauthorized access and malicious attacks of computers, servers, electronic circuits, in short, all systems with technological infrastructure. We can detail the concept of cyber security under four headings: network security, application security, information security and operational security. Figure 2 shows the schematic of this organization.

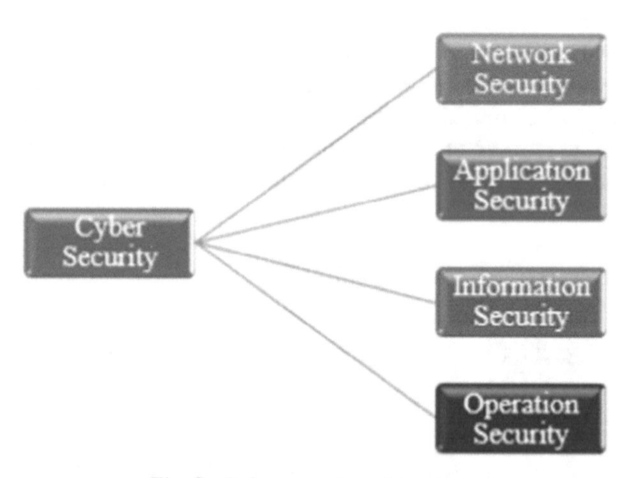

Fig. 2. Cyber security subheadings

Within the title of network security, it includes the activities of protecting computer networks that are desired to be manipulated by malware or attackers. The application security title focuses on the protection of developed software from external threats. Application security processes start with the software development process of the application. Under the heading of information security, the protection of the data of individuals, companies, or devices to be stored and the security measures during data transfer are examined. Operational security, on the other hand, covers the security of the transactions to be carried out, data processing processes and authorization procedures.

Taking precautions against cyber threats will prevent data loss and disruption of workflow for individuals and companies. It would be a correct grouping to examine cyber-attacks and threats under ten headings so far. In Fig. 3, there is a representation of cyber-attacks and threats under headings.

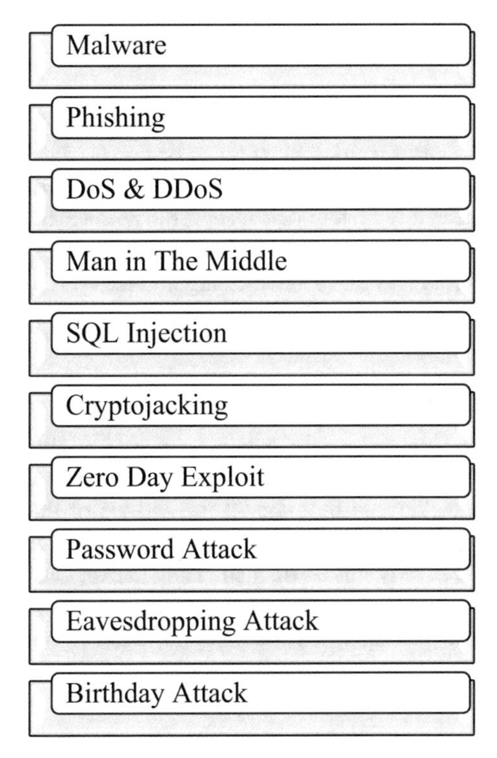

Fig. 3. Types of cyber-attacks and threats

3.1 Malware

The word malware is used as a definition that covers harmful and malicious software such as worms, viruses, Trojan horses [2]. This malicious software can change the network configurations of personal and corporate computers without the user's authority, reproduce itself in numbers within the computer, and change access permissions [3]. Since this situation creates a great vulnerability, it poses a high-level dangerous threat. To protect against such attacks, programs should be kept up-to-date, and the firewall should not be turned off.

3.2 Phishing

Such attacks are carried out by methods that are generally described as deception or traps. At the root of the attacks are pitfalls such as spoofing a page, spoofing an email, spoofing a panel. The aim is to capture people's credit cards, login usernames and passwords. Clicking on links sent via emails or messages is exposed to attack. The most effective way to protect from such attacks is to carefully check the accuracy and security of the incoming message or e-mail source.

3.3 DoS and DDoS

Denial of Services (DoS) and Distributed Denial of Services (DDoS) attack types are attacks that are carried out to restrict, slow down or prevent the operation of real-time service systems. The basis of these attacks is to make multiple and continuous requests to a web server, a database or any system that accepts the request. In this way, it is aimed to slow down, stop or out of service the system. In order not to be exposed to these attacks, script-based protection, DDOS and DoS filtering services or DoS or DDoS protected servers should be preferred.

3.4 Man in the Middle (MITM)

In this attack type, attackers can spoof a web service or a network, or hide between the user and the web service, or between the user and the network access to be used. With this move, the attacker can monitor all the actions of the user on the internet. From this moment on, there is no data security for the user. Movements performed on the monitored network may not be tracked. The attacker can modify any operation performed. In such cases, attacks such as ARP Poisoning, DNS Spoofing, Port Stealing, STP Mangling, DHCP Spoofing, ICMP Redirection, IRDP Spoofing, Route Mangling, Traffic Tunneling are frequently encountered cyber threats. How MITM attacks work is tried to be explained with the image in Fig. 4.

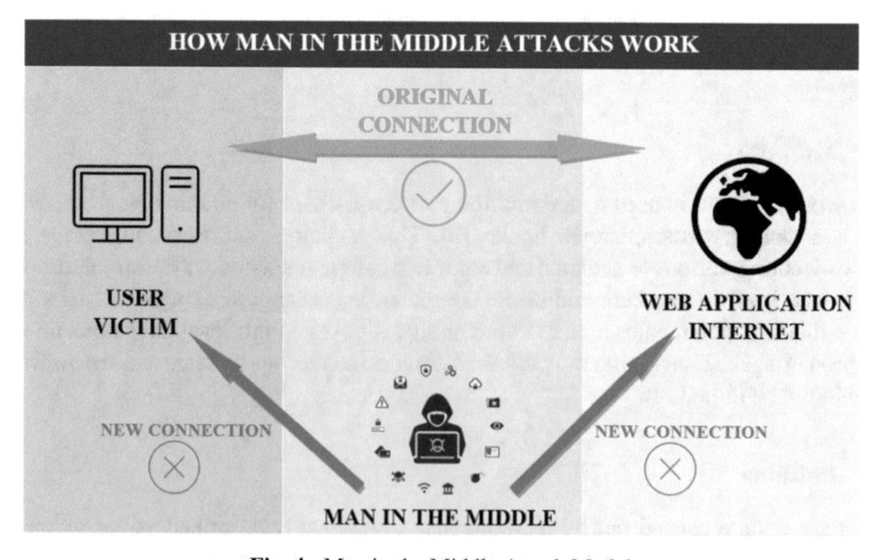

Fig. 4. Man in the Middle Attack Model

It is very difficult to detect such attacks. For this reason, it will be the right move to take the precautions to prevent being attacked as much as possible. If we briefly list these methods.

- Authentication plays an important role in verifying the message source.
- Authentication methods should be preferred to protect against fake packets in communication with the server.
- TLS certificate will make TCP stronger against MITM attack.
- Avoiding passwordless access to wireless access points.
- Prefer strong passwords.
- Paying attention to the fact that the firewall and protocols are always open and up to date
- Sites with https instead of http should be used to avoid unsafe and potentially fraudulent sites.

3.5 SQL Injection

SQL queries are used as databases in many software applications used today. Especially in web applications, accessing and tampering with the database by manipulating the query sentences sent to the remote server should be examined under this heading. In a SQL Injection attack, destructions such as deleting records in the database, changing the records, obtaining a copy, adding malicious code pieces to the system can be performed. To carry out this attack, it is necessary to find security vulnerabilities in the application that uses the database. Among the methods, there are attacks based on user input, attacks based on cookies, attacks based on http headers. The most famous attack types are In-band, Blind or Inferential and Out-of-band attacks. It is very important to keep the software versions used to be protected from such attacks. In addition, Web Application Firewall (WAF) can prevent these attacks by filtering out potentially dangerous web requests.

3.6 Crypto Jacking

In this type of attack, the attacker uses the resources of the computer or mobile device of the person he chooses as the victim to gain income from crypto mining without permission and authorization. In this type of attack, a malicious software running in the background is usually installed on the other party's device, without being noticed by the victim. The malware processes cryptocurrency mining or steals assets from the victim's crypto wallets. There are two popular ways to carry out the attack. The first of these is to upload a crypto mining code that will run secretly on the computer by clicking on a link to the victim via e-mail. The second is to infect a web page or ad content with JavaScript code that runs automatically in the browser of the victim's device. Devices exposed to this attack experience poor performance, overheating, and abnormal values in resource usage. There are several effective methods of avoiding this type of attack. Using a good protective cybersecurity program, using ad-blocking applications, disabling JavaScript are recommended protection methods.

3.7 Zero Day Exploit

In this type of attack, even if the vulnerability on the existing system is noticed by the developers, the attack is carried out by hackers without the opportunity to fix it. The attack is organized to disrupt the working order of the system or to steal data records. Before a planned patch of an improved software is revealed, a malicious code snippet called exploit is prepared by hackers. Afterwards, this code snippet is inserted into the existing application with social engineering methods or a technical process. These attacks are usually carried out on operating systems, web browsers, internet of things (IoT) systems. It is a type of attack that is very difficult to detect and is delayed due to its nature. For this reason, keeping the applications and operating systems on the devices constantly updated is of vital importance for protection. The firewall running on the devices must remain up to date and open.

3.8 Password Attack

It is a type of cyber-attack carried out on the encryption mechanism that is widely used when logging into any platform. Also known as Brute Force, this type of attack is based on the principle of trying to break the victim's password by constantly generating random values. There are simple ways to prevent this type of attack. A solution can be produced for password systems that are entered incorrectly above a certain number, by stopping or blocking measures.

3.9 Eavesdropping

Based on this attack type, there are eavesdropping processes. Cyber attackers perform eavesdropping on data transmitted through network connections used by devices such as smartphones and computers. There are two ways to do this, active listening, and passive listening. In passive listening processes, no changes are made to the listened data. In active listening, attackers can perform ancestors and modify the data they listen to. This attack type is quite like the MITM attack. Passive eavesdropping is almost impossible to detect unless there is an obvious problem with network operation that you will notice. To protect from these attacks, it is important to use strong passwords, to set different passwords for different applications, not to prefer collectively used wireless network connections, to avoid unsafe links and connections, to stay away from links that contain surveys, donations, campaigns, and advertisements.

3.10 Birthday Attack

This type of attack is a type of attack against mechanisms that ensure the authenticity of a message to be transmitted, an enhanced application or a digital signature. In hash algorithms, a fixed-length hash value is created independently of the data to be processed. There are two random content generation processes that are equivalent to the hash value generated at the base of the attack. When random fake content is obtained that will give a digest value equal to the hash value of the content of the original message, the original and the fake message are easily interchangeable, and this situation becomes undetectable.

This attack type can be thought of as a more developed and customized version of the Brute Force attack type. It depends on the higher probability of multiplication between random attack attempts and a fixed degree of permutation, as in the Birthday paradox based on development.

4 Standard- AI-Powered Cyber Attacks and Protection Solution Proposal Works

Many of the attacks are organized on Internet of Things (IoT) technologies. In addition, the number of web technologies and servers is not to be underestimated. Cyber-attacks and defense methods on IoT devices are discussed in detail by Kuzlu et al. This research [4] is shown in Table 1.

Cyber attackers are constantly changing their attacks by empowering them with artificial intelligence-based techniques [17]. In this study, the existing literature on artificial intelligence-guided cyber-attacks, inappropriate use of artificial intelligence in cyberspace and the negative effects of artificial intelligence-guided cyber-attacks has been reviewed.

Papernot et al. proposed an attack model based on a new backup training algorithm that uses synthetic data generation to generate competing samples that are misclassified by black-box DNNs. In the attack design, a remote DNN submitted by Metamind was verified by targeting and forced to misclassify 84.24% of the samples [18]. The result of all these studies is that it has the potential to cause traffic accidents in autonomous vehicles and is to show an example of misclassification of traffic signs by AI algorithms [19].

Finlayson et al.'s study shows that hostile samples can manipulate deep learning systems in three clinical areas. They have concretely demonstrated why and how attacks can be carried out on imaging technologies in the healthcare field. It is necessary to raise awareness in demonstrating existing vulnerabilities when using deep learning systems in clinical settings and to further investigate the field-specific features of medical learning systems from the machine learning community.

Another study investigated the current state-of-the-art technology of next-generation threats using AI techniques such as deep neural network for the attacker and evaluated the efficiency of AI-Medium attacker defense methods under a cloud-based system where the attacker uses an AI technique to launch an advanced attack by finding the shortest attack path. A Markovian dynamic game is proposed. CVSS metrics were used to measure the values of the zero-sum game model in decision making [20].

Siroya conducted a study on the role of AI in Cyber Security. He proposed a new system with the idea he gained because of the literature study of this new field of study. This system proposal is designed to protect against cyber threats. We can think of the constructed structure as layered. There will be a firewall in front of the server, accessing it requires a Captcha authentication. He suggested that there should be a machine learning model right after these stages. Botnet detection will be done in this machine learning area. Then either the process will be stopped, or the anti-virus software will be activated [21].

Table 1. IoT attack and methods of protecting against the attack

IoT Attack	Methods Of Protecting Against The Attack
Physical attacks	Tamper-proof hardware should be preferred. Hardware-based security connections [5] should be provided. Destroy commands should be used. Self-destruct should be carried out [6]
Man-in-the-Middle	Software updates should be followed, and software should be kept up to date. The firewall should be configured according to its service purpose. It should be noted that the preferred passwords are strong. Wi-Fi should not be preferred if you are not sure of its security [7]
Bluetooth MITM	Turn off the visibility of used devices and make them undiscoverable. Block unknown devices. Use two-factor authentication [6]. Use strong matching methods such as the Elliptic Curve Diffie-Hellman public key cryptography or the Out-of-Band method [8]
False data injection	Make sure your software is up to date. Set firewalls to correct and appropriate configuration. Do not use Wi-Fi if you are not sure of its security [7]. Use anomaly detection [8]
Botnets	Run antivirus scans at specific time intervals. Avoid suspicious email attachments or download links, run your updates regularly [9]
Mirai botnet	Run updates regularly, change default login on IoT devices [10]
Denial of service attacks	Use a DoS protection service, use an antivirus and firewall [11]
Bricker Bot	Run your software updates regularly. Use the appropriate firewall for your purpose. Configure authentication mechanisms for actions such as login [12]
Fuzzing and symbolic execution	Properly secure a login of your choice on your device and free it from threats. Limit your device's allowed entries [13]
Dataset poisoning	Scan your device periodically such as data cleaning and anomaly detection, use micro models [14]
Algorithm poisoning	Use a validation dataset to remove local models that have a large negative impact on the error rate or cause large losses [15]
Model poisoning	Protect the system in which the model resides [16]

Although AI and information systems are at the beginning of their application-based development and have many shortcomings, they are very useful for preventing cyber-attacks and threats. AI technology can be used for automation and false positive detection. They are also involved in the execution of daily tasks of computers, such as analyzing network traffic, providing access based on a criterion of rules, and detecting anomalies in the system. Cyber-attacks are becoming more common day by day and are becoming increasingly complex as an organization. AI can assist in detecting malicious cases by monitoring a broader scope of factors, identifying normal and abnormal activity patterns without looking for different traces of malicious activity. Identifying this activity through AI in the early stages of the attack will help mitigate and locate the attack. The last application of AI that will be useful is immunity. If AI can be trained like the human body's immune system, it can find and neutralize all threats much more quickly and effectively [22].

AI frequently uses supervised machine learning in today's cybersecurity field. A malicious and threat database is provided for the training of the model to be developed as a requirement of this method. Although it has been found to be successful, unsupervised machine learning method comes into play because it is weak in terms of providing the database and applicability to all threats. In unsupervised learning, threats and normal usage times are learned automatically by the model. This method also requires more human intervention from outside.

Neural networks, on the other hand, can help diagnose malicious IP traffic when applied together with clustering techniques [23]. Skin learning algorithms, on the other hand, are used to continuously identify new methods of protection against a known virus with maximum protection and minimum data loss. Most importantly, however, predictive capabilities are used to catch problems before they are triggered by calculating the behavior patterns used in the system [Chan].

5 Conclusion

Artificial intelligence technologies are in their most popular era today. Today, it is combined with almost all fields of science, contributing to different fields and seems to progress in a similar way. The issue of cyber security is also very important for individuals, institutions and even states. It will be inevitable for such two magnificent technological fields to be used together. In this study, we have compiled information that reveals how destructive cyber-attacks can be created when these two magnificent topics are used together, and how strong and reliable cyber defense techniques can be designed. There is no doubt that AI is the foundation of today's best anomaly detection systems, and as it evolves, it will become pervasive in different cybersecurity divisions.

References

1. Ghobakhloo, M.: Industry 4.0, digitization, and opportunities for sustainability. J. Clean. Prod. **252**, 119869 (2020)
2. Almomani, A., Altaher, A., Ramadass, S.: Application of adaptive neuro-fuzzy inference system for information security. J. Comput. Sci. **8**(6), 983–986 (2012)
3. Bauer, J.M., van Eeten, M.J.G.: Cybersecurity: stakeholder incentives, externalities, and policy options. Telecommun. Policy **33**(10–11), 706–719 (2009)
4. Kuzlu, M., Fair, C., Guler, O.: Role of artificial intelligence in the Internet of Things (IoT) cybersecurity. Disc. Int. Things **1**(1), 1–14 (2021)

5. Woo, S.: The right security for IoT: physical attacks and how to counter them. In: Minj, V.P., (ed.) Proft From IoT. https://iot.electronicsforu.com/headlines/the-right-security-for-iot-physical-attacks-and-how-to-counter-them/. Accessed 13 June 2019
6. Akram, H., Dimitri, K., Mohammed, M.: A comprehensive IoT attacks survey based on a building-blocked reference mode. Int. J. Adv. Comput. Sci. Appl. (2018). https://doi.org/10.14569/IJACSA.2018.090349
7. Cekerevac, Z., Dvorak, Z., Prigoda, L., Čekerevac, P.: Internet of things and the man-in-the-middle attacks–security and economic risks. MEST J. **5**, 15–25 (2017). https://doi.org/10.12709/mest.05.05.02.03
8. Mode, G., Calyam, P., Hoque, K.: False data injection attacks in Internet of Things and deep learning enabled predictive analytics (2019)
9. Porter, E.: What is a botnet? And how to protect yourself in 2020. SafetyDetectives, Safety Detectives. https://www.safetydetectives.com/blog/%20what-is-a-botnet-and-how-to-protect-yourself-in/#review-2. Accessed 28 Dec 2019
10. Hendrickson, J.: What is the Mirai botnet, and how can I protect my devices? How to geek, LifeSavvy media. https://www.howtogeek.com/408036/what-is-the-mirai-botnet-and-how-can-i-protect-my-devices/. Accessed 22 Mar 2019
11. Understanding denial of service attacks. Cybersecurity and infrastructure security agency CISA. https://www.cisa.gov/ncas/tips/ST04-015. Accessed 20 Nov 2019
12. BrickerBot Malware emerges, permanently bricks IoT devices. Trend Micro, Trend Micro Incorporated. https://www.trendmicro.com/vinfo/us/%20security/news/internet-of-things/brickerbot-malware-permanently-bricks-iot-devices. Accessed 19 Apr 2017
13. Jurn, J., Kim, T., Kim, H.: An automated vulnerability detection and remediation method for software security. Sustainability **10**, 1652 (2018). https://doi.org/10.3390/su10051652
14. Moisejevs, I.: Poisoning attacks on machine learning. towards data science, medium. https://www.towardsdatascience.com/poisoning-attac%20ks-on-machine-learning-1f247c254db. Accessed 15 July 2019
15. Fang, M., et al.: Local model poisoning attacks to Byzantine-Robust federated learning. In: Usenix security symposium.arXiv:1911.11815. Accessed 6 Apr 2020
16. Comiter M.: Attacking artifcial intelligence. Belfer Center for Science and International Afairs, Belfer Center for Science and International Afairs. http://www.belfercenter.org/sites/default/fles/2019-08/AttackingAI/AttackingAI.pdf. Accessed 25 Aug 2019
17. Guembe, B., et al.: The emerging threat of AI-driven cyber attacks: a review. Appl. Artif. Intell. 1–34 (2022)
18. Papernot, N., McDaniel, P., Goodfellow, I., Jha, S., Celik, Z.B., Swami, A.: Proceedings of the 2017 ACM on Asia Conference on Computer and Communications Security, pp. 506–519. ACM (2017)
19. Yamin, M.M., Ullah, M., Ullah, H., Katt, B.: Weaponized AI for cyber attacks. J. Inf. Secur. Appl. **57**, 102722 (2021)
20. Alavizadeh, H., Jang-Jaccard, J., Alpcan, T., Camtepe, S.A.: A Markov Game Model for AI-based Cyber Security Attack Mitigation. arXiv preprint: arXiv:2107.09258 (2021)
21. Siroya, N., Mandot, M.: Role of AI in Cyber Security. Artif. Intell. Data Min. Appr. Secur. Fram. 1–9 (2021)
22. Chan, L., et al.: Survey of AI in cybersecurity for information technology management. In: 2019 IEEE Technology & Engineering Management Conference (TEMSCON), pp. 1–8. IEEE, June 2019
23. Veiga, A.P.: Applications of Artificial Intelligence to Network Security. arXiv [cs.CR], 27 Mar 2018

A QR Code-Based Robust Color Image Watermarking Technique

Gökhan Azizoğlu[1(✉)] and Ahmet Nusret Toprak[2]

[1] Department of Computer Engineering, Sivas Cumhuriyet University, Sivas, Turkey
gazizoglu@cumhuriyet.edu.tr

[2] Department of Computer Engineering, Erciyes University, Kayseri, Turkey
antoprak@erciyes.edu.tr

Abstract. With the recent development of image manipulation and distribution software, protecting digital images' copyright and intellectual property rights has become more challenging. One of the most effective approaches that address this problem is robust image watermarking. The fundamental purpose of this study is to demonstrate that the fault tolerance and high data capacity features of QR codes can be utilized in robust image watermarking. This study hence introduces a robust QR code watermarking technique based on Discrete Wavelet Transform (DWT) and Discrete Cosine Transform (DCT) for color images. The proposed scheme first transforms the original host image from RGB to YCbCr color space. It then converts the watermark information to be hidden into a QR code and embeds it into the host image. The results of various experiments on color images reveal that using QR codes as watermarks enables effective extraction of embedded data and improves robustness to different attacks, including histogram equalization, noise insertion, and sharpening.

This study is an extended version of a paper that appeared in ICAIAME 2022 [1].

Keywords: discrete wavelet transform · digital image watermarking · data hiding · QR code watermarking

1 Introduction

Modern image manipulation and internet technologies have made digital image duplication, reproduction, modification, and distribution very easy and cost-effective. The protection of digital images has become a crucial concern because of these advances. Robust image watermarking, commonly utilized in copyright protection, integrity control, digital fingerprinting, and broadcast monitoring, provides an alternative solution to these problems. In the robust watermarking methods, it is aimed that the watermark embedded in the image withstand image processing and geometric transformation attacks [2]. For this purpose, many robust watermarking methods have been presented in the literature [3–5]. However, in these methods, attacks on the watermarked image may cause damage to the watermark information.

D. J. Hemanth et al. (Eds.): ICAIAME 2022, ECPSCI 7, pp. 445–457, 2023.
https://doi.org/10.1007/978-3-031-31956-3_38

Previous studies have revealed that using a QR code (an acronym for quick response code) can improve the robustness and capacity of image watermarking methods. The objective of QR code-based image watermarking approaches is to take advantage of the intrinsic error-correction features of the structure of the QR code, in conjunction with its great data capacity. The error-correction functionality of the QR code ensures that the QR code is correctly decoded despite errors unless the error surpasses the error correction capacity. This way, the robustness of the watermark embedded as a QR code could be increased.

Considerable research has been done in recent years to increase the security of watermark information and make image watermarking methods more robust by using the inherent structure of QR codes. In one of these studies, Li et al. (2017) introduces a robust watermarking scheme to ensure the integrity of QR codes. In their method, a QR code produced with the watermark belonging to the copyright owner was used as a cover image, and the same watermark information was embedded in the original image using Discrete Wavelet Transform (DWT) and Singular Value Decomposition (SVD) techniques [6]. Patvardhan et al. (2018) proposed a digital image watermarking technique using DWT and SVD techniques to embed a watermark in the color image images in YCbCr color space. This study demonstrated that employing QR codes as watermark information improved capacity, robustness, and imperceptibility [7]. In a method proposed by Hsu et al., Discrete Cosine Transform (DCT) is used to embed QR code information into a cover image. This method enhances robustness by employing the fault correction mechanism of the QR code [8]. In another work, Sun et al. (2021) proposed an encrypted digital image watermarking scheme based on DCT and SVD techniques. In this work, the authors demonstrated that watermark information could be extracted from a QR code even if it has been significantly distorted by various attacks, such as image processing and geometric transformation [9]. Pan et al. (2021) proposed an image watermarking scheme based on meta-heuristic algorithms and using DWT-SVD techniques to embed watermark information in QR codes. Their method aims to increase capacity while enhancing robustness and imperceptibility [10].

This study presents a blind and robust QR code-based watermarking method for color images using a combination of DWT and DCT. In the proposed method, a QR code is employed as a watermark. Also, the proposed method requires neither the original watermark nor the original image for watermark extraction. In the proposed technique, the information is first shuffled using the Arnold transform and then converted into a QR code. After converting the color host image from RGB to YCbCr, the watermark information is embedded in the coefficients, which are obtained by applying DWT and DCT on the Cr channel. The experimental results reveal that using QR codes allows lossless watermark extraction and improves the robustness against different attacks, such as histogram equalization, noise insertion, compression, and sharpening. The following are the most important contributions made by our work:

1. We introduce a blind and robust QR code-based watermarking method for color images employing DWT and DCT.
2. We used a QR code as a watermark to extract the watermark from the watermarked image without data loss.

3. The proposed method supports blind extraction, which eliminates the need for the original image or watermark.

The remaining part of this study proceeds as follows: The second section describes the details of the proposed blind and robust QR code watermarking method, the third section presents experimental results and evaluation of the proposed technique, and the final section gives the conclusion.

2 Blind Robust QR Code Watermarking Scheme

This section details the proposed QR code-based image watermarking algorithm. The following subsection explains QR Code. Section 2.2 gives brief information about the Arnold transform. Section 2.3 and 2.4 describes the proposed watermark embedding and extracting procedures, respectively.

2.1 QR Code

A QR Code is a two-dimensional (2D) barcode created in 1994 by Denso Wave, a Toyota company, for tracking inventory in the production of vehicle parts [11]. To date, QR codes have been utilized in various industries, including digital payments, data gathering, public service, logistics management, and product traceability. QR codes have some significant advantages over conventional barcodes such as large data capacity, quick readability, and high fault tolerance [9]. As can be seen in Fig. 1, there are different types of QR codes, from the first version (21 × 21) to the fortieth version (177 × 177).

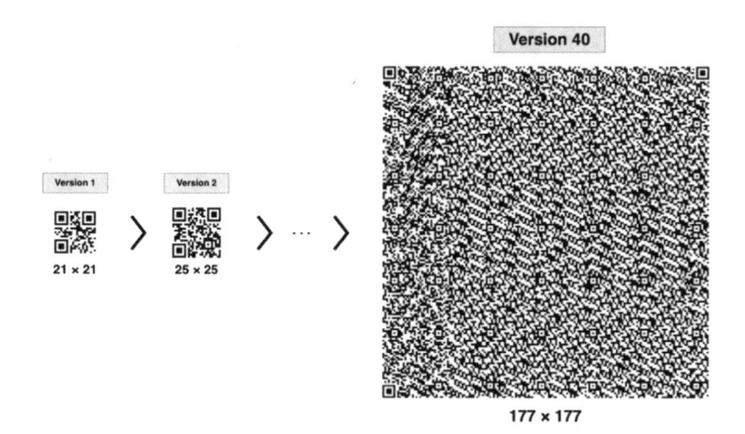

Fig. 1. There are a variety of QR code versions, starting at version 1 and going up to version 40

The QR code is capable of correcting errors so that it can be properly decoded regardless of whether it is noisy or partially destroyed. There are four levels of error corrections (ECC): low level (%7), medium level (%15), quarter level (%25), and high level (%30). The largest amount of data a QR code may store varies according to the

Table 1. The quantity of information that can be encoded in a QR code version_40 [11].

Version	Modules	ECC Level	Data Bits (mixed)	Numeric	Alpha-numeric	Binary	Kanji
40	177 × 177	L	23,648	7,089	4,296	2,953	1,817
		M	18,672	5,596	3,391	2,331	1,435
		Q	13,328	3,993	2,420	1,663	1,024
		H	10,208	3,057	1,852	1,273	784

character type and error correction level. The possible amount of data stored for different character types and error levels in version 40 is given in Table 1 [11].

Each QR code is made up of square modules that are laid up in a regular square array, as well as function patterns and an encoding region. The QR code is encircled by a border that is referred to as the quiet zone. QR codes include specific shapes named function patterns, which must be placed in particular sections of the QR code to allow scanners to recognize and direct the decoding process effectively. Figure 2 represents the general structure of the QR codes [11].

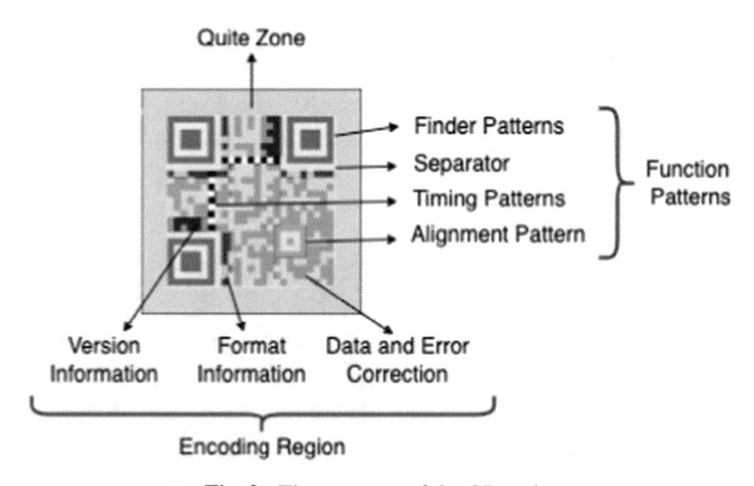

Fig. 2. The structure of the QR code.

2.2 Arnold Transform

After performing a shuffling operation, the Arnold transform is an algorithm that can restore pixels to their original state. Because of its ability to spread pixels throughout an image, the Arnold transform is widely used in several watermarking methods. This feature ensures that any faults in the watermarked image are spread throughout the image so that the watermark can be restored even though errors occur on the watermarked image.

The Arnold Transform for the N × N image is computed as follows [12]:

$$\begin{pmatrix} f' \\ g' \end{pmatrix} = \begin{pmatrix} 1 & 1 \\ 1 & 2 \end{pmatrix} \begin{pmatrix} f \\ g \end{pmatrix} mod\,(N) \tag{1}$$

where, f and g represent the positions of the original image pixels, f' and g' represent the position of the scrambled image. To maximize the effectiveness of the Arnold Transformation, the technique must be repeated multiple times. N transforms later, the image is turned to white noise [13].

2.3 QR Code Embedding Procedure

During the process of QR code watermark embedding, an original cover image I^O, size of 1024 × 1024 × 3, and a binary QR code watermark, size of 128 × 128, were used. The schematic diagram of the QR code embedding procedure is demonstrated in Fig. 3.

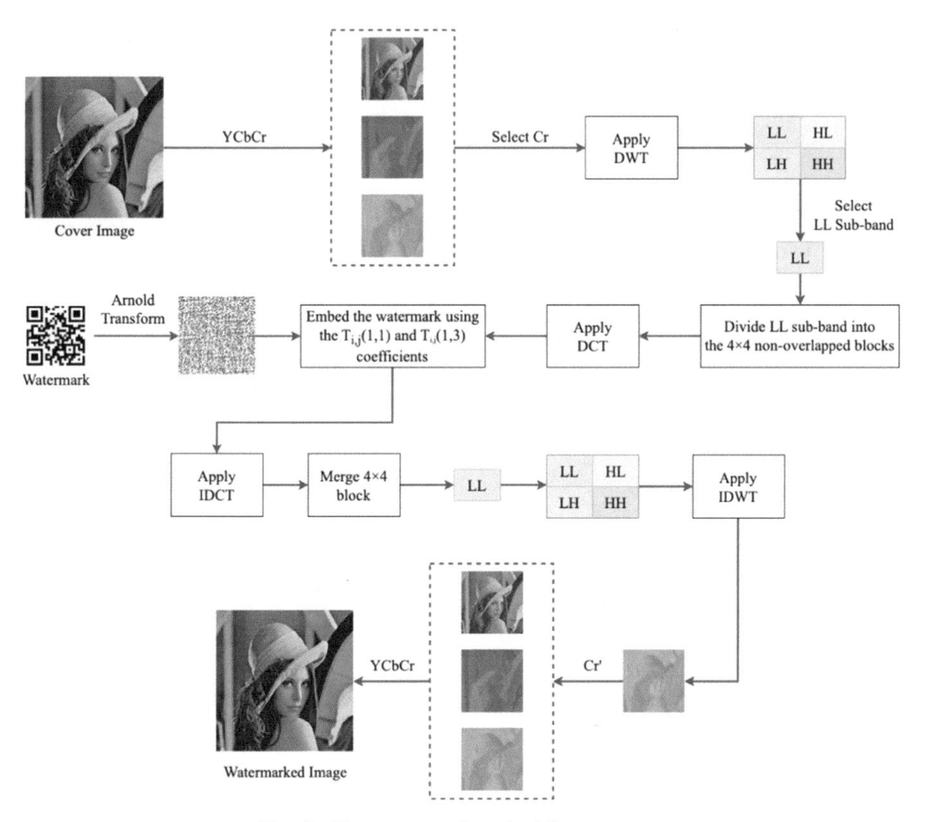

Fig. 3. The watermark embedding process.

The following steps describe the QR code embedding procedure in detail:

1. The original 1024 × 1024 × 3 cover image, I^O, is transformed into the YCbCr color space and Y, Cb, and Cr elements are obtained.
2. DWT is performed on the Cr (I^O_{Cr}), and LL_{Cr}, HL_{Cr}, LH_{Cr} and HH_{Cr} sub-bands are obtained:

$$LL_{Cr}, LH_{Cr}, HL_{Cr}, HH_{Cr} = DWT\left(I^O_{Cr}\right) \tag{2}$$

3. LL_{Cr} sub-band is divided into 4 × 4 non-overlapped blocks, $K_{i,j}$ for $i = 1,2, \ldots, 128$ and $j = 1, 2, \ldots,128$.
4. QR code watermark, size of 128 × 128, is converted to the binary, W, and then scrambled with Arnold transform, W^A.
5. For each block, $K_{i,j}$:

   5.1. The coefficient matrix $T_{i,j}$ is obtained by applied DCT to the block, $K_{i,j}$:

$$T_{i,j} = DCT\left(K_{i,j}\right) \tag{3}$$

   5.2. The watermark bit $W(i,j)$ is embedded using the $T_{i,j}(1,1)$ and $T_{i,j}(1,3)$ coefficients using the following equation:

$$T_{i,j} = \begin{cases} swap\left(T_{i,j}(1,1), T_{i,j}(1,3)\right), & \mathit{If}\ W(i,j) = 1\ \mathit{and}\ T_{i,j}(1,1) < T_{i,j}(1,3) \\ swap\left(T_{i,j}(1,1), T_{i,j}(1,3)\right), & \mathit{If}\ W(i,j) = 0\ \mathit{and}\ T_{i,j}(1,1) > T_{i,j}(1,3) \\ T_{i,j}, & \mathit{otherwise} \end{cases} \tag{4}$$

   5.3. The difference between the $T_{i,j}(1,1)$ and $T_{i,j}$. (1,3) coefficients is increased by using Eq. (5):

$$T'_{i,j} = \begin{cases} T_{i,j}(1,1) + 10, & \mathit{If}\ \left(T_{i,j}(1,1) - T_{i,j}(1,3)\right) < 10\ \mathit{and}\ T_{i,j}(1,1) > T_{i,j}(1,3) \\ T_{i,j}(1,1) - 10, & \mathit{If}\ \left(T_{i,j}(1,3) - T_{i,j}(1,1)\right) < 10\ \mathit{and}\ T_{i,j}(1,1) < T_{i,j}(1,3) \end{cases} \tag{5}$$

   5.4. The obtained coefficients are subjected to IDCT to produce the watermarked block $K'_{i,j}$:

$$K'_{i,j} = IDCT(T'_{i,j}) \tag{6}$$

6. The 4 × 4 watermarked blocks, $K'_{i,j}$, are combined to produce new sub-band LL'_{Cr}.
7. The IDWT is used to produce the watermarked Cr, I^w_{Cr}.

$$I^W_{Cr} = IDWT\left(LL'_{Cr}, LH_{Cr}, HL_{Cr}, HH_{Cr}\right) \tag{7}$$

8. Watermarked color image, I^w, is reconstructed from the color components I^O_Y, I^O_{Cb} and I^w_{Cr}.

2.4 QR Code Extracting Procedure

In the QR code extracting procedure, binary 128 × 128 QR is extracted from the 1024 × 1024 × 3 watermarked color image, I^W. Figure 4 depicts the schematic representation of the QR code extraction technique.

The following steps describe the watermark extracting process in detail:

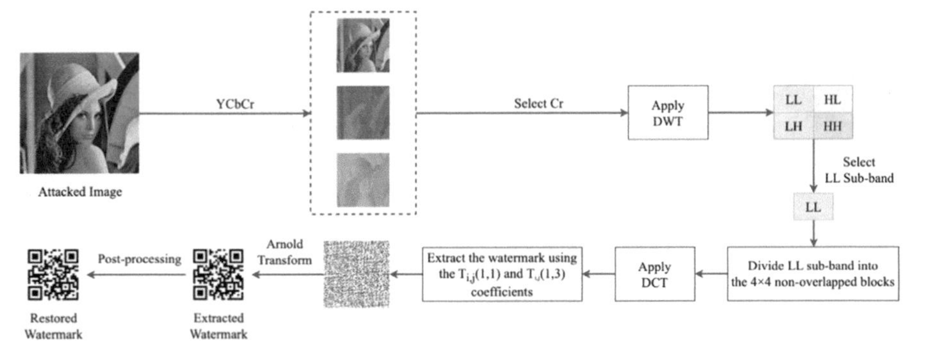

Fig. 4. The method of extracting the watermark from the attacked image

1. The 1024 × 1024 × 3 attacked image, I^A, is transformed into YCbCr color space and the Y, Cb, and Cr elements are obtained.
2. DWT is performed on the Cr (I^A_{Cr}), and LL^*_{Cr}, HL^*_{Cr}, LH^*_{Cr} and HH^*_{Cr} sub-bands are obtained:

$$LL^*_{Cr}, LH^*_{Cr}, HL^*_{Cr}, HH^*_{Cr} = DWT\left(I^A_{Cr}\right) \quad (8)$$

3. LL^*_{Cr} sub-band is divided into 4 × 4 non-overlapped blocks, $K^*_{i,j}$ for $i = 1,2, \ldots, 128$ and $j = 1, 2, \ldots, 128$.
4. For each block, $K^*_{i,j}$:

   4.1. The coefficient matrix $T^*_{i,j}$ is obtained by applied DCT to the block, $K^*_{i,j}$:

$$T^*_{i,j} = DCT\left(K^*_{i,j}\right) \quad (9)$$

   4.2. The scrambled watermark bit $W^{A*}(i,j)$ is extracted using the $T^*_{i,j}(1,1)$ and $T^*_{i,j}(1,3)$ coefficients:

$$W^{A*}(i,j) = \begin{cases} 0, & \textit{If } T_{i,j}(1,1) < T_{i,j}(1,3) \\ 1, & \textit{otherwise} \end{cases} \quad (10)$$

5. Using inverse Arnold transform, scrambled watermark image, W^{A^*}, is decoded to produce extracted binary QR Code watermark image, W^E.
6. To make the QR code more readable, postprocessing is applied, which includes noise removal and finder pattern correction.
7. Corrected QR code is decoded to obtain the hidden information.

The watermark extracted from the attacked image could be severely damaged. In this study, a process sequence is performed to the extracted watermark to increase the readability of the QR code. The post-processing includes the removal of noise on the QR code and the correction of the finder pattern. As mentioned above, the noise on the W^E is first removed. To this end, a 3 × 3 median filter is applied to remove the noise. Then, the extracted QR code's finder pattern is drawn again. The position and proportions of the finder pattern are similar for all extracted QR codes from the watermarked image. This way, the QR code's readability is improved, and watermark information can be acquired. Figure 5 shows an example of a post-processing operation. After post-processing, the QR code has become clearer and more readable, as can be seen in the figure. This application demonstrates that QR codes are well-suited for representing watermark information in watermarking methods.

Fig. 5. An illustration of a post-processing operation; a) Watermark before post-processing, b) Watermark after post-processing

3 Experimental Results

This section is structured into three different subsections. The first one presents information about the experimental setup, the second part discusses the outcomes of experiments conducted to find the imperceptibility of the method, and the last one provides the visual and quantitative results of the experiments that are conducted to examine the robustness of the method against various attacks.

3.1 Experimental Setup

The $1024 \times 1024 \times 3$ cover images *Lena*, *Baboon*, *Jet Plane*, and *Peppers* are chosen to provide an overview of the proposed method, as demonstrated in Fig. 6. The text "20220520Azizoglu" was used as a watermark with a size of 128×128, converted to version 3 and high ECC level (%30) QR code.

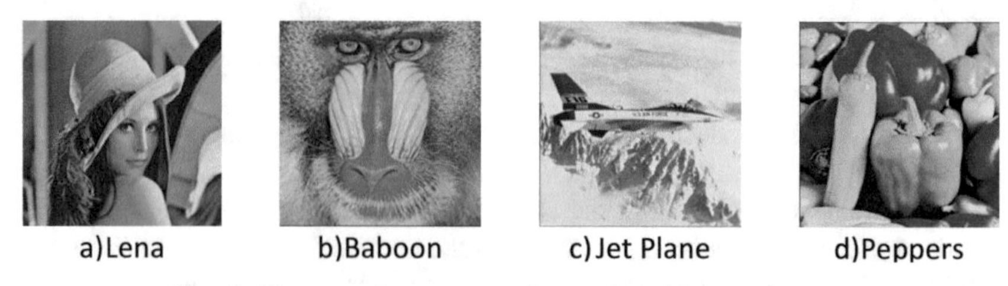

Fig. 6. The cover images are to be employed in experiments.

Fig. 7. The QR code watermark image.

3.2 Imperceptibility Analysis of the Watermarked Images

This section gives the results of the experiments carried out to evaluate the imperceptibility of the proposed QR code-based image watermarking technique. In the literature, the peak signal-to-noise ratio (PSNR) is commonly used to measure the imperceptibility of image watermarking methods. Since a large PSNR value indicates a high degree of similarity between two images, a high PSNR value between the original and watermarked image means high imperceptibility. The PSNR can be calculated as follows [14]:

$$PSNR = 10log_{10}\left(\frac{255^2}{\frac{1}{P\times Q\times R}\Sigma_{i=1}^{P}\Sigma_{j=1}^{Q}\Sigma_{k=1}^{R}(I_{i,j,k}^{O} - I_{i,j,k}^{W})^2}\right) \tag{11}$$

where, $P \times Q \times R$ represents the image size, $I_{i,j,k}^{O}$ and $I_{i,j,k}^{W}$ represent the pixels located at (i, j, k) of the original image, I^O and the watermarked image, I^W, respectively [4].

The perceptual quality of the images, as demonstrated in Fig. 8, has been preserved throughout the watermarking process. Table 2 gives the PSNR values of the test images after being watermarked using the watermark shown in Fig. 7. The average PSNR value for the QR watermark was obtained at 38.6492, which, as can be seen from the table, indicates that the imperceptibility of the watermarked image is satisfactory.

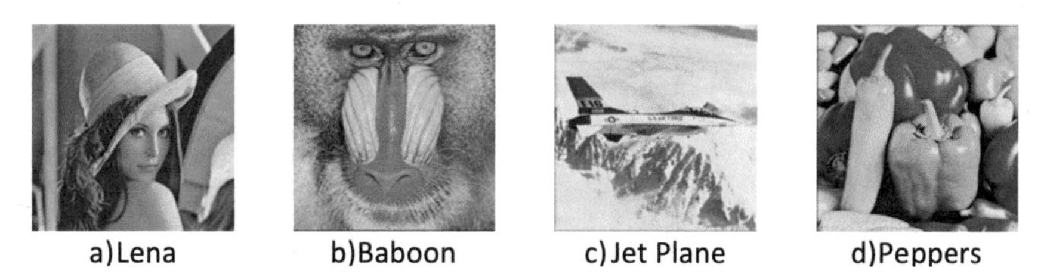

Fig. 8. The watermarked color images with QR code watermark.

3.3 Robustness and QR Code Readability Evaluation

This section summarizes the results of the experiments aimed to analyze the robustness of the proposed technique. In this study, the NC metric is performed to measure the

Table 2. The perceptual quality of the proposed method (PSNR dB).

Watermark	Lena	Baboon	Jet Plane	Peppers	Average
QR Code	41,0962	37,0071	39,3686	37,1250	38,6492

robustness. The NC is calculated as described in the following equation [7]:

$$NC = \frac{\sum_{i=1}^{P}\sum_{j=1}^{Q}\left(D_{i,j}^{E} - D_{M}^{E}\right)\left(D_{i,j}^{O} - D_{M}^{O}\right)}{\sqrt{\left(\sum_{i=1}^{P}\sum_{j=1}^{Q}\left(D_{i,j}^{E} - D_{M}^{E}\right)^{2}\right)\left(\sum_{i=1}^{P}\sum_{j=1}^{P}\left(D_{i,j}^{O} - D_{M}^{O}\right)^{2}\right)}} \tag{12}$$

where, $P \times Q$ is the watermark image size, $D_{i,j}^{E}$ and $D_{i,j}^{O}$ are pixels located at (i, j) of the extracted and original QR codes, and D_{M}^{E} and D_{M}^{O} represent mean values of the extracted and original QR codes, respectively [7].

In the experiments, the proposed technique is subjected to different types of attacks to evaluate its robustness. For this purpose, the most commonly used attacks in the literature were selected. Table 3 shows the applied attacks and their parameters.

Table 3. Details of the attacks that were utilized to evaluate the robustness of the proposed QR-based image watermarking technique.

Attack	Abbreviation	Detail
Gaussian White Noise	GWN	3 × 3 kernel size
Speckle Noise	SN	0.01 variance
Salt & Pepper Noise	SPN	0.01 density
Average Filtering	AF	3 × 3 filter size
Median Filtering	MF	3 × 3 filter size
Sharpening	SA	3 × 3 filter size
Blurring	BL	Gaussian lowpass filter of size 5 × 5
JPEG Compression	JPEG	20% Quality
Histogram Equalization	HE	–
Rotation	RO	Clockwise 25°
Resizing	RS	Bilinear resizing (1024 › 512 › 1024)

Figures 9–12 show QR code watermarks that were extracted from the Lena, Baboon, Jet Plane, and Peppers images in terms of different attacks. All extracted watermarks were post-processed, and their readability was improved. From the figures, it can be said that extracted watermarks are well preserved. Furthermore, it can be clearly seen that the post-processing process improves the readability of the QR codes and enables decoding of the content.

Table 4. The NC values of extracted QR code watermarks for different images.

Attack	Lena	Baboon	Jet Plane	Peppers
GWN	0.9843	0.9850	0.9817	0.9843
SPN	0.9703	0.9788	0.9688	0.9711
SN	0.9856	0.9857	0.9847	0.9856
AF	0.9854	0.9854	0.9853	0.9857
MF	0.9856	0.9855	0.9852	0.9858
SA	0.9856	0.9856	0.9848	0.9856
BL	0.9855	0.9857	0.9850	0.9857
RS	0.9856	0.9857	0.9852	0.9856
JPEG	0.9817	0.9850	0.9502	0.9800
HE	0.9794	0.9852	0.9831	0.9860
RO	0.9857	0.9811	0.9855	0.9850

Fig. 9. QR code watermarks extracted from *Lena* image attacked by; a) GWN, b) SPN, c) SN, d) AF, e) MF, f) SA, g) BL, h) RS, i) JPEG, j) HE, k) RO

Fig. 10. QR code watermarks extracted from Baboon image attacked by; a) GWN, b) SPN, c) SN, d) AF, e) MF, f) SA, g) BL, h) RS, i) JPEG, j) HE, k) RO

Fig. 11. QR code watermarks extracted from *Jet Plane* image attacked by; a) GWN, b) SPN, c) SN, d) AF, e) MF, f) SA, g) BL, h) RS, i) JPEG, j) HE, k) RO

Table 4 gives the NC values between the original and extracted watermarks for *Lena*, *Baboon*, *Jet Plane*, and *Peppers* images. The NC values indicate that the extracted watermarks are similar to the original ones. Moreover, the readability of QR codes has been checked by using the QR code reading feature of the camera application of the

Fig. 12. QR code watermarks extracted from *Peppers* image attacked by; a) GWN, b) SPN, c) SN, d) AF, e) MF, f) SA, g) BL, h) RS, i) JPEG, j) HE, k) RO

iPhone 12 pro device. The watermark information could be decoded from all extracted QR code watermarks.

4 Conclusion

This study set out to determine how effective the use of QR codes as watermarks is in terms of robustness and *capacity*. Therefore, a QR code-based robust and blind watermarking method using DWT and DCT for digital color images was presented in this study. In addition, several experiments were performed to demonstrate the effectiveness of the QR code-based image watermarking by embedding a watermark consisting of a QR code in well-known color test images. The acquired results have indicated that in contrast to most other robust methods, the watermark information can be extracted losslessly from the attacked watermarked image. Also, the proposed method supports blind extraction, which means it does not require the original image and watermark. In addition, experiments have revealed that the proposed QR-based image watermarking technique is remarkably robust to different kinds of attacks. Consequently, it demonstrates better practical applicability for protecting the copyright of digital assets.

References

1. Azizoglu, G., Toprak, A.N.: A Robust DWT-DCT-based QR code watermarking method. In: International Conference on Artificial Intelligence and Applied Mathematics in Engineering, ICAIAME 2022, vol.156 (2022)
2. Azizoglu, G., Toprak, A.N.: A novel reversible fragile watermarking in DWT domain for tamper localization and digital image authentication. In: 9th International Symposium on Digital Forensics and Security, ISDFS 2021 (2021)
3. Ariatmanto, D., Ernawan, F.: An improved robust image watermarking by using different embedding strengths. Multimed. Tools Appl. **79**(17–18), 12041–12067 (2020). https://doi.org/10.1007/s11042-019-08338-x
4. Abdulrahman, A.K., Ozturk, S.: A novel hybrid DCT and DWT based robust watermarking algorithm for color images. Multimed. Tools Appl. **78**(12), 17027–17049 (2019). https://doi.org/10.1007/s11042-018-7085-z
5. Liu, J., Li, J., Ma, J., Sadiq, N., Bhatti, U.A., Ai, Y.: A robust multi-watermarking algorithm for medical images based on DTCWT-DCT and Henon Map. Appl. Sci. **9**(4), 700 (2019)
6. Li, D., Gao, X., Sun, Y., Cui, L.: Research on anti-counterfeiting technology based on QR code image watermarking algorithm. Int. J. Multimed. Ubiquit. Eng. **12**(5), 57–66 (2017)
7. Patvardhan, C., Kumar, P., Vasantha Lakshmi, C.: Effective Color image watermarking scheme using YCbCr color space and QR code. Multimed. Tools Appl. **77**(10), 12655–12677 (2017). https://doi.org/10.1007/s11042-017-4909-1

8. Hsu, L.Y., Hu, H.T., Chou, H.H.: A blind robust QR code watermarking approach based on DCT. In: Proceedings - 2019 4th International Conference on Control, Robotics and Cybernetics, CRC 2019, pp. 174–178 (2019)
9. Sun, L., Liang, S., Chen, P., Chen, Y.: Encrypted digital watermarking algorithm for quick response code using discrete cosine transform and singular value decomposition. Multimed. Tools Appl. **80**(7), 10285–10300 (2020). https://doi.org/10.1007/s11042-020-10075-5
10. Pan, J.S., Sun, X.X., Chu, S.C., Abraham, A., Yan, B.: Digital watermarking with improved SMS applied for QR code. Eng. Appl. Artif. Intell. **97**, 104049 (2021)
11. Tiwari, S.: An introduction to QR code technology, pp. 39–44 (2017)
12. Chow, Y.W., Susilo, W., Tonien, J., Zong, W.: A QR code watermarking approach based on the DWT-DCT technique. In: Pieprzyk, J., Suriadi, S. (eds.) Information Security and Privacy. ACISP 2017. LNCS, vol. 10343, pp. 314–331. Springer, Cham (2017). https://doi.org/10.1007/978-3-319-59870-3_18
13. Bhatti, U.A., et al.: New watermarking algorithm utilizing quaternion Fourier transform with advanced scrambling and secure encryption. Multimed. Tools Appl. **80**(9), 13367–13387 (2021). https://doi.org/10.1007/s11042-020-10257-1
14. Gul, E., Toprak, A.N.: Contourlet and discrete cosine transform based quality guaranteed robust image watermarking method using artificial bee colony algorithm. Expert Syst. Appl. **212**, 118730 (2023)

Repairing of Wall Cracks in Historical Bridges After Severe Earthquakes Using Image Processing Technics and Additive Manufacturing Methods

Pinar Usta[1](✉), Merdan Özkahraman[2], Muzaffer Eylence[2], Bekir Aksoy[2], and Koray Özsoy[3]

[1] Technology Faculty, Civil Engineering Department, Isparta University of Applied Sciences, Isparta, Turkey
pinarusta@isparta.edu.tr
[2] Technology Faculty, Mechatronics Department, Isparta University of Applied Sciences, Isparta, Turkey
[3] Machine and Metal Technologies Department, Isparta University of Applied Sciences, Isparta OSB Vocational School, Isparta, Turkey

Abstract. Historical buildings are prone to seismic damage, due to either construction features or effects of changes over the time. Historical buildings, which have a face about the past of the regions and countries in which they are located, are an important part of cultural heritage and provide information about the conditions and characteristics of the period in which they were built. In the repair and strengthening of damaged historical buildings, it is extremely important to preserve the originality of the building and to have good workmanship in the repair. Repairs to these structures require detailed work and special craftsmanship. This increases the difficulty of repairing old masonry buildings by preserving their "original" condition. Moreover, Traditional methods for repairing damages in buildings require a certain amount of time, regardless of the scale of the damage. Recently, technologies produced using additive manufacturing methods have begun to affect the construction industry more. In civil engineering, with a additive manufacturing methods, structurally sufficient results can be obtained to meet the quality and on-site construction safety requirements of designs at various scales and from different perspectives. Historical stone bridges are one of the most important parts of historical structures. In this study, it is aimed to determine the damage conditions of historical stone bridges by using artificial intelligence methods. In addition, it has been emphasized that for the repair of damaged elements and parts of stone bridges, 3D image processing techniques and manufacturing methods can be used to produce elements suitable for stone bridges and in desired dimensions.

Keywords: Historical Buildings · image processing · additive manufacturing methods · repairing

D. J. Hemanth et al. (Eds.): ICAIAME 2022, ECPSCI 7, pp. 458–464, 2023.
https://doi.org/10.1007/978-3-031-31956-3_39

1 Introduction

Anatolian lands, which have hosted many different civilizations for centuries due to its geographical location, contain many historical structures in the process from ancient times to the present, when examined in terms of Turkish history. Bridges connecting the two sides of the lands have an important place in terms of cultural heritage in order to meet the transportation and communication needs after the shelter, which is the most basic need for people (Tektas et al, 2017).

Historical bridges have artistic and architectural significance in human societies. Natural disasters such as floods and earthquakes have threatened these historical structures (Naderi and Zekavati, 2018).

The historical heritage and historical structures are an asset to be preserved for future generations. In particular, the historical heritage building for their size and different component, material, and realization require innovative systems for their preservation (Carnì et al., 2022). Everyone is responsible for the preservation, livelihood and sustainability of historical stone bridge structures that are common to all people (Tektas et al, 2017).

Existing bridge structures are exposed to various environmental and operational loads during their service. These external loads have negative effects on bridges and accelerate structural damage. However, bridges can be encountered and damaged by extreme events such as earthquakes throughout their service life. Therefore, it is necessary to recognize the structural condition of the bridges in a timely manner and take precautions to ensure safety. Traditionally, visual inspection has played quite significant role in detecting surface defects and assessing structural condition. However, visual inspection is labor-intensive, time-consuming, and subjective, even for well-trained inspectors, so it cannot exactly track status changes in real time (Sun et al., 2020).

Especially the damages in historical buildings prevent the structures from reaching future generations as cultural heritage. In addition, in cases where intervention in historical structures is not allowed, it is often not possible to damage assessment to such structures. So, the preservation of the historical heritage structure is an important task that requires the implementation of innovative systems. In this field, interesting advantages are provided by the use of Artificial Intelligent methods (Carnì et al., 2020).

To understand what role AI techniques can play in bridge engineering, it is important to provide an overview of bridge engineering tasks. This overview will help organize current research work and point to areas that have the greatest promise to benefit from AI techniques and those areas which are in urgent need for such techniques. Studies on AI techniques for bridge engineering have focused primarily on the two most critical issues in bridge engineering: design (mostly preliminary) and maintenance (mostly inspection) (Reich, 1996).

Damage assessment (DI) in structures is very important for scientific research and engineering applications and is also the most difficult to perform. Uncertainties due to damages exist in both structural models and external inputs (Sun et al., 2020).

Numerous articles on computer applications in civil and structural engineering shows that the technology has proven its importance for the field of civil and structural engineering. Besides, Artificial Intelligence (AI) techniques have been proven very effective to undertake classification tasks in very complex problems. These techniques have exhibited huge potential of application in various engineering fields, such as civil engineering (Yardım, 2013). It is widely used in additive manufacturing technology, one of the computer-aided digital industry components, in the damage assessment of old stone bridges (Ali et al., 2022).

3D printing is an additive manufacturing method from three-dimensional model data to produce complex shaped parts (Praveena et al., 2022). Additive manufacturing technologies are divided into seven categories such as powder-based fusion, direct energy deposition, material jetting, polymerization, material extrusion, binder jetting, and sheet lamination (Alammar et al., 2022). Additive manufacturing technologies, which are components of Industry 4.0, are widely used such as engineering, medical, military, aviation, construction, architecture and fashion etc. (Rouf et. al, 2022).

In this paper, the ability to automatically detect critical events by using artificial intelligence methods to monitor historical heritage bridges has been examined.

2 Material and Methods

2.1 Material

2.1.1 Datasets

In the study, 350 damaged real bridge images were used by increasing them to 1050 images by using data augmentation methods. Figure 1 shows a sample image of a damaged stone bridge.

Fig. 1. Historical Stone bridge

2.1.2 YOLOV5 Algorithm

YOLO (You Only Look Once) is a convolutional neural network architecture that was brought to the literature by Glenn et al. in 2020 and can perform image segmentation and classification (Jiang et al. 2022). YOLOV5 is a system whose architecture is based on detecting objects in the image by dividing the images into grids and performing feature extraction from the image (Kıran et al., 2022). Unlike previous YOLO architectures, the YOLOV5 architecture was trained in image segmentation operations, and it was found that it performs these operations successfully. (Yang et al., 2020). YOLOV5 architecture is shown on Fig. 2.

Overview of YOLOv5

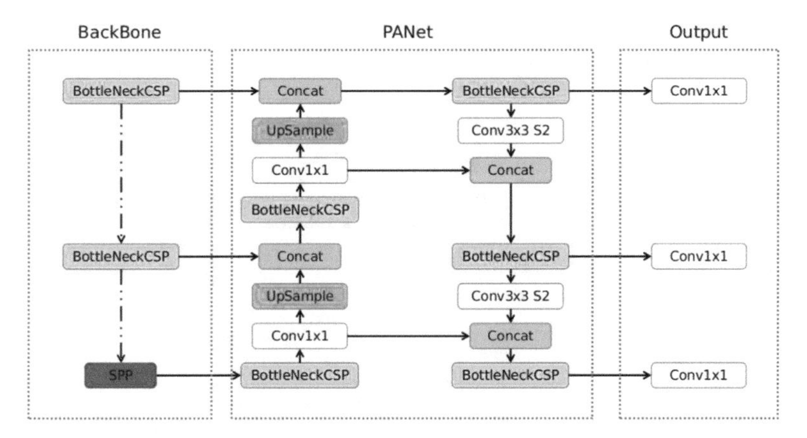

Fig. 2. YOLOV5 architecture

2.1.3 Performance Evaluation Metrics

In this study, precision, recall and mAP (mean average precision) metrics were used to test the accuracy of the model being trained. Formulas for metrics are given in Eqs. 1, 2 and 3.

$$Precision = \frac{TP}{TP + FP} \tag{1}$$

$$Recall = \frac{TP}{TP + FN} \tag{2}$$

$$mAP = \frac{1}{n(T)} \sum_{rT} AP_r \tag{3}$$

In the above equation, the line is positive; true negative; false negative; represents the number that is true, but the prediction is incorrect, and a false positive represents the number that is incorrect but classified as correct in the prediction. mAP is used as an evaluation criterion in detection algorithms. Its value ranges from 0 to 1, and the closer it is to 1, the better the result. AP given in the formula is the average precision (Li et al., 2021).

2.2 Methods

Damaged areas on the image were marked with the polygon tool using the labelimg program on the images in the Data Set. After the image tagging process was completed, the images were transferred to the cloud system using the Colab environment developed by Google. Training tools of the YOLOV5 architecture were installed in the Colab environment and the training process of the tagged data was carried out. The training phase lasted approximately 43 min. For training, 80% of the data was used as training file and 20% as test data. In order to evaluate the accuracy of the data during the training phase, 10% of the training data is reserved for the validation data. The workflow diagram of the work done on Fig. 3 is given.

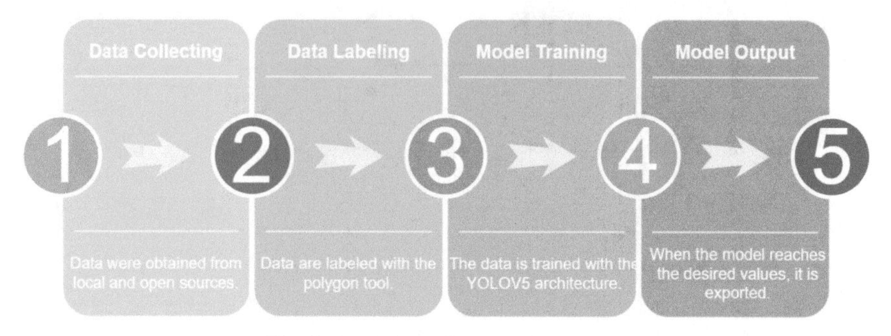

Fig. 3. Workflow diagram of the study

3 Findings and Conclusions

The training results used in the study are given in Table 1. When the table is examined, it is observed that the Precision value of the model is 0.9723, the Recall value is 0.9488, and the mAP value is 0.9789. In order to test the model speed, it has been observed that when a visual that has never been given a model before is given, it responds in 25 s and detects it with 0.97 accuracy. Figure 4 shows this image as it came out of the model.

Table 1: Model training results

Model	Precision	Recall	mAP	Time (Train)
YOLOV5	0.9723	0.9488	0.9789	25 s

By using artificial intelligence methods, the damage conditions of historical stone bridges can be determined. In addition, for the repair of damaged elements and parts of stone bridges, 3D printing can be used to produce elements suitable for the stone bridge and in desired dimensions. By using 3D image processing technique and manufacturing methods, human-induced and bad workmanship errors can be prevented, the repair process of cracks and fragmentation damages in historical buildings can be completed and accelerated. Prototype of damaged areas on old stone bridges was fabricated using a 3D printer as shown in Fig. 5.

Fig. 4. Sample test image and result

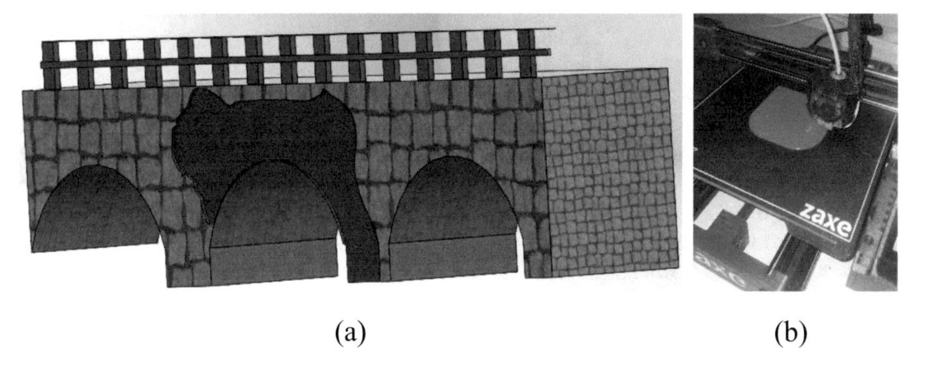

(a) (b)

Fig. 5. Prototype old stone bridge by using 3D Printer a) CAD solid model, b) manufacturing

Acknowledgments. We thank all users who shared the images used in the study on open access websites.

References

Carnì, D.L., Scuro, C., Olivito, R.S., Crocco, M.S., Lamonaca, F.: Artificial Intelligence based monitoring system for historical building preservation. In: 2020 IMEKO TC-4 International Conference on Metrology for Archaeology and Cultural Heritage, pp. 112–116 (2020)

Carnì, D.L., Scuro, C., Olivito, R.S., Lamonaca, F., Alì, G., Calà, S.: Structural health monitoring system design for historical heritage building. In: Journal of Physics: Conference Series, vol. 2204, no. 1, p. 012039. IOP Publishing, April 2022

Reich, Y.: Artificial intelligence in bridge engineering. Comput.-Aided Civil Infrastruct. Eng. **11**(6), 433–445 (1996)

Sun, L., Shang, Z., Xia, Y., Bhowmick, S., Nagarajaiah, S.: Review of bridge structural health monitoring aided by big data and artificial intelligence: from condition assessment to damage detection. J. Struct. Eng. **146**(5), 04020073 (2020)

Li, Y., Sun, L., Zhang, W., Nagarajaiah, S.: Bridge damage detection from the equivalent damage load by multitype measurements. Struct. Control. Health Monit. **28**(5), e2709 (2021)

Yardım, Y.: Applications of artificial intelligent systems in bridge engineering. In: 1st International Symposium on Computing in Informatics and Mathematics, July 2013

Naderi, M., Zekavati, M.: Assessment of seismic behavior stone bridge using a finite element method and discrete element method. Earthq. Struct **14**(4), 297–303 (2018)

Tektas, E., İlerisoy, Z.Y., Tuna, M.E.: Assesment of historical kars stone bridge in terms of sustainability of historical heritage. In: Fırat, S., Kinuthia, J., Abu-Tair, A. (eds) International Sustainable Buildings Symposium. ISBS 2017. Lecture Notes in Civil Engineering, vol. 7, pp. 291–305. Springer, Cham (2018). https://doi.org/10.1007/978-3-319-64349-6_23

Jiang, H., et al.: A survey on deep learning-based change detection from high-resolution remote sensing images. Remote Sens. **14**(7), 1552 (2022)

Kıran, E., Karasulu, B., Borandag, E.: Gemi Çeşitlerinin Derin Öğrenme Tabanlı Sınıflandırılmasında Farklı Ölçeklerdeki Görüntülerin Kullanımı. J. Intell. Syst.: Theory Appl. **5**(2), 161–167 (2022). https://doi.org/10.38016/jista.1118740

Yang, G., et al.: Face mask recognition system with YOLOV5 based on image recognition. In: 2020 IEEE 6th International Conference on Computer and Communications (ICCC), pp. 1398–1404. IEEE, December 2020

Li, M., Zhang, Z., Lei, L., Wang, X., Guo, X.: Agricultural greenhouses detection in high-resolution satellite images based on convolutional neural networks: comparison of faster R-CNN, YOLO v3 and SSD. Sensors **20**(17), 4938 (2020)

Ali, M.H., Issayev, G., Shehab, E., Sarfraz, S.: A critical review of 3D printing and digital manufacturing in construction engineering. Rapid Prototyp. J. (2022)

Praveena, B.A., Lokesh, N., Buradi, A., Santhosh, N., Praveena, B.L., Vignesh, R.: A comprehensive review of emerging additive manufacturing (3D printing technology): methods, materials, applications, challenges, trends and future potential. Mater. Today: Proc. **52**, 1309–1313 (2022)

Alammar, A., Kois, J.C., Revilla-León, M., Att, W.: Additive manufacturing technologies: current status and future perspectives. J. Prosthodont. **31**(S1), 4–12 (2022)

Rouf, S., et al.: Additive manufacturing technologies: industrial and medical applications. Sust. Oper. Comput. (2022)

Analysis Efficiency Characteristics Multiservice Telecommunication Networks Taking into the Account of Properties Self-similar Traffic

Bayram G. Ibrahimov(✉), Mehman H. Hasanov, and Ali D. Tagiyev

Department of Radioengineering and Telecommunication, Azerbaijan Technical University, H. Javid Ave. 25, AZ 1073 Baku, Azerbaijan
i.bayram@mail.ru

Abstract. Qualitative and quantitative indicators quality of functioning multiservice telecommunication networks based on ICT technology of the digital economy, using the concept NGN (Next Generation Network) and FN (Future Network) in terms of the properties self-similar traffic, are analyzed. Self-similarity of traffic has a significant impact on the quality of communication. In this case, research is mainly focused on the statistical characteristics of queues, since buffering and throughputs are the main resource provisioning strategy of communication networks. And it turns out that traditional queuing analysis based on the Poisson flow assumption cannot accurately predict system performance under conditions of self-similar traffic. Important directions for the development NGN and FN multiservice networks based on end-to-end digital technologies in the digital economy are studied. The tasks multi rate service systems, a sufficient amount information and network resources, which provide indicators quality of functioning networks, both indicators quality of service (QoS) and indicators quality of experience (QoE), are considered. Based on the model, analytical expressions are obtained for evaluating the performance indicators hardware-software complexes of a communication system, taking into account the properties self-similar traffic.

Keywords: multiservice network · QoS · self-similar traffic · hardware and software complex · QoE · channel and network resource · network performance · efficiency · probabilistic-temporal characteristics

1 Introduction

At present, the development and widespread introduction information and communication technologies - the technology building distributed communication networks in the sphere of a single information infrastructure and in all spheres human activity is manifested in global multiservice telecommunication and computer networks with packet switching [1, 2]. A multiservice telecommunication network is a communication network built in accordance with the concept of a communication network and forming a single information and telecommunications structure that supports all types of traffic

D. J. Hemanth et al. (Eds.): ICAIAME 2022, ECPSCI 7, pp. 465–472, 2023.
https://doi.org/10.1007/978-3-031-31956-3_40

(data, voice, video) and provides all types of services at any point, in any set and volume, with differentiated guaranteed quality and at prices that satisfy different categories users [1, 2].

In multiservice telecommunications networks, the transmitted useful and service traffic have a special structure, i.e. traffics have self-similarity properties [3–6]. In order to build multiservice telecommunication networks based on ICT technology, digital twins, artificial intelligence (AI), machine learning (ML) and end-to-end digital technologies, special attention should be paid to their efficiency and reliability in the transmission and processing self similar traffic [6–11].

For an accurate analysis real network traffic, it has been established [4, 9, 12–14] that the existing models useful and service traffic generated by multimedia traffic in a communication network differ from traditional Poisson models used to study the nature voice traffic. An important question that arises in this case is the creation an adequate traffic model of a multiservice network that transmits packet flows that take into account the properties traffic self-similarity.

In the telecommunications system, for a long time, researchers believed that the behavior network traffic corresponds to a Poisson process. Over time, the number studies and measurements of network flow characteristics has increased [5, 8, 15]. As a result, it was noticed that not always the flow of packets in a multiservice network can be modeled using the Poisson process. In this case, the duration of service differs from the exponential law.

Therefore, the analysis of the behavior flow of traffic packets in the study of the characteristics efficiency multiservice telecommunication networks, taking into account the properties self-similarity traffic, is extremely important.

This paper analyzes the efficiency characteristics multiservice telecommunication networks with a heterogeneous architecture when transmitting streams selfsimilar traffic packets.

2 General Statement of the Research Problem

System-technical analysis useful and service traffic showed that a large amount work [1, 2, 9] is devoted to the study of the behavior network traffic. In [3, 5, 13, 14], tele traffic models based on a self similar random process were studied, and in [4, 14, 16], the characteristics NGN and FN networks were analyzed taking into account the self similarity properties network traffic. In addition, this problem is successfully considered by many scientists from leading universities and research centers around the world [2, 7, 9, 17–20].

It should be noted that the first and important stage this work was the creation of an adequate model that characterizes the behavior network traffic and showed that the behavior network traffic is successfully modeled using the so-called self similar process. The self similarity property is associated with one of the fractal types, that is, when the scale changes, the correlation structure of the self-similar process remains unchanged [3, 12, 19].

Thus, numerous studies have shown that real network traffic is a self-similar random process [3, 5, 10, 18, 20].

In order to more accurately describe the behavior new types traffic (elastic and inelastic traffic, as well as patient and impatient traffic, etc.), it is necessary to study the performance characteristics of a multiservice network, taking into account the properties traffic self similarity [5, 14].

These services require multi-rate service systems, a sufficient amount information, channel and network resources that provide indicators of the quality of functioning multi-service networks, such as QoS (Quality of Service) indicators and QoE (Quality of Experience) quality of experience indicators [5, 10, 15].

To formalize the problem, a mathematic al model is proposed that takes into account the self-similarity useful and service traffic, which will most accurately reflect the telecommunication processes occurring in the network under study when providing multimedia services. These will allow you to obtain analytical expressions for calculating the main indicators of the effectiveness and quality of functioning NP (Network Performance) network [12, 14].

The solution of the problem under consideration requires an integrated approach to study the behavior real network traffic in order to evaluate the effectiveness multiservice telecommunication networks with a selfsimilar incoming traffic flow.

3 Analysis Selfsimilarity Useful and Service Traffic and Description of the Model

The conducted studies of the statistical properties information flows of traffic packets have shown [4, 5, 12] that traffic in communication networks is a self-similar random process at a time. The self-similar nature network traffic observed in modern communication networks is due to the fact that multiservice networks are integral and are used to transmit voice, data, video traffic, files and other types information presented in the form standardized packets.

Self-similar traffic has a significant impact on the efficiency characteristics communication networks. In particular, as established by measurements [3, 4], with an increase in the size of the capacity of the buffer storage at the input communication channel, the probability losses decrease much more slowly than exponentially, and rapidly decreases according to an exponential law with an increase in the throughput of the communication channel.

Suppose is X a semi-infinite segment of a broadly stationary real random process discrete time $t \in N \triangleq \{1, 2, ...\}$. Let us define the average value and the variance of the process X, respectively, through numerical probabilistic characteristics [3, 12]:

$$\mu = E[X_t] < \infty, \ \sigma^2 = D[X_t] < \infty. \tag{1}$$

Using (1), you can determine the indicators - the correlation coefficient of the process X:

$$r(k) = \frac{E(X_{t+k} - \mu)(X_t - \mu)}{\sigma^2}, \tag{2}$$

Then the coefficient auto covariance of the process X is defined as follows:

$$b(k) = \sigma^2 \cdot r(k), k \in Z_+ \triangleq \{1, 2, \ldots\} \tag{3}$$

Thus, from (1), (2) and (3) the process is called strictly self-similar traffic in a broad sense, taking into account the Hurst exponent H, which is expressed as follows:

$$H = 1 - 0,5\beta, 0 < \beta < 1. \tag{4}$$

Here values $\beta-$ indicates how slowly dependence decreases in X. This means that the smaller β, the stronger the dependence in X. A similar interpretation can be given to the Hurstexponent H, lying in the range $0.5 < H < 1$. In addition, the process X is called self-similar in the narrow sense with the parameter H and β, if

$$m^{1-H} \cdot X^{(m)} \stackrel{\bullet}{=} X, \tag{5}$$

It follows from the above that the fulfillment of condition (1)–(5) is necessary and sufficient for the self-similarity useful and service traffic.

It follows from the operation algorithm and traffic behavior that the considered communication network is a queuing system (QS) of the type $fMG/N_m/N_{BS}$ with some assumptions [12]. The considered self-similar traffic Y is a stream of packets with a length taken as a unit time and the distribution of service duration is arbitrary, serving terminals N_m have a common buffer store with a limited capacity N_{BS}.

Packets are generated by sources so that traffic Y is a superposition of traffic packet streams, which have numbers $n \in Z$.

Given the above, we assume that the considered selfsimilar traffic $Y(\lambda, n, t)$ is a superposition traffic packets generated by different load sources and is defined as follows [3, 12]:

$$Y_t(\lambda, n, t) = \sum_{n \in Z} U_t(t - t_n + 1), \quad t \in Z,\ i = 1, 2, 3, \ldots, n, \tag{6}$$

Given a QS type model $fMG/N_m/N_{BS}$ and problem statement, the studied efficiency multiservice networks will be evaluated for three large groups indicators: the average queue length in the system $L(\lambda, H)$, average waiting time $T_{wt}(\lambda, H)$ and containers N_{BS} with the help which it is possible to mathematically formulate the proposed approach.

4 Research and Analysis of the Probabilistic-Temporal Characteristics Communication Networks

Consideration quality issues in a communication network based on NGN and FN occurs at three levels with three indicators, such as QoS, NP and QoE. In multiservice communication networks in real time, the main characteristic of the guaranteed quality of service for traffic is the average delay time for the transmission of voice and video traffic packets [5, 6, 15].

Based on the "End to end" algorithm, the allowable average packet delay of self-similar traffic in the network is expressed as follows:

$$T_{dt}(\lambda_i, H) = \inf[(1/K_c) \cdot \sum_{j=1}^{N_m} T_{ij}(\lambda_i, H)] \le T_{dt}^{all \cdot}(\lambda_i, H), j = \overline{1, N_m}, i = \overline{1, n} \tag{7}$$

wher N_m– the number hardware-software complexes in the communication network through which the streams voice and video traffic packets pass. Naturally, for voice traffic according to ITU-T, G.114 recommendations [5, 12, 15], the acceptable value is $T_{dt}^{all.}(\lambda_i, H) \leq 200, ..., 350$ ms.

Expression (7) characterizes the total delay in the transmission packets over a communication network and is an indicator QoS, NP and QoE traffic. In addition, analysis (7) showed that the use of a limited number hardware-software N_m complexes SDN, NFV, LTE, IMS and 5G NR technologies helps to minimize the time during $T_{dt}(\lambda_i, H)$ the transmission i-th traffic.

An important characteristic of the communication network in the system $fMG/N_m/N_{BS}$ with queues is the average waiting time of service in the buffer drive of hardware-software systems. Using Little's formula, we have.

$$\text{Å}[T_{wt}(\lambda_i, H)] = \frac{N_m}{f(H)} \cdot E[L_{ql}(b_i, C_s^2)] \cdot \mu \cdot \rho(\lambda_i, H) \ , \quad i = \overline{1, n} \ , \tag{8}$$

where $f(H) = 2H$– a function that takes into account the self-similarity property of incoming traffic packets; H- is the Hurst coefficient for the traffic flow, where the value is noted $H = 0.5$ corresponds to a low degree of self-similarity, and the value $H \to 1$ corresponds to a high degree self-similarity - non-stationary traffic, traffic with a complex structure; $E[L_{ql}(b_i, C_s^2)]$– average queue length in the system, depending on network parameters b_i è $C_s^2; \mu_i$– the average service rate of the i-th traffic and is equal to $\mu_i = 1/b_i$, $i = \overline{1, n}$; E – expectation sign;

Considering the coefficient variation C_s^2 and the load factors of the incoming process, the average value queue length in the system is determined as follows:

$$E[L_{ql}(b_i, C_s^2)] = \rho(\lambda_i, H) \cdot \frac{0.5(1 + C_s^2)}{[1 - \rho(\lambda_i, H)]} \cdot \lambda \cdot b_i \cdot f(H), \quad i = \overline{1, n} \tag{9}$$

It can be seen from expressions (8) and (9) that the average waiting time at the beginning of service and the average queue length in the system are directly proportional to the network parameters N_m, μ_i and Hurstcoefficient for the traffic flow, which is an indicator QoS, NP and QoE and probabilistic-temporal characteristics of multiservice networks.

Figure 1 shows a graphical dependence of the average waiting time at the beginning service on the number hardware-software complexes SDN, NFV, LTE, IMS and 5G NR technologies N_m for a given load factor of the communication network $\rho(\lambda_i, H)$ and the rate transmission streams of traffic packets.

Graphical dependency analysis shows that with an increase in the limited number hardware-software complexes $N_m \geq (40, ..., 50) \cdot 10^3$, meeting the requirements of system reliability, helps to minimize the average waiting time for packet flows at a given network speed $V_{i.k} \leq (0.064, ..., 8.0)$, $f(H) = \{H_1 = 0.60; \ H_2 = 0.66\}$ and load factor of a multiservice communication network varies within the following limits $\rho(\lambda_i, H) = \{\rho_1 = 0.55; \ \rho_2 = 0.70\}$.

Studies have shown [3, 6, 13, 21–23] that in order to support the architecture IntServ and DiffServ in multiservice communication networks, it is necessary to use network resources efficiently.

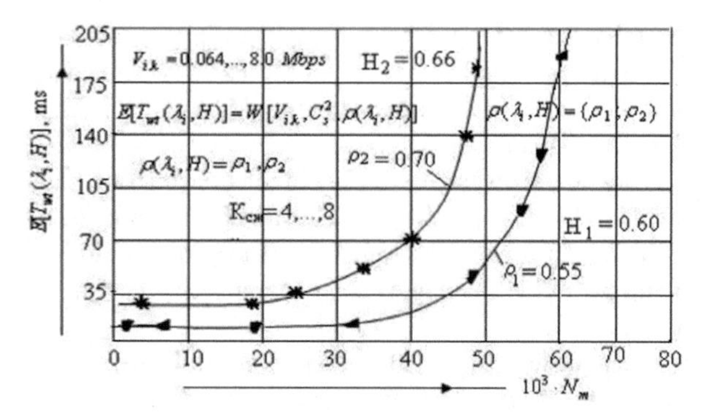

Fig. 1. Graphical dependence of the average waiting time on the number hardware-software complexes of the technology for a given network parameter

In this case, the network resource is classified into buffer and channel. The buffer resource is a queue packets organized on the routers communication network, and the channel resource, in turn, is characterized by the throughput $C_{i.\max}$ communication channels and hardware and software systems.

It is the load of the queue buffer and communication channels that significantly affects the numerical values of the key QoS, NP and QoE indicators, the average packet delay, jitter, the level packet loss and the performance multiservice telecommunication networks in general.

During the normal functioning communication networks, when there is no unlimited increase in the queue, the coefficients effective use resources of the channel $R_{ko}(\lambda_i, H)$ equipment (switches, integrated multiplexers, routers, controllers, etc.) should be less than one:

$$R_{ko}(\lambda, H) = f(H) \cdot \sum_{i=1}^{n} [\frac{\lambda_i}{N_m \cdot C_{i.\max}} \cdot L_{i.p}] \leq 1, \quad i = \overline{1, n} \tag{10}$$

The fulfillment of condition (10) makes it possible to determine the resource reserves of the channel equipment of multiservice communication networks:

$$R_r = 1 - R_{kr}(\lambda, H) \tag{11}$$

Thus, the obtained relations (10) and (11) make it possible to more accurately estimate the effective use of channel and buffer resources of the queuing system.

5 Conclusion

On the basis of the study, a new approach to the construction of a mathematical model for the analysis of the throughput characteristics and probabilistic-temporal characteristics multiservice communication networks based on the technology building distributed communication networks in the provision of multimedia services and applications is proposed.

On the basis of the model, the influence traffic self-similarity on the main characteristics of the efficiency communication network, which contributes to the improvement QoS, NP and QoE indicators, was studied.

References

1. Giordani, M., Mezzavilla, M., Zorzi, M.: Initial access in 5G mm wave cellular networks. IEEE Commun. Mag. **54**, 40–47 (2016)
2. Ibrahimov, B.G., Tagiyev, A.D.: Mathematical model for optimizing the efficiency of the functioning multiservice communication networks on base SDN & NFV technologies. In: Proceedings International Conference "Engineering Management of Communication and Technology", IEEE Conference, pp. 1–4. Springer, Cham (2021)
3. Shelukhin, O.I.: Modeling of Information Systems. Textbook for universities, p. 516. M: Hot line - Telecom (2018)
4. Ibrahimov, B.G., Alieva, A.A.: Research and analysis indicators of the quality of service multimedia traffic using fuzzy logic. In: Aliev, R.A., Kacprzyk, J., Pedrycz, W., Jamshidi, Mo., Babanli, M., Sadikoglu, F.M. (eds.) ICAFS 2020. AISC, vol. 1306, pp. 773–780. Springer, Cham (2021). https://doi.org/10.1007/978-3-030-64058-3_97
5. Galkin, A.M., Simonina, O.A., Yanovsky, G.G.: Analysis of the characteristics NGN networks, taking into account the properties traffic self-similarity, Elektrosvyaz, no. 12, pp. 23–25 (2010)
6. Ibrahimov, B.G., Humbatov, R.T., Ibrahimov, R.F.: Analysis performance multiservice telecommunication networks with using architectural concept future networks. T-Comm. **6**(12), 84–88 (2018)
7. Yusifbayli, N.A., Guliyev, H., Aliyev, A.: Voltage control system for electrical networks based on fuzzy sets. In: Aliev, R.A., Yusupbekov, N.R., Kacprzyk, J., Pedrycz, W., Sadikoglu, F.M. (eds.) 11th World Conference "Intelligent System for Industrial Automation" (WCIS-2020). WCIS 2020. Advances in Intelligent Systems and Computing, vol 1323, pp.55–63. Springer, Cham (2021). https://doi.org/10.1007/978-3-030-68004-6_8
8. Ibrahimov, B.G.: Research and estimation characteristics of terminal equipiment a link multiservice communication networks. Autom. Control. Comput. Sci. **46**(6), 54–59 (2010)
9. Kreutz D., Ramos F., Verissimo P., Rothenberg C., Azodolmolky S., Uhlig S.: Software-defined networking: a comprehensive survey. Proc. IEEE **103**(1), 14–76 (2015)
10. Ibrahimov, B.G., Namazov M.B., Quliev M.N.: Analysis performance indicators network multiservice infrastructure using innovative technologies. In: Proceedings of the 7-th International Conference on Control and Optimization with Industrial Applications, Vol. 2, 176–178 (2020)
11. Ismibayli, E.G., Islamov, I.J.: New approach to definition of potential of the electric field created by set distribution in space of electric charges. IFAC-PapersOnLine **51**(30), 410–414 (2018)
12. Ibrahimov, B.G., Ismaylova, S.R.: Research and analysis performance indicators multiservice signal networks NGN/IMS. Int. J. Eng. Sci. Res. Technol. **6**(12), 295–300 (2017)
13. Xiong, B., Yang, K., Zhao, J., Li, W., Li, K.: Performance evaluation of OpenFlow-based soft-ware-defined networks based on queueing model. Comput. Netw. **102**, 172–185 (2016)
14. Ibrahimov, B.G., Tagiyev, A.D., Aliyeva A.A.: Research and analysis of the functioning quality of hardware and software complexes automotive service systems with using logistic approach. In: Proceedings International Conference on Problems of Logıstıcs, Management and Operatıon in the East-West Transport Corrıdor, pp. 64–69 (2021)

15. Naumov, V.A., Samuilov, K.E., Samuilov, A.K.: On the total amount of resources occupied by serviced customers. Autom. Remote. Control. **77**(8), 1419–1427 (2016). https://doi.org/10.1134/S0005117916080087
16. Gasimov, V.A.: The modified method of the least significant bits for reliable information hiding in graphic files. Int. J. Inf. Secur. Sci. **8**(1), 1–10 (2019)
17. Perrot, N., Reynaud, T.: Optimal placement of controllers in a resilient SDN Architecture. In: 12th International Conference on the Design of Reliable Communication Networks, pp. 145–151 (2016)
18. Evans, J.W., Filsfils, C.: Deploying IP and MPLS QoS forMultiservice Networks: Theory & Practice, p. 456. Morgan Kaufmann Published, Cambridge (2010)
19. Tatarnikova, T.M., Volskiy, A.V.: Estimation of probabilistic-temporal characteristics of network nodes with traffic differentiation. Inf. Contr. Syst. **3**, 54–60 (2018)
20. Coker, O., Azodolmolky, S.: Software-defined Networking with Openflow–, 2nd edn., p. 246. Packt Publishing Limited, Birmingham (2017)
21. Islamov, I.J., Ismibayli, E.G., Hasanov, M.H., Gaziyev, Y.G., Ahmadova, S.R., Abdullayev, R.: Calculation of the electromagnetic field of a rectangular waveguide with chiral medium. Progr. Electromagnet. Res. **84**, 97–114 (2019)
22. Islamov, I.J., Ismibayli, E.G.: Experimental study of characteristics of microwave devices transition from rectangular waveguide to the megaphone. IFAC-PapersOnLine **51**(30), 477–479 (2018)
23. Islamov, I.J., Ismibayli, E.G., Gaziyev, Y.G., Ahmadova, S.R., Abdullayev, R.: Modeling of the electromagnetic feld of a rectangular waveguide with side holes. Prog. Electromagn. Res. **81**, 127–132 (2019)

Research and Analysis Methods for Prediction of Service Traffic Signaling Systems Using Neural Network Technologies

Bayram G. Ibrahimov[1](✉), Cemal H. Yilmaz[2], Almaz A. Aliyeva[2], and Yusif A. Sonmez[1]

[1] Department of Radio Engineering and Telecommunication, Azerbaijan Technical University, H. Javid Ave. 25, AZ 1073 Baku, Azerbaijan
i.bayram@mail.ru

[2] Department of Information Technology, Mingechaur State University, Zahid Khalilov, 23, AZ 4015 Mingechaur, Azerbaijan

Abstract. Methods for predicting the service traffic signaling systems and protocols in multiservice communication networks based on the architectural concepts Next Generation Network (NGN) and Future Networks (FN) with the introduction packet switching are analyzed. This paper discusses the effectiveness methods for predicting the signal traffic of signaling systems and protocols using the concept neural networks in the provision multimedia services and applications. Methods for introducing fuzziness into the structures fuzzy-neural networks are studied based on the provisions of the theory fuzzy sets and fuzzy logic. Membership functions are built using fuzzy production rules, a method for predicting signal traffic and operational control network characteristics is considered, and an analytical expression is obtained to assess the quality of forecasting service traffic in a communication network. A new approach is proposed to control network characteristics, Quality of Service (QoS) and Quality of Experience (QoE) multimedia traffic using fuzzy logic.

Keywords: membership function · prediction · efficiency · signal traffic · neural network · hurst coefficient · fuzzy set theory and fuzzy logic

1 Introduction

The conducted research [1, 2] showed that multiservice packet-switched communication networks are complex systems and processes when modeling service and useful traffic, which requires effective methods and tools for their study. The efficiency multiservice communication networks based on NGN and FN depends on the method of forecasting service traffic and the functioning of the network for transmitting service and useful information based on neural network technology or fuzzy-neural networks [3, 4].

An analysis of the possibility using classical methods for calculating the waiting time for service traffic signaling systems in the queue showed a significant difference between the results calculating network characteristics and real values. To improve the

D. J. Hemanth et al. (Eds.): ICAIAME 2022, ECPSCI 7, pp. 473–482, 2023.
https://doi.org/10.1007/978-3-031-31956-3_41

quality of service traffic delivery, it is necessary to take into account a number important factors that affect the calculation network characteristics, but existing algorithms do not allow dynamic tracking of their changes [5, 6].

Based on the study [2, 5, 7], it has been established that modern network methods for designing complex signaling systems and telecommunication processes do not allow calculating the main characteristics of the network - the incoming load and quality of service signaling traffic, but provide only a rough estimate of the performance of the basic components NGN and FN networks. Overcoming these contradictions will improve the quality of designing complex telecommunication systems and ensure real-time control of the characteristics of the service traffic signaling network.

Given the importance developing fuzzy models used to describe complex weakly formalized telecommunications control systems, the effectiveness methods for predicting the signal traffic systems and signaling protocols using the concept neural networks and fuzzy-neural networks in the provision multimedia services and applications is considered.

2 General Statement of the Research Problem

The system-technical analysis of the stationary characteristics of service traffic showed that the existing algorithms and methods for predicting the traffic of the beginning congestion in the signaling system of the NGN and FN networks do not allow dynamic and real-time notification approaching a given congestion threshold. Often, traditional methods processing and controlling the transmission of service traffic of signaling systems and protocols NGN and FN networks not only do not allow adequate processing data, but also do not allow taking into account the uncertainty naturally inherent in these data.

To solve the tasks, it is advisable to use fuzzy logic and the theory of fuzzy sets, which are one of the effective approaches to solving this problem [3, 5–10]. The use fuzzy logic for modeling and managing a wide class processes has proven to be effective in comparison with the use classical methods.

The task constructing a time series forecast is typical for neural networks that implement fuzzy production rules. However, the difficulty using them for forecasting service traffic lies in the choice input data representation and the choice an appropriate neural network architecture for time series forecasting [5, 11, 12]. These circumstances characterize the contradiction between effective methods for predicting the characteristics of the service traffic NGN and FN networks, and the absence in the models dynamic adjustment parameters that would make it possible to quickly respond to the approaching congestion threshold.

The main direction for solving the problem can be the use of neural network forecasting methods. Taking into account the characteristic features of this method for predicting the service traffic of systems and signaling protocols NGN and FN networks, a number of advantages can be distinguished [4, 8, 12]:

- the absence of a formal model of the predictable service traffic process;
- quick adaptation to changing conditions in the provision of multimedia services.

Considering the above, this paper considers the effectiveness of methods for dynamic prediction of service traffic of systems and signaling protocols of NGN and FN networks using the concept of neural networks that implement fuzzy production rules.

3 Construction Membership Functions Using Fuzzy Production Rules

It is known [1, 3, 8] that membership functions premises and conclusions can have different forms. Most often, for the convenience practical implementation and minimization computational costs in the implementation of algorithms for fuzzy inference and learning parameters of the rules ($L - R$), functions are used: triangular, trapezoidal, sigmoid and Gaussian membership functions. It should be noted that the implementation various components of a fuzzy production model using neural network technologies involves the use of formalisms in the form neurons and artificial neural networks, namely: multiplication input signals of signaling systems by weight coefficients, addition input service signals, nonlinear transformation of the results of this summation.

Figure 1 shows a simple implementation of the membership function using neural network technologies, which consists of two layers. The block diagram consists mainly two layers: Layer 1 consists of an adder, multiplier and integrator, and the second layer consists of an adder only.

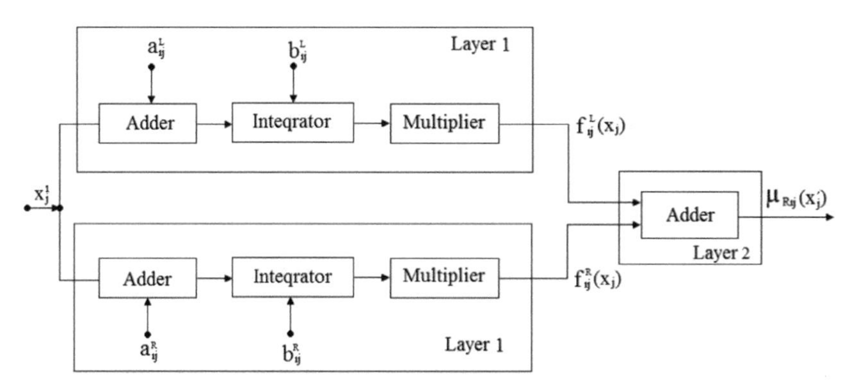

Fig. 1. Structural and functional diagram for the implementation of the membership function using a neural network

The studied Fig. 1 performs the formation of the prerequisites fuzzy production rules when predicting the signal traffic of signaling systems and protocols using the concept neural networks. In addition, the proposed scheme is an example membership implementation using a neural network [4, 5].

Based on Fig. 1, an example of a neural network method for implementing the Gaussian membership function is given when predicting the signal traffic of signaling systems and protocols [3, 5, 12]:

$$\mu_{Ad}(x_j) = \exp[-(x_j - a_{ij})/b_{ij}]^2. \tag{1}$$

It follows from expression (1) that using a two-layer neural network to predict signal traffic, two neurons of the first layer which have sigmoid activation functions and is described as follows:

$$f_{ij}^{L}(x_j) = \frac{1}{1 + \exp[-b_{ij}^{L}(x_j + a_{ij}^{L})]} \tag{2}$$

Based on (2), it is similarly written for the second layer of the sigmoid activation function:

$$f_{ij}^{R}(x_j) = \frac{1}{1 + \exp[-b_{ij}^{R}(x_j + a_{ij}^{R})]}. \tag{3}$$

From (2) and (3) it can be seen that the only neuron of the second layer with a linear activation function serves to combine these sigmoid functions, shifted relative to each other and having opposite signs:

$$\mu_{Aij}(x_j) = f_{ij}^{L}(x_j) + f_{ij}^{R}(x_j), \tag{4}$$

where a_{ij}^{L}, a_{ij}^{R} и b_{ij}^{L}, b_{ij}^{R} - are configurable membership function parameters using neural network technologies.

Considering an example of the implementation of the membership function using neural network technologies, the custom parameters can be defined as follows:

$$a_{ij} = 0.5(a_{ij}^{L} + a_{ij}^{R}), b_{ij} = b_{ij}^{L} = b_{ij}^{R}. \tag{5}$$

Thus, the obtained analytical expressions have shown that in various fuzzy production models, the formation prerequisites for fuzzy production rules for dynamic forecasting of service traffic is implemented in the same way.

4 Method for Predicting Signal Traffic and Operational Control Network Characteristics

The main difficulties in using fuzzy models for solving practical problems are usually related to the a priori determination of the components these models, since these components are often chosen subjectively, they may not be quite adequate to the simulated NGN/IMS (Internet Protocol Multimedia Subsystem) signaling system [2]. The signaling system under study is like a network platform NGN/IMS, which provides overload protection: in the event of an overload in any direction of the signaling network, the MTP (Message Transfer Unit) layer 3 protocol sends signaling messages along backup links or routes. As a result, Softswitch/IMS network operators face an urgent task - operational control network characteristics of signaling traffic.

The analysis of the existing methods operational control characteristics of the signaling system, which improve the quality of Softswitch/IMS operation, showed that the urgent task is not only to substantiate and select a method for calculating signaling traffic, but also a method for its dynamic forecasting.

The basis for predicting the load is the statistical data on the number of signaling messages served by the signaling system per unit time. In this case, the controlled parameters are: loading of the signaling link, distribution messages by length, distribution of time intervals between message arrivals [4, 5, 11]. Then, in general, the forecasting problem is described as follows. Let the values of the time series be given

$$X = (x_1, x_2, \ldots, x_t), t = 1, 2, ..., n, \tag{6}$$

where x_t – the value of the analyzed indicator signaling system, registered in the t –th time interval ($t = 1, \ldots, n$).

Using the chosen forecasting method, it is required to estimate the value of the future values of the series

$$\widehat{X} = \left(\widehat{x}_{n+1}, \widehat{x}_{n+2}, \ldots, \widehat{x}_{n+\tau}\right), \tag{7}$$

where τ – forecasting horizon and is determined by the inequality $1 \leq \tau \leq n$.

To assess the quality of the service traffic forecast, the value is introduced

$$Y(t) = \left|\widehat{X}(t) - X(t)\right|. \tag{8}$$

It has been established [4, 5] that the existing algorithms for predicting the beginning overload do not allow real-time notification approaching the specified overload threshold of the signaling link. In this section, the formulas obtained on the basis of a new approach allow us to control the QoS indicator and the QoE quality criterion for multimedia traffic using a fuzzy logic apparatus.

5 Signal Traffic Prediction Using Fuzzy Neural Networks

Studies [1, 3, 4, 6] have shown that there are various options for endowing fuzzy neural networks of this type with fuzziness. The inputs, outputs, as well as the weights of the neurons of such networks can be fuzzy. Suppose that type 1 neural fuzzy networks (inputs are fuzzy, weights are crisp and outputs are crisp) are mainly used to classify fuzzy input vectors into crisp classes [3, 4], and type 2 networks (inputs are fuzzy, weights are fuzzy) and outputs are fuzzy). In this case, they are used to implement fuzzy production models.

Here, the usual neural fuzzy network is understood as a neural network with fuzzy inputs and outputs and/or fuzzy weights, with a sigmoid activation function, all operations in which are defined based on the principle fuzzy expansion L. Zadeh [2, 6].

In Fig. 2 the structure of a conventional neural fuzzy network is shown, which is a Softswitch/IMS system, which receives service signals x_1 and x_2 of the type of significant signal units at the input. Figure 2 describes a neural fuzzy network. Assume that both signal traffic and weight signals are fuzzy numbers. Two-input neurons do not change their inputs, so their outputs are the same as their inputs.

Signaling traffic x_i inputs combined with weights w_i:

$$r_i = w_i \cdot x_i, i = 1, 2 \tag{9}$$

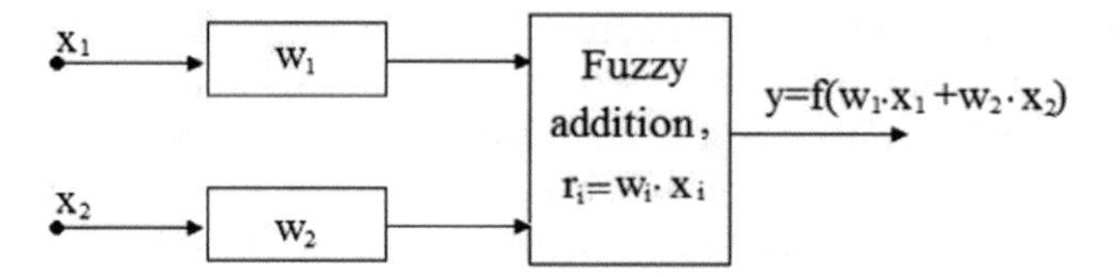

Fig. 2. Structure of a neural fuzzy network

Based on Fig. 2, the calculation r_i uses the principle of fuzzy expansion L. Zadeh. The values r_i are aggregated using the standard fuzzy extended addition [7]:

$$net = net(r) = r_1 + r_2 = w_1 \cdot x_2 + w_2 \cdot x_2 \tag{10}$$

From Fig. 2 it can be seen that the output value of the neuron is formed as a result of the transformation value *net* by the sigmoid function f:

$$y = f(net) = f(w_1 x_1 + w_2 x_2), \tag{11}$$

In (11), an important parameter is $f(net)$ and is the sigmoid membership functions of service traffic and is found as follows:

$$f(net) = [1 + \exp(-net)]^{-1}, \tag{12}$$

The membership function of the output of a neuron to the output fuzzy set is calculated based on the principle fuzzy generalization L.Zade:

$$\mu(y) = \begin{cases} (w_1 x_1 + w_2 x_2)(f^{-1}(y)), & if\ ,\ 0 \le y \le 1, \\ 0, & other \end{cases}, \tag{13}$$

where $f^{-1}(y) = \ln y - \ln(1 - y)$.

The research results showed [5, 7, 12] that ordinary regular-neural fuzzy networks are monotonic, that is, if $x_1 \subset x_1'$ and $x_2 \subset x_2'$, then expressions (11) will take the following form:

$$y = f(w_1 x_1 + w_2 x_2) \subset y' = f(w_1 x_1' + w_2 x_2') \tag{14}$$

where f - sigmoid activation function, and all operators in this expression are implemented on the basis of the principle fuzzy generalization L. Zadeh.

Thus, the latter means that neural fuzzy networks based on the principle fuzzy generalization can be universal approximations only for continuous monotonic functions. If a function is not monotonic, then there is no certainty that it will be approximated by a neural fuzzy network based on the fuzzy extension principle.

Unlike ordinary neurons, such fuzzy neurons have several outputs. The inputs these neurons are fuzzy, and the degrees belonging to fuzzy sets are formed at the outputs. The activation fuzzy neurons is a fuzzy process.

6 Description of the Fuzzy Inference Systems

Taking into account the above forecasting methods using theory fuzzy sets, fuzzy neural networks and fuzzy logic, let's consider predicting self-similar service traffic in a communication system, also in fuzzy conditions. Since traditional loading parameters require taking into account the invariant value of the Hurst coefficient, H. Therefore, in this system, there is a decision-making process that occurs in fuzzy conditions. Based on the study, it was established [2, 12] that among the important parameters in the fuzzy control system and in the neural fuzzy network is the load factor communication channels $\rho(\lambda_i, H_i)$, taking into account the self-similarity of the predicted traffic with the Hurst coefficient H_i, $i = \overline{1, k}$.

Based on the proposed approach to monitoring the QoS & QoE indicator multimedia self-similar traffic and the quality of network, we consider the method calculating the Hurst coefficient using the mathematical apparatus of fuzzy sets. Here, an important characteristic of a fuzzy set is the Membership Function. It should be noted that when using the apparatus in question, the expected value of the coefficient H_i from area $H_э$ according to the experimental results for various methods can be determined using the fuzzy integral [4]. In this case, it is a priori necessary to determine the distribution of the fuzzy density of the weights these values $g(H_i)$, $H_i \in H_e$. Here $H_i \in H_e$ means the aggregate of the expected value of the Hurst coefficient in the management QoS & QoE traffic.

Given the processes that occur using fuzzy logic, it is advisable to use the membership function $\mu_W(H_i)$ of an element H_i to a fuzzy set W. The analysis systems and decision-making under fuzzy conditions is the subject of the fundamental work L. Zade and R. Bellman [8, 12].

On the basis [8, 12], which allows changing the QoS & QoE parameters using the fuzzy logic apparatus from the domain magnitude W and H_e there is a set of ordered pairs and is described as follows:

$$W = \{\rho(\lambda_i, H_i), \ \mu_W(H_i)\}, \quad H_i \in H_e, \tag{15}$$

where $\mu_W(H_i)$– function that determines the degree membership of an element H_i to a Fuzzy Set W.

Given the algorithm for calculating the Hurst coefficient using the fuzzy logic apparatus, the membership function is expressed as follows:

$$\mu_W(H) = \sum_i \mu_W(H_i) \big/ (H_i), (H_i) \in H_e, \tag{16}$$

As a result of the study different method for calculating indicators self-similar traffic, the following values Hurst coefficient are obtained:

$$H_i = \{(i = 1, ..., 4) : H_1 = 0,68, H_2 = 0,75; H_3 = 0,86; H_4 = 0,95\}$$

Now we can determine the value of the function membership function ordered by decreasing degrees as follows:

$$Y(H_i) = \sum_{i=1}^{4} H_i \cdot \mu_W(H_i), \quad i = \overline{1, 4} \tag{17}$$

Then, according to the calculated data, numerical values $Y(H_i)$ are found as

$$Y(H_i) : h(H_i) = 0,43, \quad Y(H_2) = 0,52, \quad Y(H_3) = 0,90, \quad Y(H_4) = 0,98$$

Based on the fuzzy control technique, the numerical values of the membership function of the process under study can be determined as follows:

$$\mu_W(H_i) = 0,43H_1^{-1} + 0,52H_2^{-1} + 0,90H_3^{-1} + 0,98H_4^{-1}, \quad (18)$$

Given (18) under given conditions, fuzzy measures take the following values:

$$g(F_1) = 0,52, g(F_2) = 0,81, g(F_3) = 0,95, g(F_4) = 1,0,$$

Based on the proposed approach and calculation, the QoS & QoE indices are considered both the average service time and the probability packet loss when using the predicted value H_i compared to determining the listed indicators by the average value of the Hurst coefficient $H_i^{av} = 0,70$,where $0 < H_i^{av} \leq 1$.

7 Construction and Verification of the Degree Adequacy Apparatus Fuzzy model

It is worth noting that when constructing and checking the degree of adequacy of the fuzzy model apparatus, the Fuzzy Logic Toolbox software package Matlab was used [2, 4, 12–15]. Figure 3 shows the function belonging $\mu_W(H_i)$ to the input variables of the coefficient effective loading communication channels in multiservice networks at $H_i^{av} = 0,70$ and $C_{i.\max} = 155$ Mbps.

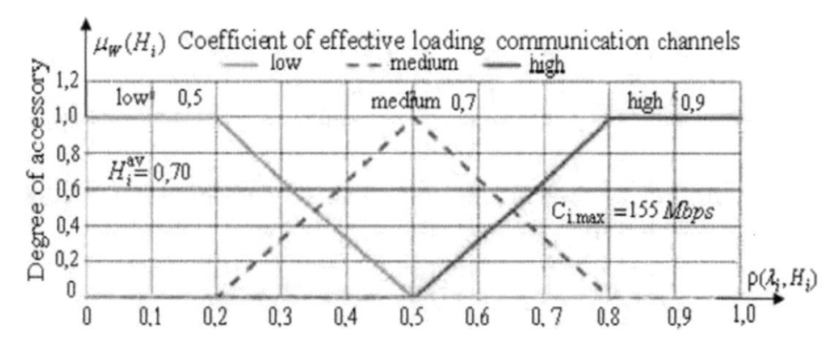

Fig. 3. Graphs membership function $\mu_W(H_i)$ input variables of the coefficient effective loading communication channels

From the graphical dependence it follows that an increase in the effective loading communication channels leads to an increase in the share lost packets, there by reducing the required QoS & QoE parameters. For various services, its noticeable change begins with the value $\rho(\lambda_i, H_i) > 0.55$, ..., 0.65 at $H_i^{av} = 0,70$ and $C_{i.\max} = 155$ Mbps.

To set membership functions in the Fuzzy Logic Toolbox, it was found that membership functions become non-linear. And this greatly complicates further research.

8 Conclusions and Future Work

As a result of the study of the construction membership functions using fuzzy production rules and a method for predicting signal traffic and operational control network characteristics, an analytical expression was obtained to assess the quality of forecasting service traffic in a communication network. The system-technical analysis showed that the studied method for predicting the network characteristics of service traffic will solve the problem ensuring the required quality of signaling systems of the NGN and FN networks when providing multimedia services using fuzzy neural network technologies based on the fuzzy extension principle. On the basis of the forecasting model, the function of belonging to the set of an element H_i is proposed W using the apparatus fuzzy logic, a graph of the function $\mu_W(H_i)$ f the input variables of the coefficient of effective traffic loading is given.

References

1. Zadeh, L.A.: Fuzzy sets. Inf. Control **8**, 338–353 (1965)
2. Ibrahimov, B.G., Ismaylova, S.R.: The effectiveness NGN/IMS networks in the establishment of a multimedia session. Am. J. Netw. Commun. **7**(1), 1–5 (2018)
3. Aliev, R.A., Gurbanov, R.S., Aliev R.R., Huseynov, O.H.: Investigation of stability of fuzzy dynamical systems. In: Proceedings of the Seventh International Conference on Applications of Fuzzy Systems and Soft Computing, pp. 158–164 (2006)
4. Ibrahimov, B.G., Alieva, A.A.: An approach to analysis of useful quality service indicator and traffic service with fuzzy logic. In: 10-th International Conference on Theory and Application of Soft Computing, Computing with Words and Perceptions–ICSCCW-2019. Advances in Intelligent Systems and Computing, vol. 1095, pp. 495–504 (2019)
5. Borisov, V.V., Kruglov, V.V., Fedulov, A.S.: Fuzzy models and networks, p. 248. Hotline - Telecom (2012)
6. Zadeh, L.: The Concept of a Linguistic Variable and its Application to Making Approximate Decisions. Mir, Moscow (1976)
7. Sokolov, D.A.: Fuzzy quality assessment system. Technol. Means Commun. **4**, 26–28 (2009)
8. Belman, R.E., Zadeh, L.A.: Decision - making in a fuzzy environment. Manag. Sci. **17**, 141–164 (1970)
9. Abrol, S., Mahajan, R.: Artificial neural network implementation on FPGA chip. Int. J. Comput. Sci. Inf. Technol. Res **3**, 11–18 (2015)
10. Yusifbayli, N., Nasibov, V.: Trends in Azerbaijan's electricity security for short-term periods. In: Aliev, R.A., Kacprzyk, J., Pedrycz, W., Jamshidi, M., Babanli, M., Sadikoglu, F.M. (eds.) ICAFS 2020. AISC, vol. 1306, pp. 565–571. Springer, Cham (2021). https://doi.org/10.1007/978-3-030-64058-3_70
11. Islamov, I.J., Hunbataliyev, E.Z., Zulfugarli, A.E.: Numerical simulation characteristics of propagation symmetric waves in microwave circular shielded waveguide with a radially inhomogeneous dielectric filling. Int. J. Microw. Wirel. Technol. **9**, 761–767 (2022)
12. Islamov, I.J., Ismibayli, E.G., Hasanov, M.H., Gaziyev, Y.G., Ahmadova, S.R., Abdullayev, R.: Calculation of the electromagnetic field of a rectangular waveguide with chiral medium. Prog. Electromagn. Res. **84**, 97–114 (2019)
13. Islamov, I.J., Ismibayli, E.G.: Experimental study of characteristics of microwave devices transition from rectangular waveguide to the megaphone. IFAC-PapersOnLine. **51**(30), 477–479 (2018)

14. Islamov, I.J., Ismibayli, E.G., Gaziyev, Y.G., Ahmadova, S.R., Abdullayev, R.: Modeling of the electromagnetic feld of a rectangular waveguide with side holes. Prog. Electromagn. Res. **81**, 127–132 (2019)
15. Ibrahimov, B.G., Alieva, A.A.: Research and analysis indicators of the quality of service multimedia traffic using fuzzy logic. In: Aliev, R.A., Kacprzyk, J., Pedrycz, W., Jamshidi, M., Babanli, M., Sadikoglu, F.M. (eds.) ICAFS 2020. AISC, vol. 1306, pp. 773–780. Springer, Cham (2021). https://doi.org/10.1007/978-3-030-64058-3_97

An Image Completion Method Using Generative Adversarial Networks

Eyyüp Yıldız[1,2](✉), Selçuk Sevgen[2], and M. Erkan Yüksel[3]

[1] Department of Computer Engineering, Faculty of Engineering-Architecture, Erzincan Binali Yıldırım University, Erzincan, Turkey
eyyup.yildiz@ogr.iu.edu.tr

[2] Department of Computer Engineering, Faculty of Engineering, İstanbul University-Cerrahpaşa, İstanbul, Turkey

[3] Department of Computer Engineering, Faculty of Engineering-Architecture, Burdur Mehmet Akif Ersoy University, Burdur, Turkey

Abstract. Image inpainting is one of the important research areas in artificial intelligence. It can be defined as the completion of the corrupted parts in an image by estimating the appropriate pixel values. After the introduction of generative adversarial networks (GAN), successful results began to be obtained in this problem compared to previous techniques. In this study, we presented an image inpainting method based on the Adversarial Learned Inference (ALI) model that jointly learns a generation network and an inference network using an adversarial process. In addition to the working principle of the ALI model, our method uses latent variable vectors in the learning phase. We analyzed our model on various datasets and obtained successful results for the image inpainting process. In our experimental studies, we observed that our method is useful for image editing, harmonizing, and image completion problems.

Keywords: Deep Learning · Generative Adversarial Networks · Adversarial Learning · Image Inpainting · Supervised Learning

1 Introduction

Successfully correcting the defects in the images is among the problems that have been studied for a long time in artificial intelligence. Until deep learning-based methods came to the fore, this problem was tried to be solved by classical machine learning methods or by manually editing the image. Machine learning methods have some drawbacks as requiring huge pre/post-processing processes and limited success. On the other hand, manually editing is dependent on human experiences, is time-consuming, and cannot be standardized. Due to such reasons, estimating the appropriate pixel values of the corrupted parts in the image could not reach the desired level for a long time. With the prominence of deep learning-based approaches, important successes have started to be achieved in the image inpainting problem. Among the deep learning methods, generative networks are the most used models in image inpainting.

D. J. Hemanth et al. (Eds.): ICAIAME 2022, ECPSCI 7, pp. 483–489, 2023.
https://doi.org/10.1007/978-3-031-31956-3_42

GAN [1], which is one of the most important architecture of the generative learning field, is based on the combination of two deep learning artificial neural networks. Instead of directly estimating the distributions of the data received as input, GAN aims to produce new data with the same distribution as possible. GAN consists of two artificial networks: generative neural network and discriminative neural network. The Generator network takes as input a noise randomly drawn from the normal distribution and tries to output it like the real image. On the other hand, the discriminator network tries to successfully classify the real images and the images produced by the generator. The training of the two networks, in which the generator tries to deceive the discriminator by producing real images as much as possible, and the discriminator tries to separate the real images from the fake images in the best way, is done by adversarial training.

In this paper, we presented an ALI [2] -based GAN model for image inpainting problems. Our model is novel in terms of the use of the ALI loss function and the use of noise at an L1 distance to generate realistic reconstructed images. In the model, real data is passed through an encoder network, and feature vectors are obtained for real data. Real images and feature vectors obtained from the encoding network are concatenated and sent to the discriminator network. The generator network consists of two neural networks: encoder and generator. First, the corrupted image is sent to a generator's encoder network, and the feature vectors of the corrupted images are obtained. These obtained feature vectors are sent to a generator network containing transposed convolution layers, and an image with appropriately filled in the corrupted part is obtained. The feature vectors obtained from the degraded image by the encoding network and the images produced by the generating network are combined and sent to the discriminator network. A discriminator is a network with convolutional layers. The discriminator tries to distinguish between real images and their feature vectors and feature vectors of distorted images and images obtained from feature vectors. The created model is trained using binary cross-entropy and L1 distance.

2 Related Works

Although different approaches are suggested for the image inpainting problem, which is defined as filling the values of missing or damaged pixels in the image in accordance with the texture in the image, the most successful approaches are the GAN-based ones. The studies carried out for this problem are finding the closest pixel values to that part in the image for the corrupted part and adapting the pixel values found with the suggested approaches to the corrupted part accordingly. With the emergence of data-oriented deep learning architectures, great progress has been made in the process of completing the corrupted parts in the image with appropriate pixel values.

Among these approaches, the first study based on GAN was published in 2016 by D. Pathak et al. The architecture is reconstructing the corrupted image with an encoder and decoder network with the L1 loss function and sending the damaged part in the created image together with the original image to a separator network to amplify its reality [3].

C. Yang et al. published an improved version of the [3], in their study. Besides the study using a neural network like the structure of [3] to complete the missing region, a style transfer network called Style Texture Network is used to increase the quality of the pixel values in the created missing region and its compatibility with the image [4].

S. Iizuka et al. proposed a neural network in which they dealt with the fact that the missing region in the image may be dependent on regions more distant from itself. In the proposed network, dilated convolution blocks, which are more useful in obtaining distant information, are placed in the middle layers of the generator instead of classical convolution blocks. Researchers have used two discriminator networks the first discriminator operates on the entire image, while the second discriminator operates only on the completed segment. The error is calculated by combining the output of the two discriminators [5].

U. Demir and G. Unal created an image completion model using a PatchGAN-based [6] structure in their study in 2018. The generator network is ResNet -based [7], and dilated convolution is used in the first and last layers. The resulting image is sent to two discriminator networks that operate on global and local patches. Another prominent difference in the study is the use of the PatchGAN network as the local discriminator [8].

In 2018, Z. Yan et al. proposed the so-called Shift-net, which is an adapted version of UNet [9], for the image inpainting problem. The model has added a shift layer to the UNet neural network structure, which includes the parts closest to the damaged part in the image. This layer has increased image completion and consistency within the image [10].

3 Method and Material

3.1 Proposed Approach

Our network is based on the ALI idea. ALI takes account of the noise distribution and its generation to infer real data distribution. ALI consists of three neural networks namely: Encoder, Generator, and Discriminator. Encoder(E) maps $x \rightarrow \tilde{z}$ via learnable parameter θ_E where x represents real data and $\tilde{z}$ represents real data distribution. On the other hand, generator(G) maps $z \rightarrow \tilde{x}$ via learnable parameter θ_G where z represents Gaussian distribution with zero mean unit variance and $\tilde{x}$ represent generated images. Discriminator takes $E(x, \tilde{z}; \theta_E)$ and $G(z, \tilde{x}; \theta_G)$ as input and try to correctly predict which network, they came from. Training the whole ALI network is done by $\mathcal{L}_{ALI}$ adversarial loss value (1). Firstly, discriminator encoder network parameters are trained to maximize $\mathcal{L}_{ALI}$. Secondly Generator network is trained by gradient flow coming from the discriminator whose parameters are hold constant, to minimize $\mathcal{L}_{ALI}$.

$$\mathcal{L}_{ALI} = \underset{G}{min}\, D^{max} \mathbb{E}_{x \in X}\left[\log(D(x, E(x)))\right] + \mathbb{E}_{z \in Z}\left[\log(1 - D(G(z), z))\right] \tag{1}$$

$$\mathcal{L}_{REC} = |(x - \tilde{x}) + (z - \tilde{z})|^1 \tag{2}$$

Figure 1 shows our proposed GAN network to solve the image inpainting problem. The network consists of three encoders and a decoder. An encoder and a decoder compose a generator network and the other two networks correspond to the encoder and discriminator respectively. All networks have multiple layers and in all encoder networks, each layer use convolution, batch normalization, and activation functions while the decoder network use transposed convolution. In the training of the network, the adversarial loss

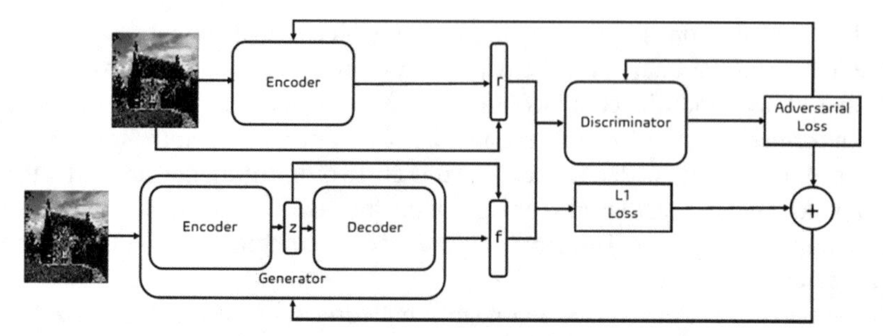

Fig. 1. Proposed network. Generator produces completed image which is concatenated with input feature vector to form f. The encoder produces feature vector which is concatenated with the input image to form r. The discriminator tries to separate its inputs correctly. All network is trained by adversarial and L1 loss

function is used to make the network generate images similar to real images. To reconstruct corrupted images, another loss function L1 distance (2) is used. Our total loss function $\mathcal{L}_{ADV}$ (3) is the weighted sum of the $\mathcal{L}_{ALI}$ and $\mathcal{L}_{REC}$ loss functions. A detailed representation of the network layers is given in Table 2. For the loss function, $\mathcal{L}_{ALI}$ and $\mathcal{L}_{REC}$ weights are set to $\gamma_{ali} = 0.25\ and\ 0.5$ and $\gamma_{rec} = 100$ respectively.

$$L_{ADV} = \gamma_{ali} L_{ALI} + \gamma_{rec} L_{REC} \tag{3}$$

3.2 Material and Implemental Details

We have used to LSUN dataset [11] to train and test our methods. LSUN Church dataset has images into ten categories, and we have used only the church images category to test our network, and we kept the size of images fixed to 64×64 and corrupted images with size $3 \times 10 \times 10$ of normally distributed randomly generated numbers. Corrupted part of images has been kept fixed in position 25–35 for left and top pixels. In Table 1, it has been shown our network details. We have used $5 \times 5, 4 \times 4$, and 1×1 convolutions (transposed convolution for generator's decoder part) with 1 or 2 strides in different layers with zero padding for all layers. We have used Batch Normalization and Leaky ReLU activation after all convolution and transposed convolution operations except for the last layers of networks. In the last layer of generator and encoder, we have used tanh activation. For the discriminator's last layer, we have used sigmoid activation. In all layers of the discriminator, dropout with a 0.2 value has been used to overcome memorizing.

The network implemented in Pytorch with Nvidia TITAN X Pascal GPU and Intel(R) Core (TM) i7-10700KF CPU @ 3.80 GHz 3.79 GHz computer. we have used the ADAM optimizer for optimization.

Table 1. The detailed structure of our network. The Generator consists of two sub-networks namely an encoder and decoder. The Discriminator consists of three sub-networks namely Image part for real and fake images, Feature part for real and fake images' feature vectors and All parts for producing real/fake results. (Op. The Operation, K. Kernel size, S. Stride number, F. Output of Layer, BN. Batch Norm. D. Dropout, Conv. Convolution, TConv. Transposed Convolution)

Generator

Encoder Part
(3x64x64 input)

Op.	K.	S.	F.	BN
Conv-5x5		2	32	Yes
Conv-4x4		2	64	Yes
Conv-4x4		1	128	Yes
Conv-4x4		2	256	Yes
Conv-4x4		1	512	Yes
Conv-1x1		1	512	Yes
Conv-1x1		1	256	No

Decoder Part
(256x1x1 input)

Op.	K.	S.	F.	BN
TConv-4x4		1	256	Yes
TConv-4x4		2	256	Yes
TConv-4x4		1	128	Yes
TConv-4x4		2	64	Yes
TConv-4x4		1	64	Yes
TConv-4x4		2	64	Yes
TConv-1x1		1	32	Yes
TConv-1x1		1	3	No

Encoder

(3x64x64 input)

Op.	K.	S.	F.	BN
Conv-5x5		2	32	Yes
Conv-4x4		2	64	Yes
Conv-4x4		1	128	Yes
Conv-4x4		2	256	Yes
Conv-4x4		1	512	Yes
Conv-1x1		1	512	Yes
Conv-1x1		1	256	No

Discriminator

Image Part
(3x64x64 input)

Op.	K.	S.	F.	BN	D.
Conv-5x5		2	32	Yes	Yes
Conv-4x4		2	64	Yes	Yes
Conv-4x4		1	128	Yes	Yes
Conv-4x4		2	256	Yes	Yes
Conv-4x4		1	512	Yes	Yes

Feature Part
(256x1x1 input)

Op.	K.	S.	F.	BN	D.
Conv-1x1		1	512	No	Yes
Conv-1x1		1	512	No	Yes

Concatenation
(Feature and Image part output)

All Part
(1024x1x1 input)

Op.	K.	S.	F.	BN	D.
Conv-1x1		1	1024	No	Yes
Conv-1x1		1	1024	No	Yes
Conv-1x1		1	1024	No	Yes

Optimizer: Adam ($\beta_1 = 0.5, \beta_2 = 0.999$)
Batch Size: 512
Epoch: 1750
Learning Rate: 3e-4

4 Discussion and Results

The church category of LSUN Dataset has approximately 126.000 church images. we started to train with the values of $\gamma_{ali} = 0.5$ and $\gamma_{rec} = 100$ for 1500 epochs. During the training between 1500 to 1750 epochs, we drop from 0.5 to 0.25 but kept y = 100, to get better image quality. The learning rate was used as 3e-4 throughout the training and was not reduced.

In Fig. 2, it can be seen the results of training for the proposed network. The network successfully predicted corrupted pixel values of images. Although the network has limited success to produce high-quality completed images. We interpret this situation

Fig. 2. Result of training dataset for proposed network. For every three-row group, from top to down; original image, corrupted image, and inpainted images.

Fig. 3. Result of test dataset for proposed network. For every three-row group, from top to down; original image, corrupted image, and inpainted images.

as the proposed method successfully estimating pixel values but needs improvement for the image inpainting problem.

Figure 3 shows the results for the test images. When we look at the difference in success between the test dataset and the training dataset, we can say that the proposed method needs to be improved. Despite 1750 iterations which is far less than most other networks in the literature, looking at the results, it can be said that the proposed model works successfully.

5 Conclusions and Future Work

Today, the image completion problem is mostly being tried to solve by using deep learning techniques. In this study, we presented a GAN-based method to complete the corrupted parts of images efficiently. We employed the Adversarial Learned Inference-based model to solve this problem. Although the method we developed works successfully on various image datasets, some aspects need improvement in producing higher-quality images. In our future work, we aim to develop a novel GAN-based framework that can automatically generate better and large-size inpainting images using subject-specific image datasets in a scalable way while maintaining the accuracy and diversity of the datasets.

References

1. Goodfellow, I.J., et al.: Generative adversarial nets. In: Advances in Neural Information Processing Systems, vol. 27 (2014). http://www.github.com/goodfeli/adversarial
2. Dumoulin, V., et al.: Adversarially Learned Inference. arXiv preprint arXiv:1606.00704 (2016). http://arxiv.org/abs/1606.00704
3. Pathak, D., Krähenbühl, P., Donahue, J., Darrell, T., Efros, A.A.: Context encoders: feature learning by inpainting. In: Proceedings of the IEEE Conference on Computer Vision and Pattern Recognition (2016)
4. Yang, C., Lu, X., Lin, Z., Shechtman, E., Wang, O., Li, H.: High-resolution image inpainting using multi-scale neural patch synthesis. In: Proceedings of the IEEE Conference on Computer Vision and Pattern Recognition (2017)
5. Iizuka, S., Simo-Serra, E., Ishikawa, H.: Globally and locally consistent image completion. ACM Trans. Graph. (ToG) **36**(4), 1–14 (2017). https://doi.org/10.1145/3072959.3073659
6. Isola, P., Zhu, J.-Y., Zhou, T., Efros, A.A.: Image-to-image translation with conditional adversarial networks. In: Proceedings of the IEEE Conference on Computer Vision and Pattern Recognition (2017). https://github.com/phillipi/pix2pix
7. He, K., Zhang, X., Ren, S., Sun, J.: Deep residual learning for image recognition. In: Proceedings of the IEEE Conference on Computer Vision and Pattern Recognition (2016). http://image-net.org/challenges/LSVRC/2015/
8. Demir, U., Unal, G.: Patch-based image inpainting with generative adversarial networks. arXiv preprint arXiv:1803.07422. (2018). http://arxiv.org/abs/1803.07422
9. Ronneberger, O., Fischer, P., Brox, T.: U-Net: convolutional networks for biomedical image segmentation. In: Navab, N., Hornegger, J., Wells, W.M., Frangi, A.F. (eds.) MICCAI 2015. LNCS, vol. 9351, pp. 234–241. Springer, Cham (2015). https://doi.org/10.1007/978-3-319-24574-4_28
10. Yan, Z., Li, X., Li, M., Zuo, W., Shan, S.: Shift-net: image inpainting via deep feature rearrangement (2018). https://github.com/Zhaoyi-Yan/Shift-Net
11. Yu, F., Seff, A., Zhang, Y., Song, S., Funkhouser, T., Xiao, J.: LSUN: construction of a large-scale image dataset using deep learning with humans in the loop. arXiv preprint arXiv:1506.03365 (2015). http://arxiv.org/abs/1506.03365

Construction 4.0 - New Possibilities, Intelligent Applications, Research Possibilities

Krzysztof Kaczorek, Nabi Ibadov(✉), and Jerzy Rosłon

Warsaw University of Technology, Warsaw, Poland
{krzysztof.kaczorek,nabi.ibadov,jerzy.roslon}@pw.edu.pl

Abstract. Negative global phenomena, such as pandemics, armed conflicts, economic crises, or natural disasters are causing significant damage to the industry. However, at the same time, they are also forcing the acceleration of the development of selected areas of the economy. In this race, the key will be to implement innovations that will give individual entities a competitive advantage. In order to increase the level of innovation in the construction sector, it is neces-sary to stop treating it as a typical industry branch and to set requirements specific to the modern technology sector. Requirements for potential ideas and solutions should be closely related to the features that characterize the modern technology sector. In the article, the authors take up the subject of the implementation of the concept of Construction 4.0 in Poland - the current status, new possibilities, barriers and limitations, applications, and future research opportunities are presented.

Keywords: Construction 4.0 · Construction Management · Innovations · Intelligent Solutions · ICT

1 Introduction

Negative global phenomena, such as pandemics, armed conflicts, economic crises, or natural disasters are causing significant damage to the industry. However, at the same time, they are also forcing the acceleration of the development of selected areas of the economy. The outbreak of the global pandemic at the turn of 2019 and 2020 made the vast majority of analyzes, specific trends and designated development plans worthless. At the beginning of 2022, optimistic forecasts finally begin to emerge: the number of vaccinated and recovering people is increasing, new variants of the Covid-19 virus are less lethal, societies have learned to function in conditions of increased sanitary requirements. The global recovery of economies is also underway. In this race, the key will be to implement innovations that will give individual entities a competitive advantage. In the article, the authors take up the subject of the implementation of the concept of Construction 4.0 in Poland - the current status, new possibilities, barriers and limitations, applications, and future research opportunities are presented.

D. J. Hemanth et al. (Eds.): ICAIAME 2022, ECPSCI 7, pp. 490–499, 2023.
https://doi.org/10.1007/978-3-031-31956-3_43

2 Sources of Competitive Advantage of Enterprises

According to M.E. Porter "competitive advantage grows fundamentally out of value a firm is able to create for its buyers that exceeds the firm's cost of creating it" [18]. It manifests itself in the correct choice of the path of building and strengthening the competitive potential of the enterprise, and in the effective use of competition instruments [24]. The result of effective implementation of the presented activities is the use of available resources and skills better than the competition, which translates into better adaptation to the requirements of a given industry, meeting the expectations of consumers, and achieving above-average profits [5].

What distinguishes competitive enterprises in the 21st century is primarily innovation, entrepreneurship, and network connections of companies in a given sector. According to [11, 18, 19], a competitive advantage can be obtained thanks to:

- continuous development and implementation of product and technological innovations in the field of organization and management, market service, etc.,
- creating new customer needs and new markets,
- a constant search for new business opportunities,
- entering new areas of activity (the so-called entrepreneurial concept of competitiveness),
- continuous restructuring which is understood mainly in terms of cost reduction (in the sphere of employment and in the product-market sphere) or, more broadly, in the re-engineering of basic business processes (the so-called restructuring concept of entrepreneurship).

3 Expectations Regarding Innovation in the Construction Sector

In order to increase the level of innovation in the construction sector, it is necessary to stop treating it as a typical industry branch and to set requirements specific to the modern technology sector. Requirements for potential ideas and solutions should be closely related to the features that characterize the modern technology sector:

- the leading role of knowledge and knowledge management,
- the leading role of research and development,
- the leading role of key competencies and key employees,
- high level of innovation and capital critical mass,
- the leading role of time pressure, race against time,
- the international or global reach of the market and creating advantages,
- the ever-shorter technology life cycle,
- the leading role of change and innovation in the development of the industry,
- the neutral or positive influence on the environment [6].

4 Innovativeness in Poland

According to the material [12] of PFR S.A. Analysis Bureau of March 2021, concerning the innovativeness of the Polish economy, it should be stated that:

- at the end of 2019, over EUR 7 billion was allocated to research and development in Poland, which was 17.1% higher than the year before, and more than 518% higher than in the year Poland joined the EU. However, it is not sufficient to provide nominal values alone as they depend on the size of the economy. A better reflection of the actual state of affairs is the comparison of outlays on R&D activity to the GDP of a given country. In nominal terms, Poland has made a very impressive growth since 2004, when about 0.5% of GDP was spent on outlays, and in 2019 it was already 1.3% (Fig. 1).

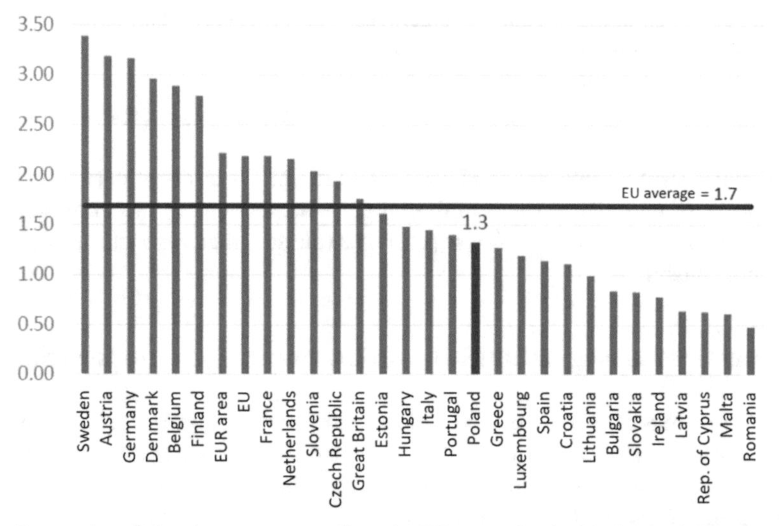

Fig. 1. Research and development expenditure in EU countries in 2019 (% of GDP) based on Eurostat data [12]

When analyzing expenditure on research and development in Poland against the EU in terms of per capita or as a% of GDP, it should be stated that Poland is currently in the second ten of the EU economies. Nevertheless, it should be emphasized that analyzing the data since 2004, Poland is beginning to approach the EU average (Fig. 2).

- In the cyclical ranking of the European Commission, the European Innovation Index, Poland scored 66.2 points, which was the fourth worst result among the EU countries and the seventh among all surveyed countries (Switzerland turned out to be the best with 177.3 points). Nevertheless, it should be noted that compared to the 2011 ranking, Poland's score increased by 12.9 points, which was a better result than the EU average of 10.5 points.

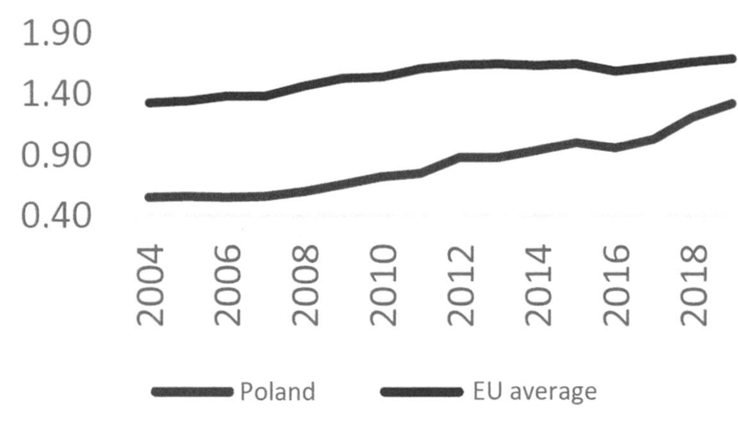

Fig. 2. Expenditure on research and development in Poland compared to the EU (% GDP) in 2004–2019 based on Eurostat data [12]

5 From the First to the Fourth Industrial Revolution - Construction 4.0

The first industrial revolution began in the 18th century by harnessing the power of steam, which made it possible to mechanize production. Muscle strength (both human and animal) was gradually being replaced by steam engines. In construction, it is primarily the dynamic development of the railway system and the construction of numerous engineering facilities (e.g. iron bridges) required by subsequent lines of railway connections. The second industrial revolution took place in the 19th century with the invention of electricity and the assembly line. Electrical installations appear in the buildings, and in 1867, Joseph Monier, a gardener of the city of Paris, patented mesh concrete pots - the first modern prefabricated elements. The third industrial revolution started in the 1970s with ubiquitous computerization, and later also automation. The design process was significantly improved, thanks to the computer software, precise analyzes were possible, which allowed for the construction of more complex structural systems. Automation of production processes translated into increased efficiency in prefabrication. Currently, the fourth industrial revolution is taking place, which is characterized by the collection, processing, and analysis of huge amounts of data. Not only intelligent systems are developed, but more and more often autonomous, which, after appropriate programming, can "learn" themselves and work with very limited human intervention. In the construction industry, the fourth industrial revolution was named Construction 4.0.

Analyzing Sects. 2–5 of this article, it should be stated that the requirements for the full implementation of Construction 4.0 are currently met in Poland. The circumstances described earlier should be supplemented with those presented in [7]:

- large and constantly growing market,
- high quality of services,
- qualified staff - both technical and managerial,
- a large presence of many foreign, international construction companies,

- high flexibility and openness of entities operating in Poland compared to foreign markets,
- increasing awareness, determination, and striving to meet the challenges of civilization, such as energy challenges or air pollution, through the construction industry.

The use of the presented advantages creates an opportunity for the Polish construction sector to use the fourth industrial revolution not only for the far-reaching development of enterprises in Poland but also for increasing the dynamics of expansion into foreign markets.

6 Development Areas Within the Construction 4.0

According to [25], the following areas of development can be mentioned under Construction 4.0:

- 3D Printing: The industrial revolution in manufacturing redefines the transition from high-quality design to a real product. Digital production plays a major role in this revolution, and 3D printing technology deserves a distinction here. Currently, it is possible to go directly from the three-dimensional model of an element to a finished 3D object with one command, using one device, without the need for retooling, with over 80 different types of materials such as steel, glass, ceramics, polymer, concrete, etc.
- Crowdfunding: The construction industry needs a steady flow of capital to operate and grow. The digital technology of crowdfunding is a new way of unlocking the needed funds. Initially, this solution was used primarily in the production, entertainment, and IT industries. However, it is taking the first steps in the construction industry with increasing confidence. With its use, one can collect funds necessary for the construction of investments, both for the private sector (e.g., residential buildings) and the public sector (e.g., infrastructure facilities).
- Robotics: This area is constantly strengthening its position in the construction industry. Previously known robots were responsible for performing a limited set of repetitive activities and were used primarily in the production of materials and components. They were most often used as part of large production lines. Thanks to the progress, the range of possibilities for robots has definitely expanded. Currently, they can be connected to various sensors, which allows them to record information about the work performed. The data can be entered into the control system, which then calibrates the robot's operating parameters, ensuring greater efficiency and precision of the process.
- Cloud computing power: Cloud computing is already a widely known and used solution. In construction, it can give design teams the opportunity to reverse the design process and start from the end. The classic chronology in design is the design of the building and the subsequent analysis of its parameters related to erection (e.g., time and cost of implementation) and operation (e.g., costs of maintaining and using the facility) [22]. Thanks to the computing power of the cloud, it becomes possible to analyze many geometric, material, economic, etc. variants and generate a solution

proposal that should be implemented in the design process in order to obtain the maximum profit, the lowest total cost of ownership, or other success indicators [13, 23].

- Crowdsourcing: This process involves the external outsourcing of tasks traditionally performed by employees of a given entity to an unidentified, usually very wide group of people. In the construction industry, this opens up an opportunity for extensive cooperation via the Internet on solving complex and complicated technical problems.
- Drones: New technologies such as drones can now be used to research, scanning, and inspection of construction sites. Inserting drone-captured images into photorealistic image applications that combine photos to create 3D models takes the real world to digital on a large scale. We are already witnessing the use of drones equipped with cameras for a variety of applications - from remote inspection of tall buildings and structures, which minimizes the risk and costs of working at heights, to the inspection of linear structures, such as pipelines or rail corridors.
- Augmented reality [14, 15, 16, 21]: This technology is used in a number of areas. The first is project planning. Thanks to the use of AR, BIM, and advanced 3D modeling, it is possible, for example, to see the visualization of the building in a real environment, on a 1: 1 scale, even before starting the implementation of works. What's more, the visualization (including the interior of the building) can also be used during the construction of the facility to correct previously made decisions, for example, related to the emergence of new circumstances in the form of collisions or delays. Another area are training activities. Scientific research unequivocally confirms that a person remembers pictures better than text. Moreover, in the case of complex machines or devices, presenting the method of operation in a descriptive manner may be difficult for the author and incomprehensible for the reader. Presenting the method of operation on a mobile device, indicating individual elements and showing how they work, significantly accelerates training and increases the level of assimilation of new information. Augmented reality has also significantly improved the process of monitoring the progress of work. Thanks to AR, inspectors have the opportunity to accurately compare the BIM model with the construction in progress. Using AR during the review of the progress of works, it is possible to take photos and create notes about possible problems which can be then integrated with the model. Moreover, AR makes it possible to compare the actually made elements with the schedule previously made in the BIM class software (BIM 4D). What's more, AR significantly reduces the risk of damage to underground installations - it is enough to provide machine operators with appropriate documentation on devices that allow viewing the area in real-time with the marked course of wires, cables, pipes, etc.
- Real-time collaboration: So-called "wearable technologies" are used to improve safety on the construction site. An example may be solutions developed by Human Condition Safety, such as smart vests, which are to help the employees wearing them perform their tasks in a better, safer, and faster way. At the same time, they serve as a kind of management cockpit operating in real-time, transmitting information on how many current employees are in areas with a higher likelihood of an accident. Another solution is the "connected helmet". Each such helmet is connected to a special system that monitors selected parameters. The main task of the device is to alert the construction site office in the event of detecting anomalies in the readings that indicate a dangerous

situation on the construction site, such as shock, impact, fall, or a significant increase or decrease in the temperature under the helmet. Along with the alert, the location of a given employee is also sent, thanks to which the reaction time is limited to a minimum, and the effectiveness of assistance increases. The helmet also facilitates work with documentation on the construction site - the system to which it is connected shows the construction management + the validity of medical examinations, health, and safety training, or the authorization to perform specialized works.
- Generative design: A human is a creative being, capable of creative work, able to process large amounts of distributed data, and think in a complex and abstract way in order to solve problems. Unfortunately, the decision-making process of a person can be burdened with habits, likes, and prejudices. The use of intelligent algorithms imitating the natural approach to design is not limited by such flaws.
- Game engines: The game engine deals with the interaction between the game elements. By translating this into the design process, it is possible to check the behavior of people using the building or the site in terms of the location of specific stores or service points. The use of such algorithms makes it possible to sell or rent retail and service space at a higher price, thus increasing the investor's profit.
- Internet of Things [27]: This is the concept of devices that can connect to the Internet or other devices using wireless networks or, more rarely, with cables. The benefits of IoT systems are already felt during investment planning, e.g. when analyzing the vicinity of the plot, when one can obtain information on the light intensity, traffic, and air pollution. Moving on to the implementation phase, construction machines begin to communicate with each other, informing each other about the current location, phase of work, or planned movements. This allows users to calculate the optimal scope of work and transport while maintaining the set priorities, minimizing the risk of a collision. The benefits of using intelligent machines increase with their number. This is due to the fact that standard methods of planning and organizing works are burdened with many unknowns. Current microprocessors can perform billions of analytical calculations per second, and the ability to stay up-to-date on the progress of work allows users to update the schedule in real-time. The concept of communicating machines applies not only to those performing work on the construction site but also takes into account those outside the place of direct work (if they are added to the IoT network). For example, one can receive up-to-date information on all deliveries of construction elements, concrete mix, or asphalt, thanks to which the user knows exactly what is happening on the arranged detours of linear investments and is able to maintain precise control over activities that so far were burdened with a significant margin of error or even impossible to predict.
- Big data [9, 3, 4, 8, 2, 10]: Big data is already quite commonly used in finance (e.g., to predict loan default rates) and retail sales (primarily to determine purchasing behavior). In the construction industry, large amounts of highly structured data are today generated mainly by BIM-class software and other design tools. This creates prospects for a new discipline, which is Construction Intelligence, i.e., the ability to forecast the future through the exploration of this data. Searching for regularities in projects and other data sources gives the possibility of identifying the entire process — from early signs of pressure in the supply chain to the best method of optimizing cash flows and the primary reason for revaluation of offers.

- Machine learning [1, 20]: this is an area of artificial intelligence devoted to algorithms that improve automatically through experience, i.e., exposure to data. Machine learning algorithms build a mathematical model from sample data, called the training set, to forecast or make decisions without being explicitly programmed by humans for that purpose. The technology allows the system to self-improve on the basis of the data provided to it. ML tasks are limited to a narrow, specific scope within which a specific system is to operate. Contrary to artificial intelligence, the machine learning process is not able to create something new, but only obtain optimal solutions to a specific problem.

7 New Requirements and Difficulties in the Implementation of the Construction 4.0

New technologies mean, apart from new opportunities, also new difficulties. According to [26], the main challenges related to the implementation of Construction 4.0 are:

- conservative mentality,
- low market openness to new ideas,
- professionals' preparation level,
- reluctance to innovate due to additional burden,
- low confidence in the adaptation of solutions developed by other countries,
- the issue of data security,
- increased demand for data analysts and IT architects,
- the need to learn new software,
- close cooperation between the construction industry and the IT industry.

The implementation of innovations can be improved and accelerated by using:

- ready-made solutions developed by foreign entities and brought to Poland for the contracts they carry out,
- experience of engineers who returned to Poland after gaining experience abroad,
- knowledge and experience of the scientific staff, especially the ones who worked on international internships with the most modern solutions on a global scale.

8 Summary and Conclusions

On the basis of the information presented in Sects. 1–7 of this publication, it should be stated that:

1. The fourth industrial revolution is becoming a reality, and in the construction industry it is known as "Construction 4.0." Its main idea is to develop innovative solutions using huge amounts of data.
2. Construction 4.0 offers a wide range of possibilities when it comes to innovative solutions. They can significantly improve the individual issues that make up the construction sector.

3. In the post-pandemic period, a race will begin related to the reconstruction of the construction sectors in individual countries. Wise use of the benefits of Construction 4.0 may not only improve the functioning of entities on the domestic market but also help in expansion into foreign markets.
4. The implementation of new solutions will require facing new challenges: primarily related to breaking the resistance of employees at various levels and close cooperation with the IT industry.
5. In the implementation of innovations under Construction 4.0, the support of entities with experience in the field of new technologies that they have acquired abroad will be extremely valuable.

References

1. Anysz, H.: Machine learning and data mining tools applied for databases of low number of records. Adv. Eng. Res. **21**(4), 346–363 (2021)
2. Anysz, H., Ibadov, N.: Neuro-fuzzy predictions of construction site completion dates. Czasopismo Techniczne **6**, 51–58 (2017)
3. Anysz, H., Rosłon, J., Foremny, A.: 7-score function for assessing the strength of association rules applied for construction risk quantifying. Appl. Sci. **12**(2), 844 (2022)
4. Anysz, H., Zbiciak, A., Ibadov, N.: The influence of input data standardization method on prediction accuracy of artificial neural networks. Procedia Eng. **153**, 66–70 (2016)
5. Faulkner, D., Bowman, C.: Strategie konkurencji, Gebethner i Ska, Warsaw, p. 29 (1996)
6. Godlewska, H., Skrzypek, E., Płonka, M.: Przewaga konkurencyjna w przedsiębiorstwie: sektor-wiedza-przestrzeń. Texter (2016)
7. Gorustowicz, M.: W kierunku Budownictwa 4.0 - perspektywy rozwoju branży budowlanej w obliczu czwartej rewolucji przemysłowej, Kwartalnik naukowy AKADEMIA ZARZĄDZANIA, vol. 4, 2nd edn., pp. 208–216 (2020)
8. Ibadov, N.: The alternative net model with the fuzzy decision node for the construction projects planning. Arch. Civ. Eng. **64**(2), 3–20 (2018)
9. Górecki, J., Bizon-Górecka, J., Michałkiewicz, K.: Execution of construction projects in the use of Big Data. Sci. Rev. Eng. Environ. Sci. **26**(2), 241–249 (2017)
10. Ibadov, N.: Construction project planning under fuzzy time constraint. Int. J. Environ. Sci. Technol. **16**(9), 4999–5006 (2018). https://doi.org/10.1007/s13762-018-1695-x
11. Ibadov, N.: Ecology of building solutions in the engineering of construction projects. Arab. J. Geosci. **13**(13), 1–9 (2020). https://doi.org/10.1007/s12517-020-05356-0
12. Kolasa, M.: Innowacyjność Polski. Chartbook, Polski Fundusz Rozwoju, Warsaw (2021)
13. Kulejewski, J., Ibadov, N., Rosłon, J., Zawistowski, J.: Cash flow optimization for renewable energy construction projects with a new approach to critical chain scheduling. Energies **14**(18), 5795 (2021)
14. Majcher, J.: Rzeczywistość rozszerzona w budownictwie (2021). http://bimcorner.com/
15. Nicał, A., Nowak, P., Rosłon, J.: Innovations in construction personnel education. In: MATEC Web of Conferences, vol. 86, p. 05005. EDP Sciences (2016)
16. Nicał, A., Nowak, P., Rosłon, J.: Szkolenie z wykorzystania technologii rzeczywistości rozszerzonej w wybranych obszarach budownictwa w UE. Przegląd Budowlany, vol. 92 (2021)
17. Pierścionek, Z.: Strategie konkurencji i rozwoju przedsiębiorstwa, pp. 209–224. Wydawnictwo Naukowe PWN, Warszawa (2006)

18. Porter, M.E.: Competitive Adventage: Creating and Sustaining Superior Performance. Free Press, New York (1985)
19. Porter, M.E.: Porter o konkurencji, p. 47. PWE, Warszawa (2001)
20. Rosłon, J.H., Kulejewski, J.E.: A hybrid approach for solving multimode resource-constrained project scheduling problem in construction. Open Eng. **9**(1), 7–13 (2019)
21. Rosłon, J., Nicał, A., Nowak, P.: ARFAT-The augmented reality formwork assembly training. In Edulearn 18. 10th International Conference on Education and New Learning Technology, Palma, 2nd–4th July 2018, Conference proceedings IATED Academy. pp. 1727–1731 (2018)
22. Zawistowski, J.: Cash-flow schedules optimization within life cycle costing (LCC). Sustainability **12**(19), 8201 (2020)
23. Sroka, B., Rosłon, J., Podolski, M., Bożejko, W., Burduk, A., Wodecki, M.: Profit optimization for multi-mode repetitive construction project with cash flows using metaheuristics. Arch. Civil Mech. Eng. **21**(2), 1–17 (2021). https://doi.org/10.1007/s43452-021-00218-2
24. Włodarczyk, T.: Źródła przewagi konkurencyjnej przedsiębiorstwa – aktualny stan dyskusji. In: Walińska, E. (ed.) Współczesne problemy finansów, rachunkowości i zarządzania (2013)
25. Branża budowlana wykorzystująca siłę cyfryzacji, Centrum Badań i Analiza Rynku, Autodesk (2021)
26. Budownictwo. Innowacje. Wizja liderów branży 2025. Centrum Badań i Analiza Rynku, Autodesk (2020)
27. Internet rzeczy. Co to jest internet rzeczy? Jak IOT zmienia budownictwo? Murator (2019)

Tone Density Based Sentiment Lexicon for Turkish

Muazzez Şule Karaşlar[1](✉), Fatih Sağlam[2], and Burkay Genç[1]

[1] Department of Computer Engineering, Hacettepe University, Ankara, Turkey
{muazzez.karaslar,burkay.genc}@hacettepe.edu.tr
[2] Department of Computer Engineering, Ufuk University, Ankara, Turkey
fatih.saglam@ufuk.edu.tr

Abstract. With the developing technology and increasing use of the internet, many sources of data have been exposed to researchers. Analysis and extraction of meaningful information from this data is a research topic under the field of natural language processing. Sentiment analysis which is a sub-field of NLP evaluates the content of data with respect to the opinion it conveys as one of positive or negative. Most sentiment analysis research is done using one of two approaches: lexicon based and machine learning based. Lexicon based approach needs a dictionary of positive and negative words which is then used to evaluate a text. Although there are an abundance of studies in English, the same cannot be claimed for Turkish. Therefore, we focus our studies in constructing a comprehensive and accurate Turkish sentiment lexicon. In this paper, we aim to develop a Turkish sentiment lexicon using a novel methodology: using statistical tone density functions computed using a very large document corpus obtained from mainstream Turkish news agencies. The lexicon obtained this way, not only assigns tone values instead of boolean polarities, but also provides sharper tones which is usually not possible with other approaches in the literature. We evaluate the performance of this lexicon in comparison with similar lexicons in the literature. Results show that the constructed sentiment lexicon in this study achieves a comparable performance and poses many potential improvement possibilities.

Keywords: Sentiment analysis · Natural Language Processing · lexicon · polarity · statistical distribution

1 Introduction

Today, with the widespread use of social media and the internet, users have access to many interpretable thoughts through the online environment. Due to this rapid increase in data in the electronic environment, the need to automatically analyze the data and extract meaningful information from the data is increasing.

Sentiment analysis, also known as Opinion Mining, is the interpretation of a subjective language element, such as speech or writing, on a particular topic. Most of the data is available in text form and using this analysis method, words and phrases can be classified as positive, negative, and neutral.

D. J. Hemanth et al. (Eds.): ICAIAME 2022, ECPSCI 7, pp. 500–514, 2023.
https://doi.org/10.1007/978-3-031-31956-3_44

Sentiment analysis which can be roughly divided into two which are machine learning and lexicon-based approach. In machine learning approach, linguistic features, machine learning models [1, 2] and algorithms are used. On the other hand, lexicon-based approach uses sentiment lexicon for identifying the sentiment tones of words and texts.

For analyzing text content by using lexicon-based approach, existence of the languages own resources is very important and necessary. Compared to other languages including Turkish, English has very rich language resources and libraries such as SentiWordNet [3], SenticNet [4], and WordNet-Affect [5]. For this purpose, translation approach from English language sources is mostly used in the literature. However, the use of language libraries obtained with this approach may prevent accurate sentiment analysis because of the different language structures.

To this end, this study aims to develop a Turkish Sentiment Lexicon using an automated pipeline. Mainstream news media sources have been used from the GDELT database to build a sentiment corpus. However, unlike the literature, lexicon is developed by a novel method: using the tone distributions based on the density functions computed over the documents.

The rest of this paper is described as follows. In the second section we discuss related works about developing sentiment lexicon and analyzing sentiment polarity of texts. The third section explains the proposed approach of developed Turkish sentiment lexicon and describes the data set. The fourth section presents evaluation and results and finally, we conclude the paper.

2 Related Works

The lexicon-based approach finds the sentiment polarity of a document by using the sentiment tone of each word in a given document. There are several well-known sentiment lexicons and most of them in English such as SentiWordNet, SenticNet, SentiStrength [6] etc. As seen in the literature, most known sentiment lexicons are in English and there is a lack in other languages sentiment lexicon including Turkish. Because of this lack, first Turkish sentiment lexicon SentiTurkNet is created by Dehkharghani [7].

While developing SentiTurkNet, they used Turkish WordNet and manually labelled synsets as positive, negative, and neutral and extracted features for synsets. Then, extracted features are used in machine learning classifiers for developing sentiment lexicon by mapping features with polarity values. They concluded that translation method for developing lexicon is not a best methodology because of the terms in one language cannot have an equivalent meaning or sentiment because of the different language structure and culture.

There are lots of study which use SentiWordNet by developing sentiment lexicon in their own language. Ucan [8] uses this lexicon for developing Turkish Sentiment Dictionary by using translation approach from English to Turkish. He used SVM (Support Vector Machine) method and movie and hotel reviews as labeled data for evaluating performance of the created lexicon. This lexicon has named as SWNetTR.

To enhance the SWNetTR lexicon, Sağlam et al. [9] conducted a study. Their study uses GDELT Database to reach Turkish news pages on the web for enriching the lexicon from 27K to 37K words. For determining the performance of this lexicon, they tested the lexicon on domain independent news texts. With this study SWNetTR lexicon expanded with around 10000 unique words and this new lexicon was named as SWNetTR-PLUS and this thesis led to our study.

For improving SWNetTR-PLUS, Fatih and his friends [10] conducted a new study, they used Turkish synonym and antonym word pairs and extended this lexicon by almost 33 percent to obtain SWNetTR++, a Turkish sentiment lexicon with 49K words. They followed a different way in their studies to expand the lexicon by using graph structure by representing words as nodes, and edges as synonym–antonym relations between words.

John et al. (2019) [11] conducted a lexicon-based sentiment analysis study on tweet data with positive, negative, and neutral ratings from sentiment140.com. In the study, sentiment classification was performed using three different lexicons, which are SentiWordNet, Hybrid Lexicon and Hybrid Lexicon followed by Sentiment Adjustment Factors named as "H+S ADF".

Classification processes in terms of polarity of the term in the sentiment lexicon may be different for domain specific words, therefore author says that the contextual polarity of the text should be considered. SentiWordNet which is a general-purpose lexicon, has accuracy of 79.80% in classification. On the other hand, in domain specific words this classification ratio decreases to the value of 68.20%. When the hybrid lexicon has applied and after sentiment adjustment factors like considering emoticon, modifiers and negation added, the accuracy of this lexicon is increased approximately 74.80%. The results show that the application of the Hybrid Lexicon Classification and Sentiment Adjustment Factors increase the accuracy of the sentiment analysis.

Yurtalan et al. [12] conducted a lexicon-based sentiment analysis study for Turkish tweets. Their sentiment lexicon includes 1181 data item with positive and negative word roots, part of speech (POS) tags, and polarity values. Author claims that literature has some Turkish sentiment lexicon with low accuracy because translation method from source language into Turkish can lose the meaning because of the morphological structure of the Turkish language. In this scope, he developed a sentiment lexicon by using POS tags of words as noun, verb, pronoun, adjective, adverb, conjunction etc. and these tags are used for determining word groups. While determining the sentiment polarity of the text, he considers the word groups and POS tags. As a result, study shows that considering the word groups on lexicon-based sentiment analysis improves the performance.

3 Method and Material

3.1 Dataset

This study uses GDELT [13] which have been explored by various authors for sentiment analysis in academic research. GDELT is a comprehensive database of web news from all over the world in over 100 languages which provides open access to metadata. We used GDELT's Global Knowledge Graph (GKG) database which consist of news articles where entities, themes, locations and tone are coded. It computes thousands of sentiment scores for every article.

All analyzes performed in this study are based on metadata from the database compiled by the GDELT project. When querying the GDELT database, only the message text metadata and the URL of the message text accessed by GDELT are provided. Since this data is not sufficient to perform a comprehensive analysis of the language used, we accessed each URL individually to confirm that the text was still accessible and to import the text into our local database so that it could be processed.

At this stage, we used the newspaper library [14] to clean the news text from HTML tags and other content unrelated to the news on the website. We took care to distribute the texts evenly over the 2-year interval so that they did not focus on a particular time or event. We selected three major news agencies (Anadolu Agency, Demiroeren News Agency, and İhlas News Agency) as news sources. As can be seen in Fig. 1, İhlas News Agency published more individual news stories than the other agencies during this period.

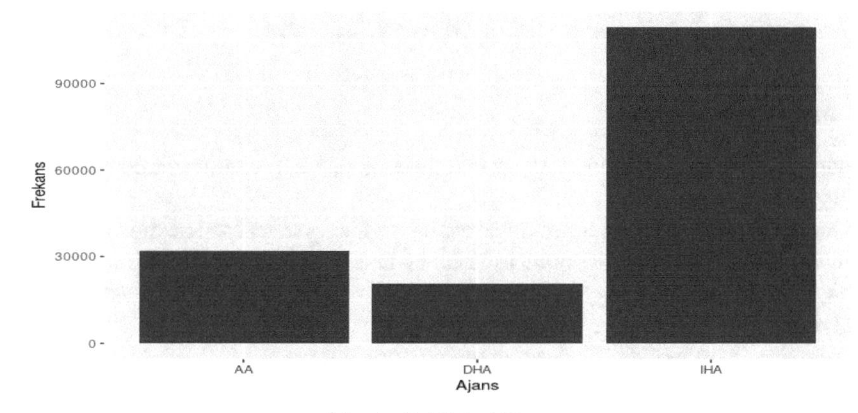

Fig. 1. Published News

The tone and polarity values from the GDELT database metadata are an important component of our study. The tone value is calculated by translating the text into English by the automatic algorithms used by the GDELT project to measure the emotional intensity of a news text. Although GDELT says that this value can range from -100 to $+100$, it has been shown that the tone values in the data we obtained were mostly between -10 and $+10$. Figure 2 shows the distribution of tone values determined by GDELT for the messages we retrieved.

Since the tone values offered by GDELT are given in a very wide range and there are almost no values at the ends of the range, we normalized the tone values in the database and reduced them to the range $[-1, +1]$.

3.1.1 Data Preprocessing

In our study, we implemented the step of removal noisy data. The newspaper library was used to clean the news text from HTML tags and other content unrelated to the news on the website as described in the dataset title. While creating sentiment corpus from documents; tokenization, linguistic operations(morphology), and removing stop words steps were applied. In some of these steps, the Zemberek library [15] was used.

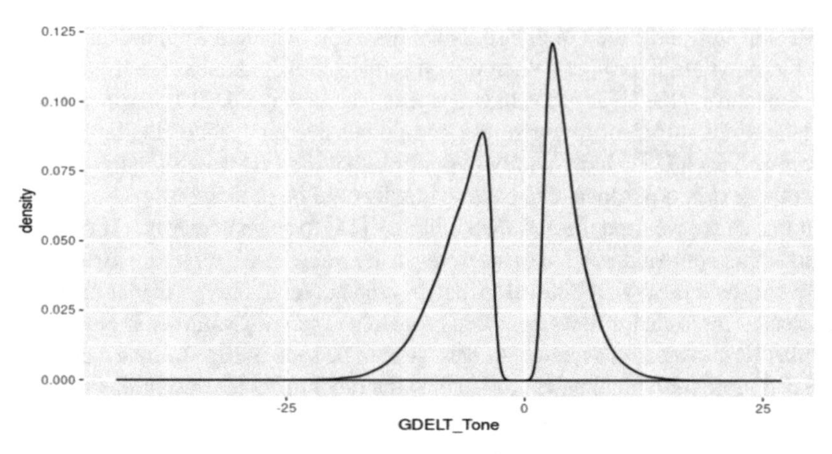

Fig. 2. GDELT Document Tone Values

3.2 Methodology

In scope of this work, sentiment tones and polarity of each word was calculated using two approaches.

One of the methods is using weighted average which is often used method in literature by finding sentiment tone, other one is based on tone distribution of terms in documents which is a novel methodology and the lexicon obtained this way, provides sharper tones which is usually not possible with other approaches in the literature.

Applying these methodologies two main metric are used:

- Frequency of words in news text
- The tone value assigned by GDELT to the news text

3.2.1 Weighted Average Approach

In weighted average approach, firstly tone vectors of terms are found and then using the Eq. 1 weighted average is calculated.

$$S_{\mathrm{T}} = \frac{\sum_{i=i}^{n}(d_i \cdot f_i)}{\sum_{i=i}^{n} f_i} \tag{1}$$

where:

n = number of documents containing term
di = sentiment tone of the document
fi = the number of times the term appears in the document
ST = sentiment tone of the term.

The tone vector of the word "velet" is shown in Table 1. The word "velet" has appeared in 3 different new documents, 4 times in total, and 2 of these news have positive

Table 1. Tone Vector of "velet" (kid)

Document	Document Tone	Term Frequency
DOC-1	−11.801	1
DOC-2	9.871	1
DOC-3	5.319	2

sentiment tone and 1 of these has negative tone. With the mathematical expression in Eq. 1, the sentiment tone value of the word "velet" is calculated as −0.185.

The document tones shown in Table 1 were normalized to be between −1 and 1, and thus the sentiment tone of the term was found between −1 and 1.

3.2.2 Tone Density Based Approach

In tone density based approach, sentiment tone is found by a novel methodology: using statistical tone distributions based on the density functions computed over the documents. This approach aims to find sharper sentiment tone for the word by calculating difference of positive and negative areas under the density function.

Total Area under the curve in probability of density function is 1 and shown in the Fig. 3.

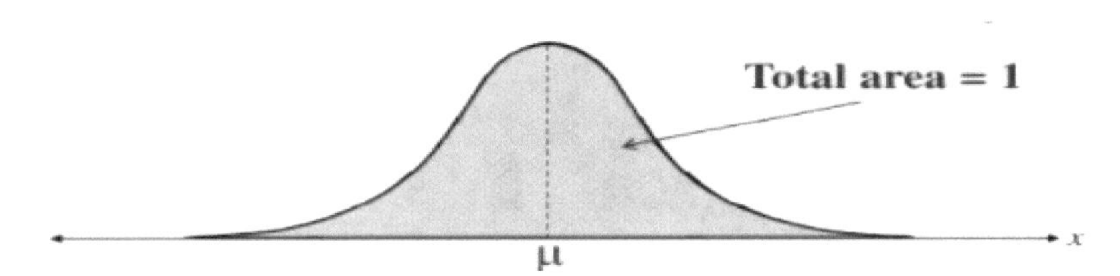

Fig. 3. Probability of density function curve

In this plot of density function in Fig. 4, calculation of negative and positive areas are based on the statistical approach described in [16].

Another way of seeing this relationship is with the following plot, where the value of the area underneath the density is mapped into a new curve that will represent the CDF.

While using this relationship, final sentiment tone of word giving the difference of positive and negative area is calculated using Eq. 2:

$$S_T = 1 - 2\mathrm{CDF}(0) \quad (2)$$

where: CDF (0) = negative area under the density function.

3.3 Sentiment Tone Results

Examples of calculated sentiment tone values using two approaches are given on the figures in this section.

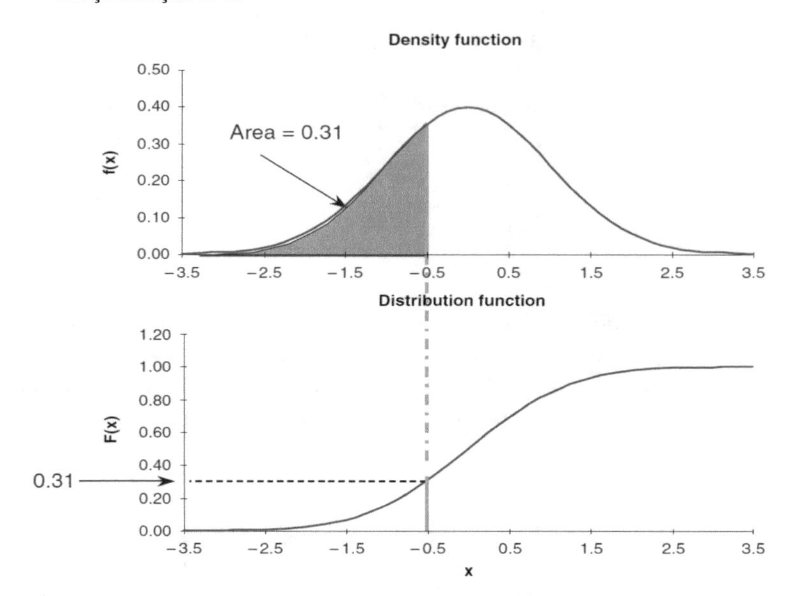

Fig. 4. The relationship between PDF and CDF

3.3.1 Positive Terms

The following Figs. 5 and 6 include words with positive sentiment tones that have a positive effect on people's minds.

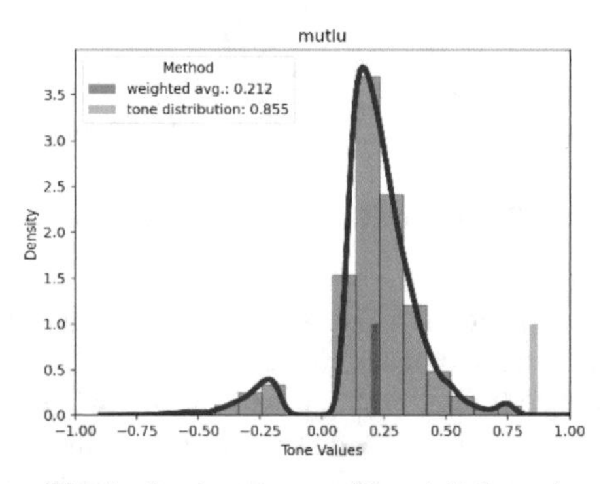

Fig. 5. Sentiment tone of "mutlu" (happy)

The figures show that the sentiment tones found in both approaches successfully polarized the words as positive. To interpret the tone values found, words with very clear positivity were selected.

The result shows that the sentiment tone found by the weighted average method did not have high positivity, while the tone found by the density based method had very strong positivity. This result was obtained because the positive area is quite dominant compared to the negative area as you can see in the figures.

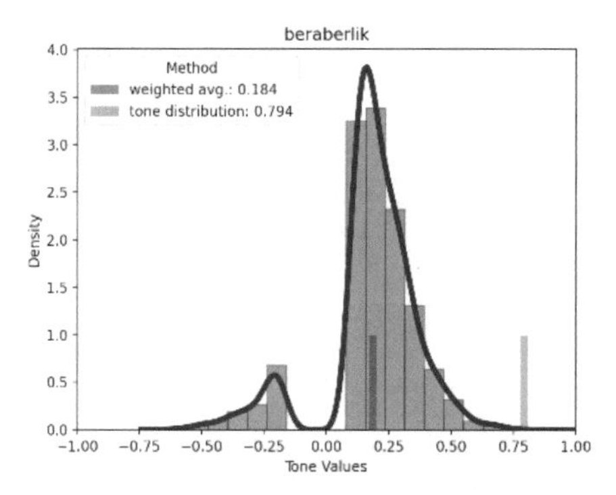

Fig. 6. Sentiment tone of "beraberlik" (togetherness)

3.3.2 NegativeTerms

The following Figs. 7 and 8 include words with negative sentiment tones that have a negative effect on people's minds.

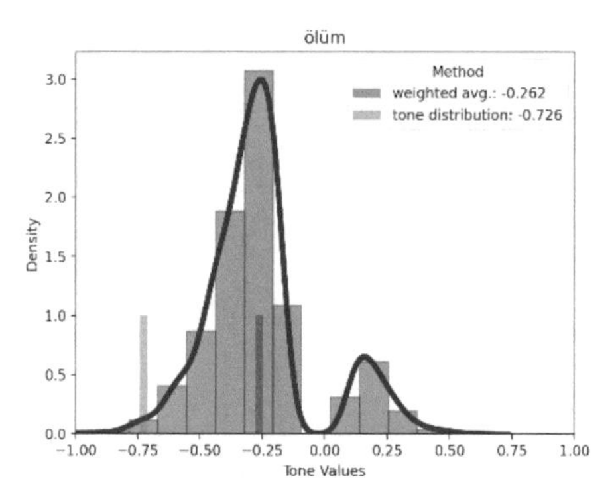

Fig. 7. Sentiment tone of "ölüm" (death)

The figures show that the sentiment tones found in both approaches successfully polarized the words as negative. To interpret the tone values found, words with very clear negativity were selected. The result shows that the sentiment tone found by the weighted average method did not have high negativity, while the tone found by the density based method had very strong negativity. This result was obtained because the negative area is quite dominant compared to the positive area as you can see in the figures.

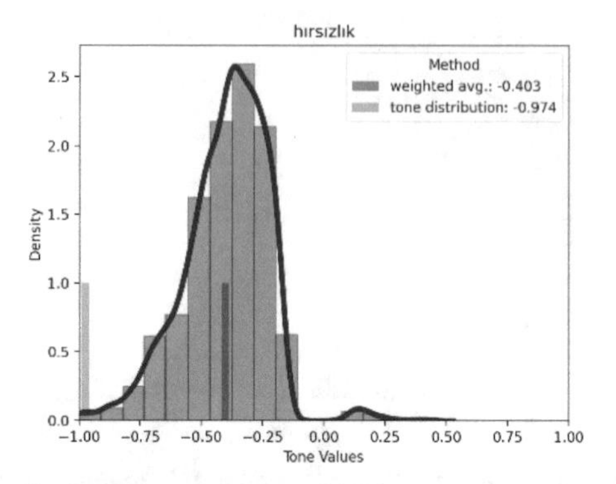

Fig. 8. Sentiment tone of "hirsizlik" (robbery)

3.3.3 Close to Neutral Terms

The following Figs. 9 and 10 include words with neutral sentiment tones that have a neutral effect on people's minds.

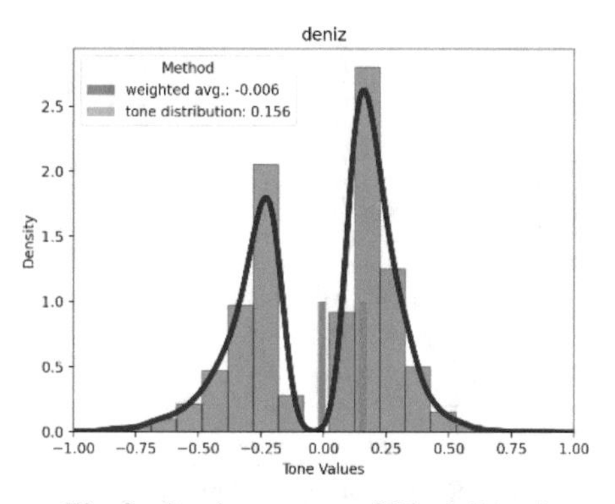

Fig. 9. Sentiment tone of "deniz" (sea)

The figures show that the sentiment tones found in both approaches are close to neutral. Since the sentiment tone value found by the weighted average method is far from finding value close to the extremes between -1 and 1, tone value is also closer to 0 than the tone density-based approach.

This result was obtained because the negative area and positive area are almost equal to each other.

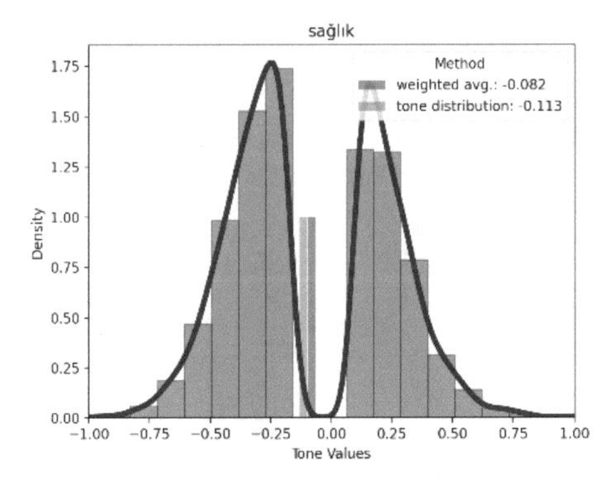

Fig. 10. Sentiment tone of "sağlik" (health)

3.4 MLTC (Manually Labeled Turkish Corpus)

For the performance evaluation of the sentiment lexicon, a labeled Turkish corpus is necessary. In order to meet this need, a total of 300 news documents were classified as positive and negative by 3 evaluators.

In the evaluation of the evaluators for a news, the final polarity of the news was determined on the basis of majority vote. This test corpus was named as MLTC-300. In Table 2, the polarity distributions of MLTC-300 is shown according to the evaluator's determinations.

Table 2. Test Set Distribution

Test Set Distribution		
HUMAN_POS	127	42%
HUMAN_NEG	173	58%
TOTAL	300	

4 Discussion and Results

The performance of the lexicons with developing in two approaches will be evaluated and also, comparison of these lexicons and also a general purpose sentiment lexicon developed by Sağlam and named SWNETTR++ with a capacity of 49K was analyzed. Results show that the constructed tone density based sentiment lexicon with novel methodology achieves a comparable performance.

The Confusion Matrix was used to see the sentiment lexicon test results. In confusion matrix, manually assigned polarity information for the "Actual" value and the polarity values determined with the help of sentiment lexicons for the "Predicted" value were taken into account.

Weighted Average Approach: First, the sentiment lexicon created by the weighted average approach was tested. According to the results, the accuracy value of the sentiment lexicon was 87.33%. The values of the confusion matrix are shown in Fig. 11. And the results obtained from the confusion matrix are shown in Table 3.

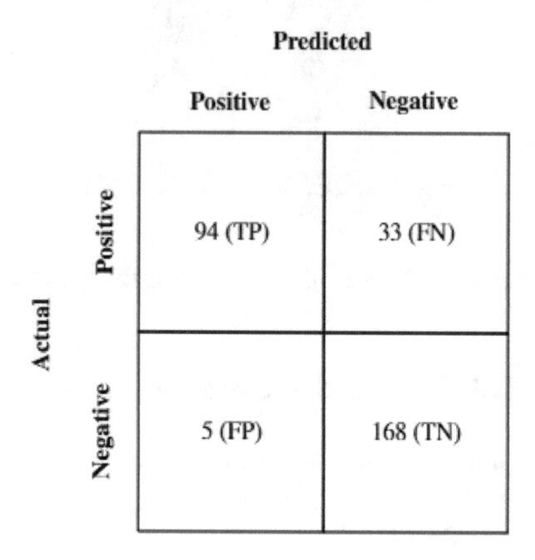

Fig. 11. Confusion Matrix for Tone Weighted Average Approach

Table 3. Performance metrics

Lexicon with Weighted Avg.	
Precision	0.9495
Recall	0.7402
F1 Score	0.8319
Accuracy	0.8733

Tone Density Based Approach: The sentiment lexicon created by the tone density based approach was tested. According to the results, the accuracy value of the sentiment lexicon was 88.67%. The values of the confusion matrix of the study are shown in Fig. 12 and the results obtained from the confusion matrix are shown in Table 4.

Merged Method: Combining the results from the two different approaches we implemented, we determined the polarity of the documents as the sign of the sentimental tone, which has a great absolute value. We called it the merged method. According to the results, the accuracy value of the merged method was 89.67%. The values of the confusion matrix of the study are shown in Fig. 13 and the results obtained from the confusion matrix are shown in Table 5.

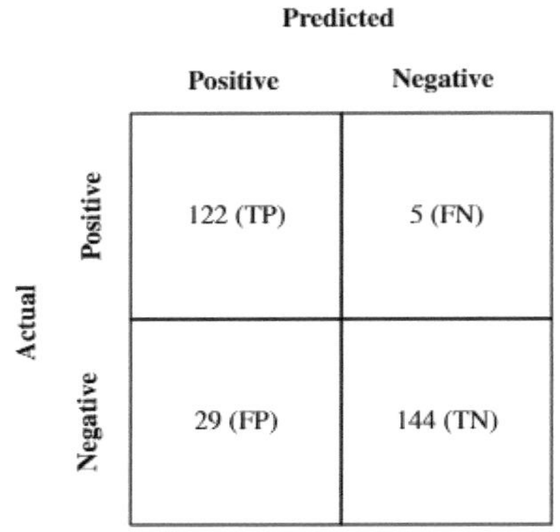

		Predicted	
		Positive	Negative
Actual	Positive	122 (TP)	5 (FN)
	Negative	29 (FP)	144 (TN)

Fig. 12. Confusion Matrix for Tone Density Based Approach

Table 4. Performance metrics

Lexicon with Tone Density	
Precision	0.8079
Recall	0.9606
F1 Score	0.8777
Accuracy	0.8867

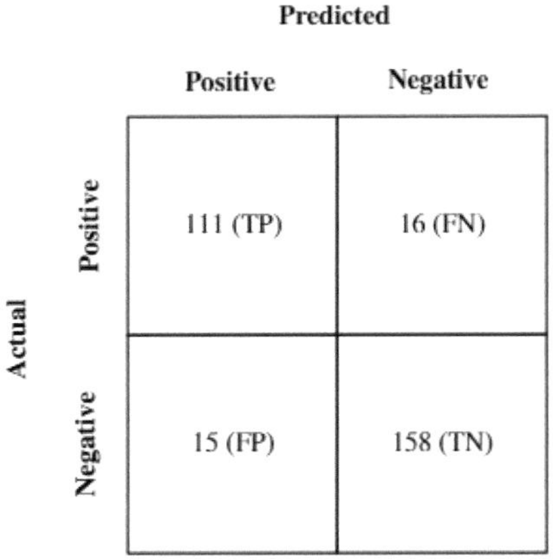

		Predicted	
		Positive	Negative
Actual	Positive	111 (TP)	16 (FN)
	Negative	15 (FP)	158 (TN)

Fig. 13. Confusion Matrix for Merged Method

SWNetTR++: The sentiment lexicon SWNetTR++ was tested. According to the results, the accuracy value of the sentiment lexicon was 89.33%. The values of the confusion matrix of the study are shown in Fig. 14 and the results obtained from the confusion matrix are shown in Table 6.

Table 5. Performance metrics

Merged Method	
Precision	0.8810
Recall	0.8740
F1 Score	0.8775
Accuracy	0.8967

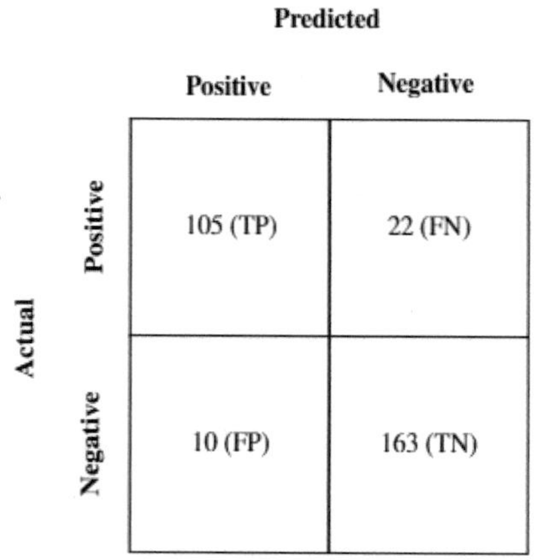

Fig. 14. Confusion Matrix for SWNetTR++

Table 6. Performance metrics

SWNetTR++	
Precision	0.9130
Recall	0.8268
F1 Score	0.8678
Accuracy	0.8933

Table 7 shows the results of the confusion matrices together. In the Recall column, which indicates how accurately positive states are predicted in this table, the most successful result was 96.06% for the lexicon based on tone density, while a low percentage was obtained with 74.02% for the weighted average lexicon. In the specificity column, which indicates how accurately the negative states are predicted, the most successful result is 97.11% for the weighted average lexicon, while the lowest result is 83.24% for the tone density based lexicon. Consequently, the weighted average lexicon was more successful in predicting negative news and the tone distribution based lexicon was more successful in predicting positive news.

Since the merged method was tested with the combination of the results of the two methods, the precision and recall values were also between the results of these two approaches, and with the balance of these values, F-score and accuracy were high.

The most successful result is shown for the weighted average lexicon in the Precision column, which is the measure of how many of the positively predicted values are actually positive. The F1 value, which is the harmonic mean of the Precision and Recall results, achieved the highest value in the tone distribution lexicon.

Table 7. Evaluation Results for 300 News

Lexicon	Accuracy	Precision	Recall	Specificity	F1 Score
WEIGHTED AVG. BASED	0.8733	0.9495	0.7402	0.9711	0.8319
TONE DENSITY BASED	0.8867	0.8079	0.9606	0.8324	0.8777
MERGED	0.8967	0.8810	0.8740	0.9133	0.8775
SWNetTR++	0.8933	0.9130	0.8268	0.9422	0.8678

5 Conclusions and Future Work

As a result of the study, we developed a Turkish sentiment lexicon using a novel methodology: using statistical tone density functions computed using a very large document corpus obtained from mainstream Turkish news agencies. We evaluate the performance of this lexicon by using manually labeled test data and compare it with similar lexicons. While accuracy value of weighted average-based lexicon is 87%, the tone density-based approach is 88% and the combination of this two gives the best accuracy with 89%. Results show that the constructed sentiment lexicon with novel methodology achieves a comparable performance and poses many potential improvement possibilities.

As a feature work, we will construct a graph (tree) based structure using the agglutinative morphology of Turkish to store and query the tone values of the words. This will further allow us to study the sentimental weights of different suffixes in the language. We will finally produce a nice and readable visualization of these graphs, which will demonstrate multiple features of the words, such as frequency in corpus, tone value, suffix strength, etc. This visualization will further allow researchers to understand the relations between words and make it possible to extend sentiment studies in the future.

References

1. Rustam, F., et al.: A performance comparison of supervised machine learning models for Covid-19 tweets sentiment analysis. Plos One **16**(2), e0245909 (2021)
2. Akba, F., Ucan, A., Sezer, E.A., Sever, H.: Assessment of feature selection metrics for sentiment analyses: Turkish movie reviews. In: 8th European Conference on Data Mining (2014)

3. Esuli, A., Sebastiani, F.: Sentiwordnet: a publicly available lexical resource for opinion mining (2006)
4. Cambria, E., Speer, R., Havasi, C., Hussain, A.: Senticnet: a publicly available semantic resource for opinion mining. In: AAAI Fall Symposium - Technical Report, pp. 14–18 (2010)
5. Strapparava, C., Valitutti, A.: Wordnet affect: an affective extension of wordnet. In: Lrec, vol. 4, no. 1083–1086 (2004)
6. Thelwall, M.: The heart and soul of the web? Sentiment strength detection in the social web with SentiStrength. In: Holyst, J. (ed.) Cyberemotions. Understanding Complex Systems, pp. 119–134. Springer, Cham (2017). https://doi.org/10.1007/978-3-319-43639-5_7
7. Dehkharghani, R., Saygin, Y., Yanikoglu, B., Oflazer, K.: SentiTurkNet: a Turkish polarity lexicon for sentiment analysis. Lang. Resour. Eval. **50**(3), 667–685 (2015). https://doi.org/10.1007/s10579-015-9307-6
8. Ucan, A., et al.: SentiWordNet for new language: automatic translation approach. In: 2016 12th International Conference on Signal-Image Technology & Internet-Based Systems (SITIS). IEEE (2016)
9. Sağlam, F., Sever, H., Genç, B.: Developing Turkish sentiment lexicon for sentiment analysis using online news media. In: 2016 IEEE/ACS 13th International Conference of Computer Systems and Applications (AICCSA). IEEE (2016)
10. Sağlam, F., Genc, B., Sever, H.: Extending a sentiment lexicon with synonym–antonym datasets: SWNetTR++. Turkish J. Electr. Eng. Comput. Sci. **27**(3), 1806–1820 (2019)
11. John, A., John, A., Sheik, R.: Context deployed sentiment analysis using hybrid lexicon. In: 2019 1st International Conference on Innovations in Information and Communication Technology (ICIICT). IEEE (2019)
12. Yurtalan, G., Koyuncu, M., Turhan, Ç.: A polarity calculation approach for lexicon-based Turkish sentiment analysis. Turkish J. Electr. Eng. Comput. Sci. **27**(2), 1325–1339 (2019)
13. GDELT. https://blog.gdeltproject.org/gdelt-translingual-translating-the-planet/. Accessed 30 Apr 2022
14. Ou-Yang, L.: Newspaper3k: Article Scraping & Curation (2016). https://github.com/codelucas/newspaper. Accessed 15 Jan 2022
15. Akin, A.A., Akin, M.D.: Zemberek, an open source NLP framework for Turkic languages. Structure **10**(2007), 1–5 (2007)
16. La Vecchia, D., Manon Felix, M.: Continuous Random Variable (2016). https://bookdown.org/tara_manon/MF_book/continuousrv.html. Accessed 08 Mar 2022

User Oriented Visualization of Very Large Spatial Data with Adaptive Voronoi Mapping (AVM)

Muhammed Tekin Ertekin(✉) and Burkay Genç

Hacettepe University, Ankara, Turkey
tekinertekin@gmail.com

Abstract. Today, with the development of location service providers and equipment providing location services, spatial data can be obtained from many different devices. This data can be used in various applications. Data are affected by both geographic and human factors, and sometimes they may show homogeneous and sometimes heterogeneous distributions. Displaying so much data in a way that users can perceive requires special methods. This study to be written aims at expressing many spatial data to the users in the most efficient way. As a result of this study, it is aimed to visualize spatial data consisting of many (e.g. millions) points globally or regionally and to do this in real time.

Keywords: visualization · spatial data · rasterization · Voronoi Diagrams · Big data

1 Introduction

Spatial data is a mathematical data set that enables an object in the real or virtual world to be expressed in the coordinate plane to which it belongs. Spatial data have been used in many disciplines such as engineering, military, industry, game technologies and mapping from past to present. Today, the explicit transfer of spatial data to the end user is done by many different methods.

With the spread of positional data generating devices such as GPS, it has become a problem to present the large number of data obtained to the users. In addition, the data of the game world in the game industry, especially in strategy-themed games, are presented on maps within the game. These data, which are generated in real environments or virtual environments such as games, are generally displayed on a point basis and this situation does not provide convenience to the user. For example, the point-based representation of a map containing tens of thousands of points to the user will not make any sense for the user. Instead, it would be more understandable to present these data to the user by processing and summarizing them on the map with a regional expression technique. In this study, we have worked on this problem.

Creating enormous spatial data on maps is a difficult task for three main reasons. First of all, visualized data points overlap with each other due to limited screen resolutions.

D. J. Hemanth et al. (Eds.): ICAIAME 2022, ECPSCI 7, pp. 515–536, 2023.
https://doi.org/10.1007/978-3-031-31956-3_45

As a result of this overlap, the points cover each other. Second, the displayed data must generally have different characteristics. For example, different colors or markers are used to represent different data. Finally, in many drawing tools and image creators, a limited amount of data can be generated, so these tools have a hard time displaying data larger than them. In this case, only a limited part of the data is represented. The resulting image is it cannot accurately represent the actual location data as it is a loss of information.

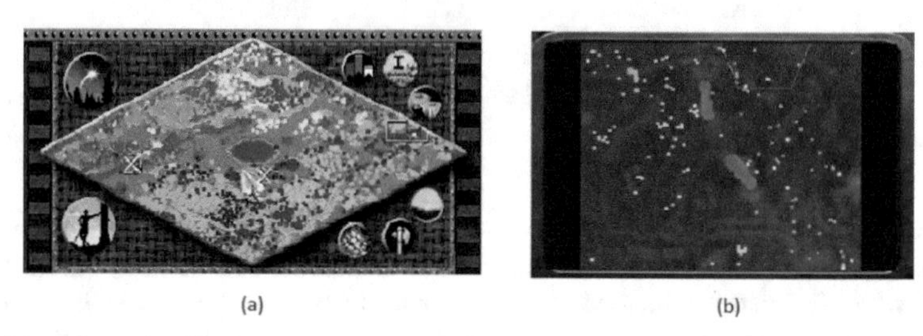

Fig. 1. Map examples in strategy games (a) Age of Empires II (b) Command and Conquer: Generals 2

Indication of positional points is used in many areas. For example, in strategy games, it is common to show the players' objects as a game map to the user, generally on the bottom right or bottom left of the screen. Examples of these maps are shown in Fig. 1. In many areas such as military, web applications, geographic information systems, displaying data on the map is important in terms of providing more meaningful data to users. Map usage within the military is shown in Fig. 2. The location of objects (warplanes, ships, etc.) in the military, especially in the air and naval forces, is shown on maps and the strategy is determined according to these data. With the development of technology, devices that make spatial data sharing instantly were added to every vehicle, even to every soldier, and thus, it was aimed to increase coordination. With the increase of technology in geographical information systems, it has become widespread to show spatial data, especially in institutions such as the municipality and the Ministry of Environment. For example, many map representations such as title deed information, building data, point data of public institutions are widely used.

Voronoi diagrams are named after Georgy Feodosevich Voronoy. He defined and worked on voronoi diagrams. He was able to express Voronoi diagrams mathematically in his work in 1908 [1].

Voronoi diagrams are a geometric definition obtained by dividing an infinite field with equal distances around previously given points. Each segment obtained from the points is called a cell. All points in a cell area are closer to the center point of that cell than any other center points. As seen in Fig. 3 and Fig. 4 [2], black dots represent center points. Colored areas are Voronoi cells corresponding to the center point.

Voronoi diagrams appear in many different areas, especially in natural life. For example, patterns on giraffes, shapes formed by textures on leaves or in the form of cracking of dried soil (Fig. 5) we can see voronoi-like shapes in nature.

These shapes, which can also be seen in a cave or on the edges of a rock, are the shapes that are familiar to the human eye and included in the previous visual memory.

Fig. 2. Military operations center map usage

Fig. 3. Voronoi diagram according to Euclidean distance measure [2]

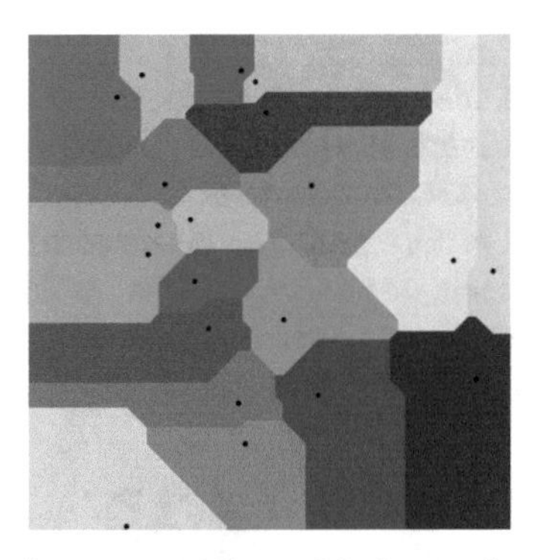

Fig. 4. Voronoi diagram according to Manhattan distance measure [2]

Basically, this is the reason for showing the mapping to the end user using voronoi diagrams in this study.

Pixelation (rasterization) is one of the first methods that comes to mind for the problem and similar problems discussed in this study. Pixelation is applied on a vector

Fig. 5. Voronoi-like shapes in nature [2]

image object or on datasets containing spatial data. The relevant data set is first divided into squares of the specified number and size. Then, the color of the data (or the area in vector images), which is the majority in this segmentation, is assigned as the color of that frame. An example of pixelation is shown in the Fig. 6.

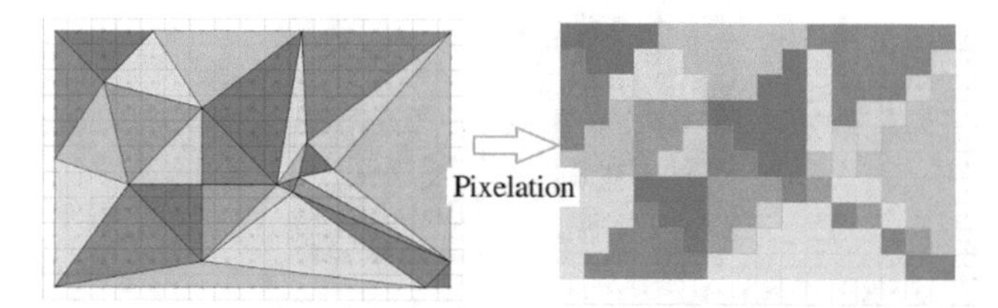

Fig. 6. Pixelation example

Michael F. Worboys did some work on pixelation in 1994 [3]. According to this study, taking an image described by figures in a graphical (usually vector) format and converting it to a raster image (a set of pixels, dots or lines that, when viewed together, forms the represented image) is called pixelization. The pixelated image can then be viewed on a computer screen, video screen, or printer [4]. This resulting map (or bitmap) can be stored in a specific file format (png, bitmap, etc.). Pixelation is frequently used in 3D model drawing applications or simple 2D image display applications.

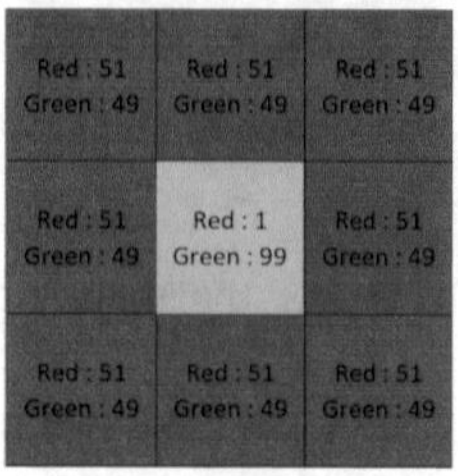

Red : 51 Green : 49	Red : 51 Green : 49	Red : 51 Green : 49
Red : 51 Green : 49	Red : 1 Green : 99	Red : 51 Green : 49
Red : 51 Green : 49	Red : 51 Green : 49	Red : 51 Green : 49

Fig. 7. Pixelation example

Although pixelation offers a simple and fast solution, it also brings some problems. First of all, pixelation causes data loss. An example of pixelation is shown in the Fig. 7. According to this example, a total of 900 data in the data set is divided into 9 different cells. When the data is examined, it is seen that there are 409 red and 491 green data. Although the dominance of green data is more than red data in the aforementioned data, the end user sees 8 red and 1 green areas. This situation presents an incorrect display to the user. For this reason, the pixelization method was not used in this study. In this study we work on overcoming common problems with adaptive voronoi mapping (AVM).

2 Related Works

In the study created by Çakmak [5], 3 solution methods are presented for the solution of our problem. The first method is like the pixelation method. Accordingly, the data set with fixed frame sizes was divided and a pixelization was applied according to these dimensions. According to the second method created in Çakmak's work, the square dimensions were created in a more adaptive way. Accordingly, if a certain entropy value cannot be achieved in a square, that square is divided into smaller squares. According to the third method created in Çakmak's work, the squares are divided according to certain rectangles, not four different squares. Accordingly, if a certain entropy value cannot be obtained in a square, that square is not divided into smaller squares, but into smaller rectangles according to entropy calculations. If a certain entropy value is provided, that square (or rectangle) is left undivided into smaller rectangles.

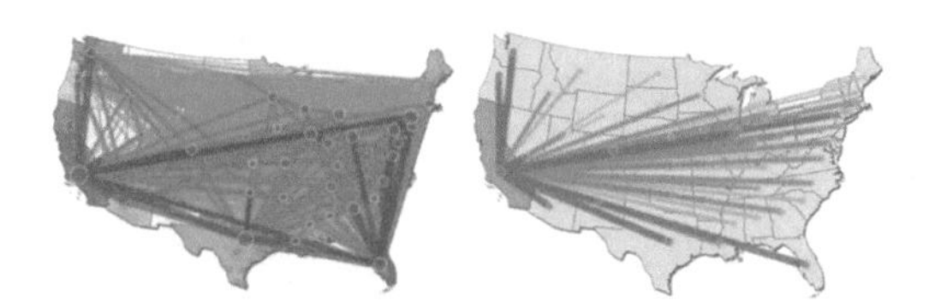

Fig. 8. American internal migration map (display of all vectors) [6]

In the study conducted by Diansheng Guo in 2009, American internal migration data were examined, and solution methods were studied on how this data could be presented to the end user [6]. The problem that is tried to be solved in this study is that eight hundred thousand vector data (migration from one city to another) cannot be presented to the user in a meaningful way. As can be seen in the Fig. 8, presenting all this data to the end user creates a serious confusion.

In the study by Diansheng Guo, these data were made more meaningful by obtaining larger nodes. Figure 9 shows the nodes used in this solution method.

Later, vectorial data was combined according to these nodes, and it was aimed to present a more meaningful visual to the end user. During this merge, the start and end points of the vectors in the combined nodes are updated and adjusted again. In the Fig. 10 the vectors are shown again as a result of joining the nodes. As can be seen from the figures, the display created by this method offers a more understandable display to the end user. In addition, the researchers developed a tool for filtering the image (age,

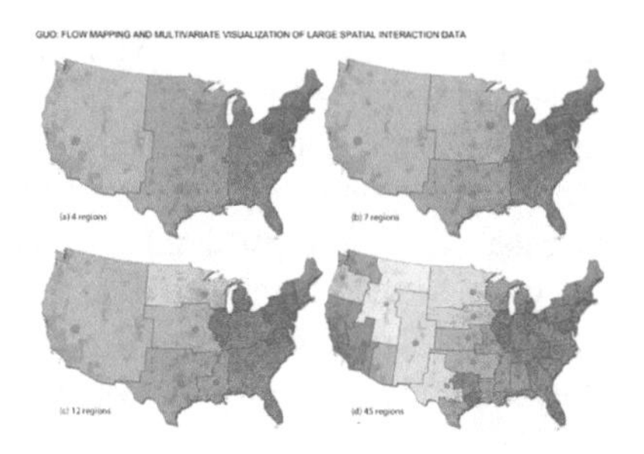

Fig. 9. American internal migration map (nodes joined) [6]

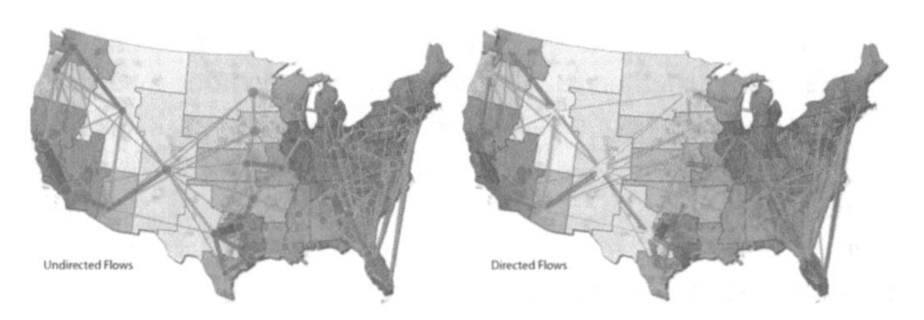

Fig. 10. American internal migration map (post-updated representation of all vectors) [6]

gender, occupation, etc.) and bringing the nodes together. As a result of this study, when the internal migration situation in America is examined, it is seen that there is an internal migration from the center to the coasts. Thus, a simplification has been achieved in the data representation.

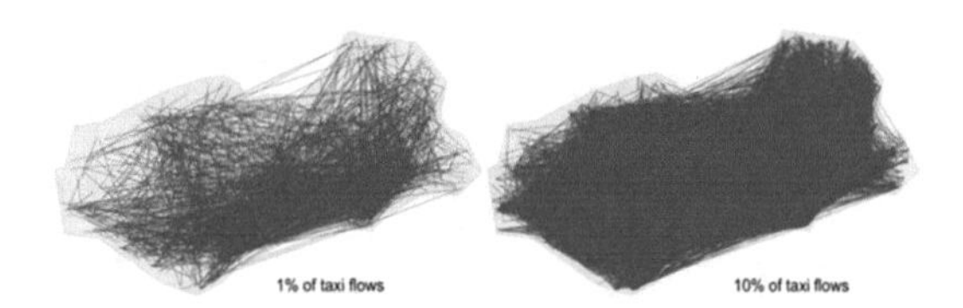

Fig. 11. Complexity of taxi data [7]

Xi Zhu and Diansheng Guo worked on an algorithm for a better representation of vector data in a way that [7] in their study in 2014 [7] similar to Diansheng Guo's solo work in 2009 [6]. The study they have done is based on the points where the taxis pick up and drop off the passengers. According to the study using the SNN (shared nearest neighbors) algorithm, the points where taxis pick up and drop off passengers are clustered. As a result of this clustering, the data confusion, of which only 1% and 10% are shown in Fig. 11, has been avoided and a simpler view has been obtained. The final state of the application is shown in Fig. 12.

In the research conducted by Piercesare Secchi, Simone Vantini and Valeria Vitelli in 2013, it was aimed to show the temperature data on the world more healthily to the end user [8]. In the study, the temperature data were grouped by considering multiple

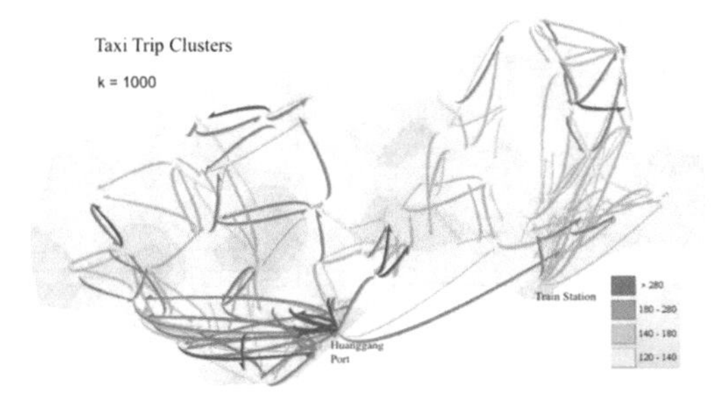

Fig. 12. Final plot created by Xi Zhu and Diansheng Guo [7]

stages and the structure formed as a result of this grouping was expressed using Voronoi diagrams. In the study, it was also shown that entropy values decrease when the number of Voronoi cells is increased.

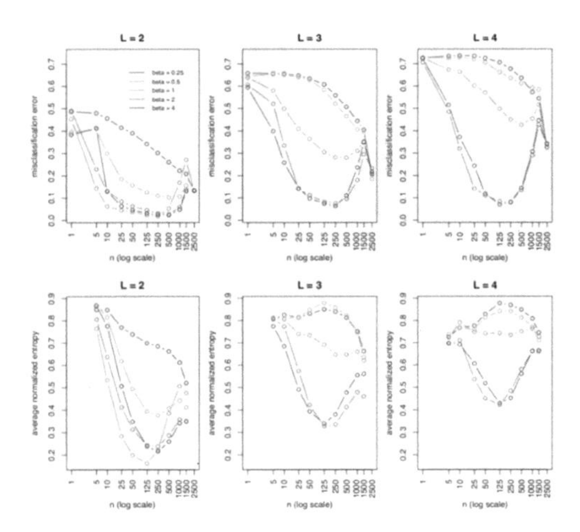

Fig. 13. Errors and entropy graph by level [8]

The change of entropies is shown in the Fig. 13. The value of n in the figure represents the number of Voronoi cells. In the study, temperature data according to months were used and a dynamic mapping method was obtained. The final view is shown in Fig. 14. The map shown in Fig. 14 was obtained by combining Voronoi diagrams.

Jiri Komzak and Marc Eisenstadt, in their study in 2001, tried to present student information on a map with spatial data on a map [9]. This study is scalable and according to the study, the fields in which the students work (undergraduate, graduate, doctorate, etc.) have been tried to be shown on the map. In the system in which scales such as continent, country and city exist, the number of students is expressed in proportion to the size of the circle. The colors in the circle show the weight of the studied areas. As a result of the study, an application was developed. Universities with data by application (data for millions of students) are shown as scalable on the map. In the system, which is

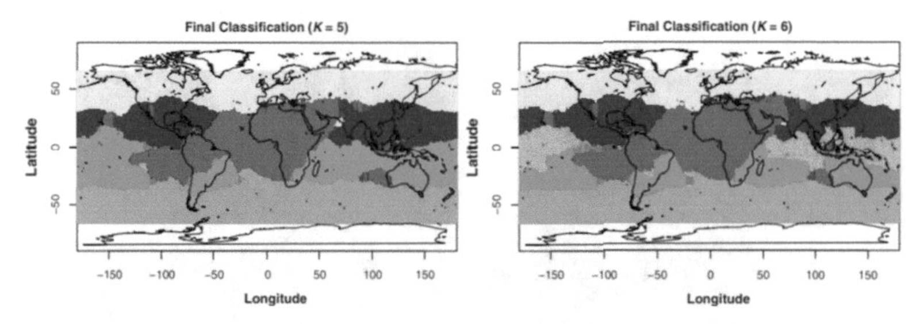

Fig. 14. Multi-colored Voronoi diagrams based on temperature data [8]

not based on instant data, images are created in the background for a certain period of time and presented to the end user statically. The application is shown in the Fig. 15.

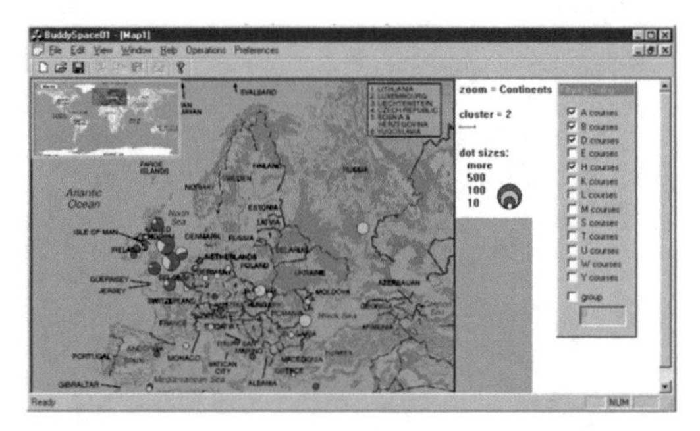

Fig. 15. Application that spatially displays student data created by Jiri Komzak and Marc Eisenstadt [9]

3 Dataset

Three data sets were used to conduct this study. The first dataset is an artificial dataset located on the plane of the World map created by Bilgehan Çakmak, who conducted a similar study before, [5]. The second data set was created by us based on the Turkey map. This set is artificially created like the set belonging to Çakmak. The third set is by us using the application programming interface of the same site, the air pollution data on the world plane at the address "https://waqi.info/" generated real data. The summary of these all data files and can be seen in Fig. 16.

First data set was previously created by Çakmak with the Matplotlib Basemap tool to be used in his Master's thesis. There are 5 different data files in the set. In the dataset B1, B2, B3, B4 and B5, the number of red and green dots is approximately equal. But the red dots are set to dominate inside some circles, but they are all over the map. Green dots are sparsely found in these circles. Are equally distributed in all areas. In this study, we needed different data sets because we were investigating the visual separation of dominant areas from other areas and how this separation could be expressed

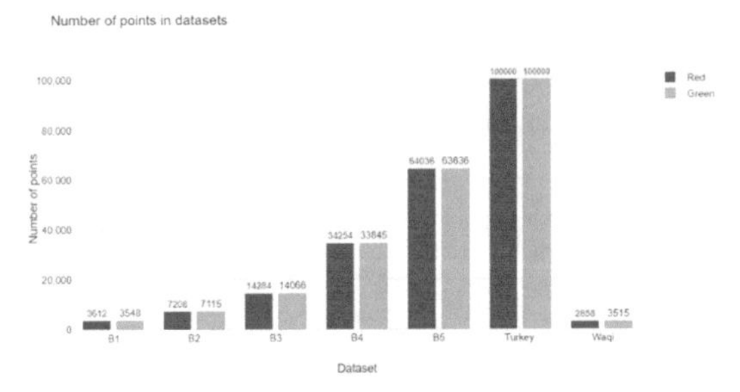

Fig. 16. Points in all dataset

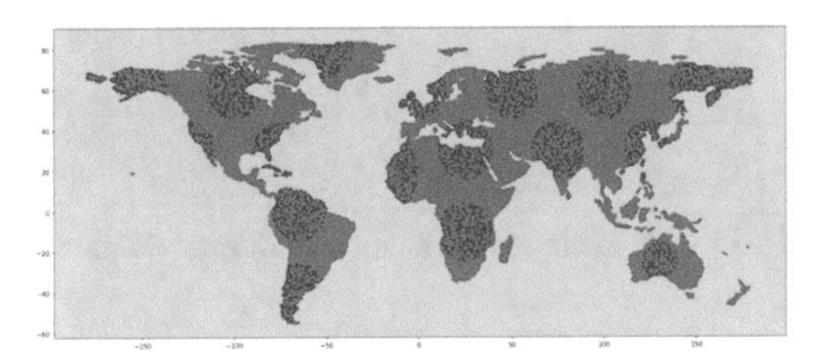

Fig. 17. Data set B5, 127672 points (64036 red - 63636 green)

mathematically. For this purpose, we used this set, which was prepared in the same direction, in 5 different sizes.

As can be clearly seen in the Fig. 17, the green dots cover the red dots. This is because the green dots are transferred to the map later than the red dots. Our main motivation in this study is to prevent such notation errors.

Second dataset created on the Turkey map was created by us. Using this map, 3 circles were selected in the center. Within these circles, the number of Type 1 Points (red) is set to be dominant, but there are red dots in every area of the map. The distribution of green dots is placed evenly in every area of the map. The ratio of red dots within the red areas seen in the Fig. 18 is approximately 73%. The rate of green spots in green areas is approximately 78%.

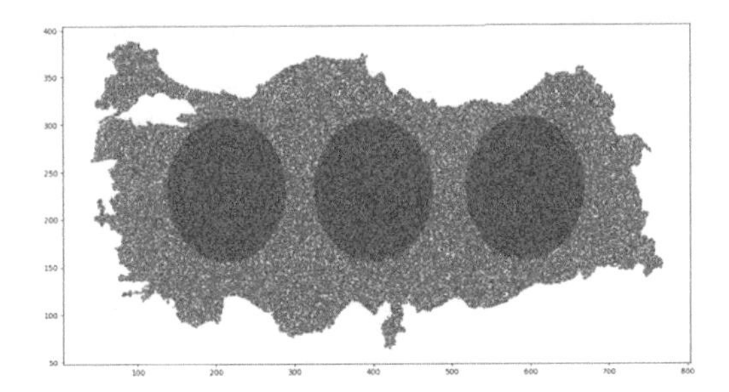

Fig. 18. Synthetic data on Turkey, red 100000, green 100000 dots

Third dataset contains the World air quality index data broadcast daily at "https://waqi.info/". Obtained on 30 July 2020 using the application programming interface of the same site at "https://aqicn.org/api/".

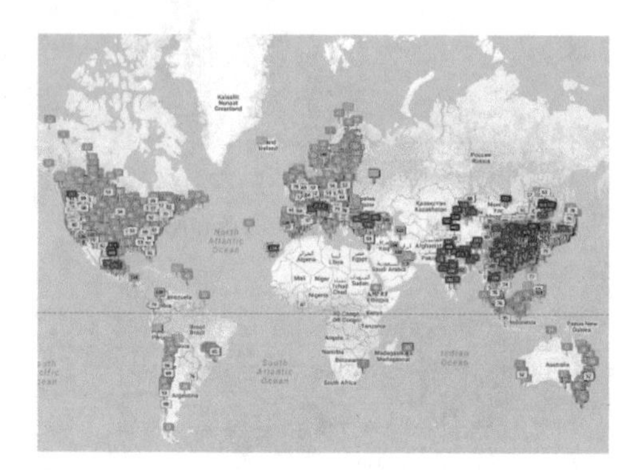

Fig. 19. World air quality index map (https://waqi.info/)

This data includes air quality data obtained from more than 6000 stations around the world.

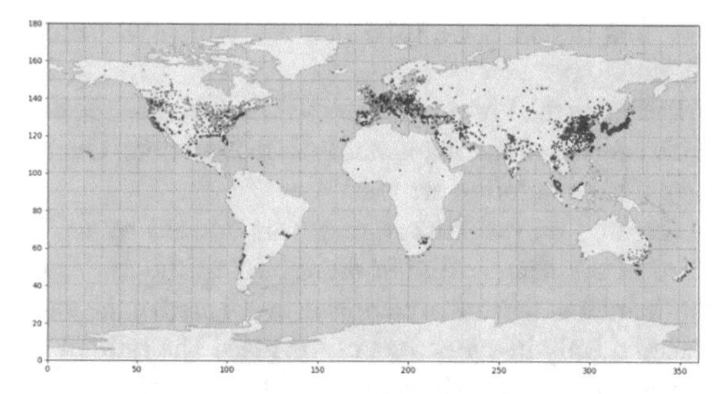

Fig. 20. World air quality index point distribution

As can be seen in the Fig. 19, this service cannot provide meaningful data to the user while showing the air quality on the map. This data has more than two classes. To fit the basic assumptions of this study, we bi-classified the data using a threshold value. The threshold value was chosen as 40 in order to create a balanced data. As seen in the Fig. 20, these points show a natural distribution between [−180, 180] on the x-axis and between [−90, 90] on the y-axis.

4 Methodology

The problem of visualizing a large number of spatial data is the problem on which this study is based. Our solution to this problem is to perform a segmentation using

Voronoi diagrams. First of all, the data set is divided into squares. These squares are then repeatedly divided into smaller squares according to the density and dominance of the data they contain. This division process continues until the allowable level. Then, the division into squares is finished. In this way, the data is placed in squares of varied sizes.

After the division into squares is completed, a random point is selected in each square. The selected points are given to the Voronoi diagram as the center point. In this way, the Voronoi diagram is obtained.

After obtaining the Voronoi diagram, it is determined which Voronoi center point is close to the points in the data set by using the Ball-tree algorithm. In this way, Voronoi cells are colored. If there is no data set point close to the Voronoi center point, it is indicated in blue in the images. Other than that, whichever group (red or green) has the majority, the Voronoi cell is colored with that color. In case of equality, green color is used. A base map to which the dataset belongs is drawn in the background in the images. Voronoi diagrams are drawn transparently on this map and a suitable display for the end user is made.

The information that this algorithm takes as input can be listed as follows:

- Dataset
- Height
- Width
- Density threshold
- Dominance threshold
- Initial frame size
- Depth

It is possible to create different maps for the same data set by using the density threshold, dominance threshold, initial frame size and depth parameters. In addition, by using these parameters, the working time can be increased or decreased according to the need. Thanks to these parameters, different qualities of mapping can be made.

In this study, an initial segmentation is made according to the first frame size given as input at the beginning of the proposed method. Accordingly, the data is read first. Then, the first squares are created by considering the height and width data of the map. The side length of these squares is determined by the first square size input. Then the points in the data set are placed inside these squares. The depth value of these first frames created is assigned as 0.

```
class Square:
        var x1, x2, y1, y2;
        var points[];
        var depth;
```

Fig. 21. Square class used

Each square class contains the values x1, x2, y1, and y2. These values represent the boundaries of the square. In addition, each frame has a depth value and a list of points. The structure of the square class is shown in the Fig. 21.

Table 1. Division of Points into Squares

Algorithm 1: Creating the early squares

```
function createSquares(var height,var width, var size,var points[]):
    squares = [];
    for x in range(math.ceil(height/size)):
        for y in range(math.ceil(width/size)):
            x1 = x * size;
            y1 = y * size;
            if(x + 1) * size > height :
                x2 = height;
            else:
                x2 = (x + 1) * size;
            if(y + 1) * size > width :
                y2 = width;
            else:
                y2 = (y + 1) * size;
            squares.append(new Square(x1,x2,y1,y2,0));
    w = math.ceil(width/size);
    for point in points:
        index = int( point[x] / size ) * w;
        index += int( point[y] / size );
        squares[index].points.append(point);
    return squares;
```

The points in the read data set are transferred into these squares after the first squares are calculated. The transfer function is shown in the Table 1. The situation that occurs as a result of the creation of the first squares is shown in Fig. 22.

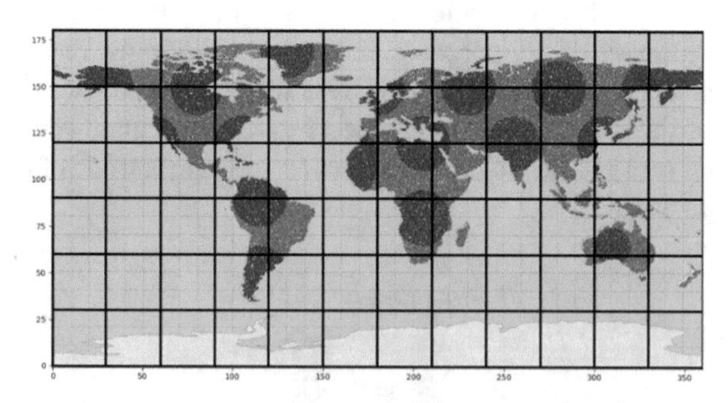

Fig. 22. Generating initial squares for the B5 dataset, square size = 30

After the initial frames are created, these frames are thrown into a queue structure. The corresponding frame is divided into four smaller squares according to the density threshold and dominance threshold inputs. The divided squares are added back to the queue. Frames crossing the density threshold and the dominance threshold are added to a new list. Also, frames that exceed the depth limit are added to the same list. The last resulting list is sent to the next function to generate Voronoi diagrams. The division function is shown in the Table 2.

Table 2. Dividing Squares into Smaller Squares

Algorithm 2: Splitting squares process

```
function divideSquares(var squares,
var depth, var density, var dominance):
    squareTemp = squares;
    lastSquares = [];
    while len(squareTemp) > 0 :
        square = squareTemp.pop();
        if len(square.points) == 0:
            lastSquares.append(square);
        else:
            red = 0;
            green = 0;
            for point in square.points:
                if points[color] == red:
                    red = red + 1;
                else:
                    green = green + 1;
            if square.depth >= depth:
                lastSquares.append(square);
            else if float(len(square.points)) /
                float((square.x2-square.x1)*(square.y2-square.y1))
                    < dominance or
                    float(math.fabs(red-green))/
                    float(red+green) < dominance:
                squareTemp.push(divideFour(square));
            else:
                lastSquares.append(square);
    return lastSquares;
```

In the Fig. 23 the situation after dividing the squares is shown. It is a reflection of the importance of the segmentation density threshold data, especially on the island of Hawaii on the left side of the figure. Accordingly, since the island density was not sufficient, the segmentation process continued. As can be seen in the coastal areas of the continents and the Mediterranean region, the coasts with data can be segmented very clearly.

Following the creation of the final squares, random points are selected within these squares. The reason for choosing random points, not the midpoints of the squares, is that Voronoi diagrams are created without any problems.

There are some problems in the representation of collinear points in Voronoi diagrams [10]. In order to avoid these problems, the center points to be used in the Voronoi diagram are randomly selected within the squares. In the Table 3, the function that selects the points and sends these points to the Voronoi diagram is expressed.

The situation resulting from the creation of the center points is shown in Fig. 24. Selected center points are indicated by large yellow dots.

These created points are sent into the Voronoi diagram as the center point. The Voronoi diagram created by selecting these center points is shown in the Fig. 25.

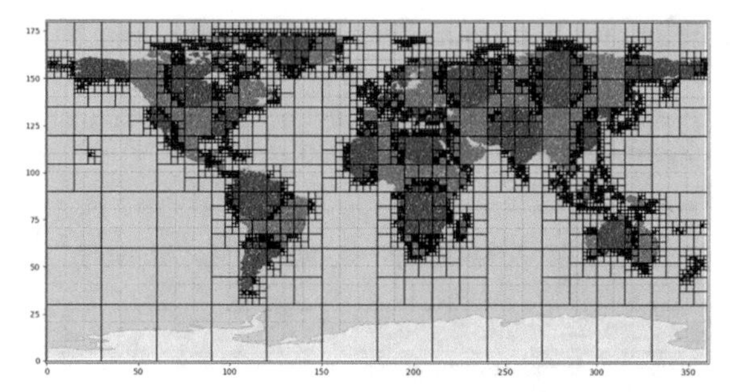

Fig. 23. For B5 dataset, split squares density threshold = 5, dominance threshold = 0.4, depth = 5

Table 3. Selecting Random Points and Generating Voronoi Diagrams

Algorithm 3: Selecting random points and generating Voronoi diagrams

```
function voronoiCreate(var squares[]):
    centerPoints = [];
    Voronoi voronoi;;
    for square in squares:
        centerPoints.append(RandomInside(square));
    voronoi = Voronoi(centerPoints);
    return voronoi;
```

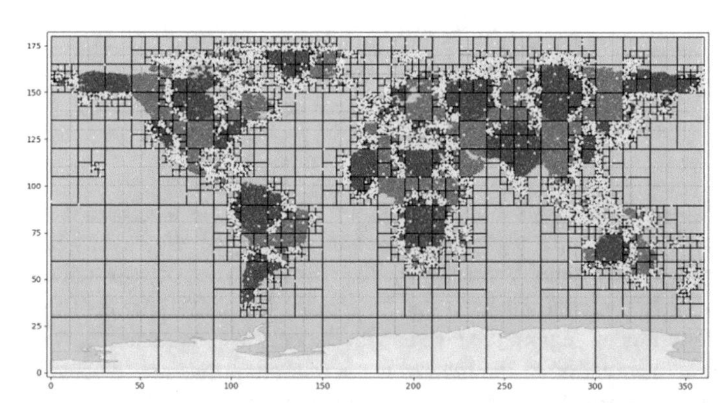

Fig. 24. Random spots selected for B5 dataset density threshold = 5, dominance threshold = 0.4, depth = 5

The points in the read data set are transferred into these squares after the first squares are calculated. The transfer function is shown in the Table 1. The situation that occurs as a result of the creation of the first squares is shown in Fig. 22. Using this tree structure, the closest a new point is within this n point to which one is found in O(log n) time complexity. Coloring Voronoi diagrams using the Ball-tree algorithm is shown in the Table 4.

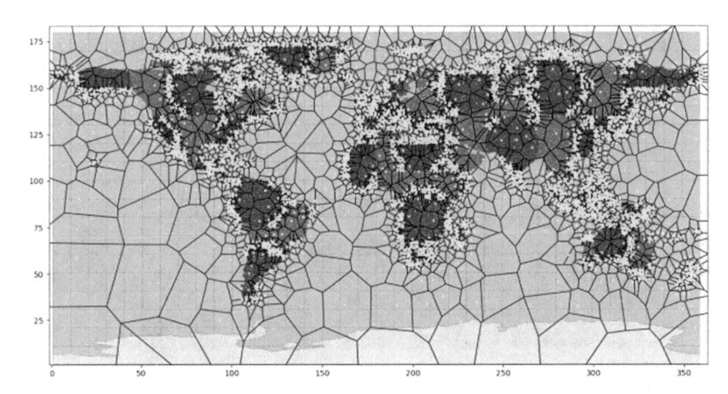

Fig. 25. Voronoi diagram for B5 dataset density threshold = 5, dominance threshold = 0.4, depth = 5

Table 4. Coloring the Voronoi Diagram with the Ball-Tree Algorithm

Algorithm 4: Coloring the Voronoi diagram with the ball-tree algorithm

```
function voronoiColoring(var voronoi, var points[]):
    colorList[] = null;
    BallTree btree = Baltree(voronoi.centerPoints);
    for point in points:
        index = btree.query(point);
        if list[index] == null:
            if point[color] == red:
                colorList[index] = -1;
            else:
                colorList[index] = 1;
        else:
            if point[color] == red:
                colorList[index] += -1;
            else:
                colorList[index] += 1;
    return colorList;
```

An empty color list is first created for coloring purposes. The size of the color list is equal to the number of Voronoi cells. Then the center points are sent into the Ball-tree algorithm. Then, the center point to which each point in the data set is closest is queried from the Ball-tree algorithm. The result of the query is saved in the color list.

For each data set point, the corresponding value in the color list is increased or decreased. The Voronoi diagram is colored according to the majority in the color list. Accordingly, if the color list item corresponding to the Voronoi cell is empty, that cell is expressed with a transparent light blue color.

If the value in the color list corresponding to the Voronoi cell is negative or 0, the cell is represented in red. If the corresponding list is positive, it is highlighted in green. If the two groups are equal, the 0 value is expressed in red because the red dots are more than the green dots in total. Colored Voronoi diagrams are shown in the Fig. 26. As a result of these processes, the classification of a large number of spatial data is completed.

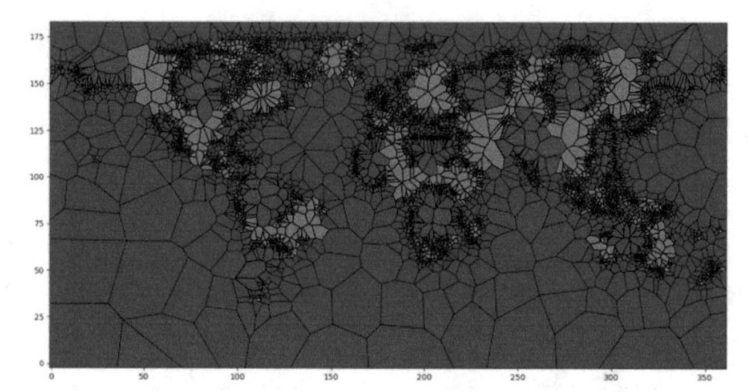

Fig. 26. Voronoi diagram (in color) generated for the B5 dataset, density threshold = 5, dominance threshold = 0.4, depth = 5

A basemap is placed behind the colored Voronoi diagram in order to display a large number of spatial data more clearly to the end user. The diagram is shown transparently so that the colored Voronoi diagram does not obscure the basemap. In the Fig. 27, the map formed as a result of the work of the whole algorithm is shown. In this display, the transparency of the red and green areas was 0.5, and the areas of the cells without data had a transparency of 0.1.

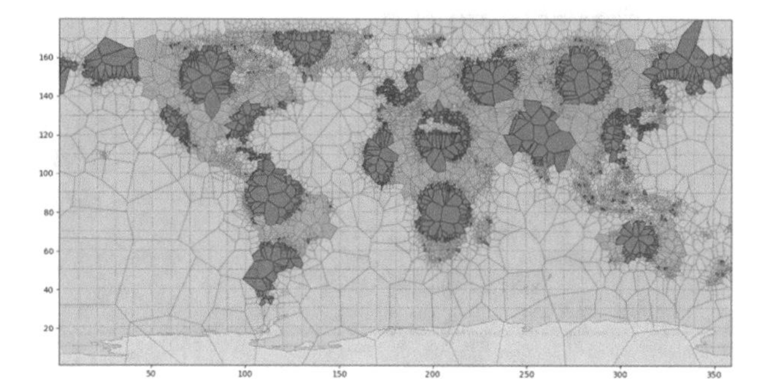

Fig. 27. Final map density threshold = 5, dominance threshold = 0.4 for the B5 dataset

The general structure of all these operations is shown in Table 5.

5 Tests and Results

In this section, we will be explaining which outputs our proposed method produces with which parameters and the numerical values of the produced outputs. We will give information about how the parameters affect the outputs visually and numerically. We will also compare our method with a similar study by Çakmak.

Our method has been tested on 7 different data sets. Among these, we will focus mostly on datasets named B1, B2, B3, B4 and B5. We used different depths so that we could analyze how our method yielded results. Accordingly, the initial square size

Table 5. Algorithm Operation with All Functions

Algorithm 5: Algorithm operation with all functions

```
function AVM(var points[],var height,var width
var size,var depth, var density, var dominance,var baseMap):
    firstSquares= createSquares(height,width,size,points);
    Squares= divideSquares(firstSquares,depth,density,dominance);
    Voronoi voronoi= voronoiCreate(Squares);
    colorList= voronoiColoring(voronoi, points);
    Display(voronoi, colorList, baseMap);
    return;
```

was chosen as 30 for each data set. In the depth parameter, 2, 3, 4, 5 and 6 values are selected. The density threshold was determined separately for each data set. We determined 5 different intensity thresholds for each dataset. Values of 0.3, 0.4, 0.5, 0.6 and 0.7 were determined at the dominance threshold. Considering all these test cases, 625 results are obtained. In order not to cause confusion, these results will be transferred to the data set, depth input, density threshold and dominance threshold, keeping other parameters constant. Visual and mathematical results will be displayed based on these parameters. In order to avoid confusion, not all of the selected parameters are shown.

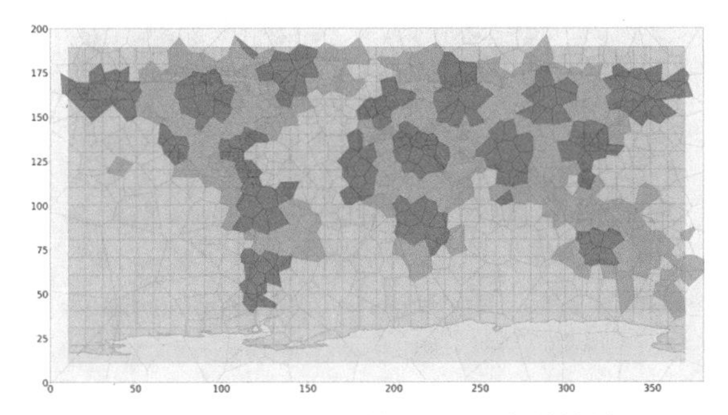

Fig. 28. Depth = 2, number of cells = 626, algorithm time = 1.189814 s, amount of ram consumed = 18.147149 megabytes

Other inputs are kept constant in order to show how the depth parameter affects the results. For this purpose, the B5 data set was chosen. The density threshold value was chosen as 5 because the same data set was used. The dominance threshold was chosen as 0.4. It is possible to obtain more detailed images by using the depth parameter. For example, when the depth parameter is selected as 2, the frames are not detailed enough and as a result, images with less detail are formed. In the Fig. 28 and 29, the images formed as a result of selecting the depth parameter as 2 and 6 are shown respectively.

The density threshold is especially important for better visualization of coasts and islands. The density threshold can be expressed as the number of dots expected to fall per unit square. Accordingly, if the density of any square falls below this threshold (for example, in areas such as an island or a coast), the square is again divided into four equal parts. If the density threshold is kept too high compared to the data set, the threshold will not be exceeded even in areas completely within the landmass. This can lead to

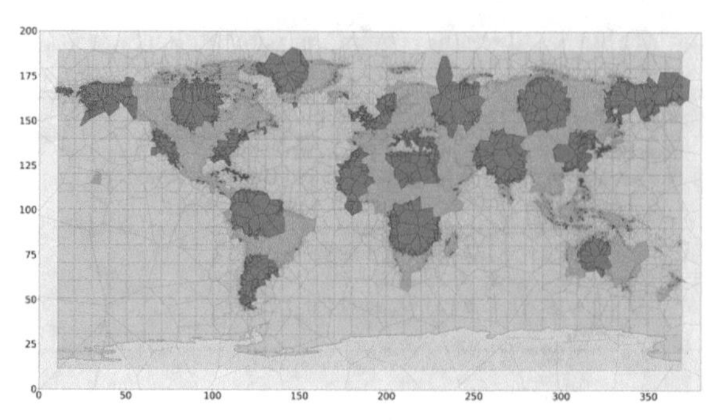

Fig. 29. Depth = 6, number of cells = 8789, algorithm time = 2.384620 s, amount of ram consumed = 27.588289 megabytes

unnecessary segmentation. While determining the density threshold parameter, it would be more accurate to determine the density of a square consisting of all landmass.

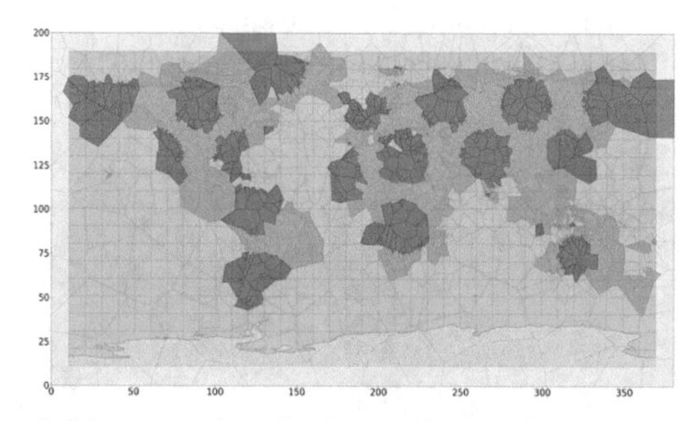

Fig. 30. Density threshold = 1, number of cells = 1607, algorithm time = 1.40253 s, amount of algorithm ram = 19.289733 megabytes

Other parameters were kept constant in order to better understand the effect of the density threshold on the created images. For this purpose, only the B5 dataset was used. Depth is set to 5. The dominance threshold was chosen as 0.4.

In the Fig. 30 and 31, the images created as a result of selecting the intensity threshold parameter as 1 and 9, respectively, are shown. Especially considering the Fig. 31, it is seen that the density threshold is kept very high and as a result of this situation, even the regions with sufficient density have to be divided. For this reason, the density threshold needs to be adjusted according to the dataset.

The dominance threshold is used to decide whether the dots in a square are sufficient for coloring. For example, if there are 70 red and 30 green dots in a certain square, the dominance value of this square is calculated as 0.4. This ratio is calculated as |red − green|/totalpoints. If the dominance threshold is chosen to be a number less than this value (for example, 0.3), this square is not divided again. If the dominance threshold is greater than this value (for example 0.5), this square is divided into smaller squares until it falls below the dominance threshold. Therefore, as a result of increasing the dominance threshold, more segmentation is made.

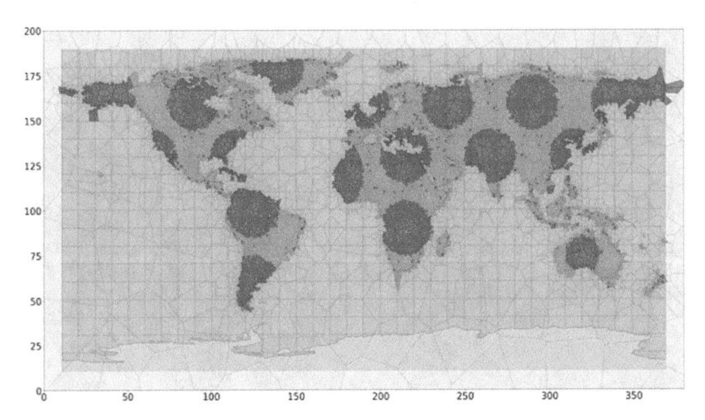

Fig. 31. Density threshold = 9, number of cells = 14585, algorithm time = 3.104696 s, amount of algorithm ram = 34.368145 megabytes

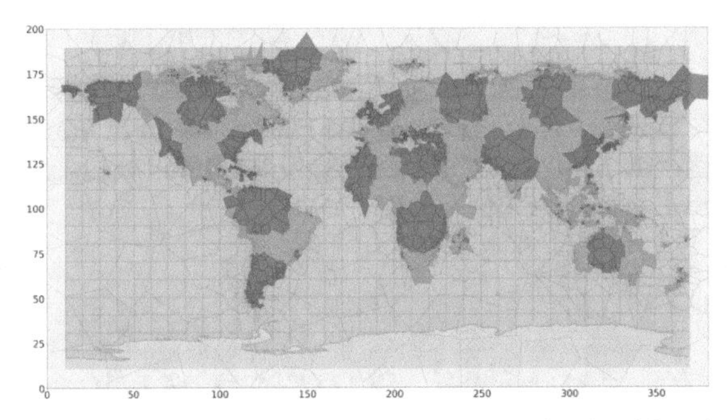

Fig. 32. Dominance threshold = 0.3, number of cells = 4883, algorithm time = 2.008595 s, amount of algorithm ram = 23.086893 megabytes

The dominance threshold separates the areas of differentiation within the data set. For example, circle-shaped dominance separations in the B5 dataset, where the green and red majority are close, can be visualized more clearly or softly by changing this parameter.

Other parameters were kept constant in order to better express the changes in the dominance threshold parameter. For this purpose, only the B5 dataset was used. Depth is set to 5. The density threshold was chosen as 5.

In the Fig. 32 and 33, the images created as a result of selecting the dominance threshold parameter as 0.3 and 0.7, respectively, are shown.

Algorithm timing is one of the fundamental analyzes that this thesis focuses on. At the beginning of this thesis, real-time or near-real-time mapping was attempted. The most time consuming parts of the proposed algorithm are dividing the squares and running the Ball-tree algorithm. Since the Ball tree algorithm consumes more time than dividing the squares, the time complexity of the algorithm is approximately the part that the Ball-tree algorithm uses to color the Voronoi diagrams.

Searching within the ball-tree algorithm has a time complexity of O(log n) for the center point n. Here n is equal to the total number of frames. Also, the total number of frames is equal to both the total number of center points and the total number of Voronoi

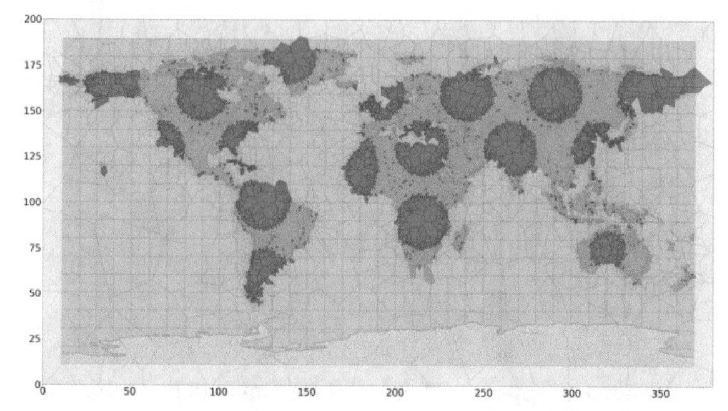

Fig. 33. Dominance threshold = 0.7, cell count = 11654, algorithm time = 2.701807 s, algorithm ram amount = 30.991777 megabytes

cells. If the data to be queried are sent to the Ball-tree algorithm one by one, the time complexity will be equal to O(m * log n). The expression m in the formula represents the total points in the data set. Thus, the time complexity for n center points and m data set points can be expressed as O(m*log n).

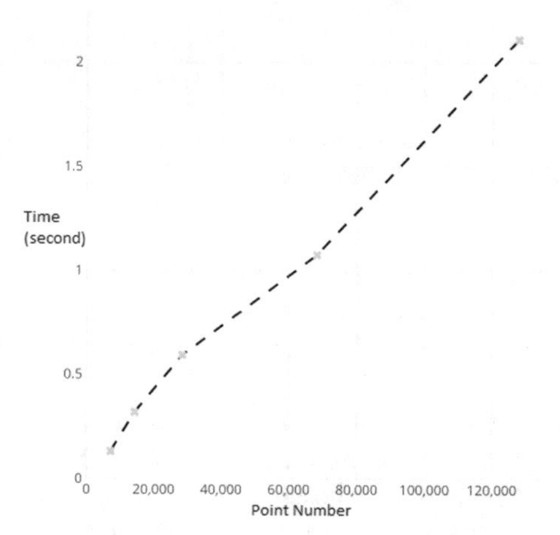

Fig. 34. Variation of algorithm runtime according to the number of points in the data set

The number of points in the data set is expressed as m. The change in this number affects the algorithm time. Accordingly, the increase in points in the data sets increases the time complexity of our algorithm. Figure 34 shows the change of algorithm time according to the change in the number of points in the data set. As it can be understood from here, there is a linear relationship between the number of points and the total running time of the algorithm.

6 Conclusion

Considering all these tests and outputs, although the method we propose does not offer high performance in terms of runtime, it has achieved significant success in terms of visuality and entropy values. Run-time can be reduced with optimized functions and better hardware. Reducing the entropy values with this method we use is important in terms of complexity management. In this study, we tried to show the spatial data to the user in a fast, clean and understandable way. We reviewed previous studies on this subject. In our work, we preferred to make representations with Voronoi diagrams because of its natural appearance and its use in computer graphics. We used the Ball-tree algorithm to find the point with the shortest distance needed in the study. We conducted studies on data sets, some of which were created by us and some of which were used before.

In our study, we used parameters such as first frame size, depth, density threshold, dominance threshold. We performed segmentation according to these parameters.

We subjected our proposed method to different tests according to the dataset and parameters used. As a result of these tests, we obtained different maps with different entropy values according to the parameters. We have shown that similar maps and entropy values occur when data sets created using the same tools and containing different points are changed. This is important for the consistency of the proposed method.

As a result of our tests, we obtained better visuals, although the run-time is longer than other methods. We also used the entropy value, which has an important place in the complexity calculation. As a result of the entropy calculations we have made, it is observed that the complexity is well managed by the proposed method.

In addition, the fact that this study is carried out using more colors instead of two types (colors) can be considered as a different study. Voronoi diagrams can be created directly rather than from squares. Again, this method may be the subject of a different research.

References

1. Voronoi, G.: Nouvelles applications des paramètres continus à la théorie des formes quadratiques. Deuxième mémoire. Recherches sur les parallélloèdres primitifs. Journal für die reine und angewandte Mathematik (Crelles J.) **1908**(134), 198–287 (1908)
2. Marko: Real-time cave destruction using 3D voronoi (2018)
3. Worboys, M.: GIS: a computer science perspective (1994)
4. Chang, K.-T.: Programming ArcObjects with VBA: A Task-Oriented Approach. CRC Press (2007)
5. Cakmak, B.: Complexity management in visualization of very large spatial data, pp. 1–68 (2017)
6. Guo, D.: Flow mapping and multivariate visualization of large spatial interaction data. IEEE Trans. Vis. Comput. Graph. **15**(6), 1041–1048 (2009)
7. Zhu, X., Guo, D.: Mapping large spatial flow data with hierarchical clustering. Trans. GIS **18**(3), 421–435 (2014)
8. Secchi, P., Vantini, S., Vitelli, V.: Bagging Voronoi classifiers for clustering spatial functional data. Int. J. Appl. Earth Obs. Geoinf. **22**, 53–64 (2013)

9. Komzak, J., Eisenstadt, M.: Visualisation of entity distribution in very large scale spatial and geographic information systems. In: KMITR-116, KMi. The Open University (2001)
10. Smorodinsky, S.: Geometric permutations and common transversals. Ph.D. dissertation. Citeseer (1998)

Comparative Analysis of Machine Learning Algorithms for Crop Mapping Based on Azersky Satellite Images

Sona Guliyeva[1,2](✉), Elman Alaskarov[1], Ismat Bakhishov[1], and Saleh Nabiyev[1]

[1] Space Agency of the Republic of Azerbaijan (Azercosmos), Baku, Azerbaijan
[2] National Aviation Academy, Baku, Azerbaijan
sona.guliyeva@azercosmos.az

Abstract. The application of satellite images using geographical information systems (GIS) for the identification and assessment of the condition of agricultural fields is constantly growing, and with this goal, the application of machine learning is increasingly expanding. In the past few years, the issue of accurately classifying agricultural crops has been actively solved. The relevance of this is justified by the need to maintain statistical data on the area of crops grown, calculate data on the food supply of the state, and also monitor the crop condition. For this, various methods are used and data of different formats are processed. The main objective of this research is to apply various machine learning algorithms to Azersky high-resolution satellite imagery for identifying crop types, as well as to evaluate the performance of the models obtained. Three machine learning classifier algorithms support vector machine (SVM), artificial neural network (ANN), and random forest (RF), were applied to identify and map crop types. As a result of this study, the accuracy of different models has been computed and the best model for the classification of crops has been chosen, depending on the provided number of training samples. Also, this paper demonstrated open-source platforms, namely Google Earth Engine (GEE) benefits for such studies, like rapid classification of crops.

Keywords: remote sensing · machine learning · deep learning · classification · crop map · Azersky · GIS · GEE

1 Introduction

In the past few years, the issue of classification of agricultural crops has been actively solved. The relevance of this is justified by the need to maintain statistical data on the areas of crops grown in the regions under consideration, calculate data on the food security of the state, and also monitor the state of crops [1]. There has been a significant rise in the interest of a wide range of organizations in using Earth observation (EO) data from space to obtain the necessary information for making timely and informed decisions in agriculture. Along with this, there are prospects and results for the development of

D. J. Hemanth et al. (Eds.): ICAIAME 2022, ECPSCI 7, pp. 537–548, 2023.
https://doi.org/10.1007/978-3-031-31956-3_46

automated methods for processing satellite data using machine learning a(ML) and artificial intelligence (AI).

Agricultural management is a major topic of interest both for economics and land management. Since crop maps show the territorial differentiation of agricultural production and economic and natural conditions for agricultural development, give an economic assessment of the resources of agricultural production. The classification of satellite images for crop mapping is also an important factor in accuracy [2]. Crop classification is one of the crucial implementation of remote sensing because this type of data can be used as input data for spatiotemporal monitoring of crops, crop yield estimation, and agricultural planning strategy development.

Satellite images as remote sensing data from space can be exploited in various fields of agriculture. One of the important and initial stages of monitoring agricultural fields in the mapping of agricultural land is the identification of various types of crops. In this case, satellite imagery with ground data is an invaluable resource that can quickly reproduce crop type mapping.

Azerbaijan has its own satellite, EO Azersky, which is operated by the Space Agency of the Republic of Azerbaijan (Azercosmos). Azersky EO satellite has the ability to receive high-resolution satellite images from anywhere in the world. Organizations from Central Asia and Eastern Europe, including Azerbaijan, use the images taken by Azersky satellite in agriculture, ecology and the environment, cadastre and cartography, emergencies and oil pollution of the sea or land. Moreover, the capabilities of the Azersky satellite play a supporting role in the socio-economic development of the country, within the framework of the State Program for the Development of the Earth Remote Sensing Service for 2019–2022 [3].

Due to the fact that crop mapping has become a mandatory requirement for most countries, and the Republic of Azerbaijan is given its importance, especially in food security. In this regard, the Geographic Information Systems Center (GIS Center) under Azercosmos is also making efforts, implementing projects for the socio-economic development of the country. Every year, the GIS center compiles a crop map for territory of country and provides the government for decision. One of the tasks in conducting this research was to improve the classification of crops using modern methods, taking into account world experience.

At present along with the improvement of ML, many classification algorithms have been applied to crop mapping, such as ISODATA, maximum likelihood classification (MLC), K-nearest neighbor (KNN), artificial neural networks (ANN), decision tree, Random Forests (RFs), and support vector machine (SVM) [4]. In this research three of these algorithms (SVM, RFs, CNN) has been applied for Azersky image classification using ArcGIS Pro 2.9.0 software.

Currently, developments are being actively carried out aimed at maximizing the integration of systems that provide maintenance of super-large distributed data archives and systems for operational data analysis. According to the researchers, the most rapidly developing system of this type is the Google Earth Engine (GEE). Users of this platform have access to both information analysis tools and the ability to run their own scripts to process any data available in the archives [5]. We have also experience of crop mapping using GEE.

The main goal of this research is to explore the prospective of Azersky images using ML algorithms to distinguish between plots of different crop types in the study area and to determine the best algorithm for crop mapping. In line with the above background and the main purpose of the study, the objectives of this study were to (1) analyze the capabilities of ML algorithms (SVM, RF and CNN) for crop mapping; (2) study the impact of the training set to the classification results; and (3) evaluate the results by accuracy assessment of classified images.

2 Material and Method

2.1 Study Area

The study area is placed in the northern part of the Goranboy region, which is located in the northeastern part of the Ganja-Dashkasan economic region, which is situated in Ganja-Gazakh geographical zone of the Republic of Azerbaijan. The relief of the Goranboy region is flat in the northeast, and mountainous in the southwest, crossed by gorges [6]. In the Ganja-Gazakh geographical zone predominate semi-desert and dry steppe plants. The average annual air temperature is 10.5–14.2 °C. The amount of precipitation varies within 275–440 mm [7].

The southern and southwestern part of Goranboy is covered by the northern slopes of the Murovdag ridge. The research area is located below the shores of the Mingachevir reservoir and stands at 304 km^2 is given in Fig. 1.

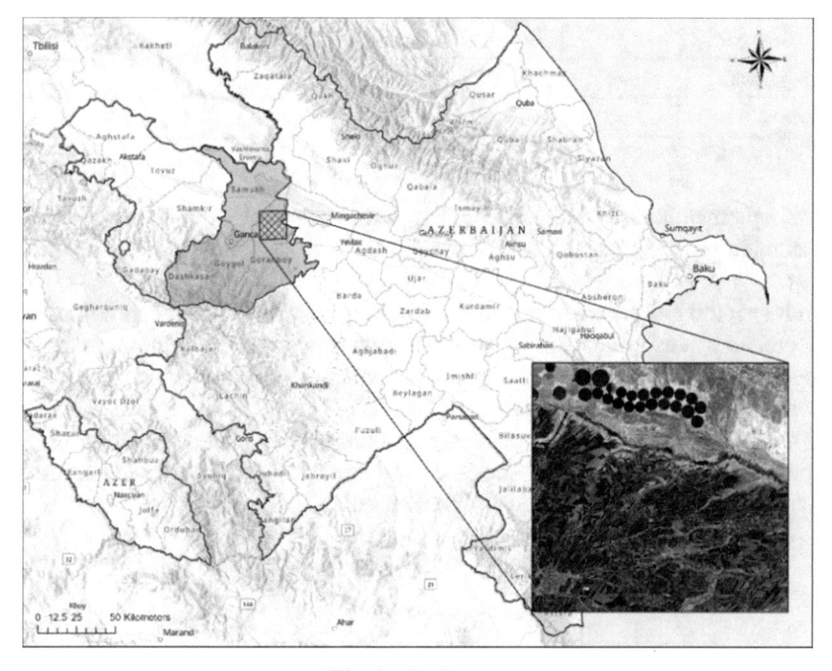

Fig. 1. Study area

The test area is dominated by agricultural arable land with the main types of crops: wheat, cotton, sunflower, corn and alfalfa.

2.2 Dataset

Remotely Sensed Data. Considering the importance of the spatial resolution of satellite images for accurate classification of crop types [1] Azersky high-resolution imagery was used in this study. Azersky is the first EO satellite of the Republic of Azerbaijan. The Azersky satellite capture four-band multispectral image with 6 m of spatial resolution and panchromatic image with 1.5 m resolution. For finer detail object characterization making the high-resolution colored image with the pansharpening of multispectral image and panchromatic image. [8]. The major technical specifications of the Azersky satellite sensor are given below in Table 1.

Table 1. Azersky satellite sensor specifications

Resolution (GSD)	Panchromatic	1.5 m
	Multispectral	6 m
Spectral bands	PAN	0.45–0.75 μm
	Blue	0.45–0.52 μm
	Green	0.53–060 μm
	Red	0.62–0.69 μm
	NIR (Near Infrared)	0.76–0.89 μm
Imaging Swath	60 km	
Altitude	694 km	

For implementing crop mapping Azersky multi-temporal satellite images have been used as main remote sensing data, obtained on April 15, May 19 and June 13, 2021 (Fig. 2).

Multispectral images are more informative in comparison with panchromatic image. Therefore, an extensive analysis of the image pixels distribution in the visible range of the spectrum makes it possible to regulate the parameters characterizing the objects of observation or the processes occurring in them [9, 11].

In this research we have used pansharpened image with spatial resolution 1.5 m and spectral resolution with 4 bands. The dates of the images were chosen according to the phenological regime of the studied agricultural fields.

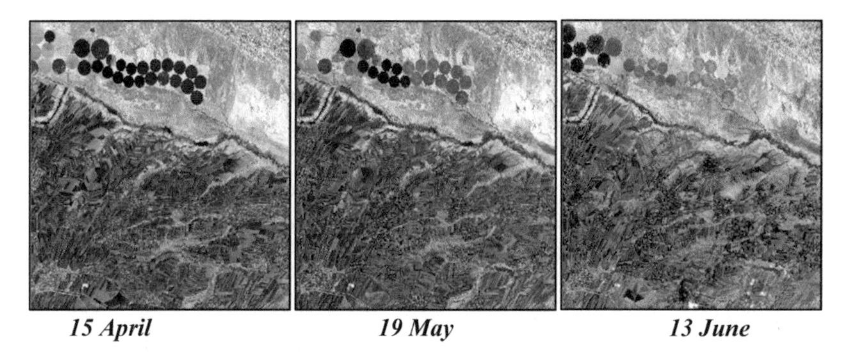

Fig. 2. Azersky satellite images

Training data. A training set with more than 200 field samples of various crops like Wheat, Cotton, Corn, etc., were collected. The details for training samples were provided in Fig. 3. A brief overview of the classification ML algorithms is provided in the ensuing section of this paper.

Fig. 3. Fragment of satellite image with training samples

As training ground data were used, which is information about the crop types. The training samples consisted of 5 main classes, namely alfalfa, corn, cotton, sunflower, and wheat. 4 classes (forest, bareland, swamp and water) created as additional classes, for identification another objects from this data.

Despite the fact that the number of samples was limited, fairly reliable classification results were obtained.

2.3 Classification Methods

Machine learning models are progressively being employed for map of agricultural crop types. Many of research has been devoted to the application of machine learning algorithms for crop mapping using satellite imagery. There are various methods are used to classify crop types and different data formats are processed. Scientists from different countries have also carried out research on the application and selection of the best image classification method for accurately identifying and classifying crop types [10]. In this study, we selected three main algorithms and tested them on the territory of Azerbaijan using Azersky satellite images.

For classification or regression issues a supervised machine learning algorithm named **Support Vector Machine (SVM)** can be used. This algorithm uses a technique to transform the data and finds an optimal boundary between the possible outputs.

This class of machine learning methods can also be used for both classification and regression recovery. The idea of the support vector machine proposed by Vapnik [12] is to construct a hyperplane that acts as a decision surface that maximally separates positive and negative examples from the training set. Translation the original vectors into a higher-dimensional space and search for a separating hyperplane with the largest gap in this space is the important point of this method.

Random Forest (RFs) is a collection of decision trees. The classification decision is made by predominance vote. Every one decision tree is built separately of the others. A subsample of the training sample is selected for each tree and the best feature is chosen for splitting them. Usually the tree is built until the selection is exhausted. The leaves of the tree should contain examples of individual class.

This method is one the continuously used ensemble learning algorithms in remote sensing for satellite image classification. For making a prediction this algorithm creates a set of decision trees, the final output of the classifier is determined by the majority voting of the trees [13].

Convolutional Neural Network (CNN), is a class of ANN. U-Net is a CNN that was developed firstly for biomedical image segmentation. For detecting and characterizing placer signals modelling framework employed classical approaches and deep learning models. The U-Net model architecture includes a feature encoding path and a decoding path. The encoding path extracts hierarchical features given labeled inputs, while the feature decoding path learns the spatial information needed to reconstruct the original inputs [14].

2.4 Image Pre-processing and Classification

The entire process of image processing and map creation was carried out based on the flowchart, shown in Fig. 4. For classifying agricultural crops, the following methods of data preprocessing have been used: bringing the size of images to a single size, removing the background that is not related to the presented class, and creating a composite from multi-temporal data.

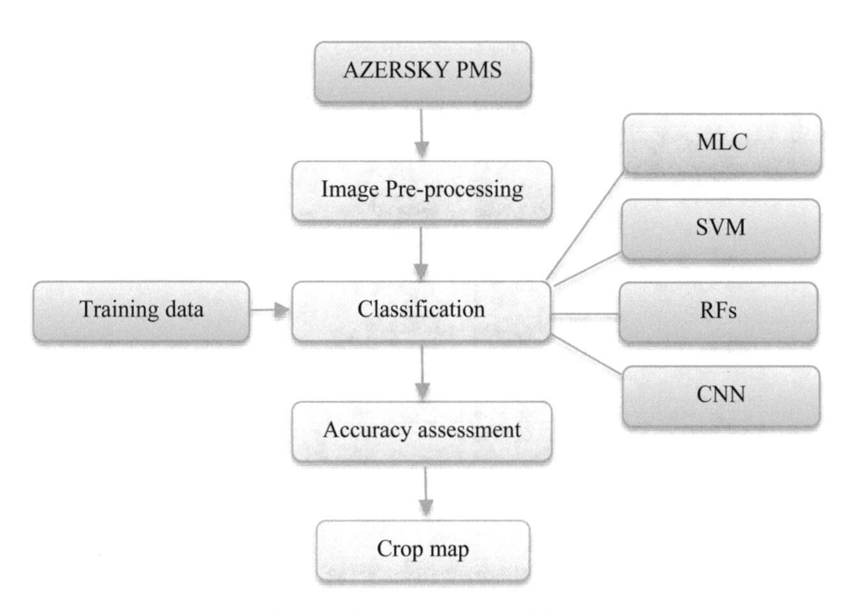

Fig. 4. Flowchart of methodology

Further, on the basis of ground data obtained in a vector format with point geometry, training samples of polygons were created that determined the types of crops and, additional classes were created for identifying other objects, like water and bareland. The previous validation of training samples has been also carried out by visual analysis on the basis of multi-temporal images, where crop types could be distinguished by their development depending on the season of the year. For example, for winter wheat, which was noted by ground data in this study area, in April the field is seen as dark green for true-color image, and in June it is already almost yellow, which indicates its maturity. Classified images by three different methods are shown in Fig. 5.

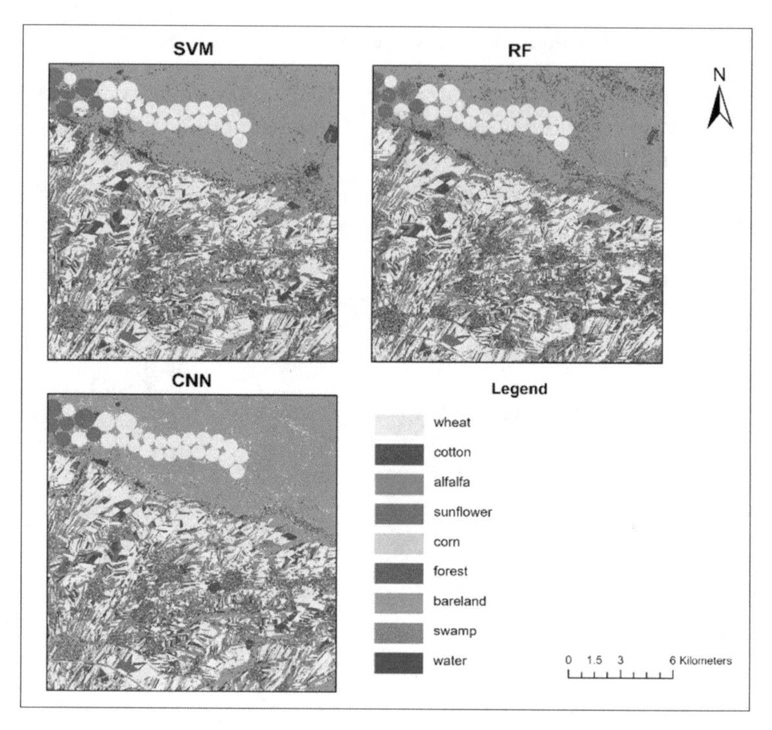

Fig. 5. Classification results

The classification maps were obtained from SVM, RFs, and U-net CNN algorithms with classes Wheat, Cotton, Alfalfa, Sunflower and Corn crops.

2.5 Accuracy Assessment

Accuracy assessment is a key component of the classification of satellite images. It compares the classified image to another data source that is considered accurate or reliable. Reliable ground data can be collected in the field; however, for this purpose time and material resources may be required. Data can be obtained by interpreting high-resolution imagery, existing classified imagery, or GIS data layers to verify accuracy.

The commonest form to estimate the accuracy of result is to generate a random points from the true data and compare their values with the classified image using a confusion matrix. This is a two-step process and may need to compare the results of different classification methods or other training sites, but perhaps there is no ground truth data, or it is based on the same imagery that was used for implementation of classification [15].

For providing accuracy assessment we have used 546 validation points, which were distributed stratified randomly. Validation points were determined based on expert knowledge. The number of training samples for the selected area was limited.

Due to the limited number of training for the territory, the RF showed the best results than CNN. In the end, 3 ML was compared with the classical maximum likelihood

(MLC) method and confirmed the effectiveness of machine learning methods compared to traditional methods of classification. Accuracy assessment of classified images implemented using the confusion matrix and the Kappa statistics were evaluated, has been in accordance with Table 2.

Table 2. Overall accuracy and Kappa ratios for classified images

	RF		SVM		CNN	
Class	U_A	P_A	U_A	P_A	U_A	P_A
wheat	1.000	1.000	0.947	0.940	0.909	0.970
cotton	0.862	0.982	0.719	0.807	0.768	0.754
alfalfa	1.000	0.825	0.804	0.719	0.851	0.702
sunflower	0.724	1.000	0.604	0.762	0.689	0.738
corn	0.857	0.581	0.783	0.581	0.609	0.452
forest	0.474	0.943	0.412	0.778	0.257	0.936
bareland	0.988	0.937	0.945	0.885	0.951	0.885
swamp	0.958	0.793	0.828	0.828	0.833	0.690
water	1.000	1.000	0.923	0.923	0.818	0.692
Overall	0.925		0.842		0.824	
Kappa	0.907		0.805		0.783	

U_A – User Accuracy, P_A – Producer Accuracy

The results demonstrate that SVM and RFs algorithms are powerful for crop type classification and achieve higher classification accuracy with less number of the training samples in the training data. In this study RFs are proven to achieve the higher classification accuracy with less number of training samples.

According to the results the identification of wheat was high because the images corresponded to the vegetation period of wheat.

3 Discussion and Results

Results taken from accuracy assessment demonstrated that the RF algorithm has a higher precision in comparison with the other algorithm in crop mapping. This algorithm has been proposed as an optimal classifier for the crop map extraction because of its higher accuracy and better consistency with the study area. Based on the classified image by RFs algorithm, a crop map has been compiled, shown in Fig. 6.

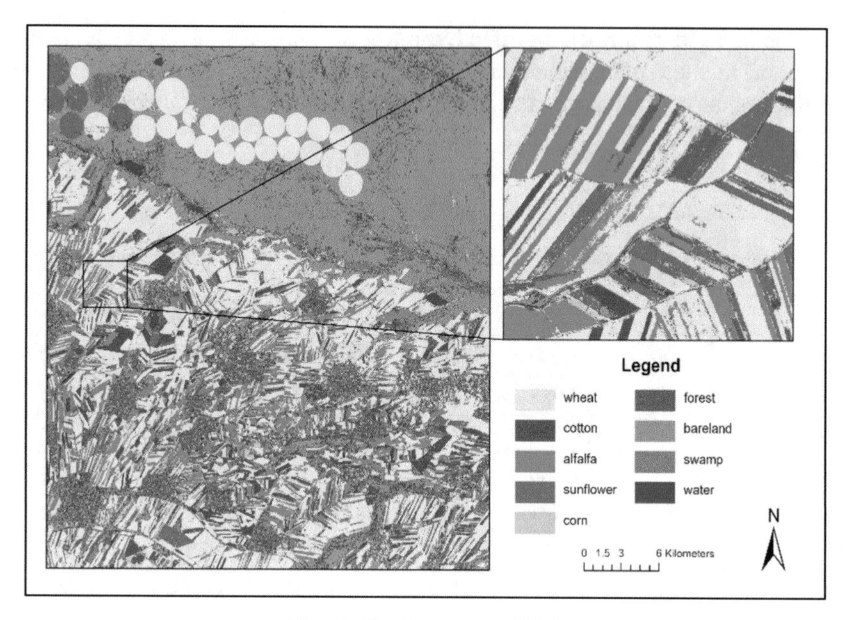

Fig. 6. Final crop map (RFs)

According to this map, the areas of the found classes were calculated (Table 3).

Table 3. Agricultural lands statistics (area)

Class	Count of pixels	km^2
wheat	46677038	87.61
cotton	13890533	26.07
alfalfa	16848847	31.62
sunflower	16658165	31.27
corn	2628029	4.93
forest	9646195	18.10
bareland	58602065	109.99
swamp	2881079	5.41
water	2504792	4.70

4 Conclusions and Future Work

The main goal of this research was to achieve a high accuracy of classification and select the method with good results. It has been used efforts to demonstrate machine learning capabilities for agricultural segments classification for a particular study area. For this reason, it was selected three machine learning methods, such as maximum likelihood, support vector machine, and random trees for achievement expected results.

According to the results of the research, for the studied types of crops, the classification accuracy, was established. The higher result has been identified during the third experiment, for Azersky image using RFs classification method.

This research investigated the performance of machine learning algorithms for the identification of crop types using high-resolution Azersky satellite images. As a result, a comparative analysis of three ML algorithms has been done and the best result has been selected for future consideration in image classification of large areas of interest.

Experimental results demonstrated the following main conclusions: (1) classification of Azersky images using ML algorithms has been done; (2) the training data are significant information in supervised classification the crop classification and for implementing classification with CNN more training samples are needed; (3) RF can produce high classification accuracy than other algorithms for the process of satellite image classification.

In conclusion, it should be noted that we have a plan for future investigation, which will focus on the considerable analysis of the crop classification with more training samples for the large-scale area using the cloud-computing.

References

1. Kononov, V.M., Asadullaev, R.G., Kuzmenko, N.I.: Algoritm podgotovki multispektralnykh sputnikovykh dannykh dlia zadachi klassifikatsii selskokhoziaistvennykh kultur. Nauchnyi rezultat. Informatsionnye tekhnologii. – T.5, №2 (2020)
2. LUO, H.X., et al.: Comparison of machine learning algorithms for mapping mango plantations based on Gaofen-1 imagery. J. Integr. Agric. **2815–2828**(19), 11 (2020)
3. State Program on Development of Earth Remote Sensing Services in the Republic of Azerbaijan for 2019–2022. https://e-qanun.az/framework/40724
4. Ustuner, M., Sanli, F.B.: Crop classification from multi-temporal PolSAR data withregularized greedy forest. Adv. Remote Sens. **1**(1), 10–15 (2021)
5. Andrii, S., Mykola, L., Nataliia, K., Alexei, N., Sergii, S.: Exploring google earth engine platform for big data processing: classification of multi-temporal satellite imagery for crop mapping. Front. Earth Sci. **5**(1–10),17 (2017)
6. İnzibati-ərazi vahidləri (2021). https://files.preslib.az/projects/azerbaijan/gl2.pdf
7. https://biodiversity.az/index.php?mod=content&id=21&lang=az
8. Whiteside, T.G., Bartolo, R.E.: Mapping aquatic vegetation in a tropical wetland using high spatial resolution multispectral satellite imagery. Remote Sens. **7**, 11664–11694 (2015). https://doi.org/10.3390/rs70911664
9. Kulik, K.N., et al.: Geoinformation monitoring of agroecosystems using remotesensing data. In: Materials of the All-Russian Scientific Conference (with international participation) "Application of RemoteSensing Measures in Agriculture", St. Petersburg, pp. 151–155 (2015)
10. Garipelly, P., Bujarampet, D., Palaka, R.: Crop classification using machine learning algorithm. In: Das, B.B., Hettiarachchi, H., Sahu, P.K., Nanda, S. (eds.) Recent Developments in Sustainable Infrastructure (ICRDSI-2020)—GEO-TRA-ENV-WRM. Lecture Notes in Civil Engineering, vol. 207, pp. 131–141. Springer, Singapore (2022). https://doi.org/10.1007/978-981-16-7509-6_11

11. Guliyeva, S.H.: Land cover/land use monitoring for agriculture features classification. Int. Arch. Photogramm. Remote Sens. Spatial Inf. Sci. XLIII-B3–2020, 61–65, (2020). https://doi.org/10.5194/isprs-archives-XLIII-B3-2020-61-2020
12. Cortes, C., Vapnik, V.: Support-vector networks. Mach. Learn. **20**, 273–297 (1995). https://doi.org/10.1007/BF00994018
13. Breiman, L.: Random forests. Mach. Learn. **45**, 5–32 (2001). https://doi.org/10.1023/A:1010933404324
14. Ronneberger, O., Fischer, P.,Brox, T.: U-Net: convolutional networks for biomedical image segmentation. In: Navab, N., Hornegger, J., Wells, W., Frangi, A. (eds.) Medical Image Computing and Computer-Assisted Intervention – MICCAI 2015. MICCAI 2015. Lecture Notes in Computer Science(), vol. 9351, pp. 234–241. Springer, Cham (2015).https://doi.org/10.1007/978-3-319-24553-9
15. https://doi.org/10.1007/978-3-319-24574-4_28, https://desktop.arcgis.com/ru/arcmap/latest/manage-data/raster-and-images/accuracy-assessment-for-image-classification.htm

Defect Detection on Steel Surface with Deep Active Learning Methods on Fewer Data

Bahadır Gölcük[(✉)] and Sevinç İlhan Omurca

Kocaeli University, 41001 Kocaeli, Turkey
bahadirgolcuk@gmail.com, silhan@kocaeli.edu.tr

Abstract. Traditional image processing and machine learning techniques are used for automated surface defect detection problems. Beyond that, deep learning techniques have proven to outperform traditional methods in several defect detection problems. Deep learning techniques are successful to cope with the challenges of detecting complex surface defects in industrial settings. Background properties, shape, sizes, lighting, color, and intensity in the detection of patterns in images can be considered among these challenges. However, another main challenge of deep learning applications in industrial surface defect detection is the insufficient amount of training data. Generally, surface defects occur rarely, and generating or labeling defective data samples are expensive processes.

This study, therefore, proposes a defect detection model which aims to free from the burden of tagging all data by training the models with less but most informative data. It is also aimed to ensure the stability of the model, increase the applicability of the object-detection models. Addressing these, an active learning method "Consistency-based Active Learning for Object Detection" (CALD) is implemented to select the most informative data. With CALD, an active learning cycle was established and the object detection model with the FasterRCNN-Resnet50 backbone was trained with only the most informational data in each cycle, and this cycle was continued until the model success reached the desired result.

Finally, it is investigated that, the accuracy of the study model by using a surface defect dataset called NEU-DET dataset which includes six typical surface defects on metal surfaces. In the first cycle of active learning, the object detection model was trained with the 500 most informative data samples which are decided with the CALD method. As the result of the first cycle, 71.5% mAP(IoU0.5) was obtained. In the next cycle, 200 more data sample was chosen and the model is trained with 700 data sample. 75,8% mAP(IoU0.5) was obtained after the second active learning cycle. In the previous studies, 100% of the data were labeled, and the highest mAP(IoU0.5) result of 77.9% was achieved. It is concluded that our study can approximately render the maximum performance level of 77.9% with fewer labeled data such as 51.8% of the data.

Keywords: Defect Detection · Deep Learning · Active Learning · Object Detection

D. J. Hemanth et al. (Eds.): ICAIAME 2022, ECPSCI 7, pp. 549–559, 2023.
https://doi.org/10.1007/978-3-031-31956-3_47

1 Introduction

The quality of the product surface is one of the important key factors for qualified industrial production processes. The surface defects of the products can substantially effect the performance indicators and the final values of the product [1]. Metallic surfaces have particular importance because of their wide range of industrial applications. Material characteristics and some industrial processing technologies can cause defects on metallic surfaces and these defects are obstacle to produce high-quality end products [2]. Therefore it is essential to study valid methods for detecting metallic surface defects. Although the detection of surface defects of products is still done by manual inspection in traditional industry, performing these processes with advanced and automated methods is a very crucial research area. At this point, traditional image processing and deep learning methods are used to develop solutions to the problems such as defect localization, detection, recognition or classification. Especially recently, the development of surface defect recognition applications using deep learning methods has become a promising research area.

Due to the automatic feature extraction capability and multi-layer nonlinear transformations, deep learning methods have great performances in challenging applications such as surface defect detection and have become popular. CNN based deep learning methods and apart from that more recently, R-CNN [3], Fast R-CNN [4], Faster R-CNN [5], SSD [6], and YOLO [7] methods which are known to be successful in object recognition tasks, have been applied to improve the defect detection.

Apart from the main advantages above, deep learning methods also provides prominent solutions for various challenges of detecting complex surface defects such as background properties, shape, size, lighting, color, and intensity in the detection of patterns in images. However, these are not the only challenges with the surface detection problem; another main challenge is insufficient amount of training data. In industrial processes, generally, surface defects occur rarely, and generating or labeling defective data samples are expensive tasks. In such states, Active learning which uses as few labeled data as possible to achieve high performance is one of promising strategies. To decide and select the informative instances for annotation play a key role for active learning.

Therefore, to overcome all these difficulties, the main motivation of the study is to develop a defect detection model that produces effective results by training with less but most informative data, instead of a model that needs large labeled data stacks in the training phase. An active learning method that designed to explore the consistency between original and augmented data - Consistency-based Active Learning for Object Detection (CALD) [8] - is implemented to achieve these objectives. With CALD, an active learning cycle was established and the object detection model with the FasterRCNN-Resnet50 backbone was trained with only the most informative data in each cycle, and the cycle was continued until the model success reached the desired result. Finally, the experiments were conducted to demonstrate the effectiveness of the proposed method on the NEU-DET [9] dataset which is a public steel surface defect dataset established by Northeastern University.

The rest of the paper is organized as follows: Sect. 2 gives related researches with Surface Defect Detection and Active Learning. In Sect. 3, the presented method is represented. Experimental setup and results are demonstrated in Sects. 4. Section 5 concludes the paper with a discussion and conclusions.

2 Related Works

2.1 Object Detection for Surface Defect Detection

Detecting surface defects is a growing topic in the industry. According to Chen et al, it is possible to group deep learning-based [10] methods used in the detection of surface defects as in Fig. 1.

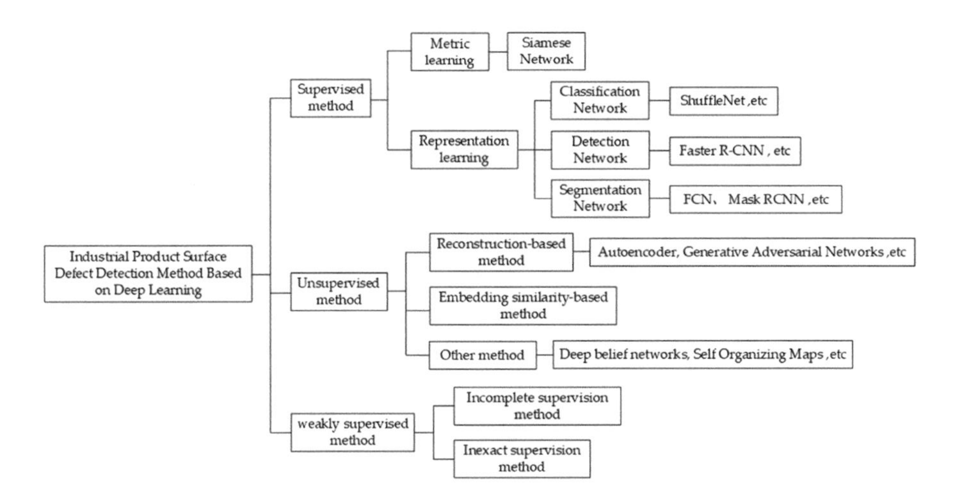

Fig. 1. Industrial product surface defect detection method based on deep learning.

In recent years, object-detection models have had a great place in product defect detection systems used in the industry. Although the classification of the defect is important, determining the region of the defect allows a more detailed analysis of the root cause of the problems. For this, many state of the art models were used and solutions were tried to be developed with these. Jiang et al. [11] focused on the defects on the shaft surfaces produced in their study and obtained successful results by using the FasterRCNN method. Lv et al. on the other hand, by using the model they developed based on the Single Shot Multi Detection model in the detection of steel defects [12]; SSD achieved higher results than YOLOv2 [13], YOLOv3 [14] and FasterRCNN models. Concentrating on the same problem, Chen et al. used MobileNetV2 [15] instead of DarkNet53 which the YOLOv3 model uses as a backbone and produced a real-time applicable model with 80.96 FPS speed and %86.96 accuracy [16]. Successful results were obtained in these supervised studies carried out to detect the error zone at a high rate, and the error zone was also detected on the products. However, in these studies, a lot of effort is required for labeling, and it is also difficult to reach the data due to the low occurrence of faulty surfaces in the field.

2.2 Active Learning

Active learning is an increasingly popular topic due to the reduction in the effort spent on data labeling and the difficulty of accessing labeled data. Furthermore, some active learning studies are related to automatic labeling. Yang et al. [17] proposed an active learning framework for auto-labeling for surface defect visual classification problem. For auto-labeling process, they implemented a novel attention-based similarity measurement network (ASMN) that measures the similarity between unlabeled and labeled samples. In general, deep learning applications require more data, so as much data as possible is obtained to solve the problem. All of these obtained data are labeled and deep learning models are trained. In the active learning strategy, with the established loop, the data containing the most information is selected according to the inquiry methods in each iteration, and the model is trained with these data.

The query part is of great importance in AL loops. At the beginning of these query methods are uncertainty-based methods [18–21]. Diversity-based approaches [22–25] and expected model change methods [26–28] among other approaches are used. The methods developed based on these approaches have been used in many detection problems. Feng et al. [29] used these approaches to detect defects in the construction site and achieved a success rate of %85. On the other hand, Koutroulis et al. [30] proposed an iterative active learning framework with CNN for chip defect classification in semiconductor wafers. They selected the most informative subsets of images based on their estimated uncertainty. Then implemented density based clustering and defined dense neighborhoods. A similar problem exists in detecting steel surface defects. In order to detect defects on steel surfaces, Lv et al. proposed the Uncertainty Sampling for Candidates method to separate the candidate images and identify the images with the most information for the model. He compared the results of this method with the results from other methods and when he labeled all the data. As a result, by labeling only %21 of the data, more successful results were achieved with less data than all the others.

Although a detection-based problem is handled in these studies and the most current active learning method used in steel surface defects, only a query method that considers and evaluates classification results has been used. The study by Yu et al. stated that bbox estimation is also one of the metrics containing the most information, and it has been determined that it plays an essential role in the selection of data. In addition, inquiry methods based on classification results are insufficient in decision-making situations when several types of errors occur in a visual. For these reasons, they developed the Consistency Based Active Learning (CALD) method in their study.

3 Methodology

In this study, we used the CALD method, which evaluates the classification and detection results together in the object detection problem and obtains better results than the active learning methods used with other object-detection models.

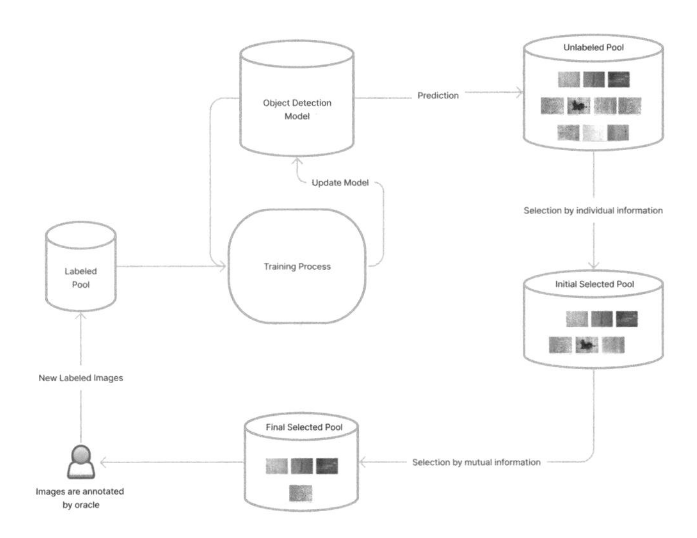

Fig. 2. Overview of CALD Method.

3.1 Overview of CALD

Consistency-based active learning cycle, with an overview diagram in Fig. 2, consists of four repositories, a labeling expert, and an object-detection model. To summarize, the active learning cycle proceeds as follows: First, an object-detection model suitable for the problem is selected. This model may be pre-trained or untrained before the cycle begins. To start the first cycle, several data are labeled and the selected model is trained with these data. As a result of the training, a new model with updated weights emerges. Unlabeled data is asked to this model one by one, and among the data asked, the data containing the most information is selected according to individual information. A second elimination method is used to compare the distributions of the classes among these selected data and organize the selected classes. This method is called mutual information. The data eliminated through this method reaches its final form and is transferred to the tagged data pool by being labeled by an expert. This cycle can be continued until the desired success rate of the model is reached. By choosing the data containing the most information among the unlabeled data at each stage, it is ensured that the model gets the best result at every stage. Detailed flow diagram in Fig. 3.

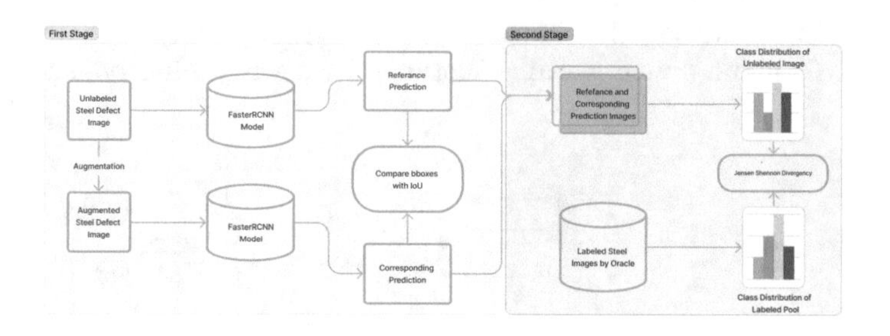

Fig. 3. Two Stages Flow Diagram

3.2 Consistency-based Individual Information

To compare the consistency of our preferred deep learning model, we need to compare the prediction results. Here, the prediction results are obtained in two ways. First, the unlabeled images (x_u) are asked directly to the model $\theta(x_u)$. Then, by applying data augmentation to these results, reference results are obtained $A(\theta(x_u)$. Bounding boxes are shown as b_k and scores as s_k in reference estimates. Second, the unlabeled images to be asked to the model are first incremented ($x_u^{'}$) and then asked the model. A comparison is made between these two model outputs. In the comparison process, bounding boxes are compared according to the IoU metric, while the score results are made according to the Jensen-Shennon (JS) divergence metric. In order to get the most reliable predicted score result, the maximum confidence is determined as a weighting factor. As a result, the consistency of an image is equal to the sum of the consistency of the bounding boxes and the scores, and is represented as:

$$m_k \& = C_k^b + C_k^s \tag{1}$$

$$C_k^b \& = IoU(b_k, b_k^{'}) \tag{2}$$

$$C_k^s \& = \frac{1}{2}[max(\varphi_n) + max(\varphi_n^{'})](1 - JS(s_k || s_k^{'}) \tag{3}$$

Obviously, $m_k[0, 2]$. The closer the result is to 0, the more inconsistent it is and the more informative this patch is. However, in some cases, when comparing the reference outputs with the corresponding outputs, it is seen that the predictions are made with low scores in unrelated regions. In this case, however, close m_k is to 0, it does not mean that it is informative. For these and similar cases, the m_k value of the informative patch is determined in two ways: (1) Keeping a certain distance from the lower bound means the paired predictions have relatively high matching degrees and high confidence. If the prediction is wrong, this patch is probably informative, because the prediction is likely the main result of the object. (2) Being far away from the upper bound, means the matching degree is worse than when m_k is on the upper bound. This indicates that the model cannot cope with the data augmentation and the prediction is most likely wrong. Thus, the consistency metric of an image is defined as:

$$M(x_u; A,) = EA[min_k|m_k|] \tag{4}$$

3.3 Mutual Information

In the active learning cycle, the class distributions may be unbalanced among the initial data selected for training the model. In this case, the tendency of the model to some data is high. In an active learning method that is selected only according to individual information, data containing information may be selected incorrectly. The idea is to compare the class distribution of each sample in the initial selected pool and the class distribution of the entire labeled pool, then select samples with large distances (i.e. with different class distributions from the labeled pool) to form the final selected pool. An algorithm is proposed using JS divergence to evaluate the distance between two class distributions. In this algorithm, firstly the softmax and class distributions of the classes in the labeled data pool are calculated, and then the class distributions of the data in the initial-selected pool are calculated and compared. Here, due to the high precision of the data from the inital-selected pool, only those with the highest reliability value of the predictions in each class are selected.

$$\Delta_L(Y_L) = Softmax([\delta_1, \delta_2, ..., \delta_m, ...]^T)\delta_m = \sum_{y_L \epsilon Y_L} I(y_L = m) \tag{5}$$

$$\Delta_U(x_U) = Softmax([\delta_1, \delta_2, ..., \delta_m, ...]^T)\delta_m = max_{s_k \epsilon \{s_k\}}\{\varphi_m | \varphi_m \epsilon s_k\} + max_{s'_j \epsilon \{s'_j\}}\{\varphi'_m | \varphi'_m \epsilon s'_j\} \tag{6}$$

4 Experiments

4.1 Dataset

In the experiments, we used public steel surface defect dataset called as NEU-DET dataset and establihed by Northeastern University. The dataset contains 1800 images of six types of steel surface defects. There are 300 images for each defect type, with a sample size of 200 × 200. The labels of steel surface defects are defined as rolled-in scale, patches, crazing, pitted surface, inclusion, and scratches. The sample images belong to defect types are shown in Fig. 3. 450 of the data were reserved to validate the model. An active learning cycle was started with 500 of the remaining 1350 and 200 were selected in each cycle and included in the system. In order to examine all the results, the active learning cycle was terminated after all 1350 data were included in the loop. At this stage, you can observe the results of each cycle and terminate the cycle when the desired success rate is achieved.

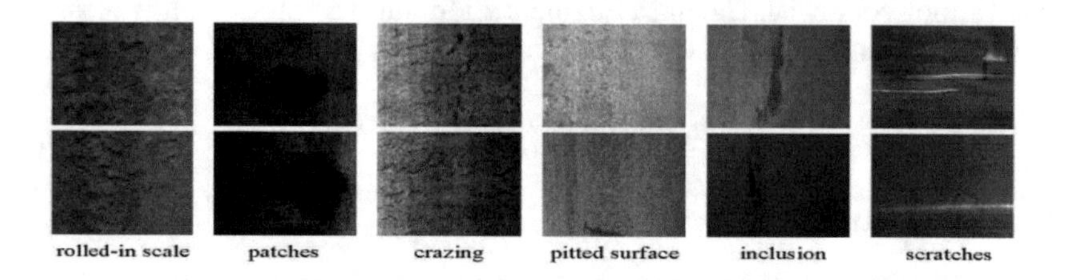

Fig. 4. Defect Types in NEU-DET Dataset.

In the active learning cycle, before creating the initial-selected pool, augmentation is applied to the data. Data augmentation methods used in the experiments are flip, smaller-resize, and rotation.

4.2 Model

FasterRCNN model with a two-stage detector structure was selected to be used in the active learning loop and the ResNet50 [31] method was used as a backbone. The learning rate parameter is defined as 0.0025 and the batch size is 4 (Fig. 5).

4.3 Experimental Results

We adopted the CALD method to the NEU-DET dataset and the obtained results are shown in Fig. 4. When the results are examined, it is seen that, when %37 of the data was labeled %71.5 mAP score was achieved, and when %51.8 of the data was labeled %75.8 mAP score was achieved. In the previous study, Chen et al. [9] trained all of the data in the NEU-DET dataset with the Faster RCNN Resnet50 model, and a %77.9 mAP score was achieved. The obtained mAP@IoU0.5 values for each cycle are shown in Table 1 and the results obtained by training all data are shown in Table 2

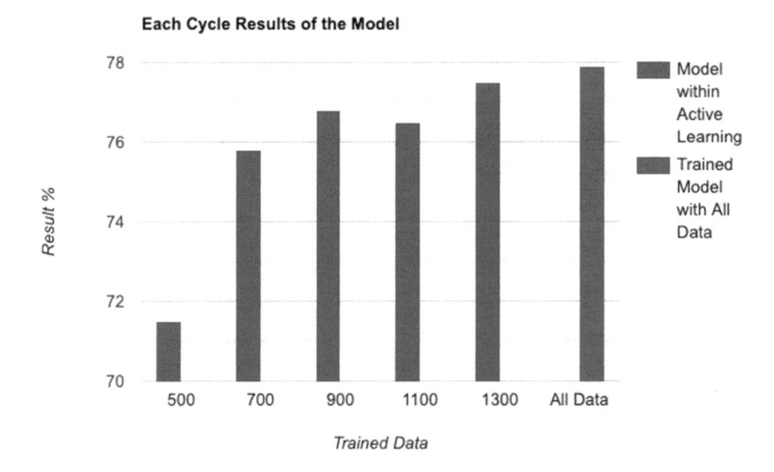

Fig. 5. Each cycle results of the model and trained model within all data.

Table 1. Details results of the each cycle

Cycle	mAP@0.5	crazing	inclusion	patches	pitted surface	rolled in scale	scratches
1	71.5	32.9	74.6	88.6	78.3	64.0	90.8
2	75.8	43.7	76.4	91.1	82.5	65.4	95.7
3	76.8	47.1	76.0	92.0	84.5	66.8	94.1
4	76.5	42.6	77.4	91.2	83.2	69.6	95.2
5	77.5	46.3	76.1	93.3	84.5	70.0	95.1

Table 2. Results of FasterRCNN Resnet50 that trained with all data

Cycle	mAP@0.5	crazing	inclusion	patches	pitted surface	rolled in scale	scratches
–	77.9	52.5	76.5	89.0	84.7	74.4	90.3

5 Conclusions

In our study, Consistency-based Active Learning for Object Detection (CALD) is presented that performs high performance surface defect detection by labelling less of the data instead of all of it. In each active learning cycle a FasterRCNN-Resnet50 backbone was trained with only the most informational data. Considering that the production data in real life contains very little surface deformation data, the proposed model makes an important contribution to detecting the surface deformations with less data.

The performance and effectiveness of the model were investigated using the called NEU-DET dataset which includes six typical surface defects on metal surfaces. When the experimental reults are analyzed, it is seen that, according to the latest active learning study with NEU-DET data set in the literature [24], both more accurate results were obtained in this study and these results were obtained in 6 error types instead of 4.

In order to improve this study, as a future direction, the effect of different detection models on the CALD active learning cycle can be investigated and the effect of the cycles can be discussed by testing them on datasets with different surface defects.

References

1. Fan, B.B., Li, W.: Application of GCB-Net based on Defect Detection Algorithm for Steel Plates, 21 April 2022, PREPRINT (Version 1) available at Research Square. https://doi.org/10.21203/rs.3.rs-1550068/v1
2. Lv, X., Duan, F., et al.: Deep active learning for surface defect detection. Sensors **20**(6), 1650 (2020). https://doi.org/10.3390/s20061650
3. Girshick, R., Donahue, J., Darrell, T., et al.: Rich feature hierarchies for accurate object detection and semantic segmentation. In: Proceedings of the IEEE Conference on Computer Vision and Pattern Recognition, pp. 580–587 (2014)
4. Girshick, R.: Fast R-CNN. In: Proceedings of the IEEE International Conference on Computer Vision, pp. 1440–1448 (2015)
5. Ren, S., He, K., Girshick, R., et al.: Faster R-CNN: towards real-time object detection with region proposal networks. In: Advances in Neural Information Processing Systems, pp. 91–99 (2015)
6. Liu, W., et al.: SSD: single shot multibox detector. In: Leibe, B., Matas, J., Sebe, N., Welling, M. (eds.) ECCV 2016. LNCS, vol. 9905, pp. 21–37. Springer, Cham (2016). https://doi.org/10.1007/978-3-319-46448-0_2
7. Redmon, J., Divvala, S., Girshick, R., et al.: You only look once: unified, real-time object detection. In: Proceedings of the IEEE Conference on Computer Vision and Pattern Recognition, pp. 779–788 (2016)
8. Yu, W., Zhu, S., Yang, T., Chen, C.: Consistency-based active learning for object detection. In: CVPR-2022 Workshop (2021). https://doi.org/10.48550/arXiv.2103.10374
9. Song, K., Yunhui, Y.: NEU Surface Defect Database, Northeastern University
10. Chen, Y., et al.: Surface defect detection methods for industrial products. MDPI Appl. Sci. **11**(16), 7657 (2021). https://doi.org/10.3390/app11167657
11. Jiang, Q., et al.: Object detection and classification of metal polishing shaft surface defects based on convolutional neural network deep learning. MDPI **10**(1), 87 (2020). https://doi.org/10.3390/app10010087
12. Lv, X., et al.: Deep metallic surface defect detection: the new benchmark and detection network. Sensors **20**(6), 1562 (2020). https://doi.org/10.3390/s20061562
13. Redmon, J., Farhadi, A.: YOLO9000: better, faster, stronger. In: Proceedings of the IEEE Conference on Computer Vision and Pattern Recognition, Honolulu, HI, USA, 21–26 July 2017, pp. 7263–7271 (2017)
14. Redmon, J.; Farhadi, A. Yolov3: An incremental improvement. arXiv 2018, arXiv:1804.02767
15. Sandler, M., et al.: MobileNetV2: inverted residuals and linear bottlenecks. In: The IEEE Conference on Computer Vision and Pattern Recognition (CVPR), pp. 4510–4520 (2018). https://doi.org/10.48550/arXiv.1801.04381
16. Chen, X., et al.: Online detection of surface defects based on improved YOLOV3. Sensors **22**, 817 (2022). https://doi.org/10.3390/s22030817

17. Yang, H., Song, K., Mao, F., Yin, Z.: Autolabeling-enhanced active learning for cost-efficient surface defect visual classification. IEEE Trans. Instrum. Meas. **70** (2021)
18. Beluch, W.H., Genewein, T., Nürnberger, A., Köhler, J.M.: The power of ensembles for active learning in image classification. In: 2018 IEEE Conference on Computer Vision and Pattern Recognition, CVPR 2018, Salt Lake City, UT, USA, 18–22 June 2018, pp. 9368–9377. IEEE Computer Society (2018)
19. Joshi, A.J., Porikli, F., Papanikolopoulos, N.: Multi-class active learning for image classification, pp. 2372–2379 (2009)
20. Lewis, D.D., Gale, W.A.: A sequential algorithm for training text classifiers, pp. 3–12 (1994)
21. Ranganathan, H., Venkateswara, H., Chakraborty, S., Panchanathan, S.: Deep Active Learning for Image Classification (2017). https://doi.org/10.1109/ICIP.2017.8297020
22. Bilgic, M., Getoor, L.: Link-based active learning. In: NIPS Workshop on Analyzing Networks and Learning with Graphs (2009)
23. Gal, Y., Islam, R., Ghahramani, Z.: Deep bayesian active learning with image data. In: Proceedings of the 34th International Conference on Machine Learning, ICML 2017, Sydney, NSW, Australia, 6–11 August 2017, pp. 1183–1192 (Proceedings of Machine Learning Research, Vol. 70). PMLR (2017)
24. Guo, Y.: Active instance sampling via matrix partition. In: Advances in Neural Information Processing Systems 23: 24th Annual Conference on Neural Information Processing Systems 2010. Proceedings of a Meeting Held, 6–9 December 2010, Vancouver, British Columbia, Canada, pp. 802–810. Curran Associates Inc. (2010)
25. Hieu Nguyen, T., Smeulders, A.: Active learning using pre-clustering. ICML **2004**, 79–79 (2004)
26. Freytag, A., Rodner, E., Denzler, J.: Selecting influential examples: active learning with expected model output changes. In: Fleet, D., Pajdla, T., Schiele, B., Tuytelaars, T. (eds.) ECCV 2014. LNCS, vol. 8692, pp. 562–577. Springer, Cham (2014). https://doi.org/10.1007/978-3-319-10593-2_37
27. Roy, N., McCallum, A.: Toward optimal active learning through monte carlo estimation of error reduction. ICML, Williamstown, pp. 441–448 (2001)
28. Settles, B., Craven, M., Ray, S.: Multiple-Instance Active Learning, pp. 1289–1296 (2007)
29. Feng, C., Liu, M.-Y., et al.: Deep Active Learning for Civil Infrastructure Defect Detection and Classification, American Society of Civil Engineers ASCE International Workshop on Computing in Civil Engineering (2017). https://doi.org/10.1061/9780784480823
30. Koutroulis, G., Santos, T., Wiedemann, M., Faistauer, C., Kern, R., Thalmann, S.: Enhanced active learning of convolutional neural networks: a case study for defect classification in the semiconductor industry. In: Proceedings of the 12th International Joint Conference on Knowledge Discovery, Knowledge Engineering and Knowledge Management (IC3K 2020) - Volume 1: KDIR, pp. 269–276 (2020)
31. He, K., Zhang, X., Ren, S., Sun, J.: Deep residual learning for image recognition (2015). https://doi.org/10.48550/arXiv.1512.03385

Determining Air Pollution Level with Machine Learning Algorithms: The Case of India

Furkan Abdurrahman Sari[1(✉)], Muhammed Ali Haşıloğlu[1], Muhammed Kürşad Uçar[2], and Hakan Güler[1]

[1] Faculty of Engineering, Civil Engineering Department, Sakarya University, 54187 Serdivan, Sakarya, Turkey
{furkansari,hguler}@sakarya.edu.tr, muhammed.hasiloglu@ogr.sakarya.edu.tr
[2] Faculty of Engineering, Electricial-Electronics, Engineering Department, Sakarya University, 54187 Serdivan, Sakarya, Turkey
mucar@sakarya.edu.tr

Abstract. Air pollution is one of the critical health problems affecting the quality of life, especially in city centers. The air quality index (AQI) is the primary parameter used to measure air pollution. This parameter is constantly measured in city centers with measuring devices that contain various sensors. Due to the high purchasing costs and the need for periodic calibration, there is a need to develop more economical technologies to calculate AQI values. This study aims to predict air pollution with minimum sensors based on machine learning. In the study, daily-based five-year air quality measurement data from India, taken from the Kaggle database, were used. Air quality data includes $PM_{2.5}$, PM_{10}, O_3, NO_2, SO_2, CO values obtained from sensors in the measuring devices, and AQI calculated with these values. A feature selection algorithm is used to reduce the sensor cost. Then, artificial intelligence-based AQI was calculated with minimum sensor data. According to the findings, artificial intelligence-based AQI calculation model performances r and $RMSE$ were determined as 0.93 and 20.57, respectively. It has been evaluated that AQI data can be calculated based on artificial intelligence with a minimum of sensors.

1 Introduction

Air pollution is a significant health problem in urban centers where the population is dense [1]. The necessities of life; heating, economic activities, and waste gases resulting from transportation are the main factors causing air pollution [2]. Humanity, who built civilization by processing and combining the materials found in nature, has created many pollutants with these activities [3]. Among the contaminants in the air, the density of particulate matter directly affects human health [4]. According to the World Health Organization (WHO) report, more than 2 million people worldwide die at an early age due to diseases caused by air pollution [5]. The air quality index (AQI) is used as an indicator of current

D. J. Hemanth et al. (Eds.): ICAIAME 2022, ECPSCI 7, pp. 560–581, 2023.
https://doi.org/10.1007/978-3-031-31956-3_48

air quality in many cities worldwide [6]. AQI is developed by the United States Environmental Protection Agency (U.S. EPA) and is made easily understandable and is made available to public access [7]. The basic AQI table prepared by the EPA is shown in Fig. 1.

Daily AQI Color	Levels of Concern	Values of Index	Description of Air Quality
Green	Good	0 to 50	Air quality is satisfactory, and air pollution poses little or no risk.
Yellow	Moderate	51 to 100	Air quality is acceptable. However, there may be a risk for some people, particularly those who are unusually sensitive to air pollution.
Orange	Unhealthy for Sensitive Groups	101 to 150	Members of sensitive groups may experience health effects. The general public is less likely to be affected.
Red	Unhealthy	151 to 200	Some members of the general public may experience health effects; members of sensitive groups may experience more serious health effects.
Purple	Very Unhealthy	201 to 300	Health alert: The risk of health effects is increased for everyone.
Maroon	Hazardous	301 and higher	Health warning of emergency conditions: everyone is more likely to be affected.

Fig. 1. Basic Air Quality Index [8]

Basically, 6 different pollutants are considered in determining the AQI index: CO, NO_2, O_3, PM_10, $PM_{2.5}$, and SO_2 [9]. There are limited concentration values for each pollutant determined epidemiologically and determined according to national air standards [10]. The concentration value of each pollutant collected from the locally located monitoring network is calculated by Equaiton 1 using the limit concentration values of the relevant contaminant in Table 1 [11].

Table 1. Break Points for AQI

AQImin - AQImax	O3		PM2.5	PM10	CO	SO2	NO2
	ppm		μg/m3		ppm	ppb	
	Period of averaging, hour						
	8	1	24	24	8	1	1
0–50	0–54	–	0–12	0–54	0.0–4.4	0–35	0–53
51–100	55–70	–	12.1–35.4	55–154	4.5–9.4	36–75	54–100
101–150	71–85	125–164	35.5–55.4	155–254	9.5–12.4	76–185	101–360
151–200	86–105	165–204	55.5–150.4	255–354	12.5–15.4	186–304	361–649
201–300	106–200	205–404	150.5–250.4	355–424	15.5–30.4	305–604	650–1249
301–500	–	405–604	250.5–500.4	425–604	30.5–50.4	605–1004	1250–2049

$$\mathrm{I_P} = \frac{I_{Hi} - I_{Lo}}{BP_{Hi} - BP_{Lo}} \times (C_P - BP_{Lo}) + I_{Lo} \tag{1}$$

Measuring airborne pollutants in city centers with an extensive monitoring system causes very high initial setup costs. In addition, after the installation, the calibration of the devices should be done at specific periods. In addition, it is not possible to measure all the pollutant particles in the air with a limited number of devices. Breathing clean air is a universal right for all people, but not all countries have the financial means to install these systems. Showing that it is possible to predict air quality with a high accuracy rate with the help of new technologies has been a source of motivation for us in our work. In the literature, there are studies to predict the change of general air quality under seasonal effects with machine learning algorithms [12]. Deep learning studies have also been carried out to determine which factors are more effective in reducing air pollution caused by traffic, which increases due to transportation activities in urban centers [13] In other studies, short and long-term air quality index estimations have been tried to show with machine learning algorithms [14,15]. In other studies, only two pattern recognition algorithms have attempted to make predictions with the help of data obtained from only two different sensors [16]. In addition, studies involving hybrid models have also been carried out [17]. There are other studies in the literature that use the open data set we use [18].

When the studies in the literature were examined, it was seen that the number of data was insufficient in some studies, and it was understood that the collected data were obtained from only a few monitoring devices. In some studies, very few estimation algorithms were used in independent estimation models during the implementation phase. In some studies where hybrid models were tested, it was observed that the number of hybridizing algorithms was insufficient. In another study in which classification algorithms were used for estimation, the feature selection algorithm was applied incorrectly, and the raw data set was applied without editing. Due to the deficiencies stated in the studies, the performance results of the models are not satisfactory. In our study, higher prediction performances will be tried to be obtained by using many models and their hybrids. The dataset will be preprocessed, and the features required for the prediction will be considered.

In this study, the raw data set was first analyzed, then a healthier data set was obtained by cleaning the empty cells with missing sensor data. Then, using Spearman correlation coefficient and Principal Component Analysis (PCA), selected features and feature groups were created. In the study, Decision Trees Ensembles (DTE), Support Vector Machine Regression Model (SVMs), Multi-Layered Feed Forward Neural Network (MLFFNN), which were widely used in previous studies, were used. In addition, hybrid versions of the models were created using three models. The performance results of the prediction models were evaluated.

2 Materials and Methods

This study has been conducted according to the flow diagram given in Fig. 2. The data is first grouped according to the correlation values and feature selection algorithm. Then, prediction models were created to be used in feature groups. At this stage, machine learning algorithms and hybrid models of these algorithms have been created. The models were trained for each feature group and in total, 84 models were acquired. Lastly, with the acquired model outputs, seven performance criteria have been calculated and the success of the models was evaluated.

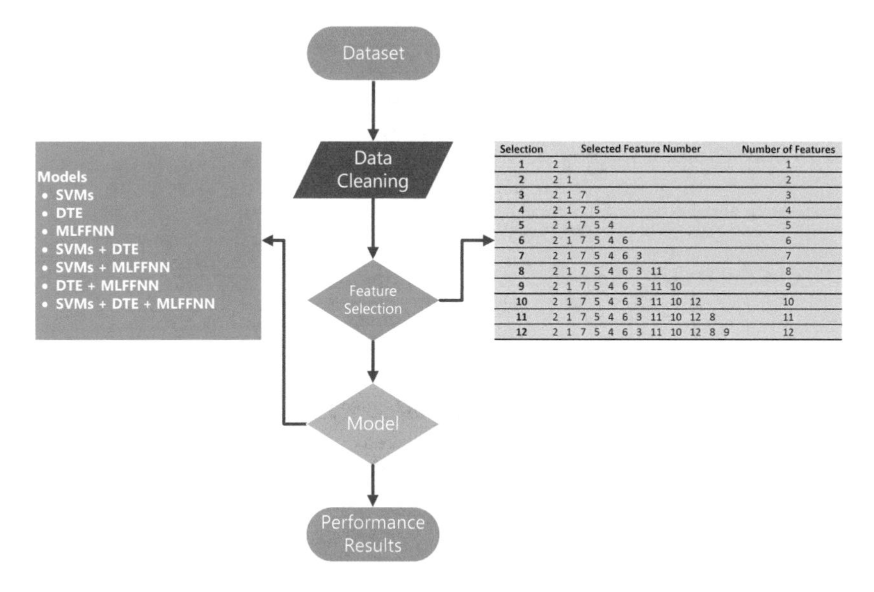

Fig. 2. Flow Diagram

2.1 Collection of Data

The data set used in this study was created with the daily air quality indexes measured and publicly published by the Government of India (Central Pollution Control Board) [19]. The data set includes an air quality index of 12 sensor data collected from various stations in 26 cities (Table 2). The values that cannot be read from the sensors were removed from the data set to improve the data quality. Then, 80% of the data was used to train the models, and the remaining 20% was used to test the models.

The statistical information of the data is presented in Table 3. In Table 4, the correlation levels of the features and the air quality index are given.

Table 2. Features

Feature Number	Feature Name
1	PM2.5
2	PM10
3	NO
4	NO2
5	NOx
6	NH3
7	CO
8	SO2
9	O3
10	Benzene
11	Toluene
12	Xylene

Table 3. Distribution of Data

Feature Number	Min	Max	Mean	Std	%95 CI		R	R2
					LB	UB		
1.00	1.09	734.56	52.48	43.10	51.65	53.31	0.90	0.803
2.00	5.77	830.10	108.49	67.88	107.18	109.80	0.92	0.839
3.00	0.02	262.00	12.20	18.88	11.84	12.57	0.37	0.139
4.00	0.01	254.78	33.19	22.91	32.75	33.63	0.55	0.302
5.00	0.02	331.50	29.91	28.19	29.37	30.46	0.56	0.313
6.00	0.10	269.93	17.10	12.74	16.85	17.35	0.41	0.172
7.00	0.00	4.74	0.70	0.44	0.69	0.71	0.61	0.374
8.00	0.10	67.26	9.91	7.93	9.75	10.06	0.27	0.073
9.00	0.03	162.33	32.22	19.94	31.84	32.61	0.21	0.043
10.00	0.00	165.41	4.43	13.42	4.17	4.69	0.33	0.107
11.00	0.00	259.03	12.12	23.77	11.66	12.58	0.34	0.116
12.00	0.00	133.60	2.86	7.05	2.73	3.00	0.29	0.085

Table 4. Relationship Between Features and AQI

Feature Number	Relationship Level with AQI
2	0.916
1	0.896
7	0.611
5	0.56
4	0.549
6	0.415
3	0.373
11	0.341
10	0.327
12	0.292
8	0.271
9	0.206

2.2 Feature Selection Algorithm

In this study, Spearman correlation coefficient and PCA methods were used as feature selection algorithm.

2.2.1 Spearman Correlation Coefficient

Correlation coefficients are a statistical measurement that provides information about the relationship between two variables. It states how the other variable reacts when we change one variable. If the variables are numerical values, Pearson and Spearman coefficients are used . It is stated as r_s for spearman correlation coefficient, d_{ii} for difference between observation order numbers, n for observation order number as in Eq. 2.

$$r_s = 1 - 6\sum_{i=1}^{n} \frac{{d_i}^2}{n \times (n^2 - 1)} \tag{2}$$

In cases where the number of observations is higher than 30 ($n > 30$), r_s's meaningfulness is evaluated by the value of t given in Eq. 3.

$$t = \frac{r_s}{\sqrt{\frac{1-{r_s}^2}{n-2}}} \tag{3}$$

The obtained t statistic is compared with the selected alpha value (in this study alpha = 0.05). In the case of $t < 0.05$, r_s is meaningful and usable [20].

The features were sorted according to the relationship levels and 12 feature groups were created (Table 5).

Table 5. Selected Features

Selection Group	Selected Features Number												Number of Features
1	2												1
2	2	1											2
3	2	1	7										3
4	2	1	7	5									4
5	2	1	7	5	4								5
6	2	1	7	5	4	6							6
7	2	1	7	5	4	6	3						7
8	2	1	7	5	4	6	3	11					8
9	2	1	7	5	4	6	3	11	10				9
10	2	1	7	5	4	6	3	11	10	12			10
11	2	1	7	5	4	6	3	11	10	12	8		11
12	2	1	7	5	4	6	3	11	10	12	8	9	12

2.2.2 Principal Component Analysis

PCA is a method that is used often to determine the components that make up the general, to reduce the number of variables, and to sort the observations objectively [20]. In this study, the main components calculated by PCA analysis are given in Table 6. From the results, it is seen that the 6 basic components can represent %95 of the entire data.

Table 6. Principal Component Analysis Results

PCA No	Coefficient	Cumulative Sum
1	71.5	71.5
2	10.7	82.2
3	4.7	86.9
4	4	90.9
5	3	93.9
6	2.7	96.6
7	1.5	98.1
8	0.8	98.9
9	0.6	99.5
10	0.3	99.8

2.3 Machine Learning

In this study, DTE, SVMs and MLFFNN algorithms were used as machine learning algorithms and hybrid structures of these algorithms were built (Table 7).

Table 7. Individual and Hybrid Models

	Machine Learning Algorithm
1	SVMs
2	DTE
3	MLFFNN
4	SVMs+DTE
5	SVMs+MLFFNN
6	DTE+MLFFNN
7	SVMs+DTE+MLFFNN

The reason for using these algorithms is that the training times are short and the accuracy rates are high [21–23]. In addition, DTE, SVMs and MLFFNN are widely used in many studies in the literature [24]. Hybrid models have been created with the basic algorithms used. An intermediate system has been created by taking common decisions from the basic models. For example, if the output of the DTE model is 6 and the output of the SVM's model is 10 for a data, the output of the DTE + SVM's hybrid model will give the value $(6+10)/2 = 8$ [25]. So, the average output of the main models will be considered as the output of the hybrid model. After all the structures were created, performance evaluations of the models were made. There were 10314 data available to use in machine learning models, 80% of them were used to train models and the remaining 20% were used to test models [26].

2.3.1 DTE

DTE is the process of combining trees to take advantage of the benefits of multiple decision trees of community methods. Single decision trees have advantages and disadvantages in many ways. A large number of decision trees can combine the advantages of single decision trees and reduce their disadvantages [23]. In the scope of this study, bagging and boosting methods were used as the methods of combining single trees and the model results that give the highest performance were presented (Fig. 3).

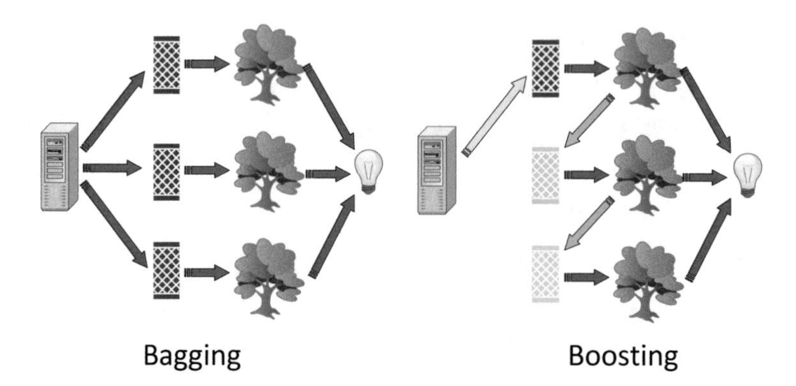

Fig. 3. Decision Tree Ensemble General Model

2.3.2 SVMs

SVMs is a supervised machine learning algorithm used for classification and regression [20,22,27,28]. This algorithm plots each sample as a point into an n-dimensional space with property values. The algorithm aims to fit the curve into the hyperplane so that it passes through the dataset. To be able to do this, support vectors are used. The maximum distance between the vectors is determined and a curve is placed between them. This curve forms the generalized solution of the data set Fig. 4.

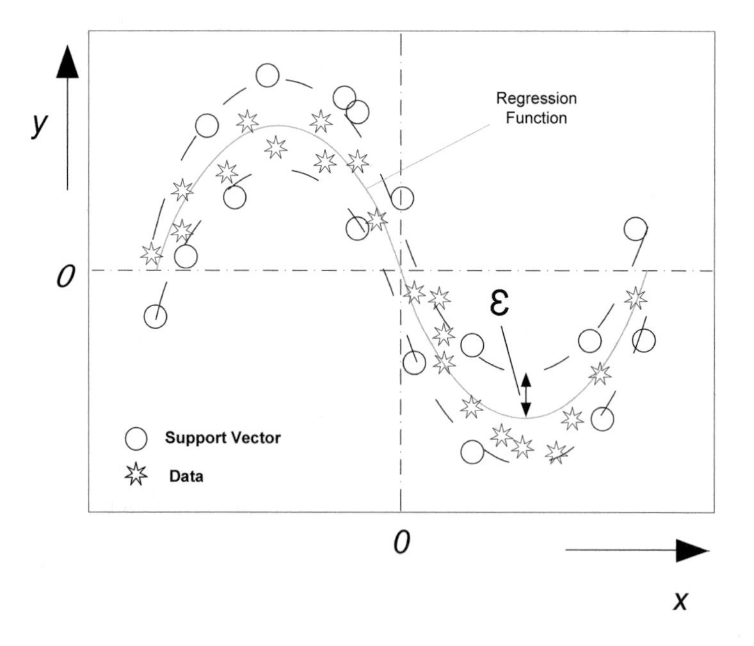

Fig. 4. SVMs General Model

2.3.3 MLFFNN

Artificial neural networks are computer architectures created by mathematically modeling the neural structures of the human brain [?]. Artificial neural networks are a system that consist of artificial nerve cells that are interconnected with each other. Multi-layered neural network is a type of feed-forward artificial neural network that consists of an input layer, an output layer, and a hidden layer between these two layers (Fig. 5) [21,29].

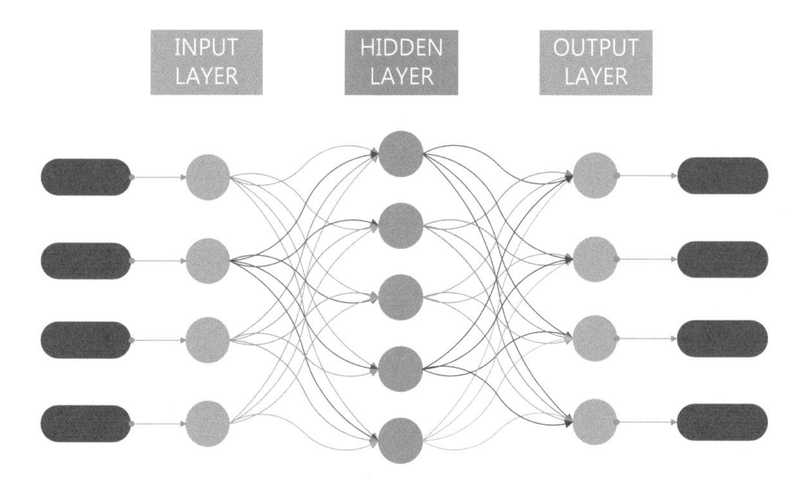

Fig. 5. MLFFNN General Network Structure

In this study, the activation function of MLFFNN, lambda, layer numbers and layer size parameters were optimized to create the best model.

2.4 Performance Evaluation Criteria

Models are trained by using 80% of the available data and then the remaining 20% of the data are used to test the models. The answers of the models to the test questions were called y, the real answers were called t, and the performance criteria were calculated. The criteria used are as follows; Mean Square Error (MSE), Root Mean Square Error ($RMSE$), Standard Error (SE), Correlation Coefficient (R), Explanatory Coefficient (R^2), Mean Absolute Difference (MAD), mean absolute percentage error ($MAPE$).

2.4.1 Raw Residues (e_i)

Raw remainder (e_i) is the difference between the actual value(t_i) and the estimation generated from the models (y_i) (Eq. 4). In the cases where e_i is close to zero means that the model works close to reality for the test data in use [20].

$$e_i = t_i - y_i \tag{4}$$

2.4.2 Mean Square Error (MSE)

The MSE, represents the average of the square of errors between the actual output and the model output (Eq. 5) [30]. Since MSE takes squares of sample errors, large differences in values give exaggerated results.

$$MSE = \frac{1}{n}\sum_{i=1}^{n} {e_i}^2 \tag{5}$$

2.4.3 Root Mean Square Error ($RMSE$)

The $RMSE$ represents the square root of the average squared errors (Eq. 6) [31]. MSE values may be too large to be easily evaluated. That is why $RMSE$ is easier to understand and evaluate than MSE.

$$RMSE = \sqrt{\frac{1}{n}\sum_{i=1}^{n} {e_i}^2} \tag{6}$$

2.4.4 Standart Error (SE)

The Standard Error indicates the compatibility of the generated models with the data [20]. The system cannot predict accurately when the correlation is less than 1. In this case, a deviance about e_i, between the estimations and actual values occur. The SE of the trained model is the standard deviation of the raw error (Eq. 7) [32].

$$SH = \sqrt{\frac{\sum_{i=1}^{n} e_i}{n-2}} \tag{7}$$

2.4.5 Error Rate ($MAD\&MAPE$)

The average absolute error is the absolute average of the difference between the model output and the actual value (Eq. 8). $MAPE$, on the other hand, indicates the percentage of the average absolute error relative to the actual output value (Eq. 9).

$$MAD = \frac{1}{n}\sum_{i=1}^{n} |t_i - y_i| \tag{8}$$

$$MAPE = \frac{1}{n}\sum_{i=1}^{n} \frac{|t_i - y_i|}{t_i} \times 100 \tag{9}$$

2.4.6 The Correlation Coefficient

The correlation coefficient can be used in numerical variables as it is used in classification problems. Since the model outputs are continuous variables, Pearson (r) or Spearman (r_s) correlation coefficients are used. When selecting these two calculation types, it is checked whether the data is normally distributed or not [20]. If the data is normally distributed, Pearson correlation coefficient will be used, otherwise the Spearman correlation coefficient (r_s) will be used.

2.4.7 The Relationship Between Correlation and Estimation

If the correlation coefficient between the two variables of interest is $|r| < 0.70$, the model estimation error rate will be high. If the correlation coefficient is $|r| > 0.90$, it means the model estimation is high, if the correlation coefficient is between 0.7 and 0.5, it means the system estimation is low.

2.4.8 Explanatory Coefficient (R^2)

R^2 is the percentage of the total change in the regression model (Eq. 10) [20]. In other words, how much of the total change in the dependent variable R^2 can be determined by the independent variable.

$$R^2 = \frac{KT_R}{KT_Y} \tag{10}$$

In the R^2 equation: Model outputs (yi), real values (t_i), number of data (n), sum of T squares (KT_T), sum of R squares (KT_R), sum of A squares (KT_A).

$$KT_T = KT_R + KT_A \tag{11}$$

$$KT_T = \sum_{i=1}^{n} (t_i - \bar{t})^2 = \sum_{i=1}^{n} t_i^2 - \frac{\left(\sum_{i=1}^{n} t_i\right)^2}{n} \tag{12}$$

$$KT_R = \sum_{i=1}^{n} (y_i - \bar{t})^2 \tag{13}$$

$$KT_A = \sum_{i=1}^{n} (t_i - y_i)^2 \tag{14}$$

2.5 Statistical Analysis

The relationship between the sensor values found in the data set and the air quality index corresponding to these values were statistically analyzed. Spearman correlation coefficients were used during the analysis. Correlation values were agreed as significant If $t < 0 : 05$ (Sect. 2.2.1).

3 Results

In this study, it is aimed to create a machine learning model that can estimate the air quality index with a high accuracy rate by reducing the number of sensor data used to calculate this index. A real data set published by the Government of India was used in the study. The data set contains 12 different characteristics and air quality indexes obtained from sensors. In this study, feature selection and sorting algorithms were used, in which existing features were evaluated in a relational manner, and 12 different data sets were created. In addition, Principal Component Analysis has been applied, which is used to reduce the number of features. In order to estimate the air quality index with these data sets, 7 different machine learning methods were used. In order to compare the estimation results of the models, performance evaluation criteria were calculated.

Even if the calculated performance evaluation criteria differ according to the feature level selected by applying the Spearman correlation coefficient, the most successful models were obtained in trainings by using all the features. However, there are no major differences between the success achieved after level 2 used and the success where all features are used in the model. It is observed that different machined learning models which were trained at the same feature selection levels, show different performance (Table 8, 9, 10 and 11).

PCA analysis, which is another method used to reduce the number of properties, was used to obtain 6 components. With this data set obtained, the models show different performances (Table 12). The most successful model obtained with 6 components was the SVMs + DTS + MLFFNN hybrid model. It was found that the models trained by applying PCA analysis were more successful than the models with feature selection.

Performance evaluation metrics of all models present in the Fig. 6. One of the criteria calculated to evaluate the performance of models is R^2 and it is expected to be close to 1, while other parameters are expected to be close to 0. A substantial increase is seem in model performance with the first feature group when the performance metrics were analyzed according to the selected features. As the number of selected features increased, the model performance continued to increase. SVMs + DTE + MLFFNN model which consists of 12th feature group showed the highest performance. However, although the MLFFNN model created with the 2nd feature group contains less features, it showed approximately the same performance values as the 12th feature group.

In this study, the relationships between the air quality index and the 12 values read from the sensors were statistically examined and the correlation values are present in the Fig. 7. The calculated t-value is less than 0.05, and the correlation coefficient varies between 0.21 and 0.92. According to these values, it is seen that there is a relationship between the data and the air quality index.

Table 8. Performance Evaluation 1

Feature Selection Level 1							
Feature Numbers	2						
Model	Performance Evaluation Criteria						
	MSE	RMSE	SH	Mape	MAD	R	R2
SVMs	943.25	30.71	30.73	15.87	18.71	0.91	0.83
DTE	919.86	30.33	30.34	16.89	19.30	0.91	0.83
MLFFNN	887.44	**29.79**	29.80	17.44	19.35	0.91	0.83
SVMs+DTE	920.50	30.34	30.35	16.29	18.88	0.91	0.83
SVMs+MLFFNN	904.14	30.07	30.08	16.49	18.85	0.91	0.83
DTE+MLFFNN	896.48	29.94	29.96	17.10	19.23	0.91	0.83
SVMs+DTE+MLFFNN	903.77	30.06	30.08	16.59	18.94	0.91	0.83
Feature Selection Level 2							
Feature Numbers	2, 1						
Model	Performance Evaluation Criteria						
	MSE	RMSE	SH	Mape	MAD	R	R2
SVMs	434.54	20.85	20.86	13.34	13.65	0.93	0.86
DTE	454.37	21.32	21.33	14.64	14.61	0.93	0.86
MLFFNN	423.19	**20.57**	20.58	14.22	13.99	0.93	0.86
SVMs+DTE	431.36	20.77	20.78	13.80	13.89	0.93	0.86
SVMs+MLFFNN	424.71	20.61	20.62	13.69	13.72	0.93	0.86
DTE+MLFFNN	429.15	20.72	20.73	14.30	14.10	0.93	0.86
SVMs+DTE+MLFFNN	425.42	20.63	20.64	13.88	13.84	0.93	0.86
Feature Selection Level 3							
Feature Numbers	2, 1, 7						
Model	Performance Evaluation Criteria						
	MSE	RMSE	SH	Mape	MAD	R	R2
SVMs	423.24	20.57	20.58	12.61	13.29	0.93	0.87
DTE	436.12	20.88	20.89	13.66	13.93	0.93	0.87
MLFFNN	404.71	**20.12**	20.13	13.42	13.58	0.93	0.87
SVMs+DTE	413.36	20.33	20.34	12.90	13.34	0.93	0.87
SVMs+MLFFNN	406.49	20.16	20.17	12.90	13.31	0.93	0.87
DTE+MLFFNN	413.09	20.32	20.33	13.46	13.66	0.93	0.87
SVMs+DTE+MLFFNN	407.52	20.19	20.20	13.04	13.38	0.93	0.87

Table 9. Performance Evaluation 2

Feature Selection Level 4							
Feature Numbers	2, 1, 7, 5						
Model	Performance Evaluation Criteria						
	MSE	RMSE	SH	Mape	MAD	R	R2
SVMs	421.50	20.53	20.54	12.42	13.24	0.94	0.88
DTE	422.42	20.55	20.56	13.34	13.71	0.94	0.87
MLFFNN	421.64	20.53	20.54	13.40	13.67	0.94	0.87
SVMs+DTE	405.53	20.14	20.15	12.64	13.19	0.94	0.88
SVMs+MLFFNN	410.66	20.26	20.27	12.73	13.24	0.94	0.88
DTE+MLFFNN	411.62	20.29	20.30	13.23	13.53	0.94	0.88
SVMs+DTE+MLFFNN	405.08	**20.13**	20.14	12.80	13.24	0.94	0.88
Feature Selection Level 5							
Feature Numbers	2, 1, 7, 5, 4						
Model	Performance Evaluation Criteria						
	MSE	RMSE	SH	Mape	MAD	R	R2
SVMs	427.19	20.67	20.68	12.39	13.30	0.94	0.88
DTE	424.71	20.61	20.62	13.53	13.78	0.94	0.87
MLFFNN	645.43	25.41	25.42	15.83	17.39	0.93	0.86
SVMs+DTE	405.21	**20.13**	20.14	12.69	13.18	0.94	0.88
SVMs+MLFFNN	452.87	21.28	21.29	13.47	14.44	0.94	0.88
DTE+MLFFNN	482.60	21.97	21.98	14.34	15.09	0.93	0.87
SVMs+DTE+MLFFNN	429.49	20.72	20.73	13.35	14.01	0.94	0.88
Feature Selection Level 6							
Feature Numbers	2, 1, 7, 5, 4, 6						
Model	Performance Evaluation Criteria						
	MSE	RMSE	SH	Mape	MAD	R	R2
SVMs	453.57	21.30	21.31	12.49	13.46	0.94	0.88
DTE	407.56	20.19	20.20	13.55	13.64	0.93	0.87
MLFFNN	402.06	20.05	20.06	13.38	13.53	0.94	0.88
SVMs+DTE	400.57	20.01	20.02	12.73	13.18	0.94	0.88
SVMs+MLFFNN	395.42	19.89	19.89	12.68	13.15	0.94	0.88
DTE+MLFFNN	391.09	19.78	19.79	13.27	13.33	0.94	0.88
SVMs+DTE+MLFFNN	387.24	**19.68**	19.69	12.79	13.09	0.94	0.88

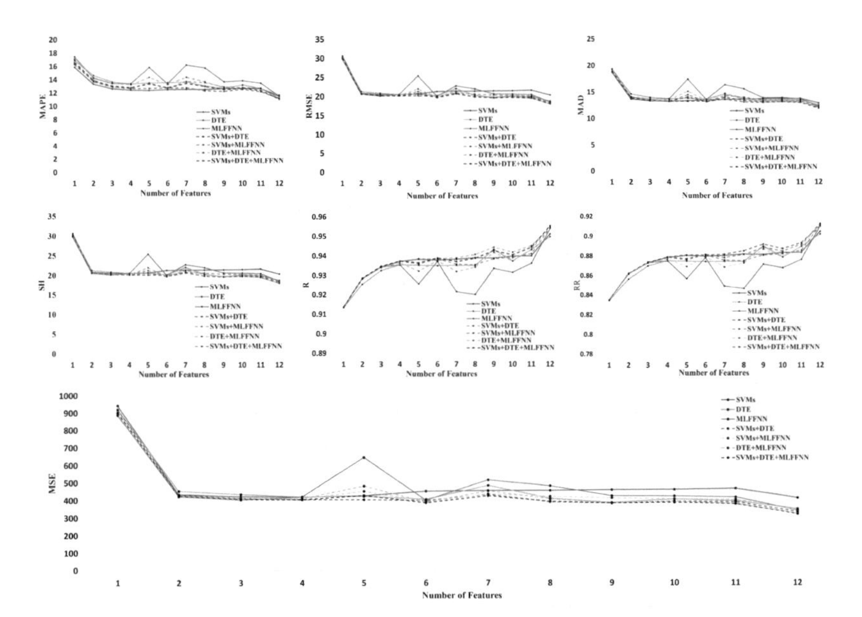

Fig. 6. Performance Metrics for All Models

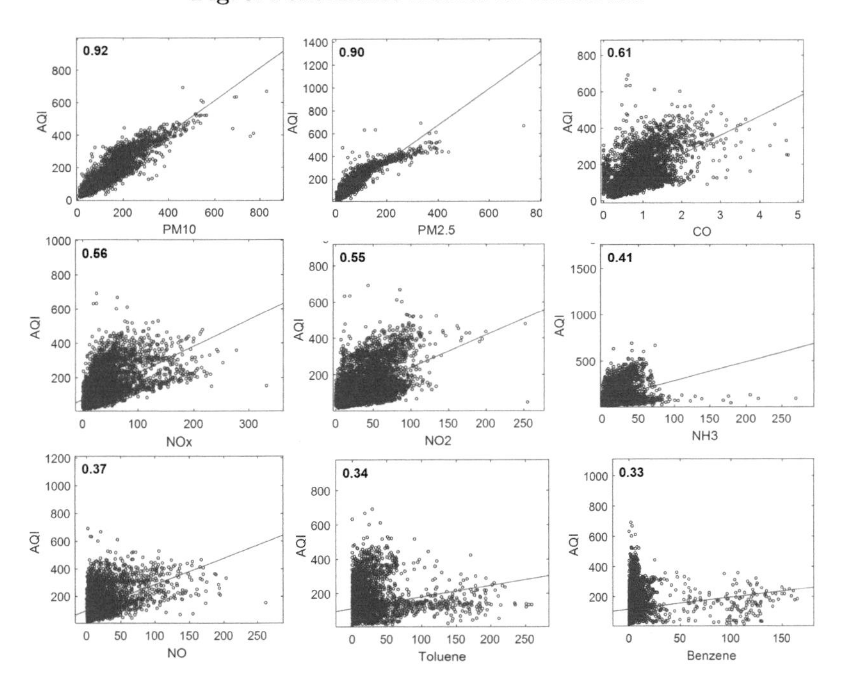

Fig. 7. Statistical Analysis of the Relationship Between AQI and Pollutants

Table 10. Performance Evaluation 3

Feature Selection Level 7							
Feature Numbers	2, 1, 7, 5, 4, 6, 3						
Model	Performance Evaluation Criteria						
	MSE	RMSE	SH	Mape	MAD	R	R2
SVMs	456.73	21.37	21.38	12.55	13.53	0.94	0.88
DTE	485.84	22.04	22.05	13.68	14.54	0.94	0.87
MLFFNN	517.47	22.75	22.76	16.21	16.33	0.92	0.85
SVMs+DTE	434.77	20.85	20.86	12.79	13.58	0.94	0.88
SVMs+MLFFNN	436.86	20.90	20.91	13.74	14.23	0.94	0.88
DTE+MLFFNN	454.88	21.33	21.34	14.36	14.70	0.93	0.87
SVMs+DTE+MLFFNN	427.33	**20.67**	20.68	13.44	13.93	0.94	0.88
Feature Selection Level 8							
Feature Numbers	2, 1, 7, 5, 4, 6, 3, 11						
Model	Performance Evaluation Criteria						
	MSE	RMSE	SH	Mape	MAD	R	R2
SVMs	457.17	21.38	21.39	12.52	13.59	0.94	0.88
DTE	412.63	20.31	20.32	13.03	13.64	0.94	0.87
MLFFNN	483.38	21.99	22.00	15.75	15.55	0.92	0.85
SVMs+DTE	392.44	**19.81**	19.82	12.31	13.05	0.94	0.88
SVMs+MLFFNN	424.27	20.60	20.61	13.56	13.95	0.94	0.88
DTE+MLFFNN	409.61	20.24	20.25	13.68	13.84	0.93	0.87
SVMs+DTE+MLFFNN	394.68	19.87	19.88	12.97	13.38	0.94	0.88
Feature Selection Level 9							
Feature Numbers	2, 1, 7, 5, 4, 6, 3, 11, 10						
Model	Performance Evaluation Criteria						
	MSE	RMSE	SH	Mape	MAD	R	R2
SVMs	460.29	21.45	21.46	12.66	13.69	0.94	0.88
DTE	387.59	19.69	19.70	12.68	13.26	0.94	0.89
MLFFNN	426.68	20.66	20.67	13.69	13.86	0.93	0.87
SVMs+DTE	384.30	**19.60**	19.61	12.24	12.96	0.94	0.89
SVMs+MLFFNN	414.32	20.35	20.36	12.91	13.46	0.94	0.88
DTE+MLFFNN	385.88	19.64	19.65	12.80	13.14	0.94	0.89
SVMs+DTE+MLFFNN	384.82	19.62	19.63	12.52	13.04	0.94	0.89

Table 11. Performance Evaluation 4

Feature Selection Level 10							
Feature Numbers	2, 1, 7, 5, 4, 6, 3, 11, 10, 12						
Model	Performance Evaluation Criteria						
	MSE	RMSE	SH	Mape	MAD	R	R2
SVMs	461.97	21.49	21.50	12.66	13.72	0.94	0.88
DTE	407.41	20.18	20.19	13.22	13.52	0.94	0.88
MLFFNN	424.95	20.61	20.62	13.84	13.89	0.93	0.87
SVMs+DTE	394.08	19.85	19.86	12.53	13.10	0.94	0.89
SVMs+MLFFNN	408.52	20.21	20.22	12.79	13.29	0.94	0.88
DTE+MLFFNN	394.79	19.87	19.88	13.17	13.35	0.94	0.88
SVMs+DTE+MLFFNN	388.36	**19.71**	19.72	12.69	13.09	0.94	0.88
Feature Selection Level 11							
Feature Numbers	2, 1, 7, 5, 4, 6, 3, 11, 10, 12, 8						
Model	Performance Evaluation Criteria						
	MSE	RMSE	SH	Mape	MAD	R	R2
SVMs	467.41	21.62	21.63	12.63	13.75	0.94	0.88
DTE	396.67	19.92	19.93	12.66	13.46	0.94	0.89
MLFFNN	418.82	20.47	20.47	13.47	13.68	0.94	0.88
SVMs+DTE	391.12	19.78	19.79	12.19	13.05	0.95	0.89
SVMs+MLFFNN	404.53	20.11	20.12	12.72	13.28	0.94	0.89
DTE+MLFFNN	381.12	19.52	19.53	12.62	13.11	0.94	0.89
SVMs+DTE+MLFFNN	380.47	**19.51**	19.52	12.36	12.96	0.94	0.89
Feature Selection Level 12							
Feature Numbers	2, 1, 7, 5, 4, 6, 3, 11, 10, 12, 8, 9						
Model	Performance Evaluation Criteria						
	MSE	RMSE	SH	Mape	MAD	R	R2
SVMs	414.06	20.35	20.36	11.47	12.87	0.95	0.90
DTE	351.88	18.76	18.77	11.66	12.48	0.95	0.90
MLFFNN	342.08	18.50	18.50	11.51	12.24	0.95	0.91
SVMs+DTE	334.43	18.29	18.30	11.13	12.09	0.95	0.91
SVMs+MLFFNN	345.44	18.59	18.60	11.18	12.13	0.95	0.91
DTE+MLFFNN	323.94	18.00	18.01	11.16	11.91	0.95	0.91
SVMs+DTE+MLFFNN	323.02	17.97	17.98	11.01	11.86	0.96	0.91

Table 12. Performance Evaluation for PCA

PCA							
PCA	6 main component						
Model	Performance Evaluation Criteria						
	MSE	RMSE	SH	Mape	MAD	R	R2
SVMs	378.59	19.46	19.47	11.72	12.86	0.95	0.90
DTE	393.14	19.83	19.84	13.12	13.58	0.95	0.89
MLFFNN	372.61	19.30	19.31	12.17	12.94	0.95	0.90
SVMs+DTE	356.79	18.89	18.90	11.94	12.64	0.95	0.90
SVMs+MLFFNN	345.76	18.59	18.60	11.57	12.43	0.95	0.90
DTE+MLFFNN	345.34	18.58	18.59	12.09	12.55	0.95	0.90
SVMs+DTE+MLFFNN	338.58	18.40	18.41	11.70	12.33	0.95	0.90

4 Discussion

In the previous sections, detailed analyses of the performances of the estimation models created by using machine learning have been described. In this section, the performances of the models based on the data set obtained in real time will be compared with the literature. With the worldwide increase in the number of academic studies and experts in the field, AQI estimation studies are also increasing [18,33]. It costs a lot to equip cities with very large electronic monitoring networks for AQI detection. It is impossible to use as much sensors as the number of pollutants in the air. Breathing in and out of a healthy air is a universal right for all living creatures. Considering the number of people who lose their lives due to bad weather conditions every year, the importance of AQI calculations is much better understood [5]. With the method presented in this study, AQI calculations can be performed with a high accuracy rate and in a much more economical way. The calibration requirement, which is a basic requirement for almost the majority of sensors, increases the operating expenses of the system. It also reduces the reliability. With the machine learning based estimation models presented in this study, it is possible to calculate AQI with a high performance and by using a minimum number of sensors. It has been observed that the number of data set is low and insufficient in some of the studies encountered in the literature [34]. In other studies, the entire data set was used when creating estimation models, but it is impossible to test the estimation performance of the model in this way. In some studies where it is said that feature selection was performed, no feature selection algorithm was actually used and only the data set was subjected to a preliminary editing process [18]. In the MLFFNN method used in this study, a $RMSE$ value of 20.57 was obtained with 2 sensor information. In a study where similar data were used in the literature, a $RMSE$ value of 24.462 was obtained with the proposed hybrid model [35]. The most prominent feature of our study is that it exhibits the same estimation success as the attribute selection algorithms (Spearman-PCA) but with much less transaction volume. Although the data set used in some of the studies in the literature seems to be long-termed, since they contain a small number of samples they are not expected to give healthy results [12]. Again, in some studies, hybrid models have been tried to be applied, however, the number of crossed machine learning algorithms has been limited to only two models (Support Vector Machine-Random Forest) [16]. In this study, we created only four hybrid models besides the individual models and achieved high accuracy rates. In this study, hybrid prediction models with different feature levels were created with the methods presented and the performances they showed were examined. Based on this, it is obvious that the study contains more detailed analysis. The $MAPE$ value of the best model proposed by Zhu has managed to reach 15.60%. The $MAPE$ value of 14.22% obtained in the model proposed in this article was obtained with the MLFFNN model which uses only two features (Table 8). It shows that the model used in this study results in a higher accuracy than similar studies in the literature. It has been observed that the performance value obtained from each model is better than the best performance of the hybrid model proposed by Zhu [35]. In this study, there is a very small performance difference between the models created by PCA analysis and the models created by using correlation coefficients. Since the number of input variables used in the models obtained by PCA analysis is large, the models with fewer features gave more economical results.

5 Conclusion

Various countries use different calculation methods in AQI calculations. However, the EPA method, the principles of which are explained in the introduction, is the gold standard. It is quite laborious that the technique requires regional epidemiological studies for each pollutant to be included in the calculation. Basically, although 6 different pollutant parameters are used in AQI calculations, other contaminants in the region are also included in the calculation. It is almost impossible to use various sensors for the measurement of each pollutant when all pollutants are considered. At the same time, it does not seem possible for all countries to have such a widespread monitoring network soon due to both high initial investment costs and ongoing maintenance costs. It is impossible to continue measurement without calibration, as the nature of sensor technology introduces the need for calibration. The importance of calibration causes even healthy measurements to be approached with suspicion from time to time. Therefore, estimating with the measurement of much fewer pollutant variables will provide significant benefits in air pollution monitoring and AQI calculations. It is a more practical and rational approach to using artificial intelligence-based systems in measures, which provide successful results with high accuracy rates in prediction models thanks to its mathematical foundations. In this study, hybrid machine learning-based models were produced by using the measurements of 12 variables collected for AQI calculations. As a result of the study, a model has been proposed that can successfully predict the air quality index with only two pollutant species data. As a result of the analysis, a system was created that can predict the air quality index with only $PM_{2.5}$ and PM_{10} measurements. The proposed forecasting model has many benefits. These can be listed as; (1) reducing the number of sensors, (2) reducing initial installation and maintenance costs, (3) reducing the infrastructure time required for measurement (4) that city centers are more sustainable living spaces. The correlation value of the model proposed in this study was recorded as 0.93. This calculated value proves that using the proposed forecasting model is possible. It also contains compatibility with studies in the literature.

Conflict of Interests. There is no conflict of interest between the authors.

Financial Support. No support was received for this research.

Ethics Committee Approval. Ethics committee approval is not required as the open source data set was used in the study.

Data Collection and Use Permission. .

References

1. Yolsal, H.: Estimation of the air quality trends in Istanbul. İktisadi İdari Bilimler Dergisi **38**, 375 (2016)
2. Khuda, K.-E.: Causes of air pollution in Bangladesh's capital city and its impacts on public health. Nat. Environ. Pollut. Technol. **19**, 1483–1490 (2020)
3. Yilmaz, F.: Hava kirliliği, bileşenleri ve sağlık. FSM İlmî Araştırmalar İnsan ve Toplum Bilimleri Dergisi 231–250 (2021)
4. Pope, C.A., Dockery, D.W.: Health effects of fine particulate air pollution: lines that connect. J. Air Waste Manag. Assoc. **56**, 709–742 (2006)
5. Who air quality guidelines for particulate matter, ozone, nitrogen dioxide and sulfur dioxide (2006)
6. Maji, S., Ahmed, S., Ghosh, S., Garg, S.K.: Evaluation of air quality index for air quality data interpretation in Delhi, India (2020)
7. History of air pollution | US EPA
8. AQI basics | airnow.gov
9. Kuznetsova, I.N., Tkacheva, Yu.V., Shalygina, I.Yu., Lezina, E.A.: Calculation of air quality index and assessment of its informativeness for Russia based on monitoring data for Moscow. Russ. Meteorol. Hydrol. **46**, 530–538 (2021)
10. Plaia, A., Ruggieri, M.: Air quality indices: a review. Rev. Environ. Sci. Biotechnol. **10**, 165–179 (2011)
11. How is the AQI calculated? | US EPA
12. Wood, D.A.: Local integrated air quality predictions from meteorology (2015 to 2020) with machine and deep learning assisted by data mining. Sustain. Anal. Model. **2**, 100002 (2022)
13. Du, W., et al.: Deciphering urban traffic impacts on air quality by deep learning and emission inventory. J. Environ. Sci. **124**, 745–757 (2023)

14. Lightstone, S.D., Moshary, F., Gross, B.: Comparing CMAQ forecasts with a neural network forecast model for PM2.5 in New York. Atmosphere **8**(161), 2017 (2017)
15. Cabaneros, S.M., Calautit, J.K., Hughes, B.R.: A review of artificial neural network models for ambient air pollution prediction. Environ. Modell. Softw. **119**, 285–304 (2019)
16. Liu, H., Li, Q., Yu, D., Gu, Y.: Air quality index and air pollutant concentration prediction based on machine learning algorithms. Appl. Sci. **9**, 4069 (2019)
17. Cabaneros, S.M.S., Calautit, J.K.S., Hughes, B.R.: Hybrid artificial neural network models for effective prediction and mitigation of urban roadside NO2 pollution. Energy Procedia **142**, 3524–3530 (2017)
18. Kekulanadara, K.M.O.V.K., Kumara, B.T.G.S., Kuhaneswaran, B.: Machine learning approach for predicting air quality index. In: 2021 International Conference on Decision Aid Sciences and Application, DASA 2021, pp. 622–626 (2021)
19. CPCB | central pollution control board
20. Mishra, S., Datta-Gupta, A.: Applied statistical modeling and data analytics: a practical guide for the petroleum geosciences (2018)
21. Bebis, G., Georgiopoulos, M.: Feed-forward neural networks. IEEE Potent. **13**, 27–31 (1994)
22. Vogt, M., Kecman, V.: Active-set methods for support vector machines. In: Wang, L. (ed.) Support Vector Machines: Theory and Applications. Studies in Fuzziness and Soft Computing, vol. 177, pp. 133–158. Springer, Heidelberg (2005). https://doi.org/10.1007/10984697_6
23. Breiman, L., Friedman, J.H., Olshen, R.A., Stone, C.J.: Classification and regression trees. Classif. Regression Trees 1–358 (2017)
24. Abiri, O., Twala, B.: Modelling the flow stress of alloy 316l using a multi-layered feed forward neural network with Bayesian regularization. In: IOP Conference Series: Materials Science and Engineering, vol. 225, p. 9 (2017)
25. Uçar, M.K., Uçar, Z., Köksal, F., Daldal, N.: Estimation of body fat percentage using hybrid machine learning algorithms. Meas.: J. Int. Meas. Confederation, **167** (2021)
26. Tanabe, K.: Pareto's 80/20 rule and the gaussian distribution. Phys. A Stat. Mech. Appl. **510**, 635–640 (2018)
27. Flake, G.W., Lawrence, S.: Efficient SVM regression training with SMO. Mach. Learn. **46**, 271–290 (2002)
28. Cherkassky, V., Ma, Y.: Practical selection of SVM parameters and noise estimation for SVM regression. Neural Netw. **17**, 113–126 (2004)
29. Polat, K., Koc, K.O.: Detection of skin diseases from dermoscopy image using the combination of convolutional neural network and one-versus-all (2020)
30. On the mean squared error of an estimator | introduction to probability | supplemental resources | mit opencourseware
31. Root mean square error (RMSE) | cros
32. Standard error | what it is, why it matters, and how to calculate
33. Rahimpour, A., Amanollahi, J., Tzanis, C.G.: Air quality data series estimation based on machine learning approaches for urban environments. Air Qual. Atmos. Health **14**, 191–201 (2021)
34. Ausati, S., Amanollahi, J.: Assessing the accuracy of ANFIS, EEMD-GRNN, PCR, and MLR models in predicting PM2.5. Atmos. Environ. **142**, 465–474 (2016)
35. Zhu, S., Lian, X., Liu, H., Hu, J., Wang, Y., Che, J.: Daily air quality index forecasting with hybrid models: a case in China. Environ. Pollut. **231**, 1232–1244 (2017)

A New Epidemic Model with Direct and Indirect Transmission with Delay and Diffusion

Fatiha Najm[1], Radouane Yafia[1(✉)], Ahmed Aghriche[2], and M. A. Aziz Alaoui[3]

[1] Department of Mathematics, Faculty of Sciences, Ibn Tofail University, Campus Universitaire, BP 133, Kénitra, Morocco
{fatiha.najm,radouane.yafia}@uit.ac.ma

[2] Department of Mathematics, Faculty of Sciences, Ibn Tofail University, Campus Universitaire, Kénitra, Morocco

[3] Normandie Univ, Le Havre, ULH, LMAH, 76600, FR-CNRS-3335, ISCN, 25 rue Ph. Lebon, 76600 Le Havre, France
aziz.alaoui@univ-lehavre.fr

Abstract. The aim of this paper is to study the dynamics of a delayed reaction-diffusion epidemic model with direct and indirect transmission modes. It's well known that, direct contact transmission occurs when there is physical contact between an infected person and a susceptible person and indirect contact transmission occurs when there is no direct human-to-human contact. We study the model in terms of existence, positivity, and boundedness of solutions for a reaction-diffusion system with homogeneous Neumann boundary conditions. The asymptotic behaviours of the disease-free equilibrium and endemic equilibrium are proved with respect to the basic reproduction number R_0. Some numerical simulations are carried out to illustrate our theoretical results. We deduce that, there is no effect of diffusion terms and time delays on the asymptotic behaviours of equilibria and there is no occurrence of Turing instability or Hopf bifurcation.

Keywords: Epidemic model · latency periods · DDE · basic reproduction number R_0 · stability · reaction-diffusion

1 Introduction and Mathematical Model

Viruses are responsible for causing many diseases, including: AIDS, Common cold, Ebola, Genital herpes, Influenza, Measles, Chickenpox, shingles and recently Coronavirus disease 2019 (COVID-19).... [5,7,10,17]. These diseases are transmitted via two main routes: direct and indirect transmission. Direct transmission is caused when there is physical contact between an infectious person and a susceptible person. While indirect transmission involves contact between a person and a contaminated object. This is often a result of unclean hands contaminating an object or environment liken. The microorganism remains on this

D. J. Hemanth et al. (Eds.): ICAIAME 2022, ECPSCI 7, pp. 582–594, 2023.
https://doi.org/10.1007/978-3-031-31956-3_49

surface to be picked up by the next person who touches it. Transmission occurs when droplets containing microorganisms generated during coughing, sneezing and talking are propelled through the air. These microorganisms land on another person, entering that new person's system through contact with his/her conjunctivae, nasal mucosa or mouth. These microorganisms are relatively large and travel only short distances (up to 6 ft/2 m). However these infectious droplets may linger on surfaces for long periods of time [10,17] (Table 1). To model this situation, we consider a delayed reaction diffusion system which takes into account the effect of direct and indirect transmission and the effect of latency periods caused by the two modes of transmission. The system is given as follows:

Table 1. Table of Survival Times of Microorganisms on Surfaces

Organism	Survival Time
Adenovirus	Up to 3 months
Clostridium difficile	Up to 5 months
Coronovirus	3 h
E. coli	Up to 16 months
Influenza	1–2 days
MRSA	Up to 7 months
M. tuberculosis	Up to 4 months
Norovirus	Up to 7 days
RSV	Up to 6 h

$$\begin{cases} \frac{\partial S(t,x,y)}{\partial t} = d_S \triangle S + \Lambda - \beta_s S(t,x,y)I(t,x,y) - \beta_W S(t,x,y)W(t,x,y) - \mu_s S(t,x,y) \\ \frac{\partial I(t,x,y)}{\partial t} = d_I \triangle I + \beta_s S(t,x,y)I(t,x,y) + \beta_W S(t,x,y)W(t,x,y) - (\gamma + \mu_I)I(t,x,y) \\ \frac{\partial R(t,x,y)}{\partial t} = d_R \triangle R + \gamma I(t,x,y) - \mu_R R(t,x,y) \\ \frac{\partial W(t,x,y)}{\partial t} = \mu_W I(t,x,y) - \varepsilon W(t,x,y) \\ S(0,x,y) = S_0(x,y) \geq 0, I(0,x,y) = I_0(x,y) \geq 0, \\ R(0,x,y) = R_0(x,y) \geq 0, W(0,x,y) = W_0(x,y) \geq 0. \end{cases} \tag{1}$$

We study system (1) with the following homogenous Neumann boundary conditions:

$$\frac{\partial S}{\partial \eta} = \frac{\partial I}{\partial \eta} = \frac{\partial R}{\partial \eta} = \frac{\partial W}{\partial \eta} = 0 \quad \text{on} \quad \partial\Omega \times (0, +\infty), \tag{2}$$

where $S(t,x,y)$, $I(t,x,y)$ and $R(t,x,y)$ are the total number of susceptible, infectious and recovered subpopulation at location (x,y) and time t, respectively. $W(x,y,t)$ is the virus concentration caused by humans resulting by coughing and sneezing. Ω is a bounded domain in $\mathbb{R}^2$ with smooth boundaries $\partial\Omega$, with Neumann boundary conditions. η is the unit outer normal to $\partial\Omega$. The diffusion symbol $\triangle = \frac{\partial^2}{\partial x^2} + \frac{\partial^2}{\partial y^2}$ denotes the Laplacian operator and the positive constants d_S, d_I and d_R are the diffusion coefficients of susceptible, infectious and recovered respectively. All parameters of the model are positive and defined in Table 2.

Table 2. Parameters definition

Parameters	Epidemiological interpretation
Λ	Birth rate parameter of S population
β_s	Direct transmission rate
β_W	Indirect transmission rate
μ_s	Death rate of susceptible population
γ	Recovery rate
μ_I	Death rate of infectious population
μ_R	Death rate of recovered population
μ_W	Shedding coefficients
$\frac{1}{\varepsilon}$	Lifetime of the virus
τ	Latency period
ν	Time needed for a population S to become infectious by coughing and sneezing

As the equation of recovered population R depends only on infectious population I, it suffices to study the following reduced system

$$\begin{cases} \frac{\partial S(t,x,y)}{\partial t} = d_S \triangle S + \Lambda - \beta_s S(t,x,y)I(t,x,y) - \beta_W S(t,x,y)W(t,x,y) - \mu_s S(t,x,y) \\ \frac{\partial I(t,x,y)}{\partial t} = d_I \triangle I + \beta_s S(t,x,y)I(t,x,y) + \beta_W S(t,x,y)W(t,x,y) - (\gamma + \mu_I)I(t,x,y) \\ \frac{\partial W(t,x,y)}{\partial t} = \mu_W I(t,x,y) - \varepsilon W(t,x,y) \\ S(0,x,y) = S_0(x,y) \geq 0, I(0,x,y) = I_0(x,y) \geq 0, W(0,x,y) = W_0(x,y) \geq 0. \end{cases} \tag{3}$$

with the following homogenous Neumann boundary conditions:

$$\frac{\partial S}{\partial \eta} = \frac{\partial I}{\partial \eta} = \frac{\partial W}{\partial \eta} = 0 \quad \text{on} \quad \partial\Omega \times (0, +\infty). \tag{4}$$

The rest of the paper is organized as follows. The next section deals with the global existence, positivity, and boundedness of solutions of problem (3). In Sect. 3, we discuss the stability analysis of equilibria. In Sect. 4, we present the numerical simulation to illustrate our result.

2 Existence, Positivity, and Boundedness of Solutions

In this section, we establish the global existence, positivity, and boundedness of solutions of problem (3) because this model describes the population. Hence, the population should remain nonnegative and bounded.

Proposition 1. *Each solution of system* (3) *initiating from a non-negative initial condition remains non-negative and bounded for all* $t \geq 0$.

Proof. System (3) can be written abstractly in the Banach space $X = C(\bar{\Omega}) \times C(\bar{\Omega})$ of the form

$$\begin{cases} u'(t) = Au(t) + F(u(t)), t > 0 \\ u(0) = u_0 \in X, \end{cases} \tag{5}$$

where $u = \begin{pmatrix} S \\ I \\ W \end{pmatrix}$, $u_0 = \begin{pmatrix} S_0 \\ I_0 \\ W_0 \end{pmatrix}$ and $Au(t) = \begin{pmatrix} d_S \bigtriangleup S \\ d_I \bigtriangleup I \\ 0 \end{pmatrix}$, and

$$F(u(t)) = \begin{pmatrix} \Lambda - \beta_s SI - \beta_W SW - \mu_s S \\ \beta_s SI + \beta_W SW - (\gamma + \mu_I) I \\ \mu_W I - \varepsilon W \end{pmatrix}. \tag{6}$$

It is clear that F is locally Lipschitz in X. From [13], we deduce that system (3) admits a unique local solution on $[0, T_{max})$, where T_{max} is the maximal existence time for solution of system (3).

In addition, system (3) can be written in the form

$$\begin{cases} \frac{\partial S(t,x,y)}{\partial t} - d_S \bigtriangleup S = F_1(S, I, W) \\ \frac{\partial I(t,x,y)}{\partial t} - d_I \bigtriangleup I = F_2(S, I, W) \\ \frac{\partial W(t,x,y)}{\partial t} = F_3(S, I, W) \\ S(0, x, y) = S_0(x, y) \geq 0, I(0, x, y) = I_0(x, y) \geq 0, W(0, x, y) = W_0(x, y) \geq 0. \end{cases} \tag{7}$$

Is easy to see that the functions F_1, F_2 and F_3 are continuously differentiable satisfying $F_1(0, I, W) = \Lambda \geq 0$, $F_2(S, 0, W) = \beta_W SW \geq 0$ and $F_3(S, I, 0) = \mu_W I \geq 0$ for all $S, I, W \geq 0$. Since initial conditions of system (3) are non negative, we deduce the positivity of the local solution (see [15]).

Now, we show the boundedness of solution. From (3) we have

$$\begin{aligned} \frac{\partial S}{\partial t} - d_S \bigtriangleup S &\leq \Lambda - \mu_S S, \\ \frac{\partial S}{\partial \eta} &= 0, \\ S(x, y, 0) = S_0(x, y) &\leq \| S_0 \|_\infty = \max_{(x,y) \in \bar{\Omega}} S_0(x, y), \end{aligned}$$

By the comparison principle [14], we have $S(x, y, t) \leq S_1(t)$, where $S_1(t) = S_0(x, y) e^{-\mu_S t} + \frac{\Lambda}{\mu}(1 - e^{-\mu_S t})$ is the solution of the problem

$$\begin{cases} \frac{dS_1}{dt} = \Lambda - \mu_S S_1 \\ S_1(0) = \| S_1 \|_\infty . \end{cases} \tag{8}$$

Since $S_1(t) \leq \max \left\{ \frac{\Lambda}{\mu_S}, \| S_1 \|_\infty \right\}$ for $t \in [0, T_\infty)$, we have that

$S(x, y, t) \leq \max \left\{ \frac{\Lambda}{\mu_S}, \| S_1 \|_\infty \right\}$, $\forall (x, y, t) \in \bar{\Omega} \times [0, T_{max})$.

From Theorem 2 given by Alikakosin [4], to establish the L^∞ uniform boundedness of $I(x,y,t)$,it is sufficient to show the L^1 uniform boundedness of $I(x,y,t)$.

Since $\frac{\partial S}{\partial \eta} = \frac{\partial I}{\partial \eta} = \frac{\partial W}{\partial \eta} = 0$ and $\dfrac{\partial(S+I+W)}{\partial t} - \triangle(d_S S + d_I I) \leq \Lambda - \delta(S + I + W)$, where $\delta = \min\{\mu_S, \gamma + \mu_I, +\mu_W, \varepsilon\}$ we get

$$\frac{\partial}{\partial t}\left(\int_\Omega (S+I+W)dxdy\right) \leq mes(\Omega)\Lambda - \delta \int_\Omega (S+I+W)dxdy.$$

Hence,

$$\int_\Omega (S+I+W)dxdy \leq mes(\Omega) \max\left\{\frac{\Lambda}{\delta}, \| S_0 + I_0 + W_0 \|_\infty\right\},$$

which implies that

$$\sup_{t\geq 0} \int_\Omega I(x,y,t)dxdy \leq K := mes(\Omega) \max\left\{\frac{\Lambda}{\delta}, \| S_0 + I_0 + W_0 \|_\infty\right\}.$$

Using [[4], Theorem 3.1], we deduce that there exists a positive constant K^* that depends on K and on $\| S_0 + I_0 + W_0 \|_\infty$ such that

$$\sup_{t\geq 0} \| I(.,.,t) \|_\infty \leq K^*.$$

From the above, we have proved that S, I and W are L^∞ bounded on $\bar{\Omega} \times [0, T_{max})$. Therefore, it follows from the standard theory for semilinear parabolic systems,(see [9]) that $T_{max} = +\infty$. This completes the proof of the proposition. □

3 The Effect of Diffusion

Using the results presented in [1–3,11–15], it is easy to get that the basic reproduction number of disease in the absence of spatial dependence is given by

$$R_0 = \frac{\Lambda(\varepsilon\beta_S + \beta_W \mu_W)}{\mu_s \varepsilon(\gamma + \mu_I)}, \tag{9}$$

which describes the average number of secondary infections produced by a single infectious individual during the entire infectious period.

The equilibrium points of model (3) are obtained by solving the algebraic system obtained by cancelling all derivatives of $S(t), I(t)$ and $W(t)$. Thus:

1. The disease free equilibrium (DFE) is: $E_0 = (S_0, 0, 0) = (\frac{\Lambda}{\mu_S}, 0, 0)$.
2. The endemic equilibrium is: $E^* = (S^*, I^*, W^*)$, with its components given by

$$S^* = \frac{\varepsilon(\gamma+\mu_I)}{\varepsilon\beta_S + \beta_W\mu_W}, \quad W^* = \frac{\mu_W}{\varepsilon} I^* \quad \text{and} \quad I^* = \frac{\Lambda}{\gamma+\mu_I} - \frac{\varepsilon\mu_S}{\varepsilon\beta_S + \beta_W\mu_W}. \tag{10}$$

The endemic equilibrium E^* is positive if the basic reproduction number R_0 of model (3) is greater than 1.

Proposition 2

- *If $R_0 \leq 1$, the system (3) has only the disease free equilibrium $E_0 = (S_0, 0, 0)$.*
- *If $R_0 > 1$, in addition to E_0 the system (3) has a unique endemic equilibrium $E^* = (S^*, I^*, W^*)$ which is a positive.*

The objective of this section is to discuss the local and global stability of equilibria.

3.1 Local Stability of Equilibria

First, we linearize the dynamical system (3) around arbitrary spatially homogeneous fixed point $\bar{E}(\bar{S}, \bar{I}, \bar{W})$ for small space-and time-dependent fluctuations and expand them in Fourier space. For this, let

$$\begin{aligned} S(\overrightarrow{x}, t) &\sim \bar{S} e^{\lambda t} e^{i \overrightarrow{k} . \overrightarrow{x}}, \\ I(\overrightarrow{x}, t) &\sim \bar{I} e^{\lambda t} e^{i \overrightarrow{k} . \overrightarrow{x}}, \end{aligned} \tag{11}$$

where $\overrightarrow{x} = (x, y)$ and $\overrightarrow{k} . \overrightarrow{k} := < \overrightarrow{k}, \overrightarrow{k} > := k^2$; $\overrightarrow{k}$ and λ are the wavenumber vector and frequency, respectively. Then we can obtain the corresponding characteristic equation as follows:

$$det(J - k^2 D - \lambda I_2) = 0, \tag{12}$$

where I_3 is the identity matrix, $D = diag(d_S, d_I, 0)$ is the diffusion matrix and J is the Jacobian matrix of (3) without diffusion ($d_S = d_I = 0$) at $\bar{E}$ which is given by

$$J = \begin{pmatrix} -\beta_S \bar{I} - \beta_W \bar{W} - \mu_S & -\beta_S \bar{S} & -\beta_W \bar{S} \\ \beta_S \bar{I} + \beta_W \bar{W} & \beta_S \bar{S} - \gamma - \mu_I & \beta_W \bar{S} \\ 0 & \mu_W & -\varepsilon \end{pmatrix}. \tag{13}$$

The characterization of the local stability of disease-free equilibrium E_0 is given by the following result.

Theorem 1. *The disease-free equilibrium E_0 is locally asymptotically stable if $R_0 \leq 1$ and it is unstable if $R_0 > 1$.*

Proof. Evaluating (12) at E_0, we have

$$(\lambda + \mu_S + k^2 d_S)\big(\lambda^2 + a_1(k)\lambda + a_0(k)\big) = 0, \tag{14}$$

where

$$\begin{aligned} a_1(k) &= -\beta_S S_0 + \gamma + \mu_I + k^2 d_I + \varepsilon \\ a_0(k) &= \varepsilon(-\beta_S S_0 + \gamma + \mu_I) - \mu_W \beta_W S_0 + \varepsilon k^2 d_I \\ &= \varepsilon(\gamma + \mu_I)(1 - R_0) + \varepsilon k^2 d_I. \end{aligned}$$

Clearly, $\lambda_1(k) = -(\mu_S + k^2 d_S) < 0$ is a root of (14). Note that if $R_0 \leq 1$, then the coefficients $a_1(k)$ and $a_0(k)$ are all strictly positive for all k, so the characteristic Eq. (14) does not admit a real strictly positive root. Hence E_0 is locally asymptotically stable if $R_0 \leq 1$. If $R_0 > 1$, $a_0(0)$ is negative, so E_0 is unstable. □

Next, we focus on the local stability of the endemic equilibrium E^*.

Theorem 2. *The endemic equilibrium E^* is locally asymptotically stable if $R_0 > 1$.*

Proof. Evaluating (12) at $E^*(S^*, I^*, W^*)$, we have

$$\lambda^3 + \alpha_2(k)\lambda^2 + \alpha_1(k)\lambda + \alpha_0(k) = 0, \tag{15}$$

where

$$\begin{aligned}
\alpha_2(k) &= \varepsilon + \beta_W \frac{S^* W^*}{I^*} + \frac{\Lambda}{S^*} + k^2(d_S + d_I)\\
\alpha_1(k) &= \varepsilon k^2 d_I + \left(\frac{\Lambda}{S^*} + k^2 d_S\right)\left(\beta_W \frac{S^* W^*}{I^*} + k^2 d_I + \varepsilon\right) + \beta_S(\gamma + \mu_I)I^*\\
\alpha_0(k) &= \varepsilon k^2 d_I \left(\frac{\Lambda}{S^*} + k^2 d_S\right) + (\gamma + \mu_I)(\beta_S \varepsilon + \mu_W \beta_W)I^*.
\end{aligned}$$

We have $\alpha_2(k), \alpha_1(k)$ and $\alpha_0(k)$ are strictly positive for all k, then E^* is locally asymptotically stable. □

3.2 Global Stability of E^*

Proposition 3. *If $R_0 > 1$, the endemic equilibrium E^* is globally asymptotically stable.*

Proof. Let us considering the following function

$$V(t) = S(t) - S^* - \int_{S^*}^{S(t)} \frac{S^*}{Z} dZ + I^* \Phi\left(\frac{I(t)}{I^*}\right) + \frac{\beta_W S^* W^*}{\varepsilon} \Phi\left(\frac{W(t)}{W^*}\right), \tag{16}$$

where $\Phi(Z) = Z - 1 - \ln Z > 0$, for $Z > 0$. It is obvious that Φ attains its strict global minimum at 1 and $\Phi(1) = 0$. Then $\Phi(Z) > 0$ and the functional V is nonnegative.

As V is Lyapunov function of (3) without diffusion (see [11]), then

$$H = \int_{\Omega} V(u(t, X))dX$$

is a Lyapunov function for system (3), where $u(t, X) = \begin{pmatrix} S(t, X) \\ I(t, X) \\ W(t, X) \end{pmatrix}$ and $X = \begin{pmatrix} x \\ y \end{pmatrix}$.

By a direct computation and Green formula, the time derivative of H satisfy the following property

$$\begin{aligned}\frac{dH}{dt} &= \int_{\Omega} \dot{V}(u(t,X))dX - S^* \int_{\Omega} \frac{|\nabla S(t,X)|^2}{S^2(t,X)} dX \\ &\quad - I^* \int_{\Omega} \frac{|\nabla I(t,X)|^2}{I^2(t,X)} dX - \frac{\beta_W S^* W^*}{\varepsilon} \int_{\Omega} \frac{|\nabla W(t,X)|^2}{W^2(t,X)} dX \\ &\le 0\end{aligned}$$

Then, the endemic equilibrium E^* is globally asymptotically stable. □

4 The Effect of Time Delay and Diffusion

In this section we study the effect of time delay and diffusion on the behaviour of equilibria [4,6,16,18] of the following delayed reaction-diffusion system,

$$\begin{cases} \frac{\partial S(t,x,y)}{\partial t} = d_S \triangle S + \Lambda - \beta_s S(t,x,y) I(t-\tau,x,y) - \beta_W S(t,x,y) W(t,x,y) - \mu_s S(t,x,y) \\ \frac{\partial I(t,x,y)}{\partial t} = d_I \triangle I + \beta_s S(t,x,y) I(t-\tau,x,y) + \beta_W S(t,x,y) W(t,x,y) - (\gamma + \mu_I) I(t,x,y) \\ \frac{\partial W(t,x,y)}{\partial t} = \mu_W I(x,y,t) - \varepsilon W(x,y,t) \\ \frac{\partial S}{\partial \eta} = \frac{\partial I}{\partial \eta} = \frac{\partial W}{\partial \eta} = 0 \quad \text{on} \quad \partial\Omega \\ S(0,x,y) = S_0(x,y) \ge 0,\, I(\theta,x,y) = \Phi(x,y,\theta) \ge 0,\, W(0,x,y) = W_0(x,y) \ge 0,\, (x,y) \in \Omega,\, \theta \in [-\tau,0], \end{cases} \tag{17}$$

where $\phi(.,.,\theta) \in C([-\tau,0],\mathbb{R}^2)$.

4.1 Stability of E_0

Note that the operator $-\triangle$ with the homogeneous Neumann boundary condition on Ω has the eigenvalues $0 = \mu_0 < \mu_1 < \mu_2 < ... < \mu_n < ...$ and $\lim_{n\to+\infty} \mu_n = \infty$. Let $S(\mu_n)$ be the eigenspace corresponding to μ_n with multiplicity $m_n \ge 1$. Let $\Phi_{nj}(1 \le j \le m_n)$ be the normalized eigenfunctions corresponding to μ_n. Then, the set $\{\Phi_{ji} : i \ge 0, 1 \le j \le m_n\}$ forms a complete orthonormal basis [8,12]. The linearization of system (17) at the disease free equilibrium E_0 can be expressed by:

$$\begin{pmatrix} S_t \\ I_t \\ W_t \end{pmatrix} = L \begin{pmatrix} S \\ I \\ W \end{pmatrix} = D \begin{pmatrix} \Delta S \\ \Delta I \\ \Delta W \end{pmatrix} + J_{11} \begin{pmatrix} S \\ I \\ W \end{pmatrix} + J_{22} \begin{pmatrix} S(t-\tau,x,y) \\ I(t-\tau,x,y) \\ W(t-\tau,x,y) \end{pmatrix}, \tag{18}$$

with

$$D = \begin{pmatrix} d_S & 0 & 0 \\ 0 & d_I & 0 \\ 0 & 0 & 0 \end{pmatrix}, \quad J_{11} = \begin{pmatrix} -\mu_S & 0 & -\beta_W S_0 \\ 0 & -(\gamma+\mu_I) & \beta_W S_0 \\ 0 & \mu_W & -\varepsilon \end{pmatrix}, \quad J_{22} = \begin{pmatrix} 0 & -\beta_S S_0 e^{-\lambda\tau} & 0 \\ 0 & \beta_S S_0 e^{-\lambda\tau} & 0 \\ 0 & 0 & 0 \end{pmatrix}.$$

Note that λ is an eigenvalue of L if and only if λ is an eigenvalue of the matrix $J_n = -\mu_n D + J_{11} + J_{22}$ for some $n \geq 0$. Therefore, the stability is translated into the distribution of roots of the following characteristic equation:

$$(\lambda + \mu_S + d_S\mu_n)[\lambda^2 + \alpha_{1n}\lambda + \alpha_{0n} - \beta_S S_0(\lambda + \varepsilon)e^{-\lambda\tau}] = 0, \tag{19}$$

Clearly, $\lambda_1(n) = -(\mu_S + d_S\mu_n) < 0$ is a root of (19), so the study of stability of E_0 is reduced to the study of the roots of

$$\lambda^2 + \alpha_{1n}\lambda + \alpha_{0n} - \beta_S S_0(\lambda + \varepsilon)e^{-\lambda\tau} = 0, \tag{20}$$

where

$$\begin{aligned}\alpha_{1n} &= \varepsilon + d_I\mu_n + \gamma + \mu_I,\\ \alpha_{0n} &= \varepsilon(d_I\mu_n + \gamma + \mu_I) - \mu_W\beta_W S_0.\end{aligned}$$

Assume that $\lambda = i\omega(\omega > 0)$ is a root of (20), then we have

$$-\omega^2 + \alpha_{1n}i\omega + \alpha_{0n} - \beta_S S_0(i\omega + \varepsilon)(\cos\omega\tau - i\sin\omega\tau) = 0. \tag{21}$$

Separating the real and imaginary parts, we have

$$\begin{cases}\omega^2 - \alpha_{0n} &= -\varepsilon\beta_S S_0\cos\omega\tau - \beta_S S_0\omega\sin\omega\tau,\\ -\alpha_{1n}\omega &= -\beta_S S_0\omega\cos\omega\tau + \varepsilon\beta_S S_0\sin\omega\tau.\end{cases} \tag{22}$$

Adding up the squares of both the equations, we obtain

$$\omega^4 + A_{1n}\omega^2 + A_{0n} = 0. \tag{23}$$

Let $z = \omega^2$, Eq. (23) becomes

$$h(z) = z^2 + A_{1n}z + A_{0n} = 0, \tag{24}$$

where

$$\begin{aligned}A_{1n} &= \alpha_{1n}^2 - 2\alpha_{0n} - (\beta_S S_0)^2,\\ A_{0n} &= \alpha_{0n}^2 - (\varepsilon\beta_S S_0)^2.\end{aligned}$$

For $n = 0$, we have

$$\begin{aligned}A_{00} &= \alpha_{00}^2 - (\varepsilon\beta_S S_0)^2,\\ &= \varepsilon(\gamma + \mu_I)(1 - R_0)(1 + R_0).\end{aligned}$$

Note that $A_{00} < 0$ if $R_0 > 1$. Hence h is continue and $\lim_{z\to+\infty} h(z) = +\infty$, then h crosses x-axis in some positive value.

Proposition 4. *The disease-free equilibrium E_0 is locally asymptotically stable if $R_0 \leq 1$ and it is unstable if $R_0 > 1$.*

4.2 Stabililty of E^*

By linearizing system (17) at E^*, we get the following corresponding characteristic equation

$$\lambda^3 + B_{2n}\lambda^2 + B_{1n}\lambda + B_{0n} + (\beta_S S^* \lambda^2 + C_{1n}\lambda + C_{0n})e^{-\lambda\tau} = 0, \qquad (25)$$

where

$$\begin{aligned}
B_{2n} &= d_S\mu_n + \beta_S I^* + \beta_W W^* + \mu_S + \varepsilon + d_I + \gamma + \mu_I,\\
B_{1n} &= \varepsilon(d_I + \gamma + \mu_I) - \mu_W\beta_W S^* + (d_S\mu_n + \beta_S I^* + \beta_W W^* + \mu_S)(\varepsilon + d_I + \gamma + \mu_I),\\
B_{0n} &= (d_S\mu_n + \beta_S I^* + \beta_W W^* + \mu_S)\varepsilon(d_I + \gamma + \mu_I) - \mu_W\beta_W S^*(d_S\mu_n + \mu_S),\\
C_{1n} &= \varepsilon\beta_S S^* + d_S\mu_n + \beta_S I^* + \beta_W W^* + \mu_S + (\beta_S I^* + \beta_W W^*)\beta_S S^*,\\
C_{0n} &= \varepsilon\beta_S S^*\big(d_S\mu_n + \mu_S + 2(\beta_S I^* + \beta_W W^*)\big).
\end{aligned}$$

Assume that $\lambda = i\omega(\omega > 0)$ is a root of (25), then we have

$$-i\omega^3 - B_{2n}\omega^2 + B_{1n}i\omega + B_{0n} + (-\beta_S S^*\omega^2 + C_{1n}i\omega + C_{0n})(\cos\omega\tau - i\sin\omega\tau) = 0. \qquad (26)$$

Separating the real and imaginary parts, we have

$$\begin{cases} B_{2n}\omega^2 - B_{0n} = (-\beta_S S^*\omega^2 + C_{0n})\cos\omega\tau + C_{1n}\omega\sin\omega\tau,\\ \omega^3 - B_{1n}\omega \quad = (\beta_S S^*\omega^2 - C_{0n})\sin\omega\tau + C_{1n}\omega\cos\omega\tau. \end{cases} \qquad (27)$$

Adding up the squares of both the equations, we obtain

$$\omega^6 + D_{2n}\omega^4 + D_{1n}\omega^2 + D_{0n} = 0. \qquad (28)$$

Let $z = \omega^2$, Eq. (28) becomes

$$g(z) = z^3 + D_{2n}z^2 + D_{1n}z + D_{0n} = 0, \qquad (29)$$

where

$$\begin{aligned}
D_{2n} &= B_{2n}^2 - 2B_{1n} - \beta_S S^*,\\
D_{1n} &= B_{1n}^2 - 2B_{0n}B_{2n} + 2\beta_S S^* C_{0n} - C_{1n}^2,\\
D_{0n} &= B_{0n}^2 - C_{0n}^2.
\end{aligned}$$

Let the hypothesis:

$$(H_n) : D_{2n} > 0, D_{0n} > 0 \quad \text{and} \quad D_{2n}D_{1n} - D_{0n} > 0. \qquad (30)$$

Proposition 5. *If $R_0 > 1$ and (H_n) is satisfied for some $n \in \mathbb{N}$, then endemic equilibrium E_1 is asymptotically stable for all $\tau > 0$.*

Proof. The proof is deduced from the Routh-Hurwitz stability criterion [6] (Figs. 1, 2, 3 and 4). □

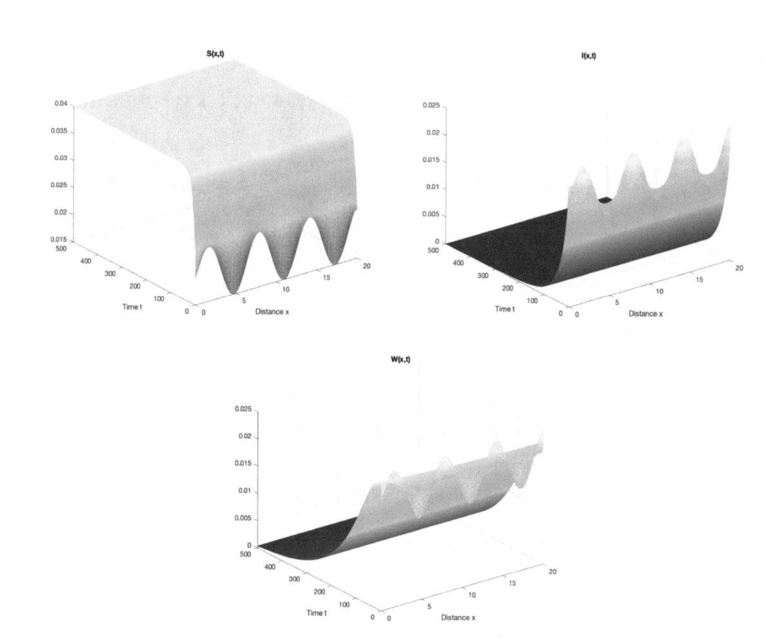

Fig. 1. Spatio-Temporal evolution of S, I and W for $\tau = 0$ and $\Lambda = 0.04$, $\beta_S = 0.01$, $\beta_W = 0.02$, $\mu_S = 0.1$, $\mu_I = 0.01$, $\mu_W = 0.02$, $\gamma = 0.2$ and $\epsilon = 0.01$, we have the stability of disease free equilibrium E_0 for $R_0 = 0.0667$

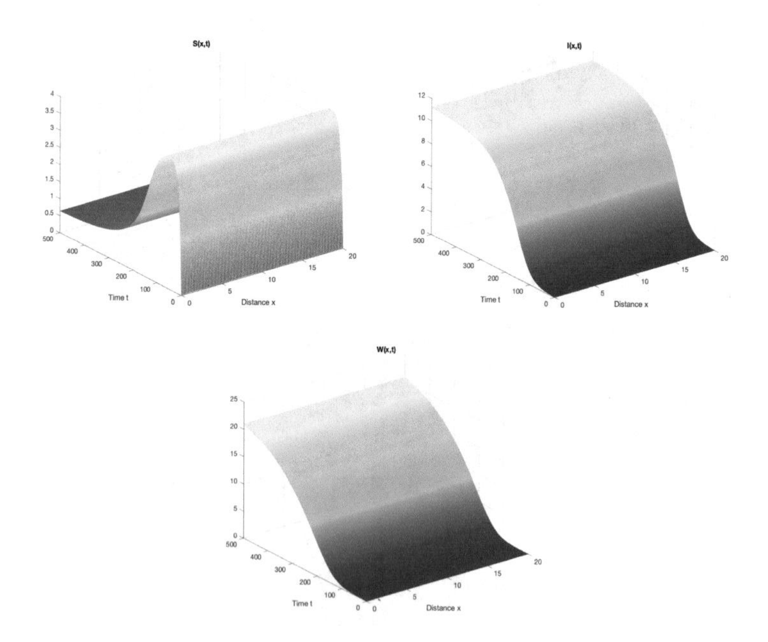

Fig. 2. Spatio-Temporal evolution of S, I and W for $\tau = 0$ and $\Lambda = 0.4$, $\beta_S = 0.01$, $\beta_W = 0.02$, $\mu_S = 0.1$, $\mu_I = 0.01$, $\mu_W = 0.02$, $\gamma = 0.2$ and $\epsilon = 0.01$, we have the stability of endemic equilibrium E^* for $R_0 = 6.6667$

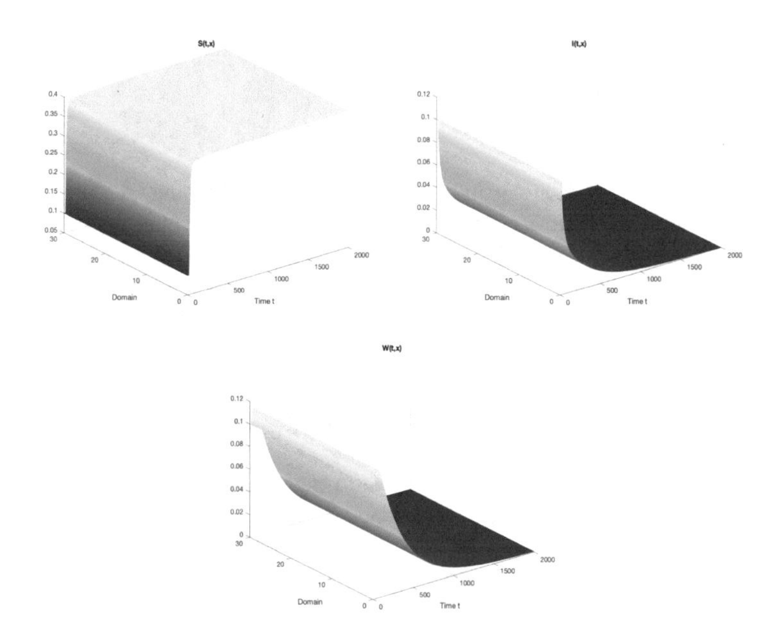

Fig. 3. Spatio-Temporal evolution of S, I and W for $\tau = 20.12$ and $\Lambda = 0.04$, $\beta_S = 0.01$, $\beta_W = 0.02$, $\mu_S = 0.1$, $\mu_I = 0.01$, $\mu_W = 0.02$, $\gamma = 0.2$ and $\epsilon = 0.01$, we have the stability of disease free equilibrium E_0 for $R_0 = 0.0667$

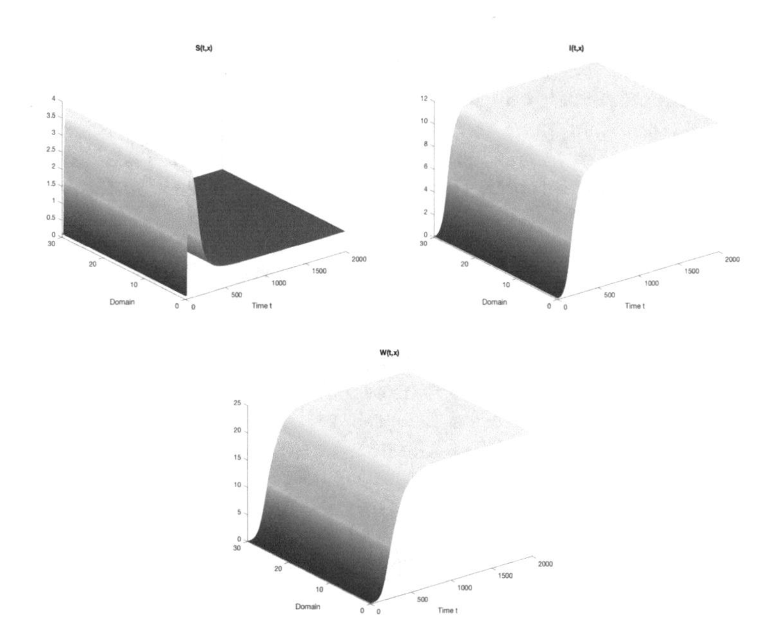

Fig. 4. Spatio-Temporal evolution of S, I and W for $\tau = 20.12$ and $\Lambda = 0.4$, $\beta_S = 0.01$, $\beta_W = 0.02$, $\mu_S = 0.1$, $\mu_I = 0.01$, $\mu_W = 0.02$, $\gamma = 0.2$ and $\epsilon = 0.01$, we have the stability of endemic equilibrium E^* for $R_0 = 6.6667$

Funding. This research was supported by CNRST (Cov/2020/102).

References

1. Abid, W., Yafia, R., Aziz Alaoui, M.A., Bouhafa, H., Abichou, A.: Instability and pattern formation in three-species food chain model via holling type II functional response on a circular domain. Int. J. Bifurcation Chaos **25**(6), 1550092 (2015)
2. Abid, W., Yafia, R., Aziz Alaoui, M.A., Bouhafa, H., Abichou, A.: Global dynamics on a circular domain of a diffusion predator-prey model with modified Leslie-Gower and Beddington-DeAngelis functional type. Evol. Equ. Control Theory **4**(2), 115–129 (2015)
3. Abid, W., Yafia, R., Aziz Alaoui, M.A., Aghriche, A.: Turing instability and Hopf bifurcation in a modified Leslie-Gower predator-prey model with cross-diffusion. Int. J. Bifurcation Chaos **28**(7), 1850089 (2018)
4. Alikakos, N.D.: An application of the invariance principle to reaction-diffusion equations. J. Differ. Equ. **33**(2), 201–225 (1979)
5. Aziz-Alaoui, M.A., Najm, F., Yafia, R.: SIARD model and effect of lockdown on the dynamics of COVID-19 disease with non total immunity. Math. Model. Nat. Phenomena **16**, 2021025 (2021)
6. Chebotarev, N.G., Meiman, N.N.: The Routh-Hurwitz problem for polynomials and entire functions. Trudy Mat. Inst., Steklov (1949)
7. COVID-19 Coronavirus Pandemic. https://www.worldometers.info/coronavirus/repro. Accessed 26 Mar 2020
8. Faria, T.: Stability and bifurcation for a delayed predator-prey model and the effect of diffusion. J. Math. Anal. Appl. **254**(2), 433–463 (2001)
9. Henry, D.: Geometric Theory of Semilinear Parabolic Equations. Lecture Notes in Mathematics, vol. 840. Springer, Germany (1993)
10. Kramer, A., Assadian, O.: Survival of microorganisms on inanimate surfaces. In: Borkow, G. (ed.) Use of Biocidal Surfaces for Reduction of Healthcare Acquired Infections, pp. 7–26. Springer, Cham (2014). https://doi.org/10.1007/978-3-319-08057-4_2
11. Najm, F., Yafia, R., Aziz-Alaoui, M.A., Aghriche, A.: COVID-19 Disease Model with Reservoir of Infection: Cleaning Surfaces and Wearing Masks Strategies, submitted
12. Ouyang, Q.: Pattern Formation in Reaction Diffusion Systems, Shanghai Scientific and Technological Education Publishing House (2000)
13. Pazy, A.: Semigroups of Linear Operators and Applications to Partial Differential Equations. Applied Mathematical Sciences, vol. 44. Springer, New York (1983)
14. Protter, M.H., Weinberger, H.F.: Maximum Principles in Differential Equations. Prentice Hall, Englewood Cliffs (1967)
15. Smoller, J.: Shock Waves and Reaction-Diffusion Equations. Springer, Berlin (1983)
16. Talibi Alaoui, H., Yafia, R.: Stability and Hopf bifurcation in an approachable haematopoietic stem cells model. Math. Biosci. **206**(2), 176–184 (2007)
17. Wißmann, J.E., et al.: Persistence of pathogens on inanimate surfaces: a narrative review. Microorganisms **9**, 343 (2021). https://doi.org/10.3390/microorganisms9020343
18. Yafia, R.: Hopf bifurcation in a delayed model for tumor-immune system competition with negative immune response. Discret. Dyn. Nat. Soc. 95296 (2006)

Machine Learning-Based Biometric Authentication with Photoplethysmography Signal

Bahadır Çokçetin[(✉)], Derya Kandaz, and Muhammed Kürşad Uçar

Sakarya University, Faculty of Engineering, Electrical-Electronics Engineering, 54187 Serdivan, Sakarya, Turkey
bahadir.cokcetin@dpu.edu.tr, {deryakandaz,mucar}@sakarya.edu.tr

Abstract. Biometric identification determines the reality of an individual's physiological or behavioral characteristics with the current identity. Performing this process with traditional methods, including fingerprint, iris, and voice, creates a security problem for a society connected to an electronic environment. The photoplethysmography (PPG) signal has ease of use, high security, and less attack proneness among these biometric features. This study aims to make biometric identification with machine learning methods using PPG signals in line with these requirements. The study includes three PPG records from 219 individuals retrieved from the open-source database. The received PPG signal is free of artifacts and noise. Twenty-five features were extracted from the filtered PPG signal. The system's performance was improved by calculating the correlation of these features with the Spearman coefficient. The data set was trained using different artificial intelligence methods, and the accuracy rate was measured as 100%. According to the study results, it is evaluated that PPG signals can be used in practice with high efficiency in biometric identification.

Keywords: Biometric Authentication · Authentication · Biomedical Signal Processing · Machine Learning · Photoplethysmography

1 Introduction

Biometric recognition systems work to recognize only the physical or behavioral characteristics of individuals that distinguish them from other individuals [1].

The use of authentication models such as the face, finger, iris recognition, voice, and other physiological and behavioral authentication technologies and traditional security methods such as passwords and PIN have recently been the subject of extensive research [2].

Physiological signals used in biometric systems are electrocardiograms (EKG), electroencephalograms (EEG), phonocardiogram (PCG), and finally, photoplethysmography (PPG) signals. These signals other than PPG can be

D. J. Hemanth et al. (Eds.): ICAIAME 2022, ECPSCI 7, pp. 595–606, 2023.
https://doi.org/10.1007/978-3-031-31956-3_50

obtained with special equipment placed in the body and appropriate environmental conditions. These conditions make it inefficient in wearability and easy availability for daily use.

PPG signals are obtained from pulse oximeters, which emit light to the skin and measure the change of light intensity transmitted or reflected through the skin. The periodicity of the reflected/transmitted light usually corresponds to the heart rhythm used for heart rate estimation. This makes PPG sensors popular for embedding in wearable devices (e.g., smartwatches). PPG signals can be received from various locations such as earlobes, fingertips, or wrist [3].

Among physiological signals, PPG signals are more prominent than other biometric signals in terms of accessibility, ease of use, reliability, and originality [4]. However, reception is susceptible to motion artifacts under normal daily living conditions, which degrades signal accuracy and impairs identification robustness.

Since PPG signals can be obtained with wearable technologies, they are difficult to imitate because they are not dependent on a certain time interval, suitable for daily use, personal and are received from a live person, that is, reliable. With these features, it is very suitable for biometric identification [3].

This study aims to provide a high accuracy rate of PPG signals with the steps of cross-validation, feature extraction and hybrid artificial intelligence methods.

In studies with PPG signals, results were obtained using different artificial intelligence methods [5–7]. Various filtering methods are used for signal preprocessing. In some studies, mean and Butterworth filters are obtained by using [5], or bandpass filter [8]. Several statistical feature extraction methods, such as addition and geometric mean, were used for feature extraction. We can divide PPG signal-based biometric recognition into algorithmic approaches, and computationally intelligent approaches [4]. Accuracy rates in related studies were not as high as in this study. In recent years, time-independent methods have come to the fore [9,10].

All possibilities were investigated with the cross-validation process in this study. Testing and training processes have been strengthened with feature extraction. A high accuracy rate has been achieved by testing different artificial intelligence algorithms.

The most distinctive feature of the study is that it achieves the highest accuracy rate for each feature.

2 Materials and Methods

The work process applied in the study is summarized in Fig. 1. According to the flow, the signals were first filtered, and then labeled matrices were created. Features were extracted from the signals and classified selectively.

2.1 Data Collection

Data obtained from open-source data-sharing platform [11]. There are 657 sampled recorded PPG signals from 219 people in the data set, three records for each person. It also includes physical labels such as gender, age, height, weight, and disease labels such as hypertension and diabetes. The sampling frequency is 1000 Hz. First, these signals were combined, and the data matrix was created.

2.2 Signal Preprocessing

657 PPG signals were labeled with each other by cross-validation. Familiar signals belonging to the same person are labeled as 1 (Table 1), while foreign signals belonging to different people are labeled as 0 (Table 2). Due to the combination, matrices of different sizes are formed. For this reason, data balancing continued. One thousand three hundred fourteen records belonging to each group were taken.

By comparison, it was seen that the data labeled as a familiar signal had 1314 rows, while the data labeled as an unfamiliar signal had 214839 rows. Using the systematic sampling theorem, data with different label values are balanced.

The sampling frequency of the PPG signal is 1000 Hz. The PPG signal was cleaned with the help of digital filters. For noise cleaning, 0.05 Hz–0.20 Hz IIR - Chebyshev Type II Band-Pass Filter was applied.

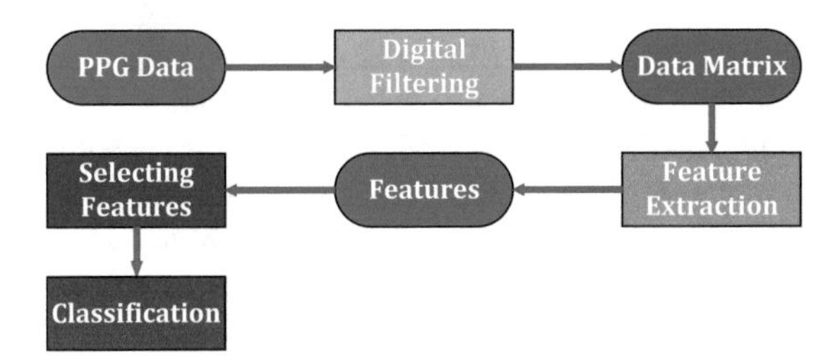

Fig. 1. Application flowchart

Table 1. Label 1 sample matrix

Number of Comparisons	First Person		Second Person		Label
	Person Number	Person Feature Number	Person Number	Person Feature Number	
1	1	1	1	1	1
2	1	1	1	2	1
3	1	1	1	3	1
4	1	2	1	2	1
5	1	2	1	3	1
6	1	3	1	3	1
7	2	1	2	1	1
8	2	1	2	2	1
9	2	1	2	3	1
10	2	2	2	2	1
...	...	...	...	...	...
1314	219	3	219	3	1

2.3 Feature Extraction

Twenty-five statistical features were extracted from the signal that was balanced and cleared from noise by filtering (Table 3). These features are descriptive statistical parameters that are frequently used in statistics. Descriptive statistical parameters show the data set's prevalence, location, skewness, and kurtosis information. The purpose of statistical-based feature extraction is to reveal the information of the PPG signal from different perspectives. In this way, it is aimed at preventing the loss of information.

The feature extraction process for the generated 1–0 matrices was applied. For 1 tag, there are 2 individual records for a row. Twenty-five features were extracted from each individual. By taking different sums and products of each extracted feature, 75 features were obtained. In this way, a feature matrix of 1314×75 will be created. The exact process was carried out within the 0 tags.

2.4 Feature Selection Algorithm - Spearman Correlation Coefficients

After the feature selection process, the relevant features were selected for the performance success of the methods to be applied. Spearman correlation coefficients were used for selected features. The relationship between familiar signal comparison tags and unfamiliar signal comparison tags was determined by Spearman correlation coefficients (r_s).

d_i i in the Spearman Correlation Coefficient (r_s) equation represents the parameter difference for the data. n represents the number of data (Eq. 1) [12] (Table 4).

Table 2. Label 0 sample matrix

Number of Comparisons	First Person		Second Person		Label
	Person Number	Person Feature Number	Person Number	Person Feature Number	
1	1	1	2	1	0
2	1	1	2	2	0
3	1	1	2	3	0
4	1	2	2	1	0
5	1	2	2	2	0
6	1	2	2	3	0
7	1	3	2	1	0
8	1	3	2	2	0
	...	...	...	...	0
...	1	3	219	3	0
...	2	1	3	1	0
...	2	1	3	2	0
...	2	1	3	3	0
...	2	2	3	1	0
...	2	2	3	2	0
...	2	2	3	3	0
...	2	3	3	1	0
...	2	3	3	2	0
...	2	3	3	3	0
...	...	...	...	...	0
...	...	...	219	3	0
...	...	...	...	...	0
214839	218	3	219	3	0

$$r_s = 1 - 6\sum_{i=1}^{n} \frac{{d_i}^2}{n \times (n^2 - 1)} \tag{1}$$

2.5 Artificial Intelligence Methods

The Classification Learner tool on the Matlab program was used for the data sets we created for each feature with the feature selection algorithm. The results for each feature are given. Table 6 results were 100% accurate. More than 30 methods were used while creating the models.

Table 3. Representation of Features mathematical and code

Nu	Feature	Equation
1	Kurtosis	$x_{kur} = \frac{\sum_{i=1}^{n}(x(i)-\overline{x})^4}{(n-1)S^4}$
2	Skewness	$x_{ske} = \frac{\sum_{i=1}^{n}(x_i-\overline{x})^3}{(n-1)S^3}$
3	* IQR	$IQR = iqr(x)$
4	CV	$CV = (S/\overline{x})100$
5	Geometric Mean	$G = \sqrt[n]{x_1 + \cdots + x_n}$
6	Harmonic Mean	$H = n/(\frac{1}{x_1} + \cdots + \frac{1}{x_n})$
7	Activity - Hjort Parameters	$A = S^2$
8	Mobility - Hjort Parameters	$M = S_1^2/S^2$
9	Complexity - Hjort Parameters	$C = \sqrt{(S_2^2/S_1^2)^2 - (S_1^2/S^2)^2}$
10	* Maximum	$x_{max} = max(x_i)$
11	Median	$\tilde{x} = \begin{cases} x_{\frac{n+1}{2}} & : x \text{ odd} \\ \frac{1}{2}(x_{\frac{n}{2}} + x_{\frac{n}{2}+1}) & : x \text{ even} \end{cases}$
12	* Mean Absolute Deviation	$MAD = mad(x)$
13	* Minimum	$x_{min} = min(x_i)$
14	* Central Moments	$CM = moment(x, 10)$
15	Mean	$\overline{x} = \frac{1}{n}\sum_{i=1}^{n} = \frac{1}{n}(x_1 + \cdots + x_n)$
16	Average Curve Length	$CL = \frac{1}{n}\sum_{i=2}^{n}\lvert x_i - x_{i-1}\rvert$
17	Average Energy	$E = \frac{1}{n}\sum_{i=1}^{n} x_i^2$
18	Root Mean Squared	$X_{rms} = \sqrt{\frac{1}{n}\sum_{i=1}^{n}\lvert x_i\rvert^2}$
19	Standard Error	$S_{\overline{x}} = S/\sqrt{n}$
20	Standard Deviation	$S = \sqrt{\frac{1}{n}\sum_{i=1}^{n}(x_i - \overline{x})}$
21	Shape Factor	$SF = X_{rms}/\left(\frac{1}{n}\sum_{i=1}^{n}\sqrt{\lvert x_i\rvert}\right)$
22	* Singular Value Decomposition	$SVD = svd(x)$
23	* 25% Trimmed Mean	$T25 = trimmean(x, 25)$
24	* 50% Trimmed Mean	$T50 = trimmean(x, 50)$
25	Average Teager Energy	$TE = \frac{1}{n}\sum_{i=3}^{n}(x_{i-1}^2 - x_i x_{i-2})$

*** The property was computed using MATLAB**
IQR Interquartile Range, CV Coefficient of Variation
S^2: **variance of the signal** x
S_1^2: **Variance of the 1st derivative of the signal** x
S_2^2: **Variance of the 2nd derivative of the signal** x

Table 4. Feature Values

Feature Number	Correlation Coefficient	Feature Number	Correlation Coefficient	Feature Number	Correlation Coefficient
1	0.0386	26	0.0052	51	0.0252
2	0.0230	27	0.0176	52	0.0965
3	0.0496	28	0.1230	53	0.0158
4	0.0436	29	0.1043	54	0.0189
5	0.0452	30	0.1351	55	0.0437
6	0.0382	31	0.1135	56	0.0366
7	0.0936	32	0.1249	57	0.0095
8	0.0811	33	0.0779	58	0.0242
9	0.0736	34	0.0912	59	0.0081
10	0.0703	35	0.1383	60	0.0633
11	0.0300	36	0.0832	61	0.0270
12	0.0519	37	0.1191	62	0.0115
13	0.0157	38	0.0103	63	0.0233
14	0.1692	39	0.1154	64	0.0057
15	0.0510	40	0.1630	65	0.0495
16	0.0865	41	0.0454	66	0.1584
17	0.0553	42	0.1604	67	0.0521
18	0.0538	43	0.1606	68	0.0521
19	0.0533	44	0.1177	69	0.0095
20	0.0533	45	0.1177	70	0.0095
21	0.0625	46	0.1653	71	0.0615
22	0.0538	47	0.1606	72	0.0521
23	0.0457	48	0.1234	73	0.0439
24	0.0349	49	0.0979	74	0.0329
25	0.0920	50	0.2049	75	0.0229
				76	1

2.6 Performance Evaluation Criteria

The following performance evaluation criteria were used to evaluate the model performances in the study. These; True Positive Rate (TPR), Positive Predictive Value (PPV), False Positive Rate (FNR), and False Discovery Rate (FDR) are performance ratings [13,14]

To test the model performances, the data set is divided into two as 80% training and 20% testing (Table 5).

Table 5. Training and test dataset distribution

Dataset	Train (80%)	Test (20%)	Total
Label 0	1051	263	1314
Label 1	1051	263	1314

3 Results

The study aims to realize a biometric recognition system with PPG. For this purpose, a different model from the literature has been proposed. According to the model, the records taken from the individual are compared with the records in the database. The system produces 1 tag if it is the same individual and 0 if it is a different individual. Twenty-five features are extracted from the PPG signal of the individual. Twenty-five features are extracted from the other individual in the database. Total, difference and multiplication operations are performed for each feature and the number of features becomes 75. Then classification is done by machine learning. The feature selection process was carried out before classification. 11 datasets were created with different amounts of selections (Table 6).

Models were created using more than 30 classifiers for the 11 datasets created in the study (Table 6). 100% accuracy rate was obtained in all datasets. It has been observed that different classifiers provide a better fit in each model. This is an indication that classifiers adapt differently to different datasets.

When the models proposed in this study are compared with the studies in the literature, it is seen that the performance is quite high (Table 7).

4 Discussions

Although PPG-based biometric systems in the literature have achieved high accuracy rates, this type of biometric recognition technology is relatively new and, therefore, can be applied in uncontrolled and unrestricted conditions. It requires more work to obtain accurate and robust recognition approaches. Some of the problems in PPG-based biometric recognition can be listed as follows [4]. (1) The time range used is limited. The stability of PPG signals over long periods (years or decades) is not yet known. (2) Studies in the literature have been studied on the distinguishability of PPG samples for datasets consisting of a certain number of users. It has not yet been studied for unlimited or larger user groups. (3) There are no studies that take into account performing activities of daily living, for example, in harsh conditions such as sports activities. (4) In the literature, there are only preliminary studies on continuous authentication methods [1]. There are no open datasets available for such studies.

Table 6. Performance Results

Feature Number	Methods	Methods Used for Performance Evaluation	TPR	PPV	FNR	FDR
4	Method1	Fine Tree	100	100	0	0
	Method2	Medium Tree	100	100	0	0
	Method3	Coarse Tree	100	100	0	0
8	Method1	Linear SVM	100	100	0	0
	Method2	Quadratic SVM	100	100	0	0
	Method3	Cubic SVM	100	100	0	0
11	Method1	Logistic Regression	100	100	0	0
	Method2	Fine KNN	100	100	0	0
	Method3	Bagged Trees	100	100	0	0
15	Method1	Narrow Neural Network	100	100	0	0
	Method2	Wide Neural Network	100	100	0	0
	Method3	Bilayered Neural Network	100	100	0	0
19	Method1	Trilayered Neural Network	100	100	0	0
	Method2	Medium Neural Network	100	100	0	0
	Method3	SVM Kernel	100	100	0	0
23	Method1	Logistic Regression	100	100	0	0
	Method2	Cubic SVM	100	100	0	0
	Method3	Fine KNN	100	100	0	0
27	Method1	Medium Tree	100	100	0	0
	Method2	Linear SVM	100	100	0	0
	Method3	Bagged Trees	100	100	0	0
30	Method1	Fine Tree	100	100	0	0
	Method2	Quadratic SVM	100	100	0	0
	Method3	Wide Neural Network	100	100	0	0
34	Method1	Linear SVM	100	100	0	0
	Method2	Bagged Trees	100	100	0	0
	Method3	Bilayered Neural Network	100	100	0	0
38	Method1	Coarse Tree	100	100	0	0
	Method2	Coarse Gaussian SVM	100	100	0	0
	Method3	Trilayered Neural Network	100	100	0	0
76	Method1	Fine Tree	100	100	0	0
	Method2	Bagged Trees	100	100	0	0
	Method3	Wide Neural Network	100	100	0	0

Table 7. Literature performance comparison chart

Nu	References	Year	Methods	Accuracy
1	[15]	2014	KNN	94.44%
2	[6]	2016	KNN	98.1%
7	[5]	2016	DBM, RBM	96.1%
3	[7]	2018	CNN	96.0%
8	[16]	2018	CNN	83.2%
4	[9]	2019	RNN, CNN	96.0%
6	[17]	2020	GAN for domain adaption	96.85%
5	[18]	2021	ANN	100.0% and 98.07%
9	[10]	2021	CNN	98.0%
10	Proposed Method	2022	Hybrid	100.0%

DBN = Deep Belief Networks, RBM = Restricted Boltzman Machines
GAN = Generative Adversarial Networks, CNN = Convolutional Neural Network
ANN = Artificial Neural Network, kNN = K-Nearest Neighbours

5 Conclusion

According to the findings, it is evaluated that the PPG signal can be used in biometric recognition with high accuracy.

This study provides a solution with a high validation rate for PPG-based biometric recognition systems. In the future, the work can be developed in different ways. (1) It can be tested in motion-related noisy environments under harsh conditions such as sporting activities. (2) Time-independent models can be created with multiple sessions. (3) By using larger data sets, new models can be created by working on data with different age and gender characteristics.

Conflict of Interests. There is no conflict of interest between the authors.

Financial Support. No support was received for this research.

Ethics Committee Approval. Ethics committee approval is not required as the open source data set was used in the study.

Data Collection and Use Permission. Data obtained from open-source data-sharing platform [11].

References

1. Bonissi, A., Labati, R.D., Perico, L., Sassi, R., Scotti, F., Sparagino, L.: A preliminary study on continuous authentication methods for photoplethysmographic biometrics. In: 2013 IEEE Workshop on Biometric Measurements and Systems for Security and Medical Applications, BioMS 2013 - Proceedings, pp. 28–33 (2013)
2. Shahid, H., Aymin, A., Remete, A.N., Aziz, S., Khan, M.U.: A survey on AI-based ECG, PPG, and PCG signals based biometric authentication system. In: 2021 International Conference on Computing, Electronic and Electrical Engineering, ICE Cube 2021 - Proceedings (2021)
3. Gu, Y.Y., Zhang, Y., Zhang, Y.T.: A novel biometric approach in human verification by photoplethysmographic signals. In: Proceedings of the IEEE/EMBS Region 8 International Conference on Information Technology Applications in Biomedicine, ITAB, 13–14 January 2003 (2003)
4. Labati, R.D., Piuri, V., Rundo, F., Scotti, F.: Photoplethysmographic biometrics: a comprehensive survey. Pattern Recogn. Lett. **156**, 119–125 (2022)
5. Jindal, V., Birjandtalab, J., Pouyan, M.B., Nourani, M.: An adaptive deep learning approach for PPG-based identification. In: Proceedings of the Annual International Conference of the IEEE Engineering in Medicine and Biology Society, EMBS, pp. 6401–6404 (2016)
6. Choudhary, T., Manikandan, M.S.: Robust photoplethysmographic (PPG) based biometric authentication for wireless body area networks and m-health applications. In: 2016 22nd National Conference on Communication, NCC (2016)
7. Everson, L., et al.: BiometricNet: deep learning based biometric identification using wrist-worn PPG. In: Proceedings - IEEE International Symposium on Circuits and Systems, 4 May 2018 (2018)
8. Donida Labati, R., Piuri, V., Rundo, F., Scotti, F., Spampinato, C.: Biometric recognition of PPG cardiac signals using transformed spectrogram images. In: Del Bimbo, A., et al. (eds.) ICPR 2021. LNCS, vol. 12668, pp. 244–257. Springer, Cham (2021). https://doi.org/10.1007/978-3-030-68793-9_17
9. Biswas, D., et al.: CorNET: deep learning framework for PPG-based heart rate estimation and biometric identification in ambulant environment. IEEE Trans. Biomed. Circ. Syst. **13**, 282–291 (2019)
10. Hwang, D.Y., Taha, B., Lee, D.S., Hatzinakos, D.: Evaluation of the time stability and uniqueness in PPG-based biometric system. IEEE Trans. Inf. Forensics Secur. **16**, 116–130 (2021)
11. Haque, C.A., Kwon, T.H., Kim, K.D.: Cuffless blood pressure estimation based on Monte Carlo simulation using photoplethysmography signals. Sensors **22**, 1175 (2022)
12. Alpar, R.: Uygulamali Çok de ĞİŞkenlİ İstatİstİksel yÖntemler
13. Canbek, G., Temizel, T.T., Sagiroglu, S., Baykal, N.: Binary classification performance measures/metrics: a comprehensive visualized roadmap to gain new insights. In: 2nd International Conference on Computer Science and Engineering, UBMK 2017, pp. 821–826 (2017)
14. Jiao, Y., Du, P.: Performance measures in evaluating machine learning based bioinformatics predictors for classifications. Quant. Biol. **4**, 320–330 (2016)
15. Kavsaoğlu, A.R., Polat, K., Bozkurt, M.R.: A novel feature ranking algorithm for biometric recognition with PPG signals. Comput. Biol. Med. **49**, 1–14 (2014)

16. Luque, J., Cortès, G., Segura, C., Maravilla, A., Esteban, J., Fabregat, J.: End-to-end photoplethysmography (PPG) based biometric authentication by using convolutional neural networks. In: European Signal Processing Conference September 2018, pp. 538–542 (2018)
17. Lee, E., Ho, A., Wang, Y.T., Huang, C.H., Lee, C.Y.: Cross-domain adaptation for biometric identification using photoplethysmogram. In: ICASSP, IEEE International Conference on Acoustics, Speech and Signal Processing - Proceedings, May 2020, pp. 1289–1293 (2020)
18. Siam, A.I., Elazm, A.A., El-Bahnasawy, N.A., El Banby, G.M., Abd El-Samie, F.E.: PPG-based human identification using Mel-frequency cepstral coefficients and neural networks. Multimedia Tools Appl. **80**(17), 26001–26019 (2021). https://doi.org/10.1007/s11042-021-10781-8

DDOS Intrusion Detection with Machine Learning Models: N-BaIoT Data Set

Celil Okur[1], Abdullah Orman[2], and Murat Dener[1](✉)

[1] Information Security Engineering, Graduate School of Natural and Applied Sciences, Gazi University, Ankara, Turkey
{celil.okur,muratdener}@gazi.edu.tr

[2] Department of Computer Technologies, Vocational School of Technical Sciences, Ankara Yıldırım Beyazıt University, Ankara, Turkey
aorman@ybu.edu.tr

Abstract. Internet of Things (IoT) technology is described as a system of different devices that can work synchronously on the same network. This synchronous working makes life easier, but it also brings some problems. The synchronous operation of the system is important for synchronous attacks such as Distributed Denial of Service (DDOS) made by hackers. When we look at the DDOS attacks carried out in recent years, the role of IoT systems is clearly seen. In this study, SimpleHome_XCS7_1003_WHT_Security_Camera dataset, one of the cyber attack (N-BaIoT) datasets made in 2016, was examined. While examining the data set, machine learning models used in many fields in recent years have been used. In an environment that includes work, attack and normal network traffic; It detects which traffic is attack and which traffic is normal using machine learning models. Contrary to classical attack detection methods, machine learning methods, which have become popular in recent years and detecting with high accuracy, have been used in this study. In addition, the aim of this study is to categorize the data set containing attack traffic data and normal traffic data with high accuracy using machine learning models.The data set was examined separately with supervised and unsupervised learning models. The data set was examined with 23 different machine learning models, namely Naive Bayes, Naive Bayes Updateable, LDA, Logistic, QDA, IBK, OneR, PART, ZeroR, RepTree, Hoeffdinf Tree, J48, J48 Consolidate, J48 Graft, JRIP, Random Forest, Random Tree, Canopy, EM, Farthestr First, Filtered Clusterer, Simple K-Means, Make Density Clusterer. While examining the data set, not all machine learning models were used. The models used are the ones that are frequently preferred in the studies in the literature and give results with a high percentage of accuracy. It is one of the most comprehensive studies carried out with machine learning models in the literature in terms of using 23 different models. As a result of the study, the highest percentage of correct detection was achieved with the Random Forest model with 99.92%.

Keywords: IoT Botnets · N-BaIoT DDOS Attacks · Machine Learning · Cyber Security

D. J. Hemanth et al. (Eds.): ICAIAME 2022, ECPSCI 7, pp. 607–619, 2023.
https://doi.org/10.1007/978-3-031-31956-3_51

1 Introduction

IoT devices have entered everyone's life in recent years. For individual use in cars, homes; As for social use, it is used in health, industry, military, transportation and social areas. It is estimated that the value of IoT devices in the world market will reach between 4 and 11 trillion dollars by 2025 [1]. This means that the number of IoT devices will increase rapidly. Accordingly, it is predicted that the number of attacks with these devices will increase. Synchronous operation of IoT systems and secure transmission of data are important for IoT systems. The communication and synchronization of the devices that make up the system takes place in the form of user-device, device-device, device-devices [2]. The number of devices in IoT systems reaches high levels according to the need. This situation can be manipulated by hackers. The synchronous use of thousands of devices in the same system or in different systems by the attacker reveals DDOS attacks that are widely heard today. In 2016, Mirai botnet attack was carried out using 100,000 devices [3]. For the attacked systems, there is no difference between attacking a small IoT device and attacking a well-equipped computer. Due to the increase in the number of IoT devices used, their unconscious use, and limited resources, these systems become the target of attackers. In addition, the IoT botnet (Gafgyt, Mirai) source codes that are easily accessible on the Internet make the attackers' job even easier. While the purpose of some of the attacks on IoT devices is to use these devices as slave devices in the background without damaging the devices in the system, the purpose of the other part is to seize the personal data processed and flowing on the devices. In this study, the N-BaIoT dataset realized with IoT devices was examined with machine learning models, and a distinction was made between offensive and normal network traffic according to its type. In the study, Mirai botnet attack traffic among DDOS attacks, which has been frequently encountered in recent years, has been examined. The aim of the study is to categorize the traffic with high accuracy when examining the attack traffic. Contrary to classical methods, machine learning models were used in the study to detect traffic with a high percentage of accuracy. When the studies in the literature are examined, the fact that the study is carried out with 23 different machine learning models shows the scope of the study. In addition, the fact that the highest percentage of accuracy obtained is higher than other studies of the relevant data set shows the contribution of the study to the literature. In the second part of the study, information about machine learning is given. In the third chapter, the application of the data set obtained from the IoT devices with machine learning models is mentioned. In the last part of the study, the results of the study are given.

Regarding the subject of the article and the research done, the remaining content is arranged as follows: In the second part, machine learning algorithm applications made with FPGA are mentioned and general literature review is explained. Following this, general information about Artificial Neural Networks and FPGAs is given and the findings obtained from the applications are given and a general discussion about them is given in the third section. Subsequently, the final chapter and the article ended with explanations of the results and some possible future studies.

2 Machine Learning Models

Due to the rapid change in technology in recent years, user interaction with technology has increased. Each interaction of users causes the production of different types of data. Purpose-oriented results are obtained by processing these data. Fast and high accuracy results are produced by using different algorithms, models and technologies with raw data. The use of machine learning technology, which is used for data processing, is increasing day by day in every field. In this study, it is aimed to categorize data with a high percentage of accuracy using different models of machine learning. Processing data with machine learning consists of various stages. In the first stage, the data to be processed is collected raw from the device, platform or application. In this state, the data cannot be used directly in the learning model. In the second stage, the data is made into a form that the learning model can understand by subjecting it to simplification and transformation processes. The data that is ready to be processed is divided into two groups as for training and for testing, according to the developer's preference. In the next stage, learning is carried out by training the learning model with the processed data. In the last stage, in order to understand the level of learning, the model is tested either with the data used in the training, with the data reserved for testing, or with a new data set, according to the developer's preference. As a result of this test, whether the learning is successful or not is seen with the accuracy value and scores. The selection of the appropriate learning model is important for successful results.

Machine learning models are made in two ways, supervised and unsupervised. In supervised learning, the data set is divided into different classes according to the characteristics of the algorithm used. Before the supervised learning is performed with the data set, the data set is labeled to indicate which class it belongs to. The model performs its learning with these labels and classifies the previously labeled data. If the label and the class are compatible, a successful learning is considered to have taken place. In short, in supervised learning, it is determined be-forehand which class the data belongs to. As a result, it is expected from the machine in which class it will be. The machine performs the classification process by using these labels. Supervised learning types are listed as Logistic Regression, Decision Trees, Linear Regression, RandomForest, Naïve Bayes, ZeroR, OneR, JRip. In unsupervised learning, the data is given as an input to the model without labels. The model groups the data completely according to the parameters it has determined. The similarity rate of the data samples in the same group (cluster) is higher than the data samples in the other cluster [4]. Unsupervised learning types are listed as Canopy, Coweb, EM, Filtered Clusterer, Simple K-Means [5].

3 Examination of the Dataset

In this part of the study, the network traffic data set of the security camera SimpleHome_XCS7_1003_WHT, which includes normal and attack traffic, is examined. The data set is included under the N-BaIoT data set in the literature [5]. Network traffic data cannot be used directly in learning models. Data are subjected to transformations and simplifications as they are made machine understandable. Although the data set used in this study was prepared by performing the necessary preprocessing, in this study the data was subjected to some preprocessing again. The dataset consists of 115 feature dimensions and 850,827 rows. The dataset includes a security camera network traffic data from IoT devices. The dataset belongs to the N-BaIoT dataset group. The distribution of IoT datasets used in the literature by years is listed in Table 1.

Table 1. Data set by years

Reference No	Data Set	Year
[6]	N-BaIoT	2018
[7]	MedBIoT	2020
[8]	Kitsune	2019
[9]	BoT_IoT	2018
[10]	TON_IoT	2019
[11]	IoTbening&attack trace	2019
[12]	IoT network intrusion	2019
[13]	IoT-23	2020

N-BaIoT data set kit consists of 9 different data sets, including Danmini Door-bell, Ecobee Thermostat, Ennio Doorbell, Philips B120N10 Baby Monitor, Provision PT 737E Security Camera, Provision PT 838 Security Camera, Samsung SNH 1011 N Webcam, SimpleHome XCS7 1002 WHT Security Camera, and SimpleHome XCS7 1003 WHT Security Camera. In the study, Simple XCS7 1003 security camera dataset was examined. The analyzed data set; mirai and gafgyt consist of three types of data sets as harmful and benign (normal). Malicious datasets consist of scan, junk, tcp, udp, udpplain, ack, syn, combo, data types. These 8 types of malicious data sets were combined in the python programming language and saved as .csv. The large row and feature sizes prevent the weka program from working properly [14]. Therefore, using the Correlation Attribute Eval tool, which is one of the tools of the Weka platform, the features of the data set are hierarchically ordered according to their effect on the results, and the first 10 features are discussed. While the classification part of the application is being performed, the data was processed by choosing the tenfold cross validation in the test part. The ten fold cross validation selected in this section; means that the data is divided into 10 equal parts and these data pieces are used both in the training phase and in the testing phase [15]. The study was examined in this way with 23 models. However, low results were obtained in terms of success percentage. Similarly, when low accuracy results were obtained in different studies in the literature, it was seen that the developers performed data reduction until the data distribution was equal [16]. In the studies in the literature, it has been stated that the data reduction process is done by randomly selecting high dimensional data types in the data set until the data sizes are equalized. The size of the mirai, gafgyt attack data examined in this study is larger than the size of the data containing normal traffic. Mirai and gafgyt data are reduced in size until they are equal to normal data. The results obtained after data reduction have reached very high accuracy percentages compared to previous results. In order to equalize the amount of data types, the gafgyt mirai attack data is reduced to equal the normal traffic data set. In the literature, the highest percentage of accuracy was achieved among the studies conducted with this data set. Table 2 shows the first 10 features used after the feature ranking process hierarchically.

Table 2. Features used in the study

No	Top 10 features out of 115 features
1	HpHp_L0.01_weight
2	HpHp_L0.01_covariance
3	HpHp_L0.1_covariance
4	HpHp_L0.01_pcc
5	HpHp_L0.1_pcc
6	HH_L0.1_pcc
7	HH_L0.01_pcc
8	HpHp_L0.01_magnitude
9	HpHp_L0.1_magnitude
10	HH_L0.01_covariance

In Table 2, using the Weka Correlation Attribute Eval tool, 115 features are ranked according to their effect weights. After this stage, the data is ready to be analyzed with Weka. The results of the data set examined with 23 models are presented in Table 3 and 4. The accuracy percentages of the supervised learning models specified in the tables are the results directly obtained from the Weka program. In unsupervised learning models, the program presents the percentage of miscalculation to the user. When this result of the program is subtracted from 100%, the accuracy percentage emerges. The Actual (Gaf, Mirai, Normal) rows in Table 4 show the actual distribution amount of the attack data, and the predict (Gaf, Mirai, Normal) columns show the data distribution amount formed by the machine separating the attack data according to its types after learning.

The data set was analyzed with different models and the accuracy percentages of all models are shown in Table 3. In Table 4, the distribution of the data according to the models as a result of the learning is given.

Table 3. Learning results

Machine learning models	Accuracy percentages
Naive Bayes	60.85%
Naive Bayes Updateable	57.56%
LDA	66.65%
Logistic	77.93%

(*continued*)

Table 3. (*continued*)

Machine learning models	Accuracy percentages
QDA	57.93%
IBK	99.88%
OneR	97.98%
PART	79.76%
ZeroR	40.55%
RepTree	99.81%
Hoeffding Tree	94.12%
J48	79.72%
J48 Consolidate	79.73%
J48 Graft	78.13%
JRIP	99.84%
Random Forest	99.92%
Random Tree	99.87%
Canopy	39.12%
EM	55.24%
FarthestFirst	39.28%
Filtered Clusterer	55.07%
Simple K-Means	55.07%
Make Density Clusterer	56.24%

The learning success of the models are shown in Fig. 1. When the graph is examined the while the model with the highest learning success is Random Forest with 99.92%, the lowest one is FarthestFirst with 39.28.

The data set examined in the application; It was searched in Web of Science, IEEExplorer, Science Direct, ERIC databases and the results obtained in the studies conducted with the data set are given in Table 5. In the literature, there are many studies conducted with the entire N-BaIoT data set. However, 7 studies were reached in the databases related to the data set examined in this study.

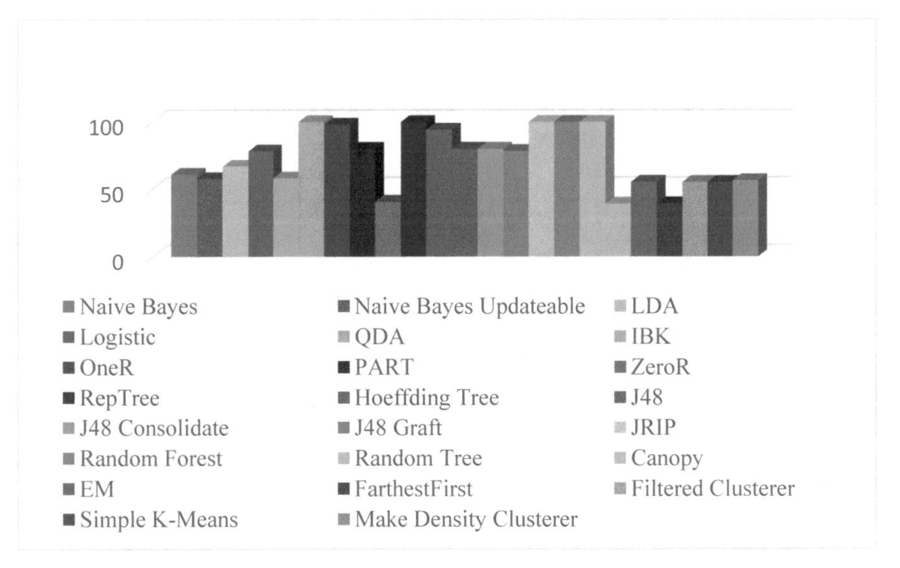

Fig. 1. Accuracy results of the models

Table 4. Data distribution results according to models as a result of learning

	Predicted Gaf	Predicted Mirai	Predicted Benign	Models
Actual Gaf	3	26991	89	Naive Bayes
	11580	726	807	Naive Bayes Updateable
	17159	121	77	LDA
	11269	15729	84	Logistic
	6	26994	82	QDA
	27074	0	8	IBK
	26978	20	84	OneR
	11332	15730	20	PART
	0	27082	0	ZeroR
	27040	1	41	REPTree
	23059	3744	279	Hoeffding Tree
	11324	15730	28	J48
	11327	15729	26	J48 Consolidate
	11268	15729	85	J48 Graft
	27054	7	21	JRip

(*continued*)

Table 4. *(continued)*

	Predicted Gaf	Predicted Mirai	Predicted Benign	Models
	27069	0	13	Random Forest
	27066	0	16	Radom Tree
	3	25998	20	Canopy
	42	25934	45	EM
	11	26010	0	FarthestFirst
	44	25949	28	Filtered Clusterer
	44	25949	28	Simple K Means
	42	25934	45	MakeDensity Based Clusterer
Actual Mirai	91	31708	0	Naive Bayes
	17	310040	573	Naive Bayes Updateable
	99033	113707	0	LDA
	61	31669	69	Logistic
	40	31761	28	QDA
	0	31780	19	IBK
	0	31799	0	OneR
	1	31772	26	PART
	0	31799	0	ZeroR
	1	31779	19	REPTree
	5	31682	112	Hoeffding Tree
	2	31767	30	J48
	2	31768	29	J48 Consolidate
	12	31682	105	J48 Graft
	0	31786	13	JRip
	0	31786	13	Random Forest
	0	31786	13	Radom Tree
	0	328039	2	Canopy
	0	28039	2	EM
	0	28041	0	FarthestFirst
	0	28039	2	Filtered Clusterer
	0	28039	2	Simple K Means

(continued)

Table 4. (*continued*)

	Predicted Gaf	Predicted Mirai	Predicted Benign	Models
	0	28039	2	MakeDensity Based Clusterer
Actual Benign	4838	1261	13429	Naive Bayes
	18	3441	508674	Naive Bayes Updateable
	5944	11420	2164	LDA
	1223	137	18168	Logistic
	4569	1266	13693	QDA
	8	59	19461	IBK
	195	1282	18051	OneR
	10	82	19436	PART
	0	19528	0	ZeroR
	25	58	19445	REPTree
	65	400	19063	Hoeffding Tree
	13	94	19421	J48
	8	93	19427	J48 Consolidate
	1118	93	18317	J48 Graft
	9	69	19450	JRip
	3	50	19475	Random Forest
	15	53	19460	Radom Tree
	585	18195	748	Canopy
	1029	6008	12491	EM
	57	19358	113	FarthestFirst
	1001	6160	12367	Filtered Clusterer
	1001	6160	12367	Simple K Means
	1029	6008	12491	MakeDensity Based Clusterer

Studies with all models in the study were examined and Mirai, Gafgyt and Normal (Benign) Roc curves of the Random Forest study, which achieved the highest accuracy rate of 99.92%, are shown in Fig. 1, Fig. 2, and Fig. 3. The area under the Roc curve shows the success of the relevant curve. This part of the study is very important. When literature is reviewed for this dataset, this studys' result is highest precision percentage. Therefore this result is valuable (Fig. 4).

ROC graphs are given in Fig. 1, Fig. 2 and Fig. 3 according to traffic types. The success of classification in ROC graphs is proportional to the formation of the curve

Table 5. Results of studies in the literature

Reference No	Applied model	Results
[16]	Random Forest	87.50% (Accuracy)
[17]	LSTM	99.91% (Accuracy)
[18]	CNN	85% (F1 Score)
[19]	LSTM-CNN	89.64% (Accuracy)
[20]	LSTM	94.80% (Accuray)
[21]	ANN	99.00%(Accuracy)
[22]	ANN	92.80% (Accuracy)

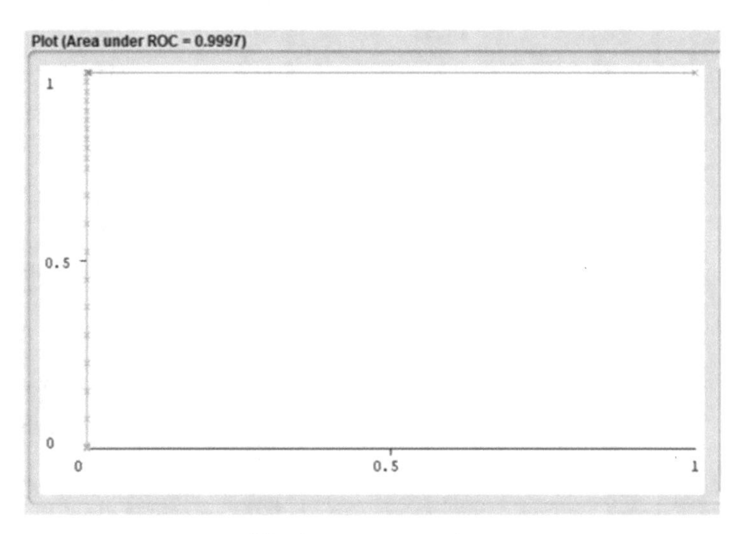

Fig. 2. Mirai ROC Chart

of the ROC graph in the upper left part. The formation of the ROC curve in the upper left part means that the size of the area under it is maximum. In this case, it shows the classification success of the model. For this reason, the fact that the curve is formed in the upper left section means that the area will be maximum, so the accuracy percentage will be maximum. In addition, X axis represents False Positive and Y axis True Positive value in ROC graph. In this respect, the closer the values of the X axis are to the origin, that is, the smaller the False Positive value, the more successful the model. On the other hand, the larger the values of the Y axis, that is, the larger the True Positive value, the more successful the model is. In short, when the X axis has the smallest values and the Y axis the largest values, the curve of the graph is formed in the upper left part. When the resulting graph is examined, it is seen that the area takes values close to the maximum. This shows that the model is successful.

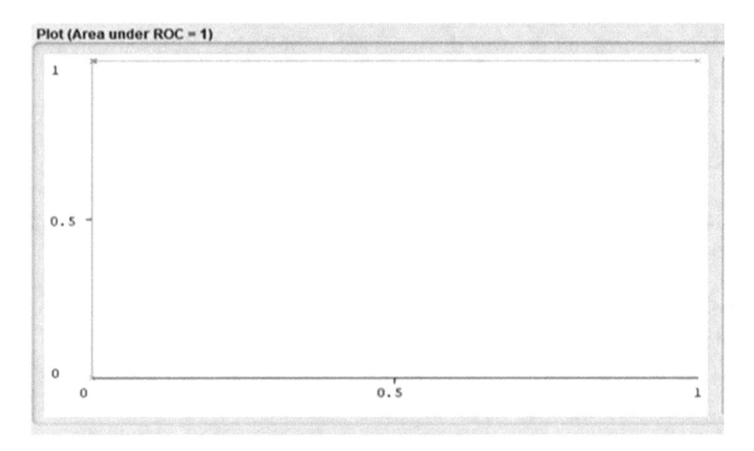

Fig. 3. Gafgyt ROC Chart

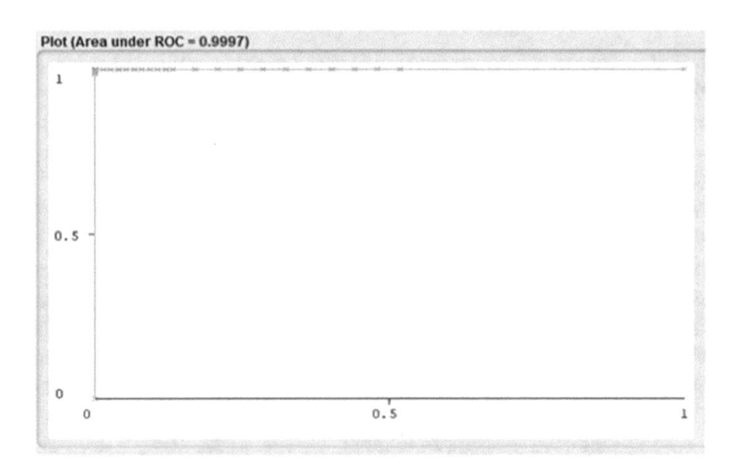

Fig.4. Normal (Benign) ROC Chart

4 Conclusions

In this study, the IoT botnet attack dataset, which has been used by cyber attackers in recent years, especially in DDOS attacks, has been examined. The aim of the study is to categorize the data set containing normal and attack network traffic as normal and attack network traffic with high accuracy by examining the data set with machine learning models. The dataset is the network traffic of the security camera, Simple-Home_XCS7_1003_WHT_Security_Camera. The dataset consists of 850,827 rows of 115 dimensional features. The large data size both prevents the Weka platform from working properly and learns with a low percentage of accuracy. Therefore, the feature and size reduction process used in the literature was applied on the data set [23]. While learning was carried out, supervised and unsupervised learning models were used. With the Weka platform, the data set was examined with 23 different machine learning models and the Random Forest model was the most successful algorithm with an accuracy rate of 99.92%.

References

1. Angrishi, K.: Turning internet of things (IoT) into internet of vulnerabilities (IoV): IoT botnets. ArXiv:1702.03681, pp. 1–17 (2017)
2. Shaikh, E., Mohiuddin, I., Manzoor, A.: Internet of things (IoT): security and privacy threats. In: 2019 2nd International Conference on Computer Applications & Information Security (ICCAIS), pp.1–6(2019). https://doi.org/10.1109/CAIS.2019.8769539
3. Doshi, R., Apthorpe, N., Feamster, N.: Machine learning DDoS detection for consumer internet of things devices. In: 2018 IEEE Symposium on Security and Privacy Workshops (2018)
4. Al Shorman, A., Faris, H., Aljarah, I.: Unsupervised intelligent system based on one class support vector machine and Grey Wolf optimization for IoT botnet detection. J. Ambient. Intell. Humaniz. Comput. **11**(7), 2809–2825 (2019). https://doi.org/10.1007/s12652-019-01387-y
5. Karaçalı, B.: Improved quasi-supervised learning by expectation-maximization. In: 2013 21st Signal Processing and Communications Applications Conference (SIU), Haspolat, pp. 1–4 (2013). https://doi.org/10.1109/SIU.2013.6531366
6. Meidan, Y., et al.: N-baiot—network-based detection of IoT botnet attacks using deep autoencoders. IEEE Pervasive Comput. **17**(3), 12–22 (2018)
7. Guerra-Manzanares, A., Medina-Galindo, J., Bahsi, H., Nõmm, S.: MedBIoT: generation of an IoT botnet dataset in a medium-sized IoT network. In: ICISSP, pp. 207–218 (2020)
8. Mirsky, Y., Doitshman, T., Elovici, Y., Shabtai, A.: Kitsune: an ensemble of autoen-coders for online network intrusion detection. In: 25th Annual Network and Distributed System Security Symposium, NDSS 2018, San Diego, California, USA, 18–21 February 2018 (2018)
9. Koroniotis, N., Moustafa, N., Sitnikova, E., Turnbull, B.: Towards the development of realistic botnet dataset in the internet of things for network forensic analytics: Bot-IoT dataset. Futur. Gener. Comput. Syst. **100**, 779–796 (2019)
10. Alsaedi, A., Moustafa, N., Tari, Z., Mahmood, A., Anwar A.: Ton_iot telemetry dataset: a new generation dataset of IoT and IIoT for data-driven intrusion detection systems. IEEE Access **8**, 165 130–165 150 (2020)
11. Hamza, A., Gharakheili, H.H., Benson, T.A., Sivaraman, V.: Detecting volumetric attacks on lot devices via SDN-based monitoring of mud activity. In: Proceedings of the 2019 ACM Symposium on SDN Research, pp. 36–48 (2019)
12. Kang H., Ahn, D.H., Lee, G.M., Yoo, J.D., Park, K.H., Kim, H.K.: IoT net-work intrusion dataset. IEEE Dataport (2019). https://dx.doi.org/10.21227/q70p-q449
13. Parmisano, A., Garcia, S., Erquiaga, M.J.: A labeled dataset with malicious and benign IoT network traffic. Stratosphere Laboratory (2020). https://www.stratosphereips.org/datasets-iot23
14. Bahşi, H., Nõmm, S., La Torre, F.B.: Dimensionality reduction for machine learning based IoT botnet detection. In: 15th International Conference on Control, Automation, Robotics and Vision (ICARCV), Singapore, pp. 1857–1862 (2018). https://doi.org/10.1109/ICARCV.2018.8581205
15. Durna, M.B.: Cross validation nedir? Nasıl çalışır? (2020). https://medium.com/bili%C5%9Fim-hareketi/cross-validation-nedir-nas%C4%B1l-%C3%A7al%C4%B1%C5%9F%C4%B1r-4ec4736e5142. Accessed 09 Nov 2020

16. Joshi, S., Abdelfattah, E.: Efficiency of different machine learning algorithms on the multivariate classification of IoT botnet attacks. In: 2020 11th IEEE Annual Ubiquitous Computing, Electronics & Mobile Communication Conference (UEMCON), pp. 0517–0521 (2020). https://doi.org/10.1109/UEMCON51285.2020.9298095
17. Hasan, T., Adnan, A., Giannetsos, T., Malik, J.: Orchestrating SDN control plane towards enhanced IoT security. In: 2020 6th IEEE Conference on Network Softwarization (NetSoft), pp. 457–464 (2020). https://doi.org/10.1109/NetSoft48620.2020.9165424
18. Kim, J., Shim, M., Hong, S., Shin, Y., Choi, E.: Intelligent detection of IoT botnets using machine learning and deep learning. Appl. Sci. **10**(19), 7009 (2020). https://doi.org/10.3390/app10197009. Accessed 8 Oct 2020
19. Alkahtani, H., Aldhyani, T.: Botnet attack detection by using CNN-LSTM model for internet of things applications. Secur. Commun. Netw. **2021**, 1–23 (2021). https://doi.org/10.1155/2021/3806459. Accessed 10 Sept 2021
20. De La Torre Parra, G., Rad, P., Choo, K., Beebe, N.: Detecting Internet of Things attacks using distributed deep learning. J. Netw. Comput. Appl. **163**, 102662 (2020). https://doi.org/10.1016/j.jnca.2020.102662. Accessed 1 Aug 2020
21. Sumaiya Thaseen, P.V.I., Reddy Gadekallu, T., Aboudaif, M.K., Abouel Nasr, E.: Robust attack detection approach for iiot using ensemble classifier. Comput. Mater. Continua **66**(3), 2457–2470 (2021). https://doi.org/10.32604/cmc.2021.013852. Accessed 2021
22. Palla, T., Tayeb, S.: Intelligent mirai malware detection for IoT nodes. Electronics **10**(11), 1241 (2021). https://doi.org/10.3390/electronics10111241. Accessed June 2021
23. Bagui, S., Wang, X., Bagui, S.: Machine learning based intrusion detection for IoT bot-net. Int. J. Mach. Learn. Comput. **11**(6) (2021)

Research and Analysis of the Efficiency Processing Systems Information Streams of Telematic Services

Bayram G. Ibrahimov[1](✉), Gulnar G. Gurbanova[1], Zafar A. Ismayilov[1], Manafedin B. Namazov[2], and Asif A. Ganbayev[2]

[1] Department of Radioengineering and Telecommunication, Azerbaijan Technical University, H. Javid ave. 25, AZ 1073 Baku, Azerbaijan
i.bayram@mail.ru, Gulnar.Gurbanova@aztu.edu.az

[2] Department of Computer Engineering and Information Technologies, Baku Engineering University, Hasan Aliyev str. 120, AZ 0101 Khirdalan, Absheron, Baku, Azerbaijan
aqanbayev@beu.edu.az

Abstract. The performance indicators of the functioning systems digital processing information flows in multiservice communication networks are analyzed. In this article, the subject research is the efficiency functioning systems for processing information flows telematic services based on the architectural concept of subsequent communication networks NGN (Next Generation Networks) using modern computer technology. As a criterion for the effectiveness of the functioning system, the performance information flow processing systems was chosen, which characterizes the following complex indicators such as the throughput of the system hardware and software in the exchange various types messages, the probabilistic-temporal characteristics useful and service traffic, the reliability telematic services system, and protection information from unauthorized access subscriber and network communication lines. The purpose this work is to study and analyze the effectiveness of the functioning systems for processing information flows telematic services using NGN technologies. As a result, the study proposed a mathematical model (MM), taking into account the nature of the processed telematics traffic and the presence of a variety network resources, on the basis which the performance indicators of the information processing system were analyzed. Analytical expressions are obtained to evaluate the indicators quality of service, information security and fault tolerance of the system functioning during the transmission various multimedia traffics. On the basis numerical calculations were carried out and graphical dependences of the probability trouble-free operation system on the failure rate software and hardware nodes and the dependence of the average length packet queue on the value of the system load factor were constructed. The given graphic dependencies showed that the presence network resources leads to a significant improvement in the performance characteristics systems for processing information flows telematic services.

Keywords: telematic services · information flow processing systems · reliability · performance · fault tolerance · quality of service · security risk · average queue length

D. J. Hemanth et al. (Eds.): ICAIAME 2022, ECPSCI 7, pp. 620–629, 2023.
https://doi.org/10.1007/978-3-031-31956-3_52

1 Introduction

At the present stage development, network infrastructures of the digital economy and the creation strategic plans "Digitalization Roadmap" require the rational organization and management resources of transport enterprises, ensuring the efficiency of the functioning systems for processing information flows telematic services [1, 2].

As a criterion for the effectiveness of the functioning system, the performance information flow processing systems was chosen, which characterizes the following complex indicators:

- throughput of the system hardware and software in the exchange of various messages;
- probabilistic-temporal characteristics useful and service transport traffic;
- reliability of functioning of the system of system telematic services;
- protection of information from unauthorized access of subscriber communication lines.

Therefore, the tasks analyzing the above performance indicators, the functioning of the transmission system and the processing information flows of telematic services using promising basic technologies and protocols of the NGN concept, are the most relevant.

The papers [3–5] analyze methods for improving the characteristics of the throughput of a hardware and software system and algorithms for protecting information from unauthorized access.

In [4, 6–8], the quality of functioning information processing systems telematic services using the NGN architectural concept and technologies was studied, and their main network and channel indicators were identified.

However, the task studying complex performance indicators of the functioning systems for processing information flows telematic services in the exchange useful and service transport traffic has not yet been fully resolved.

In this paper, we consider the solution of the problem formulated above - the study and evaluation of the effectiveness functioning systems for processing information flows telematic services using NGN technologies.

2 General Statement of the Problem Research and Construction of a Mathematical Model

To evaluate the performance indicators of the functioning of the system for processing information flows of telematic services, it is necessary to develop a mathematical model (MM), taking into account the nature of the transmitted traffic and the presence of many network resources [2, 9–11].

It has been established [1, 4, 6, 12] that the studied message processing systems are a queuing system (QS), which, according to the Basharin-Kendall encoding, correspond to a common type $M/G/1/N_{bs}$ with one serving terminal server.

Let us assume that the incoming packet flow into the queuing system has Poisson distribution laws with the parameter λ_i, and the service duration of i-th traffic has an arbitrary distribution function $B(t)$ with moments b_i.

We assume that with a critical load $\rho_i \le 1, i = \overline{1, n}$ in the system, the number of waiting places is limited to N_{bs}. Then, the studied MM is a single-channel queuing system of general type $M/G/1/N_{bs}$ with limited queues.

The performance of the system under study is determined by the number of streams of traffic packets transmitted per unit of time under conditions of given reliability and information security, which are functionally described by the following relationship:

$$P_i(\lambda_i) = F\left[K_i, N_{sq}(\lambda_i), I_i, C_i\right], i = \overline{1, n}, \tag{1}$$

where $N_{sq}(\lambda_i)$– is the average waiting time for a packet in the service queue in the system, taking into account the incoming flow rate λ_i when processing the i-th traffic, $i = \overline{1, n}$;

$P_i(\lambda_i)$– performance of the system of hardware and software, taking into account the rate of receipt of the incoming stream λ_i when processing the i-th traffic, $i = \overline{1, n}$;

K_i - factor of maintaining fault tolerance of hardware and software when processing the i-th traffic flow, $i = \overline{1, n}$;

I_i - coefficient of information security of the operation of hardware and software when processing the i-th traffic flow, $i = \overline{1, n}$;

C_i - capacity of the system of hardware and software in the transmission of the i-th traffic flow ($C_i = C_{i.max}$), $i = \overline{1, n}$.

The mathematical formulation of the task of the proposed MM for assessing the performance indicators of systems for processing information flows of telematic services is described by the following functional transformation of traffic:

$$E_{ft}(\lambda)\{t_{in}\} \rightarrow \{\lambda\} \left| \begin{array}{l} K_i \ge K_{i.allow.}, i = \overline{1, n} \\ N_{sq}(\lambda_i) \le N_{sq.allow.}(\lambda_i) \\ C_i \le C_{i.allow.}, i = \overline{1, n} \\ I_i \le I_{i.allow.}, i = \overline{1, n} \end{array} \right., \tag{2}$$

under the following restrictions

$$T_{\det.}(\lambda_i) \le T_{\det.}^{allow.}(\lambda_i), P_{fail.} \le P_{fail.}^{allow.}, i = \overline{1, n}, \tag{3}$$

where $T_{\det.}(\lambda_i)$– is the packet delay time from the moment it arrives to the moment it is sent, taking into account the rate arrival of the incoming stream λ_i;

$P_{fail.}$– probability of denial of service to a packet of incoming traffic;

$T_{\det.}^{allow.}(\lambda_i)$, $P_{fail.}^{allow.}$, $C_{sq.allow.}$, $K_{i.allow.}$ and $I_{i.allow.}$– respectively, the allowable value of the value packet delay time, the probability failure, the average waiting time of the packet in the queue for services in the system, throughputs, the coefficient fault tolerance and the coefficient information security of the functioning of the software and hardware transport services during the transmission of the i-th packet stream, $i = \overline{1, n}$.

Expressions (1), (2) and (3) define the essence of the new approach under consideration, taking into account the intensity transport traffic, on the basis which the MM is proposed for analyzing the performance indicators of the systems for processing information flows of telematic services.

3 Investigation of the Fault Tolerance Functioning Message Processing System

In [6], it was revealed that one of the important performance criteria of information processing systems is the coefficient of maintaining the fault tolerance of the system operation K_i, $i = \overline{1, n}$, which determine the reliability indicators:

$$K_i = F[P_{RS}(\Lambda, t), P(\Lambda), P_{i.PFO}(t, \Lambda)], i = \overline{1, n}. \tag{4}$$

Based on the study [2, 6], it was found that the fault tolerance indicators are determined by the choice of means to ensure the reliability of active functional units both switches and controllers using NGN network protocols, firewalls, and groups network servers. These are important and functional compositions of the hardware and software of the information flow processing system.

Consider one of the options for assessing the reliability of a system consisting of a single network server that serves the i-th stream of traffic packets. Then the assessment of the reliability of the system is carried out using the well-known mathematical apparatus [6, 7]:

$$P_{RS}(\Lambda, t) = \Pi_{i=1}^{n}[P_{i,PFO}(t, \Lambda)], \ i = \overline{1, n}, \tag{5}$$

where $P_{i.PFO}(t, \Lambda)$– is the probability of failure-free operation of the system, $i = \overline{1, n}$; Λ–failure rate of network switching nodes, (1/s).

Expression (5) characterizes the fault tolerance of the system operation with the failure rate Λ of the system hardware and software.

Considering the number of switching nodes N_n and the availability factor, the probability of failure of hardware and software while maintaining the functioning of the system is expressed as follows [7, 8, 10]:

$$P(\Lambda) = 1 - [1 - K_a(\Lambda)]^{N_n}, \tag{6}$$

where $K_a(\Lambda)$– are the availability factors of the system hardware and software and is expressed as

$$K_a(\Lambda) = \Lambda_r/(\Lambda + \Lambda_r), \tag{7}$$

where Λ_r–is the intensity of system hardware and software restorations.

Formula (6) and (7) characterizes one of the most important indicators of the coefficient of system fault tolerance K_i, $i = \overline{1, n}$ for given load factors of hardware and software nodes when processing information flows of useful $\lambda_{i.u}$ and serve $\lambda_{i.s}$ of traffic, which is found by the expression:

$$\rho_i = (\lambda_{i.u} + \lambda_{i.s})/(\mu_i \cdot C_i) \leq 1, i = \overline{1, n}, \tag{8}$$

where μ_i–is the service rate of the i-th traffic flow and is equal to $\mu_i^{-1} = b_i$.

Expressions (8) also determine the coefficient of efficient use of the system hardware and software when servicing information flows of the i-th traffic of automotive services.

4 Numerical Analysis Results

Figure 1 shows a graphical dependence of the probability failure-free operation on the failure rate hardware and software nodes for a given system load factor.

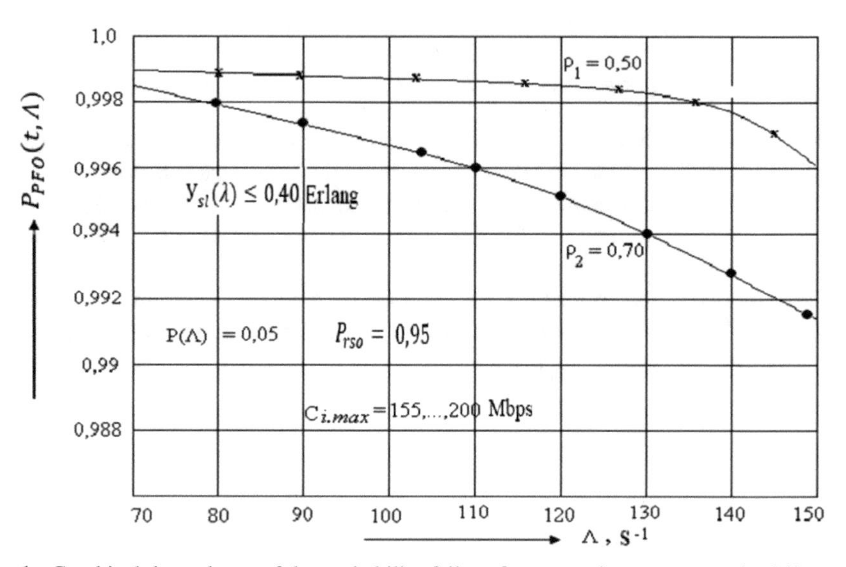

Fig. 1. Graphical dependence of the probability failure-free operation system on the failure rate hardware and software nodes

Thus, from The analysis of the graphic dependence shows that in the case of an increase in the failure rate, with the load intensity of the service channel $Y_{sl}(\lambda) \leq 0, 40$ Erlang, the probability failure $P(\Lambda) = 0, 95$ the restoration of the system operability $P_{rso} = 0, 95$ and $C_{i.\max} = (155, \ldots, 200)$ Mbps.

The probability system uptime is greatly reduced. Its noticeable change begins with the values $\Lambda \geq 125s^{-1}$.

Expression (5) and the given graphical dependencies

$P_{PFO}(t, \Lambda) = F[\rho_1, \Lambda, \rho_2]$ it follows that with an increase in the failure rate of the complexes, it leads to a deterioration in the fault tolerance of the system operation at $\rho_i \leq 0, 65$, and thus thereby worsens the protection information from unauthorized access subscriber and network communication lines in general.

5 Analysis and Protection of Information in the Message Processing System

It is known [6, 7] that the spread of information technology has given a lot of new opportunities in the communication system, business, science and everyday life. However, along with this, methods of fraud, theft and substitution of information have also developed.

In a system with an increase in the number of information technologies, the need to ensure information security in the transmission of useful and service traffic is growing [2, 5, 6, 9].

Taking into account possible unauthorized access and cyber threats, the security factor for the functioning of software and hardware $I_i^s(\lambda_i)$ when servicing the i-th flow of traffic packets is found by the expression:

$$I_i^s(\lambda_i) = \left[1 - I_r^{hf}(\lambda_i)\right] \leq 1, i = \overline{1, n}, \quad (9)$$

where $I_i^{hf}(\lambda_i)$– is the hazard coefficient of the functioning of the service system for the i-th flow of traffic packets with intensity λ_i, $i = \overline{1, n}$.

Expressions (9) evaluate the level of information protection against the probability of a threat to information security in case of unauthorized access to the system.

6 Research on the Effectiveness of the Use Innovative Technology Resources

At the present stage development and implementation of the architectural concept NGN and FN (Future Network), it is expected that the family SDN (Software Defined Networking), NFV (Network Functions Virtualization), 5G/IMT-2020 and IMS (Internet Protocol Multimedia Subsystem) technologies, united under the general name technologies for building distributed communication networks, will become a standard solution for multiservice communication networks. These systems and technologies provide high bandwidth, low network interface latency and accelerate the launch of new multimedia and telematics services and applications [4, 12–14].

It should be noted that the preliminary requirements for SDN, NFV, 5G/IMT-2020 and IMS technologies were put forward by the 3-rd Generation Partnership Project (3GPP), a non-profit consortium that develops specifications for multiservice communication networks.

In this paper [13, 14], a set mathematical model of a communication system is investigated and analyzed, which characterizes the effectiveness multiservice communication networks built in accordance with the concept NGN and FN in terms reserving a part information, channel and network resources for the case stationary transceiver terminal devices and mobile hardware and software.

It is assumed here that after the receipt of the interrupt signal, the resource requirement for continuing the session service changes, while an increase in the resource requirement corresponds to the appearance, and decrease - the disappearance of the mobile blocker on the line of sight between the base station BS and the terminal equipment.

In this case, we believe that due to the use of the technology of building distributed communication networks - SDN, NFV, 5G/IMT-2020 and IMS, multiservice communication networks have a resource of volume ΔF_k, Hz.

The reservation mechanism [4, 13–18] assumes that only a part of the resource, namely $\eta_k \cdot \Delta F$, is available for requests to establish a new session, while the entire volume of the resource is available for sessions already accepted into the system. Reserving a share $(1 - \eta_k)$ of the resource for sessions accepted for servicing is intended to protect them from being dropped if the line-of-sight blocking between the base station and terminal equipment occurs in a state when the entire public resource is busy, that is, $\eta_k \leq \eta_{k.tot.}$.

From the above, it can be seen that a communication network with resources ΔF_k is a queuing system. Now let's consider a queuing system with intensity λ, with a finite number of channels N_k, hardware and software systems, and a resource of limited volume ΔF.

We assume that the system serves two types of claims - new and repeated streams non-uniform traffic packets. The entire free resource from the total volume of resource B is available to the re-claim. If the volume of the requested resource for the re-claim is greater than the free resource, the claim is discarded without completing service. Following [1], to derive the stationary distribution in analytical form, we introduce the Markov process

$$Y(t) = W[M(t),\ \eta(t)], \tag{10}$$

where $M(t)$– the number active sessions in the system, and $\eta(t)$– the total amount of occupied resources.

The state space for a process $Y(t)$ is as follows:

$$Y = \bigcup\nolimits_{n=1}^{N_k} Y_n, \quad Y_n = \{(n,\ r)\ :\ 0 \leq r \leq \Delta F_k,\ P_r^{(n)} > 0\}, \tag{11}$$

where $P_r^{(n)}$– denotes n– multiple convolution of probabilities P_r, $r \geq 0$, and interpreted as the probability that n– requests occupy resource r units.

Considering the above, the proposed mechanism will work more efficiently in congestion conditions and with sessions that have high requirements for data transfer speed when providing telematic services, which increases its demand for communication systems using SDN, NFV, 5G / IMT-2020 and IMS technologies.

7 Estimation of the Average Residence Time of a Flow Traffic Packets in the System

Based on the analysis of the $M/G/1/N_{bs}$ model, it was revealed that one of the key characteristics for evaluating system performance is the average waiting time for a packet in the queue $E\big[N_{sq}(\lambda_i)\big]$ for services in the telematics services system [6, 9].

Taking into account the Polyachek-Khinchin formula, the average waiting time for a packet in the service queue in the system telematics services, taking into account the incoming flow rate λ_i, when processing the i-th traffic, is determined by the following expression [2]:

$$E\big[N_{sq}(\lambda_i)\big] = \frac{\lambda_{i.u} + \lambda_{i.s}}{C_i \cdot \mu_i} \cdot \frac{(1 + C_v^2)}{2(1 - \rho_i)} \cdot E[t_{trans}],\ i = \overline{1, n}, \tag{12}$$

where C_v^2– is the square of the time packet transmission variation coefficient and determines the structural complexity of the traffic, and is calculated as

$$C_v^2 = \sigma_v^2 / m_v^2, \tag{13}$$

m_v^2, σ_v^2– the value of the mathematical expectation of the packet transit time and its dispersion, respectively;

$E[t_{trans}]$– average packet transmission time through the switching nodes of the system, without taking into account the waiting time in the queue.

Expression (12) determines one of the important probabilistic and temporal characteristics of the transmitted useful and service traffic and is an indicator of QoS (QoS - Quality of Service) of telematics services. In addition, in (10) an important indicator is the load factor serving the software and hardware of the communication system and is expressed as

$$\rho_i = E[t_{trans}]/\lambda_i. \tag{14}$$

Figure 2 shows a graphical dependence of the average packet waiting time in the QS queue on the load factor of the serving software and hardware complex for a given switching node availability factor and system throughput $C_{i.\ \max}$.

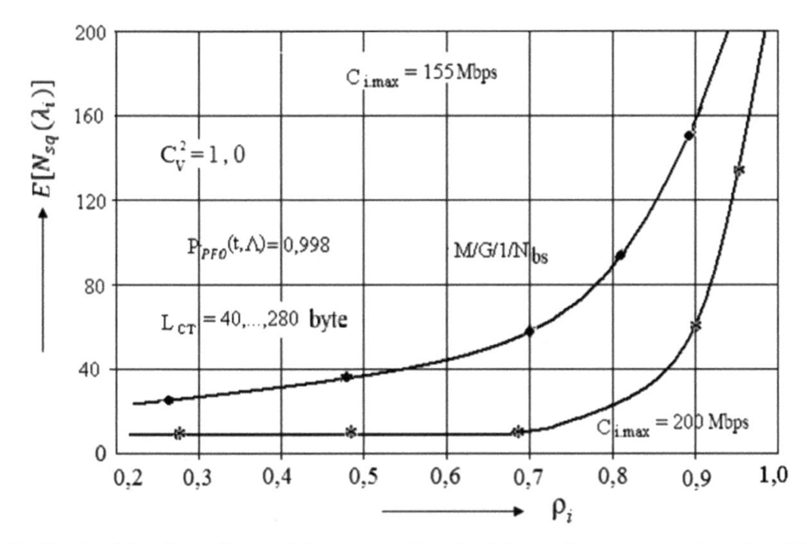

Fig. 2. Graph of the dependence of the average length of the packet queue on the value QS load factor

The analysis of the graphic dependence shows that with an increase in the load factor of QS $\rho \geq (0, 63, ..., 0, 85)$, which meets the requirement of fault-tolerance of functioning and information security of the system at $Y_{loda.} = (0, 20, ..., 0, 40) Earl.$, $C_s^2 - 1, 0$ and $C_{i.\ \max} \leq (155, \ldots 200)$ Mbps, the average queue length in the network node decreases. Its noticeable change begins with values $\rho \geq 0, 65$.

8 Analysis of the Reliability and Information Security Indicator of Telematic Services

The following important characteristics were selected as a criterion indicators of system reliability and information security: average time between failures, attack threats with the intensity occurrence λ_a, intensity elimination μ_a of the real threat attack, and the parameter failure flow Λ_{no}.

Based on the reliability model, the failure flow parameter λ_{no} for the stationary section is found by the following expression [4, 13]:

$$\lambda_{no} = \sum_{i \in Q_{pc}} P_i \cdot \sum_{i \in Q_{oc}} \Lambda_{ij}, \tag{15}$$

where $Q_{pc}-$ many health conditions of telematics services based on FN;

$Q_{oc}-$ many of telematics services failure conditions and equal $Q_{oc} = 1 - Q_{pc}$;

$\Lambda_{ij}-$ the intensity of the transition from $i-$ th working state, the probability of finding the system P_i in which $j-$ th inoperative state.

In order to evaluate information security parameters, the following important characteristics of the threat of attack were selected, which are expressed by the following formulas [4, 12]:

$$\lambda_a = 1/E\left[T_{ya}\right], \ \mu_a = \lambda_a \cdot P_a \cdot (1 - P_a), \tag{16}$$

where $E[T_{ya}]$ average network failure time; P_a- the probability of readiness for safe operation of the protected information system when changing the security parameters of the protection system λ_a and μ_a communication networks.

Based on (15) and (16), it is possible to determine the stationary coefficient K_k of system availability during the safe operation hardware-software complexes networks:

$$K_k = E[T_{ya}]/\lambda_{no}. \tag{17}$$

Expression (17) is one of the important integral characteristics of system telematic services based on the architectural concept FN, which is necessary so that the information protection system does not introduce large service delays.

Thus, on the basis of the proposed model, analytical expressions have been obtained that make it possible to evaluate the performance indicators of the information processing system.

9 Conclusions

As a result of the study, MMs were proposed, taking into account the presence many network resources, the nature useful and service traffic of telematic services, on the basis of which the performance indicators of the information processing system based on the NGN and FN architectural concept using modern technologies were analyzed. On the basis of the model, the obtained analytical expressions and the given graphical dependencies showed that the presence network resources leads to a significant improvement in the performance characteristics information flow processing systems.

References

1. Kleinrock, L.: Queueing Systems, vol. I: Theory, p. 432. Wiley-Interscience, New York, NY (1975)
2. Ibrahimov, B.G., Hashimov, E.Q.: Analysis and selection performance indicators multiservice communication networks based on the concept NGN and FN. Computer and Information Systems and Technologies, Kharkiv, April, 96–98 (2021)
3. Seitz, N.: ITU-T QoS standards for IP-based networks. IEEE Commun. Mag. **41**(6), 82–89 (2003)
4. Ibrahimov, B.G., Humbatov, R.T., Ibrahimov, R.F.: Performance multiservice telecommunication networks based on the architectural concept future networks using SDN technology. T-Comm **12** (12), 84–88 (2018)
5. Efimushkin, V.A., Kozachenko, Y.M., Ledovskikh, T.V., Shcherbakova, E.N.: The future image of the unified telecommunications network and the Russian Federation. Electrosvyaz (10), 18–27 (2018)
6. Ibrahimov, B.G.: Research and estimation characteristics of terminal equipiment a link multiservice communication networks. Autom. Control. Comput. Sci. **46**(6), 54–59 (2010)
7. Aliev, T.I.: Fundamentals of System Design, p. 120. ITMO University, St. Petersburg (2015)
8. Bankov Dmitry, Khorov Evgeny, Lyakhov Andrey. . On the limits of LoRaWAN channel access. In: Engineering and Telecommunication (EnT), International Conference on/IEEE, pp. 10–14 (2016)
9. Ibrahimov, B.G., Alieva A.A. (2021). Research and Analysis Indicators of the Quality of Service Multimedia Traffic Using Fuzzy Logic. In: Aliev R.A., Kacprzyk J., Pedrycz W., Jamshidi M., Babanli M., Sadikoglu F.M. (eds) 14th International Conference on Theory and Application of Fuzzy Systems and Soft Computing, ICAFS-2020. Advances in Intelligent Systems and Computing, vol 1306. Springer, Cham, 773–780
10. Bogatyrev, V.A., Bogatyrev, A.V.: Reliability of functioning real-time cluster systems with fragmentation and redundant query servicing. Inf. Technol. **22**(6), 409–416 (2016)
11. Xiong, B., Yang, K., Zhao, J., Li, W., Li, K.: Performance evaluation of openflow-based soft-ware-defined networks based on queueing model. Comput. Netw. **102**, 172–185 (2016)
12. Kartashevskiy, I., Buranova, M.: Calculation of packet jitter for correlated traffic. In: International Conference on Internet of Things, Smart Spaces, and Next Generation Networks and Systems. NEW2AN 2019, vol. 11660, pp. 610–620 (2019)
13. Ibrahimov B.G., Hasanov A.H. (2021). Research of the quality of functioning multiservice communication networks when establishing a multimedia session. Computer and Information systems and technologies. Kharkiv, april, 55
14. Moltchanov, D., et al.: Improving session continuity with bandwidth reservation in mmwave communications. IEEE Wirel. Commun. Lett. **8**(1), 105–108 (2019)
15. Islamov, I.J., Hunbataliyev, E.Z., Zulfugarli, A.E.: Numerical simulation characteristics of propagation symmetric waves in microwave circular shielded waveguide with a radially inhomogeneous dielectric filling. Camb. Univ. Press Int. J. Microw. Wirel. Technol. **9**, 761–767 (2022)
16. Islamov, I.J., Ismibayli, E.G., Hasanov, M.H., Gaziyev, Y.G., Ahmadova, S.R., Abdullayev, R.: Calculation of the electromagnetic field of a rectangular waveguide with chiral medium. Progress Electromagn. Res. **84**, 97–114 (2019)
17. Islamov, I.J., Ismibayli, E.G.: Experimental study of characteristics of microwave devices transition from rectangular waveguide to the megaphone. IFAC-PapersOnLine. **51**(30), 477–479 (2018)
18. Islamov, I.J., Ismibayli, E.G., Gaziyev, Y.G., Ahmadova, S.R., Abdullayev, R.: Modeling of the electromagnetic feld of a rectangular waveguide with side holes. Prog. Electromagn. Res. **81**, 127–132 (2019)

An Artificial Intelligence-Based Air Quality Health Index Determination: A Case Study in Sakarya

Salman Ahmed Nur(✉), Refik Alemdar, Ufuk Süğürtin, Adem Taşın, and Muhammed Kürşad Uçar

Electrical and Electronics Engineering Department, Sakarya University, Sakarya, Turkey
salmaanahmednuur@gmail.com, mucar@sakarya.edu.tr

Abstract. Long-term exposure to air pollutants has been shown to increase the risk of respiratory disease, heart disease, and lung cancer among healthy individuals. Numerical air quality indices have been developed to characterize air pollution levels. Air Quality Health Index (AQHI) is one of these indices and it is calculated from the relationship between pollutant particles ($PM_{2.5}$), oxygen (O_3), and nitrogen dioxide (NO_2) in the environment and short-term mortality rates. The purpose of this study is to build regression models to determine the AQHI of Sakarya without any calculation. The data including $PM_{2.5}$, O_3, and NO_2 parameters were taken from the website of the Ministry of Environment, Urbanization and Climate Change in Türkiye. In this study, which features are required for a good regression performance are revealed using the Spearman feature selection algorithm. To evaluate the performance of the models, the mean-square error (MSE) and root-mean-square error (RMSE) were used. The results showed that the GPR-based model performed better in determining the AQHI (MSE = 4.48e−06, RMSE = 0.00211). It was concluded that artificial intelligence-based methods can be used in the determination of AQHI without calculation.

Keywords: Artificial intelligence · Prediction · Regression · Machine learning · Air quality health index

1 Introduction

Today, humanity is struggling with many environmental crises such as air pollution, hazardous waste, global warming, and depletion of resources [1]. Air pollution is the dangerous level of pollutants in the form of gas, water vapor, smoke, and dust that harm all living things in the atmosphere [2]. Air pollution can be human-caused such as transportation, heating, and industry, or naturally sourced, such as desert dust, volcanic activities, and forest fires [3]. As a result of low-quality fossil fuels for heating purposes and industrial activities, the concentration of pollutants in the air is gradually increasing. A rapidly growing industry, population growth, and continuing urbanization have increased air pollution. Air pollution has a crucial place on the quality of life in countries or cities which are industrialized. Approximately 91% of the world's population lives in

D. J. Hemanth et al. (Eds.): ICAIAME 2022, ECPSCI 7, pp. 630–639, 2023.
https://doi.org/10.1007/978-3-031-31956-3_53

areas that exceed the air quality limits set by the World Health Organization (WHO) [4]. People living in lower-income, middle, and less-developed countries also experience air pollution problems. Exposure to air pollution has serious health consequences. According to the WHO, an estimated 4.2 million deaths per year from heart disease, stroke, lung cancer, and chronic respiratory diseases are caused by air pollution [5]. Millions of people die from diseases caused by air pollution [6].

In order to prevent this air pollution in cities, air pollution must be measured from specific points, and necessary measures should be taken according to these measurement results [7].

2 The Air Quality Health Index (AQHI)

There are air quality monitoring stations in cities at various points for monitoring air quality [8]. Installing these stations is costly and may require expensive maintenance afterward. Therefore, the importance of air pollution forecasting studies has increased. The values to be measured for air quality are determined by international and national evaluations [9]. Accordingly, air pollution is determined according to the amounts of sulphur dioxide (SO_2), nitrogen dioxide (NO_2), ozone (O_3), carbon monoxide (CO) gases, and dust particles (PM_{10}). Traditional mathematical models are used to extract the desired results from these measurements.

In the literature, various machine learning algorithms were used to estimate air pollution and successful prediction results were obtained based on the concentrations of essential air pollutants such as particulate matter (PM_{10}), nitrogen oxide (NO), sulphur dioxide (SO_2) with a size between 2.5 μm and 10 μm. One study, measurement data of air pollutant gases such as nitrogen dioxide (NO_2), SO_2, ozone (O_3), carbon monoxide (CO), and PM_{10} were discussed for Adana province, and the air quality index was predicted with various machine learning algorithms [3]. In the air pollution estimation study for Erzincan province [10], daily records of PM_{10} and SO_2 pollutants between 2016–2018 were used, and estimation was made with KNN, one of the machine learning methods. In their study, Veljanovska and Dimoski (2018) compared the success results of classification algorithms of k-nearest neighbour, decision tree, neural networks, and support vector machines in determining the air quality index of the city of Skopje [11]. Countries have used the Air Quality Index (AQI) to mitigate the potential health consequences of exposure to outdoor air pollution. The Air Quality Index (AQI), reported in many countries worldwide, is a simple and easy-to-understand numerical scale with a 0–500 range. The AQI value concentration is calculated by comparing air pollutant criteria. The pollutants determine air quality with the highest AQI score among all these criteria [12]. The pollutant, on the other hand, that deviates the greatest from the reference standard determines the AQI [13]. The Air Quality Index (AQI) has received much criticism for not accurately reflecting the cumulative health consequences of many air contaminants. One of these criticisms is the disregard of people at high health risk in the low concentration of pollutants, where the value of AQI is classified as "good." In addition, AQI is inadequate to show the risk level of short-term exposure to air pollution.

For these reasons, the Air Quality Health Index (AQHI) was developed and implemented by Canada in order to deal with these possible disadvantages of AQI [14]. Later,

Table 1. The AQHI categories

Health Risk	AQHI Category
Low	1 to 3
Moderate	4 to 6
High	7 to 10
Very high	Above 10

Data from Environment Canada [19].

the same index was adapted to Switzerland and the Chinese cities of Shanghai [15], Guangzhou [16], Tianjin [17], and Hong Kong [18]. This index was calculated by estimating time series analysis of the deadly short-term risks of selected air pollutants, including particulate matter (PM_{10} or $PM_{2.5}$), nitrogen dioxide (NO_2), and ozone (O_3).

AQHI is an index in which the possible risk effects of air pollution are transmitted in health messages classified on a scale of 1 and 10. The following table (Table 1) shows the messages that correspond to these values.

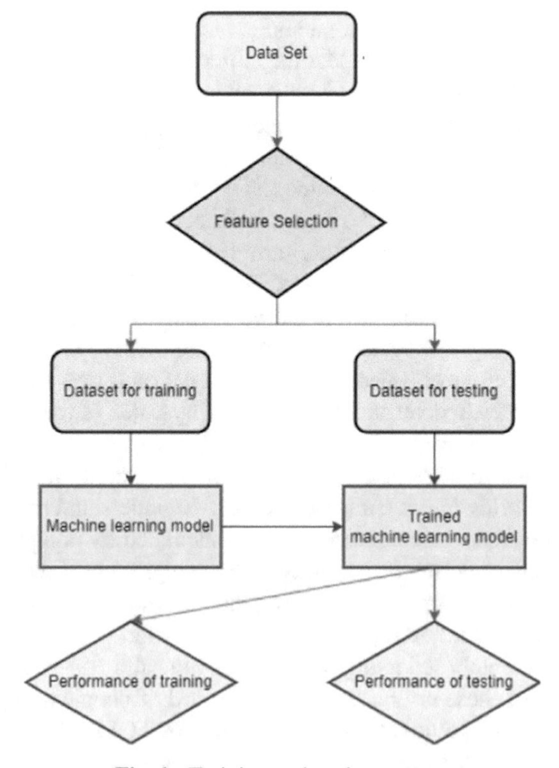

Fig. 1. Training and testing process

3 Method and Material

The processing steps of machine learning algorithms are given in Fig. 1 for determining air quality health index.

3.1 Data Set

In this study, the dataset obtained from the Air Monitoring Station sensor data of the Ministry of Environment, Urbanization and Climate Change in Türkiye was used [20].

The dataset consists of all days of the years of 2019, 2020 and 2021 and the first 15 days of January 2022. It has 1110 instances with each consists of 11 features (PM10, PM2.5, SO_2, NO_2, NO_X, NO, O_3, Air Temperature, Wind Speed, Relative Humidity, Air Pressure) and one label (AQHI). The data set is shown in Table 2.

Table 2. The Data set used for machine learning

Subject No	Features											Label
	PM_{10}	$PM_{2.5}$	SO2	NO_2	NO_X	NO	O_3	Air Temp.	Wind Speed	Relative Hum.	Air Pres.	AQHI
1	89,5	63,36	22,48	34,74	68,61	22,11	13,82	5,46	0,95	68,65	1019,21	6,68
2	114,29	85,91	15,48	38,01	93,43	36,17	3,76	6,75	0,63	75,67	1009,51	7,54
3	61,2	54,41	9,67	33,85	62,3	18,54	11,31	5,83	0,7	99,42	1007,8	6,04
4	31,29	28,79	5,18	26,33	52,49	17,03	23,73	4,17	1,94	100	1010,52	4,82
5	31,38	27,92	6,63	29,73	76,21	30,31	21,22	3,95	0,81	95,23	1012,86	4,94
6	51,52	49,07	9,89	30,19	65,71	23,16	15,5	3,25	1,23	100	1006,11	5,69
7	32,54	30,54	8,43	33,74	72,51	25,31	19,78	2,5	1,26	100	1013,01	5,34
8	25,93	24,83	5,38	30,14	53,54	16,04	25,09	1,43	1,57	95,78	1014,91	5,03
9	63,91	47,43	17,42	37,13	73,44	24,81	19,13	7,5	2,3	77,78	1012,7	6,40
10	42,71	17,98	14,65	24,29	35,98	7,67	38,03	9,72	3,25	47,5	1008,09	4,89
…	…	…	…	…	…	…	…	…	…	…	…	…
…	…	…	…	…	…	…	…	…	…	…	…	…
…	…	…	…	…	…	…	…	…	…	…	…	…
1110	53,89	24,86	19,59	41,09	92,79	33,74	31,75	4,28	0,99	66,33	1020,93	6,33

3.2 Machine Learning Algorithms

Regression algorithms were used because the label used in this study consists of numeric values [21]. Regression algorithms are often used in many prediction applications. Many regression algorithms are available, such as Regression Trees (RT) [22], Gaussian Process Regression (GPR) [23], and Support Vector Machine Regression (SVM) [22]. Within the scope of this study, the above algorithms were used.

3.3 Sampling and Training/Testing Process

The division of data into training and testing is one of the most important processes in the design model. There are two points here.

1. What should be the training and testing split percentages?
2. What data will we choose?

The difference between the training and test rates is 50% for significant correlations between the features and the label. Because the test results would be negatively impacted, using less than 50% of the training set is not suggested [24]. The non-formal preference is to choose 80% of the training and 20% of the test.

Data selection can be made in simple sampling and systematic sampling. Simple sampling can also be used to test even-numbered data, for example, to train single-numbered data. However, random numbers can be generated, and data corresponding to these numbers can be preferred.

However, it is crucial to select the same data, especially in applications where performances between methods are compared. In this application, a systematic sampling method was used.

3.4 Performance Criteria for Regression

The attempt to explain the relationship between variables with mathematical equations can be carried out by Regression [25]. This process is done in the hidden layer with machine learning algorithms, not equations in machine learning. After a variable enters the machine, the results are obtained from the machine.

Some parameters are calculated in the interpretation of the results of Regression. The most used performance evaluation criteria in regression problems are MSE and RMSE.

3.4.1 MSE

We can determine how closely a regression line resembles a set of points using the mean squared error (MSE). The term "MSE" stands for "mean squared error" (Eq. 1). The e_i the equation expresses the difference between actual values (t_i) and estimated values (y_i) and calculated ($e_i = t_i - y_i$). Here n is the total number of observations.

$$MSE = \frac{1}{n}\sum_{i=1}^{n} e_i^2 \tag{1}$$

3.4.2 RMSE

The square root of MSE is referred to as RMSE. The error rate reduces as the MSE and RMSE approaches zero (Eq. 2).

$$RMSE = \sqrt{\frac{1}{n}\sum_{i=1}^{n} e_i^2} \tag{2}$$

4 Results and Discussion

We carried out the training, testing, and performance testing steps with the codes we wrote in Matlab. The effect of changes in test percentage on test performance was examined, and the results are given in Table 3.

MSE and RMSE were used as performance criteria. In addition, processing time has been added. As can be seen from the results, the algorithm with the longest computation time is SVM whereas RT has the fastest computation time.

Comparing the simulation results, the GPR (MSE = 4.48e−06, RMSE = 0.00211) gives the best performance.

In addition, the regression graphs of the test percentage value at which each method performed best (marked in the table) are shown in the graph (Fig. 2(a), (b), (c)).

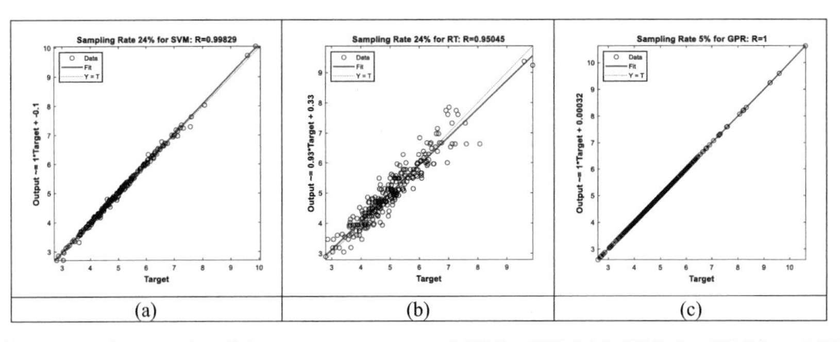

Fig. 2. Regression graphs of the test percentage at 24% for SVM (a), 24% for RT (b) and 5% for GPR (c)

Table 3. The effect of changes in test percentage on test performance

Test percentage	SVM			RT			GPR		
	MSE	RMSE	Elapsed Time (s)	MSE	RMSE	Elapsed Time (s)	MSE	RMSE	Elapsed Time (s)
5%	0.01035	0.10173	39.90	0.27788	0.52714	0.0330	4.08e-06	0.00202	7.37
10%	0.00973	0.09866	42.10	0.17908	0.42318	0.0277	7.47e-06	0.00273	3.89
16%	0.00501	0.07078	38.99	0.17091	0.41341	0.0311	4.69e-06	0.00216	3.58
20%	0.00750	0.08665	35.76	0.19069	0.43668	0.0355	5.61e-06	0.00236	3.19
24%	0.00461	0.06792	32.70	0.10820	0.32893	0.0368	4.48e-06	0.00211	3.19
33%	0.00542	0.07368	28.28	0.17254	0.41538	0.0266	4.35e-06	0.00208	2.57

After finding that the best performance was achieved at 24% test percentage, it was decided that the test percentage to be 24%. The next step is to look at how feature selection affects test results.

The Spearman feature selection algorithm was applied. The Spearman algorithm has selected the features which are mentioned in Table 4. A total of seven features were chosen by the Spearman algorithm.

The new data set consisting of the features selected by the algorithm is divided into 24% test and 76% training by systematic sampling. The results obtained were compared with those obtained when using all the features.

The model has given the same performance (when reduced features with the Spearman feature selection algorithm) as when used all features. The reduction of data in the data matrix has also contributed to reducing computation time.

The model performance generated by manually adding one feature at a time is given in the following table (Table 6).

The best performance is obtained when using the features selected by the spearman algorithm, according to a comparison of Tables 5 and 6's results.

Table 4. The Features that Spearman algorithm selected

Sub-ject No	Features											Label
	PM_{10}	$PM_{2.5}$	SO_2	NO_2	NO_X	NO	O_3	Air Temp.	Wind Speed	Rela-tive Hum.	Air Pres.	AQHI
1	89,5	63,36	22,48	34,74	68,61	22,11	13,82	5,46	0,95	68,65	1019,21	6,68
2	114,29	85,91	15,48	38,01	93,43	36,17	3,76	6,75	0,63	75,67	1009,51	7,54
3	61,2	54,41	9,67	33,85	62,3	18,54	11,31	5,83	0,7	99,42	1007,8	6,04
4	31,29	28,79	5,18	26,33	52,49	17,03	23,73	4,17	1,94	100	1010,52	4,82
5	31,38	27,92	6,63	29,73	76,21	30,31	21,22	3,95	0,81	95,23	1012,86	4,94
6	51,52	49,07	9,89	30,19	65,71	23,16	15,5	3,25	1,23	100	1006,11	5,69
7	32,54	30,54	8,43	33,74	72,51	25,31	19,78	2,5	1,26	100	1013,01	5,34
8	25,93	24,83	5,38	30,14	53,54	16,04	25,09	1,43	1,57	95,78	1014,91	5,03
9	63,91	47,43	17,42	37,13	73,44	24,81	19,13	7,5	2,3	77,78	1012,7	6,40
10	42,71	17,98	14,65	24,29	35,98	7,67	38,03	9,72	3,25	47,5	1008,09	4,89
…	…	…	…	…	…	…	…	…	…	…	…	…
1110	53,89	24,86	19,59	41,09	92,79	33,74	31,75	4,28	0,99	66,33	1020,93	6,33

Table 5. Elapsed times for all features vs Spearman

Sampling Rate	SVM			RT			GPR		
	MSE	**RMSE**	**Elapsed Time (s)**	**MSE**	**RMSE**	**Elapsed Time (s)**	**MSE**	**RMSE**	**Elapsed Time (s)**
All-Feature	0.00461	0.06792	32.70	0.10820	0.32893	0.0368	4.48e-06	0.00211	3.19
Spearman	0.00461	0.06792	33.20	0.1082	0.3289	0.1492	4.48e-06	0.00211	3.49

Table 6. The effect of manual feature selection on test perfomance

Number of features	SVM		RT		GPR	
	MSE	RMSE	MSE	RMSE	MSE	RMSE
1	0.6705	0.8188	1.1837	1.0880	0.7470	0.8643
2	0.6640	0.8148	1.1570	1.0756	0.7444	0.8627
3	0.6633	0.8144	1.0126	1.0063	0.7449	0.8631
4	0.4847	0.6962	0.7615	0.8726	0.3904	0.6248
5	0.3911	0.6253	0.4973	0.7052	0.2519	0.5019
6	0.3942	0.6279	0.3347	0.5785	0.2213	0.4704
7	0.0049	0.0704	0.1139	0.3375	3.54e−06	0.0018
8	0.0043	0.0661	0.1168	0.3417	3.61e−06	0.0019
9	0.0044	0.0668	0.1220	0.3493	3.75e−06	0.0019
10	0.0041	0.0642	0.1056	0.3250	3.88e−06	0.0019
11	0.0046	0.0679	0.1081	0.3289	4.48e−06	0.0021

5 Conclusions and Future Work

The aim of this study is to build regression models that determine the AQHI of Sakarya without any calculation. For this, it was examined three regression algorithms which are SVM,RT and GPR.The results show that The best performance is obtained by GPR algorithm but the fastest model is achieved with RT. It is concluded that an adequate prediction performance can be achieved when used the first 6 features and 10th feature that spearman algorithm selected. The paper concludes that it can be reduced the sensor cost by using only seven features. Accordingly, a data collection mechanism that uses only these sensors is sufficient for the determining AQHI.

References

1. Vitousek, P.M.: Beyond global warming: ecology and global change. Ecology **75**(7), 1861–1876 (1994). https://doi.org/10.2307/1941591
2. Atacak, I., Arici, N., Guner, D.: Modelling and evaluating air quality with fuzzy logic algorithm-Ankara-Cebeci sample. Int. J. Intell. Syst. Appl. Eng. **5**(4), 263–268 (2017). https://doi.org/10.18201/ijisae.2017533902
3. Irmak, M.E., Aydilek, I.B.: Hava kalite indeksinin tahmin başarısının artırılması için topluluk regresyon algoritmalarının kullanılması. Acad. Platf. J. Eng. Sci. **7**(3), 507–514 (2019)
4. Ambient air pollution: a global assessment of exposure and burden of disease. https://apps.who.int/iris/handle/10665/250141?locale-attribute=en&mbid=synd_yahoolife. Accessed 06 May 2022
5. Ierodiakonou, D., et al.: Ambient air pollution. J. Allergy Clin. Immunol. **137**(2), 390–399 (2016). https://doi.org/10.1016/j.jaci.2015.05.028
6. Zhang, Q., et al.: Transboundary health impacts of transported global air pollution and international trade. Nature **543**(7647), 705–709 (2017). https://doi.org/10.1038/nature21712

7. Tecer, L.H.: Hava kirliliği ve sağlığımız. Bilim ve Aklın Aydınlığında Eğitim **135**, 15–29 (2011)
8. SİM (Sürekli İzleme Merkezi) | T.C. Çevre, Şehircilik ve İklim Değişikliği Bakanlığı. http://sim.csb.gov.tr/SERVICES/airquality. Accessed 06 May 2022
9. Yangyang, X., Bin, Z., Lin, Z., Rong, L.: Spatiotemporal variations of PM2.5 and PM10 concentrations between 31 Chinese cities and their relationships with SO2, NO2, CO and O3. Particuology **20**, 141–149 (2015). https://doi.org/10.1016/J.PARTIC.2015.01.003
10. Altunkaynak, A., Başakın, E.E.: Dalgacık K-EN yakın komşuluk yöntemi ile hava kirliliği tahmini (2020)
11. Veljanovska, K., Dimoski, A.: Air quality index prediction using simple machine learning algorithms. Int. J. Emerg. Trends Technol. Comput. Sci. **7**(1), 25–30 (2018)
12. Kumari, S., Jain, M.K.: A critical review on air quality index. Environ. Pollut. 87–102 (2018)
13. Suman: Air quality indices: a review of methods to interpret air quality status. Mater. Today Proc. **34**, 863–868 (2021). https://doi.org/10.1016/j.matpr.2020.07.141
14. Stieb, D.M., Burnett, R.T., Smith-Doiron, M., Brion, O., Shin, H.H., Economou, V.: A new multipollutant, no-threshold air quality health index based on short-term associations observed in daily time-series analyses. J. Air Waste Manag. Assoc. **58**(3), 435–450 (2008)
15. Chen, H.: Review of air quality index and air quality health index. desLibris (2013). https://policycommons.net/artifacts/1209997/review-of-air-quality-index-and-air-quality-health-index/
16. Li, X., et al.: The construction and validity analysis of AQHI based on mortality risk: a case study in Guangzhou, China. Environ. Pollut. **220**, 487–494 (2017). https://doi.org/10.1016/j.envpol.2016.09.091
17. Zeng, Q., Fan, L., Ni, Y., Li, G., Gu, Q.: Construction of AQHI based on the exposure relationship between air pollution and YLL in northern China. Sci. Total Environ. **710**, 136264 (2020). https://doi.org/10.1016/j.scitotenv.2019.136264
18. Wong, T.W., Tam, W.W.S., Yu, I.T.S., Lau, A.K.H., Pang, S.W., Wong, A.H.S.: Developing a risk-based air quality health index. Atmos. Environ. **76**, 52–58 (2013). https://doi.org/10.1016/j.atmosenv.2012.06.071
19. Health Canada. Understanding Air Quality Health Index messages (2015). https://www.canada.ca/en/environment-climate-change/services/air-quality-health-index/understanding-messages.html. Accessed 06 May 2022
20. Hava Kalitesi Veri Bankası | T.C. Çevre, Şehircilik ve İklim Değişikliği Bakanlığı. https://sim.csb.gov.tr/STN/STN_Report/DataBank. Accessed 16 May 2022
21. Uçar, M.K., Nour, M., Sindi, H., Polat, K.: The effect of training and testing process on machine learning in biomedical datasets. Math. Probl. Eng. **2020**, 2836236 (2020). https://doi.org/10.1155/2020/2836236
22. Aldrich, C., Auret, L.: Unsupervised Process Monitoring and Fault Diagnosis with Machine Learning Methods, vol. 16, no. 3. Springer, Cham (2013)
23. Kang, M., Jameson, N.J.: Machine learning: fundamentals. In: Prognostics and Health Management of Electronics: Fundamentals, Machine Learning, and the Internet of Things, pp. 85–109 (2018)
24. Henrique, B.M., Sobreiro, V.A., Kimura, H.: Literature review: machine learning techniques applied to financial market prediction. Expert Syst. Appl. **124**, 226–251 (2019)
25. Alpar, R.: Applied Statistic and Validation-Reliability. Detay Publishing Ankara (2010)

Anomaly Detection in Sliding Windows Using Dissimilarity Metrics in Time Series Data

Ekin Can Erkuş[1,2(✉)] and Vilda Purutçuoğlu[2,3]

[1] Huawei Turkey R&D Center, Intelligent Applications Department, İstanbul, Turkey
ekincanerkus@hotmail.com, ekin.can.erkus1@huawei.com
[2] Biomedical Engineering, Middle East Technical University, Ankara, Turkey
vpurutcu@metu.edu.tr
[3] Department of Statistics, Middle East Technical University, Ankara, Turkey

Abstract. Anomaly detection in time series data is a useful approach for classifying different types of time series data. A working real-time anomaly detection algorithm and application may provide an opportunity for an immediate-like response to the anomalies. Dissimilarity metrics provide a statistical approach to the difference between two time-series data parts and can be used to differentiate the different data behavior from baseline data. Moreover, machine learning classifiers are partially useful in discriminating the outlying features obtained by the data samples. Hence, this study proposes a new approach to detect anomalies in real-time by a unique implementation of dissimilarity metrics with machine learning classification techniques. Furthermore, a self-learning sliding windows technique is implemented to provide a subject-specific anomaly detection approach. The proposed approach is tested in an experimental setup and with real data. Although an additional parameter optimization process is required, the results provide anomaly detection in real-time with perfect accuracy for some datasets.

Keywords: Sliding windows · clustering · machine learning · data generation · feature extraction · anomaly detection · dissimilarity

1 Introduction

Anomalies in time series refer to the disturbances in the data behavior and may hinder some of the valuable information from the data [1]. Motion artifacts, non-systematic noises, missing values, or a mixture of more than two different data sources can be considered examples of bad anomalous conditions. They generally alter the data characteristics for some time intervals and reduce the efficiency of the analyses [2]. Therefore, they are aimed to be detected and eliminated before the main analyses. There are numerous amounts of research in anomaly and outlier detection and many approaches to detect anomalies in time series data [3–7].

Anomalies may occur in a single interval, irregular or quasi-periodical intervals in the data. Anomalies, especially quasi-periodic anomalies, may have different probabilistic distributions from the base data [8]. Hence, detecting the anomalies requires prior

D. J. Hemanth et al. (Eds.): ICAIAME 2022, ECPSCI 7, pp. 640–651, 2023.
https://doi.org/10.1007/978-3-031-31956-3_54

information about the data to be used. Therefore, the first step to detect the anomalies in the data may be to reveal the behavior of the base data. Thus, the anomalies are easier to find, excluding the base data model. Many studies in the literature aim to detect anomalies directly from the given data, but, some of them specifically use the distributional information of the baseline data [9–11].

Quasi-periodic anomalies particularly occur in quasi-periodic data, and are found in many fields, specifically biological and mechanical systems and seasonal measurements [12]. Some methods in the literature are proposed to reveal quasi-periodic anomalies in time series data by using different approaches, such as frequency domain transformation [9], modeling [13], or sliding windows [14]. Although most of the studies prove that their methods perform better than others, their performances rely on the selected datasets, application setups, and parameters.

Real-time anomaly detection is a specific and challenging topic in time series analysis. The applications of real-time anomaly detection include patient monitoring [15], traffic control [16], radar tracking or defense systems [17]. Hence, real-time processing of time series requires relatively faster algorithms to match the sampling speed of the data collection device. The number of features extracted from a data interval changes the data processing speed [18]. Therefore, the minimum number of features is required for an acceptable classification accuracy for the anomalous data parts.

This study aims to provide preliminary results about the usability of the dissimilarity metrics for the upcoming studies regarding anomaly detection in time series. Another aim is to observe the anomaly detection performances of dynamic time warping (DTW), specifically, in quasi-periodic data for future studies. Furthermore, testing a new approach to the computation of the grand average in sliding windows is set as another goal. Three synthetic datasets are initially generated to achieve the goals, each representing a different data behavior. Moreover, they are duplicated by generating noisier versions of each.

Moreover, periodic anomalous data intervals are added to each of them. The datasets are processed using sliding windows to mimic a real-time data analysis. The details of the generation of the synthetic datasets, their analyses using dissimilarity metrics, and the evaluation steps of their performances are explained in the Methods section. The results of the analyses and their interpretation are given in the Results and Discussion section.

2 Methods

2.1 Synthetic Data Generation

Three different synthesized datasets are used to test the dissimilarity-based anomaly detection approaches. Each dataset is a single time series data with 1000000 samples, consisting of a baseline-event data train with 1000 samples in each condition. Therefore, there are 500 baseline and 500 event conditions, and each is concatenated with another. Moreover, each dataset is generated in 2 independent versions, representing the relatively low noisy with a signal-to-noise ratio (SNR) of 10dB and the high noisy form with an SNR value of 0dB. The noise is added to each sample by using another Gaussian distribution and aimed to provide a more realistic aspect to the generated datasets.

The event data are simulated with different distributional parameters than the baseline data in order to induce the anomaly-like structures in time series data. Moreover, each dataset has a different characteristic of the anomalies in the time series.

The first dataset contains the baseline data produced by a Gaussian normal distribution with the mean value μ of 0 and the variance value σ^2 of 1, i.e., white noise. On the other hand, distribution of the event data is generated again by using the Gaussian distribution with a variance of 1 but a mean parameter value of 3. Hence, such a dataset is used to mimic the anomalous structures with higher mean values than the baseline data, while the overall data behavior is not quasi-periodic. A representation of the generated dataset can be found in Fig. 1, where the groups of lower amplitude samples refer to the baseline, and others refer to the anomalous conditions.

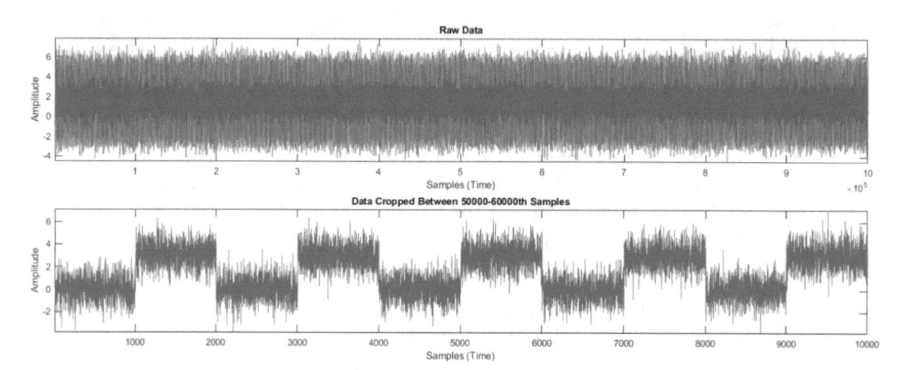

Fig. 1. Snapshot of the complete generated dataset 1 (upper) and zoomed in between 50000th and 60000th samples (lower). Baseline: Gaussian with $\mu = 0$, $\sigma^2 = 1$. Anomalous: Gaussian with $\mu = 3$, $\sigma^2 = 1$.

The second dataset is prepared to test the high amplitude anomalous conditions in quasi-periodic data. Hence, the generated samples are from a sinusoidal with a preset amplitude and sampling rate (Fs) value. Here, the baseline data samples are generated by using the sampling rate of 5 Hz, and their oscillation amplitude is set around 1, including the added noise. On the other hand, the event data samples are simulated with the same sampling rate values, but, with an oscillation amplitude of 3. A snapshot of the generated dataset can be found in Fig. 2, where the segmentations of the baseline data can be distinguished with lower amplitudes.

The final dataset is similar to the previous dataset in terms of quasi-periodicity, but this time, the sampling rate differs for the anomalous condition from the baseline data. Here, the oscillation amplitudes are set as 1, but the sampling rate is selected as 15Hz for the anomalous conditions. Therefore, the higher oscillatory periodic behavior of the anomalous data is aimed to be tested. An example representation of the dataset can be found in Fig. 3, where the anomalous data has a distinguishably higher oscillation rate, although the amplitudes are similar.

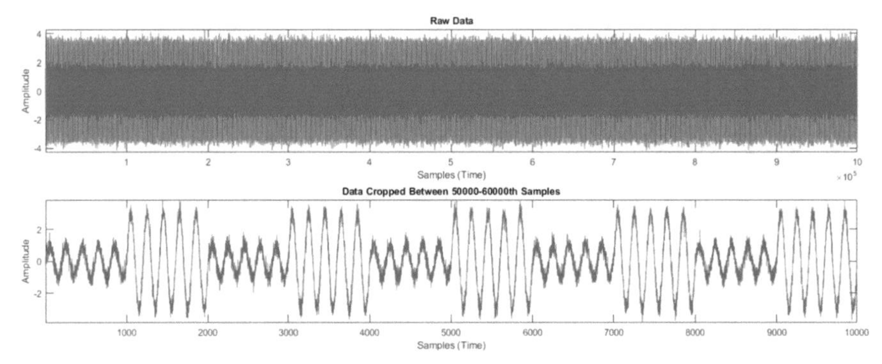

Fig. 2. Snapshot of the complete generated dataset 2 (upper) and zoomed in between 50000th and 60000th samples (lower). Baseline: Sinusoidal with amplitude = 1, and sampling rate = 5 Hz. Anomalous: Sinusoidal with amplitude = 3, and sampling rate = 5 Hz.

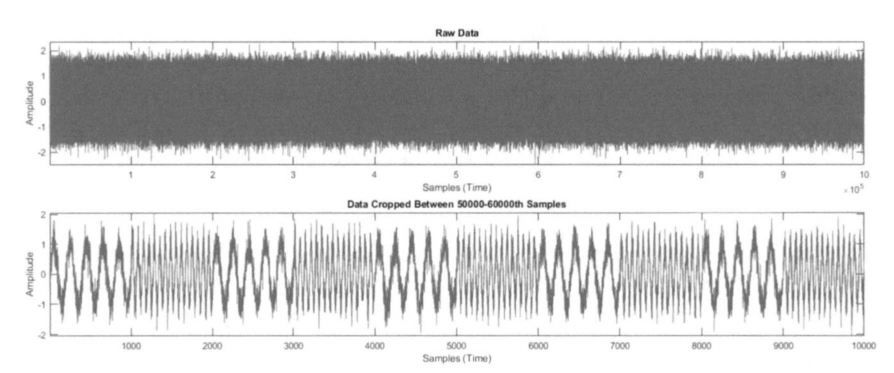

Fig. 3. Snapshot of the complete generated dataset 3 (upper) and zoomed in between 50000th and 60000th samples (lower). Baseline: Sinusoidal with amplitude = 1, and sampling rate = 5 Hz. Anomalous: Sinusoidal with amplitude = 1, and sampling rate = 15 Hz.

2.2 Experimental Setup and Sliding Windows

The sliding windows method is used to mimic the real-time processing of time series data. Indeed, the window and parameters of the analyses should fit the real-time specifications of the device to be used. But in this study, such limitations are omitted. Hence, to simulate a real-time operation, the parameters for the sliding windows, such as the window size and the slide size, should be defined. The window size and the slide size are illustrated in Fig. 4.

The window size and the slide size are expected to change the performances of the analyses. Therefore, an optimization process is required to find the best values for them for the given dataset. However, to keep the simplicity in this study, both the window and slide sizes are kept constant. The window size is set as 1000 to match the sizes of the conditions and 2000 to cover both condition sizes in independent trials, while the slide size is set as 500, 1000, and 2000 independently for different cases of analyses.

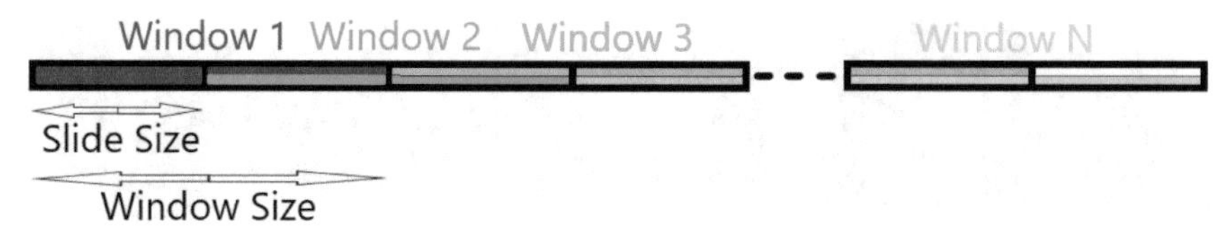

Fig. 4. An illustration of the window size and the slide size.

The processing algorithm takes the cropped sliding windows, one per iteration, applies the operations for analyses such as feature extraction and decision making, and proceeds to the next window. The samples in the next window are determined by the window size and the slide size.

2.3 Feature Extraction

The feature extraction process is the crucial step before evaluating the windows since the features represent different aspects of the data windows. This study focuses on the dissimilarity metrics and aims to apply them uniquely in real-time time series analysis.

In the operation step of the iterating windows, the maximum value in the window is found. Then, a temporary window is formed by centering the maximum value and shifting the sides of the window respectively. The next step is the computation the sample-wise grand average of the windows. The grand average is updated with each new window. In other words, the new grand average is updated by the sample-wise mean values of the corresponding samples of the windows that have been iterated so far. Instead of storing all window values and obtaining the sample-wise mean, the grand average is iteratively computed and assigned a weight value of the window that is currently being iterated to improve the computational efficiency in the software environment. Then, it is averaged with the next grand average of the window with the weight of 1, considering their weights as can be found in Eq. 1.

$$GA_i = \frac{\left[GA_{i-1} * (i-1)\right] + W_i}{i}. \tag{1}$$

In Eq. 1, i stands for the number of the current window on iteration, GA_i represents the grand average of the windows that have been iterated so far, and W_i is the samples of the current window. Note that the addition operation is performed sample-wise between the vectors.

On the other hand, along with the grand average computation as in Eq. 1, the feature extraction step mainly relies on the dissimilarities between the ith grand average and the ith window, which are computed sample-wise. The anomaly detection hypothesis in this study assumes such dissimilarity increases if an anomalous structure exists in the respective window. Three dissimilarity metrics, Euclidean, Chebyshev, and dynamic time warping (DTW) distances are selected to obtain the dissimilarity feature values from each window.

Euclidean distance is the most commonly used distance measurement between two samples. It provides linear and one-to-one dissimilarity between the sample pairs of two vectors with the same length. The Euclidean dissimilarity may also be referred to as the L_2 norm, and its formula can be found in Eq. 2.

$$d_{euc}(GA_i, W_i) = \sqrt{\sum_{j=1}^{n} \left(GA_{ij} - W_{ij}\right)^2}. \tag{2}$$

Here, j stands for the number of the sample in the vector with the window size of n, GA_i is the grand average, and W_i is the window.

Chebyshev distance is selected as the second dissimilarity feature, and it may also be referred to as the L_∞ norm. Chebyshev distance simply measures the maximum of the sample-wise differences between GA_i and W_i. Chebyshev distance relies on one-to-one matching between the samples and can be calculated by using Eq. 3.

$$d_{che}(GA_i, W_i) = \max_j \left|GA_{ij} - W_{ij}\right|. \tag{3}$$

The final dissimilarity metric used in this study is dynamic time warping (DTW). DTW provides a non-linear matching between the samples of GA_i and W_i through an optimization process that involves several constraints. Due to its computational complexity, DTW is most of the time the slowest compared to Euclidean and Chebyshev. A rough formulation of DTW dissimilarity between GA_i and W_i can be found in Eq. 4.

$$d_{dtw}(GA_i, W_i) = \left|GA_{ij} - W_{ik}\right| + \min_j \begin{cases} D(j-1, k) \\ D(j-1, k-1) \\ D(k-1) \end{cases}. \tag{4}$$

Here, j and k refer to the sample numbers, and D is the optimization function. More information about DTW and its optimization process can be found in [19, 20].

2.4 Clustering and Decision Making

The final step in the algorithm is to determine whether a window is anomalous or not. Here, the extracted dissimilarity metric values in each window are used to provide a decision about the current window. An unsupervised clustering algorithm, k-medoids, is used to classify the streaming values of the features out of each window that is processed. K-medoids are also referred to as partitioning around medoids (PAM) algorithm.

The k-medoids algorithm starts by defining the value number of clusters parameter, k, which is used as the number of feature values arbitrarily selected by the algorithm at the initial phase, also can be called seeds. Since the number of classes to classify is two in this study, the k value is selected as two. The remaining feature values are assigned to one of such initially selected values by their distances. The distance is computed for each of the remaining feature values and measures the difference between the initially selected feature values. The operation is repeated by selecting another arbitrary k value until the difference is minimized. Then, considering the assigned seeds, the feature values are labeled to the classes accordingly. Hence, returning to the main hypothesis, the windows marked as anomalous is assumed to have a different cluster of values than the rest.

The clustering approach in this study follows the subject-wise learning approach, where the classification model is dynamically trained and updated to reduce the classification error rate. Since there is no prior information about the data, the error rate is expected to be high. Therefore, a small portion from the beginning of the data is selected as the training set. After some empirical results performed with 0.5, 1, 2, 3, 5, 10%, the training set size is selected as 2% of the total data length. Such 2% yields the overall appropriate results with the minimum training set. In other words, the values of 0.5 and 1 resulted in bad performances, whereas the performance does not significantly increase for the larger training set percentage values. The training portion of the data is marked from the start of the data where the algorithm operates. But the decisions are not considered for the respective windows.

The unsupervised classification is processed through the overall data. Thus, it results in the decision of each window during the operation. As more data are processed, the accuracy for detecting the anomalous windows increases. After the data is processed completely, the overall accuracy is computed by dividing the correctly found anomalous windows by the total number of anomalous windows for the simulated datasets. The results for different conditions and datasets are reported in the Results and Discussion Section.

3 Results and Discussion

The overall accuracy of detecting the anomalous windows per dataset is reported in this section. A simple MATLAB figure is prepared to monitor the data processing in real-time. So that the user can see the cropped window, cumulative data, some recent data, the cumulative feature values, the grand average so far to the current window, and the overall decisions on each window so far. A snapshot of such figures can be found in Fig. 5.

Firstly, the effect of window size is investigated by comparing the accuracies between the window size values of 1000 and 2000. The window size of 1000 is selected to match the anomalous condition length in the simulated datasets exactly. Hence, each window with a slide size of 1000 represents a unique window in the data. Such window alternates between the baseline and anomalous intervals in each iteration. On the other hand, the window size of 2000 is selected to obtain windows composed of baseline and anomalous parts with equal length. The slide size of 1000 and the window size of 2000 alternate between the baseline and anomalous for half of the window. Therefore, each window except the first and the last is processed twice. The accuracy results of the experiment under different window size for each simulated data are presented in Table 1.

According to the results in Table 1, the anomalies are detected with better accuracy in the sinusoidal dataset with different amplitude values for the matching window size of 1000 than the window size of 2000. Moreover, the matching window size can perfectly detect the anomalous windows in that sinusoidal dataset. However, for the rest of the datasets, a window size of 2000 provided better results than the matching window. The reason might be the fitting patterns and forming harmonics in the grand average for the sinusoidal dataset with different amplitude values, as shown in Fig. 6.

The second experiment compares the slide sizes for the same window size. The matching window size with 1000 samples is selected to make it standard. The slide

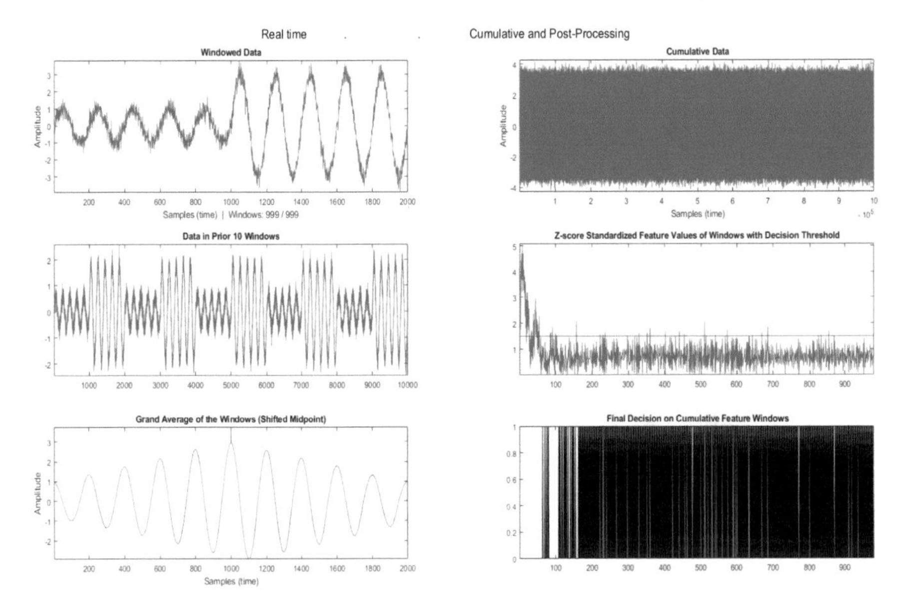

Fig. 5. A snapshot of the real-time operation from MATLAB figure.

Table 1. Comparison of accuracies for window size via Euclidean dissimilarity metric, when the slide size is selected as 1000. 'B' refers to baseline, and 'A' stands for anomalous windows, 'SNR' refers to signal-to-noise ratio, 'Fs' is the sampling rate.

Dataset generation conditions		Window size	
		1000	2000
B: Gaussian. $\mu = 0, \sigma^2 = 1$ and A: Gaussian. $\mu = 3, \sigma^2 = 1$	SNR = 10	0.7661	0.7534
	SNR = 0	0.6835	0.7787
B: Sinusoidal. Amplitude = 1, Sampling rate = 5 and A: Sinusoidal. Amplitude = 3, Fs = 5	SNR = 10	1.000	0.8991
	SNR = 0	0.9980	0.7105
B: Sinusoidal. Amplitude = 1, Sampling rate = 5 and A: Sinusoidal. Amplitude = 1, Fs = 15	SNR = 10	0.7189	0.8226
	SNR = 0	0.5305	0.5749

sizes are selected as 500, 1000, and 2000. Here, the slide size of 500 samples makes the samples, except in the first and the last windows to be investigated twice, while alternating between the perfectly matching and the half matching of the correct anomalous windows. 1000 is to match the windows perfectly, and 2000 is used as a control set, where it only captures the baseline parts. The accuracy results of the experiment under distinct slide size can be found in Table 2.

Table 2 proves the importance of the selection of the correct slide size for the analysis. Considering the sinusoidal data with different amplitude values for data conditions, matching the patterns of the grand average with the window is critical. The classification

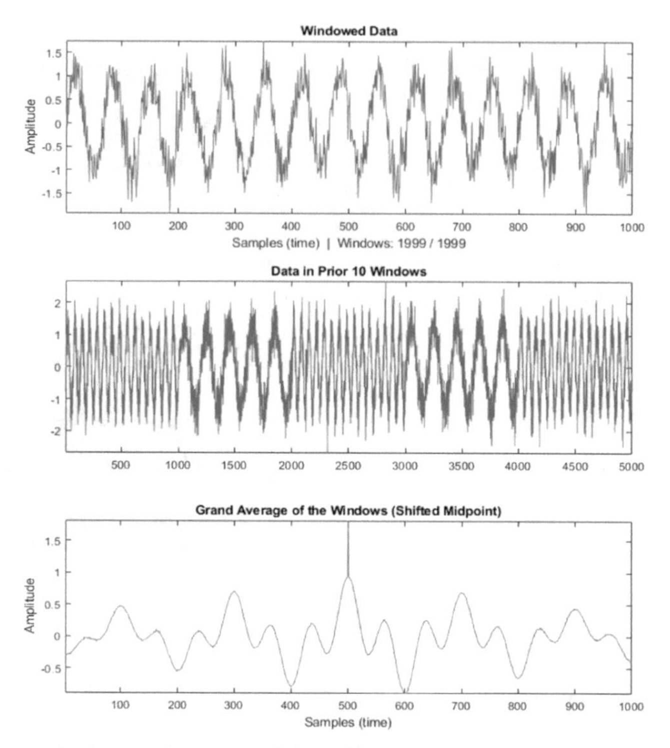

Fig. 6. A pattern in the grand average of the analysis for the sinusoidal dataset with different amplitudes for the data conditions.

Table 2. Comparison of accuracies for slide size via Euclidean dissimilarity metric, when the window size is selected as 1000. 'B' refers to baseline, and 'A' stands for anomalous windows, 'SNR' refers to signal-to-noise ratio, 'Fs' is the sampling rate.

Dataset generation conditions		Slide size		
		500	1000	2000
B: Gaussian. $\mu = 0, \sigma^2 = 1$ and A: Gaussian. $\mu = 3, \sigma^2 = 1$	SNR = 10	0.7918	0.7661	0.5020
	SNR = 0	0.7037	0.6835	0.5183
B: Sinusoidal. Amplitude = 1, Sampling rate = 5 and A: Sinusoidal. Amplitude = 3, Fs = 5	SNR = 10	0.5665	1.000	0.5102
	SNR = 0	0.5395	0.9980	0.5061
B: Sinusoidal. Amplitude = 1, Sampling rate = 5 and A: Sinusoidal. Amplitude = 1, Fs = 15	SNR = 10	0.7506	0.7189	0.5000
	SNR = 0	0.5485	0.5305	0.5082

accuracy falls to 56% for half of the matching slide size from a perfect classification for the matching.

The next experiment compares the performances of different dissimilarity metrics for a fixed window of 1000 and slide size of 1000 samples. The classification accuracy results of the dissimilarity metrics to discriminate the anomalous windows can be found in Table 3.

Table 3. Comparison of accuracies for dissimilarity metrics for the window size of 1000 and the slide size of 1000. 'B' refers to baseline, and 'A' stands for anomalous windows, 'SNR' refers to signal-to-noise ratio, 'Fs' is the sampling rate.

Dataset generation conditions		Dissimilarity metric		
		Euclidean	Chebyshev	Dynamic time warping (DTW)
B: Gaussian. Mean = 0, σ^2 = 1 and A: Gaussian. Mean = 3, $\sigma^2 = 1$	SNR = 10	0.7661	0.6354	0.5479
	SNR = 0	0.6835	0.6415	0.5794
B: Sinusoidal. Amplitude = 1, Sampling rate = 5 and A: Sinusoidal. Amplitude = 3, Fs = 5	SNR = 10	1.000	1.000	0.9980
	SNR = 0	0.9980	0.9084	0.9919
B: Sinusoidal. Amplitude = 1, Sampling rate = 5 and A: Sinusoidal. Amplitude = 1, Fs = 15	SNR = 10	0.7189	0.6863	1.000
	SNR = 0	0.5305	0.5092	0.8870

According to the accuracy values of dissimilarity metrics in Table 3, the Euclidean distance outperforms the Chebyshev distance in every situation. On the other hand, DTW is significantly preferable for the sinusoidal datasets with perfect classification accuracy, while other metrics have up to 71% classification accuracy. DTW's high performance with the periodic data is the non-linear matching of the periodic peaks of the current window and the grand average.

Moreover, considering the results in Tables 1, 2, and 3, the noise reduces the performance of revealing the anomalous windows in the datasets. Hence, before the anomaly detection process, an operation of the noise reduction may be used to improve the performance. However, since it increases the computational complexity, it may not be used for real-time analyses.

Finally, the findings are compared based on the computational times of the dissimilarity metrics. The computational times are given in seconds and represent the overall average computational times, including the graphical representation for all experiments for the window size of 1000 and 2000, independently. The results of the computational times in terms of seconds are presented in Table 4.

The outcomes of the computational time in Table 2 suggest to use the Chebyshev dissimilarity for the fastest computational requirements. But, DTW can be proposed to be used in the real-time analysis only if the real-time operation is matching or not lagging the data collection/sampling process. On the other hand, considering the higher accuracy values of DTW for the quasi-periodic data, it can still be usable with a few computational-cost performance modifications.

Table 4. Comparison of average computational times (in s) of dissimilarity metrics for the number of windows.

Dissimilarity metric	Window size	
	1000	2000
Euclidean	**167.54**	**178.57**
Chebyshev	**164.41**	**171.05**
Dynamic time warping	**198.90**	**216.73**

4 Conclusions and Future Work

This study proposes a real-time anomaly detection approach by using dissimilarity metrics in a univariate manner to observe the base performances with the lowest computational times. An experimental setup by using sliding windows is used to perform the analyses on a non-real-time system. Several datasets with different data behaviors are produced by concatenating the baseline and anomalous windows, which are generated under a variety of data distributions. The experiments observe the effects of the window size, slide size, and dissimilarity metric on detection of the anomalous windows.

The study results prove the usability of the dissimilarity metrics between the windows and the iterative grand averages for anomaly detection purposes. The analyses are preferred to be performed by using a single feature in order to improve the computational times to fit the real-time analysis speed of the time series. Moreover, the results also suggest the application of the dynamic time warping (DTW) when analyzing quasi-periodic data while suggesting not to use DTW for real-time applications if the computational time is slower than the sampling time of the device which the data are collected.

This study also shows the importance of matching slide size with the anomaly pattern. Finding an optimal window size and slide size is a major concept in time series analysis with moving windows. The optimization of the window size and slide size is an extension of this study. Moreover, repeating the analysis on several other simulated datasets with various conditions may also provide the wide usability of the proposed sliding windows method with the dissimilarity metrics. On the other hand, a repetition of this study by using multimodal classification is indeed a major topic for another study.

References

1. Lane, T., Brodley, C.E.: Temporal sequence learning and data reduction for anomaly detection. ACM Trans. Inf. Syst. Secur. **2**(3), 295–331 (1999). https://doi.org/10.1145/322510.322526
2. Ariyaluran Habeeb, R.A., Nasaruddin, F., Gani, A., Targio Hashem, I.A., Ahmed, E., Imran, M.: Real-time big data processing for anomaly detection: a Survey. Int. J. Inf. Manage. **45**, 289–307 (2019). https://doi.org/10.1016/J.IJINFOMGT.2018.08.006
3. Al-Anbuky, A., et al.: A survey of outlier detection techniques in IoT: review and classification. J. Sens. Actuat. Netw. **11**(1), 4 (2022). https://doi.org/10.3390/JSAN11010004
4. Chandola, V., Banerjee, A., Kumar, V.; Anomaly detection: a survey. ACM Comput. Surv. (2009). https://doi.org/10.1145/1541880.1541882

5. Chandola, V., Banerjee, A., Kumar, V.: Outlier detection: a survey. ACM Comput. Surv. (2007)
6. Chandola, V., Banerjee, A., Kumar, V.: Anomaly detection for discrete sequences: a survey. IEEE Trans. Knowl. Data Eng. **24**(5), 823–839 (2012). https://doi.org/10.1109/TKDE.2010.235
7. Gupta, M., Gao, J., Aggarwal, C.C., Han, J.: Outlier detection for temporal data: a survey. IEEE Trans. Knowl. Data Eng. (2014). https://doi.org/10.1109/TKDE.2013.184
8. Ren, H., Ye, Z., Li, Z.: Anomaly detection based on a dynamic Markov model. Inf. Sci. (Ny) **411**, 52–65 (2017). https://doi.org/10.1016/J.INS.2017.05.021
9. Erkuş, E.C., Purutçuoğlu, V.: Outlier detection and quasi-periodicity optimization algorithm: Frequency domain based outlier detection (FOD). Eur. J. Oper. Res. **291**(2), 560–574 (2021). https://doi.org/10.1016/J.EJOR.2020.01.014
10. Xiuyao, S., Mingxi, W., Jermaine, C., Ranka, S.: Conditional anomaly detection. IEEE Trans. Knowl. Data Eng. **19**(5), 631–644 (2007). https://doi.org/10.1109/TKDE.2007.1009
11. Zenati, H., Romain, M., Foo, C.S., Lecouat, B., Chandrasekhar, V.: Adversarially learned anomaly detection. In: Proceedings - IEEE International Conference on Data Mining, ICDM, vol. 2018-November, pp. 727–736 (2018). https://doi.org/10.1109/ICDM.2018.00088
12. Mcpherson, S.R.: Event based measurement and analysis of internet network traffic (2011)
13. Chakraborty, G., Kamiyama, T., Takahashi, H., Kinoshita, T.: An efficient anomaly detection in quasi-periodic time series data—a case study with ECG. In: Rojas, I., Pomares, H., Valenzuela, O. (eds.) ITISE 2017. CS, pp. 147–157. Springer, Cham (2018). https://doi.org/10.1007/978-3-319-96944-2_10
14. Malhotra, P., Ramakrishnan, A., Anand, G., Vig, L., Agarwal, P., Shroff, G.: LSTM-based encoder-decoder for multi-sensor anomaly detection (2016). https://doi.org/10.48550/arxiv.1607.00148
15. Amarbayasgalan, T., Pham, V.H., Theera-Umpon, N., Ryu, K.H.: Unsupervised anomaly detection approach for time-series in multi-domains using deep reconstruction error. Symmetry **12**(8), 1251 (2020). https://doi.org/10.3390/SYM12081251
16. Zuo, F., Gao, J., Yang, D., Ozbay, K.: A novel methodology of time dependent mean field based multilayer unsupervised anomaly detection using traffic surveillance videos. In: 2019 IEEE Intelligent Transportation Systems Conference, ITSC 2019, pp. 376–381 (2019). https://doi.org/10.1109/ITSC.2019.8917034
17. Riveiro, M., Pallotta, G., Vespe, M.: Maritime anomaly detection: a review. Wiley Interdiscip. Rev. Data Min. Knowl. Discov. **8**(5), e1266 (2018). https://doi.org/10.1002/WIDM.1266
18. Nanduri, A., Sherry, L.: Anomaly detection in aircraft data using recurrent neural networks (RNN). In: ICNS 2016 Security an Integrated CNS System to Meet Future Challenges (2016). https://doi.org/10.1109/ICNSURV.2016.7486356
19. Müller, M.: Dynamic time warping. Inf. Retr. Music Motion 69–84 (2007). https://doi.org/10.1007/978-3-540-74048-3_4
20. Senin, P.: Dynamic time warping algorithm review (2008)

Operating a Mobile Robot as a Blockchain-Powered ROS Peer: TurtleBot Application

Mehmed Oğuz Şen[1](✉), Fatih Okumuş[2], and Adnan Fatih Kocamaz[1]

[1] Department of Computer Engineering, Inonu University, Malatya, Turkey
{oguz.sen,fatih.kocamaz}@inonu.edu.tr

[2] Department of Software Engineering, Inonu University, Malatya, Turkey
fatih.okumus@inonu.edu.tr

Abstract. Decentralized control and management of multiple mobile robots is a promising approach to eliminate the disadvantages of centralized systems such as maintenance of a server, single point of failure and redundant repetitive task execution. Dynamic communication between robots in this system can be achieved by the decentralized nature of blockchain systems. In this paper, the integration of the Hyperledger Fabric blockchain platform into the Robot Operating System (ROS) for execution on an embedded system is presented with an example of an objective test robot that performs tasks delivered via the Fabric network. The Fabric network has a modular design in which components can be adjusted according to the functional requirements of a network of mobile robots. Studies on Fabric applications for systems of multiple mobile robots mostly focus on the theoretical approaches supported with simulations; therefore, this study focuses on the applicability of ROS integrated Fabric on a real robot as a contribution to the literature. A Fabric network with two peers is set up and transmission of management commands to the test robot is performed via sockets by a client application running on the ROS peer. During the experiments, the Fabric network operates on actual computing devices, not on Docker containers. Experimental results show that our test robot operates as a ROS peer in a Fabric network and completes the tasks assigned to it via Fabric. Several different mobile robots can be operated by using the same infrastructure.

1 Introduction

In most of the current mobile robot applications, management of a multi-robot system with mobile robots is usually done by using central computing machinery (typically a server). This approach makes maintenance of the server compulsorily, if the connection between robots and the server is lost, the whole system is affected by this. As an alternative approach, decentralized management eliminates the need for a central server and gives robots the ability to achieve consensus in the case of an emergency.

In a dynamic network of mobile robots, adding a new robot becomes problematic in terms of communication when static communication is preferred between robots. Components of communication (protocols, procedures, hardware etc.) need to be adjusted

D. J. Hemanth et al. (Eds.): ICAIAME 2022, ECPSCI 7, pp. 652–661, 2023.
https://doi.org/10.1007/978-3-031-31956-3_55

depending on the number of robots in the network. Blockchain systems are designed for providing a secure and decentralized solution to communication problems in distributed networks of mobile robots. They provide a smooth communication infrastructure which operates independently from the number of robots and thus sustains robust transactions between robots.

However, the concept of controlling multi-robot systems with mobile robots, storing acquired data and providing communication by using a blockchain system has been presented so far in simulation environments. In this paper, we perform the execution of a Hyperledger Fabric blockchain platform on a ROS enabled embedded system using TurtleBot as a real-world robotic device. Integration of TurtleBot into the Hyperledger Fabric network is done by operating TurtleBot with a ROS peer as a Fabric peer. Our application is intended to be a demonstration of a mobile robot operation in a Fabric network and can be extended to multiple mobile robots in the same manner. Mentioning its advantages, we also explain why Hyperledger Fabric is a more feasible solution than Ethereum as a blockchain platform.

1.1 Blockchain

The initial motivation behind blockchain systems was expressed as overseeing the foundation of a self-governing monetary system that operates independently of the current financial system and its components such as the SWIFT monetary network, the founders of Bitcoin [1] blockchain system. With the foundation of the Ethereum blockchain network [2], applications of blockchain expanded into several fields such as finance, healthcare and supply chain by the notion of secure and reliable data management in distributed systems without the need for a centralized authority. Blockchain research is ongoing with the advances in quantum resilient cryptographic hash function definitions, consensus algorithms, network topology and integration of IoT and other useful systems.

1.2 Blockchain and Mobile Robots

Mobile robots with autonomous behaviours are one of the key research areas in robotics with ongoing research in artificial intelligence and related concepts. Current studies on mobile robots have already fruited in many disciplines such as military [3], search and rescue [4], molecular robotics [5], industrial logistics [6] and health [7]. Moreover, the modern world witnessed the necessity of autonomous robots in pandemics like COVID-19, swarms of mobile robots have the potential of aiding humanity in the treatment of COVID-19 patients and in dealing with the pandemic [7, 8].

Proof of Work (PoW) consensus algorithm is used in Bitcoin and Ethereum blockchain networks since these networks do not have a node hierarchy and anyone can connect to them without the need for permission thus a consensus must be achieved meanwhile preventing malicious tampering [9]. The downside of PoW is that a race condition between nodes consuming more and more energy is inevitable since only a single node is rewarded as an incentive for contributing a new block to the ledger. There doesn't exist a hardcoded hash rate limit which a node can achieve, each node can eagerly enhance its hash power (GPUs, ASIC devices etc.) to receive this valuable prize [9]. This makes a robot to join and operate in a public blockchain network unreasonable because a robot's computing device is designed for controlling the robot, acquiring data from its sensor suite and thus cannot compete with the powerhouse nodes in the network. Therefore, private or authorized blockchain systems with a lightweight consensus algorithm preventing race conditions must be preferred for the decentralized operation of mobile robots.

1.3 Hyperledger Fabric

Observing the potential of blockchain applications, Linux Foundation founded the Hyperledger consortium in 2015 to oversee and carry out open-source blockchain projects. Being one of the members, IBM proposed Hyperledger Fabric as a private and authorized blockchain platform for enterprise-level decentralized application development [10]. Fabric provides a modular system for developers, where ledger data can be stored in multiple formats and other deterministic consensus algorithms can be set to be used by peers. Channels can be created for members connected to a Fabric network to join and a separate ledger is initiated in each channel to store transactions between members of it. This allows members to conduct a private communication medium whenever they want.

In this study, we focus on Fabric's configuration for an application on real robots such as TurtleBot. Ethereum is quite popular as a blockchain platform with many decentralized applications. Using Solidity, developers can implement custom smart contracts in Ethereum. On the other hand, Fabric provides different language options in application development. As a protection for overloading the network with spamming, transactions occurring in the main Ethereum network consume ETH cryptocurrency. Hence Ethereum developers need to consider a trade-off between transaction performance and its cost when implementing a smart contract [9]. Operating a private network eliminates the need of using real ETH but nodes must be given fake ETH when necessary. If a node is out of ETH, it will not be able to commit a transaction. Transaction execution in Fabric networks is independent of any cryptocurrency and since a node must have an identity to join a Fabric network, privacy and security requirements of a system of multiple mobile robots can be met and resilience against Sybil attacks can be achieved with the infrastructure of Fabric.

1.4 Robot Operating System (ROS)

Mobile robots participating in a Fabric network can communicate and share information securely; however, dealing with different mobile robots is not an easy task, unless a middleware is used for communication at the hardware level. Since robots can be equipped with different hardware components, developing a separate application for each robot can be unfeasible when all mobile robots in a multi-robot system are different from each other. As an open-source middleware, Robot Operating System (ROS) solves this problem by managing low-level device control and giving access to robot hardware components and is a popular choice in robotics applications. ROS has a large library which supports most of the popular mobile robots sold worldwide. Integrating ROS with Fabric on TurtleBot as a real mobile robot, utilization of an authorized blockchain system to a real mobile robot managed by a robot development platform was presented for the first time in this study.

The next sections are organized as follows: A literature review of blockchain applications in mobile robot systems is given in Sect. 2. Details of a ROS integrated Fabric system including TurtleBot as a peer are explained in Sect. 3. Results of the experiment conducted with TurtleBot connected to the Fabric network are shown in Sect. 4 and finally, a conclusion and directions for future work are given in Sect. 5.

2 Related Work

Several applications of mobile robots with blockchain for managing both a single robot and multi-robot systems including swarm robots [11] have been proposed, with the majority of studies conducted in the literature dated in 2020 and 2021. They can be classified as consensus proposals [12], path planning [13], cooperative working [14], information sharing [15], collective decision making [16], robot partitioning [17], task allocation [18, 19]. An approach for cooperation and collaboration in ad hoc heterogeneous multi-robot systems with Ethereum was proposed and a discussion about its utilization in smart cities was given by Queralta and Westerlund [20].

Reviewing the studies in the literature shows that Ethereum is the preferred blockchain network in most of the applications and there exists no application using Hyperledger Fabric as a blockchain system for the operation of the swarm and multi-robot systems with mobile robots. Hence it can be stated that there is still much to do about blockchain applications in mobile robots and many ideas and concepts awaiting to be thoroughly studied can emerge in future. This study is expected to make a remarkable contribution to the literature and to be a reference for future studies about the utilization of blockchain systems for swarm robotics and multi-robot systems.

3 Method

3.1 Overview of Hyperledger Fabric

In Hyperledger Fabric, blockchain data is stored as a ledger database. Data modelling is done while defining the class of the ledger object. A ledger is composed of blockchain and world state database components. A typical Fabric network consists of organizations with peer nodes and nodes providing orderer and certificate authority (CA) services. A peer node is responsible for communicating with applications connected to the blockchain network and executing smart contracts which can query or update the ledger stored in itself. If a ledger update is performed, each peer in the same channel updates their ledger with the latest added blocks. One or several peer nodes can be associated with an organization, where the administrator node of the organization can accept or decline smart contracts for running on their organization peers. An orderer is responsible for maintaining a distributed ordering service which orders the transaction requests of committing data to the ledger. A certificate authority (CA) is responsible for issuing certificates for user and organization identification on the blockchain network.

A chaincode is defined as a package of smart contract code to be deployed to a blockchain network. The Fabric has official Go, Java and Javascript APIs to develop chaincodes and interact with blockchain network gateways. An illustration of the Fabric network with a single mobile robot peer can be seen in Fig. 1.

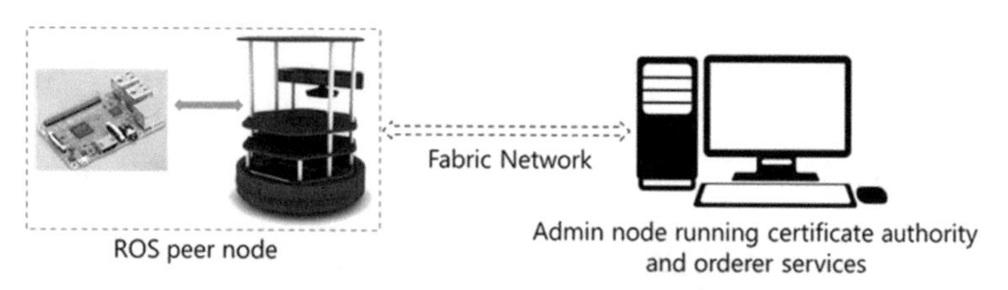

Fig. 1. Overview of a Fabric network with two nodes (1 ROS peer node and 1 admin node).

3.2 Hardware

Hardware components of a TurtleBot2 [21] peer are composed of two parts, TurtleBot2 itself and the computer (which can be a single board computer or a laptop for practical use) used for remote connection and controlling the robot. Kobuki TurtleBot2 is preferred to run the Hyperledger Fabric blockchain network on a real robot. Kobuki TurtleBot2 is a ROS compliant, low-cost, personal robot kit with open-source software. Its sensing capabilities can be enhanced by plugging in external sensors like RFID, Asus Xtion PRO LIVE, and Microsoft Kinect for Xbox. A portable computer (like a netbook) mounted on its base is required to send commands to the robot for movement around the environment. Kobuki TurtleBot2 that we used also contains an RFID reader and four antennas as extrasensory devices as shown in Fig. 2; however, the application presented in this study can be executed without using them.

Fig. 2. Kobuki TurtleBot2 used in the experiments. The robot was equipped with an RFID reader and antennas in addition to its sensor suite to detect RFID tags for positioning and real-time navigation in indoor logistics applications [22].

3.3 Blockchain and ROS Integration

To operate a mobile robot as a peer in a Fabric network, the robot needs to execute Fabric peer binaries on its computing devices. Since TurtleBot2 is a ROS compliant mobile robot, its firmware is set to receive instructions to be executed from a computer connected to itself via USB. In our case, a laptop placed on top of TurtleBot 2 is connected to the robot as its computing device and it runs ROS modules to command the robot. It can also run Fabric peer binaries concurrently so executing ROS modules and peer binaries in separate Linux terminals at the same time provides the integration of Fabric and ROS.

Although the sample applications presented in official Fabric documentation uses Docker to define peer nodes as Docker containers and to create a blockchain network with Docker Compose, it is only suitable for testing the correct execution of chaincodes. To form a blockchain network without using Docker containers, a peer in a Fabric network can be defined by installing Fabric API, setting the parameters and executing the Fabric peer binaries on any PC running a Linux distribution or a node in a Kubernetes cluster. Likewise, an administrator node can be defined by the execution of Fabric certificate authority and orderer binaries. These executables can be run in the background as services which can be enabled and started on systemd daemon, a software responsible for system and service management for Linux operating systems, at peer and admin nodes respectively or can be run as processes in separate terminals to monitor them easily.

For concurrent execution of ROS and Fabric binaries, ROS Kinetic was installed on the laptop with Ubuntu 16.04 to execute ROS modules. Installing Ubuntu version 16.04 is essential because ROS TurtleBot2 packages don't work with later versions (released after Kinetic) of ROS and Ubuntu. To integrate Fabric into a ROS peer, we executed the Fabric peer binary file in a separate terminal. We also stored IP addresses of peer and orderer nodes in the "etc/.hosts" file, so both nodes can know each other's address on the network. Since there is no official release of a Fabric Python SDK; it is not feasible to use the same programming language for both Fabric and ROS API, so a Java client application for connecting to the Fabric network and a Python application for ROS modules must be implemented and executed separately. Moreover, these two applications need to interact with each other. To resolve this matter, we implemented methods in both applications for interprocess communication via sockets as shown in Fig. 3. Both applications instantiate socket objects which communicate via port 6020. The port number is chosen arbitrarily and can be adjusted to any value between 0 and 65353 except for reserved port numbers for avoiding conflict.

3.4 ROS Integrated Blockchain Network Setup

After integration of Fabric and ROS as ROS peers, we set up a Fabric network with 1 ROS peer node and 1 admin node executing certificate authority and orderer services. Since real computers (A laptop for ROS peer node and a PC for admin node) are used as network nodes, our blockchain network is formed without using Docker Compose.

A PC with Linux operating system (Ubuntu Server 20.04) as the admin node was connected to the same network as the laptop. As described above, orderer and certificate authority binaries were executed in separate terminals at the admin node. Likewise, the

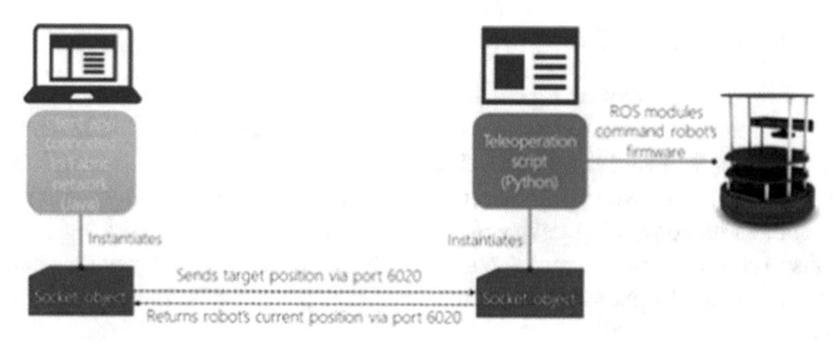

Fig. 3. Illustration of the communication between the Java client application connected to Fabric network and Python teleoperation application via sockets. Both sides instantiate socket objects which communicate via port 6020. Socket object instantiated by Java application sends commands related to the task information obtained from Fabric network and in addition to receiving data from other side, socket object instantiated by Python application returns data acquired from robot such as position information and other sensory data. The teleoperation application is responsible for delivering commands to the robot.

peer service was started in a separate terminal at ROS peer (a laptop executing Ubuntu 16.04 Linux distribution).

After setting up and launching the blockchain network, a basic smart contract written in Java was deployed on the network as a chaincode package for testing it. Then the chaincode package including our smart contract code for robot and task definitions was deployed on the network and started on the ROS peer. We implemented a decentralized client application in Java to connect to the Fabric network and invoke methods for task assignments to TurtleBot2 as illustrated in Fig. 4.

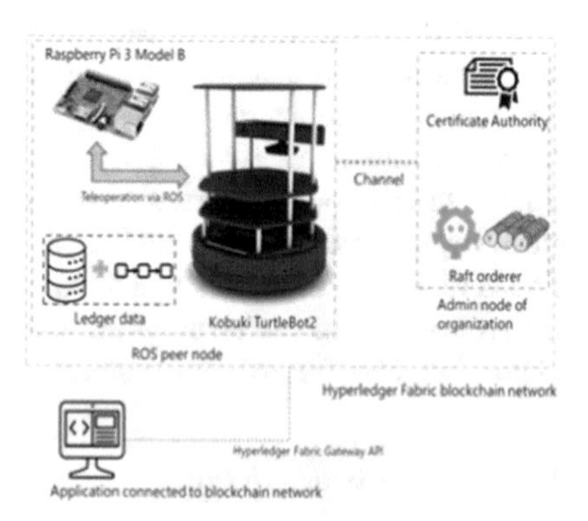

Fig. 4. Illustration of the Fabric network components in detail. A Java application is implemented for task assignment to TurtleBot2 and querying ledger data (position information etc.).

4 Experimental Results

TurtleBot2 was initially placed in the start position and tasked to reach the goal position shown in Fig. 5. Invocation of the method for task execution in the smart contract deployed on the blockchain network prompted TurtleBot2 to move along the path and reach the goal.

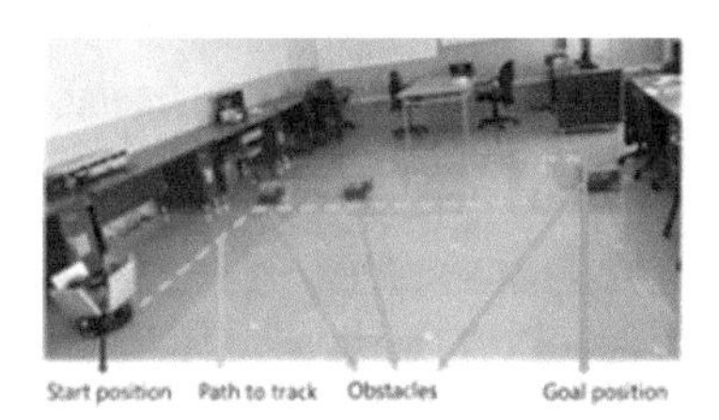

Fig. 5. Task start and goal positions for TurtleBot2. The robot is expected to follow the path depicted in dashed lines to avoid obstacles and reach the goal position.

The movement of TurtleBot2 along the expected path can be seen in Fig. 6. After following the path and avoiding the obstacles, TurtleBot2 finally reached the goal position as shown in Fig. 6(d).

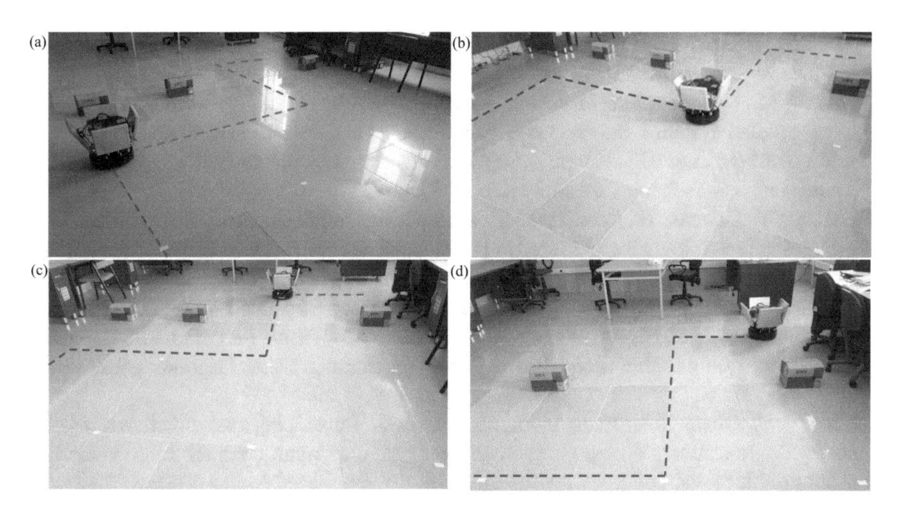

Fig. 6. TurtleBot2 moves along the path and avoids the first obstacle (a) and second obstacle (b) on its way. After passing between obstacles in (c), the robot arrives at the goal position and completes its task (d). Red dashed lines indicate the distance covered by the robot and blue dashed lines indicate the remaining path to the goal.

5 Conclusion and Future Work

In this study, the application of Hyperledger Fabric as a blockchain system to Kobuki TurtleBot2 operated by ROS is demonstrated. After giving a summary of related work, an overview of the blockchain network presented in this paper is given. Experiments are conducted in a real-world environment using Kobuki TurtleBot2. Considering the results, it can be seen that a single robot connected to the Fabric network as a ROS peer could be operated decentralized.

For future work, we plan to operate decentralized management of a swarm of mobile robots. Therefore, a decentralized control over real robots would be possible for missions meant to be carried out by robot swarms. We also plan to apply our proposed blockchain network structure to swarms of heterogeneous robots where robots with different computational capabilities may be assigned different roles in a blockchain network for efficiency and robustness in transactions.

References

1. Nakamoto, S.: Bitcoin: A peer-to-peer electronic cash system. Decentralized Bus. Rev. 21260 (2008)
2. Buterin, V.: A next-generation smart contract and decentralized application platform. ethereum project white paper. Technical report (2014). https://ethereum.org/en/whitepaper/. Accessed 1 May 2022
3. Sangeetha, M., Srinivasan, K.: Swarm robotics: a new framework of military robots. In Journal of Physics: Conference Series, vol. 1717, no. 1, p. 012017. IOP Publishing (2021)
4. Dadgar, M., Couceiro, M.S., Hamzeh, A.: RbRDPSO: Repulsion-based RDPSO for robotic target searching. Iran. J. Sci. Technol. Trans. Electr. Eng. **44**(1), 551–563 (2020)
5. Kabir, A.M.R., Inoue, D., Kakugo, A.: Molecular swarm robots: recent progress and future challenges. Sci. Technol. Adv. Mater. **21**(1), 323–332 (2020)
6. Okumuş, F., Dönmez, E., Kocamaz, A.F.: A cloudware architecture for collaboration of multiple agvs in indoor logistics: case study in fabric manufacturing enterprises. Electronics **9**(12), 2023 (2020)
7. Holland, J., et al.: Service robots in the healthcare sector. Robotics **10**(1), 47 (2021)
8. Farkh, R., Marouani, H., Al Jaloud, K., Alhuwaimel, S., Quasim, M.T., Fouad, Y.: Intelligent autonomous-robot control for medical applications. Comput. Mater. Continua, **68**, 2189–2203 2021
9. Zarir, A.A., Oliva, G.A., Jiang, Z.M., Hassan, A.E.: Developing cost-effective blockchain-powered applications: a case study of the gas usage of smart contract transactions in the ethereum blockchain platform. ACM Trans. Softw. Eng. Methodol. (TOSEM) **30**(3), 1–38 (2021)
10. Androulaki, E., et al.: hyperledger fabric: a distributed operating system for permissioned blockchains. In: Proceedings of the Thirteenth EuroSys Conference, pp. 1–15 (2018)
11. Strobel, V., Castelló Ferrer, E., Dorigo, M.: Blockchain technology secures robot swarms: a comparison of consensus protocols and their resilience to byzantine robots. Front. Robot. AI, **7**, 54 (2020)
12. Liu, J., Xie, M., Chen, S., Ma, C., Gong, Q.: An improved DPoS consensus mechanism in blockchain based on PLTS for the smart autonomous multi-robot system. Inf. Sci. **575**, 528-541 (2020)

13. Singh, P.K., Singh, R., Nandi, S.K., Ghafoor, K.Z., Rawat, D.B., Nandi, S.: An efficient blockchain-based approach for cooperative decision making in swarm robotics. Internet Technol. Lett. **3**(1), e140 (2020)
14. Mokhtar, A., Murphy, N., Bruton, J.: Blockchain-based multi-robot path planning. In: 2019 IEEE 5th World Forum on Internet of Things (WF-IoT), pp. 584–589. IEEE (2019)
15. Karthik, S., Chandhar, N.P., Akil, M., Chander, S., Amogh, J., Aditya, R.: Bee-Bots: a blockchain based decentralised swarm robotic system. In: 2020 6th International Conference on Control, Automation and Robotics (ICCAR), pp. 145–150. IEEE 2020
16. Nishida, Y., Kaneko, K., Sharma, S., & Sakurai, K., 2018. Suppressing chain size of blockchain-based information sharing for swarm robotic systems. In 2018 Sixth International Symposium on Computing and Networking Workshops (CANDARW) (pp. 524–528). IEEE
17. Nguyen, T.T., Hatua, A., Sung, A.H.: Blockchain approach to solve collective decision making problems for swarm robotics. In: Prieto, J., Das, A.K., Ferretti, S., Pinto, A., Corchado, J.M. (eds.) BLOCKCHAIN 2019. AISC, vol. 1010, pp. 118–125. Springer, Cham (2020). https://doi.org/10.1007/978-3-030-23813-1_15
18. Tran, J.A., Ramachandran, G.S., Shah, P.M., Danilov, C.B., Santiago, R.A., Krishnamachari, B.: SwarmDAG: a partition tolerant distributed ledger protocol for swarm robotics. Ledger **4**(Supp 1), 25–31 (2019)
19. Basegio, T.L., Michelin, R.A., Zorzo, A.F., Bordini, R.H.: A decentralised approach to task allocation using blockchain. In: El Fallah-Seghrouchni, A., Ricci, A., Son, T.C. (eds.) EMAS 2017. LNCS (LNAI), vol. 10738, pp. 75–91. Springer, Cham (2018). https://doi.org/10.1007/978-3-319-91899-0_5
20. Grey, J., Godage, I., Seneviratne, O.: Swarm contracts: smart contracts in robotic swarms with varying agent behavior. In: 2020 IEEE International Conference on Blockchain (Blockchain), pp. 265–272. IEEE (2020)
21. Queralta, J.P., Westerlund, T.: Blockchain powered collaboration in heterogeneous swarms of robots. arXiv preprint arXiv:1912.01711.(2019
22. TurtleBot2 open-source robot development kit. https://www.turtlebot.com/turtlebot2. Accessed 1 May 2022
23. Okumuş, F.: Cloud-based autonomous robot management for indoor logistics activities (Thesis No. 656926) [Doctoral dissertation, Inonu University]. Council of Higher Education Thesis Center. (in Turkish) 2020

Opportunities and Prospects for the Application of Intelligent Robotic Devices in the Agricultural Sector

A. Mustafayeva[1](✉), E. Israfilova[1], E. Aliyev[2], E. KHalilov[1], and G. Bakhsiyeva[1]

[1] Department of Information Technologies, Mingachevir State University, Mingachevir, Azerbaijan
{aida.mustafayeva,elmira.israfilova,elnur.xalilov, gunel.baxshiyeva}@mdu.edu.az

[2] Internal Control Department, Mingachevir State University, Mingachevir, Azerbaijan
elchin.aliyev@mdu.edu.az

Abstract. In developed countries of the world economy, various types of robotic devices are widely used in the processing of data directly related to the monitoring of land, crops and products of the agricultural sector, economic assets and other resources. The growing interest in this field in Azerbaijan also began with the reforms and state programs implemented by the country's leadership on the sustainable development of various aspects of the digital economy, the study of international experience, the promotion of startup ideas and projects. In general, the problems facing agriculture in our country are productivity and labor; flows in two directions, including technical and technological issues. The presented article explores the possibilities and prospects of using robotic devices with artificial intelligence in solving the problems covered by both directions. The ability of the proposed intelligent robotic devices to walk and fly closer to the ground will provide more accuracy, flexibility and mobility based on the data collected and experience gained without damaging the product in the field. The creation and application of such devices will give impetus to the sustainable development of digital agriculture in our country.

Keywords: robotic device · artificial intelligence · drone · ground monitoring · neural network

1 Related Works

The application of mobile robotics in agriculture allows solving problems such as increasing labor productivity, efficient use of equipment and machinery, staff shortages, 24/7 working hours and so on. It is known that mobile robots are a branch of system engineering and robotics. Technical solutions of artificial intelligence give more effective results in ensuring the autonomous operation of these devices.

Mobile robots have penetrated both the control, analysis and planning of agriculture, as well as the direct mechanism of action. Drones of various purposes and functions are

D. J. Hemanth et al. (Eds.): ICAIAME 2022, ECPSCI 7, pp. 662–670, 2023.
https://doi.org/10.1007/978-3-031-31956-3_56

widely used in agriculture around the world. The integration and control of drones and robotic devices with the use of artificial intelligence is an urgent issue for the sustainable development of agriculture. In the near future, most of the development of this technology will fall on the agricultural sector. In this regard, the main purpose of the study is to develop scientific and theoretical provisions and recommendations aimed at increasing the efficiency of agricultural robotization.

Pages [1–4] examine the benefits of using artificial intelligence and robotics in agriculture. It is known that the agricultural sector suffers from plant diseases, pest infestations, water shortages, weeds and many other problems. These problems cause significant crop losses and economic losses. At the same time, agricultural experiments lead to environmental hazards (destruction of animals due to drugs, air pollution, etc.). The application of intelligent technologies in robotics and mechatronic systems allows solving these problems effectively.

[7] describes a mathematical model and control algorithm for a six-rotor drone. The control algorithm under consideration is non-trivial and is relatively difficult to adjust. It is advisable to test the mathematical model before commissioning, as an incorrectly tuned control algorithm can damage the device. The mathematical model is based on classical Newton's laws. It is known that the classical methodology does not fully meet the required quality indicators in uncertain conditions. This raises the issue of the application of artificial intelligence.

In [10–15], the application of unmanned aerial vehicles is widely analyzed for its functionality, both in industry as a whole and in its individual sectors. The existing problems in the agricultural sector are reflected, and ways to solve these problems with the help of unmanned aerial vehicles are proposed. In the scientific work, a scheme reflecting the functional interface of automated unmanned aerial vehicles was created. An electronic communication mechanism has been formed between the drone as well as the technical components of the automation equipment. At the end, an explanation of the software used to control the drone was given.

Thus, the analysis suggests that the use of artificial intelligence, multifunctional drones with combined, hybrid architecture in various areas of the agricultural sector is relevant and one of the main factors influencing the development of digital agriculture.

2 Method and Material

2.1 Statement of the Problem

There are two main directions of the problems facing agriculture in our country during the Fourth Industrial Revolution:

1. Productivity and labor issues.
2. Techniques and technology issues.

Problems covering productivity and labor include:

- Decrease in employment due to profitability (decrease in income) (closure of jobs).
- The problem of attracting labor force due to the seasonal nature of agricultural activities. This problem is especially acute for entrepreneurs.

- Increase in labor costs (increase in fuel, fertilizer costs and minimum wage).
- Maximum losses in fruit and vegetable harvesting.
- Lack of water resources.
- The soil is loamy, stony and saline.
- Negative impact of weeds on agriculture and crop production.
- Proliferation of areas affected by pests and fungi.

Problems related to equipment and technology can be noted as follows:

- Expensive technical equipment; operation of this equipment on low-quality and low-productivity land plots.
- Lack of control over problems that may arise during the cultivation and ripening of crops in large areas (fire, driving in the fields and exposure to insects, etc.).

In modern conditions, additional opportunities are needed to maintain the working mechanism of agriculture, train new staff and support and develop related industries. It is impossible to imagine the innovative development of agriculture without the application of robotics. For this purpose, it is necessary and sufficient to design multifunctional, robotic hybrid devices with artificial intelligence (DRON & ROBOT & ARTIFICIAL INTELLIGENCE) that solve these problems.

Private companies around the world use BIG DATA analytics to generate ideas for agriculture, to implement those ideas, and to create different methods for artificial intelligence and even IoT applications. The application of robotics and artificial intelligence solutions analyzes information on air, soil, irrigation and plant health, allowing farmers to grow a better product, regardless of the season, with a minimum of time, money and labor.

The functionality of agro-robotic devices with artificial intelligence is related to the collection and processing of data through cameras and sensors placed on the platform. Based on the processed data, it is planned to send a number of necessary notifications to farmers before and after planting:

- Soil, saline and stony soil before planting.
- Determination of water supply and water consumption in plant stems.
- Identification of primary sources of pests.
- Direct interventions without reaching the point of economic damage.
- Organization of additional measures for plant nutrition and protection (for example, the application of digital recipes to farmers in the diagnosis and treatment of plant and soil diseases using artificial intelligence).
- Lack of organic and inorganic substances in the soil.
- Identification of weeds.
- Determination of notifications about ripening and readiness of the product, etc.

Analyzing the capabilities of the proposed multifunctional robotic devices, the use of electromagnetic pulse waves as a perspective direction is also considered. This technology is intended to be used, for example, to harvest crops from walnut trees with low losses, as well as to remove pests from the field.

The research area includes agricultural production areas such as cotton growing, grain growing, vegetable growing, rice growing, and fruit growing (Fig. 1).

Fig. 1. Areas of the agricultural sector

Processed data include soil condition (analysis of soil by categories, analysis of soil ecological condition), fertilizer shortages, irrigation, plant health and production productivity. The collected data will provide both autonomous and remote control of robotic devices, ensuring optimal decision-making based on the application of neural networks, which is an area of artificial intelligence.

2.2 Analysis of the Application of Intelligent Robotic Devices and Definition of Functional Scheme

Robotic agricultural devices have the following functional capabilities:

- Implementation of remote irrigation.
- Determination of water resources in the area.
- Ensuring harvest.
- Soil monitoring.
- Identification of areas attacked by pests and areas inhabited by toxigenic species of fungi.
- Mobile (GPS) communication capabilities.

Based on the listed functional capabilities, the proposed device will allow to improve the following indicators:

- Increase productivity;
- Time saving;
- Easy to use;
- Mapping of sown areas;
- Accelerate the payback period of the investment.

The functional diagram of the intelligent robotic device is shown in Fig. 2.

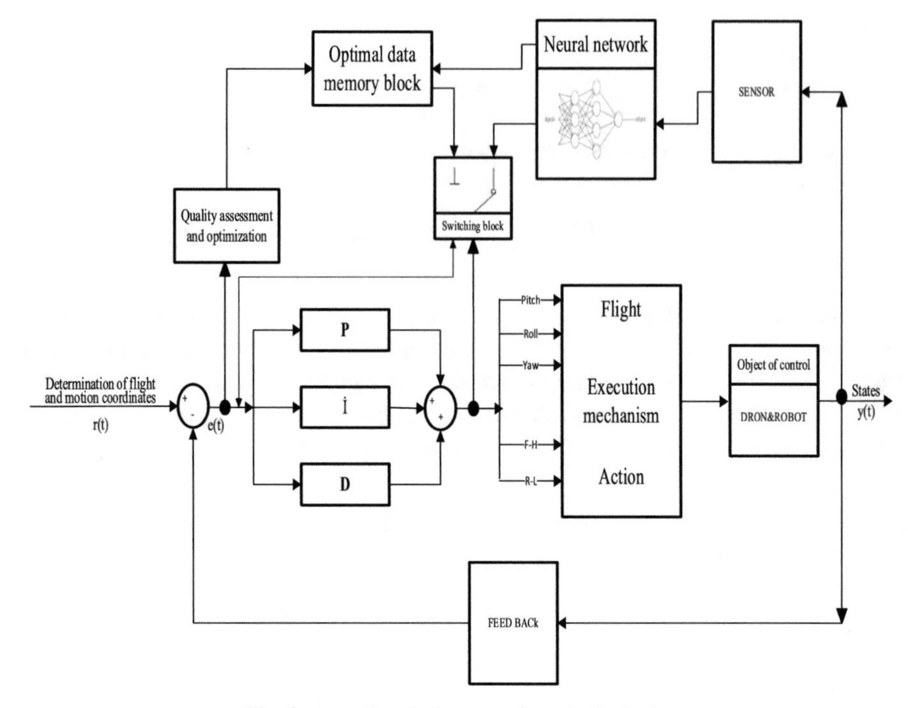

Fig. 2. Functional diagram of a robotic device

The functionality of the functional blocks shown in Fig. 2 can be summarized as follows. (1) The position variables (by flight and motion) in the OXYZ coordinate system of the control object are described in Fig. 3.

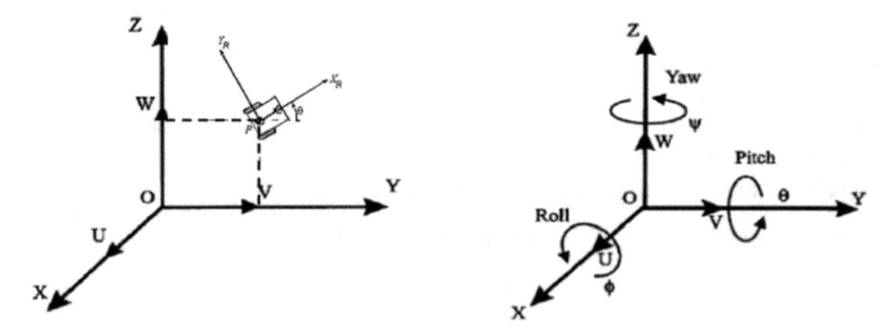

Fig. 3. Status variables of a robotic device

The status variables of the control object are defined as follows.

To move:

$$\begin{cases} \ddot{x}_w(t) = a_1\ddot{x}_w(t) - a_2\dot{\phi}(t)\dot{y}_w(t) + \overline{u}_1(t) \\ \ddot{y}_w(t) = a_1\dot{\phi}_w(t)\dot{x}_w(t) - a_1\dot{y}_w(t) + \overline{u}_2(t) \\ \ddot{\phi}_w(t) = a_3\dot{\phi}(t) + \overline{u}_3(t). \end{cases} \quad (1)$$

Here $\dot{x}_w(t)$, $\dot{y}_w(t)$ – *t is the state variable in the coordinate system $O_w(t)$, $X_w(t)$, $Y_w(t)$* are state variables in the coordinate system; $\phi(t)$ – is the angle of rotation of the controlled object at t, the value of which is calculated by taking the positive direction of the coordinate axis; $\overline{u}_1(t)$, $\overline{u}_2(t)$, $\overline{u}_3(t)$ – control effects; a_1, a_2, a_3 – depending on the physical and geometric parameters of the mobile robot moving in all directions, the coefficients are determined as follows:

$$a_1 = -2J/\left(mr^2 + 2I_w\right),$$
$$a_2 = 2I_w/\left(mr^2 + 2I_w\right)$$
$$a_3 = -4cL^2/4I_wL^2 + I_vr^2,$$

here, m, L, r, I_v, c, I_w – are the parameters of a moving mobile robot.

For the flight:

$$\begin{aligned} \ddot{x} &= (\sin\phi\sin\varphi + \cos\phi\sin\theta\cos\varphi) + u_1(t) \\ \ddot{y} &= (-\cos\phi\sin\varphi + \sin\phi\sin\theta\cos\varphi) + u_2(t) \\ \ddot{z} &= -g + (\cos\theta\cos\varphi) + u_3(t). \end{aligned} \quad (2)$$

Here, $\ddot{x}(t)$, $\ddot{y}(t)$, $\ddot{z}(t)$ – The state variables in the OXYZ coordinate system of the object to be controlled at t; $\varphi(t)$, $\phi(t)$, $\theta(t)$– there are angles of rotation and rotation of the controlled object at t, the value of which is calculated from the positive direction of the axis OXYZ; $u_1(t)$, $u_2(t)$, $u_3(t)$ (in accordance with *yaw*, *roll* and *pitch*) – there are control effects, which are defined as follows:

$$U_i = \sum_{i=1}^{3} k\omega_i^2. \quad (3)$$

In this case, the angular velocity of the three state variables will be calculated as follows:

$$\begin{aligned} U_Y &= k(-\omega_1^2 + \omega_2^2 - \omega_3^2 + \omega_4^2) \\ U_R &= k(-\omega_1^2 + \omega_3^2) \\ U_P &= k(\omega_2^2 - \omega_4^2). \end{aligned} \quad (4)$$

In block (3) (Fig. 2), based on the PID control law, the appropriate regulatory regulator monitors the values of flight and motion variables and performs a comparative analysis through feedback.

$$
\begin{aligned}
U_Y &= K_{P\phi}(\phi_r - \phi) + K_{I\phi}\int(\phi_r - \phi) + K_{D\phi}\frac{d}{dt}(\phi_r - \phi) \\
U_R &= K_{P\varphi}(\varphi_r - \varphi) + K_{I\varphi}\int(\varphi_r - \varphi) + K_{D\varphi}\frac{d}{dt}(\varphi_r - \varphi) \\
U_P &= K_{P\theta}(\theta_r - \theta) + K_{I\theta}\int(\theta_r - \theta) + K_{D\theta}\frac{d}{dt}(\theta_r - \theta).
\end{aligned}
\tag{5}
$$

K_{Pm} $(m = 1, 2, \ldots, M)$ the values of the regulator coefficients (8) are determined by means of the switch key block. (6) depending on the purpose, the neural control system (7) determines the type of problem based on the information received from the sensor and makes a decision to solve the problem. As can be seen in the figure, the signals from the sensors are converted into pulses - the control parameters of the memory generators at the appropriate signal level. This range of requirements characterizes the short-term action plan for the current moment of the robotic device. According to this plan, a number of control measures are formed in the actuators of the robot. For example, when three types of data when PID is used as a control law: K_P- satellite data $(P = k_pe(t))$, k_I – observation data $(I = K_I \int e(t)dt)$, k_D – experience data $(D = k_D(d/dt)e(t)dt)$ toplayan blok yaradılır.

The quality indicators of the robotic device are evaluated in block (4) and $k_p^{opt} = k_{pj}, j \in \lceil 1, m \rceil$ is selected by the best of the regulator's generated tuning parameters (5) and the optimal data is transferred to the memory block $(k_P^{opt}, k_I^{opt}, k_D^{opt}, D = D^*)$.

It should be noted that as the number of experiments increases, it becomes more difficult to find a local optimal solution. That is, it is equally difficult to determine the optimal values of the Jopt collected in the memory block (5), as well as the optimal value of the settings of the regulator. For this purpose, it is very important to determine the value of the discrete step of the generators to determine the global optimum using neural network technology. (8) The switch block sets the value of the regulator's tuning parameters by switching from experimental mode to real mode. In this mode, it is possible to analyze the effectiveness of the control system.

2.3 Prospects for the Application of Intelligent Robotic Devices in the Agricultural Sector

Intelligent robotic technologies allow the agricultural sector to solve the above-mentioned problems by analyzing the data collected by sensors or computer vision systems with machine learning algorithms. In addition, these technologies can provide an analytical analysis of market demand, forecasting product prices, as well as determining the optimal time for planting and harvesting (Fig. 4).

In general, the application of artificial intelligence technology determines the development of digital agriculture in the near future on three trends:

1. Effective use of agricultural machinery and robotic devices: There are different types of agricultural machinery and robotic devices (flying, walking, etc.) depending on the purpose and functionality (eg, spraying, irrigation, soil monitoring, mapping, etc.).

2. Soil and crop monitoring: determination of saline, watery and stony soils before planting, analysis of water resources and water consumption in plants, analysis of organic and inorganic substances, detection of primary sources of pests and weeds, assessment of plant vegetation, etc. covers issues such as.
3. Analytical analyzes that perform forecasting: comparison of data received from cameras and satellites, their analysis is carried out on the basis of the use of robotic devices with artificial intelligence, resulting in forecasting and optimal decisions.

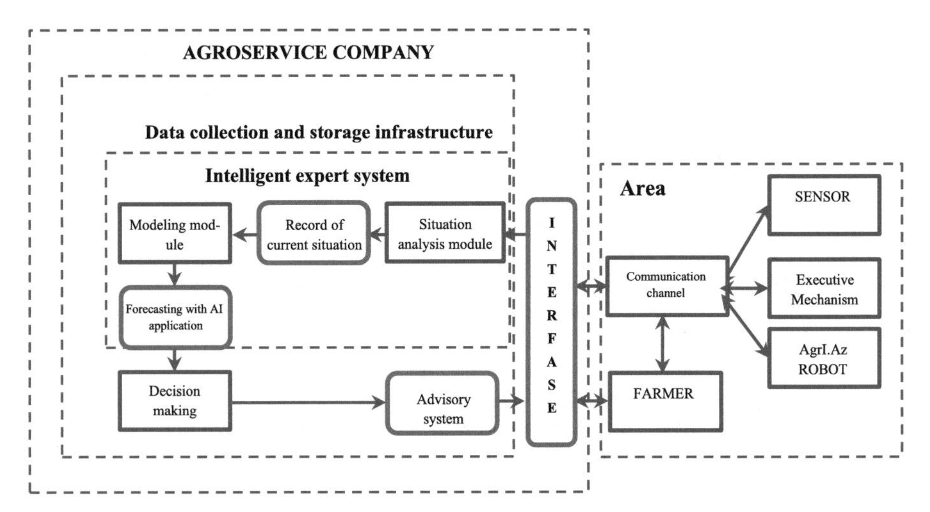

Fig. 4. Block diagram of an intelligent robotic system on a farm

There are great opportunities to improve agriculture. Sustainable and accurate agriculture in our country is one of the important factors influencing the development of the digital economy and requires special attention.

3 Discussion and Results

Thus, the proposed robotic device may at first glance be similar in terms of the principle of operation of similar technologies available around the world. However, the main idea here is to create a hybrid architecture based on the synthesis of drones, robots and artificial intelligence. The introduction of digital technologies in the agricultural sector will allow, on the one hand, to reduce the use of external resources, and on the other hand, to maximize the use of local factors of production.

4 Conclusions and Future Work

The proposed robotic device monitors soil, fertilizer shortages, water resources, and forecasts of seasonal crop growth and productivity levels in the above sectors to address problems (pests, weeds, drought, and s.) aims to implement multifunctional, robotic devices with artificial intelligence for timely detection and elimination. In addition, the use of electromagnetic pulse waves in multifunctional robotic devices can have a positive effect on productivity by increasing the yield in perennial tall trees. This will lead to a loss-free collection of the product in the short term, reducing the cost.

Unlike the objects studied in [1–10], the proposed robotic device has the ability to walk and fly. Data is obtained and processed through cameras and sensors on the platform. These data can be used to determine whether the soil is saline, waterlogged and stony before planting, to analyze water resources and water consumption in plant stems, to identify primary sources of pests, to intervene without reaching economic losses, and to implement additional measures for plant nutrition and protection, such as artificial intelligence. Warning of diagnosis and treatment in the form of digital prescriptions, etc. fulfillment is possible.

References

1. Aliyev, R.A., Jafarov, S.M., Babayev, M.C., Huseynov, B.G.: Principles of Construction and Design of Intelligent Systems, 368 p. Nargiz, Baku (2005)
2. Mustafayeva, A.M.: Prospects for the development of artificial intelligence technologies. In: International Scientific Conference Sustainable Development Strategy: Global Trends, National Experiences and New Goals. Mingachevir State University, 10–11 December 2021, pp. 47–53. https://www.mdu.edu.az/images/pdf/KONFRANS_2021_1.pdf
3. Akgül, M., Yurtseven, H., Demir, M., Akay, A., Gülci, S., Öztürk T.: Numerical elevation model in high sensitivity with unmanned aerial vehicles production and possibilities of use in forestry. Istanbul Univ. For. Fac. Mag. (J. Fac. For. Istanbul Univ.) 104–118 (2016)
4. Shet, A., Shekar, P.: Artificial Intelligence and Robotics ın the Field of Agriculture. https://www.researchgate.net/publication/347438971
5. Bechar, A., Vigneault, C.: Agricultural robots for field operations: concepts and components. Biosyst. Eng. **149**, 94–111 (2016). [CrossRef]
6. DJI 2017: Smarter agriculture package. http://www.precisionhawk.com/agriculture
7. Lenniy, D.: Artificial ıntelligence in agriculture: rooting out the seed of doubt. 6(104) (2021). ScienceDaily. https://intellias.com/artificial-intelligence-in-agriculture
8. Liang, X., Fang, Y., Sun, N., Lin, H.: Nonlinear hierarchical control for unmanned quadrotor transportation systems. IEEE Trans. Ind. Electron. **65**, 3395–3405 (2017)
9. Sylvester, G. (ed.): E-agriculture in action: drones for agriculture. Food and Agriculture Organization of the United Nations and International Telecommunication Union Bangkok (2018). http://www.fao.org/3/i8494en/i8494en.pdf
10. Wang, J.: The design of 'Six rotor UAV model of Agriculture' (2014). (Translated from the Chinese). http://www.taodocs.com/p-23157432.html
11. Yin, W., Zhang, J.N., Xiao, Y.M.Y.P., He, Y.H.: Application of the Internet of things in agriculture. In: E-Agriculture in Action, pp. 81–85. FAO and ITA, Bangkok (2017)
12. Zhang, Z., Cui, T.S., Liu, X.F., Zhang, Z.C., Feng, Z.Y.: The design and implementation of 'the use of UVA in agricultural ground surveillance system'. J. Agric. Mech. Res. **39**(11), 64–68 (2017). https://doi.org/10.13427/j.cnki.njyi.2017.11.011
13. https://www.lifewire.com/drones-could-help-farmers-raise-more-food-5220130
14. https://www.agrobotuav.com/agriculture-machinery-equipment/20-liters-agriculturaldrone-for-pesticide.html
15. https://www.copter.bg/en/dji-agras-agriculture-drones/1352-dji-agras-t20-comboagricul ture-drone-with-4-batteries-and-charger.html
16. Gabuev, K.O., Gongalo, V.O., Kucherenko, N.A., Shipko, A.I.: Automatic control system of the unmanned aerial vehicle. http://creativecommons.org/licenses/by/4.0/

Detection of Diabetic Macular Edema Disease with Segmentation of OCT Images

Saliha Yeşilyurt[1,3](✉), Altan Göktaş[2], Alper Baştürk[3], Bahriye Akay[3], Derviş Karaboğa[3], and Özkan Ufuk Nalbantoglu[3]

[1] Sivas Cumhuriyet University, Sivas, Turkey
salihayesilyurt@cumhuriyet.edu.tr
[2] Kayseri Acıbadem Hospital, Kayseri, Turkey
[3] Erciyes University, Kayseri, Turkey
{ab,bahriye,karaboga,nalbantoglu}@erciyes.edu.tr

Abstract. Diabetic macular edema (DME) is a condition in which the blood vessels of the retina become disrupted, and fluid accumulates between the retinal layers due to long-term hyperglycemia. It is a complication unnoticeable in the early stages of diabetes but can cause visual impairment and blindness, affecting millions of people with diabetes. Therefore, monitoring of retinal morphology and fluid accumulation is required properly to protect diabetic patients from blindness. In this study, deep learning methods were used to segment the retinal layers and fluid, which is a crucial step in diagnosing eye diseases. The U-Net and DeepLabV3+ model was trained with different backbones on OCT B-scan images obtained from 10 patients. According to the experimental results, the best Dice score for retinal layers was obtained with different ResNet backbones of the U-Net model. For fluid segmentation, the best Dice score (65.94%) was obtained with the ResNet101 backbone architecture of the DeepLabV3+ model.

Keywords: Optical coherence tomography · deep learning · diabetic macular edema · U-Net · DeepLabV3

1 Introduction

Optical coherence tomography (OCT) is a high-speed, non-invasive imaging modality that uses near-infrared light to create a cross-sectional retina view. This cross-sectional information is used extensively in ophthalmology to aid in the early detection and cure of retinal diseases such as diabetic macular edema (DME) [1].

DME is a disease in which the blood vessels of the retina are damaged, and fluid accumulates between the retinal layers due to long-term hyperglycemia. It is one of the most common causes of vision loss. About 100 million people worldwide show signs of diabetes-related macular edema [2].

Semantic segmentation is considered a pixel classification problem. It provides a complete understanding of the image compared to object recognition methods since it gives a more detailed view of objects. Recently, convolutional neural networks (CNN)

D. J. Hemanth et al. (Eds.): ICAIAME 2022, ECPSCI 7, pp. 671–679, 2023.
https://doi.org/10.1007/978-3-031-31956-3_57

have generally been utilized in medical image segmentation tasks. New methods based on deep convolutional neural networks (DCNN) outperform traditional methods, such as edge detection filters and mathematical methods [3].

The segmentation of retinal layers is an important step, and an expert's manual annotation of the layers is subjective and time-consuming. Moreover, OCT images suffer from artifacts that degrade image quality and affect the accuracy of image analysis. In small OCT images, the shallow contrast of adjacent tissue layers results in blurring layer boundaries. In addition, the presence of pathologies leads to changes in retinal morphology. These problems spur the development of methods to segment retinal layers and fluids in OCT images [4, 5].

Regarding the article's subject and the research done, the rest of this content is organized as follows: In the second part, segmentation methods using retinal OCT images are mentioned, and a general literature review is explained. In the third section, general information about the problem is given, and the proposed method is detailed. Following this, experimental results and metrics are given in the fourth section. Subsequently, the final chapter and the article ended with concluding remarks.

2 Related Works

In recent years, many retinal layer segmentation methods based on CNN have been developed in the literature since deep learning architectures have achieved superior performance in medical imaging processes.

In their work, Chiu et al. (2015) developed a 7-layer OCT segmentation applying kernel regression-based classification to determine the DME and OCT layer boundaries. Then, they obtained a Dice coefficient of 0.78 by combining this method with an approach utilizing Graph Theory and Dynamic Programming (GTDP) [6].

Fang et al. (2017) presented a hybrid ConvNet and graph-based technique for segmentation of the 9-layer borders of the retina. This technique employs CNN to extract useful features of retinal layer boundaries. Although this method can automate layer segmentation to some extent, the computational cost is high and sensitive to the location of the boundaries [7].

ReLayNet, proposed by Roy et al. (2017), uses a deep architecture of a fully convolutional-based network (FCN) for end-to-end segmentation of retinal layers, which significantly reduces the computational cost compared to patch-based approaches. ReLayNet is based on U-Net [8] widely used in image segmentation. It uses an encoder to learn contextual features and a decoder for semantic segmentation [9].

Pekala et al. (2019) presented an FCN model inspired by the U-Net and DenseNet models, together with Gaussian process-based regression. Although the presented method performs favorably compared to human error, its segmentation accuracy is low [5].

Liu et al. (2019) developed a semi-supervised method (SGNet) employing an adversarial learning approach with a segmentation network modified from ReLayNet and a discriminator network. The developed method automatically segments nine retinal layers and fluid regions in OCT B-scans [10].

Wang et al. (2021) developed a boundary-aware U-Net (BAU-Net) based on encoder and decoder architecture for retinal layers segmentation in OCT images [4].

In this study, experiments on the segmentation of the retinal layer and Fluid, which is an essential step in the diagnosis of ocular disease, are presented. DeepLabV3+, one of the up-to-date deep learning models, and U-Net, frequently used in biomedical images, are utilized. U-Net and DeepLabV3+ models are trained with six backbones. The obtained results are more successful for the Fluid class in terms of mIoU and Dice score than the literature.

3 Method and Material

3.1 Problem Definition

For a retinal OCT image, each x = (r, c) pixel is assigned a class of {0, 1, …, C − 1} for C = 10. The classes are shown in Fig. 1; RaR, RbR, Fluid and 7 retinal layers.

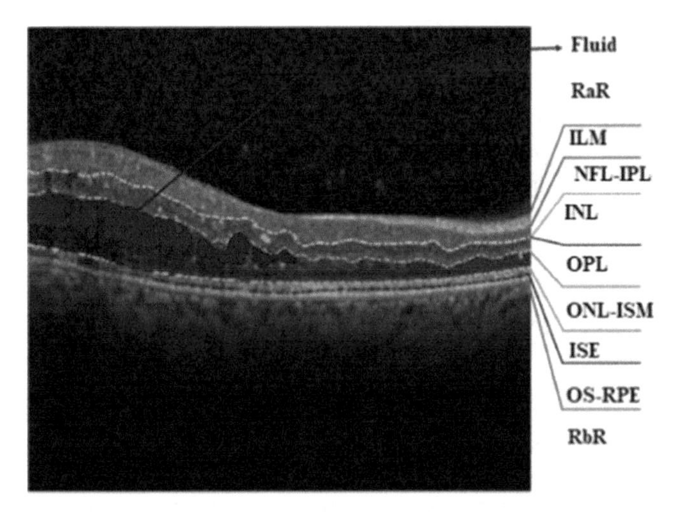

Fig. 1. Accumulated fluid and retinal layers in a sample OCT-B image [6].

3.2 Dataset

The Duke dataset consists of 10 OCT volumes from 10 patients with DME. 110 annotated SD-OCT B-scan images which are size 496 × 768, are obtained from 10 OCT volumes (11 B-scans per patient). These 110 B-scans are annotated for retinal layers and fluid regions by two expert clinicians [6]. A B-Scan is a 2D image, and multiple B-scans compose a 3D representation (a volume). Figure 2 illustrates the structure and coordinate system of such an OCT volume.

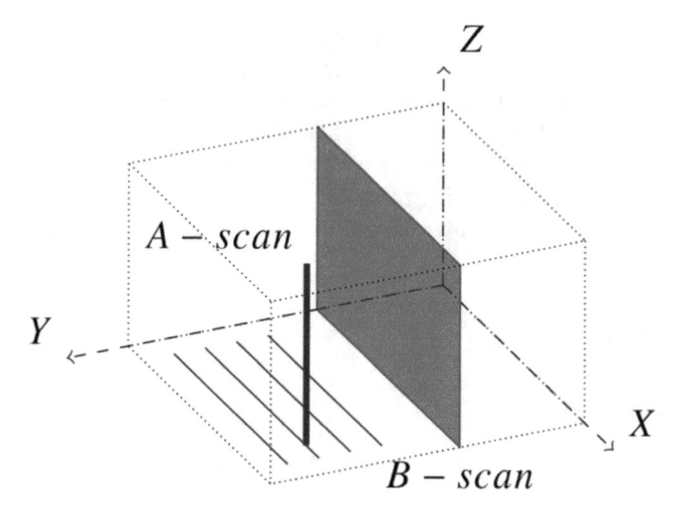

Fig. 2. An OCT volume structure and coordinate system [11].

3.3 Data Preprocessing

The size of a B-scan varies depending on the volume in which it originates. All scans were scaled to an image size of 480 × 480 so that the network's input is the same size. Then, the RGB image was converted to a gray level, and image pixel values were normalized in the range of [0 1]. Finally, annotated images are one-hot encoded, and data was prepared for training.

3.4 Model Architecture

This study utilized architecture, as in Fig. 3, inspired by encoder-decoder models. Comparisons were made between 12 different architectures using two decoder models and six backbones. In doing so, backbones (resnet50, resnet101, resnext50_32x4d, mobilenet_v2, efficientnet-b0, timm-mobilenetv3_small_minimal_100) and models (DeepLabV3+ and U-Net) in the segmentation_models [12] library were used.

U-Net [8] is a convolutional neural network architecture frequently used for biomedical image segmentation. It contains two parts, the contracting path, and the expanding path. The contracting path is an encoder, while the expanding path is a decoder. State-of-the-art models used for semantic segmentation are generally based on an encoder-decoder architecture such as U-Net. However, the presence of objects of various sizes in the image leads to problems classifying small objects. To address this problem, Chen et al. [13] presented the DeepLab neural network architecture. DeepLab was developed to increase the receptive field of feature maps effectively. It utilizes Atrous Spatial Pyramid Pooling (ASPP) module and dilation convolution with different rates. ASPP captures image context at multiple scales, as well as objects. For this reason, ASPP provides quality identifiers for objects of various sizes. In this study, DeepLabV3+, the extended version of DeepLab, was used. Experiments confirm the effectiveness of these models in semantic segmentation tasks.

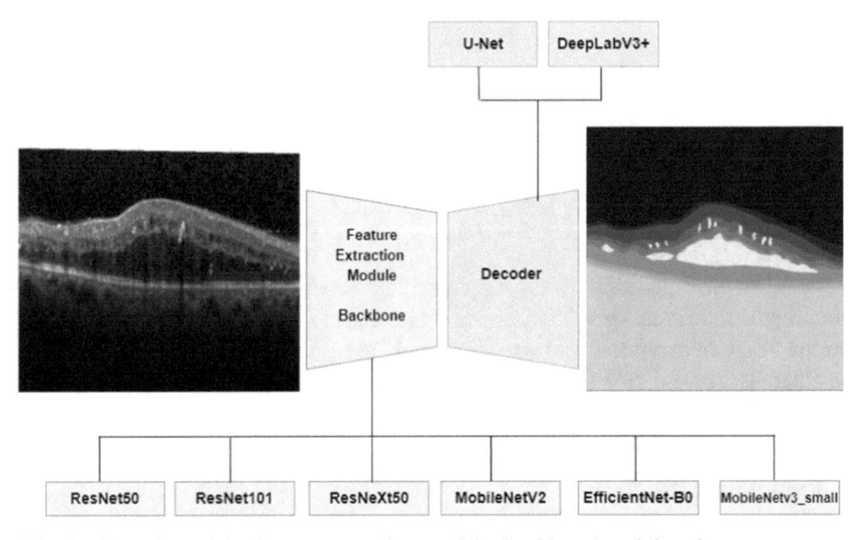

Fig. 3. Overview of the feature extraction module (backbone) and decoder components

Six different backbones were employed to extract the features from the image. These backbones are pre-trained models using ImageNet dataset. In devices with limited resources, such as mobile devices, backbones with few parameters are chosen to observe the accuracy of the medical image segmentation task.

Training models were composed of the models mentioned above (decoders) and backbones. K- fold cross-validation was performed to eliminate bias in the results and improve model prediction. The dataset (110 images for ten patients, 11 images per patient) was split into five equally sized non-overlapping subjects. Eight subjects (88 images) were used for the training, and the remaining two subjects (22 images) for the testing. This process was repeated with a 5-fold cross-validation so that each subject is included in the test set.

BCEWithLogitsLoss, with its default parameters, was used as the loss function to take advantage of the log-sum-exp trick. Additionally, the model weights were updated the RMSProp optimizer with the parameters weight_decay = 1e−8 and momentum = 0.9. Besides, the learning rate lr = 1e−5, batch_size = 8, and epoch = 180 were set. The best model was saved for each cross-validation training, then used during the corresponding testing.

4 Experiments

4.1 Evaluation Metrics

Several evaluation metrics are utilized in evaluating segmentation performance. Generally, positive-based metrics are used in the segmentation, so the metrics by which the overlap area is characterized are essential. IoU (Intersection over Union) and Dice Score are employed since these metrics are commonly employed to evaluate segmentation performance.

$$mIoU = \frac{1}{k+1}\sum\nolimits_{i=0}^{k}\frac{\sum_{i=1}^{l}TP_i}{\sum_{i=1}^{l}(TP_i + FP_i + FN_i)} \quad (1)$$

$$mDice = \frac{1}{k+1}\sum\nolimits_{i=0}^{k}\frac{2*\sum_{i=1}^{l}TP_i}{2*\sum_{i=1}^{l}TP_i + \sum_{i=1}^{l}(FP_i + FN_i)} \quad (2)$$

The IoU metric quantifies the overlap ratio between the ground truth and the predicted value over the total area. Dice Score is closely related to the IoU and positively correlated. Assuming that there are k + 1 classes, IoU and Dice Score are calculated separately for each class. The calculation method of the average IoU (mIoU) is shown in Eq. (1), and the average Dice (mDice) score is shown in Eq. (2).

4.2 Experiment Results

In this study, deep neural networks (DNNs) were obtained using six different backbones with U-Net and DeepLabV3+ decoders. Feature maps extracted from these backbones were given as input to the expanding path of the U-Net decoder and as input to the Atrous Spatial Pyramid Pool (ASPP) module of the DeepLabV3+ decoder. Then, the decoder side was used to improve the quality of these extracted feature maps. Thus, automatic segmentation of OCT images accomplished.

Table 1. Semantic segmentation results (%) of 5-fold cross-validation. The best performance is shown in bold (PN: parameter number, M: million, TI: test image)

Decoder	Backbone	PN	mDice	mIoU	TI/s
U-Net	ResNet50	23M	**86.77**	76.90	2.25
	ResNet101	42M	86.69	**75.77**	2.51
	ResNeXt50	22M	86.63	79.90	2.34
	MobileNetV2	2M	85.30	74.56	2.08
	EfficientNet-B0	4M	83.99	75.21	2.06
	MobileNetV3-small	0.43M	81.93	72.61	1.82
DeepLabV3+	ResNet50	23M	84.92	76.24	2.61
	ResNet101	42M	85.26	76.68	2.87
	ResNeXt50	22M	84.80	76.07	3.18
	MobileNetV2	2M	82.25	72.78	1.88
	EfficientNet-B0	4M	83.79	74.75	2.05
	MobileNetV3-small	0.43M	79.64	69.60	1.82

The semantic segmentation results of U-Net and DeepLabV3+ decoders with six different backbones are shown in Table 1. As the performance evaluation and comparison metric, the mean of averages is used from obtained for each class after 5-fold cross-validation. According to the segmentation results, the best outcome is obtained from ResNet50-U-Net (86.77% - mDice) and ResNeXt50-U-Net (mIoU - 79.90%). The best Dice score (65.94%) for the Fluid class is achieved with the ResNet101-DeepLabV3+ architecture.

Table 2. Comparison of the proposed method (ResNet101-DeepLabV3+) with the existing methods in the literature.

	U-Net	ReLayNet	BRUNet	BAU-Net	ResNet101 – DeepLabV3+
ILM	85.5	86.3	84.3	87.3	86.44
NFL-IPL	91.4	91.5	86.0	93.7	89.45
INL	79.8	78.7	72.7	81.8	77.40
OPL	80.9	76.2	71.5	82.6	75.09
ONL-ISM	89.0	90.5	81.1	90.6	88.88
ISE	86.5	82.3	82.3	89.4	86.61
OS-RPE	90.7	90.4	79.2	90.8	83.25
Fluid	45.2	49.5	–	52.7	**65.94**

Table 2 presents the results of the dice score of all classes on the Duke dataset and compares the ResNet101-DeepLabV3+ architecture with other methods in the literature. Additionally, our method achieves the best performance of 65.94% for the Fluid class, with improvements of 16.44% and 13.24% over ReLayNet and BaU-Net, respectively. As shown in Table 2, all the methods in the literature have lower performance in the Fluid region compared to the layer regions. The reason is that, as in b and e of Fig. 4, the shape of the Fluid is irregular, its number is various, and several small, accumulated fluid regions exist. Besides, the low contrast between the accumulated Fluid and the background complicates the Fluid regions' segmentation task. In this regard, architectures using U-Net as the decoder showed the best performance for the segmentation of retinal layers, while the best results for the segmentation of Fluid have obtained with the DeepLabV3+ decoder. When the architectures in this study are compared with each other, the fact that the architectures using the DeepLabV3+ decoder generally achieve the best result in the Fluid class is due to the success of DeepLabV3+ in classifying small objects compared to U-Net.

The average inference time for a test image is 6.43 s for the BAU-Net method. In contrast, for all conducted experiments in this study, the inference time of a test image ranged from 1.82 to 3.18 s, as shown in Table 1. In this regard, the experiments meet the clinical requirements for OCT retinal image segmentation.

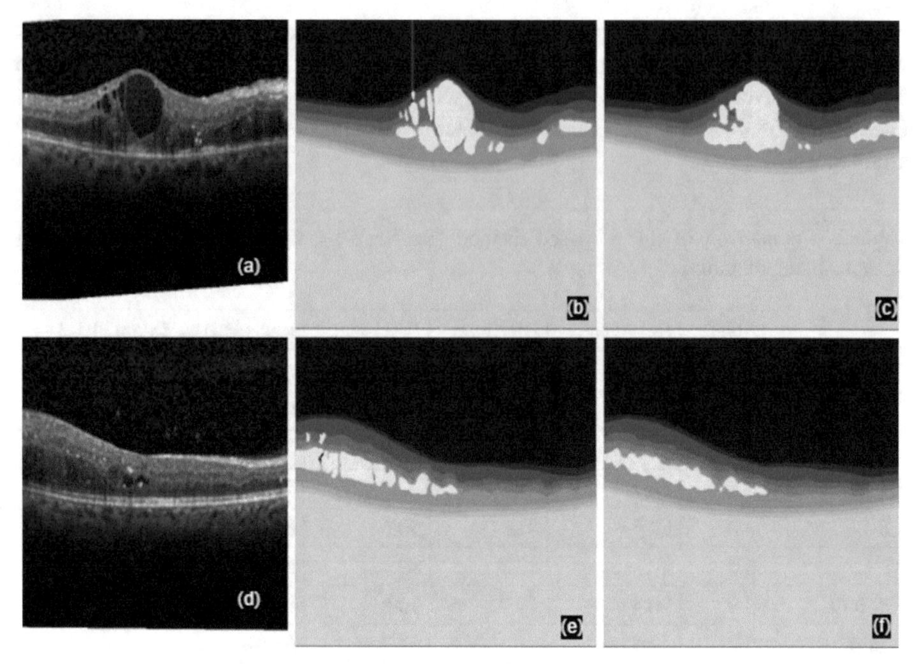

Fig. 4. Examples of B-scans from datasets and corresponding manual annotations. (a), (d) is input images, (b), (e) is annotated images, (c), (f) is the segmented prediction by ResNet101-DeepLabV3+ architecture.

5 Conclusions and Discussions

As a result of the study, it is shown that using encoder-decoder architecture in the automatic segmentation of retinal layers and fluids in OCT B scans has significant positive effects. For this, a series of experiments were carried out. In these experiments, U-Net, a popular model for biomedical images, and DeepLabV3+, one of the up-to-date deep learning models, were used with different backbones. The obtained models were validated with 5-fold cross-validation on the Duke DME dataset and compared with methods available in the literature. The best Dice score for retina layers was obtained with the different Resnet backbones of the U-Net model. In contrast, the fluid segmentation was obtained with the ResNet101 backbone architecture of the DeepLabV3+ model. The experimental outcomes in the Duke dataset show that the proposed ResNet101-DeepLabV3+ method outperforms the methods in the literature in segmenting the fluids. Moreover, the inference time of a test image in the proposed method is 2.87 s, which meets clinical needs. In this regard, the research results are essential to help in the early diagnosis and treatment of retinal diseases and are convenient for real-life applications.

References

1. Drexler, W., Fujimoto, J.G.: State-of-the-art retinal optical coherence tomography. Prog. Retin. Eye Res. **27**(1), 45–88 (2008). https://doi.org/10.1016/J.PRETEYERES.2007.07.005

2. Duphare, C., Desai, K., Gupta, P., Patel, B.C.: Diabetic macular Edema. StatPearls (2021). https://www.ncbi.nlm.nih.gov/books/NBK554384/
3. Hesamian, M.H., Jia, W., He, X., Kennedy, P.: Deep learning techniques for medical image segmentation: achievements and challenges. J. Digit. Imaging **32**(4), 582–596 (2019). https://doi.org/10.1007/S10278-019-00227-X/TABLES/2
4. Wang, B., Wei, W., Qiu, S., Wang, S., Li, D., He, H.: Boundary aware U-net for retinal layers segmentation in optical coherence tomography images. IEEE J. Biomed. Health Inform. **25**(8), 3029–3040 (2021). https://doi.org/10.1109/JBHI.2021.3066208
5. Pekala, M., Joshi, N., Liu, T.Y.A., Bressler, N.M., DeBuc, D.C., Burlina, P.: Deep learning based retinal OCT segmentation. Comput. Biol. Med. **114**(November 2018), 103445 (2019). https://doi.org/10.1016/j.compbiomed.2019.103445
6. Chiu, S.J., Allingham, M.J., Mettu, P.S., Cousins, S.W., Izatt, J.A., Farsiu, S.: Kernel regression based segmentation of optical coherence tomography images with diabetic macular edema. Biomed. Opt. Express **6**(4), 1172 (2015). https://doi.org/10.1364/BOE.6.001172
7. Fang, L., Cunefare, D., Wang, C., Guymer, R.H., Li, S., Farsiu, S.: Automatic segmentation of nine retinal layer boundaries in OCT images of non-exudative AMD patients using deep learning and graph search. Biomed. Opt. Express **8**(5), 2732–2744 (2017). https://doi.org/10.1364/BOE.8.002732
8. Ronneberger, O., Fischer, P., Brox, T.: U-net: convolutional networks for biomedical image segmentation. In: Navab, N., Hornegger, J., Wells, W.M., Frangi, A.F. (eds.) MICCAI 2015. LNCS, vol. 9351, pp. 234–241. Springer, Cham (2015). https://doi.org/10.1007/978-3-319-24574-4_28
9. Roy, A.G., et al.: ReLayNet: retinal layer and fluid segmentation of macular optical coherence tomography using fully convolutional networks. Biomed. Opt. Express **8**(8), 3627 (2017). https://doi.org/10.1364/BOE.8.003627
10. Liu, X., et al.: Semi-supervised automatic segmentation of layer and fluid region in retinal optical coherence tomography images using adversarial learning. IEEE Access **7**, 3046–3061 (2019). https://doi.org/10.1109/ACCESS.2018.2889321
11. Montuoro, A., Waldstein, S.M., Gerendas, B.S., Schmidt-erfurth, U., Bogunovi, H.: Joint retinal layer and fluid segmentation in OCT scans of eyes with severe macular edema using unsupervised representation and auto-context. Biomed. Opt. Express **8**(3), 1874–1888 (2017). https://doi.org/10.1364/BOE.8.001874
12. Yakubovskiy, P.: Segmentation models pytorch. GitHub repository. GitHub (2020)
13. Chen, L.C., Papandreou, G., Kokkinos, I., Murphy, K., Yuille, A.L.: DeepLab: semantic image segmentation with deep convolutional nets, atrous convolution, and fully connected CRFs. IEEE Trans. Pattern Anal. Mach. Intell. **40**(4), 834–848 (2018). https://doi.org/10.1109/TPAMI.2017.2699184

Innovative Photodetector for LIDAR Systems

K. Huseynzada[1,2(✉)], A. Sadigov[2], and J. Naghiyev[2]

[1] Mingachevir State University, Mingachevir AZ 4500, Azerbaijan
Khayala.huseynzada@gmail.com
[2] Innovation and Digital Development Agency of MDDT, Baku AZ 1073, Azerbaijan

Abstract. Optical navigation is widely used in automated robotics systems, smart cars, aviation. Every year the improvement of optical photo sensors expands their fields of application. Such systems or Lidar (Light Detection and Ranging) systems have different designs depending on the application. But the principle of operation is the same, light photons of a certain wavelength are emitted onto the object, the reflected light is detected by a photodetector, which in turn transmits a digital signal to the electronics, through which special software saves this signal as a point in the overall picture. Visualization of the entire object requires a large number of recorded photosignals at different angles of light, the more points there are, the better the visualization will be. Therefore, the speed and recovery time of the photodetector plays a key role in the process of obtaining a 3D image. The paper presents the concept of a new silicon photodetector. The calculation results show that the proposed photodetector has a recovery time 10 times better than its counterparts. The developed photodetector has high speed, low noise level and high resolution. The improvement of these parameters enables the developed photodetectors to become an indispensable component for lidar systems.

Keywords: LIDAR · MAPD · SiPM · MAPT · PDE · photon detection efficiency

1 Introduction

The very first known variation of modern lidar systems arose in nature millions of years ago. The bat uses a guidance system called sonar in modern times (SONAR, ultrasonic locator). They emit short pulses of ultrasound through their nostrils and catch the return signal with their ears, which look like two antennas. This feature provides the bat with a three-dimensional view of its surroundings, allowing it to avoid obstacles and find prey. The signals are short (50–100 ms) ultrasonic bursts with a constant frequency of tens of kilohertz. Bats are able to detect a wire obstacle at a distance of 17 m. The detection range depends on the wire diameter. A wire with a diameter of 0.4 mm will be found from a distance of 4 m, and a wire with a diameter of 0.08 mm from 50 cm. People began to develop such systems at the beginning of the 20th century. Christian Hülsmeier's "telemobiloscope", invented in 1904, was the first variation of the radar sensor. This device used radio waves that were outside the audible range. It consisted of an antenna, a receiver and a transmitter. Initially, it was used to detect metal objects, in particular ships at sea, in order to prevent their collision. This prototype radar had a

D. J. Hemanth et al. (Eds.): ICAIAME 2022, ECPSCI 7, pp. 680–690, 2023.
https://doi.org/10.1007/978-3-031-31956-3_58

range of 3,000 m, much less than current radars. When an object was detected, there was a sound signal from a bell that stopped sounding when the object disappeared from the field of view of the device. Radars emit pulsed signals. Distance is measured by the time it takes for an impulse to travel the distance to and from the target. It is also possible to measure the Doppler frequency shift to measure the speed of an object.

LiDAR provides the most efficient approach to fast, high resolution 3D mapping. Lidars require only one detectable photon per ranging, as opposed to hundreds or thousands of detectable photons when ranging with traditional rangefinders. The higher efficiency of lidar provides more opportunities for 3D mapping by:

- larger swath;
- better spatial resolution;
- less data collection time;
- greater density of returned range values.

In LiDAR systems, sensitivity to single photons is combined with photodetectors that have a sensitivity recovery time of nanoseconds and operate in sunlight conditions. Due to these advantages and the use of wavelengths from the NIR range, lidars are able to measure through quasi-transparent obstacles such as vegetation, fog, thin clouds, haze, and others. In addition, when using a green (~532 nm) laser as a working wavelength, additional opportunities for underwater measurements (bathymetry) appear.

Thus, LiDAR systems, among other things, can be used in such critical applications as:

- optical sounding and cartography;
- automation and unmanned systems;
- gas monitoring;
- perimeter security.

The operating principle of a lidar is simple. The object (surface) is illuminated with a short light pulse, and the time is measured after which the signal returns to the source. When you shine a flashlight on an object (surface), you see light reflecting off the object and returning to your retina. Light travels very fast - about 300,000 km per second, or 0.3 m per nanosecond - so turning on the light seems instantaneous. However, it is not. The light returns with some delay, which depends on the distance to the object. The distance traveled by a photon on its way to the object and back can be calculated using the formula:

$$\text{distance} = (\text{speed of light} \times \text{time of flight})/2$$

The equipment needed to measure this small amount of time must be extremely fast. This has only become possible with advances in modern electronics and computing technology. Lidar fires fast short pulses of laser radiation at an object (surface) with a frequency of up to 150,000 pulses per second. A sensor on the instrument measures the amount of time it takes for the pulse to return. Light travels at a constant and known speed, so lidar can calculate the distance between it and the target with high accuracy.

There are two types of lidar measurement methods: direct measurement method (also known as non-coherent), and coherent detection.

Coherent systems are best suited for Doppler or phase sensitive measurements and typically use optical heterodyne detection. This allows them to operate at much lower power, but at the same time, the design of the photodetector circuit is much more complicated. There are two main categories of pulsed lidars: micropulse and high-energy systems.

Micropulse lidars operate on more powerful computer technology with greater computing capabilities.

These lasers are of lower power and are classified as "eye safe", allowing them to be used with little or no special precautions.

High pulse energy lidars are mainly used in atmospheric research, where they are often used to measure various atmospheric parameters such as cloud height, layering and density, cloud particle properties, temperature, pressure, wind, humidity, and concentration of gases in the atmosphere.

Most lidar systems use four main components:

Lasers are classified by wavelength. Lasers with a wavelength of 600–1000 nm are more often used for non-scientific purposes, but since they can be focused and easily absorbed by the eye, the maximum power must be limited to make them "eye-safe". 1550 nm lasers are a common alternative because they are not aimed at the eye and are "eye safe" at higher power levels. Another advantage of the 1550 nm wavelengths is that they are not visible in night vision devices and are therefore well suited for military applications. Diode-pumped solid-state lasers with a wavelength of 1064 nm are often used in modern lidar systems. Bathymetric systems use diode-pumped lasers with a wavelength of 532 nm, which penetrates water with much less attenuation than 1064 nm. Better resolution can be achieved using shorter pulses, provided the receiver detector and electronics have enough bandwidth to handle the increase in data traffic.

Scanners and Optics. The speed at which images can be acquired by lidar depends on the scanning speed. A variety of scanning methods are used for various purposes, using oscillating flat mirrors, rotating polygonal mirrors, MEMS mirrors.

Navigation Systems. When LiDAR is installed on a mobile platform, such as a satellite, aircraft, or vehicle, the absolute location and orientation of the sensor must be determined in order to store useful data. Global positioning systems (GPS) provide accurate geographic information regarding the position of the sensor and the inertial measurement module provides information about the orientation of the sensor.

Photodetectors and Receiving Electronics. A photodetector is a device that reads and records the signal returned to the system. There are two main types of photodetectors, solid state detectors such as silicon avalanche photodiodes and photomultipliers.

In modern LIDAR systems, avalanche photodiodes are mainly used, which, compared with photomultipliers, have a number of advantages, such as compactness (3×3 mm), low power consumption and operating voltage (55 V), high resistance to magnetic fields and shock resistance, high photon detection efficiency.

The paper presents the concept of a new photodetector based on silicon. The photodetector is a development of recent years and is a consistent design of a line of photodetectors such as Micropixel Avalanche Photodiodes (MAPD). An Azerbaijani scientist Z.Sadygov first proposed MAPD, and patented basic designs [1–3]. The last decade, the different design of MAPD photodiodes were investigated [4–14, 15].

2 Literature Review

Avalanche photodiodes (APDs) are semiconductor photodetectors with internal photocurrent amplification. Structurally, in avalanche photodiodes, between the light absorption region (p-region) and the n-region of the semiconductor p-p-n structure, there is an additional p-semiconductor layer, that is, the APD structure has the form p-p-p-n. At a high reverse bias voltage, carriers drifting in the p-region acquire kinetic energy sufficient for impact ionization of atoms in the semiconductor crystal lattice. Due to the large, about 105 V/cm, electric field strength near the boundary of p- and n-semiconductors, the primary electron-hole pair formed during the absorption of one quantum can create tens or hundreds of secondary pairs. As a result of the avalanche multiplication of the number of carriers, the photocurrent in the APD, in comparison with the photocurrent in the p-I-n photodiode, increases by a factor of 100,000, which contributes to an increase in the sensitivity of such a photodetector by more than an order of magnitude. The main disadvantage of APD is the relatively large noise caused by temperature fluctuations in the value of the avalanche multiplication factor. The value of the reverse bias voltage in modern APDs lies in the range of 30–200 V and is set with high accuracy, for example, about 0.1 V. The operating frequency band of the APD reaches 80 GHz. And the length of the regeneration section of the transmission line.

An avalanche photodiode is a semiconductor photodetector in which an increase in quantum efficiency is realized due to internal amplification due to avalanche multiplication in a reverse biased p-n junction.

For the avalanche multiplication to exist, two conditions must be met:

1) The electric field of the space charge region must be large enough for the electron to gain energy over the mean free path greater than the band gap:
2) The width of the space charge region must be significantly greater than the mean free path

Micropixel avalanche photodiodes (MAPD) are devices of a new type for detecting light flashes of low intensity (at the level of single photons). The main advantages of these detectors, as compared to traditional devices (PMTs), is that the MAPD is a more compact device, insensitive to the magnetic field, it can have a higher photo detection efficiency, while combining lower cost. The use of such photodiodes makes it possible to create compact radiation detectors that can be used to develop devices of a new generation for a wide range of applications. Let us note some areas of their application: fundamental physical research; medical diagnostic equipment: X-ray tomograph, PET, single photon emission tomograph; registration of radiation (individual dosimeters, radiation reconnaissance, monitoring), etc. MAPD is a set of microcells - pixels. Each pixel operates in the so-called "Geiger" mode and is capable of detecting single photons. "Geiger" mode provides a very high gain up to 10^6. However, the pixel works as a photon counter in yes/no mode and is not able to register the intensity of the incident radiation. The main idea in creating an MAPD is to combine a set of such counters (cells, pixels) into a matrix on a common silicon substrate. If the intensity of the light flash is low, then the probability of the occurrence of two or more photoelectrons in each individual pixel is small, as a result of which the signal formed on the total load is proportional to the number of photons incident on the device. Thus, the MAPD as a whole is a device capable of detecting light intensity (the number of photons) with a dynamic range corresponding to the number of pixels of the photodetector. However, optical crosstalk and the effect after pulses (after pulsing effect), manifested at high coefficients ($\approx 10^6$) of the avalanche signal amplification, as well as a large specific capacitance and the associated insufficient speed with a large sensitive area, significantly limit the wide application MAPD with surface pixels. All these shortcomings significantly worsen the amplitude and time resolution of the device.

Optical crosstalk is due to the fact that during the avalanche multiplication of charge carriers in pixels, light radiation in the visible and infrared ranges is emitted. These photons hit neighboring pixels and start an avalanche process there, creating additional noise. There are a number of ways to combat optical crosstalk, such as reducing the gain, creating reflective or absorbing grooves around the perimeter of pixels. However, it is not possible to completely suppress the effect of optical crosstalk.

After the pulses are associated with the capture of charge carriers in the avalanche region of the MAPD. Charge carriers are captured in deep traps and can stay there for a long time (~100 ns). Captured charge carriers are randomly released and start a new avalanche process in the same pixel after the initial trigger. This is how the impulses appear after a certain delay in relation to the primary signal impulses. The amplitude after the pulses increases monotonically with increasing delay time, since the pixel discharged by the primary avalanche process has time to charge up to a certain level. Naturally, these random pulses (after the pulses) are summed up with the primary signal, worsening the operating parameters of the device.

Despite the fact that MAPDs (or SiPMs) have a gain of $\approx 10^5 \div 10^6$, they are significantly inferior in terms of the size of the photosensitive area to traditional electrovacuum photomultipliers. This is due to the high specific capacitance (~30 pF/mm^2) of the MAPD. For example, an MAPD with a working area of 1 cm^2 will have a capacitance of about 3000 pF, which is unacceptable for high-speed detectors, because the duration of the output signals at a load of 50 Ω will be about $3000 * 50 = 150$ ns.

3 Concept of Photodetector

In the framework of this work, we have developed and experimentally implemented a new avalanche photodetector, with the help of which it is possible to solve the above three problems of the well-known design of MAPDs (or SiPMs), namely, to significantly reduce the probability of optical crosstalk after pulses, and also to significantly increase the area of the device and improve its performance. The main idea behind the new photodetector is to reduce the avalanche gain in pixels and use an individual amplifying element for each pixel to obtain a sufficiently high gain. Such an individual amplifying element is a bipolar (or unipolar) microtransistor, the base (gate) of which is connected to the pixel. A microtransistor, such as an npn type, may be formed directly over a small area of a p-type pixel fabricated on an n-type substrate. For example, a microtransistor with a size of 2 μm * 2 μm can occupy no more than 0.5% of the area of a micropixel with a size of 15 μm * 15 μm.

The design of the developed photodetector is a micropixel avalanche phototransistor - MAPT (Fig. 1). It contains an array of micropixels with individual quenching resistors and an array of microtransistors with individual ballast resistors. The device has two independent outputs for picking up a signal: a normal output ("common out"), as in the well-known MAPD (or SiPM) and a fast output ("fast out"). All micropixels are connected with individual quenching resistors to one common metal bus, and all microtransistors are connected through individual ballast resistors to another metal bus. The avalanche gain region is spatially separated from the transistor gain region to prevent positive feedback. The fact is that when the electrons injected from the emitter enter the avalanche region, they will be strengthened by an avalanche with the creation of holes that will accumulate at the emitter, causing a new injection of electrons. This process leads to the generation of sinusoidal oscillations that are not related to the detection of light signals.

The proposed design is shown in Fig. 1. The semiconductor detector includes a semiconductor layer 1, where a matrix of semiconductor regions 2 was formed, forming potential barriers in the form of a p-n junction. Each of these areas has an individual microresistor 3 connecting it to a common metal bus 4. The microresistors and the conductive bus are isolated from the semiconductor layer by a dielectric layer 5. Individual emitters 6 are formed on the surface of these semiconductor regions, which are connected to another independent metal bus 7 through microresistors 8, The device has pin 9 with the semiconductor layer.

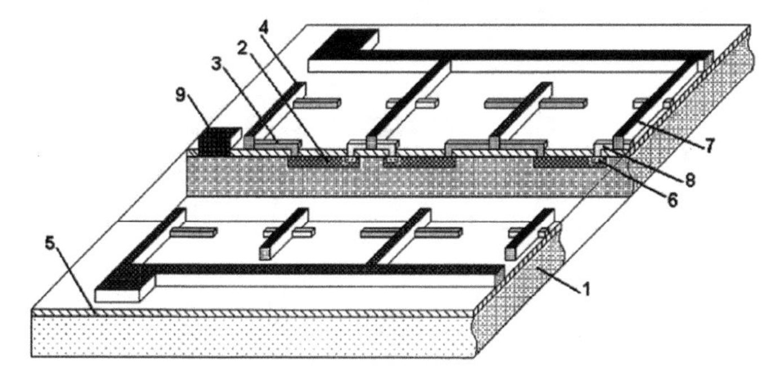

Fig. 1. Schematic view of Micropixel Avalanche Phototransistor

The operation of the proposed MAPT concept is based on the operation of a single pixel of the MAPD structure in binary mode. When a photon hits a pixel, a photoelectron is formed, which, through the field, forms up to 105 photoelectrons, thereby causing a voltage drop. This voltage (>0.8 V) is enough to open the emitter-base junction of the microtransistor, causing a large current to flow through the microtransistor circuit.

To expand the use of MAPD in various applications and devices, as well as in a number of nuclear physics detectors, medical hybrid tomographs of a new generation and in lidar systems that require high-speed photon registration, capable of measuring the arrival time of charged particles, gamma rays or photon beams with an accuracy of at least 100 ps, the design needs to be improved. For such photodiodes, an important parameter is the rate of rise of the leading edge and the recovery time of the photodiode itself. Therefore, a design change was required, which led to the creation of a new photodiode - the Micropixel Avalanche Phototransistor (MAPT) (Fig. 2).

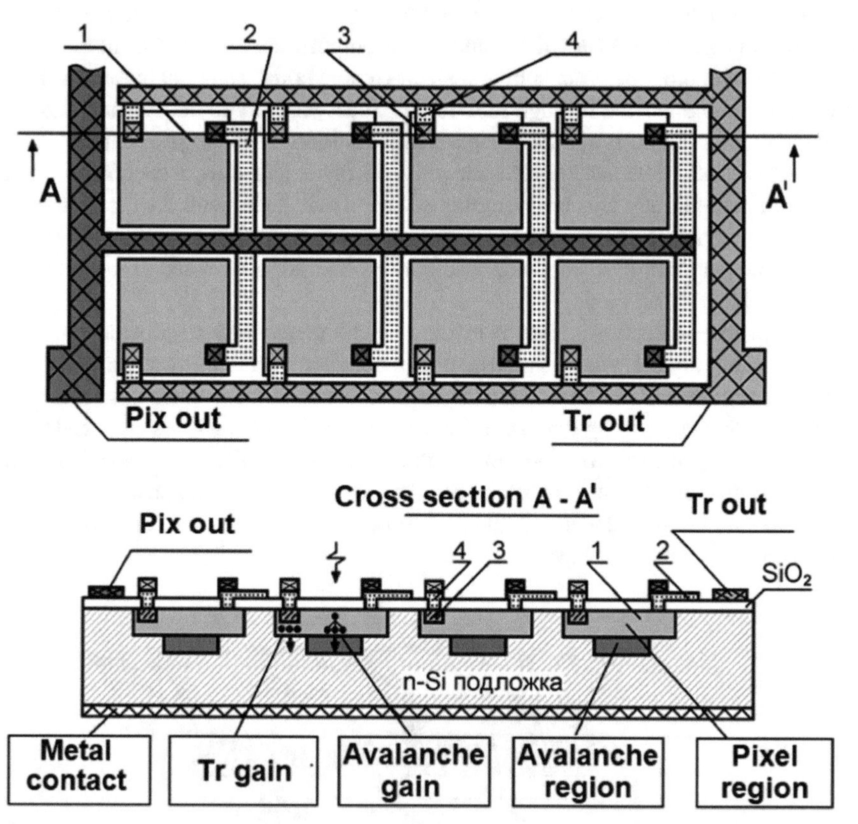

Fig. 2. Design of a micropixel avalanche phototransistor fabricated on an n-type silicon substrate. 1-Micropixel, 2-Micropixel Resistor, 3-Microtransistor, 4-Microtransistor Resistor.

Methods for picking up a fast signal from MAPD (or SiPM) pixels using an independent electrode are known in the scientific literature. For example, special micro-capacitors are used for this, formed on a part of the pixel surface. All micro-capacitors are connected to an additional metal bus, which has an independent output for removing the fast signal ("fast out"). In addition, the device has a second independent signal pickup bus connected to the pixels through individual quenching resistors ("common out"). Figure 3 shows the electrical equivalent circuit of such a device from SensL (www.

sensl.com). For comparison, the electrical equivalent circuit of the MAPT device developed by us is also shown there. It can be seen that in order to obtain a fast signal in the SiPM photodetector from SensL, a real photosignal incident on a pixel quenching resistor is differentiated. This leads to a significant drop in the amplitude (as well as the charge) of the received signal from the "Fast out" output. In the MAPT device developed by us, the signal amplitude is not attenuated, but, on the contrary, significantly increases.

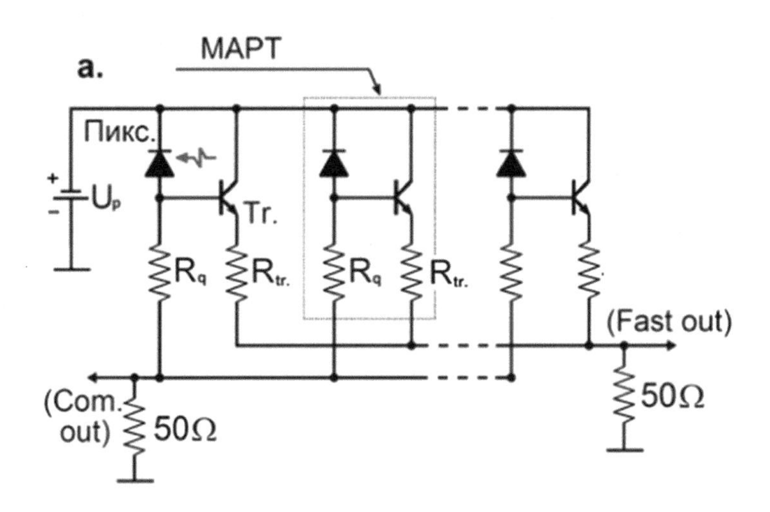

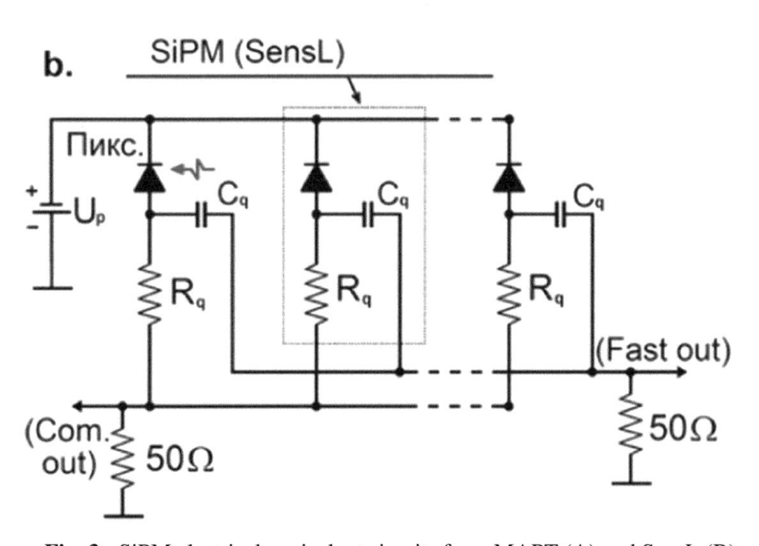

Fig. 3. SiPM electrical equivalent circuits from MAPT (A) and SensL (B).

The experimental data presented in Fig. 4 show a significant advantage of MAPT over SiPM from SensL. The comparison of the two instruments was carried out as follows. Appropriate bias voltages were applied to the devices, at which the gain in both devices reached M $\approx 1.6 * 10^6$. Such a signal is represented by curve 1 in Fig. 4. This signal was taken from the normal output (see Fig. 3, "Common out" output). At the same time, the signal was also taken from the fast exit ("Fast out"). Both signals were amplified by two identical amplifiers with a bandwidth of 80 MHz and a gain of 120, then fed to a two-beam oscilloscope with a bandwidth of 200 MHz. In this case, the edges of the signals were determined by the bandwidth of the amplifier, but this does not impair the accuracy of the comparison. The amplitude of the charge signal taken from the fast output of the MAPT exceeds (9.7 pC/1.2 pC) $\approx$ 8 times the amplitude of the corresponding signal taken from the SensL SiPM (Fig. 5).

The amplitude of the signal taken from the fast output of the developed MAPT can be significantly increased by increasing the overvoltage and reducing the resistance of the ballast resistor R(tr.). In this case, the growth rate (mV/ns) of the leading edge of the signal will increase significantly, which will improve the accuracy of timing in time-of-flight devices.

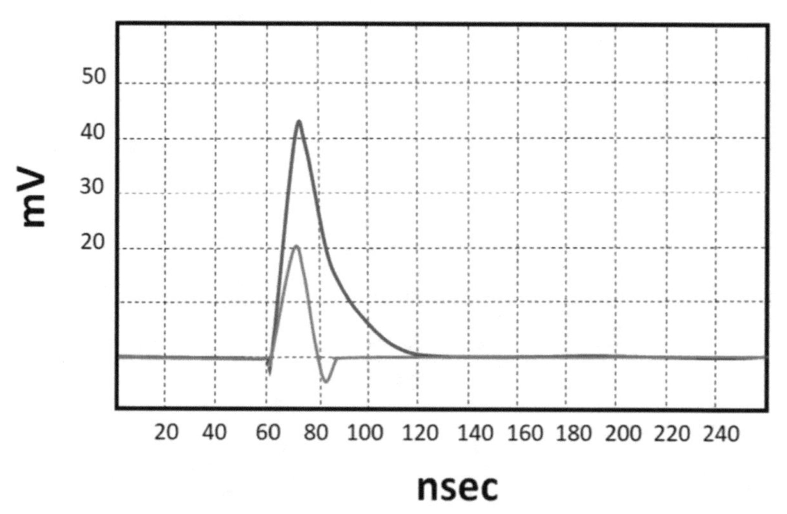

Fig. 4. Oscillogram of photodiodes. The red curve is the photo signal from the MAPT, the blue curve is the photo signal of the photodiode from the SensL company

The results of comparing the signal taken from the fast output ("Fast out") of both devices are presented in Table 1.

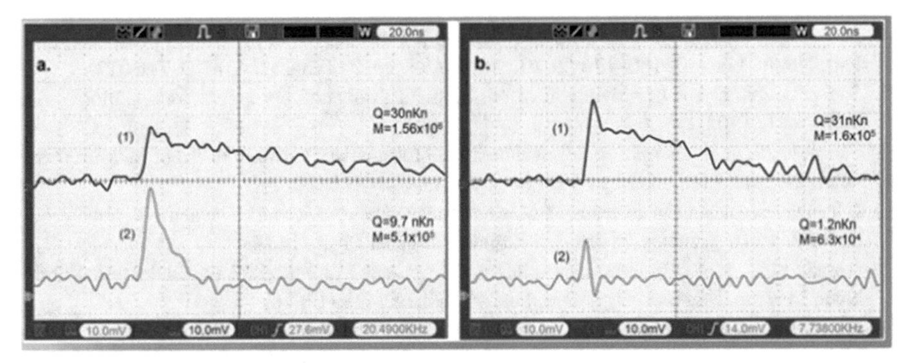

Fig. 5. Oscillograms of single-electron signals obtained from MAPT (a) and SiPM devices from SensL (b).

Table 1. Results of comparing the parameters of the developed MAPT with a foreign analogue from the company "SensL" (Ireland).

Parameter	SiPM	
	"SensL"	MAPT
Amplitude of signal	1,2 mV	2,8 mV
Gain	0,63 * 105	5,1 * 105
Rise rate of signal edge	2,4 * 10^5 V/s	5 * 10^5 V/s
Sensitive area	3 mm * 3 mm	3 mm * 3 mm

4 Results

A new design of a micropixel avalanche photodiode with microtransistors has been developed. In terms of performance parameters, the proposed photodiode exceeds its counterparts by 8 times. Also, the new design is freed from such shortcomings as afterpulses and optical crosstalk.

Acknowledgment. This Project has received funding from the European Union's Horizon 2021-SE-01 research and innovation programme under the Marie Sklodowska-Curie grant agreement 101086178.

References

1. Sadygov, Z.Y.-O., et al.: Photodetector array and method of manufacture. U.S. Patent No. 9,917,118 (2018)
2. Садыгов, З.Я., Садыгов, А.З.: Полупроводниковый лавинный фотоприемник, Пат. 2650417 РФ. Бюл. № 11. 9 с (2018)
3. Sadygov, Z.Y.-O., Sadygov, A.: Multi-pixel avalanche transistor. U.S. Patent No. 9,252,317, 2 February 2016

4. Sadygov, Z., Sadigov, A., Khorev, S.: Silicon photomultipliers: status and prospects. Phys. Part. Nuclei Lett. **17**, 160–176 (2020). https://doi.org/10.1134/S154747712002017X
5. Sadigova, N., et al.: Improvement of buried pixel avalanche photodetectors. Colloquium-J. **5**(92), 8–11 (2021)
6. Sadigov, A., et al.: A micropixel avalanche phototransistor for time of flight measurements. Nucl. Instrum. Methods Phys. Res. Sect. A **845**, 621–622 (2017)
7. Jafarova, E., et al.: On features of barrier capacitance of micropixel avalanche photodiodes on different frequencies. Mater. Sci. Condens. Matter Phys. (2014)
8. Ahmadov, F., et al.: On iterative model of performance of micropixel avalanche photodiodes. Nucl. Instrum. Methods Phys. Res. Sect. A **912**, 287–289 (2018)
9. Holík, M., et al.: Miniaturized read-out interface "Spectrig MAPD" dedicated for silicon photomultipliers. Nucl. Instrum. Methods Phys. Res. Sect. A. **978**, 164440 (2020)
10. Nuriyev, S., et al.: Performance of a new generation of micropixel avalanche photodiodes with high pixel density and high photon detection efficiency. Nucl. Instrum. Methods Phys. Res. Sect. A **912**, 320–322 (2018)
11. Akbarov, R.A., et al.: Fast neutron detectors with silicon photomultiplier readouts. Nucl. Instrum. Methods Phys. Res. Sect. A **936**, 549–551 (2019)
12. Ahmadov, F., et al.: New gamma detector modules based on micropixel avalanche photodiode. J. Instrum. **12**(01), C01003 (2017)
13. Ahmadov, F., et al.: A new physical model of Geiger-mode avalanche photodiodes. J. Instrum. **15**(01), C01009 (2020)
14. Ahmadov, F., et al.: Investigation of parameters of new MAPD-3NM silicon photomultipliers. J. Instrum. **17**(01), C01001 (2022)

FastText Word Embedding Model in Aspect-Level Sentiment Analysis of Airline Customer Reviews for Agglutinative Languages: A Case Study for Turkish

Akın Özçift(✉)

Department of Software Engineering, Hasan Ferdi Turgutlu Technology Faculty, Manisa Celal Bayar University, Manisa, Turkey
akin.ozcift@mcbu.edu.tr

Abstract. Extraction of knowledge from user generated data is important to improve the quality of services. Particularly, user reviews influences other consumers to make decisions about buying a product or using a service such as Airlines, Hotels, Telecommunication etc. Mining polarity of users about a service or a product from this valuable data is the research of Sentiment Analysis (SA). Recent studies focus on extraction sentiment of various aspects of products or services. This article proposes an efficient method for Aspect-Based Sentiment Analysis (ABSA) of Turkish airline reviews learning word representations. These representations are used to train a classifier with word embedding model, i.e., fastText (FT), which is used in ABSA task evaluations. Morphologically rich or agglutinative languages such as Turkish require preprocessing and feature engineering to comprehend both syntactic and semantic features. FT may be a solution of this problem that automates classification task with overcoming language preprocessing and feature engineering tasks to some extent. In this article, we apply Skip-gram (SG) word embedding based FT classifier to identify aspect categories and sentiment polarities of newly collected Turkish airline reviews data. The results of the experiments show that the proposed approach simplifies language preprocessing with acceptable accuracies in terms of F1-score.

Keywords: Aspect-Level Sentiment Analysis · Machine Learning · Word Embedding · Customer Review · Shallow Neural Network · FastText

1 Introduction

User generated content is a valuable resource to understand user decision mechanisms while preferring a service or a product [1]. Customers express their opinions about any product or any service using countless number of social media platforms as reviews. Researches have shown that consumer reviews about particular item influence corresponding customer choices. In other words, user purchasing decisions are affected with product or service reviews [2]. There is a continuous research for the extraction of sentiments from user reviews to improve service or product quality [3]. SA, analysis of user

D. J. Hemanth et al. (Eds.): ICAIAME 2022, ECPSCI 7, pp. 691–702, 2023.
https://doi.org/10.1007/978-3-031-31956-3_59

opinions, is defined as the investigation of polarity of user reviews as negative or positive. However, even a single sentence related to an item may practically contain conflicting opinions regarding different *aspects* of the same item [4]. For example, an airway review sentence may express positive attitude for an on time flight and negative opinion about comfort of the airplane at the same time. Therefore overall sentiment extraction of a document or sentence may become inefficient [5]. Comparison of two approaches is summarized in Fig. 1 for airline review context.

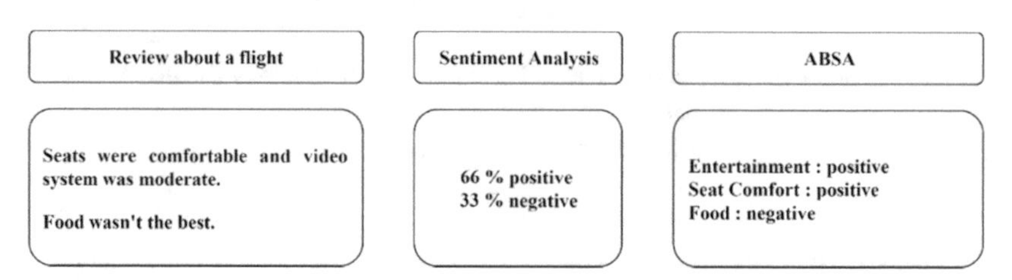

Fig. 1. Comparison of Sentiment Analysis and ABSA evaluation methods

An effective solution to the evaluation of this variation is the research field of Aspect-Based Sentiment Analysis (ABSA). In this manner, ABSA has two main tasks: (i) first identifying the aspect of the target text and then (ii) investigating polarity of that particular aspect [5].

Recent ABSA reviews show that most of the research is on English language and there are infrequent studies on the other languages [6]. In particular, though Turkish is a largely spoken language, the number of the studies in this field are very rare even quantitatively. In this context, Turkish, an agglutinative and a morphologically rich language, has research opportunities either in the creation of new datasets or in the application of state of the art algorithms.

Practically, any document should be represented in form of a vector model before a language processing task. Conventional bag of words (BOW) modelling represents text as independent word units and hence language processing tasks have problem of losing semantic or syntactic relations between words with data sparsity. Furthermore, traditional BOW approaches also requires tokenization, stop word filtering, stemming, lemmatization and word weighting steps to obtain an efficient model [7–9]. To remedy these problems word embedding models were developed. A word2vec extension embedding model, i.e. fastText, learns vectors as n-gram of characters as opposed to traditional word embedding models. This approach help classifiers understand suffixes or prefixes that makes it an appropriate solution for morphologically rich languages with suffix/prefix based language properties [10]. In this manner, fastText is eligible to catch morphological and semantic features of Turkish with the elimination of heavy pre-processing tasks.

The main contributions of this study is as follows: (i) a new dataset for Turkish ABSA analysis, (ii) development of the word embedding model based on fastText classifier for Turkish (iii) one of the first Turkish studies in airline reviews domain and (iv) a comprehensive ABSA literature overview.

We organize the remaining of the article as: In Sect. 2, we present the literature work for ABSA studies. Section 3 includes statistical properties of data and the data collection strategy. In Sect. 4, we present the details of the proposed algorithm and Grid-Search based parameter optimization for improving model's accuracy. While Sect. 5 handles the corresponding experimental results, Sect. 6 concludes overall research.

2 Related Works

In this section, we review SA studies in general and then we focus on ABSA searches particularly in Turkish language.

2.1 Sentiment Analysis (SA) Survey

In general, SA systems are categorized as (i) Machine Learning Strategies, and (ii) Lexicon based approaches [8]. In the first study, a Support Vector Machine (SVM) algorithm was used to identify sentiments of Turkish movie reviews [9]. Another ML based SA study was conducted in [10] for Turkish political news with SVM and Naïve Bayes (NB) algorithms. Two works from literature making use of various ML algorithms to obtain sentiments of Turkish Tweets are given in [11–13] respectively. Recent works of [6, 15–19] are studies based on well-known feature filtering tasks such as Information Gain (IG), Gini Index (GI), Chi Square (CH) and ML algorithms such as NB, SVM, Decision Trees (DT), Artificial Neural Networks (ANN). Lexicon-based techniques, includes studies of [20–25] which are applied to Tweets, movie reviews and news with the use of lexicons and ML algorithms to obtain polarities of respective texts. Though the mentioned studies makes use of various BOW and pre-processing approaches, the two studies applied embedding approaches to finance related Tweets [26] and customer reviews [27] for Random Forests (RF) and extreme gradient boosting (xgboost) classifiers.

2.2 Aspect Based Sentiment Analysis (ABSA) Survey

From Natural Language Processing (NLP) perspective, ABSA research can be categorized as supervised, unsupervised (Topic Modelling), frequency based and syntax-based approaches [18]. Primarily, while supervised methods use datasets annotated for aspects categories, unsupervised methods obtains aspect categories with the use of Latent Dirichlet Allocation (LDA) based algorithms [6].

The first Turkish ABSA studies from literature used for wireless carriers reviews [19] and hotel reviews [20]. And the other studies making use of POS-tagging and LDA techniques are given in [21–23]. Another recent study using a Conditional Random Fields (CRF) approach for Turkish restaurant reviews was assessed in [35].

Being one of the principal motivations of this study, there are only a few works on Turkish ABSA in the literature and therefore we selected airline reviews for ABSA domain based on fastText embedding modelling.

2.3 Word Embedding Models for Sentiment Analysis (SA) and Aspect Based Sentiment Analysis (ABSA)

Complex language analysis steps for Turkish may effectively be eased with word embedding approaches [36, 37]. Particular ABSA studies in the literature using embedding approaches are as follows: In their recent work, Hoang et al., have generated word embedding model for English restaurant reviews on top of Restricted Boltzmann Machine (RBM) [24]. Another work that make use of aspect based word embedding model was evaluated for English customer reviews with Convolutional Neural Networks (CNNs) [25]. Another work for word2vec-CNN couple was conducted for English movie reviews [26]. Other than English, another work based on Arabic word embedding with CNN was evaluated for restaurant, movie and book reviews [27]. In a recent study for Russian, the authors used fastText word embedding model for their ABSA task [28]. Aspects of Spanish Tweets were identified based on word2vec and SVM in [29].

In the following section, we explain the methodology used for Turkish aspect identification for airline customer domain.

3 Methodology

3.1 Word Embedding Approaches

Machine Learning algorithms require representation of text in terms of vectors or recently as word embedding's. More precisely, the goal is to represent words in a way that as much as information as possible to be encoded in a low dimensional vector space. Hence, continuous CBOW (Continuous BOW), SG and Glove embedding models were developed [30, 31]. In this respect, word2vec and Glove are open-source distributions based on the embedding model research [32].

CBOW architecture is a feedforward neural network language model predicts a word from its surrounding context minimizing the loss function in Eq. 1.

$$E = -\log\left(p\left(\vec{w}_t|\vec{W}_t\right)\right) \quad (1)$$

where w_t is the word to be predicted and $W_t = w_{t-n}, \ldots, w_t, \ldots, w_{t+n}$ denotes the sequence of the words in that context. The SG model is a little bit different from CBOW in that its goal is the prediction of the surrounding words of the target word instead of prediction of the target word itself. Glove, another widely used word embedding model, combines global matrix factorization and local context window techniques resting on a linear regression model while creating the corresponding neural model [33].

The main difference of fastText from word2vec-Glove pair is that the former represents a word as character n-grams. This architecture makes the model more suitable for morphologically rich or agglutinative languages having prefix-suffix nature. The architecture of fastText word embedding is explained in the next section.

3.2 FastText Word Embedding Model

The difficulty in learning word representations in morphologically rich languages was stated in [34] and sub-word or character-level information extension to continuous word embedding models was proposed as an alternative.

The architecture of the sub-word model is inspired from continuous SG model and it is briefly explained as follows: Traditional SG, gets a word list W with $w \in \{1, \ldots, W\}$ and the goal in this framework is to learn vectors for each word while maximizing the function in Eq. 2.

$$\sum_{t=1}^{T} \sum_{c \in C_t} logp(w_c|w_t) \tag{2}$$

The problem is to obtain context words w_c for given w_t depending on a score function $s(w_t, w_c) = \boldsymbol{u}_{w_t}^T \boldsymbol{v}_{w_c}$ which is calculated through scalar product of the word vector $\boldsymbol{u}$ and context word vector $\boldsymbol{v}$. One problem with this approach is that while it can learn embeddings for words occurring in the vocabulary, they cannot generate vector representation for out-of-vocabulary terms. On the other hand, sub-word information based fastText n-gram character embedding model solves this issue. In other words, fastText generates embedding of words as character n-grams and the embedding of a word is computed as the sum of its constituent n-grams. The scoring function which represents a word as sum of its n-grams is defined in Eq. 3.

$$s(w, c) = \sum_{g \in Ģ_w} \boldsymbol{z}_g^T \boldsymbol{v}_c \tag{3}$$

In this equation, a word is denoted as $Ģ_w \subset \{1,\ldots,G\}$in terms of its n-grams of size G. This model is qualified to share representations across words and it therefore allows to learn rich word forms particularly in morphologically rich languages [34].

In particular, fastText was developed to comprehend an embedding architecture and a shallow neural network classifier making it a compatible solution for morphologically rich language tasks. The architecture of fastText word embedding model is given in Fig. 2.

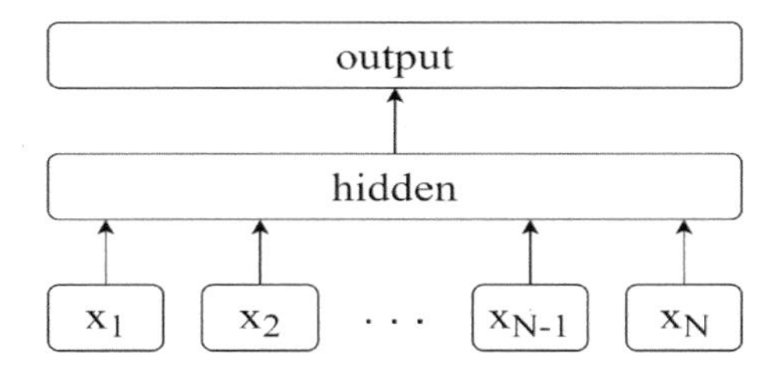

Fig. 2. FastText model for a sentence with *n-gram* features of *N*.

The n-gram features, i.e. $x_1, \ldots, x_N$, are embedded and averaged to generate the hidden variable. More details and explanations of parameters of the model is given in [35].

4 The Proposed Approach

The outline of the overall steps for the proposed approach is given in Fig. 3 for airline customer ABSA analysis.

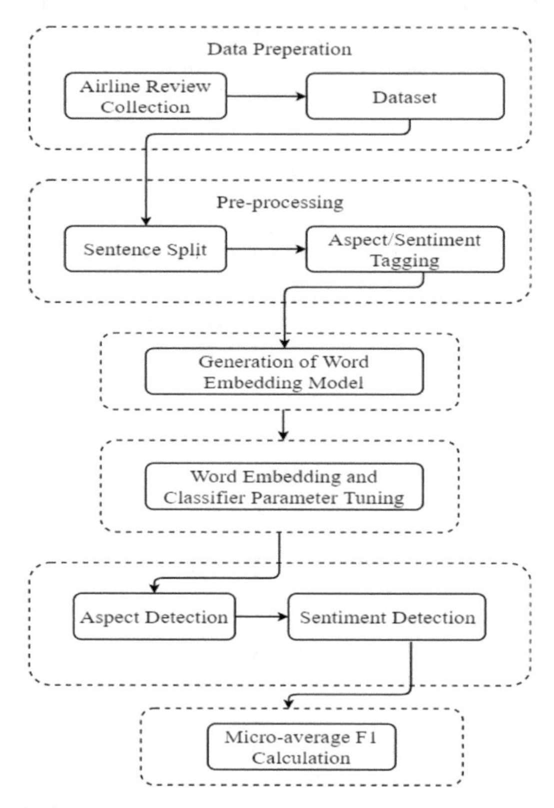

Fig. 3. Outline of the proposed ABSA polarity detection pipeline

5 Experiments and Results

5.1 Dataset Preparation and Language Pre-processing

In aspect-level sentiment analysis there are two different subtasks as aspect-category sentiment analysis (ACSA) and aspect-term sentiment analysis (ATSA). While the first task deals to obtain the aspect category of a text, the latter identifies the specific aspect terms themselves. In our study, we conduct ACSA to predict sentiment of an airline review regard to predefined aspect categories.

In this context, we randomly collected 1980 Turkish reviews about an international airline company from a travelling agency web-site. The properties of dataset is given in

Table 1. Composition of dataset

Polarity	Number	Proportion
Positive	1544	77.97%
Negative	436	22.03%
Total	1980	100.0%

Table 1 and it is randomly split into train, validation and test sets with 70%, 10% and 20% percentages in the evaluation of aspect and polarity models.

In this step, we pre-defined seven aspect categories as *cabin staff*, *food*, *entertainment*, *cleanliness*, *seat comfort*, *on time flight* and *value*. The first task in data processing step is to determine the aspect of each airline review. This tagging process was realized manually with a majority vote scheme. In more clear terms, three experts from NLP domain voted on each review and the majority of votes were used to decide final aspect label and its polarity. The dataset construction for aspect identification task is summarized in Fig. 4.

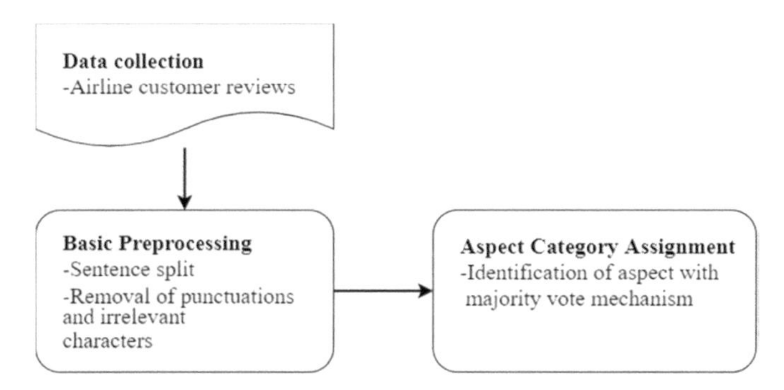

Fig. 4. The aspect sentiment dataset construction pipeline

5.2 Preparing Word Embedding Model

The evaluation of experiments first requires domain specific word embedding models to be prepared. The second requirement is to derive the optimal parameters of fastText to obtain the best accuracy:

i) We prepared the word embedding model and trained the classifier with default parameters of epoch size (ep), learning rate (lr), word n-gram (ng), dimension (dim) and loss function (ls).In this context, "ep" is the number of times that the algorithm evaluates each data point and this size may be varied from 5 to 100's. "lr" defines how fast the model to be updated in training phase and "lr" may be changed from 0.1 to 1.0. "ng" value is in the boundary of characters to represent words while obtaining model. The boundary of "ng" is 1 and 5 and by default 1 is unigram model. For example word 'comfort' is represented as <com, omf, mfo, for, ort> for ng = 3. "dim" parameter defines the size of the vectors and it can be changed typically from 50 to 300. In fastText, there are two loss functions as hierarchical softmax (hs) and standart softmax. The difference between the two is that hs approximates standart softmax function faster.

ii) The parameters of classifiers, i.e. "ep", "lr", "ng", "dim" and "ls" obtained models were optimized with the use of a grid-search strategy that is explained in Sect. 5.3.

5.3 Hyper-parameter Optimization with Grid-Search

Hyper-parameter optimization is the selection of parameter set so that the ML algorithm is to solve the problem with the highest precision. Since fastText model has parameters affecting the performance of the language task, obtaining the optimal set of parameters require a mechanism such as grid-search. In this context, the parameters of "ep", "lr", "ng" and "dim" are searched to calculate to optimize f-measure score [36].

Table 2. Grid search boundaries of each parameter

Parameter	Boundary
epoch size (ep)	5–200
n-gram (ng)	1–5
learning rate (lr)	0–1
dimension (dim)	50–300
loss function (ls)	softmax

The search boundaries of each parameter is given in Table 2. With the use of parameters, for 70%, 10% and 20% train, validation, test splits, the average F1-score performance of the model changes from 78.00% to the maximum 82.75. The maximum averaged F1 value, 82.75%, is obtained for ep = 5, ng = 2, lr = 1, dim = 50 with ls to be softmax.

5.4 The Experimental Protocol and Evaluation Metrics

The following experiments are evaluated with the dataset split into train, validation and test sets as 70%, 10% and 20% percentages respectively. Then optimal parameters for fastText word-embedding model and the corresponding shallow neural network fastText classifier were obtained with the use of grid search mechanism. The optimized model is evaluated with Precision (Pr), Recall (Re) and F1 scores (F1) for each aspect group. In particular, Re, also known as sensitivity, is calculated as the ratio of the number of true positives divided by the sum of the true positives and the false negatives. Furthermore, Pr is also known as positive predictive value and it is the ratio of the number of true positives divided by the sum of the true positives and false positives. F1 is a widely used evaluation metric in many domains of data science. The main advantage of F1 is that it evaluates the performance of a ML method as the weighted average of Pr and Re. The metric measures prediction performance of a method more realistically particularly for an imbalanced class distribution [37, 38]. In this context, the ABSA prediction is a multinomial classification problem with 7 aspect classes, where we considered Pr, Re and F1 for each class, and average of F1 for overall prediction performance of the proposed approach.

5.5 Experimental Results

With the use of performance metrics explained above, we repeated all experiments ten times and we calculated each ABSA score on average. The results of the experiments are given in Table 3.

Table 3. Experimental Results of the proposed method for ABSA detection

Aspect Class	Precison (%)	Recall (%)	F1
Food	95.24	90.91	93.02
Entertainment	90.00	81.82	85.71
Cabin staff	86.49	72.73	79.01
On time flight	86.96	72.73	79.21
Cleanliness	86.11	70.45	77.50
Seat comfort	86.84	75.00	80.49
Value	89.74	79.55	84.34
Average	88.76	77.59	82.75

The aspect-sentiment identification of fastText word embedding model is summarized in Fig. 5.

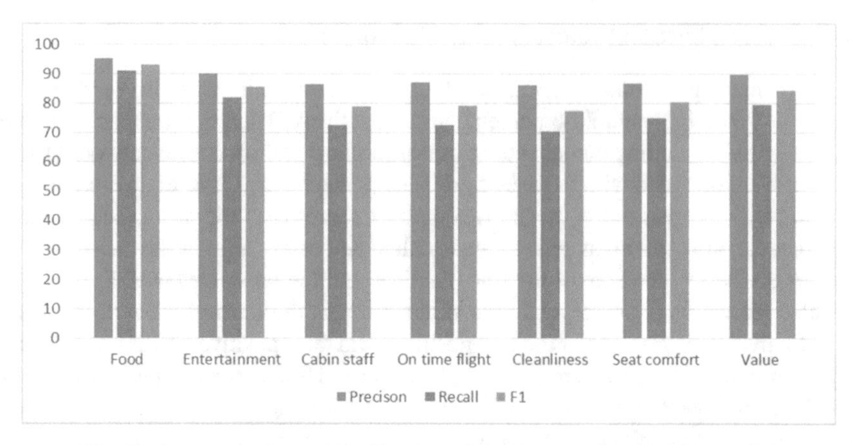

Fig. 5. Aspect-sentiment identification of fastText word embedding model

6 Conclusions and Future Work

In this study, a word-embedding model for (ABSA) of airline customer reviews in Turkish is presented. In particular, agglutinative languages such as Turkish requires extensive language processing steps before any language tasks. Principally, ABSA is a more refined sentiment analysis, it involves substantial morphological and syntactic pipelines to obtain efficient results. The research conducted in this study proposes the usage of fastText word embedding model in ABSA particularly for morphologically rich languages such as Turkish. In this work, two main tasks, i.e. aspect category identification and corresponding sentiment polarity detection, are researched. The obtained results in terms of Recall, Precision and F1 indicate the efficiency of the proposed method. Another finding is that fastText is tested to be reliable for multi-class problems such as ABSA.

References

1. Yang, B., Liu, Y., Liang, Y., Tang, M.: Exploiting user experience from online customer reviews for product design. Int. J. Inf. Manag. **46**, 173–186 (2019)
2. Do, H.H., Prasad, P.W.C., Maag, A., Alsadoon, A.: Deep learning for aspect-based sentiment analysis: a comparative review. Expert Syst. Appl. **118**, 272–299 (2019)
3. Medhat, W., Hassan, A., Korashy, H.: Sentiment analysis algorithms and applications: a survey. Ain Shams Eng. J. **5**(4), 1093–1113 (2014). https://doi.org/10.1016/J.ASEJ.2014.04.011
4. Al-Smadi, M., Al-Ayyoub, M., Jararweh, Y., Qawasmeh, O.: Enhancing aspect-based sentiment analysis of Arabic hotels' reviews using morphological, syntactic and semantic features. Inf. Process. Manag. **56**(2), 308–319 (2019)
5. Song, M., Park, H., Shin, K.S.: Attention-based long short-term memory network using sentiment lexicon embedding for aspect-level sentiment analysis in Korean. Inf. Process. Manag. **56**(3), 637–653 (2019)
6. Schouten, K., Frasincar, F.: Survey on aspect-level sentiment analysis. IEEE Trans. Knowl. Data Eng. **28**(3), 813–830 (2016)

7. Kincl, T., Novak, M., Piribil, J.: Improving sentiment analysis performance on morphologically rich languages: language and domain independent approach. Comput. Speech Lang. **56**, 36–51 (2019)
8. Abirami, A.M., Gayathri, V.: A survey on sentiment analysis methods and approach. In: 2016 Eighth International Conference on Advanced Computing (ICoAC), pp. 72–76 (2017)
9. Eroğul, U.: Sentiment analysis in Turkish. Middle East Technical University (2009)
10. Kaya, M., Fidan, G., Toroslu, I.H.: Sentiment analysis of Turkish political news. In: 2012 IEEE/WIC/ACM International Conferences on Web Intelligence and Intelligent Agent Technology, vol. 1, pp. 174–180 (2012)
11. Amasyalı, M.F.: Active learning for Turkish sentiment analysis. In: 2013 IEEE INISTA, pp. 1–4 (2013)
12. Çoban, Ö., Tümüklü Özyer, G.: The impact of term weighting method on Twitter sentiment analysis. Pamukkale Univ. J. Eng. Sci. **24**(2), 283–291 (2018)
13. Akba, F., Uçan, A., Sezer, E.A., Sever, H.: Assessment of feature selection metrics for sentiment analyses: Turkish movie reviews. In: 8th European Conference Data Mining, vol. 191, pp. 180–184 (2014)
14. Yelmen, I., Zontul, M., Kaynar, O., Sonmez, F.: A novel hybrid approach for sentiment classification of Turkish tweets for GSM operators. Int. J. Circ. Syst. Sig. Process. **12**, 637–645 (2018)
15. Parlar, T., Özel, S.A.: A new feature selection method for sentiment analysis of Turkish reviews. In: 2016 International Symposium on INnovations in Intelligent SysTems and Applications (INISTA), pp. 1–6 (2016)
16. Coban, O., Ozyildirim, B.M., Ozel, S.A.: An empirical study of the extreme learning machine for Twitter sentiment analysis. Int. J. Intell. Syst. Appl. Eng. **6**(3), 178–184 (2018)
17. Yıldırım, E., Çetin, F.S., Eryiğit, G., Temel, T.: The impact of NLP on Turkish sentiment analysis. Türkiye Bilişim Vakfı Bilgisayar Bilimleri ve Mühendisliği Dergisi **7**(1), 43–51 (2015)
18. Dey, S.: Aspect extraction and sentiment classification of mobile apps using app-store reviews. arXiv preprint arXiv:1712.03430 (2017)
19. Akbaş, E.: Aspect based opinion mining on Turkish tweets. İhsan Dogramacı Bilkent University (2012)
20. Omurca, S.I., Ekinci, E., Türkmen, H.: An annotated corpus for Turkish sentiment analysis at sentence level. In: 2017 International Artificial Intelligence and Data Processing Symposium (IDAP), pp. 1–5 (2017)
21. Türkmen, H., Omurca, S.I., Ekinci, E.: An aspect based sentiment analysis on Turkish hotel reviews. Girne Am. Univ. J. Soc. Appl. Sci. **6**, 9–15 (2016)
22. Ekinci, E., Omurca, S.I.: An aspect-sentiment pair extraction approach based on latent Dirichlet allocation for Turkish. Int. J. Intell. Syst. Appl. Eng. **6**(3), 209–213 (2018)
23. Ekinci, E., Omurca, S.İ.: Product aspect extraction with topic model. Türkiye Bilişim Vakfı Bilgisayar Bilimleri ve Mühendisliği Dergisi **9**(1), 51–58 (2016)
24. Nguyen-Hoang, B.-D., Ha, Q.-V., Nghiem, M.-Q.: Aspect-based sentiment analysis using word embedding restricted Boltzmann machines. In: Nguyen, H.T.T., Snasel, V. (eds.) CSoNet 2016. LNCS, vol. 9795, pp. 285–297. Springer, Cham (2016). https://doi.org/10.1007/978-3-319-42345-6_25
25. Du, H., Xu, X., Cheng, X., Wu, D., Liu, Y., Yu, Z.: Aspect-specific sentimental word embedding sentiment analysis of online reviews. In: 25th International Conference Companion on World Wide Web, pp. 29–30 (2016)
26. Rezaeinia, S.M., Ghodsi, A., Rahmani, R.: Improving the accuracy of pre-trained word embeddings for sentiment analysis. arXiv preprint arXiv:1711.08609 (2017)

27. Dahou, A., Elaziz, M.A., Zhou, J., Xiong, S.: Arabic sentiment classification using convolutional neural network and differential evolution algorithm. Comput. Intell. Neurosci. **2019**, 1–16 (2019)
28. Romanov, V., Khusainova, A.: Evaluation of morphological embeddings for the Russian language. In: 2019 3rd International Conference on Natural Language Processing and Information Retrieval, pp. 144–148 (2019)
29. Fiestas, F.T., de Patrones, G.D.R., Cabezudo, M.A.S.: C100TPUCP at tass 2017: word embedding experiments for aspect-based sentiment analysis in Spanish tweets. In: TASS 2017: Workshop on Semantic Analysis at SEPLN, pp. 85–90 (2017)
30. Pennington, J., Socher, R., Manning, C.D.: GloVe: global vectors for word representation. In: 2014 Conference on Empirical Methods in Natural Language Processing (EMNLP), pp. 1532–1543 (2014)
31. Van Landeghem, J.: A survey of word embedding literature. Academia (2016)
32. Camacho-Collados, J., Pilehvar, M.T.: From word to sense embeddings: a survey on vector representations of meaning. J. Artif. Intell. Res. **63**, 743–788 (2018)
33. Joulin, A., Grave, E., Bojanowski, P., Mikolov, T.: Bag of tricks for efficient text classification. arXiv preprint arXiv:1607.01759 (2016)
34. Bojanowski, P., Grave, E., Joulin, A., Mikolov, T.: Enriching word vectors with subword information. Trans. Assoc. Comput. Linguist. **5**, 135–146 (2017)
35. Bhattacharjee, J.: FastText Quick Start Guide: Get Started with Facebook's Library for Text Representation and Classification. Packt Publishing Ltd. (2018)
36. Özçift, A., Akarsu, K., Yumuk, F., Söylemez, C.: Advancing natural language processing (NLP) applications of morphologically rich languages with bidirectional encoder representations from transformers (BERT): an empirical case study for Turkish. Automatika **62**(2), 226–238 (2021)
37. Bozuyla, M., Özçift, A.: Developing a fake news identification model with advanced deep language transformers for Turkish COVID-19 misinformation data. Turk. J. Electr. Eng. Comput. Sci. **30**(3), 908–926 (2022)
38. Borandag, E., Ozcift, A., Kilinc, D., Yucalar, F.: Majority vote feature selection algorithm in software fault prediction. Comput. Sci. Inf. Syst. **16**(2), 515–539 (2019)

Prediction of Electric Energy in Hydroelectric Plants by Machine Learning Methods: The Example of Mingachevir Dam

Almaz Aliyeva[1(✉)], Mevlüt Ersoy[2], and M. Erol Keskin[3]

[1] Mingachevir State University, Mingachevir, AZ 4500, Azerbaijan
almaz.aliyeva@mdu.edu.az
[2] Department of Computer Engineering, Engineering Faculty, Suleyman Demirel University, 32200 Isparta, Turkey
[3] Department of Civil Engineering, Engineering Faculty, Suleyman Demirel University, 32200 Isparta, Turkey

Abstract. Hydroelectric power plants are important energy production areas. It is a clean energy source in many parts of the world and is generally a preferred energy production source in regions with high precipitation. With the effect of global warming, studies should be carried out according to the energy target according to the measures to be taken according to the forecasts of the energy produced in hydroelectric power plants in the coming years. In many studies, prediction models have been developed for hydroelectric power plants from different perspectives. In this study, the amount of energy that can be produced annually according to the maximum and minimum water level of the water entering and leaving the power plant for many years has been estimated by machine learning models. The predictions obtained from the models were 89% successful.

Keywords: Time Series Forecasting · Support Vector Regression · xGBoost · Energy Production

1 Introduction

Between 1997 and 2020, worldwide energy consumption has increased by approximately 60%. The technical production potential of Asian countries is approximately 6800 TWatt/year. In the survey studies conducted in some countries in the world, hydroelectric energy has been shown among the economic development parameters in the future. Hydroelectric power plants are an important energy source for the discovery of clean and sustainable energy sources in Azerbaijan. The amount of energy produced in hydroelectric power plants varies from year to year. The water levels in the power plant, the amount of water entering the electricity generation point are effective on the amount of energy produced. Changes in water level during the year; There may be many reasons such as humidity of the air, sunshine duration, cloud ratio in the air, precipitation. For this reason, the high number of parameters in the solution of these time series problems affected by hydrological events makes the solution of the problem difficult.

D. J. Hemanth et al. (Eds.): ICAIAME 2022, ECPSCI 7, pp. 703–712, 2023.
https://doi.org/10.1007/978-3-031-31956-3_60

Different approaches to energy production estimations in hydroelectric power plants are discussed. These approaches are related to the estimation of the water level in the reservoir, the estimation of the reservoir flow rate, the estimation of the water entering and leaving the dam reservoir, and the estimation of the energy production capacities. There are studies dealing with different calculation methods regarding the water estimates in the reservoir. In these studies, time series processing algorithms [1], empirical orthogonal functions [2], error correction-based estimation [3], multivariate approaches [4] or ensemble-based algorithms [5], fuzzy logic-based algorithms were used. At the same time, Machine Learning (ML) approaches have been successfully applied to these problems. The applied algorithms were generally obtained according to the data obtained from the dams in different countries.

One of the first approaches to use ML algorithms for water level estimation in reservoirs is which compares the performance of artificial neural networks and neuro-fuzzy approaches to the short-term water level problem in two German rivers from hydrological upstream data. Adaptive neuro-fuzzy inference algorithms are also discussed in [6] and [7] for water level estimation in reservoirs after typhoon events. In [8], the performance of different ML algorithms such as neural networks, support vector regression and deep learning algorithms is evaluated in a reservoir operation problem (mainly input and output prediction) at different time scales at Gezhouba Dam. Yangtze river in China. In [9] a recurrent neural network was proposed to predict the flow of a reservoir from a distributed hydrological model. In [10], different ML regressions were applied to a water level estimation problem using in situ measurements of stream flow and precipitation at a hydroelectric power station in Galicia, Spain. Various hybrid ML approaches have also been proposed for reservoir water availability problems, such as hybridizing a neural network with a genetic algorithm [11] to improve network training. This hybrid approach has been tested in the dam water level estimation problem downstream of the Han River in China [11]. The authors propose a support vector regression (SVR) algorithm hybridized with a fruit fly algorithm for optimization of SVR parameters. In this case, Iran was tested in the river flow estimation problem in the Urmia Lake basin.

In these studies, the success rates of xgBoost and SVR methods were aimed for the prediction of energy production in hydroelectric power plants. Hydroelectric power plants are not linear on long-term scales for many factors. The applied machine learning models were chosen because they gave successful results in the predictions made with non-linear input data. In addition, they are models with high success rates thanks to the wide-range use of parameters. At the same time, ML methods are frequently used for long-term forecasts in reservoir management.

The main purpose of this study is to analyze the performance in the forecasting of long-term energy generation at the hydroelectric power plant at Mingachevir Dam.

In the second part of the study, explanations were made about the use of machine learning algorithms xGBoost and SVR algorithms. In the third chapter, hyperparameters and data set of machine learning algorithms are mentioned. The results obtained in the fourth section were analyzed and a general evaluation was made in the fifth section.

2 Materials and Methods

2.1 XGBoost Algorithm

Ensemble learning algorithms is a machine learning based on divide and discover operations in solving regression and classification problems [12]. Several models achieve a result by performing a specific task. The results are then combined to produce an accuracy-efficient model. One of these models is extermly Gradient Boosting, in short, the xGBoost algorithm [13]. An important aspect of XGBoost models is their scalability. Good for multiple scenarios with fewer resource requirements than existing forecasting models, parallel and distributed computing in XGBoost accelerates model learning and enables faster model discovery. Xgboost uses a collection of classification and regression trees (CARTs) to fit the data from the training dataset to the examples [14]. These ensemble trees are divided into binary trees and a decision rule is created for each. Then, a score is made at each leaf node and an output is created by summing the scores in the output. Assuming that the data sample used for training is (x_i, y_i), where $x_i \in \mathbb{R}^n$, $y_i \in \mathbb{R}$. Here, x_i denotes an n-dimensional feature vector and y_i represents the output with respect to x_i. Accordingly, the XGBoost model is defined as Eq. 1.

$$y_i^{(t)} = \sum\nolimits_{k=1}^{t} f_k(x_i) = y_i^{(t-1)} + f_t(x_i) \quad (1)$$

where, $y_i^{(t)}$, after iteration t., i. is the estimated value obtained according to the sample. $y_i^{(t-1)}$, estimation result of previous t − 1 tree and $f_t(x_i)$, tth tree is the model output result. In order to determine the optimization direction in the XGBoost algorithm, the objective function is determined as follows (Eq. 2).

$$OF = \sum\nolimits_{i=1}^{n} l(\hat{y}_i, y_i) + \sum\nolimits_{i=0}^{t} \Omega(f_i) \quad (2)$$

where, $\sum_{i=1}^{n} l(\hat{y}_i, y_i)$ is the loss function determined according to a special type of problem. In this study, the logloss loss function is preferred because the estimation problem is carried out.

2.2 Support Vector Regression

Support Vector Regression is a machine learning algorithm in which the training process is performed with the observed data [15]. SVR ensures that the deviations of the training results due to experimental measurement errors are minimized. This shows better generalization and prediction performance on test data and data observed over the future. SVR projects the input values into a higher dimensional feature space for linear modeling of nonlinear training data samples. Accordingly, the function format of SVR can be expressed as Eq. 3 [16].

$$y(x) = \alpha\delta(x) + \beta \quad (3)$$

$\alpha \in \mathbb{R}$: slope vector

$\delta(x) : \mathbb{R}^n \rightarrow \mathbb{R}^m$; $(m > n)$: high-dimensional feature space
$\beta \in \mathbb{R}$: inner product operator

The training process is carried out with the following objective function of the linear model created by reducing it to high-dimensional space (Eq. 4).

$$\min_{\alpha \epsilon R^m, \beta \epsilon R} J(\alpha, \beta) = \frac{||\alpha||^2}{2} + C \sum_{i=1}^{N} L_\varepsilon(y_i, f(x_i)) \frac{||\alpha||^2}{2} \tag{4}$$

$\frac{||\alpha||^2}{2}$: Square of weight vector norm
$L_\varepsilon(y_i, f(x_i))$: Experimental Error (Loss) Function
C: Regulation constant

The C hyperparameter is used to control training data errors and margins based on values of weights or experimental errors. Choosing the C value large or small is related to whether the errors in the data set are large or small. For this reason, the rate of errors in the data set should be considered in the selection of C. Therefore, there are no rules regarding the choice of the C value. In Eq. (5), the epsilon insensitive Laplace loss function used to identify support vectors can be seen.

$$L_\varepsilon(y_i, f(x_i)) = \begin{cases} 0 & for|y_i - f(x_i)| < \varepsilon \\ |y_i - f(x_i)| - \varepsilon & for|y_i - f(x_i)| \geq \varepsilon \end{cases} \tag{5}$$

In Eq. (6), a_i and a_i^* are named as the Lagrange multipliers (V. Utkin, 2019).

$$f(x) = \sum_{i=1}^{N} (a_i - a_i^*) K(x_i, x_j) + b \tag{6}$$

$K(x_i, x_j)$: Kernel function

The SVR model is to determine nonlinear classes using kernel functions. The value of Kernel is equal to the inner products of x_i and x_j vectors in $\phi(x_i)$ and $\phi(x_j)$ feature spaces. Typical kernel functions are Linear, Sigmoid, Polynomial, and Radial Basis Function (RBF) functions. RBF can be seen in Eq. (7).

$$K(x_i, x_j) = \exp\left(-\gamma \|x_i - x_j\|^2\right), \forall \gamma \in R^+ \tag{7}$$

x_i, x_j: vectors in and feature spaces
γ: positive real number

Kernel Parameters should be carefully chosen indirectly to the structure of the higher dimensional space. In this study, the RBF kernel was chosen as the kernel function because the training data were similar in terms of Euclidean distance.

2.3 Efficiency Criteria

Statistical metrics shown in the table were used to evaluate and compare the prediction performances of the models used in this study (Table 1).

Table 1. Efficiency criteria used in the study

Indicator	Definition	Expression
RMSE	Root-mean-square error	$\sqrt{\frac{1}{T}\sum_{t=1}^{T}\left(y_{obs}-\hat{y}_{pre}\right)^2}$
MAE	Mean absolute error	$\frac{1}{T}\sum_{t=1}^{T}\left\|y_{obs}-\hat{y}_{pre}\right\|$
r	Correlation Coefficient	$\frac{p\left(\sum_{i=1}^{p}\hat{y}_{pre}y_{obs}\right)-\left(\sum_{i=1}^{p}\hat{y}_{pre}\right)\left(\sum_{i=1}^{p}y_{obs}\right)}{\sqrt{\left(p\sum_{i=1}^{p}y_{obs}^2-\left(\sum_{i=1}^{p}y_{obs}\right)^2\right)\left(\left(p\sum_{i=1}^{p}\hat{y}_{pre}^2-\left(\sum_{i=1}^{p}\hat{y}_{pre}\right)^2\right)\right)}}$

3 Case Study

In this study, total monthly rainfall data measured from the Mingachevir Dam. The Mingachevir Dam is located in the northwest of Azerbaijan with the highest altitude in the country and with the most flood disasters than the surrounding cities in this region. In Fig. 1, the annual average of energy production in Mingachevir Dam of 67-years (January 1954–December 2021) time series measured. Geographical data and statistical characteristics of Mingachevir Dam can be seen in Table 2. In order to do a time series analysis, the training data should consist of the test data with the observations that were made beforehand. [18]. For this reason, energy production data is divided into two groups as training data (1954 to 2000) and test data (2000–2021). The results are obtained by using the Python Programming Language (Table 2).

Table 2. Geographic information of the Mingachevir Dam where the data used between 1954 and 2021 is obtained

Latitude (N)	Longitude (E)	Surface Area (km^2)	Min (GWh)	Max (GWh)	Mean (GWh)
40°47′24″	47°1′42″	605	69.24	1840.52	977.89

First of all, Min Max scaling has been made on the data available in the design of the system. At the end of this process, the data are normalized to the 0–1 interval according to Eq. 8 in order to be evaluated.

$$x_{scaled}=\frac{x-x_{min}}{x_{max}-x_{min}} \tag{8}$$

3.1 Input Variable Determination

As the input data, the maximum and minimum water level obtained from the dam was determined as the amount of water entering and leaving the reservoir. The energy production measurements obtained as KWh as output data were converted into GWh units and evaluated (Table 3).

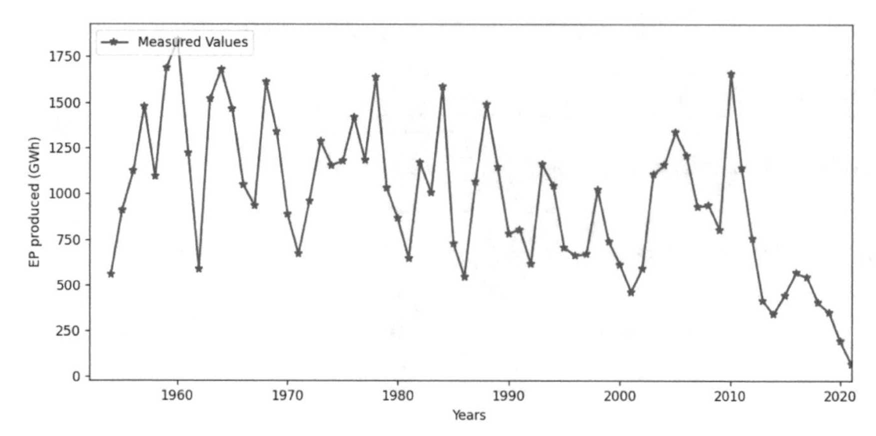

Fig. 1. Annual energy production amounts measured by Mingechewir Dam between 1954–2021

Table 3. Input features

Years	Max Water Level	Min Water Level	Water Entering the Reservoir	Water From the Reservoir	Energy Production		
					KWh	MWh	GWh
1954	69,21	55,80	12648,00	6924,00	557700000,00	557700,00	557,70
1955	70,45	62,36	14775,00	8965,00	906914840,00	906914,84	906,91
1956	82,30	69,82	14139,00	7236,00	1122000000,00	1122000,00	1122,00
1957	80,09	76,15	9965,30	12102,30	1479135520,00	1479135,52	1479,14
...							
2018	74,80	69,82	9621,00	8646,00	404549000,00	404549,00	404,55
2019	73,35	68,21	8068,00	8259,00	351960893,00	351960,89	351,96
2020	71,37	67,83	6707,00	5987,00	195039913,00	195039,91	195,04
2021	73,86	68,80	3336,00	2289,00	69243212,00	69243,21	69,24

3.2 Parameters of Models

The fact that the energy production data at hydroelectric power plants differ by years and the energy production trends of the Mingachevir Dam make it difficult to predict. Therefore, it is a difficult problem to adjust the parameter value of ML algorithms used in estimation. The determined parameter values were determined using the trial-and-error method.

Table 4 shows the estimated parameter values for the SVR and XGBoost algorithms. In the determination of these parameter values, the hyperparameter values that make the best correlation coefficient were sought. The RBF model was determined as the Core model for the SVR algorithm. The training process was designed according to these parameters. In the XGBoost model, the Max_Depth, Leaf and n_estimators parameters were determined as an experiment, and the training model was created.

Table 4. Intervals of Hyperparameter values according to models

Models	Parameters	Parameter Interval
SVR	C	[10000]
	ε	[0.01]
	γ	[0.001]
	Kernel	{RBF}
xGBoost	max_depth	3
	min_samples_leaf	3
	Alpha	1.7
	Learning_rate	0.1
	n_estimators	100

4 Results and Discussions

This section describes the results for SVR and xGBoost performance for EP in hydroelectric power plants estimated during the training and testing phases. The training phases of the SVR and xGBoost models to predict the amount of EP for the next two years were performed using EP time series input data from 62 years. The intervals determined for the hyperparameters of the SVR and xGBoost models according to the search models are given in Table 4.

According to Table 5, the best estimation results were obtained for the 21-year EP prediction of Mingechewir Dam according to the SVR model for EP measurement. During the training phase in Mingechewir Dam, the RMSE value of the SVR model was calculated as 0.161, while it was calculated as 0.036 in the xGBoost model. During the testing phase, the RMSE value of the SVR model was calculated as 1.116, and as 1.29 in the xgBoost method. In this case, it was concluded that the estimated values were obtained closer to the test data in the SVR algorithm. Considering the correlation values of the estimated values, it can be said that there is a 1% difference for both algorithms.

Figure 2 and Fig. 3 are presented to evaluate the performance of machine learning methods used in an integrated manner. The graphs compare the test phase results of the SVR and xgBoost models. According to the trend of 21-year EP estimates, the performances of all methods are satisfactory. At the same time, in this estimation model, as in other estimation performances, estimations were not at a sufficient level in unexpectedly decreasing measurement values. However, it is clear that the forecast models capture the annual EP trends.

Table 5. Performance values according to initial values of SVR and xGBoost

SVR						xGBoost					
Training			Testing			Training			Testing		
RMSE	R	MAE	RMSE	R	MAE	RMSE	R	MAE	RMSE	R	MAE
0.161	0.917	0.121	1.116	0.896	0.470	0.036	0.997	0.028	1.29	0.88	0.514

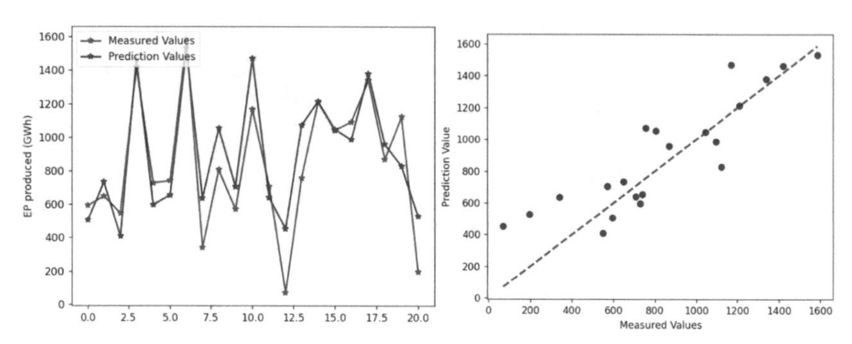

Fig. 2. Comparison of regression and forecasting values of the XGBoost method in the test phase for Mingechewir Dam

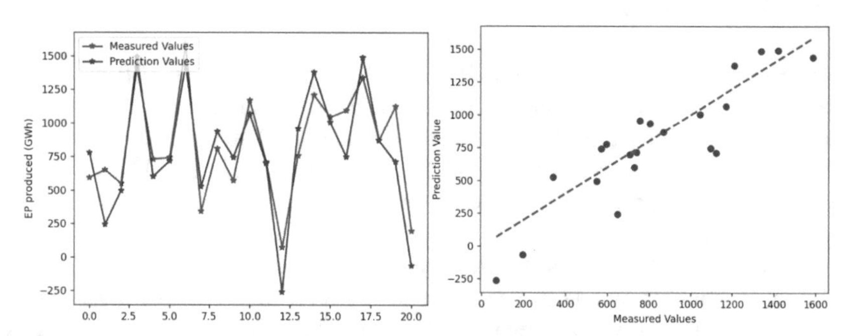

Fig. 3. Comparison of regression and forecasting values of the SVR method in the test phase for Mingechewir Dam

5 Conclusion

In this study, xgBoost and SVR models are used to forecast the EP amounts of the Mingachevir Dam that are located in the Eastern north of Azerbaijan. Annual EP data from 1954 to 2021 are given as inputs to the models. Forward-looking annual EP forecasting was made with the models. Initial parameters are static determined while creating the SVR-ABC model. The amount of energy produced was estimated by using the Maximum and Minimum Water level and the amount of water entering and leaving the Dam reservoir as input data to the models. When these values are taken into consideration, it is determined that the SVR model can provide better predictions than the xGBoost model.

In future studies, predictions will be made with different machine learning algorithms and deep learning algorithms.

References

1. Castillo-Botón, C., et al.: Analysis and prediction of dammed water level in a hydropower reservoir using machine learning and persistence-based techniques. Water **12** (2020). https://doi.org/10.3390/W12061528
2. Unal, Y.S., Deniz, A., Toros, H., Incecik, S.: Temporal and spatial patterns of precipitation variability for annual, wet, and dry seasons in Turkey. Int. J. Climatol. **32**, 392–405 (2012). https://doi.org/10.1002/joc.2274
3. Aksoy, B., Selbaş, R.: Estimation of wind turbine energy production value by using machine learning algorithms and development of implementation program. Energy Sources Part A Recovery Utilization Environ. Eff. **43**, 692–704 (2021). https://doi.org/10.1080/15567036.2019.1631410
4. Cao, E., et al.: A hybrid feature selection-multidimensional LSTM framework for deformation prediction of super high arch dams. KSCE J. Civ. Eng. **26**, 4603–4616 (2022). https://doi.org/10.1007/s12205-022-1553-8
5. Madrid, E.A., Antonio, N.: Short-term electricity load forecasting with machine learning. Inf. **12**, 1–21 (2021). https://doi.org/10.3390/info12020050
6. Chang, F.J., Chang, Y.T.: Adaptive neuro-fuzzy inference system for prediction of water level in reservoir. Adv. Water Resour. **29**, 1 (2006). https://doi.org/10.1016/j.advwatres.2005.04.015
7. Wang, A.P., Liao, H.Y., Chang, T.H.: Adaptive neuro-fuzzy inference system on downstream water level forecasting. In: Proceedings - 5th Fifth International Conference on Fuzzy Systems and Knowledge Discovery. FSKD 2008. vol. 3, pp. 503–507 (2008). https://doi.org/10.1109/FSKD.2008.671
8. Yang, S., Yang, D., Chen, J., Zhao, B.: Real-time reservoir operation using recurrent neural networks and inflow forecast from a distributed hydrological model. J. Hydrol. **579**, 124229 (2019). https://doi.org/10.1016/j.jhydrol.2019.124229
9. Deng, L., Yang, P., Liu, W.: An improved genetic algorithm. In: 2019 IEEE 5th International Conference on Computer Communications. ICCC 2019, pp. 47–51 (2019). https://doi.org/10.1109/ICCC47050.2019.9064374
10. Samadianfard, S., Jarhan, S., Salwana, E., Mosavi, A., Shamshirband, S., Akib, S.: Support vector regression integrated with fruit fly optimization algorithm for river flow forecasting in Lake Urmia Basin. Water **11** (2019). https://doi.org/10.3390/w11091934
11. Castelletti, A., Pianosi, F., Quach, X., Soncini-Sessa, R.: Assessing water reservoirs management and development in northern Vietnam. Hydrol. Earth Syst. Sci. **16**, 189–199 (2012). https://doi.org/10.5194/hess-16-189-2012
12. Wang, L., et al.: Efficient reliability analysis of earth dam slope stability using extreme gradient boosting method. Acta Geotech. **15**, 3135–3150 (2020). https://doi.org/10.1007/s11440-020-00962-4
13. Chornovol, O., Kondratenko, G., Sidenko, I., Kondratenko, Y.: Intelligent forecasting system for NPP's energy production. In: Proceedings of the 2020 IEEE 3rd International Conference on Data Stream Mining & Processing. DSMP 2020, pp. 102–107 (2020). https://doi.org/10.1109/DSMP47368.2020.9204275
14. Yu, X., Wang, Y., Wu, L., Chen, G., Wang, L., Qin, H.: Comparison of support vector regression and extreme gradient boosting for decomposition-based data-driven 10-day streamflow forecasting. J. Hydrol. **582**, 124293 (2020). https://doi.org/10.1016/j.jhydrol.2019.124293

15. Boser, B.E., Laboratories, T.B., Guyon, I.M., Vapnik, V.N.: A training algorithm for optimal margin Classifiers. Symp. A Q. J. Mod. Foreign Lit. (1992)
16. Samsudin, R., Shabri, A., Saad, P.: A comparison of time series forecasting using support vector machine and artificial neural network model. J. Appl. Sci. **10**, 950–958 (2010)
17. Utkin, L.V.: An imprecise extension of SVM-based machine learning models. Neurocomputing **331**, 18–32 (2019). https://doi.org/10.1016/j.neucom.2018.11.053
18. Hyndman, R.J., Athanasopoulos, G.: Forecasting: principles and practice. OTexts.com (2014)

Technology and Software for Traffic Flow Management

Kahramanov Sebuhi Abdul(✉)

Mingachevir State University, Mingachevir, AZ 4500, Azerbaijan
q.sebuhi_2018@mail.ru

Abstract. The research touched upon topical issues related to the development and application of the "Safe bus" and "Smart bus stop" approaches as one of the areas of implementation of the "smart city" concept. The article explores the essence of the concept of "smart city" using the opportunities of ICT to increase the efficiency of public transport, digitalization and development of this area. The main technical and technological features of the "Safe Bus" and "Smart Bus" systems are described. The article also highlights the benefits of using a monitoring system (NVMS) to safely manage and improve the transport system in tourist areas. At the end of the study, a modern model was formed using various technologies to provide public transport. In this model, the forms of traffic flow control using modern technologies are shown, and the final result obtained from various sources and planning is noted.

Keywords: Smart city · Safe bus · Clever will stop · transportation · digital · monitoring · technology · flow

1 Introduction

Traffic planning and modeling software not only provides a good idea of the level of freight and passenger traffic, but also provides information about the high level of transport services. Increasing the utilization rate of transport services and the orientation of society to social indicators is associated with increasing the efficiency of transport networks [1].

The development of transport management technologies is associated with the formation and improvement of population mobility forecasting and the selection of an appropriate route for travel. The intensity of the transport network depends not only on the parameters of the road network and temporary weather conditions, but also on many factors that correspond to the social and economic situation of the people at the same time [6].

Transport models formed using modern information technologies and multifunctional software are based on the required software. It is based on the spatial characteristics of the city and region, along with all known information on transport demand and supply, and calculates the probability of the greatest spread of both transport and passenger flows over the network [5]. Accurate and correct calculations then form the

D. J. Hemanth et al. (Eds.): ICAIAME 2022, ECPSCI 7, pp. 713–722, 2023.
https://doi.org/10.1007/978-3-031-31956-3_61

basis of forecasts that will ensure the development of the city in accordance with modern requirements and form an analytical basis for making decisions on the development of the city's transport network.

2 Method

One of the areas driving the development of the transport industry is the vehicle monitoring system (VMS). The monitoring system applied to the transport is designed for displacement management (fleet management and compliance with transport routes, route optimization), regular monitoring of vehicles (location of the vehicles, as well as its speed and design faults) and remote monitoring of vehicles allows you to apply a control system.

The constructive structure and working principle of the monitoring system is shown in Fig. 1.

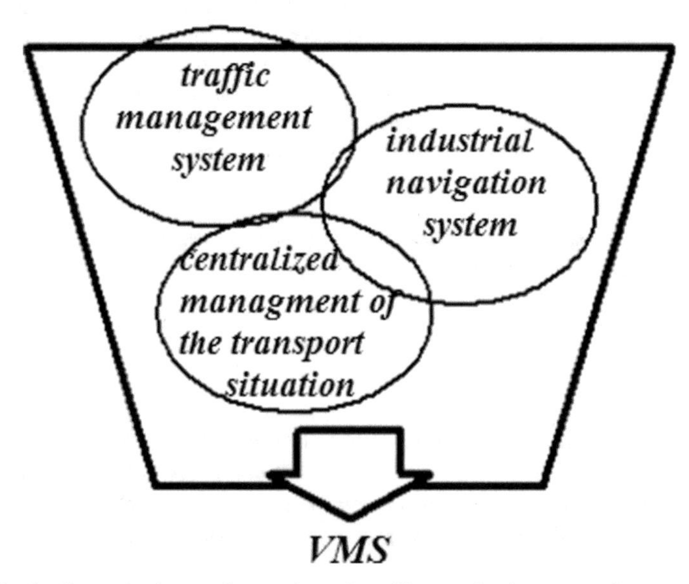

Fig. 1. General scheme of operation of satellite monitoring system in transport

Such monitoring is carried out to locate an object equipped with navigation and communication facilities [8]. Today, in order to ensure safety and deliver various types of information to passengers in the vehicle, there is an on-board set, video surveillance device, passenger information, etc. A "Safe Bus" (Fig. 2) is used, as well as a "Smart Bus Stop" (Fig. 3), which significantly improves passenger awareness and the quality of public transport services.

1 - rear view panel;
2 - smoke and temperature sensors in the cabin;

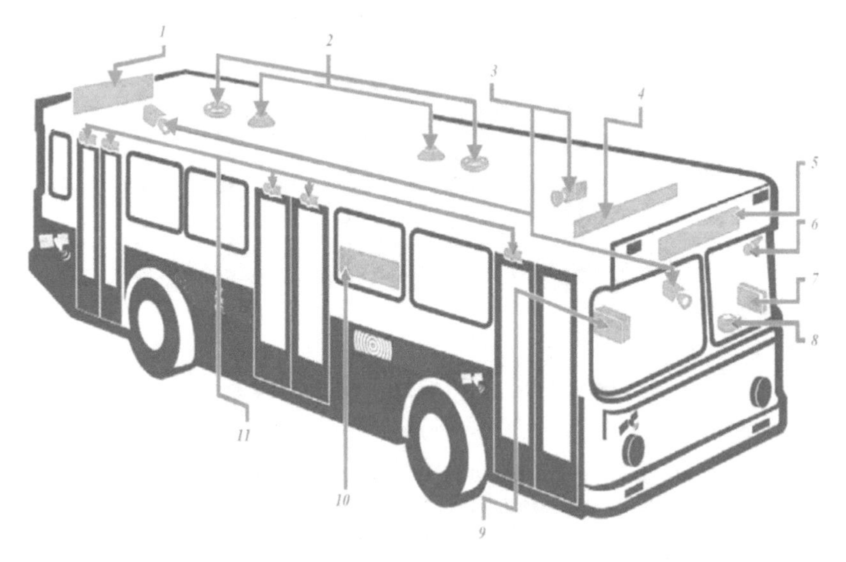

Fig. 2. Composition of the "Safe Bus" system:

3 - surveillance cameras;
4 - "linear running" panel;
5 - front view board;
6 - video recording system microphone;
7 - board navigation system and communication device;
8 - secret "alarm button";
9 - passenger flow control sensors;
10 - side view board;
11 - Passenger flow sensors.

In addition to the traditional information about the expected arrival of trolleybuses, trams and buses, there is also a 24-h video surveillance cameras, voice communication with the control center, mobile police and emergency call system in case of traffic accidents or other emergencies. The system automatically transmits the actual arrival time to the LED information board in order to inform passengers in time and quickly about the arrival time of trolleybuses, trams and buses, taking into account the actual movement during peak hours (traffic jams).

1 - information board with routes;
2 - information board with parking names;
3 - control unit indicating the time of arrival of the vehicle and having an information board;
4 - sensor screen.

In addition to managing transport systems and processes, the use of navigation systems covers the following key and important objectives in terms of national interests:

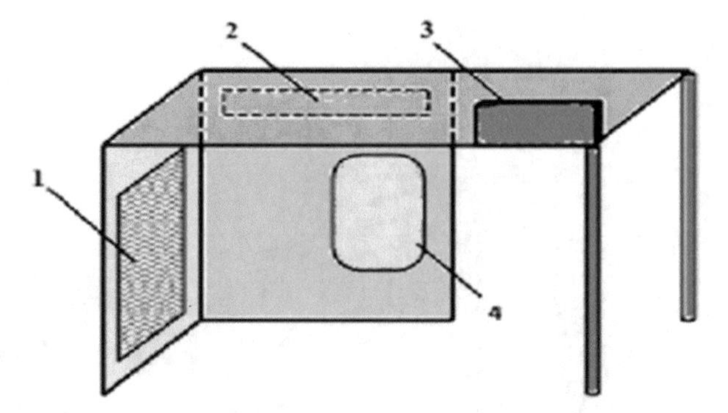

Fig. 3. Information board at the parking lot:

- Automatic identification of areas where traffic accidents and emergencies occur, and information security of transportation (primarily dangerous goods) in the Ministry of Internal Affairs and the Ministry of Emergency Situations, as well as operative interaction with ambulances;
- Establishment of a system that automatically detects the location of the vehicle, is able to solve traffic flow tasks in real time, automatically receives an "SOS" signal from the driver and immediately contacts the special operations services of the Ministry of Internal Affairs and Emergency Situations;
- Ensuring on-line driving and relocation of vehicles during emergency response.

3 Finds and Discussion

The application of new technologies in the proper organization of traffic flows ensures on-line management of traffic flows and the selection of the best option for the current situation [7].

PTV VISION, Extend Sim, GPSS World, AnyLogic, Business Studio and other software products are used in this field. However, the most well-known and widely used model is the PTV VISION software, which is widely used in Germany.

The mechanism used in traffic management includes a number of measures aimed at coordinating traffic management activities to achieve high results [2]. Such control systems must be fast, large-scale, accurate and reliable.

The complexity of the system is the organization of work in a single information environment, which has a distributed form that continues in each section and has the ability to interfere with processes.

The speed of the management system is assessed by the ability to adjust the accounting policy and management of the transport system in accordance with the changing requirements of enterprises and legislation in the transport sector, as well as to change the direction of the main work.

The scale of the system is based on finding new research methods, finding resources to expand production and other opportunities, and a perfect approach to traffic management within a region or megalopolis.

Accuracy and reliability is the ability to conduct research and take appropriate action to determine the right direction in traffic flow management, as well as to obtain and apply information from reliable sources to assess the real situation.

Traffic tables and graphs provide a comparative analysis of the main features of the simulation program system for traffic flow management [8]. In the process of modeling simulation programs, it became clear that simulation software packages with a high enough performance have a high cost [3].

PTV VISION Visum is a software package designed for computers that allows you to display all types of personal and public transport in a single model at the macro level. The main part of the whole system is PTV VISION Visum software, which is a top level of transport network management (Fig. 4).

Fig. 4. PTV VISION Visum module

When using PTV VISION software, the parameters of the road network, traffic intensity and speed, transport hubs: population density, level of development of industry and trade, recreation areas, etc. are formed to form traffic flows. You need to create a database that includes.

Visum module is a transport information system that combines information features as a base:

- formation of existing and at the same time predictable traffic flow management;
- construction of network models suitable for road and public transport of any size;
- proper assessment of the profitability of routes, including the management of public transport, which includes the effective establishment of coordination in the schedule;
- analysis and accurate assessment of traffic flows for different modes of transport;
- providing traffic forecasts based on the principles of "if something happens…"
- formation of an appropriate platform for communication between transport and information systems.

The distribution of individual traffic flows is based on determining the most cost-effective route, balance process, training methods, and road use fees.

The flow of products is based on the distribution of flows, the transport system, the route network and well-designed schedules. The city transport department and transport companies first plan the activities for the carriers and then organize the control work.

Especially built-in PTV VISION software modules help reduces public transport planning costs as much as possible. Expenses and income can be visually demonstrated and analyzed. Efficiency parameters are calculated in detail according to network data and transport schedules.

The main advantage of PTV VISION software is the ability to predict the identified activities to ensure traffic. These include the construction of new and modern streets, their reconstruction, intersections at different levels, possible changes that may occur during the driving of vehicles, the construction of individual settlements in the city, pre-planning of the consequences of emergencies, etc. allows you to accelerate and model the development of the transport network.

The construction of a macroscopic model of transport networks can be divided into three main stages:

1. Development stage of transport infrastructure. This stage includes the intersection of nodes and segments and the development of a transport infrastructure model that is used as an object to show these routes. The nodes are connected by segments, thus creating a network. In order to determine the length of a road in particular, a network is usually built on the basis of a spatial plan or map showing the scale. The following characteristics are defined for each object of a segment type: the maximum allowable speed, the ability to allow movement and the type of transport system applied from this segment. At the junctions, possible directions for the next movement are determined. One-way traffic is mainly modeled by stopping traffic in one of the existing directions in the transport network.
2. Zoning (zoning) implies the division of the studied object, ie the zones where the traffic flows, on the territory. Regions can have completely different geometric shapes, in which case the flow will start from its geometric center. Each zone can also be displayed as a dot. With the help of connections, the center of the selected area is connected to the transport infrastructure, and thus the entry and exit points of the area are organized. There can be a large number of such points in each region, and it is necessary to determine the total weight of each point, which in turn will determine the percentage of traffic flow entering that point.
3. Formulation and development of on-demand model. It should be noted that the model, formed and used on the basis of demand, has a specific purpose and determines the number of movements between the regions by choosing a special vehicle. Based on the formation of a macro model of the network, the demand model is presented in the form of a single matrix that determines the volume of traffic between regions.

When using one of the main methods to identify transport proposals and analyze proposals at the same time - PTV VISION Visum-based software package, the parameters of assessing the load of network objects or the quality of connections between transport areas, use the method of transport distribution based on optimization of transport is.

The redistribution rule is based on the identification of algorithms that determine the appropriate routes between the source and the target. The search for any rule is carried out after the selection of a rule that forms a related requirement for routing. Route combinations contain information necessary to calculate parameters such as time, distance traveled, and number of traffic. Visum defines different distribution rules for all types of transport. The applied search rules and requirements differ from the distribution rules. The basis of redistribution is the loading costs of the network objects used (nodes, segments, connections, bends, routes).

The correct calculation of the intensity in the field, in particular the calculation of the volume of transit traffic in an automated mode, can be done using the 6 rules of the Visum module. In this case, we are talking about the static redistribution of time in the top five without any specific modeling. The sixth rule works against the background of a corresponding dynamic model, ie taking into account the values of changes in parameters over time.

The sequential redistribution rule divides the messaging matrix into several percentage-matrix matrices. The messages of this part of the matrices are redistributed step by step in the network. In this case, the resistance obtained from the previous step must be taken into account in order to find the right way.

The redistribution of balance is carried out in accordance with Wardrop's first principle: "Each person in the flow of traffic chooses his own route so that in the end the travel time on all alternative routes is equal, but each transition to another route increases individual time on the road." As a primary solution, a state of equilibrium is created in the case of a multi-stage iteration, consistent with the sequential redistribution of flows. In the internal repetition phase, the messaging routes are initially balanced by the moving vehicles. In the next iteration, it is checked whether there are new routes that are less resistant to the current state of the network.

The system typically reflects the "identification process" of participants as they move through the network. Based on the principle of "either all or none of them", drivers must take into account the information they received during the last trip to search for a new road.

The Tribute method, developed as a rule by the French research society INRETS, is used to model networks, especially on roads where taxes are applied. Unlike classical rules based on a temporary time value, the Tribute benefits from time values found without random distribution. At the same time, when looking for ways, several alternative methods are used, taking into account two balance criteria - time and cost. Road use taxes are formed for a segment of the transport system as a certain price or for a sequence of segments between nodes (non-linear payment system).

Random redistribution is based in part on incomplete data, given that the parameters (duration, length and cost of travel) are subjectively perceived by transport participants as important for road selection. In addition, the choice of roads that are not shown in the model at all depends on the personal qualities of the transport participants. Both approaches go hand in hand, and in practice, if these paths are chosen and the selection is strictly followed, the Wardrop principle will not be overloaded, as the parameters are relatively close to objective values. Therefore, in a random distribution, many alternative routes are first calculated, and then the demand is divided according to the separation model (for example, the traffic distribution model described in Visum is the Logit model).

Dynamically, the redistribution of traffic flows is carried out with the precise construction of the time axis, in contrast to the above rules. The redistribution time is divided into separate time intervals, respectively, and the load and resistance are determined separately for each such time interval found. At each departure interval, the demand for current information is distributed according to the separation model, as in the case of random redistribution. Such a structure leads to a temporary re-saturation of the network (increase in intensity). The choice of routes during the day, as well as the time of departure, if necessary, varies from time to time [4].

In modern transport science, a matrix (language) is a messaging matrix that explains the state of the transport system and the traffic flows within it from the point of view of distribution. The main elements of the messaging matrices are the number of trips between pairs of transport networks, based on the functional division of the simulated area, provided for in the city master plan or other normative planning documents.

In the modern era, as a rule, there are a number of different problems that can or cannot be solved due to the extremely high costs that prevent the collection of all statistical messages necessary for modeling based on standard methods [9]. From this point of view, the issue is to develop the flexibility of mathematical models in transportation processes, as well as the work to adapt the initial data to the compositional features at a higher level. According to the above, in order to find the message matrix, a mathematical model is first formed.

The results of the calibrated transport model can be presented in two ways: they are matrices with different characteristics for different zone pairs. Such features include the average transport speed between zones, the average duration of movement, and so on may be indicators such as. Message matrices for private and public transport are compiled using the obtained cost matrices and urban population statistics. For this purpose, the MULI software module is used, or if the number of blocks used is more than 30, then Excel software is used for calculation.

- The main statistics used to form the messaging matrix are as follows:
- The number of people in each of the selected blocks;
- Number of able-bodied population;
- The number of jobs suitable for able-bodied people;
- The exact number of employees in the service sector.

As a result, the appropriate messaging matrices for private and public transport are being replaced by PTV VISION, ensuring the movement of the population and the distribution of individual transport.

4 Conclusion

Convenience is even more important for a city dweller, for example, to continue the route by easily passing from one vehicle to another, and at the same time to know the exact time of arrival of the vehicle. In this case, all urban transport must be integrated into a single system. This is impossible without the creation of intellectual platforms. The concept of a planned transport system is recognized as the most effective approach in shaping urban transport. The development of an intelligent transport network, and in particular "smart parking", seems to be a very promising direction. Increased use of public transport will reduce the number of private cars, accidents and environmental pollution on the roads. As a result of the study, a model was proposed for the management of public transport, the effective modernization of the fare system, the optimization of passenger flows, ensuring traffic flows and safe movement in the city. The task of improving transport safety in large cities is always relevant, and the introduction of a monitoring system can be an effective tool to increase the safety of the transport complex.

References

1. Amato, G., Carrara, F., Falchi, F., Gennaro, C., Vairo, C.: Car parking occupancy detection using smart camera networks and deep learning. In: Proceedings - IEEE Symposium on Computers and Communications, pp. 1212–1217, August 2016
2. Chen, Z., Xia, J.C., Irawan, B.: Development of fuzzy logic forecast models for location-based parking finding services. Math. Probl. Eng. **2013**, 473471 (2013)
3. Sumalee, A., Ho, H.W.: Smarter and more connected: future intelligent transportation system IATSS Res. **42**(2), 67–71 (2018)
4. Ismagilova, E., Hughes, L., Rana, N.P., Dwivedi, Y.K.: Security, privacy and risks within smart cities: literature review and development of a smart city interaction framework. Inf. Syst. Front. **22**(4), 393–414 (2022). https://doi.org/10.1007/s10796-020-10044-1
5. Vlasov, V.M., Efimenko, D.B., Bogumil, V.N.: Information Technologies in Automobile Transport, 256 p.Publishing Center "Academy", Moscow (2014)
6. Pervukhin, D.A., Afanaseva, O.V., Ilyushin, Y.V.: Information Networks and Telecommunications, 267 p. SPB: Publishing house "Satish" (2015)
7. Samuilov, K.E.: Networks and Telecommunications, 359 p. Publishing house "Yurait", Moscow (2015)
8. Gorev, A.E.: Information Technologies in Transport. Electronic Identification of Vehicles and Transport Equipment: Study Guide, 96 p. Publishing House SPBGACU, St. Petersburg (2010)
9. Shangin, V.F.: Information Security and Protection of Information, 702 p. DMK Press, Moscow (2014)
10. Waksman, S.A. (ed.): Information Technologies in the Management of Urban Public Passenger Transport (Tasks, Experience, Problems), 250 p. AMB Publishing House, Yekaterinburg (2012)
11. Chrpa, L., Magazzeni, D., McCabe, K., McCluskey, T.L., Vallati, M.: Automated planning for urban traffic control: strategic vehicle routing to respect air quality limitations. Intell. Artif. **10**, 113–128 (2016)
12. Djahel, S., Doolan, R., Muntean, G.M., Murphy, J.: A communications-oriented perspective on traffic management systems for smart cities: challenges and innovative approaches. IEEE Commun. Surv. Tutor. **17**, 125–151 (2015)
13. Zhang, X., Onieva, E., Perallos, A., Osaba, E., Lee, V.: Hierarchical fuzzy rule-based system optimized with genetic algorithms for short term traffic congestion prediction. Transp. Res. C Emerg. Technol. **43**, 127–142 (2014)
14. Habtie, A.B., Abraham, A., Midekso, D.: Artificial neural network based real-time urban road traffic state estimation framework. In: Abraham, A., Falcon, R., Koeppen, M. (eds.) Computational Intelligence in Wireless Sensor Networks. SCI, vol. 676, pp. 73–97. Springer, Cham (2017). https://doi.org/10.1007/978-3-319-47715-2_4

15. Papageorgiou, M. (ed.): Springer, Berlin (1983). https://doi.org/10.1007/BFb0044049
16. Ardekani, S.A., Herman, R.: Urban network-wide variables and their relations. Transp. Sci. **21**(1) (1987)
17. Mauro, V.: Road network control. In: Papageorgious, M. (ed.) Concise Encyclopedia of Traffic and Transportation Systems. Advanced in Systems, Control in Information Engineering, pp. 361–366. Pergamon Press (1991)
18. Shvetsov, V.I.: Mathematical modeling of traffic flows. Avtomat i Telemekh **11**, 3–46 (2003)
19. Diveev, A.I., Sofronova, E.A.: Synthesis of intelligent control of traffic flows in urban roads based on the logical network operator method. In: Proceedings of European Control Conference 2013, pp. 3512–3517 (2013)
20. Diveev, A.I., Sofronova, E.A., Mikhalev, V.A.: Model predictive control for urban traffic flows. In: Proceedings of 2016 IEEE International Conference on Systems, Man and Cybernetics 2016, pp. 3051–3056 (2016)

Comparision of Deep Learning Methods for Detecting COVID-19 in X-Ray Images

Hakan Yüksel(✉)

Department of Computer Technologies, Technical Sciences Vocational School, Isparta University of Applied Science, Isparta, Turkey
hakanyuksel@isparta.edu.tr

Abstract. COVID-19 is a contagious virus that can pass from person to person and causes respiratory tract infection. This virus spread rapidly in a short time and turned into an epidemic and became a global pandemic. Early diagnosis, detection and initiation of the treatment process are of great importance against viruses with a high probability of transmission. For the discovery of COVID-19, it's possible to descry it from oral slaver or lung images. RT-PCR (Reverse Transcription-Polymerase Chain Reaction) is frequently used to confirm the opinion of Covid-19. However, the process of concluding this test takes approximately 4 to 6 h. In addition, it facilitates the rapid spread of the disease due to the fact that it does not give accurate results at every stage of the disease and the long time interval in the process maintenance. The use of radiological methods (X-Ray and Computed Tomography) to make an early diagnosis both reduces contact-related procedures and facilitates rapid results. Last years, many deep learning approaches are presented on the realization of disease detection on radiological images. In this study, a deep literacy grounded approach is used for accurate and rapid-fire diagnosis of COVID-19 from radiological images. Among the deep literacy approaches, Convolutional Neural Networks and Deep Neural Networks and deep literacy ways that can classify lung X-Ray images as COVID-19, normal and viral pneumonia cases were compared. Experiments done with CNN and DNN techniques were performed on the same dataset, and the performance evaluation of each technique was carried out.

Keywords: Deep Learning · Convolutional Neural Networks · Deep Neural Networks · X-Ray · COVID-19

1 Introduction

Covid-19 emerged in the last months of 2019 in China and was seen in the whole world in a short time and is defined as a pandemic by the World Health Organization because of the global epidemic [1, 2]. As of the second quarter of 2022, there are more than five hundred million cases and over 6 million virus-related deaths, according to the WHO. The virus, which already has a high contagious rate, increases the rate of contagion with the mutations it has passed, making it difficult to control the epidemic.

D. J. Hemanth et al. (Eds.): ICAIAME 2022, ECPSCI 7, pp. 723–739, 2023.
https://doi.org/10.1007/978-3-031-31956-3_62

Examples of coronavirus are severe and middle acute respiratory syndrome. This new disease is accepted as an epidemic and is named as acute respiratory syndrome-coronavirus-2 [3, 4].

Covid-19 does not exactly show common symptoms for everyone after its virus is transmitted to humans. Although it is possible to develop the disease without showing any symptoms, the most seen symptoms are high fever, breath shortness and dry cough. In cases where the complaint develops more oppressively, it has been observed that it results in order failure, severe respiratory failure, pneumonia, and indeed death. It's known that Covid-19 is spread through driblets and contact. It's transmitted by gobbling these patches or by physical contact with people carrying the contagion, similar as hand or mucous contact or breathing [5].

Malaria medicine, pain relievers, antibiotics, cough medicine, antipyretic are generally applied for treatment of the complaint after infection. It is decided to attend patients looking at their illness, complaints and severity levels [6, 7]. Therefore, in order to slow down or even stop the rapidly increasing number of infected cases and to save our lives from its negative effects in all areas, it is necessary to detect and isolate those who carry the virus quickly and accurately.

One of the most important steps in curbing the spread of the COVID-19 epidemic is the effective screening of people suspected of being infected to hinder further spread of the disease in the population. RT-PCR is the most commonly used way for diagnosis, since mRNA can be detected sensitively from a small number of tissue cells today [8]. In other words, RT-PCR detection is obtained from sputum and other lower respiratory tract secretions, nasopharyngeal swabs and stool and blood examples. Although RT-PCR testing is a standard to screen for COVID-19, the testing of RT-PCR is a very complex, time-consuming and labor-intensive process. Not all healthcare organizations have such testing requirements. Since the beginning of the epidemic, the number of RT-PCR detection reagents has been very limited and costly. The high rate of false negative results of the testing of the RT-PCR in the early stages of the disease shows that the method is not suitable for early COVID-19 diagnosis and cure [9].

Despite development of passive and mRNA-based vaccines in the course of the epidemic, the emergence of different variants of the virus (omicron, delta, etc.), possible side effects, effectiveness against the virus, how many doses to use, cross-country sharing, logistics and anti-vaccine debates are discussed in public. Therefore, although countries follow different methods in the corona virus epidemic, the valid and commonly accepted approach is disease detection and isolation. Therefore, accurate diagnosis of the disease is critical.

Any technological device which enables high-accuracy and quick and diagnosis of COVID-19 infection is beneficial for healthcare professionals. RT-PCR test, DPI test, saliva test, antibody test, lung tissue biopsy and medical imaging methods are used in the COVID-19 treatment. Although the viral nucleic acid test is a basic method that gives consistent results in the diagnosis of Covid-19, factors such as the availability and quality of laboratories prevent this test from being widely performed [10]. Since the effect of the disease on the lungs is quite high, it is thought that the diagnosis made with the use of CT and X-ray will be both consistent, fast and the disadvantages of viral nucleic acid tests will be eliminated.

Due to the high perceptivity of the lung in CT reviews performed on cases diagnosed with Covid-19, irregular murk or ground glass nebulosity are frequently observed in the lung, so it has been used as an important reciprocal index in Covid-19 scanning.

The radiological images of the chest area contain important details about lung diseases. The details in these images are very effective in detecting the COVID-19 disease. COVID-19 shows similar findings with other lung diseases on chest images. For this reason, examinations on radiological images should be done with precision. Therefore, in last years, the methods of Deep Learning are used in scanning on radiological images. Approaches based on these methods enable fast, reliable and automated analysis [11].

2 Related Works

In the literature, it is seen that there are many studies using deep learning-based approaches in order to detect COVID-19 through X-Ray images. Approaches differ in terms of the models they use.

Li et al. (2020), designed a three-dimensional DL model from chest CT images to detect COVID-19 disease. They stated that this model can accurately detect COVID-19 and distinguish it from other lung diseases and pneumonia. They achieved 90% sensitivity using the COVNet model [12].

El Asnaou and Chawki (2021), studied image-based COVID-19 case recognition using deep learning techniques to classify the pneumonia with X-Ray and CT images. They applied several deep learning architectures such as DenseNet201, MobileNet V2, VGG19, VGG16, Inception ResNet V2, Resnet50 and Inception V3 to X-Ray and tomography scan images. According to the results, the most accurate result was obtained from the ResNet50 model with 96.61%. Inception ResNet V2 outperformed other architectures with 92.18% accuracy [13].

Apostolopoulos et al. (2020), detected the COVID-19 with CNN, using X-Ray images (3905) as two-class and seven-class. They stated that the deep learning model had 99.18% accuracy rate and the sensitivity of it was found 97.36%. 99.42% specificity rate was found in the diagnosis of the disease in the two-class case. In the seven-class case, they achieved 87.66% accuracy [14].

Zheng et al. (2020), proposed CNN method using 3D Computed Tomography (DeCoVNet) for detecting COVID-19. They reached 90% accuracy and 91% sensitivity in the study using deep learning algorithm [15].

Wang et al. (2020), COVID-Net which called to proposed a deep learning network structure for the diagnosis of covid-19 disease. They classified viral, bacterial and normal pneumonia and COVID-19 tags with X-Ray images. In the study, residual network structures were preferred for high performance and easy trainability. In comparison with VGG-19 and ResNet-50, diagnostic success with an overall accuracy rate of 93.3% was achieved [16].

Horry et al. (2020), made a covid-19 classification with CNN models. In this experimental study, they used x-ray, ultrasound and CT images. Transfer learning-based VGG16, VGG19, ResNet50, InceptionV3, InceptionResNet, NasNetLarge, DenseNet networks were applied on the data. N-CLAHE method was preferred to reduce noise.

The results were obtained in the VGG19 model for X-Ray, ultrasound and CT images, and classification was made with 86%, 100% and 84% accuracy rates, respectively [17].

Ozturk et al. (2020) proposed the DarkCovidNet model which is based on the DarkNet-19 architecture and made binary and multiple classification. As a result of experimental studies, 87.02% and 98.08% accuracy rates were obtained in binary and multiple classification, respectively [18]. Therefore, the use of X-Ray and CT like medical imaging methods has gained importance once again for the diagnosis of COVID-19 in the beginning of the disease.

Rahimzadeh et al. (2020), used a new dataset of CT (nearly 48260) to detect COVID-19. They trained the images in the dataset using the ImageNet dataset. An accuracy of 98.49% was achieved by classifying CT images with ResNet50V2, a 50-layer network [19].

Mangal et al. (2020), suggested a model in order to diagnose the COVID-19 with the use of deep convolutional neural networks. They set up a prefabricated CheXNet model with a 121-layer DenseNet followed by an associated layer. In the four-class classification, they achieved 87.2% accuracy in their results [20].

Alom et al. (2020), proposed a method detect COVID-19 patients with deep learning methods. They conducted a study using CT images and X-ray. In this study, a new method has been proposed to determine the infected regions on CT images and X-ray. Together with the results of the experiments obtained, they achieved 98.78% accuracy from CT images and the accuracy rate from X-ray images is found 84.67% and high-accuracy results were obtained in the detection of the infected area, especially in CT images. [21].

Salman et al. (2020) made a study using artificial intelligence methods in order to detect COVID-19 disease. They applied CNN model on 260 X-Ray images to diagnose SARS, ARDS, COVID-19, MERS, and Normal X-Ray disease. They found COVID-19 virus presence with 100% accuracy rate [22].

Jaiswal et al. (2020) DenseNet201 which used deep transfer learning based in their work for the classification of patients infected with COVID-19. As a result of the classification, it separated as positive (+) or negative (−) and 97% rate for the accuracy rate [23].

Butt et al. (2020) applied deep learning methods using datasets obtained from CT chest scans in order to obtain COVID-19 quickly and reliably. In this study, which was performed using the Convolutional Neural Network model, 98.2% sensitivity and 92.2% specificity values were obtained. They found that the results obtained using the deep learning model showed higher accuracy than the RT-PCR test used in Covid-19 detection [24].

Wang et al. (2020), suggested new proposal method (COVID-Net), including a deep learning network structure in order to diagnose COVID-19 disease. Radiological images, an overall accuracy rate of 93.3% was achieved in comparison with VGG-19 and ResNet-50. It is stated that there are more consistent results against these two networks in order to diagnose COVID-19 [16].

Pathak et al. (2020), used a ResNet32 method applying a transfer learning approach for COVID-19 diagnosis. Experiments were carried out using the dataset containing Covid-19 and normal values by examining 852 CT images. The model proposed in their experimental studies provided 96.22% training rate 93.01% testing rate accuracy [25].

Song et al. (2021), a new deep learning-based CT diagnostic system in order to diagnose COVID-19. They named the model as DRE-Net, which they stated had the ability to distinguish COVID-19 patients accurately from bacterial pneumonia patients. They identified COVID-19 patients with 86% accuracy and 79% sensitivity rate [26].

Sert et al. (2021), using the ResNet-50 model, classified COVID-19 and normal CT data. They compared CT scan images with a deep learning model combined with merging techniques to diagnose COVID-19. As a result, they stated that the Resnet-50 model outperformed all other models [27].

Karthik et al. (2021) focused on classification and detection of infected regions in order to diagnose Covid-19. In feature extraction, the smart filter results to the classification are given as input again by applying deep learning. The performance of the model developed as a result of the experiments was determined as 92.42% accuracy [28].

Monshi et al. (2021), focused on performance improvement in their EfficentNet-50 based solution named CovidXRayNet. The performance of hyper parameters and data augmentation methods has been improved. In the classifications made according to the COVID-19, Normal, and other pneumonia labels on the data, it provided a success increase of 11.93% and 4.97% in the VGG-19 and ResNet models, respectively. The developed CovidXRayNet model classifies with 95.82% accuracy [29].

Karakanis and Leonridis (2021), proposed dual and multiple classification for the diagnosis of Covid-19. The results obtained are compared with the ResNet8 network. The proposed method outperformed the ResNet8 network with an accuracy rate of 99.7% in binary classification and 98.3% in multi classification [30].

In this study, the opinion of cases with COVID-19, who can classify lung X-ray images as COVID-19, viral and normal pneumonia cases, by using Deep Neural Networks and Convolutional Neural Networks ways, extensively used deep literacy approaches, and discovery is targeted. An open access dataset containing X-ray images was used in order to search proposed models' performances. The work is planned as follows. Section 2, the material and system are bandied, and at this stage, the data set applied in the experimental study and the models used are mentioned. Section 3, the experimental studies are explained and the results attained from these trials are presented. In the last part, the evaluation of the results attained and commentary for farther studies are added.

3 Method and Material

3.1 Material

According to the literature, there are open access datasets for training machine learning models. Within the study, the popular Kaggle database containing current and real lung X-ray images was used [9, 31]. Viral Pneumonia, normal and COVID-19 patients' example lung X-Ray images in this dataset are shown in Fig. 1. In this data, there are 3886 images belonging to three classes, 1341 are normal, 1345 are viral pneumonia and 1200 of which are COVID-19 (Fig. 2).

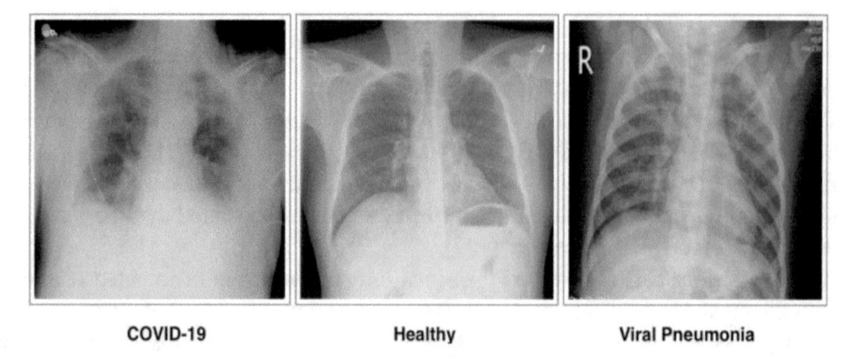

Fig. 1. Normal, viral pneumonia and covid-19 images in X-Ray images

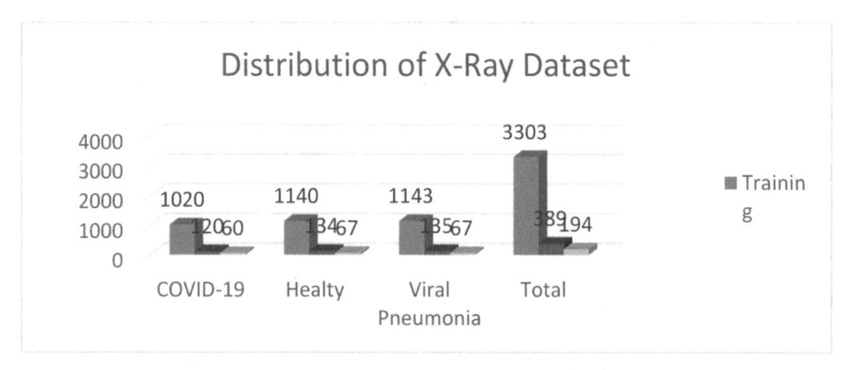

Fig. 2. The categorical distribution of the dataset for normal, viral pneumonia and covid-19

3.2 Method

Invisible image features can be revealed with Deep learning techniques from original ones [32]. A deep convolutional neural network is an intuitive and powerful network architecture in deep learning. It is frequently used in order to recognize pattern and classify images. In this study, performance evaluation and comparison of CNN and DNN technique, which can classify lung X-Ray images as normal, viral pneumonia and COVID-19 patients on the same dataset (Fig. 3).

3.2.1 CNN Model

It is one of the most widely used neural networks due to its success in image data. CNN has proven to be extremely useful in feature extraction and learning and has been used in many studies [33]. In last years, with the developing deep learning technology, more efficient ESA (DCNN) models such as VGG, ResNet, DenseNet, EfficientNet have been recently proposed. These deep ESAs perform well in image classification tasks, making it possible for computers to outperform humans at visual classification. A basic CNN architecture consists of convolution, flattened linear unit, pooling and fully connected

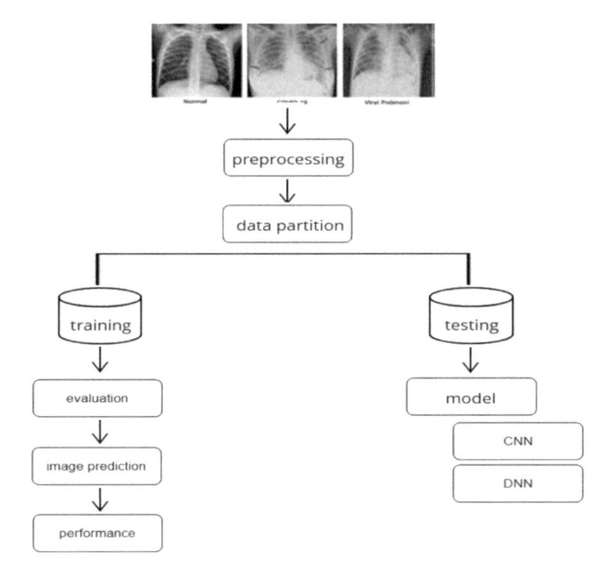

Fig. 3. Architecture used in COVID-19 detection

layer [34]. A basic CNN architecture used X-Ray Image to detect COVID-19. Figure 4 is given such as.

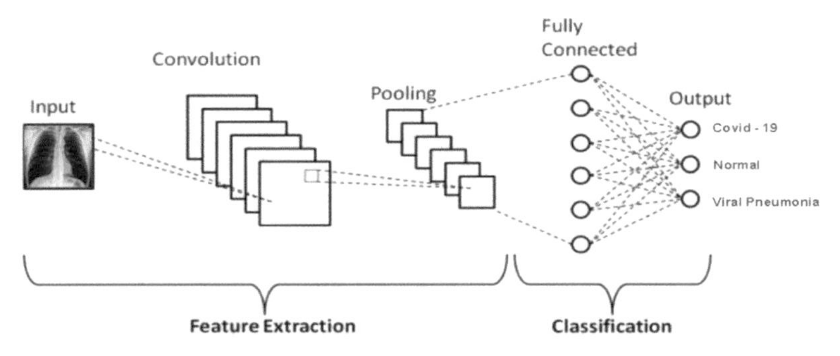

Fig. 4. CNN architecture

Convolutional Layer: Its end is to use a sludge matrix to the input with the convolution process, which forms the base of convolutional neural networks, and to use the results for the coming layer. Convolution is performed by covering a sludge over the image matrix. The sludge matrix is shifted by the specified step and the convolution operation is used The result then's given as input for the coming layer, if not the last layer. However, it represents the affair image, if this is the last layer. As seen in Eq. (1), the sludge portions (f) are calculated by multiplying the equal sized windows (w) in the image and taking their sum [35–37].

$$w(x, y) * f(x, y) = \sum_{i=-a}^{a} \sum_{j=-b}^{b} w(i, j) f(x + i, y + j) \quad (1)$$

Pooling Layer (PL): It is the layer where size reduction, which is an important function in CNN architectures, is performed. With the defined filters, a pooled feature map is created by operations such as taking the maximum value (maximum pooling), taking the average value (average pooling) on the pixel values in the image. Thus, it reduces the computational load for the next layer. Pixel values are lost in this process, but these losses create less computational load for the next layers.

Flattening Layer: Convolution and pooling can be applied multiple times in succession. It is the layer where the Flattened Linear Unit (ReLU) activation function is applied to the output data of the convolutional layers. The matrix obtained after these processes must be flattened to be used in the fully connected layer.

Fully Connected Layer: The fully connected layer is usually found at the end of the ESA and is used in classification operations. This layer is used to optimize class scores. After the fully connected layer, the final estimation is performed with the help of the activation function, which is preferred specific to the classification problem. Softmax activation function is used to release or dilute nodes below a certain threshold value.

$$softmax(xi) = \frac{e^{xi}}{\sum_{k=1}^{K} e^{k}} \quad (2)$$

Classification: ESA is the last layer for the classification process is performed. The affair values of this layer depend on the number of objects to be honored and are equal to the number of classes. ReLU activation function is used to avoid linearity in the network, the loftiest probability value gives the class prognosticated by the model.

The CNN model suggested within the study consists of a flattened layer, input layer, four (complication) complication, 4 pooling layers, three completely connected layers, and a classifier (affair). The proposed CNN model is shown in Fig. 5.

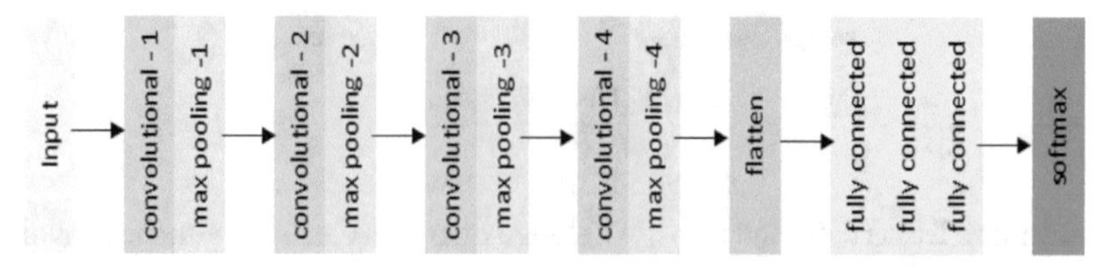

Fig. 5. Suggested CNN architecture

ReLU activation function is used after all convolution layers. After each convolution and ReLU operation, the number of step shifts (stride = 2) was taken and maximum pooling was done with a filter size of 3 × 3. After this stage, the multidimensional feature set is flattened and transformed into a one-dimensional index shape and transmitted to the next fully connected layers as input. In the training process, the number of revolutions

(Epoch) = 50, the batch size (Batch size) = 64 and the learning rate (Learning Rate) = 0.001.

ReLU and Softmax are activation functions that are widely used in the literature. In neural networks, the generalization performance of the network may vary according to the activation function. In this context, the ReLU activation function is often preferred in deep neural networks, and the fast learning of the network also provides high classification success [38]. In this study, ReLU activation function is in order to avoid linearity in the network.

3.2.2 DNN Model

DNNs are a neural network with more than two layers. The difference between DNN deep learning model and the CNN deep learning model is that in DNN deep learning model, hidden layers' number is decreased by one after the pooling layer after the first layer, and this situation is done again with loops. After the number of hidden layers is finished, smoothing is done and a model with a much larger number of layers is obtained than the CNN model. As seen in Fig. 3.12, each DNN layer performs operations similar to the steps of a CNN algorithm, such as convolution, pooling, and normalization. This deep learning model has the same structure as other neural networks, except that the number of hidden layers is higher. In Fig. 6, DNN architecture, there are 8 layers and every one of them contains one derivative of the convolution, maximum pooling and normalization stages. But the last three layers are completely interconnected [39].

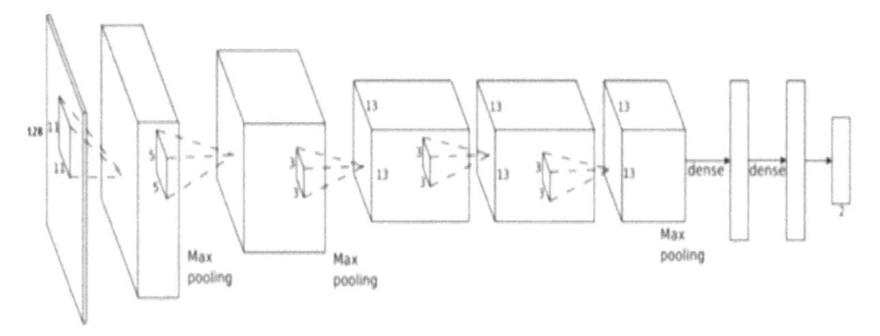

Fig. 6. Deep learning architecture

Incoming data to the deep neural network; Classification is made through three layers: input, hidden and output, and the x-ray lung image is determined to be COVID-19. In Fig. 7, the deep neural network architecture is given.

3.2.3 Performance Criteria

In general, for performance evaluation in classification algorithms, the actual and estimated values of the classes are compared with the confusion matrix. Receiver Operating Characteristics (ROC) is a method frequently used in bioinformatics to measure classification performance [40].

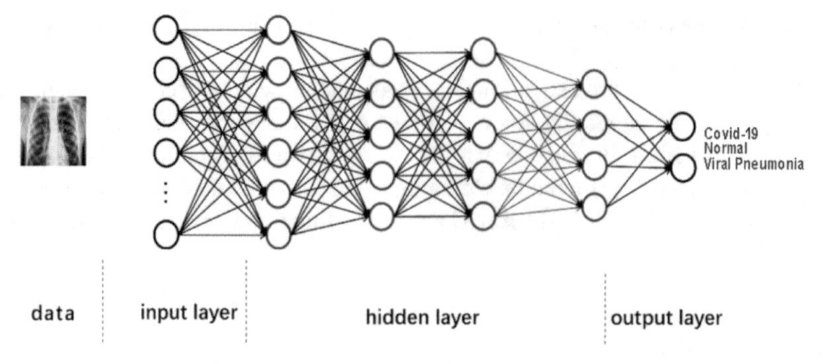

Fig. 7. Proposed deep learning architecture

In order to compare and evaluate the models used in the study, accuracy, precision, sensitivity, specificity and F1-score criteria were used as performance criteria. The confusion matrix given in Table 1 is used to calculate the performance criteria for a prediction model. The calculation of performance measures from this matrix is presented in Eq. 3–6.

Table 1. Confusion Matrix

Actual class	Predicted Class	
	Actually Positive (1)	Actually Negative (0)
Positive	True Positives (TPs)	False Positives (FPs)
Negative	False Negatives (FNs)	True Negatives (TNs)

$$Accuracy_{class\_i} = \frac{TP_{class\_i} + TN_{class\_i}}{TP_{class\_i} + TN_{class\_i} + FP_{class\_i} + FN_{class\_i}} \tag{3}$$

$$Precision_{class\_i} = \frac{TP_{class\_i}}{TP_{class\_i} + FP_{class\_i}} \tag{4}$$

$$Sensitivity_{class\_i} = \frac{TP_{class\_i}}{TP_{class\_i} + FN_{class\_i}} \tag{5}$$

$$F1\_score_{class\_i} = 2\frac{Precision_{class\_i} * Sensitivity_{class\_i}}{Precision_{class\_i} + Sensitivity_{class\_i}} \tag{6}$$

4 Experimental Studies

The comparison and classification studies of deep neural network and Convolutional neural network were made on the sample lung X-Ray images of 3886 normal, viral pneumonia and Covid-19 patients in the open source dataset. The performance of the networks was evaluated using four performance measures such as accuracy, precision, F1-score and responsiveness (recall). In the study, the experiments were done on a desktop computer with these features;

- CPU, Intel Core (TM) i7 9300H,
- Graphics Card, NVIDIA GeForce GTX 1650 (4 GB)
- Memory, 8 GB.

Proposed models are processed using Python 3.7 along with the relevant libraries. Before testing, all images were uniformly resized to 256 * 256 before being transferred to the model. A 256 * 256-pixel black background has been added to the images in order to standardize the images with different pixel ratios, to prevent distortion and to achieve a complete transformation. In the experimental study, X-Ray image data was reserved for 85% training, 10% validation and 5% testing.

4.1 Results from the CNN Model

"ReLU" activation function is used in convolutional layers and "Softmax" activation function is applied in output layer. Adam optimization algorithm was preferred in the optimization of the network. The data is categorized into three clusters named testing, training and validation. In Fig. 8, a sample performance evaluation obtained from the confusion matrix is made.

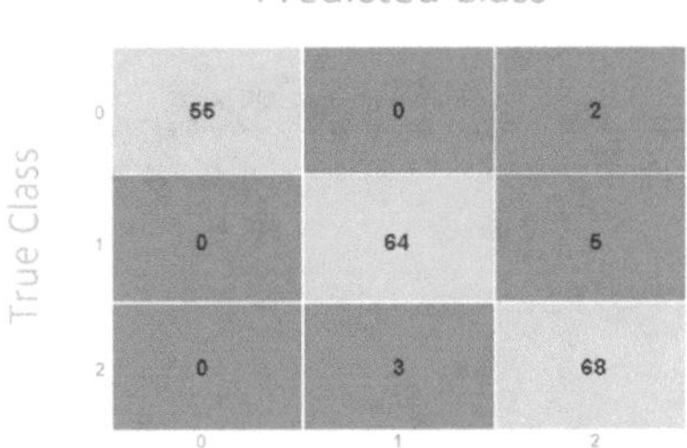

Fig. 8. Confusion Matrix of CNN model

Table 2 indicates the results of accuracy, precision, sensitivity and F1-Score criteria in 3 different classes that emerged with the CNN model.

Table 2. CNN Model Performance Rates

CNN	Accuracy	Precision	Sensitivity	F1-Score
Normal	97.4%	97.6%	98.2%	98.1%
Viral Pneumonia	96.1%	96.8%	96.8%	97.1%
COVID-19	96.9%	97.1%	97.4%	97.3%

The figure shows the accuracy of training/test and the loss graph of training/test (50 iterations) obtained with the Deep ESA model proposed in this study. Both accuracy curves of test and training indicate an increasing slope when the number of iterations increases. As shown in Fig. 8, the training process and the learning of the network are at a good learning rate. Figure 9 shows that the loss curve emphasizes the error rate decrease. While there is a decrease of each iteration in loss value, the accuracy rate increases with training set and learning occurs.

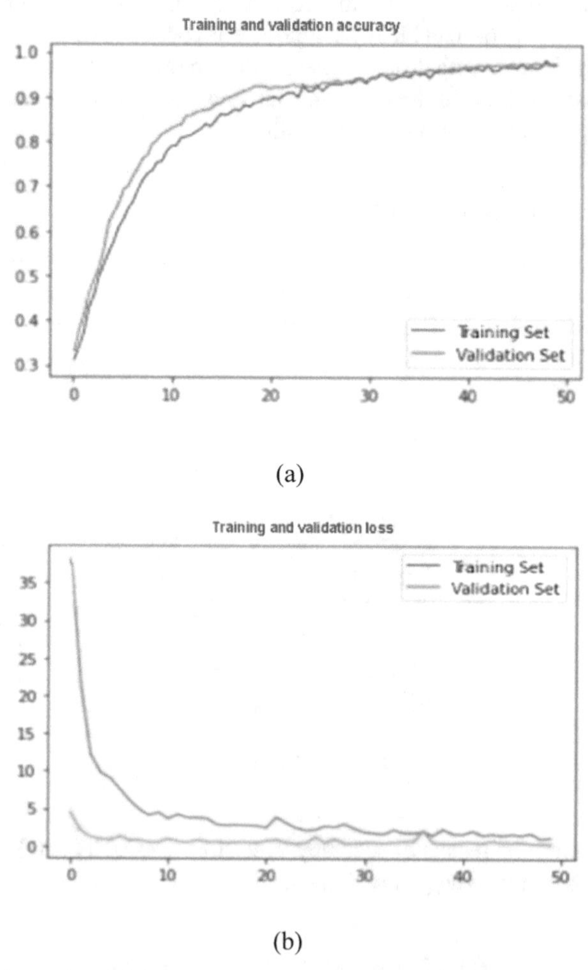

Fig. 9. (a) Training and validation accuracy (b) Training and validation loss graphs in the CNN model

4.2 Results from the DNN Model

Training, testing and verification processes were carried out in 3 categories of X-Ray images. In the DNN model, which is one of the deep learning methods, the data set is divided into 3 groups such as 85% training, 10% testing and 5% validation rates. An exemplary performance evaluation obtained from the confusion matrix in Fig. 10 was made.

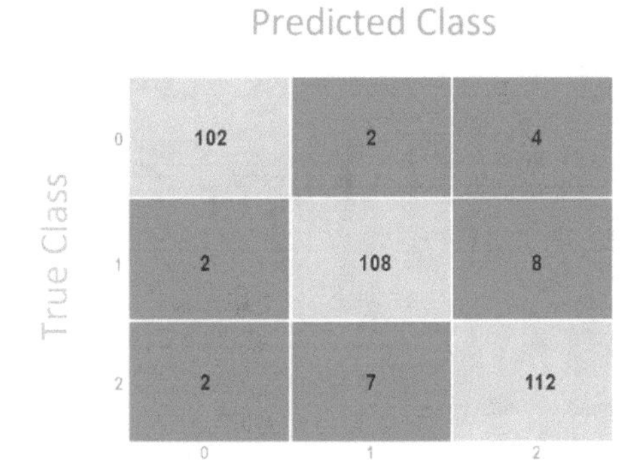

Fig. 10. Confusion Matrix of CNN model

Table 3 shows the results of accuracy, precision, sensitivity, and F1-Score criteria in 3 different classes that emerged with the DNN model.

Table 3. DNN Model Performance Rates

DNN	Accuracy	Precision	Sensitivity	F1-Score
Normal	95.8%	96.1%	96.2%	96.25%
Viral Pneumonia	94.5%	94.8%	94.9%	95.1%
COVID-19	96.6%	96.8%	97.0%	97.1%

Figure 11 shows the training and validation accuracy and training and validation loss graphs (for 50 iterations) obtained with the DNN model proposed in this study.

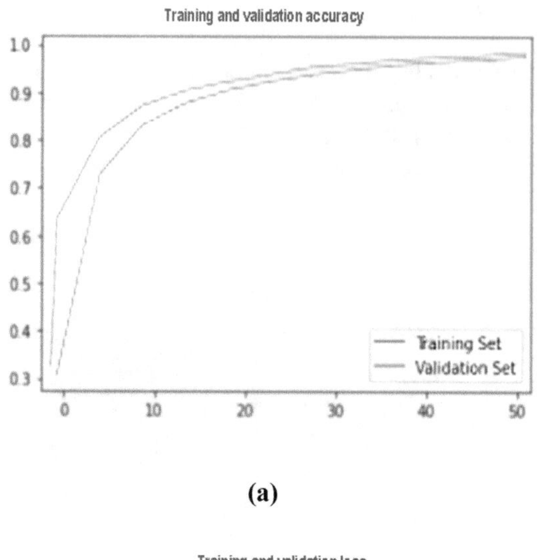

(a)

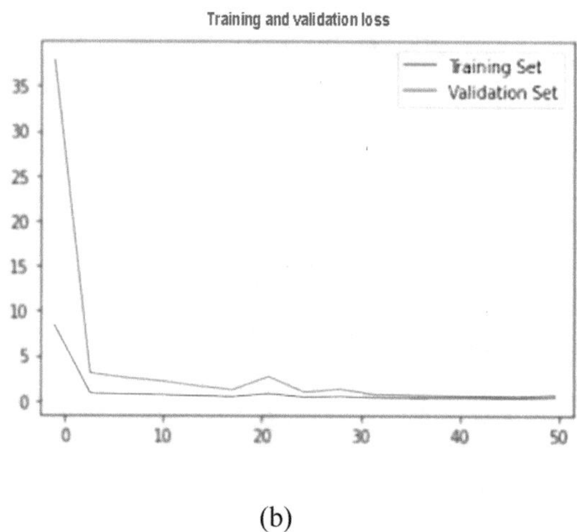

(b)

Fig. 11. (a) Training and validation accuracy (b) Training and Validation loss graphs in the CNN model

4.3 Comparison of the Model

When the findings obtained from the study are examined (Table 4), it is seen that the highest performance in terms of diagnosis-diagnosis is obtained with CNN, one of the deep learning algorithms.

Finally, 2 (two) different tests were carried out using the proposed CNN and DNN models and real data. The experimental results obtained from these tests are given in Table 5.

According to the experimental results (Table 5), the best performance and the lowest error rate were obtained with the CNN technique.

Table 4. Comparison of CNN and DNN Algorithm results

Model	Training	Training Loss	Testing	Testing Loss
CNN	99.74	0.9	98.55	4.7
DNN	98.25	1.85	97.6	6.2

Table 5. CNN and DNN sample experimental results

No	CNN	CNN Loss	DNN	DNN Loss
1	96.95	3.05	85.04	14.96
2	98.27	1.73	94.67	5.33

5 Discussion and Conclusion

COVID-19 pandemic's rapid spread in the world and its negative effects on people clearly show that the issue should be examined in detail and that positive cases should be detected in the early stages and rapid and correct intervention should be made. Lung X-ray images usage in early diagnosis of pneumonia and COVID-19. But some viral pneumonia (pneumonia) images look like COVID-19 images and have similar common features. So it is hard for radiologists to differentiate similar lung diseases from COVID-19. The fact that the symptoms of COVID-19 are similar to viral pneumonia can cause to misdiagnoses in this context.

In this study were presented which CNN and DNN model, and experimental studies were carried out on open access lung X-ray images. In this dataset, it has a total of 3886 images of three classes: COVID-19, Normal and Viral Pneumonia. For this purpose, a deep CNN and DNN model was applied in this study, which can classify lung X-ray images as viral pneumonia (pneumonia), normal and COVID-19 patients. This study was carried out by separating different amounts of data for training, validation and testing.

The results showed that the CNN Model is the more performance values than the DNN method. In addition, the performance of the model is directly proportional to the number of data. It reveals that the CNN model can be easily used in the health system by being further developed with hybrid approaches.

As mentioned in the literature studies, many sub-models of different techniques have been tried on lung images within the scope of COVID-19. When the studies are examined, it is seen that the accuracy rate is above 90% in many studies [13, 14, 18, 21, 29, 31]. The larger the sample of the data set, the higher the accuracy rate. This shows how important medical images are in disease detection and follow-up, and with the developing technology, both faster and more accurate results are achieved.

References

1. Sohrabi, C., et al.: World Health Organization declares global emergency: a review of the 2019 novel coronavirus (COVID-19). Int. J. Surg. **76**, 71–76 (2020)

2. Knight, T.E.: Severe acute respiratory syndrome coronavirus 2 and coronavirus disease 2019: a clinical overview and primer. Biopreservation Biobanking **18**(6), 492–502 (2020)
3. Chen, T.M., Rui, J., Wang, Q.P., Zhao, Z.Y., Cui, J.A., Yin, L.: A mathematical model for simulating the phase-based transmissibility of a novel coronavirus. Infect. Dis. Poverty **9**(1), 1–8 (2020)
4. World Health Organization. Naming the coronavirus disease (COVID-19) and the virus that causes it (2021). https://www.who.int/emergencies/diseases/novel-coronavirus-2019/technical-guidance/naming-the-coronavirus-disease-(covid-2019)-and-the-virus-that-causes-it
5. Toğaçar, M., Ergen, B., Cömert, Z.: COVID-19 detection using deep learning models to exploit Social Mimic Optimization and structured chest X-ray images using fuzzy color and stacking approaches. Comput. Biol. Med. **121**, 103805 (2020)
6. Brunese, L., Mercaldo, F., Reginelli, A., Santone, A.: Explainable deep learning for pulmonary disease and coronavirus COVID-19 detection from X-rays. Comput. Methods Programs Biomed. **196**, 105608 (2020)
7. Pereira, R.M., Bertolini, D., Teixeira, L.O., Silla, C.N., Jr., Costa, Y.M.: COVID-19 identification in chest X-ray images on flat and hierarchical classification scenarios. Comput. Methods Programs Biomed. **194**, 105532 (2020)
8. Bustin, S.A.: Absolute quantification of mRNA using real-time reverse transcription polymerase chain reaction assays. J. Mol. Endocrinol. **25**(2), 169–193 (2000)
9. Chowdhury, M.E., et al.: Can AI help in screening viral and COVID-19 pneumonia? IEEE Access **8**, 132665–132676 (2020)
10. Wang, S., et al.: A deep learning algorithm using CT images to screen for Corona Virus Disease (COVID-19). Eur. Radiol. **31**(8), 6096–6104 (2021)
11. Panwar, H., Gupta, P.K., Siddiqui, M.K., Morales-Menendez, R., Singh, V.: Application of deep learning for fast detection of COVID-19 in X-Rays using nCOVnet. Chaos Solitons Fractals **138**, 109944 (2020)
12. Li, L., et al.: Artificial intelligence distinguishes COVID-19 from community acquired pneumonia on chest CT. Radiology (2020)
13. El Asnaoui, K., Chawki, Y., Idri, A.: Automated methods for detection and classification pneumonia based on X-ray images using deep learning. In: Maleh, Y., Baddi, Y., Alazab, M., Tawalbeh, L., Romdhani, I. (eds.) Artificial Intelligence and Blockchain for Future Cybersecurity Applications. SBD, vol. 90, pp. 257–284. Springer, Cham (2021). https://doi.org/10.1007/978-3-030-74575-2_14
14. Apostolopoulos, I.D., Aznaouridis, S.I., Tzani, M.A.: Extracting possibly representative COVID-19 biomarkers from X-ray images with deep learning approach and image data related to pulmonary diseases. J. Med. Biol. Eng. **40**(3), 462–469 (2020)
15. Zheng, C., et al.: Deep learning-based detection for COVID-19 from chest CT using weak label. MedRxiv, 2020.03.12.20027185. https://doi.org/10.1101/2020.03.12.20027185
16. Wang, L., Lin, Z.Q., Wong, A.: Covid-net: a tailored deep convolutional neural network design for detection of Covid-19 cases from chest X-ray images. Sci. Rep. **10**(1), 1–12 (2020)
17. Horry, M.J., et al.: COVID-19 detection through transfer learning using multimodal imaging data. IEEE Access **8**, 149808–149824 (2020)
18. Ozturk, T., Talo, M., Yildirim, E.A., Baloglu, U.B., Yildirim, O., Acharya, U.R.: Automated detection of COVID-19 cases using deep neural networks with X-ray images. Comput. Biol. Med. **121**, 103792 (2020)
19. Rahimzadeh, M., Attar, A.: A modified deep convolutional neural network for detecting COVID-19 and pneumonia from chest X-ray images based on the concatenation of Xception and ResNet50V2. Inform. Med. Unlocked **19**, 100360 (2020)
20. Mangal, A., et al.: CovidAID: COVID-19 detection using chest X-ray. arXiv preprint arXiv:2004.09803 (2020)

21. Alom, M.Z., et al. COVID_MTNet: COVID-19 detection with multi-task deep learning approaches. arXiv preprint arXiv:2004.03747 (2020)
22. Salman, F.M., Abu-Naser, S.S., Alajrami, E., Abu-Nasser, B.S., Alashqar, B.A.: Covid-19 detection using artificial intelligence (2020)
23. Jaiswal, A., Gianchandani, N., Singh, D., Kumar, V., Kaur, M.: Classification of the COVID-19 infected patients using DenseNet201 based deep transfer learning. J. Biomol. Struct. Dyn. **39**(15), 5682–5689 (2021)
24. Butt, C., Gill, J., Chun, D., Babu, B.A.: Retracted article: Deep learning system to screen coronavirus disease 2019 pneumonia. Appl. Intell. (2020). https://doi.org/10.1007/s10489-020-01714-3. Epub ahead of print. PMCID: PMC7175452
25. Pathak, Y., Shukla, P.K., Tiwari, A., Stalin, S., Singh, S.: Deep transfer learning based classification model for COVID-19 disease. Irbm (2020)
26. Song, Y., et al.: Deep learning enables accurate diagnosis of novel coronavirus (COVID-19) with CT images. IEEE/ACM Trans. Comput. Biol. Bioinf. **18**(6), 2775–2780 (2021)
27. Serte, S., Demirel, H.: Deep learning for diagnosis of COVID-19 using 3D CT scans. Comput. Biol. Med. **132**, 104306 (2021)
28. Karthik, R., Menaka, R., Hariharan, M.: Learning distinctive filters for COVID-19 detection from chest X-ray using shuffled residual CNN. Appl. Soft Comput. **99**, 106744 (2021)
29. Monshi, M.M.A., Poon, J., Chung, V., Monshi, F.M.: CovidXrayNet: optimizing data augmentation and CNN hyperparameters for improved COVID-19 detection from CXR. Comput. Biol. Med. **133**, 104375 (2021)
30. Karakanis, S., Leontidis, G.: Lightweight deep learning models for detecting COVID-19 from chest X-ray images. Comput. Biol. Med. **130**, 104181 (2021)
31. Chowdhury, M.E., et al.: Covid-19 Chest X-Ray Database (2020). https://www.kaggle.com/tawsifurrahman/covid19-radiography-database
32. Çinare, O., Yağlanoğlu, M.: Determination of Covid-19 possible cases by using deep learning techniques. Sakarya Univ. J. Sci. **25**(1), 1–11 (2021)
33. Liu, J.: Review of deep learning-based approaches for COVID-19 detection. In: 2021 2nd International Conference on Computing and Data Science (CDS), Stanford, CA, USA, pp. 366–371. IEEE (2021)
34. Yılmaz, A., Kaya, U.: Derin Öğrenme (2019)
35. Ciresan, D.C., Meier, U., Masci, J., Gambardella, L.M., Schmidhuber, J.: Flexible, high performance convolutional neural networks for image classification. In: Twenty-Second International Joint Conference on Artificial Intelligence (2011)
36. Meunier, L.C.V., Chandy, D.A.: Design of convolution neural network for facial emotion recognition. In: 2019 2nd International Conference on Signal Processing and Communication (ICSPC), pp. 376–379. IEEE (2019)
37. Umer, M., Ashraf, I., Ullah, S., Mehmood, A., Choi, G.S.: COVINet: a convolutional neural network approach for predicting COVID-19 from chest X-ray images. J. Ambient. Intell. Humaniz. Comput. **13**(1), 535–547 (2022)
38. Yıldız, O.: Melanoma detection from dermoscopy images with deep learning methods: a comprehensive study. J. Faculty Eng. Archit. Gazi Univ. **34**(4), 2241–2260 (2019)
39. Cichy, R.M., Khosla, A., Pantazis, D., Torralba, A., Oliva, A.: Comparison of deep neural networks to spatio-temporal cortical dynamics of human visual object recognition reveals hierarchical correspondence. Sci. Rep. **6**(1), 1–13 (2016)
40. Lasko, T.A., Bhagwat, J.G., Zou, K.H., Ohno-Machado, L.: The use of receiver operating characteristic curves in biomedical informatics. J. Biomed. Inform. **38**(5), 404–415 (2005)

Rule-Based Cardiovascular Disease Diagnosis

Ayşe Ünlü[1,2], Derya Kandaz[1,2](✉), Gültekin Çağil[1,2], and Muhammed Kürşad Uçar[1,2]

[1] Industrial Engineering, Faculty of Engineering, Sakarya University, Sakarya, Turkey
ayse.unlu1@ogr.sakarya.edu.tr, {deryakandaz,cagil,mucar}@sakarya.edu.tr
[2] Electrical-Electronics Engineering, Faculty of Engineering, Sakarya University, Sakarya, Turkey

Abstract. Cardiovascular disease is defined as the effect of damage to the heart and vessels in the body. The prevalence of this disease and the fact that it causes death day by day show that early diagnosis is vital. Diagnosis of cardiovascular diseases in the clinic takes a long time due to its complexity. In this context, it is aimed to diagnose cardiovascular disease in individuals more accurately and in a short time by using a rule-based decision tree algorithm in the study. The study was carried out on seventy thousand data sets taken from the Kaggle database. First, using the Spearman correlation, age, gender, weight, height, blood pressure, sugar, cholesterol, smoking, alcohol and exercise characteristics that cause cardiovascular disease were compared with the label and the relationship between them was determined. Then, a rule-based decision tree algorithm was developed. Diagnostic algorithms have been extracted as tree-based clinically usable constructs. According to the study findings, a rule-based diagnostic algorithm was developed with an accuracy of approximately 73%. According to the results of the method used, it is thought that the diagnosis algorithm of cardiovascular diseases can be applied in the clinic.

Keywords: Cardiovascular Disease · Prediction · Rule-Based Decision Trees · Spearman correlation

1 Introduction

According to the World Health Organization (WHO), cardiovascular disease (CVD) is a significant public health problem and the leading cause of death globally. According to the estimation results, 17.9 million people die from this disease every year [1,2]. Cardiovascular disease arises from the dysfunction of the heart and vessels and causes angina pectoris, myocardial infarction, coronary heart disease, heart failure, arrhythmia, and disorders of the circulatory system generally associated with atherosclerosis [3]. In patients at risk of cardiovascular disease, high blood pressure, excess weight, and obesity can be seen together with

D. J. Hemanth et al. (Eds.): ICAIAME 2022, ECPSCI 7, pp. 740–750, 2023.
https://doi.org/10.1007/978-3-031-31956-3_63

glucose and lipids [4]. Most types of this disease can be prevented by eliminating or applying regularly and correctly the behavioral risk factors such as tobacco use, unhealthy food consumption, excessive alcohol use, and a sedentary life [5].

Artificial intelligence is defined as a technology that improves itself gradually with the data it collects by imitating human intelligence in decision making, problem-solving and learning processes [6–8]. Today, with the developments in artificial intelligence, it is possible to create meaningful, reliable and quality data that can identify confidential and valuable information, thanks to the big data created in clinical environments. Artificial intelligence studies use various methods such as machine learning techniques, regression problems, classification, image recognition, image and signal processing, medical diagnosis, and prediction. Machine learning methods applied in the health sector help doctors in the early detection, detection and prediction of diseases. Ganesh.

The number of people suffering from cardiovascular diseases worldwide is increasing day by day. The most effective measure for the control of heart disease in people is to predict the disease before getting the disease [9]. For this reason, researchers have been working on different approaches and algorithms for years. In the Table 1, the studies done in the literature on the subject are examined.

2 Materials and Methods

This study used cardiovascular disease data from the Kaggle data site. For the study, the flow chart in the Fig. 1 was followed. First of all, the Spearman feature extraction algorithm was applied to the data, and the features that had the most effect on causing the disease were determined. Then, the data were classified with the decision tree algorithm and analyzed according to the performance evaluation criteria. All calculations made in the study were carried out using the MATLAB program.

2.1 Dataset Overview

The data set used in the study was obtained from the examination data of people with cardiovascular disease. The disease consists of 11 features. The data set containing the information of 70000 patients labeled individuals as healthy (0) and sick (1). Of the people in the data set, 34979 were patients, and 35021 were healthy. The Table 2 shows the names, definitions, data types and defined categorical names of the attributes in the data set.

Table 1. Representation of Features mathematical and code

References	Year	Name of the study	Methods Used
[10]	2020	Veri Madenciliği İle Kalp Hastalığı Teşhis	Naive Bayes, Karar Ağacı, Rastgele Orman, Çoklu Algılayıcılar, k-en yakın komşu , Lojistik Regresyon, Destek vektör makinesi
[11]	2020	Predicting Heart Disease at Early Stages using Machine Learning: A Survey	Yapay sinir ağı , Karar ağacı , Rastgele orman , Destek vektör makinesi , Bayes, k-en yakın komşu algoritması
[12]	2021	Genetik Algoritma Yaklaşımıyla Öznitelik Seçimi Kullanılarak Makine Öğrenmesi Algoritmaları ile Kalp Hastalığı Tahmini	Genetik Algoritma, K-En Yakın Komşu , Lojistik Regresyon, Karar Ağacı , Rastgele Orman, Naive Bayes ve Destek Vektör Makinesi
[13]	2021	Heart Disease Prediction using Hybrid machine Learning Model	Karar Ağacı yaklaşık , Rastgele orman , Hibrit model
[14]	2021	Relief Özellik Seçim Yöntem Tabanlı Önerilen Hibrit Model ile Kalp Hastalığı Teşhisi	Relief özellik çıkarımı yöntemi, Logistik regresyon, Karar ağaçları, YSA, K-en yakın komşular , Bayes , Destek vektör makineleri
[15]	2020	Yapay Zeka Yöntemleri Kullanılarak Kalp Hastalığının Tespiti	Random Forest yöntemi ve Parçacık Sürü Optimizasyonu
[16]	2020	Cardiovascular Disease Forecast using Machine Learning Paradigms	Lojistik regresyon, Karar ağacı, SVM ve Naive bayes
[17]	2021	Early Prediction of Cardiovascular Diseases Using Feature Selection and Machine Learning Techniques	Rastgele Orman ile özellik çıkarımı , Destek Vektör Makinesi, Lojistik Regresyon, K-neaest Neighbors, Karar ağacı ve XGB

2.2 Spearman Feature Selection Algorithm

Feature selection is a commonly used data preprocessing step to determine the most important and most valuable features before creating a predictive model. It is crucial to determine the risk factors associated with the disease in health science. It is aimed to obtain faster and better computational results with these specified attributes and to eliminate unnecessary attributes that do not affect the result [18]. For this reason, the Spearman correlation coefficient-based feature

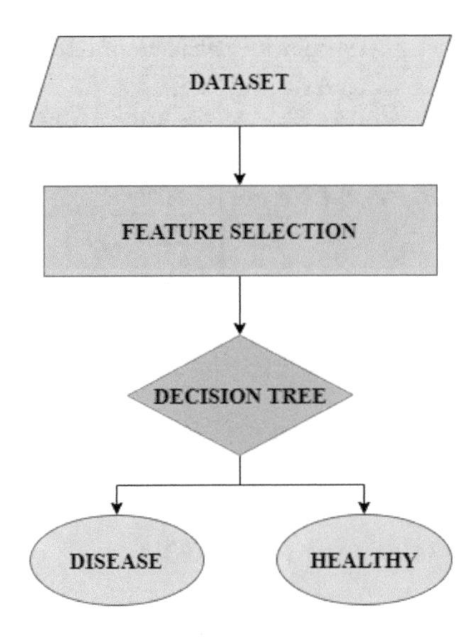

Fig. 1. Akış diyagramı

Table 2. Distribution of features

Attribute Name	Definition	Data Type	Data
Age	Patient's Age(day)	Numerical	day
Height	Patient Height (cm)	Numerical	cm
Weight	Patient's Weight (kg)	Numerical	kg
Gender	Patient's Gender	Discrete	1: Woman 2: Man
Systolic Blood Pressure	Patient's Systolic Blood Pressure (ap hi)	Numerical	
Diastolic Blood Pressure	Patient's Diastolic Blood Pressure(ap lo)	Numerical	
Cholesterol	Patient's Cholesterol Level	Discrete	1: Normal 2: Above Normal 3: Very Above Normal
Glucose	Patient's glucose level	Discrete	1: Normal 2: Above Normal 3: Very Above Normal
Smoking	Patient's smoking status	Discrete	0: Not Using 1: Using
Alcohol Consumption	Patient's alcohol use status	Discrete	0: Not Using 1: Using
Physical Activity	The patient's physical activity status	Discrete	0: Not Using 1: Using
Cardiovascular Disease Status	Patient cardiovascular disease status	Discrete	0: Healthy 1: Patient

selection algorithm was used in the study. As a result of the calculations, the coefficients of the features are given in the Table 3.

Table 3. Spearman Coefficients of Attributes

Attribute Name	Spearman Coefficient
Systolic Blood Pressure	0.4519
Diastolic Blood Pressure	0.3626
Age	0.2344
Cholesterol	0.2151
Gender	0.1827
Glucose	0.0915
Physical Activity	0.0357
Smoking	0.0155
Weight	0.0124
Height	0.0081
Alcohol Consumption	0.0073

2.3 Decision Trees

The decision tree is a machine learning algorithm among the essential artificial intelligence techniques used for classification and prediction. The most important reason why the decision tree has been widely studied in the literature in recent years is that the rules used in the creation of tree structures are understandable and straightforward [?]. Decision trees are algorithms that create the tree structure starting from the top to classify the data. This structure consists of root nodes, branches, and leaves. The outgoing branches of a node, all possible results of the test in the node, and leaves are considered where classification occurs [6]. The decision result or decision model can be graphically represented as a decision tree (Table 2) [19] (Figs. 2 and 3).

Fifty percent of the data was used in model creation, and the other fifty percent in testing the model in order to perform classification with decision trees in the data set (Table 3). For each data group, the training data set was created with the help of the systematic sampling method. The rest were used in the testing phase. The performance evaluation criteria of the model built on the test data were tested.

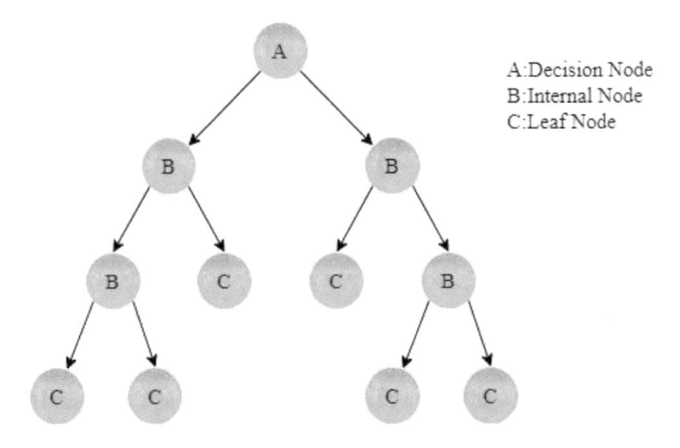

Fig. 2. Decision Tree Structure

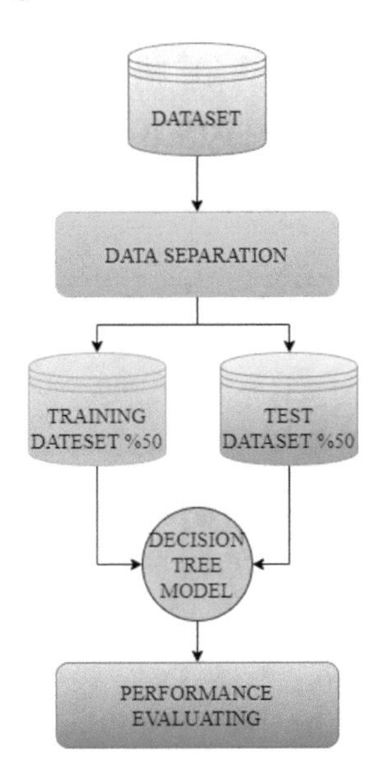

Fig. 3. Splitting the dataset into training and test data

2.4 Performance Evaluation Criteria

The confusion matrix was used to calculate the success of the models created in the study (Table 4). In order to evaluate the performance over true positive (TP), false positive (FP), false negative (FN), and true negative (TN) values in the matrix, the accuracy rates in the test set, sensitivity, and specificity of the classes, AUC and finally F measurement values were calculated [20–22].

Table 4. Confusion matrix

		Actual Results	
		Positive	Negative
Estimation Results	Positive	TP	FP
	Negative	FN	TN

3 Results

The study aimed to predict cardiovascular disease, and a decision tree algorithm was used in line with this goal. The diagnosis of the disease was made on the data set of 70000. Five data and attributes from the data in the data set are shown in Table 5. Firstly, using Spearman correlation, attributes, such as age, gender, and weight, were compared with the label, and the relationship between them was determined. It was concluded that systolic blood pressure was the highest with the label, and alcohol consumption was the lowest. Fifty percent of the data set is allocated as training data and 50% as test data through systematic sampling. Then, a rule-based decision tree algorithm was developed, and the model's performance was calculated. The estimation result of the model was found by taking the arithmetic average of the Accuracy, Sensitivity, Specificity, F score, and AUC results. Performance values vary between 0.71 and 0.73. According to these performance results, the method used can be applied in the clinic.

Table 5. Confusion matrix

id	age	gender	height	weight	ap hi	ap lo	cholesterol	gluc	smoke	alco	active	cardio
0	18393	2	168	62	110	80	1	1	0	0	1	0
1	20228	1	156	85	140	90	3	1	0	0	1	1
2	18857	1	165	64	130	70	3	1	0	0	0	1
3	17623	2	169	82	150	100	1	1	0	0	1	1
4	17474	1	156	56	100	60	1	1	0	0	0	0
8	21914	1	151	67	120	80	2	2	0	0	0	0

In future studies, performance values can be increased by using different artificial intelligence methods (Table 6).

Table 6. Confusion matrix

Accuracy rate	73.151
Sensitivity	0.738
Specificity	0.725
F score	0.731
AUC value	0.731

Table 7. Application Results

Split 1

FN	Accuracy Rate	Sensivity	Specificity	F-Score	AUC	Mean
1	71.73	0.80	0.63	0.71	0.72	0.71
2	71.73	0.80	0.63	0.71	0.72	0.71
3	71.73	0.80	0.63	0.71	0.72	0.71
4	71.73	0.80	0.63	0.71	0.72	0.71
5	71.73	0.80	0.63	0.71	0.72	0.71
6	71.73	0.80	0.63	0.71	0.72	0.71
7	71.73	0.80	0.63	0.71	0.72	0.71
8	71.73	0.80	0.63	0.71	0.72	0.71
9	71.73	0.80	0.63	0.71	0.72	0.71
10	71.73	0.80	0.63	0.71	0.72	0.71
11	71.73	0.80	0.63	0.71	0.72	0.71

Split 2

FN	Accuracy Rate	Sensivity	Specificity	F-Score	AUC	Mean
1	71.73	0.80	0.63	0.71	0.72	0.71
2	71.73	0.80	0.63	0.71	0.72	0.71
3	71.73	0.80	0.63	0.71	0.72	0.71
4	71.73	0.80	0.63	0.71	0.72	0.71
5	71.73	0.80	0.63	0.71	0.72	0.71
6	71.73	0.80	0.63	0.71	0.72	0.71
7	71.73	0.80	0.63	0.71	0.72	0.71
8	71.73	0.80	0.63	0.71	0.72	0.71
9	71.73	0.80	0.63	0.71	0.72	0.71
10	71.73	0.80	0.63	0.71	0.72	0.71
11	71.73	0.80	0.63	0.71	0.72	0.71

Split 3

FN	Accuracy Rate	Sensivity	Specificity	F-Score	AUC	Mean
1	71.73	0.80	0.63	0.71	0.72	0.71
2	71.73	0.80	0.63	0.71	0.72	0.71
3	71.73	0.80	0.63	0.71	0.72	0.71
4	71.73	0.80	0.63	0.71	0.72	0.71
5	71.73	0.80	0.63	0.71	0.72	0.71
6	71.73	0.80	0.63	0.71	0.72	0.71
7	71.73	0.80	0.63	0.71	0.72	0.71
8	71.73	0.80	0.63	0.71	0.72	0.71
9	71.73	0.80	0.63	0.71	0.72	0.71
10	71.73	0.80	0.63	0.71	0.72	0.71
11	71.73	0.80	0.63	0.71	0.72	0.71

Split 4

FN	Accuracy Rate	Sensivity	Specificity	F-Score	AUC	Mean
1	71.69	0.80	0.63	0.71	0.72	0.71
2	71.73	0.80	0.63	0.71	0.72	0.71
3	72.37	0.75	0.70	0.72	0.72	0.72
4	71.95	0.79	0.65	0.71	0.72	0.72
5	71.95	0.79	0.65	0.71	0.72	0.72
6	71.95	0.79	0.65	0.71	0.72	0.72
7	71.95	0.79	0.65	0.71	0.72	0.72
8	71.95	0.79	0.65	0.71	0.72	0.72
9	71.95	0.79	0.65	0.71	0.72	0.72
10	71.95	0.79	0.65	0.71	0.72	0.72
11	71.95	0.79	0.65	0.71	0.72	0.72

Split 5

FN	Accuracy Rate	Sensivity	Specificity	F-Score	AUC	Mean
1	71.69	0.80	0.63	0.71	0.72	0.71
2	71.73	0.80	0.63	0.71	0.72	0.71
3	72.37	0.75	0.70	0.72	0.72	0.72
4	72.81	0.78	0.68	0.72	0.73	0.73
5	72.81	0.78	0.68	0.72	0.73	0.73
6	72.81	0.78	0.68	0.72	0.73	0.73
7	72.81	0.78	0.68	0.72	0.73	0.73
8	72.81	0.78	0.68	0.72	0.73	0.73
9	72.81	0.78	0.68	0.72	0.73	0.73
10	72.81	0.78	0.68	0.72	0.73	0.73
11	72.81	0.78	0.68	0.72	0.73	0.73

Split 6

FN	Accuracy Rate	Sensivity	Specificity	F-Score	AUC	Mean
1	71.69	0.80	0.63	0.71	0.72	0.71
2	71.73	0.80	0.63	0.71	0.72	0.71
3	72.37	0.75	0.70	0.72	0.72	0.72
4	72.81	0.78	0.68	0.72	0.73	0.73
5	72.81	0.78	0.68	0.72	0.73	0.73
6	72.81	0.78	0.68	0.72	0.73	0.73
7	72.81	0.78	0.68	0.72	0.73	0.73
8	72.81	0.78	0.68	0.72	0.73	0.73
9	72.81	0.78	0.68	0.72	0.73	0.73
10	72.81	0.78	0.68	0.72	0.73	0.73
11	72.81	0.78	0.68	0.72	0.73	0.73

Split 7

FN	Accuracy rate	Sensivity	Specificity	F-Score	AUC	Mean
1	71.69	0.80	0.63	0.71	0.72	0.71
2	71.73	0.80	0.63	0.71	0.72	0.71
3	72.37	0.75	0.70	0.72	0.72	0.72
4	72.81	0.78	0.68	0.72	0.73	0.73
5	72.81	0.78	0.68	0.72	0.73	0.73
6	72.81	0.78	0.68	0.72	0.73	0.73
7	72.81	0.78	0.68	0.72	0.73	0.73
8	72.81	0.78	0.68	0.72	0.73	0.73
9	72.81	0.78	0.68	0.72	0.73	0.73
10	72.81	0.78	0.68	0.72	0.73	0.73
11	72.81	0.78	0.68	0.72	0.73	0.73

Split 8

FN	Accuracy Rate	Sensivity	Specificity	F-Score	AUC	Mean
1	71.78	0.80	0.63	0.71	0.72	0.71
2	71.77	0.80	0.63	0.71	0.72	0.71
3	72.37	0.75	0.70	0.72	0.72	0.72
4	72.81	0.78	0.68	0.72	0.73	0.73
5	72.81	0.78	0.68	0.72	0.73	0.73
6	72.81	0.78	0.68	0.72	0.73	0.73
7	72.81	0.78	0.68	0.72	0.73	0.73
8	72.81	0.78	0.68	0.72	0.73	0.73
9	72.81	0.78	0.68	0.72	0.73	0.73
10	72.81	0.78	0.68	0.72	0.73	0.73
11	72.81	0.78	0.68	0.72	0.73	0.73

Table 8. Application Results

Split 9

FN	Accuracy Rate	Sensivity	Specificity	F-Score	AUC	Mean
1	71.78	0.80	0.63	0.71	0.72	0.71
2	71.77	0.80	0.63	0.71	0.72	0.71
3	72.37	0.75	0.70	0.72	0.72	0.72
4	73.05	0.73	0.73	0.73	0.73	0.73
5	73.05	0.73	0.73	0.73	0.73	0.73
6	73.05	0.73	0.73	0.73	0.73	0.73
7	73.05	0.73	0.73	0.73	0.73	0.73
8	73.05	0.73	0.73	0.73	0.73	0.73
9	73.05	0.73	0.73	0.73	0.73	0.73
10	73.05	0.73	0.73	0.73	0.73	0.73
11	73.05	0.73	0.73	0.73	0.73	0.73

Split 10

FN	Accuracy Rate	Sensivity	Specificity	F-Score	AUC	Mean
1	71.75	0.80	0.63	0.71	0.72	0.71
2	71.74	0.80	0.63	0.71	0.72	0.71
3	72.37	0.75	0.70	0.72	0.72	0.72
4	73.05	0.73	0.73	0.73	0.73	0.73
5	73.05	0.73	0.73	0.73	0.73	0.73
6	73.15	0.74	0.73	0.73	0.73	0.73
7	73.15	0.74	0.73	0.73	0.73	0.73
8	73.15	0.74	0.73	0.73	0.73	0.73
9	73.15	0.74	0.73	0.73	0.73	0.73
10	73.15	0.74	0.73	0.73	0.73	0.73
11	73.15	0.74	0.73	0.73	0.73	0.73

Split 11

FN	Accuracy Rate	Sensivity	Specificity	F-Score	AUC	Mean
1	71.74	0.80	0.63	0.71	0.72	0.71
2	71.74	0.80	0.63	0.71	0.72	0.71
3	72.37	0.75	0.70	0.72	0.72	0.72
4	73.05	0.73	0.73	0.73	0.73	0.73
5	73.05	0.73	0.73	0.73	0.73	0.73
6	73.15	0.74	0.73	0.73	0.73	0.73
7	73.15	0.74	0.73	0.73	0.73	0.73
8	73.15	0.74	0.73	0.73	0.73	0.73
9	73.15	0.74	0.73	0.73	0.73	0.73
10	73.15	0.74	0.73	0.73	0.73	0.73
11	73.15	0.74	0.73	0.73	0.73	0.73

Split 12

FN	Accuracy Rate	Sensivity	Specificity	F-Score	AUC	Mean
1	71.74	0.80	0.63	0.71	0.72	0.71
2	71.74	0.80	0.63	0.71	0.72	0.71
3	72.37	0.75	0.70	0.72	0.72	0.72
4	73.05	0.73	0.73	0.73	0.73	0.73
5	73.05	0.73	0.73	0.73	0.73	0.73
6	73.15	0.74	0.73	0.73	0.73	0.73
7	73.15	0.74	0.73	0.73	0.73	0.73
8	73.15	0.74	0.73	0.73	0.73	0.73
9	73.15	0.74	0.73	0.73	0.73	0.73
10	73.15	0.74	0.73	0.73	0.73	0.73
11	73.15	0.74	0.73	0.73	0.73	0.73

Split 13

FN	Accuracy Rate	Sensivity	Specificity	F-Score	AUC	Mean
1	71.72	0.81	0.63	0.71	0.72	0.71
2	71.74	0.80	0.63	0.71	0.72	0.71
3	72.35	0.75	0.70	0.72	0.72	0.72
4	73.02	0.73	0.73	0.73	0.73	0.73
5	73.02	0.73	0.73	0.73	0.73	0.73
6	73.12	0.74	0.72	0.73	0.73	0.73
7	73.12	0.74	0.72	0.73	0.73	0.73
8	73.12	0.74	0.72	0.73	0.73	0.73
9	73.12	0.74	0.72	0.73	0.73	0.73
10	73.12	0.74	0.72	0.73	0.73	0.73
11	73.12	0.74	0.72	0.73	0.73	0.73

Split 14

FN	Accuracy Rate	Sensivity	Specificity	F-Score	AUC	Mean
1	71.73	0.81	0.63	0.71	0.72	0.71
2	71.74	0.80	0.63	0.71	0.72	0.71
3	72.33	0.75	0.70	0.72	0.72	0.72
4	73.01	0.73	0.73	0.73	0.73	0.73
5	73.02	0.73	0.73	0.73	0.73	0.73
6	73.12	0.74	0.72	0.73	0.73	0.73
7	73.12	0.74	0.72	0.73	0.73	0.73
8	73.12	0.74	0.72	0.73	0.73	0.73
9	73.12	0.74	0.72	0.73	0.73	0.73
10	73.12	0.74	0.72	0.73	0.73	0.73
11	73.12	0.74	0.72	0.73	0.73	0.73

Split 15

FN	Accuracy Rate	Sensivity	Specificity	F-Score	AUC	Mean
1	71.73	0.81	0.63	0.71	0.72	0.71
2	71.74	0.80	0.63	0.71	0.72	0.71
3	72.33	0.75	0.70	0.72	0.72	0.72
4	73.01	0.73	0.73	0.73	0.73	0.73
5	73.02	0.73	0.73	0.73	0.73	0.73
6	73.12	0.74	0.72	0.73	0.73	0.73
7	73.12	0.74	0.72	0.73	0.73	0.73
8	73.12	0.74	0.72	0.73	0.73	0.73
9	73.12	0.74	0.72	0.73	0.73	0.73
10	73.12	0.74	0.72	0.73	0.73	0.73
11	73.12	0.74	0.72	0.73	0.73	0.73

4 Conclusion

Diagnosis and treatment of cardiovascular diseases are of vital importance for humans. Unfortunately, health clinics cannot provide adequate equipment for diagnosing people in today's world. For this reason, the need for new technologies is increasing day by day. In this study, a machine learning-based decision tree algorithm was developed based on predicting sick individuals. The study aims to eliminate the difficulties of clinicians in diagnosing patients and provide solutions with the algorithm developed for the problems that arise with this difficulty. The algorithm works in the minimum and maximum performance range of approximately 71% and 73%, respectively. The decision tree was evaluated according to the number of splits, but it was determined that the split did not affect performance. Accordingly, the decision tree took its most straightforward form for one split, and the performance ratio remained at the desired level. The decision tree's complexity has been reduced to the simplest structure, making it easier for users to understand. Predicting cardiovascular diseases with a high-performance rate is performed with a rule-based algorithm.

5 Acknowledgement

Conflict of Interests. There is no conflict of interest between the authors.

Financial Support. No financial support was received for this study.

Ethics Committee Approval. Ethics committee approval is not required as the open source data set was used in the study.

Data Collection and Use Permission. The data was obtained from the IEEE Dataport open source site.

References

1. Zhao, Y., Wood, E.P., Mirin, N., Cook, S.H., Chunara, R.: Social determinants in machine learning cardiovascular disease prediction models: a systematic review. Am. J. Prev. Med. **61**(4), 596–605 (2021)
2. Voloshynskyi, O., Vysotska, V., Bublyk, M.: Cardiovascular disease prediction based on machine learning technology. In: 2021 IEEE 16th International Conference on Computer Sciences and Information Technologies (CSIT), vol. 1, pp. 69–75. IEEE (2021)
3. Yang, L., et al.: Study of cardiovascular disease prediction model based on random forest in eastern China. Sci. Rep. **10**(1), 5245 (2020)
4. Choi, Y.Y., et al.: Cardiovascular disease prediction model in patients with hypertension using deep learning: analysis of the National Health Insurance Service Database from Republic of Korea. CardioMetabolic Syndr. J. **1**(2), 145–154 (2021)
5. Martins, B., Ferreira, D., Neto, C., Abelha, A., Machado, J.: Data mining for cardiovascular disease prediction. J. Med. Syst. **45**(1), 1–8 (2021). https://doi.org/10.1007/s10916-020-01682-8
6. Ali, M., et al.: Meme kanserinin teşhis edilmesinde karar ağacı ve knn algoritmalarının karşılaştırmalı başarım analizi. Acad. Perspect. Procedia **2**, 544–552 (2019)
7. Yazar, S., et al.: Araştirmamakales i / research article
8. Faizal, A.S.M., Thevarajah, T.M., Khor, S.M., Chang, S.W.: A review of risk prediction models in cardiovascular disease: conventional approach vs. artificial intelligent approach. Comput. Methods Programs Biomed. **207**, 106190 (2021)

9. Nawaz, M.S., Shoaib, B., Ashraf, M.A.: Intelligent cardiovascular disease prediction empowered with gradient descent optimization. Heliyon **7**(5), e06948 (2021)
10. Makalesi, A., Taşcı, M.E., Şamlı, R.: Veri madenciliği İle kalp hastalığı teşhisi **. Eur. J. Sci. Technol. 88–95 (2020). Special Issue
11. Katarya, R., Srinivas, P.: Predicting heart disease at early stages using machine learning: a survey. In: 2020 International Conference on Electronics and Sustainable Communication Systems (ICESC), pp. 302–305. IEEE (2020)
12. Vatansever, B., Aydin, H., Çetinkaya, A.: Genetik algoritma yaklaşımıyla Öznitelik seçimi kullanılarak makine Öğrenmesi algoritmaları ile kalp hastalığı tahmini. J. Sci. Technol. Eng. Res. **2**, 67–80 (2021)
13. Kavitha, M., Gnaneswar, G., Dinesh, R., Sai, Y.R., Suraj, R.S.: Heart disease prediction using hybrid machine learning model. In: 2021 6th International Conference on Inventive Computation Technologies (ICICT), pp. 1329–1333. IEEE (2021)
14. Makalesi, A., Yılmaz, A., Sümer, E.: Ek sayı. Eur. J. Sci. Technol. **1**, 609–615 (2021)
15. Ekrem, Ö., Salman, O.K.M., Aksoy, B., İnan, S.A.: Yapay zekÂ yÖntemlerİ kullanilarak kalp hastaliginin tespİtİ. Mühendislik Bilimleri ve Tasarım Dergisi **8**, 241–254 (2020)
16. Islam, S., Jahan, N., Khatun, M.E.: Cardiovascular disease forecast using machine learning paradigms. In: 2020 Fourth International Conference on Computing Methodologies and Communication (ICCMC), pp. 487–490. IEEE (2020)
17. Rashme, T.Y., Islam, L., Jahan, S., Prova, A.A.: Early prediction of cardiovascular diseases using feature selection and machine learning techniques. In: 2021 6th International Conference on Communication and Electronics Systems (ICCES), pp. 1554–1559. IEEE (2021)
18. Hasan, N., Bao, Y.: Comparing different feature selection algorithms for cardiovascular disease prediction. Health Technol. **11**(1), 49–62 (2020). https://doi.org/10.1007/s12553-020-00499-2
19. Li, R., Yang, S., Xie, W.: Cardiovascular disease prediction model based on logistic regression and Euclidean distance. In: 2021 4th International Conference on Advanced Electronic Materials, Computers and Software Engineering (AEMCSE), pp. 711–715. IEEE (2021)
20. Shanbehzadeh, M., Nopour, R., Kazemi-Arpanahi, H.: Using decision tree algorithms for estimating ICU admission of COVID-19 patients. Inform. Med. Unlocked **30**, 100919 (2022)
21. Javaid, H., Manor, R., Kumarnsit, E., Chatpun, S.: Decision tree in working memory task effectively characterizes EEG signals in healthy aging adults. IRBM **43**(6), 705–714 (2022)
22. Bashir, S., Almazroi, A.A., Ashfaq, S., Almazroi, A.A., Khan, F.H.: A knowledge-based clinical decision support system utilizing an intelligent ensemble voting scheme for improved cardiovascular disease prediction. IEEE Access **9**, 130805–130822 (2021)

Author Index

A
Abdiyeva-Aliyeva, Gunay 149
Abdul, Kahramanov Sebuhi 713
Aghriche, Ahmed 582
Akay, Bahriye 671
Akgun, Devrim 130
Aksoy, Bekir 458
Alaskarov, Elman 537
Alemdar, Refik 630
Aliyev, E. 662
Aliyeva, Almaz 703
Aliyeva, Almaz A. 473
Aydın, Muhammed Ali 288, 360
Aydın, Zafer 337
Ayyildiz, Ertugrul 186
Aziz Alaoui, M. A. 582
Azizoğlu, Gökhan 445

B
Babayev, Igbal 348
Babayev, Jahid 348
Bakhishov, Ismat 537
Bakhsiyeva, G. 662
Balık, Hasan Hüseyin 360
Baştürk, Alper 671
Bölücü, Necva 102
Bulut, Hasan 218
Bushuiev, Denis 348
Bushuieva, Victoria 348
Bushuyev, Sergey 348
Bushuyeva, Natalia 348
Buyukozkan, Kadir 186

C
Çağil, Gültekin 740
Canbay, Pelin 102
Çelik, Kayhan 176
Çelik, Mehmetcan 298
Çelikten, Azer 218
Çokçetn, Bahadır 595
Çuha, Fatma Aybike 407

D
Dener, Murat 607
Dizdaroğlu, Bekir 197
Duysak, Hüseyin 396

E
Erdem, O. Ayhan 415
Erkuş, Ekin Can 640
Erol Keskin, M. 703
Ersoy, Mevlüt 434, 703
Ertekin, Muhammed Tekin 515
Eylence, Muzaffer 458
Eyupoglu, Can 82

G
Ganbayev, Asif A. 620
Gasimov, Vagif 374
Genç, Burkay 56, 320, 500, 515
Göktaş, Altan 671
Gölcük, Bahadır 549
Görmez, Yasin 337
Güler, Hakan 560
Guliyeva, Sona 537
Günay, Melih 254
Gurarslan, Gurhan 30
Gurbanova, Gulnar G. 620
Gürfidan, Remzi 434
Gürkaş-Aydin, Zeynep 274
Gürtürk, Uğur 274

H
Hangun, Batuhan 264
Hasanov, Mehman H. 160, 465
Haşıloğlu, Muhammed Ali 560
Hematyar, Mehran 149
Hunbataliyev, Elmar Z. 16, 160

D. J. Hemanth et al. (Eds.): ICAIAME 2022, ECPSCI 7, pp. 751–753, 2023.
https://doi.org/10.1007/978-3-031-31956-3

Huseynov, Vildan 244
Huseynzada, K. 680

I
Ibadov, Nabi 490
Ibrahimov, Bayram G. 465, 473, 620
Imamoglu, Gul 186
Islamov, Islam J. 160
Ismayilov, Zafar A. 620
Ismoilov, Shodijon 30
Israfilova, E. 662

J
Jafarov, Toghrul 244

K
Kaczorek, Krzysztof 490
Kandaz, Derya 595, 740
Kara, Mustafa 360
Karaboğa, Derviş 671
Karacan, Hacer 39
Karaşlar, Muazzez Şule 500
Keskin, Sıddıka Nilay 73
KHalilov, E. 662
Kibar, Turan 56
Kilim, Oğuzhan 434
Kocamaz, Adnan Fatih 652
Kocyigit, Hulya 109
Kolukısa, Burak 337

M
Mammadli, Asgar 244
Mammadov, Gurban 140
Mammadov, Rahim 140
Manafova, Ilaha 244
Mertoğlu, Uğur 320
Merzeh, Hisham Raad Jafer 360
Murat, Mirac 186
Mustafayeva, A. 662

N
Nabiyev, Saleh 537
Naghiyev, J. 680
Najm, Fatiha 582
Nalbantoglu, Özkan Ufuk 671
Namazov, Manafedin B. 620
Nazari, Hamidullah 130
Nur, Salman Ahmed 630

O
Okumuş, Fatih 652
Okur, Celil 607
Omurca, Sevinç İlhan 549
Onan, Aytuğ 218
Orman, Abdullah 607
Ozcan, Alper 1
Özçift, Akın 691
Özkahraman, Merdan 458
Özsoy, Koray 458
Ozturk, Nurullah 254
Ozturk, Oktay 1, 264

P
Peker, Haldun Alpaslan 407
Piriev, Sahib 383
Purutçuoğlu, Vilda 640

R
Rahimova, Elena 140
Rosłon, Jerzy 490

S
Sadigov, A. 680
Sağlam, Fatih 500
Sahin, Aleyna 186
Sahin, Ismet Can 82
Şahin, Savaş 298, 423
Sari, Furkan Abdurrahman 560
Şen, Mehmed Oğuz 652
Şenokur, Cemre 309
Seven, Engin 288
Sevgen, Selçuk 483
Sevri, Mehmet 39
Söğüt, Esra 415
Sonmez, Yusif A. 473
Soydemir, Mehmet Uğur 298, 423
Süğürtin, Ufuk 630
Suleymanov, Umid 244
Sungur, Ahmetcan 73
Süzen, Ahmet Ali 122

T
Tagiyev, Ali D. 465
Tanağardıgil, İbrahim 298
Tanoğlu, Gamze 30
Taşın, Adem 630
Tirsi, Arslan 423

Toprak, Ahmet Nusret 445
Turguner, Cansin 288

U
Uçar, Muhammed Kürşad 560, 595, 630, 740
Ünlü, Ayşe 740
Usta, Pinar 458

Y
Yafia, Radouane 582
Yazıcı, Mehmet Fatih 73
Yerlikaya, Mehmet 396
Yeşilyurt, Saliha 671
Yıldırım, Gamze 230
Yıldız, Doğan 197
Yıldız, Eyyüp 483
Yıldız, Gülcan 197
Yilmaz, Cemal H. 473
Yüksel, Hakan 723
Yüksel, M. Erkan 483
Yüzbaşı, Şuayip 230

Z
Zalluhoğlu, Cemil 309

Printed in the United States
by Baker & Taylor Publisher Services